REVIEWS in MINERALOGY and GEOCHEMISTRY

Volume 49   2002

APPLICATIONS OF SYNCHROTRON RADIATION IN LOW-TEMPERATURE GEOCHEMISTRY AND ENVIRONMENTAL SCIENCES

Editors:

Paul A. Fenter

Environmental Research Division
Argonne National Laboratory
Argonne, Illinois

Mark L. Rivers

Department of Geophysical Sciences &
Consortium for Advanced Radiation Sources
The University of Chicago
Chicago, Illinois

Neil C. Sturchio

Dept. of Earth and Environmental Sciences
University of Illinois at Chicago
Chicago, Illinois

Stephen R. Sutton

Department of Geophysical Sciences &
Consortium for Advanced Radiation Sources
The University of Chicago
Chicago, Illinois

FRONT COVER: The figure on the front cover is the color version of Figure 1 for chapter 7. The foreground sketch illustrates the various types of interactions of adsorbate molecules or atoms at a mineral-water interface. The photo in the background shows how mineral surface processes can be used to attenuate metal mobility and bioavailability in the environment by *in situ* remediation techniques. The vegetation shown was established at the Barren Jales gold mine spoil (Portugal), and was successful on the treated As spoil with *Holcus lanatus* the 1st year and *Pinus pinaster* the 2nd year.

BACK COVER: An image of Earth is superimposed with a schematic of a synchrotron light source. The electron orbit in the synchrotron, shown in white, gives off two types of synchrotron radiation: "insertion device radiation" (bold arrows) from the periodic wiggles in the straight sections and the broad fan of "bending magnet radiation" from the circular arcs. (Image of Earth, courtesy of NASA and the National Space Science Data Center (NSSDC), was taken by Apollo 17 astronauts as they left Earth orbit en route to the Moon on Dec. 7, 1972).

Series Editors: **Jodi J. Rosso & Paul H. Ribbe**

GEOCHEMICAL SOCIETY
MINERALOGICAL SOCIETY of AMERICA

Reviews in Mineralogy
and Geochemistry

(Formerly: Reviews in Mineralogy)

ISSN 1529-6466

Volume 49

Applications of Synchrotron Radiation in Low-Temperature Geochemistry and Environmental Sciences

ISBN 0-939950-54-5

** This series of review volumes is published jointly under the banner of the Mineralogical Society of America and the Geochemical Society. The newly titled *Reviews in Mineralogy and Geochemistry* has been numbered contiguously with the previous series, *Reviews in Mineralogy*.

Additional copies of this volume as well as others in this series may be obtained at moderate cost from:

The Mineralogical Society of America
1015 Eighteenth Street, NW, Suite 601
Washington, DC 20036 U.S.A.

DEDICATION

Dr. William C. Luth has had a long and distinguished career in research, education and in the government. He was a leader in experimental petrology and in training graduate students at Stanford University. His efforts at Sandia National Laboratory and at the Department of Energy's headquarters resulted in the initiation and long-term support of many of the cutting edge research projects whose results form the foundations of these short courses. Bill's broad interest in understanding fundamental geochemical processes and their applications to national problems is a continuous thread through both his university and government career. He retired in 1996, but his efforts to foster excellent basic research, and to promote the development of advanced analytical capabilities gave a unique focus to the basic research portfolio in Geosciences at the Department of Energy. He has been, and continues to be, a friend and mentor to many of us. It is appropriate to celebrate his career in education and government service with this series of courses in cutting-edge geochemistry that have particular focus on Department of Energy-related science, at a time when he can still enjoy the recognition of his contributions.

APPLICATIONS OF SYNCHROTRON RADIATION IN LOW-TEMPERATURE GEOCHEMISTRY AND ENVIRONMENTAL SCIENCE

49   *Reviews in Mineralogy and Geochemistry*   **49**

FOREWORD

The review chapters in this volume were the basis for a short course on the applications of synchrotron radiation in low-temperature geochemistry and environmental science sponsored by the U. S. Department of Energy (DOE), Argonne National Laboratory (ANL), Geochemical Society (GS) and Mineralogical Society of America (MSA). The chapter authors were speakers at the short course which was held Dec. 4-5, 2002 in Monterey, California. A session of the same title ("Synchrotron Applications to Low-Temperature Geochemistry and Environmental Science") was held during the American Geophysical Union Fall Meeting (Dec. 6-10, 2002) in San Francisco following the GS/MSA short course.

This volume is the eleventh of a series of review volumes published jointly under the banner of MSA and GS.

The editors of this volume deserve a round of applause for the tremendous effort made to get this volume done on time. I also want to thank my infinitely patient and understanding family, Kevin, Ethan and Natalie. Without their willing sacrifice, I couldn't have done the job required of me.

Jodi J. Rosso, Series Editor
West Richland, Washington

October 10, 2002

PREFACE AND ACKNOWLEDGMENTS

This volume was produced in response to the need for a comprehensive introduction to the continually evolving state of the art of synchrotron radiation applications in low-temperature geochemistry and environmental science. It owes much to the hard work and imagination of the devoted cadre of sleep-deprived individuals who blazed a trail that many others are beginning to follow. Synchrotron radiation methods have opened new scientific vistas in the earth and environmental sciences, and progress in this direction will undoubtedly continue.

The organization of this volume is as follows. Chapter 1 (Brown and Sturchio) gives a fairly comprehensive overview of synchrotron radiation applications in low temperature geochemistry and environmental science. The presentation is organized by synchrotron methods and scientific issues. It also has an extensive reference list that should prove valuable as a starting point for further research. Chapter 2 (Sham and Rivers) describes the ways that synchrotron radiation is generated, including a history of synchrotrons and a discussion of aspects of synchrotron radiation that are important to the experimentalist.

The remaining chapters of the volume are organized into two groups. Chapters 3 through 6 describe specific synchrotron methods that are most useful for single-crystal surface and mineral-fluid interface studies. Chapters 7 through 9 describe methods that can be used more generally for investigating complex polyphase fine-grained or amorphous materials, including soils, rocks, and organic matter.

Chapter 3 (Fenter) presents the elementary theory of synchrotron X-ray reflectivity along with examples of recent applications, with emphasis on in situ studies of mineral-fluid interfaces. Chapter 4 (Bedzyk and Cheng) summarizes the theory of X-ray standing waves (XSW), the various methods for using XSW in surface and interfaces studies, and gives a brief review of recent applications in geochemistry and mineralogy. Chapter 5 (Waychunas) covers the theory and applications of grazing-incidence X-ray absorption and emission spectroscopy, with recent examples of studies at mineral surfaces. Chapter 6 (Hirschmugl) describes the theory and applications of synchrotron infrared microspectroscopy.

Chapter 7 (Manceau, Marcus, and Tamura) gives background and examples of the combined application of synchrotron X-ray microfluorescence, microdiffraction, and microabsorption spectroscopy in characterizing the distribution and speciation of metals in soils and sediments. Chapter 8 (Sutton, Newville, Rivers, Lanzirotti, Eng, and Bertsch) demonstrates a wide variety of applications of synchrotron X-ray microspectroscopy and microtomography in characterizing earth and environmental materials and processes. Finally, Chapter 9 (Myneni) presents a review of the principles and applications of soft X-ray microspectroscopic studies of natural organic materials.

All of these chapters review the state of the art of synchrotron radiation applications in low temperature geochemistry and environmental science, and offer speculations on future developments. The reader of this volume will acquire an appreciation of the theory and applications of synchrotron radiation in low temperature geochemistry and environmental science, as well as the significant advances that have been made in this area in the past two decades (especially since the advent of the third-generation synchrotron sources). We hope that this volume will inspire new users to "see the light" and pursue their research using the potent tool of synchrotron radiation.

The editors of this volume wish to acknowledge with gratitude the efforts of the following individuals: Jodi Rosso (Series Editor of *RiMG* for The Geochemical Society) for her patience and aplomb in doing the copy editing and layout of this volume in record time (despite the tardiness of the authors); Alex Speer (Mineralogical Society of America) for helpful guidance at every stage of planning this short course; previous short course organizers Charlie Alpers (USGS) and Randy Cygan (Sandia National Laboratories) for sharing their experiences; the anonymous reviewers who generously gave their time to help the authors improve their chapters; and Nick Woodward (Office of Basic Energy Sciences, U. S. Department of Energy) and Chris Reilly (Director, Environmental Research Division, Argonne National Laboratory) for their support and encouragement in this endeavor.

Paul Fenter
Mark Rivers
Neil Sturchio
Steve Sutton

Chicago, Illinois
September 25, 2002

THE AUTHORS

CHAPTER ONE

Gordon Brown *completed his Ph.D. degree in 1970 at Virginia Polytechnic Institute and State University, where he focused on X-ray crystallographic studies of olivines and chemical bonding in framework silicates. Following a year-long post-doctoral position at SUNY Stony Brook, where he developed high-temperature single-crystal X-ray diffraction methods, carried out crystallographic studies of returned Lunar samples from the Apollo missions, and studied Al/Si order-disorder in feldspars using neutron diffraction methods, he became an Assistant Professor of Geological & Geophysical Sciences at Princeton University. In 1973, he moved to Stanford University, where he is now the D.W. Kirby Professor of Earth Sciences and Professor and Chair*

of the Stanford Synchrotron Radiation Laboratory Faculty. Brown carried out one of the first X-ray absorption spectroscopy studies of earth materials (Fe-bearing minerals and silicate glasses) at the Stanford Synchrotron Radiation Project in 1977, and has been an active user of the intense X-rays from synchrotrons for the past 25 years. While at Stanford, Brown's research focus has been on the structure and properties of silicate glasses and melts, the response of mineral structures to high temperature, and most recently the environmental chemistry of metal and metalloid ions at mineral/water interfaces. When not teaching, working with his students and post-docs on various research projects, or working on chapters for RiMG volumes, he enjoys fly fishing, taking hikes with his Jack Russell Terrier Timmy (see photo), and horseback riding with his wife Nancy and his Appaloosa named Jenny.

Neil C. Sturchio *received his PhD in Earth and Planetary Sciences from Washington University in 1983, for studying the petrology and geochemistry of metamorphic rocks in the Eastern Desert of Egypt. After a two-year postdoc, he joined the staff at Argonne National Laboratory for 15 years to pursue studies of trace element and isotopic behavior in rock-water systems, leading to a program of synchrotron radiation studies of mineral-fluid interfaces at the National Synchrotron Light Source and the Advanced Photon Source. Sturchio is currently Professor of Geochemistry and Head of the Department of Earth & Environmental Sciences at the University of Illinois at Chicago.*

CHAPTER TWO

T.K. (Tsun-Kong) Sham *received his PhD in Chemistry from the University of Western Ontario for the studies of Mössbauer spectroscopy. He was on the staff of the Chemistry Department at Brookhaven National Laboratory for ten years before returning to the University of Western Ontario in 1988 and is presently a Professor in Chemistry and the Scientific Director of the Canadian Synchrotron Radiation Facility at the Synchrotron Radiation Center, University of Wisconsin-Madison. He has been involved in synchrotron research since 1975, is a scientific member of the SRI-CAT at the Advanced Photon Source and a Senior Scientific Consultant for the Canadian Light Source (University of Saskatchewan, Saskatoon, Canada).*

Mark Rivers *received his PhD in Geology and Geophysics from the University of California at Berkeley for studies of the ultrasonic properties of silicate melts. He is currently a Senior Scientist in the Department of Geophysical Sciences and Consortium for Advanced Radiation Sources (CARS) at the University of Chicago. He has been involved in synchrotron radiation research since 1984, is Co-Project Leader for the GeoSoilEnviroCARS beam lines (Sector 13) at the Advanced Photon Source (Argonne National Laboratory, IL, USA). His current research interests include the application of computed microtomography to problems in earth and environmental sciences.*

CHAPTER THREE

Paul Fenter *attended the Rensselaer Polytechnic Institute from 1980-1984 where he received a B.S. in Physics, and attended the University of Pennsylvania from 1984-1990 graduating with a Ph.D. in Physics. From 1990-1997 he was a post-doctoral fellow and then a research staff member at Princeton University. He joined the staff at Argonne National Laboratory in 1997 where he uses synchrotron X-ray scattering techniques to probe geochemical processes at mineral-water interfaces.*

CHAPTER FOUR

Michael Bedzyk *received his Ph.D. in Physics in 1982 from the State University of New York at Albany. Part of his graduate research took place at Bell Labs in Murray Hill. From 1982-1984 he was a Research Associate at the Hamburg Synchrotron Laboratory, Deutsches Elektronen-Synchrotron in Germany. After which he joined the Cornell High Energy Synchrotron Source (CHESS) as a Staff Scientist from 1985-1991. During that time he was also an Adjunct Associate Professor in the Materials Science Department at Cornell University. In 1991 he joined the faculty at Northwestern University as an Associate Professor in the Materials Science Department and became a Professor in the Materials Science and the Physics Departments at Northwestern in 2000. Dr. Bedzyk was the recipient*
of the Warren Award for Diffraction Physics in 1994 from the American Crystallographic Association and is a Fellow of the American Physical Society. His research interests center around the use of synchrotron X-ray radiation for atomic-scale analysis of surfaces, interfaces, thin-films and nanostructures. He has played a key role in pioneering the use of X-ray standing waves as a probe for investigating layered structures. His work includes structural studies of adsorbate surface phases on single crystal surfaces, semiconductor strained-layer heteroepitaxy, ferroelectric thin-films, the diffuse double layer that forms at electrified liquid / solid interfaces, and ultra-thin organic films.

Likwan Cheng *received a M.E. degree (1988) from Cornell University and a Ph.D. (1998) from Northwestern University. From 1988-1991, he was a staff member of the Laboratory for Planetary Studies at Cornell University, where he synthesized an analog of cometary material that has since become a widely used property standard in the astronomy community. He is currently a postdoctoral associate in Argonne National Laboratory's Environmental Research Division, where he has studied mineral-water interactions using X-rays.*

CHAPTER FIVE

Glenn Waychunas *received his Ph.D. in Geochemistry from UCLA in 1979, having joined the staff of the Center for Materials Research (CMR) at Stanford University the previous year as a senior scientist in charge of the X-ray laboratories. At CMR his main work involved X-ray crystallography and X-ray characterization of epitaxial thin films, in the course of which he developed a new type of diffraction instrument with colleague Paul Flinn. He also developed a long-standing collaboration with Gordon Brown's group in the school of Earth Sciences that led to a close association with SSRL and synchrotron studies of minerals. This collaboration produced many of the earliest mineralogical studies (EXAFS, XANES, polarized studies, soft X-ray spectroscopy, X-ray diffraction) using synchrotron sources. A mineralogical spectroscopist, Waychunas has also made contributions to Mössbauer, optical and luminescence spectroscopy. He has authored or co-authored seven chapters for the RIMG series, and twice had stints as American Mineralogist associate editor. He has been a fellow of the MSA for 15 years. He moved to the Lawrence Berkeley National Laboratory in 1997 as staff scientist in the Earth Sciences Division where he is currently concentrating on soft X-ray mineral interface studies, the structure of water and ionic complexes near interfaces, and nanomineralogy. He is a collaborator in the Molecular Environmental Science beamline at the Advanced Light Source (ALS), and one of the instigators and early collaborators in the GSECARS sector at the Advanced Photon Source.*

CHAPTER SIX

Carol Hirschmugl *is an Associate Professor of Physics, University of Wisconsin-Milwaukee. Hirschmugl has pioneered the use of the brightest infrared sources in the world at Brookhaven National Laboratory (Long Island, New York) and the Lawrence Berkeley National Laboratory (Berkeley, California). After graduating with a PhD in Applied Physics at Yale University, she received an Alexander von Humboldt Stiftung to do research in Berlin, Germany and the University of California President's Postdoctoral Fellowship to work at the Lawrence Berkeley National Laboratory. She has been a faculty member at the University of WI-Milwaukee for five years. Her present research interests include examinations of low energy dynamics and structure at oxide/water interfaces, and infrared spectromicroscopy to study physiological effects of environmental stimuli to single cells of phytoplankton.*

CHAPTER SEVEN

Alain Manceau *(right) received his Ph. D. in 1984 from the University of Paris 7. The same year he joined the CNRS as a researcher, becoming Director of Research in 1994. In 1992 he moved to the University Joseph Fourier in Grenoble where he established the team of Environmental Geochemistry, a component of the LGIT at the Observatoire des*  *Sciences de l'Univers. His research interests focus on environmental mineralogy and biogeochemistry of metal contaminants and trace elements using X-ray structural techniques. In the mid-80s, he initiated a new research program on the structure and surface reactivity of poorly crystallized Fe oxides. In the early 90s, this program was extended to Mn oxides, and specifically to minerals of the birnessite family. In the mid-90s, he pioneered the application of synchrotron techniques to determination of the speciation of heavy metals in natural systems. In the last two years, he was a key developer of an X-ray microprobe at the Advanced Light Source of the Lawrence Berkeley National Laboratory dedicated to the study of complex environmental materials. He is also co-lead PI of the French Absorption spectroscopy beamline in Material and Environmental sciences (FAME) at the European Synchrotron Radiation Facility (ESRF) in Grenoble.*

Matthew Marcus *(center) received his Ph.D. in 1978 from Harvard and joined Bell laboratories the same year. At Bell Labs, he worked on a variety of problems in materials science, with an emphasis on structure. Some of these problems include the structure of the liquid crystal blue phase, the precipitation kinetics of Cu in Al films (probed by EXAFS), the relation between local structure and luminescence of Er in silica and silicon, and the structure and vibrations of nanoparticles of Au and CdSe. His contributions to EXAFS technique include methods for preparing samples and improved methods for fitting sets of data taken at different temperatures. In 1998, he left Bell Labs to work for KLA-Tencor, helping develop a new kind of PEEM-related electron microscope for wafer inspection. In 2001 he took on the position of Beamline Scientist at the Advanced Light Source, where he collaborates on environmental and materials problems using an X-ray microprobe.*

Nobumichi Tamura *(left) obtained his Ph.D. in 1993 at the Institut National Polytechnique de Grenoble (INPG) for his work on the structure of quasicrystals and crystalline approximant phases. In 1998 he moved to Oak Ridge National Laboratory to contribute to the development of a new synchrotron-based X-ray microfocus technique capable of resolving strain and texture in thin films with submicrometer spatial resolution. He applied this technique in the field of microelectronics. He is currently staff scientist at the Lawrence Berkeley National Laboratory, where he leads the X-ray microdiffraction project at the Advanced Light Source. His research interest is presently focused on the study of mechanical properties of thin films at mesoscopic scale using synchrotron radiation.*

CHAPTER EIGHT

Steve Sutton *received his PhD in Earth and Planetary Sciences from Washington University in St. Louis, MO for studies of the thermoluminescence properties of shocked rocks from Meteor Crater, Arizona, including the determination of the absolute age of the impact event. He is currently a Senior Scientist in the Department of Geophysical Sciences and Consortium for Advanced Radiation Sources (CARS) at the University of Chicago. He has been involved in synchrotron radiation research since 1985, is Co-Project Leader for the GeoSoilEnviroCARS beam lines (Sector 13) at the Advanced Photon Source (Argonne National Laboratory, IL, USA) and Spokesperson for the X-ray Microprobe beam line (X26A) at the National Synchrotron Light Source (Brookhaven National Laboratory, NY, USA).*

Paul M. Bertsch *is a Professor of Soil Physical Chemistry and Mineralogy and Director of the Savannah River Ecology Laboratory of The University of Georgia. His research is focused on delineating the molecular form of contaminants (chemical speciation) in complex environmental samples and relating this to transportability and bioavailability. He is a participating research team member of the X-26A, microprobe beamline at the National Synchrotron Light Source, and a design team member of Sector 13, the Geo/Soil/Enviro CARS Collaborative Access Team, at the Advanced Photon Source. He served as Vice Chair of the Board of Governors of the Consortium for Advanced Radiation Sources at the University of Chicago from 1994-1999, and was recently appointed to the Scientific Advisory Committee for the Advanced Photon Source.*

Matt Newville *took his PhD in Physics from the University of Washington in Seattle, WA for the development of advanced analysis tools for X-ray Absorption Fine-Structure and the application of these to study the thermodynamic properties of metallic alloys. He is currently a Staff Scientist at the Consortium for Advanced Radiation Sources (CARS) at the University of Chicago.*

Antonio Lanzirotti *received his PhD in Geochemistry from the State University of New York at Stony Brook for studies of U-Pb isotope systematics and trace element behavior in metamorphic systems, in particular towards the development of high precision U-Pb chronometers for low U/Pb metamorphic minerals. He is currently a beamline scientist with the Consortium for Advanced Radiation Sources (CARS) at the University of Chicago. He has been involved in synchrotron radiation research since 1992, is the beamline local contact for the X-ray Microprobe beam line (X26A) at the National Synchrotron Light Source.*

Peter Eng *received his PhD in Condensed Matter Physics from State University of New York at Stony Brook, for the study of lead monolayer superlattice ordering kinetics on the nickel (001) surface using high resolution surface X-ray diffraction. He then spent three years as a postdoctoral member of the technical staff at AT&T Bell Laboratories, Murray Hill NJ, where he conducted research of the structure, kinetics, and phase transitions of surfaces and interfaces, using X-ray diffraction. He is currently as Senior Research Scientist in the James Franck Institute and Consortium for Advanced Radiation Sources (CARS) at the University of Chicago. He has been involved in synchrotron radiation research since 1986 and during the past seven years has been involved in the design,*

construction and operation of the GeoSoilEnviroCARS (GSE)CARS laboratory at the Advanced Photon Source. Here he has applied lower dimensional X-ray diffraction, X-ray absorption fine structure, X-ray fluorescence microprobe, microtomography and microcrystallography to study problems in earth and environmental science.

CHAPTER NINE

Satish C. B. Myneni *(Ph.D, Ohio State University) is an Assistant Professor in the Department of Geosciences at Princeton University and is also affiliated with the Departments of Chemistry and Civil and Environmental Engineering. He conducts his research activities at the Lawrence Berkeley National Laboratory where he is a Faculty Scientist. He earlier worked in the Berkeley Lab as a post-doctoral researcher and as Geological Scientist and developed soft X-ray spectroscopy and spectromicroscopy methods for examining light elements in aqueous solutions and on surfaces. His research interests are to explore ion solvation and complexation, and the chemistry of natural organic molecules at environmental interfaces.*

Applications of Synchrotron Radiation in Low-Temperature Geochemistry and Environmental Science

Table of Contents

1 **An Overview of Synchrotron Radiation Applications to Low Temperature Geochemistry and Environmental Science**

Gordon E. Brown, Jr. and Neil C. Sturchio

2   A Brief Overview of Synchrotron Radiation

T. K. Sham and Mark L. Rivers

3   X-ray Reflectivity as a Probe of Mineral-Fluid Interfaces: A User Guide

Paul A. Fenter

4   X-ray Standing Wave Studies of Minerals and Mineral Surfaces: Principles and Applications

Michael J. Bedzyk and Likwan Cheng

5   Grazing-incidence X-ray Absorption and Emission Spectroscopy

Glenn A. Waychunas

6   Applications of Storage Ring Infrared Spectromicroscopy and Reflection-Absorption Spectroscopy to Geochemistry and Environmental Science

Carol J. Hirschmugl

7

Quantitative Speciation of Heavy Metals
in Soils and Sediments
by Synchrotron X-ray Techniques

Alain Manceau, Matthew A. Marcus,
and Nobumichi Tamura

8

Microfluorescence and Microtomography
Analyses of Heterogeneous
Earth and Environmental Materials

Stephen R. Sutton, Paul M. Bertsch,
Matthew Newville, Mark Rivers,
Antonio Lanzirotti and Peter Eng

9   Soft X-ray Spectroscopy and Spectromicroscopy Studies of Organic Molecules in the Environment

Satish C. B. Myneni

1

An Overview of
Synchrotron Radiation Applications to
Low Temperature Geochemistry
and Environmental Science

Gordon E. Brown, Jr.[1,2] and Neil C. Sturchio[3]

[1]*Department of Geological and Environmental Sciences*
[2]*Stanford Synchrotron Radiation Laboratory*
Stanford University
Stanford, California, 94305-2115, U.S.A.

[3]*Department of Earth and Environmental Sciences*
University of Illinois at Chicago
845 West Taylor Street
Chicago, Illinois, 60607-7059, U.S.A.

INTRODUCTION

The availability of synchrotron radiation (SR) to the scientific community has literally revolutionized the way X-ray science is done in many disciplines, including low temperature geochemistry and environmental science. The key reason is that SR provides continuum vacuum ultraviolet (VUV) and X-ray radiation five to ten orders of magnitude brighter than that from standard sealed or rotating anode X-ray tubes (Winick 1987; Altarelli et al. 1998). Although SR was first observed indirectly by John Blewitt in 1945 (Blewitt 1946) and directly by Floyd Haber in 1946 at the General Electric 100-MeV Betatron in Schenectady, NY (see Elder et al. 1947; Baldwin 1975), it took 10 to 15 years before the first systematic applications of SR, which involved spectroscopic studies of the VUV absorption of selected elements (Tomboulian and Hartman 1956) using the 300-MeV synchrotron at Cornell University and of rare gases (Madden and Codling 1963) using the National Bureau of Standards SURF I synchrotron. As of September 2002, there are about 75 storage ring-based SR sources in operation, in construction, funded, or in advanced planning in 23 countries, with 10 fully dedicated SR storage ring facilities in the U.S.. A listing of these sources can be obtained at the following web site: *http://www-ssrl.slac.stanford.edu/sr_sources.html.*

The first SR experiments relevant to low temperature geochemistry and environmental science, although not performed on earth or environmental materials, were X-ray absorption fine structure (XAFS) spectroscopy measurements on amorphous and crystalline germanium oxide conducted on the SPEAR storage ring at the Stanford Synchrotron Radiation Project in 1971 by Dale Sayers, Farrel Lytle, and Edward Stern (Sayers et al. 1971). Prior to the availability of SR in the hard X-ray energy range (> 5 keV), XAFS spectroscopy measurements were impractical because of the high X-ray flux required and the need for a continuously tunable range of X-ray energies extending up to 1000 eV above the absorption threshold of the element of interest. In the 30 years since these SR-based XAFS measurements, this method has been applied to materials ranging from metalloproteins (e.g., Hasnain and Hodgson, 1999) and catalysts (e.g., Lytle et al. 1980; Sinfelt et al. 1984; Maire et al. 1986; Koningsberger et al. 1999) to silicate liquids at high temperatures (e.g., Brown et al. 1995a), cation complexes in aqueous and hydrothermal solutions (e.g., Fontaine et al. 1978; Hoffman et al. 2001), and complex environmental samples containing heavy metal and metalloid contaminants and pollutants

1529-6466/00/0049-0001$15.00

(e.g., Cotter-Howells et al. 1994; Pickering et al. 1995; Manceau et al. 1996, 2000a; Morris et al. 1996; Hesterberg et al. 1997a; Sayers et al. 1997; Foster et al. 1998; O'Day et al. 1998, 1999, 2000a; Morin et al. 1999, 2001; Ostergren et al. 1999; Welter et al. 1999).

In addition to XAFS spectroscopy methods, SR-based micro-X-ray fluorescence spectroscopy (Bertsch and Hunter 2001) and X-ray diffraction methods (Parise 1999) are now commonly used in studies of earth and environmental materials. Less commonly used SR methods include VUV and soft X-ray photoemission spectroscopy, X-ray standing waves, X-ray reflectivity measurements, small angle X-ray scattering, X-ray microtomography, X-ray microscopy measurements, photoelectron diffraction, and synchrotron infrared spectroscopy. In spite of their limited use to date, however, each of these methods is beginning to provide useful, and in many cases, unique information about earth and environmental materials and their interactions with aqueous fluids, atmospheric gases, organic matter, microbial organisms, plant roots, and environmental contaminant ions (e.g., Brown et al. 1999a; Bertsch and Seaman 1999; Bertsch and Hunter 2001). The potential of many of these methods, particularly those that require high brightness (μXAFS, μXRD, X-ray reflectivity, X-ray standing waves, X-ray microtomography, and X-ray microscopy), could not be fully realized prior to the availability of third-generation synchrotron radiation sources in the mid-1990's.

The purposes of this chapter are to (1) discuss some of the major issues in low temperature geochemistry and environmental science that can be addressed using SR methods, (2) provide an overview of a number of these methods, (3) highlight some of the applications of SR to these fields, (4) provide an extensive list of references to the literature on these applications, and (5) forecast future SR applications in low temperature geochemistry and environmental science, including those that might make use of the first X-ray free electron laser, which is currently under development in the U.S. and Germany. Because of the importance of aqueous solutions and solid/solution interfaces in these fields, we have also included sections on these two topics that provide brief overviews of current knowledge, particularly molecular-scale information derived from SR studies.

Combining low temperature geochemistry and environmental science in this short course volume is appropriate because both areas involve similar materials, similar processes, and similar ranges of earth surface conditions: $T = -50°$ to $350°C$, $P =$ one to several hundred bars, $pH = 0$ to 14, $Eh = -0.8$ to $+1.2$ volts (i.e., the upper and lower stability limits of water, which are dependent on pH). Although low temperature geochemistry and certain areas of environmental science can be thought of as parts of a continuum in terms of chemical and biological processes, one can distinguish between low temperature geochemistry and environmental science as follows: the study of the chemistry and reactivity of pristine natural environments, materials, and processes at or near the earth's surface is *low temperature geochemistry*, whereas once contaminants or pollutants are introduced into a geological system, the disciplinary area becomes *environmental chemistry/geochemistry*. Contaminants or pollutants may be either anthropogenic (such as chlorinated solvents or nuclear waste) or naturally occurring (such as radon gas or As in groundwaters derived from earth materials). This is a somewhat arbitrary distinction, however, if one considers the concept of biosphere, which is the intersection of geosphere, hydrosphere, and atmosphere where life exists. Defined in this way, the biosphere encompasses all of the earth surface conditions, materials, and processes that define low temperature geochemistry and environmental science. The term *Molecular Environmental Science* is now widely used to describe molecular-level studies of chemical and biological processes affecting the speciation and behavior of contaminants and pollutants in the biosphere.

Rather than presenting a comprehensive review here, we shall focus on selected applications that illustrate the utility of SR methods in addressing problems relevant to low temperature geochemistry and environmental science. More comprehensive reviews of the applications of selected SR methods to these fields, as well as the methods themselves, are given in the remaining chapters in this RiMG volume. The interested reader is also directed to other sources for recent reviews of related areas, including applications of (1) SR to clay science (Schulze et al. 1999); (2) SR to soil and plant science (Schulze and Bertsch 1995); (3) XAFS spectroscopy to the reaction of metal cations in aqueous solutions with mineral surfaces, organic matter, microbial organisms, and marine biomass (Brown and Parks 2001); (4) XAFS spectroscopy to materials and environmental science (Conradson 1998); (5) synchrotron X-ray diffraction methods to single crystal and powdered earth materials (Parise 1999); (6) synchrotron X-ray microprobe methods to earth and environmental materials (Sutton and Rivers 1999; Bertsch and Hunter 2001); and (7) synchrotron radiation infrared (SR-IR) spectroscopy and microspectroscopy methods to hydrous minerals (Lu et al. 1999) and fluid/volatile inclusions in minerals (Guilhaumou et al. 1998) (see Hirschmugl 2002 and in this volume for reviews of recent applications of SR-IR spectroscopy to surfaces and interfaces and environmental and geochemical samples).

IMPORTANT ISSUES IN LOW TEMPERATURE GEOCHEMISTRY AND ENVIRONMENTAL SCIENCE

Water and atmospheric or soil gases are ubiquitous in low temperature geochemical and environmental systems

As a result, experimental studies of materials and processes relevant to both types of systems are best conducted under *in situ* conditions at appropriate T, P, pH, and Eh values, generally with bulk water, water vapor, or other relevant gases present. Although ex-situ studies are often necessary and provide valuable base-line information, they may result in sample changes and don't mimic conditions in natural systems. For example, an ex-situ measurement, such as one might make using a transmission electron microscope (TEM), requires drying the sample before preparing a thin section of a portion of it by ultra-microtoming, Ar-ion thinning, or fracturing, and placing the thinned, dried sample in the ultra-high vacuum (UHV) chamber of the TEM. A potential problem in all ex-situ UHV work is that sample preparation and the UHV environment may alter the sample irreversibly. This is particularly true, e.g., for hydrous phases such as ferrihydrite or alunite coatings in a streambed draining from a mine tailings site that may contain sorbed heavy metal contaminant ions. It is also true of contaminated soil samples that contain hydrous mineral phases as well as organic matter and microbial organisms, where such drying may cause the speciation of the contaminant ion to change. An advantage of some SR methods, such as XAFS spectroscopy, is that little if any sample preparation is required, and measurements can be made in many cases under *in situ* conditions using appropriate controlled-atmosphere, controlled-temperature fluid or flow-through cells with X-ray transparent windows. In many cases, the information provided by SR methods cannot be obtained using non-SR techniques.

Knowledge of the molecular-level speciation of trace elements and environmental contaminants and pollutants is critical for understanding their behavior

Of key importance in studies of samples and processes relevant to low temperature geochemistry and environmental science is determination of the phases present, phase associations, and the concentration and chemical speciation of inorganic and organic contaminants and pollutants or of trace elements in the case of pristine systems.

Pollutants are defined as elements or compounds whose presence in a natural system produces undesirable physical, chemical, or biological characteristics, whereas *contaminants* cause deviations from the normal composition of the environment but do not cause detrimental effects (Alloway and Ayres 1993). The term *toxin* is sometimes used in place of pollutant. The term *speciation*, as used here, refers to (1) the identity of the element (or elements), (2) its (their) oxidation state, (3) its (their) physical state (i.e., phase association; presence in a liquid, gaseous, or solid phase (amorphous or crystalline), colloidal particle, animal or plant cell, or biofilm; presence as a surface coating or thin film on a solid, as a sorption complex (monomeric or polymeric) on a solid, colloidal particle, or an organic substance; etc.), (4) its empirical formula, and (5) its detailed molecular structure. Knowledge of the chemical speciation of contaminants and pollutants is essential for predicting their environmental properties, i.e., their stability, mobility, toxicity, and potential bioavailability to humans and other organisms (Stern 1987; Brown et al. 1999b), as well as for designing effective remediation strategies.

A good example of the importance of speciation in determining the environmental properties of an inorganic pollutant is chromium, which is redox sensitive and can occur in 6+, 3+, 2+, and 0 oxidation states in earth and environmental materials depending upon Eh and pH. Chromium is essential to human health because of its importance in glucose metabolism (Kieffer 1992). The maximum recommended intake of Cr for adult males and females is 20-35 μg day^{-1} (NAS/NRC 2002). Above this threshold, ingested chromium may have serious health consequences including tumors, ulcers, and cancer induced by alteration of the DNA template (Snow and Lu 1989; Lewis 1991). In the chromate $(Cr(VI)O_4^{2-})$ form, chromium is known to be teratogenic (Abbasi and Soni 1984), mutagenic (Paschin et al. 1983; Beyersmann et al. 1984), and carcinogenic (Gauglhofer and Bianchi 1992). In addition, the more oxidized and mobile Cr(VI) form has been found to be more toxic than Cr(III) to various microbiota (Babich and Stotzky 1983). The most oxidized form of chromium occurs in aqueous solutions as a tetrahedral oxoanion $(CrO_4^{2-}, HCrO_4^{-}, or Cr_2O_7^{2-}, depending on pH and Cr concentration; Baes and Mesmer 1986) with a Cr-O bond length of ≈ 1.6 Å. Under less oxidizing conditions, chromium occurs primarily in the 3+ oxidation state as a hexaaquo complex or hydrolysis product (Baes and Mesmer 1986), with a Cr-O bond length of ≈ 2 Å. In this form, chromium is not particularly toxic, does not have the detrimental health effects of Cr(VI), and typically forms insoluble solids that effectively sequester chromium from the biosphere. These differences in molecular-level speciation of Cr(VI) vs. Cr(III) are responsible for the differences in environmental properties of these two species (Fig. 1). Figure 1 also shows that the X-ray absorption near edge structure (XANES) spectra of these two species provide clear and easily measurable fingerprints of Cr(VI) and Cr(III).

Another example of the effect of speciation on toxicity is provided by the various congeners of tetrachlorodibenzo-*p*-dioxin (TCDD). Even though the congeners have the same composition, they exhibit toxicity differences of over four orders in magnitude (Bunce 1991).

Earth and environmental materials are complex

Environmental samples such as soils consist of multiple solid phases, which can be crystalline, poorly crystalline (or amorphous), or a mixture of the two. The solid phases may be coated in part by other crystalline or amorphous phases (e.g., hydrous aluminum, iron, or manganese oxides), natural organic matter, fungi, or microbial biofilms, and are always in contact with an aqueous solution and/or a gas, as discussed above. The contaminant or pollutant species in environmental samples are often at low concentrations (e.g., parts per million, ppm, or parts per billion, ppb), and are

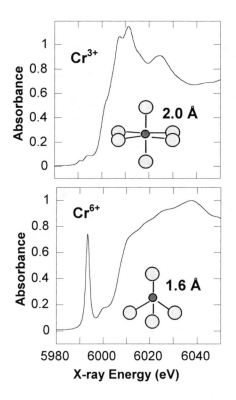

Figure 1. NEXAFS spectra of Cr in the trivalent (top) and hexavalent (bottom) chemical forms. The inset molecules show that Cr(III) is coordinated by six oxygens at an average Cr(III)-O distance of 2.0 Å and that Cr(VI) is coordinated by four oxygens at an average Cr(VI)-O distance of 1.6 Å. These Cr K-edge XANES data were collected on SSRL beam line 4-3. (after Peterson et al. 1996).

heterogeneously distributed among different phases. Figure 2 illustrates some of this complexity. SR methods can provide unique information on the identity of minor phases (amorphous or crystalline), the speciation of sorbed metal ions and organics, and their sub-micron spatial distribution and phase association.

Figure 2 also illustrates the need for a multidisciplinary/interdisciplinary approach to the molecular-level problems posed by complex environmental systems. Contributions from geochemists, inorganic and organic chemists, microbiologists, mineralogists, plant pathologists, soil chemists, surface scientists, synchrotron scientists, and toxicologists, among others, are needed to unravel these complexities (e.g., de Stasio et al. 2001).

Complementary characterization methods are necessary

Because of the complexity of environmental and low temperature geochemical samples, it is highly unlikely that any single characterization method can provide the information about chemical speciation, spatial distribution, and physical association needed to understand the processes leading to their formation or alteration or, in the case of contaminant or pollutant species, remediation methods required to isolate these species from the biosphere or transform them into environmentally benign forms. The use of multiple complementary analytical methods is often necessary to obtain this information. For example, characterization of the speciation, concentration, distribution, and phase association of Zn in complex mining wastes and associated Zn-contaminated soils requires a combination of methods, including inductively coupled plasma-mass spectrometry analysis, Zn K-edge XAFS spectroscopy, μXRF mapping of Zn and related

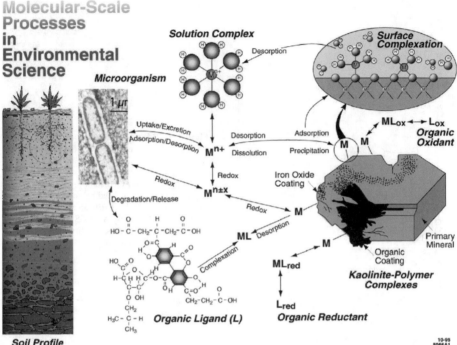

Molecular-Scale Processes in Environmental Science

Solution Complex

Surface Complexation

Microorganism

Soil Profile

Figure 2. Schematic illustration of some of the molecular-scale processes that affect the fates of environmental contaminants. Such processes range from dissolution of mineral particles in soils, which can release natural contaminants into pore waters, to the binding or sorption of metals (M) and organic compounds (L) to mineral surfaces, which can effectively immobilize contaminants and reduce their bioavailability. Some contaminant elements such as chromium, arsenic, or selenium can undergo oxidation or reduction when they interact with mineral surfaces and organic compounds. In addition, microorganisms and plants can have a profound influence on chemical reactions occurring at the earth's surface. For example, microorganisms often play a major role in the degradation of organic contaminants and in the oxidation and reduction of heavy metals. The root-soil interface (rhizosphere - see circled area in soil profile) is an area of particularly intense chemical and biological activity where organic acids, sugars, and other organic compounds are exuded by live plant roots. pH can be as much as 2 units lower, and microbial counts can be 10 to 50 fold higher at the root surface than in the bulk soil a few millimeters away. Thus, mineral weathering and the solubility of mineral elements and anthropogenic contaminants are generally greater in surface soils, where plant and microbial activity are higher, than in deeper parts of the soil and geologic column. An understanding of the molecular-scale mechanisms and rates of these processes requires a knowledge of the types, spatial distribution, and reactivity of contaminant species, the nature of mineral surfaces (including coatings that affect their reactivity) and colloidal particles, the types and distribution of organic compounds and microorganisms in soils and pore waters, and the effect of a host of variables on the rates of dissolution, adsorption, desorption, precipitation, degradation, and redox reactions. (after Fig. 1 in Brown et al. 1999b)

elements such as Mn and Fe with which it might be associated, bulk X-ray diffraction (XRD) and μXRD, electron microprobe analysis, transmission, analytical, and scanning electron microscopy, and sequential chemical extractions (see, e.g., Webb et al. 2000; Isaure et al. 2002; Juillot et al. 2002; Manceau et al., this volume).

Other examples of the usefulness of complementary characterization methods are the studies by Persson et al. (1991), which used a combination of large angle X-ray scattering, XAFS, and vibrational spectroscopy to define the structure of Cu(I)-halide complexes in a variety of aqueous and non-aqueous solutions, and by Allen et al. (1995a) and Clark et al. (1996), which used a combination of EXAFS, multinuclear NMR, and Raman spectroscopies and X-ray diffraction to determine the structure of uranyl and neptunyl carbonate complexes in near-neutral pH aqueous solutions. In addition, the study of Teng et al. (2001) used a combination of atomic force microscopy and X-ray reflectivity to study the dissolution of orthoclase surfaces as a function of pH.

Parallel studies of real and model systems are essential

Studies of simplified model systems are required because of the difficulty of gaining a clear understanding of the complex chemical and biological processes that operate in natural systems. Parallel studies of natural systems are required to identify (1) the relevant solid and amorphous phases present; (2) the concentrations, phase associations, and molecular-level speciation of contaminant or pollutant species; (3) the presence or absence of natural organic matter and microbial organisms; (4) redox conditions and the relevant reductants and/or oxidants present; (5) the presence of inorganic and/or organic coatings and biofilms on mineral particles; and (6) the composition of natural waters and gases in contact with these particles. For example, the interaction of aqueous metal ions with mineral particles in soil samples is immensely complex and cannot typically be understood at a fundamental chemical or biological level unless the problem is addressed in a series of parallel experimental studies involving both the complex natural system and simplified analog systems in which the number of variables and phases is controlled. The complexity of the model systems can be increased in successive experimental studies, and as understanding is gained, can ultimately approach the complexity of the natural system. However, without appropriate model system studies for comparison, it is unlikely that chemisorbed species of aqueous metal ions could be identified in natural samples (e.g., Foster et al. 1998; Morin et al. 1999; Ostergren et al. 1999). Without parallel studies of natural systems, the appropriate model systems may not be chosen for studies of relevant processes such as sorption reactions (Bertsch and Seaman 1999). This approach implies that some studies will be more fundamental in nature, while others will be more applied.

The nature of the solid/water interface and of sorbed species must be known

Most low temperature geochemical and environmental reactions occur at solid/water and solid/gas interfaces (Stumm et al. 1987). A highly schematic view of a solid/aqueous solution interface is shown in Figure 3, illustrating the electrical double layer that develops in response to the charge on the mineral surface and to the interactions of the charged surface with interfacial water and the counter-ions and co-ions that occur in the EDL. Such interfaces are more complex in natural systems because of the presence of inorganic ligands such as sulfate or carbonate, natural organic matter, xenobiotic organics, microbial biofilms, fungi, plant roots, and poorly crystalline coatings (Brown et al., 1999a). The major scientific objective of SR studies of mineral/water interfaces is to better understand the influence of molecular-scale processes on macroscopic geochemical cycles and mass transfer in the natural environment. Important specific questions relevant to mineral/water interface reactions that can be addressed using SR methods include the following: (1) how do the structure and composition of mineral surfaces differ from the bulk when mineral particles are in contact with aqueous solutions or a humid atmosphere? (2) what are the compositions and structures of adsorbates on mineral surfaces and how are adsorbates bonded to the mineral surface? (3) do adsorbates of a particular element differ from solution species of the same element because of effects of the solid surface and/or changes in the properties of water near the interface relative to bulk water? (4) what are the types and density of reactive sites on mineral surfaces and

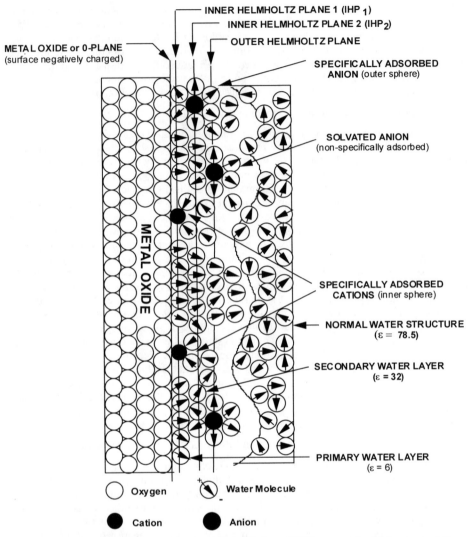

Figure 3. Highly schematic view of the electrical double layer (EDL) at a metal oxide/aqueous solution interface showing (1) hydrated cations specifically adsorbed as inner-sphere complexes on the negatively charged mineral surface (pH > pH$_{pzc}$ of the metal oxide); (2) hydrated anions specifically and non-specifically adsorbed as outer-sphere complexes; (3) the various planes associated with the Gouy-Chapman-Grahame-Stern model of the EDL; and (4) the variation in water structure and dielectric constant (ε) of water as a function of distance from the interface. (from Brown and Parks 2001, with permission)

what are their responses to changes in pH, ionic strength, and other variables? (5) how do the structure and composition of the electrical double layer (EDL) change with changing solution conditions? (6) how do organic coatings or microbial biofilms, proximity to plant roots or fungi, or the presence of inorganic or organic ligands change the structure and properties of the EDL at mineral/solution interfaces and the intrinsic reactivity of the mineral surface? (7) how do adsorbed species affect the surface charge of mineral particles and how does this, in turn, affect colloid stability? XAFS spectroscopy, X-ray standing wave methods, X-ray reflectivity, X-ray microscopy, and SR-based photoemission spectroscopy can provide unique information on chemical processes at solid/water, solid/gas, solid/microbial biofilm, and solid/organic matter interfaces (e.g., Brown et al. 1999a; Brown and Parks 2001; Bedzyk and Cheng, this volume; Fenter, this volume).

Nanoscale earth and environmental materials play a major role in interfacial chemical reactions

Enormous quantities of nanometer (10^{-9} m)-scale particles occur in natural waters, aquifers, soils, sediments, mine tailings, industrial effluents, and the atmosphere. For example, it is estimated that an average cubic meter of remote continental air contains 300 million particles having diameters between 0.1 and 0.01 µm (Cadle 1966). Natural nanoparticles, which can form via inorganic and biological pathways (e.g., Labrenz et al. 2000), can be thought of as part of a continuum of atomic clusters, ranging in size from hydrated ions in aqueous solution to bulk mineral particles and aggregates of particles. There is growing recognition that nanoparticles of earth and environmental materials play major roles in low temperature geochemical and environmental processes (Banfield and Zhang, 2001). This is true because such nanoparticles collectively have immense surface areas relative to mm-sized and larger particles of similar mass, and chemical reactivity is proportional to surface area. There is also growing evidence that the structure and properties of nanoparticles change as particle size decreases (Gleiter 1992; Siegel 1996). For example, McHale et al. (1997) predicted on the basis of the molecular dynamics simulations of Blonski and Garofalini (1993) and microcalorimetric measurements that γ-Al_2O_3 rather than α-Al_2O_3 becomes the energetically stable polymorph of alumina when particle surface area exceeds 125 m^2g^{-1}. Other studies have shown that TiO_2 nanoparticles exhibit differences in photocatalytic reduction of redox-sensitive metals like Cu(II) and Hg(II) in the presence and absence of surface adsorbers like alanine, thiolactic acid, and ascorbic acid relative to bulk TiO_2 (Chen et al. 1997a,b, 1999; Rajh et al. 1999). These differences in catalytic activity have been attributed to the different surface structures of nanoparticulate vs. bulk TiO_2. Using EXAFS spectroscopy, Chen et al. (1997a,b) found shorter Ti-O bonds and increasing disorder around Ti with decreasing size of the TiO_2 nanoparticles. They suggested that the unique surface chemistry exhibited by nanoparticulate TiO_2 is related to the increasing number of coordinatively unsaturated surface Ti sites with decreasing nanoparticle size. In a similar EXAFS study of nanoparticulate α-Fe_2O_3, Chen et al. (2002) found that surface Fe sites are undercoordinated relative to Fe in the bulk structure and are restructured to octahedral sites when the nanoparticles are reacted with enediol ligands.

In addition to SR characterization studies of the structures of nanoparticles such as those above, there have been Fe K-edge EXAFS studies of the formation and structural evolution of hematite nanoparticles from aqueous solutions using Fe K-edge EXAFS (Combes et al. 1986, 1989a,b, 1990) and Fe K-EXAFS and SAXS studies of the evolution of ferrous and ferric oxyhydroxide nanoparticles in anoxic and oxic solutions, respectively, containing SiO_4 ligands, which were found to limit the sizes of clusters formed (Doelsch et al. 2000, 2002; Masion et al. 2001). There have also been EXAFS and WAXS studies of growth inhibition of nanoparticles of ferric hydroxide by arsenate (Waychunas et al. 1993, 1995, 1996) and by Ni(II) and Pb(II) (Ford et al. 1999). Whereas Ni(II) was found to be incorporated into the ferrihydrite nanoparticles, Pb(II) and arsenate were found to be dominantly adsorbed to the particle surfaces.

Colloidal particles, which are operationally defined as particles having at least one dimension in the submicrometer size range, include the size range of nanoparticles

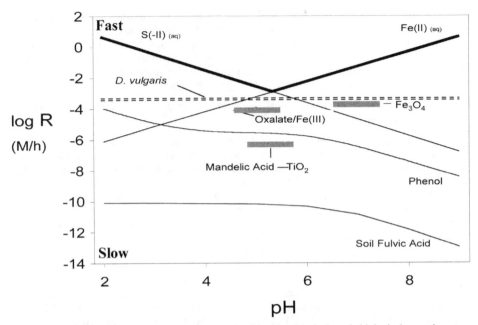

Figure 4. Rate comparison for chromate reduction by chemical and biological constituents, demonstrating the dominance (bold line) of chemical pathways at all typical pH values of natural systems. For calculated initial rates, initial [Cr(VI)] = 100 µM, [Fe(II)] = 30 µM, [S(−II)] = 10 µM. (from Fendorf et al. 2000, with permission)

(Hunter 1993). Colloidal particles are widely believed to play an important role in the transport of heavy metal and organic contaminants in the environment (e.g., Grolimund et al. 1996; Buffle et al. 1998). For example, Kersting et al. (1999) concluded that colloid-facilitated transport of Pu is responsible for its subsurface movement from the Nevada Test Site to a site 1.3 km south over a relatively short time period ($\approx$ 40 years), based on the unique $^{240}Pu/^{239}Pu$ isotope ratios of the samples. In addition, Kaplan et al. (1994) have found that colloid-facilitated transport of Ra, Th, U, Pu, Am, and Cm in an acidic plume beneath the Savanah River Site, Aiken, South Carolina, helps explain the faster than anticipated transport of these actinides. This study also found that Pu and Th are most strongly sorbed on the colloidal particles, whereas Am, Cm, and Ra are more weakly sorbed. EXAFS spectroscopy and SR-based XRF are now shedding light on how Pu and other actinide ions sorb on colloidal mineral particles and undergo redox transformations in some cases (Kaplan et al. 1994; Allen et al. 2001). Another example of colloid-facilitated transport comes from recent laboratory column experiments of Hg-mine wastes, which showed that significant amounts of Hg-containing colloidal material from calcine piles associated with Hg mines in the California Coast Range can be generated by a change in aqueous solution composition, such as occurs when the first autumn rains infiltrate these piles (Lowry et al. 2002; Shaw et al. 2002). In these latter studies, EXAFS spectroscopy, coupled with transmission electron microscopy characterization of the column-generated colloids, showed that the colloidal particles are primarily HgS rather than iron oxides on which Hg(II) is adsorbed.

The field of *nanogeoscience* has recently emerged as a subdiscipline of the geosciences in response to the recognition that natural and anthropogenic nanoparticles, including colloids, play a major role in geochemical and environmental processes (Banfield and Navrotsky 2001). Synchrotron radiation methods, particularly EXAFS spectroscopy and X-ray microscopy methods, can make unique contributions to our understanding of this class of earth materials.

Reaction pathways and kinetics of environmental and low temperature geochemical reactions are of critical importance

The fastest reaction pathway will typically control the fate of a contaminant or pollutant species in a natural system that has a number of possible reaction pathways with different kinetics. A good example of this point is the study by Fendorf et al. (2000) of reduction rates of Cr(VI) in various anaerobic model systems containing inorganic and organic/biological reductants. This study showed that dissolved sulfide, at low to neutral pH, and dissolved Fe(II), at neutral to high pH, are the most rapid reductants of Cr(VI) relative to a host of other abiotic and biotic reductants, including magnetite, *D. vulgaris*, oxalate/Fe(III), phenol, soil fulvic acid, and mandelic acid/TiO_2 (Fig. 4). Thus detailed laboratory studies are needed to evaluate the kinetics of multiple potential reaction pathways to determine which pathway(s) are most important in a natural system.

Another important consideration is the metastability of many mineral phases in the near-surface environment. Kinetics of low-temperature mineral transformations and their dependence on the chemical conditions of coexisting fluids are not well known, despite the important influence these transformations may have on particle size, surface area, and the transport behavior of associated trace elements. Some of these issues can be addressed effectively using SR methods, but the long time scale of some such processes cannot be observed in typical experimental time scales.

Computational chemistry can play an important role in studies of interfacial reactions

It is not likely that experimental studies can reveal the full mechanisms of chemical reactions at mineral/solution or mineral/gas interfaces using currently available methods, due in part to the fact that most spectroscopic and scattering methods are often too slow to identify reaction intermediates. Modern computational chemistry has an important role to play in predicting transition state complexes and rate-limiting steps in a variety of reactions of importance in environmental chemistry and low temperature geochemistry. Although most environmental and geochemical reactions are too complex for computational chemistry approaches using adequate levels of theory, some relatively simple interfacial reactions can now be successfully addressed using a combination of theory and experiment. This point has been illustrated by studies of the interaction of water with MgO surfaces (e.g., Liu et al. 1998a; Giordano et al. 1998; Odelius 1999), which are discussed in more detail in a later section of this chapter. Another example of the application of computational chemistry to interfaces is the MD simulation study of water at the mica (001)/water interface by Park and Sposito (2002), the results of which are consistent with SR X-ray reflectivity studies of interfacial water on mica (001) surfaces (Cheng et al. 2001a). An example of the usefulness of computational chemistry in interpreting spectroscopic data on aqueous species is the combined EXAFS, FTIR, and quantum chemical study of Ga(III)-acetate complexes in aqueous solutions (Clausen et al. 2002).

Inorganic and organic ligands and coatings can have a major impact on the sorption behavior of metal ions as well as on mineral dissolution

Common inorganic ligands can enhance or inhibit the sorption of metal ions from aqueous solutions onto mineral surfaces depending upon whether or not they result in the formation of ternary surface complexes or stable solution species (Fig. 5). For example, the common ligand sulfate enhances the sorption of Pb(II) on goethite surfaces over the

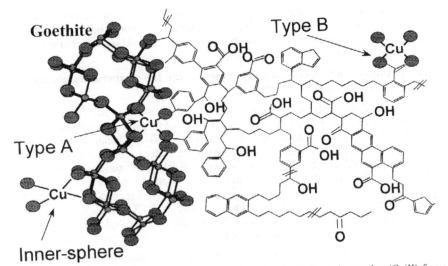

Figure 5. Cu(II)-humic complex on the goethite surface, showing a Type A complex (Cu(II) forms bridge between humic substance and the goethite surface) and a Type B complex (humic substance forms a bridge between Cu(II) and the goethite surface). Also shown is an inner-sphere bidentate Cu(II) complex on the goethite surface. XAFS data consistent with this model were collected on NSLS beam line X-11A. [Reprinted from Alcacio et al. (2001), Fig. 1, with permission from Elsevier Science]

pH range 5 to 7 due to the reduction of the positive surface charge of goethite at these pH values and the formation of a stable ternary Pb(II)/SO$_4^{2-}$/goethite surface complex (Ostergren et al. 2000a). In contrast, the presence of chloride ligands inhibits the sorption of Hg(II) on goethite at similar pH values because of the formation of a stable Hg(II)-chloride aqueous complex (Kim et al. 2002a). Another example is the effect of carbonate ligands on U(VI) sorption on minerals such as kaolinite and goethite. Above pH 7, where the stable triscarbonato solution complex of U(VI) becomes the dominant U(VI) solution species, U(VI) sorption on these surfaces is inhibited because of the stability of the solution complex (Redden et al. 1998; Thompson et al. 1998). It is interesting to note that recent XAFS and electrophoretic mobility studies of the sorption of U(VI) on powdered hematite in the presence of ambient air showed that stable ternary U(VI) carbonato complexes formed on the surface of hematite (Bargar et al. 1999a, 2000a). These results indicate that the effects of such inorganic ligands must be evaluated for specific sets of conditions and specific sorbate/sorbent systems in order to evaluate their impact on metal ion sorption reactions. XAFS spectroscopy, coupled with attenuated total reflectance (ATR)-FTIR measurements, is useful in revealing the presence of ternary surface complexes in sorption systems containing metal ions and complexing ligands.

    Organic ligands can have effects similar to inorganic ligands on aqueous metal ion sorption onto mineral surfaces. For example, bipyridine enhances the sorption of Cu(II) on γ-alumina, but it inhibits the sorption of Cu(II) on amorphous silica under certain conditions (Bourg et al. 1979; Cheah et al. 1997). In addition, citrate (Redden et al. 2001) and malonic acid (Lenhart et al. 2001) can enhance the sorption of U(VI) and Pb(II), respectively, onto goethite through the formation of ternary surface complexes, as revealed by EXAFS and FTIR spectroscopy. Oxalate and other organic ligands are known to enhance the dissolution of metal oxides (Stone and Morgan 1987; Stumm and Furrer 1987); however, the exact reaction mechanisms are not known. X-ray reflectivity studies by Teng et al. (2001) have provided important details about feldspar dissolution as a function of pH in the presence and absence of oxalate. Further mechanistic information on the effect of organic ligands on mineral dissolution and growth was provided by the X-ray reflectivity study of the adsorption of the growth inhibitor HEDP on barite (Fenter et al. 2001a). Future studies of mineral dissolution and growth reactions using soft X-ray XAFS spectroscopy at the C K-edge in the presence of water vapor may provide some of this information as well (see Myneni, this volume).

Natural organic matter has long been thought to form coatings on mineral surfaces that can result in significant changes in mineral surface charge and reactivity to metal ions in solution (e.g., Neihof and Loeb 1972, 1974; Hunter and Liss 1979; Hunter 1980; Balistrieri et al. 1981; Davis and Gloor 1981; Tipping 1981; Tipping and Cooke 1982; Davis 1984; Mayer 1994a,b; Au et al. 1999). More recent evidence, however, suggests that such coatings comprise only a small fraction (< 15%) of an effective monolayer on mineral particles in ocean and estuarine waters (Ransom et al. 1997; Mayer 1999) and in soils (Mayer and Xing 2001). Nonetheless, organic coatings when present in discrete regions on mineral surfaces may change the surface charge and reactivity of mineral surfaces and may affect sorption reactions of aqueous metal ions. X-ray standing wave, X-ray reflectivity, and X-ray spectromicroscopy methods can be used to investigate these phenomena. For example, Fenter and Sturchio (1999) demonstrated how X-ray reflectivity could be used to provide a detailed characterization of the growth and structure of stearate monolayers on calcite.

Biota can have a major effect on the speciation and sorption of contaminant and pollutant species

Microbial organisms and fungi such as lichens can enhance dissolution of mineral surfaces (Banfield and Hamers 1997; Banfield et al. 1999), thus changing the types and densities of surface functional groups, which in turn can affect the reactivity of the surface. They can also generate nanoparticulate precipitates (Banfield and Zhang 2001). Important questions are whether microbial biofilm coatings on mineral surfaces block reactive sites or provide alternative functional groups that can bind metal ions. X-ray standing wave and XAFS methods can uniquely address these questions (e.g., Templeton et al. 1999, 2001; Hennig et al. 2001; Kelly et al. 2001, 2002; Webb et al. 2001; Panak et al. 2002).

Molecular-scale mechanisms of bioremediation and phytoremediation of environmental toxins are typically not known

Many microorganisms and plants are capable of transforming toxic chemical species into less toxic forms (e.g., Lytle et al. 1996, 1998; Hunter et al. 1997). Some plants are particularly useful for remediation of contaminants in soils and natural waters because they hyperaccumulate specific toxins (e.g., Van der Lelie et al. 2001; Fuhrmann et al. 2002). In most cases, however, the molecular-scale mechanisms of these transformations or of hyperaccumulation are not known. This is a fertile research field for geochemists and mineralogists that will require multidisciplinary studies and will benefit from SR-based microspectroscopy methods.

Scaling of molecular-level observations to mesoscopic and macroscopic phenomena: the "Holy Grail" of low temperature geochemistry and environmental science

It would be extremely useful if laboratory measurements at a variety of spatial scales, including the nanometer scale, could be accurately extrapolated to the field scale so that the consequences of a field-scale pollution event could be assessed and predicted in quantitative detail as a function of time and location. Conversely, it would be useful if field-scale geochemical data could be conveniently interpreted in terms of microscopic processes. Neither of these options is usually possible, however, because of missing information at intermediate spatial scales or lack of knowledge about the microscopic forms and processes affecting the stability of the pollutant. For example, pollutant transport is controlled by many variables, including the advective flow of polluted groundwater, which in turn is controlled in part by subsurface hydraulic gradients and the connectivity of fractures and pore spaces in rocks and soils (see Fig. 6). A few of the variables and processes that affect pollutant transport are (1) the types and distribution of solid phases and surface coatings present in the subsurface and how they react with (or sorb) dissolved pollutant species, (2) the transformation of redox-sensitive pollutant species into more (or less) reactive and mobile forms through reactions with abiotic and biotic reductants and/or oxidants in the subsurface, (3) the stability of colloidal particles and how they sorb pollutant ions, and (4) the dissolution of mineral phases and the precipitation of new phases that could enhance or impede groundwater flow and/or release or sequester pollutant ions (Fig. 2). While one can study these variables and processes in the laboratory, the extent to which they occur in the subsurface is not typically known. Even after field sampling, flow-through column experiments designed to simulate reactive transport for a given geologic or soil column, and detailed characterization studies of the reaction products, it is impossible to construct a sufficiently detailed three-dimensional model of the subsurface that accounts for all of the geology, hydrology, geochemistry, and microbiology. Perhaps a more attainable "Holy Grail" is the use of molecular-scale information obtained from experimental and analytical studies of natural systems and relevant model systems to identify the important reactants and products, their spatial distribution and phase association, and the most likely reaction pathways and kinetics for different pollutant species.

An important consideration in any study that uses nanoscale analytical/ characterization methods such as SR-microspectroscopy to define these variables is the statistical significance of the result relative to the field scale. Care must be taken to ensure that the cumulative sampling volumes probed using such techniques are sufficient to be representative of the field scale, particularly for highly heterogeneous systems. Moreover, it is also true that field scale geochemical sampling is often too limited to be representative of the average distribution of a given pollutant or phase.

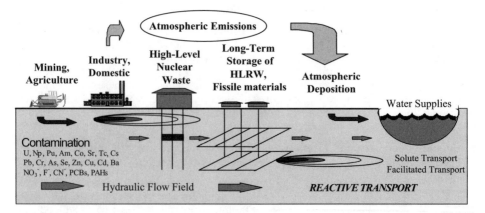

Figure 6. Field-scale schematic view of sources and transport of environmental contaminants and pollutants.

AN OVERVIEW OF SYNCHROTRON RADIATION METHODS USEFUL IN LOW TEMPERATURE GEOCHEMISTRY AND ENVIRONMENTAL SCIENCE

The following SR methods are useful in characterization of environmental and low temperature geochemical samples. More thorough reviews of some of these SR methods are given in the other chapters in this RiMG volume.

X-ray absorption fine structure (XAFS) spectroscopy

The most widely used SR method in low temperature geochemistry and environmental science is XAFS spectroscopy. Early applications of XAFS spectroscopy to these fields as well as to mineralogy are reviewed by Brown et al. (1988). XAFS spectroscopy measures the absorption of X-rays by a selected element in a sample at and above that element's characteristic absorption edge energy (see Fig. 1). Such measurements are typically feasible only with X-rays from a synchrotron radiation source, particularly if the element of interest is at low concentration (< 1 wt.%). Absorption edges correspond to excitation of deep core electrons to valence or continuum levels of an atom and occur at unique energies for each element. The X-ray absorption near edge structure (XANES) spectrum is sensitive to the valence state of the atom and the geometry and types of surrounding atoms (Fig. 1), whereas the extended X-ray absorption fine structure (EXAFS) spectrum provides quantitative information on the distances between absorbing atoms and surrounding atoms (out to a radial distance of about 5 Å), the atomic number of surrounding atoms, and their relative disorder (both positional and thermal). Because XAFS spectroscopy is a local probe, it does not require the long-range order of a crystalline sample and is one of the few methods that can provide quantitative structural information about the local environments around cations and anions in amorphous materials, such as silicate glasses (e.g., Calas et al, 1987, Farges et al. 1991, 1992) and silicate melts (e.g., Brown et al. 1995a; Farges and Brown 1996; Farges et al. 1996a), or in aqueous solutions (e.g., Fontaine et al. 1978; Lagarde et al. 1980; Farges et al. 1993a), hydrothermal/supercritical fluids (e.g., Anderson et al. 1995; Mosselmans et al. 1996; Seward et al. 1999; Hoffmann et al. 1999; Mayanovic et al. 2002), or complex environmental samples (see O'Day 1999 and Manceau et al. in this volume for reviews). X-ray or neutron scattering can provide this information in cases where the pair correlations of interest are separable from all other pair correlations contributing to a radial distribution function. SR-based differential anomalous scattering (see discussion below) can probe the medium-range structure around selected elements in amorphous and poorly crystalline materials by combining the element selectivity of XAFS spectroscopy with the high k-range of X-ray scattering. Also, because XAFS spectroscopy is element specific, it can be quite sensitive to the surface of a sample in which the element of interest occurs only at or near the surface. This feature allows one to determine the geometry, composition, and mode of attachment of a range of adsorbate ions at mineral/water interfaces (Hayes et al. 1987; Brown, 1990; Manceau et al. 1992a; Charlet and Manceau 1993; Brown et al. 1995b, 1999c).

What types of samples can be investigated using XAFS spectroscopy? XAFS experiments can be done on different types of samples (powdered vs. single crystals, glasses, liquids, gases, combinations of solids and aqueous solutions, solids and gases, etc.) under a variety of conditions (ambient T and P, low T, high T, high P, high P and T, in the presence of water and gases), which makes the method particularly useful for studying environmental and low temperature geochemical processes under appropriate *in situ* conditions. XAFS data for low element concentration samples can be collected on insertion device beamlines and analyzed for elements with atomic number Z > 19 (K) at concentrations as low as the low ppm range, assuming a favorable matrix that does not reduce S/N because of fluorescence interference from other elements. XAFS spectra of elements with $Z \geq 4$ (Be) can also be collected, although low-Z elements ($Z = 4$ (Be) – 19 (K)), which have K-edge energies of 111.5 to 3,608 eV, present complications relative to high-Z elements ($Z \geq 20$) due to the need for a UHV sample chamber, vacuum-compatible detectors, and special UHV beamlines for elements with $Z \leq 14$. For elements with 15 (P) $\leq Z \leq$ 19 (K), He-filled beam paths and He-filled sample chambers are typically required. These sample chamber and beam line constraints, coupled with the lower fluorescence yield from low-Z elements, means that their concentrations must be higher than high-Z elements in order to make XAFS data collection feasible. The K-edge energies of Ca to Cd (4,038 to 26,711 eV) and the L-edge energies of In to U (3,730 to 17,166 eV) are sufficiently high such that no special sample chambers or beam paths are typically needed for XAFS studies of these elements, unless a controlled reaction environment is desired, and standard hard X-ray beamlines can be used.

How fast can XAFS data be collected? XAFS spectra can be collected rapidly enough to allow some chemical reactions to be followed in real time. Using conventional data collection methods on a bending magnet beamline, for example, one can collect a XANES spectrum in about one to ten minutes, depending on the energy resolution required and the element concentration, and a single EXAFS spectrum (to $k_{max} = 12 \text{ Å}^{-1}$) in 10-20 minutes, assuming that the concentration of the element of interest is relatively high (> 1 wt %). The time scale for XANES spectra can be reduced to less than a minute using insertion-device beamlines such as a high-flux wiggler magnet line or a high-brightness undulator magnet line, so reactions slower than a minute or two can be studied by conventional XANES spectroscopy. Extending EXAFS data collection to higher k-values, which is needed to clearly resolve second and more distant neighbors in many cases, requires additional data acquisition time. In addition, for samples with dilute concentration levels of an element (< 1000 ppm), averaging of multiple EXAFS scans is necessary to achieve a signal-to-noise level required for extracting accurate interatomic distances and coordination numbers.

Two studies illustrate the utility of XAFS spectroscopy measurements in following chemical reaction pathways in real time. One involved the reduction of Se(VI) on green rust surfaces (Myneni et al. 1997a) (Fig. 7) and showed that this abiotic pathway for Se(VI) reduction is as rapid as the most rapid biotic pathways. The other (Brenner et al. 2002) followed the reductive dissolution and biomineralization of iron hydroxide under dynamic flow conditions. More rapid reaction dynamics (on the order of μseconds), including phase transitions, can be studied in certain cases using Quick EXAFS or QEXAFS (e.g., Frahm et al. 1993; Wong et al. 1995; Mansour et al. 1998; Bornebusch et al. 1999; Sole et al. 1999), in which the crystal monochromator moves continuously rather than in the normal step scanning fashion. Fast reactions can also be studied using dispersive EXAFS (e.g., Flank et al. 1982; Itié et al. 1989; Fontaine et al. 1992; Allen et al. 1993, 1995a; Pascarelli et al. 1999) in which a white beam of SR hits a bent crystal monochromator, which causes the beam to be focused onto a sample, then is dispersed onto a linear position-sensitive detector, resulting in a "single shot" XAFS spectrum. These more rapid measurements are limited, however, to elements at relatively high concentrations (typically > 1 wt.%) due to the lower signal-to-noise in rapidly acquired XAFS spectra for low element concentrations (fluorescence yield detection for dispersive EXAFS is under development and will allow measurements on dilute samples). The term "quick X-ray absorption spectroscopy" has also been used by Gaillard et al. (2001) to represent a fast EXAFS data analysis procedure, not fast data collection.

XANES spectroelectrochemistry. A technique known as XANES spectro-electrochemistry has been developed by Antonio et al. (1997) and Soderholm et al.

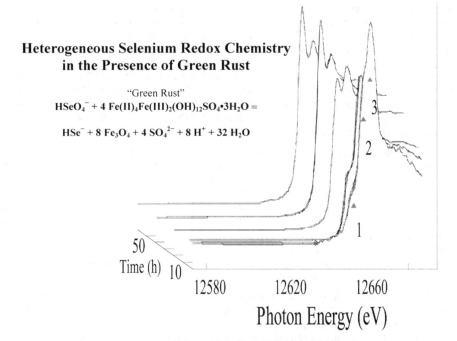

Heterogeneous Selenium Redox Chemistry in the Presence of Green Rust

"Green Rust"

$HSeO_4^- + 4\ Fe(II)_4Fe(III)_2(OH)_{12}SO_4 \cdot 3H_2O =$

$HSe^- + 8\ Fe_3O_4 + 4\ SO_4^{2-} + 8\ H^+ + 32\ H_2O$

3

2

1

50

Time (h) 10

12580   12620   12660

Photon Energy (eV)

Figure 7. Se K-edge XANES spectra as a function of reaction time of the reaction products of aqueous selenate ($HSeO_4^-$) with green rust at pH 7.0. The numbers 1, 2, and 3 next to the vertical arrows indicate Se(0), Se(IV), and Se(VI), respectively. With increasing time, Se(0) becomes predominant, with some Se(IV) also present. Data were taken on SSRL beam line 6-2. (after Myneni et al. 1997)

(1999) and makes use of a special electrochemical sample cell, which allows XANES and EXAFS spectroscopy measurements to be made on redox-sensitive metal ion species in aqueous solutions, such as Pu, as a function of Eh.

Detection methods for XAFS spectroscopy. XAFS experiments can be done with different types of detection schemes, including (1) transmission (for concentrated samples) in which the X-ray beam passes through the sample and the flux of the transmitted beam is typically measured using a gas-filled ion chamber, (2) fluorescence X-ray yield (for dilute samples) in which the fluorescent X-rays resulting from core electron excitations are measured using a gas-filled ion chamber (Lytle et al. 1984), a solid-state multi-element detector (Cramer and Scott 1981; Waychunas and Brown 1994), or a pin-diode detector (Gauthier et al., 1999), (3) electron yield (for enhanced surface sensitivity) in which the ejected electrons are measured using a positively biased metal grid or foil, or (4) X-ray excited optical luminescence (XEOL) in which a photodiode array is used for detection of luminescence (Soderholm et al. 1998).

Geometry of XAFS experiments. XAFS spectroscopy can also be done using different types of geometries, including transmission geometry, in which the incident X-ray beam strikes the sample at 90°, or fluorescence-yield geometry, in which the beam strikes the sample at ≈ 45° in order to maximize the solid angle at which the fluorescent X-ray beam intersects the detector. Self-absorption effects can be minimized in fluorescence-yield measurements by arranging the sample at a right angle to the incident X-ray beam (Tröger et al. 1992) and can be corrected for using the procedure outlined by Pfalzer et al. (1999). It is also possible to employ grazing-incidence geometry in which the incident X-ray beam strikes a flat single crystal sample at or below the critical angle of the sample, resulting in total external reflection of the X-ray beam, which greatly enhances the surface sensitivity of the method (Heald et al. 1988). A number of grazing-incidence or "glancing-angle" XAFS studies of geochemical and environmental samples have been carried out over the past few years (e.g., Towle et al. 1995a, 1999a,b; Bargar et al. 1996, 1997c; Farquhar et al. 1996, 1997; Pattrick et al. 1998, 1999; England et al. 1999a,b; Fitts et al. 1999a; Grolimund et al. 1999; Trainor et al. 1999, 2002a; Waychunas et al. 1999). The term "REFLEXAFS" is also used by some workers. Grazing-incidence XAFS studies of earth and environmental materials are discussed in detail by Waychunas (this volume).

The highly polarized nature of a synchrotron X-ray beam in the plane of the storage ring can be exploited to determine the orientation of bonds for a molecule on a mineral surface or the orientation of polyhedra at the edges of mineral grains with sheet structures (polarized XAFS spectroscopy) (e.g., Manceau et al. 1988b, 1990a, 1998, 1999a, this volume; Waychunas and Brown 1990; Hazemann et al. 1992; Li et al. 1995c; Hudson et al. 1996; Schlegel et al. 1998, 1999a,b, 2001a; Dähn et al. 2001, 2002a). X-ray absorption is enhanced when the electric vector of the X-ray beam is parallel to specific bonds between atoms and is weak when the e-vector is perpendicular to bonds. Materials with anisotropic structures, such as micas and clays, are more sensitive to polarized XAFS measurements than isotropic structures.

Soft X-ray/VUV XAFS measurements. In addition to intermediate to hard X-ray energies (> 5 keV), which the majority of XAFS studies to date have used, XAFS spectroscopy can be done at the K-edges of low atomic number elements such as C, N, O, F, Na, Mg, Al, Si, P, and S (284-2472 eV) or the L-edges of first-row transition elements ($\approx$ 450-1200 eV) using VUV or "soft" X-ray energies (see Myneni, this volume). Recent O K-edge EXAFS studies of water have shown that EXAFS is sensitive to H backscatterers (Wilson et al. 2000) as will be discussed later in the section on water. Such spectra must be collected using a UHV beam line and a UHV sample chamber because of the strong attenuation of low energy photons by air or gases such as He. This type of experiment adds an additional level of complication relative to hard X-ray XAFS experiments because of the UHV sample chamber and the limits it places on sample handling, manipulation, and *in situ* treatments, such as heating or reaction with aqueous solutions. However, use of differentially pumped UHV sample chambers and of photon in-photon out methods (e.g., Ogletree et al. 2000) allows one to collect XANES spectra at low X-ray energies on wet samples. For example, carbon K-edge μXANES spectra on fulvic acids in aqueous solution using the ALS transmission X-ray microscope beamline 8.0 has yielded unique information on the different types of C-containing functional groups as a function of spatial location (Myneni et al. 1998). In addition, the use of a UHV preparation chamber attached to a UHV analysis chamber allows samples to be transferred to the prep chamber after initial characterization, reacted with water vapor, other gases, or cations in aqueous solutions, returned to the analysis chamber, and characterized using UHV surface science methods, including XAFS spectroscopy, after reaction (e.g., Kendelewicz et al. 2000a,b). The acronyms SEXAFS, which stands for *surface* EXAFS, and NEXAFS, which stands for near edge X-ray absorption fine structure, are often used to describe these types of low energy XAFS spectra. These techniques are particularly surface sensitive when measured in partial electron or Auger electron yield, because of the very low escape depths of these electrons; they are more bulk sensitive when measured in total electron yield.

Examples of bulk-sensitive NEXAFS spectroscopy studies include investigations of the coordination environments of oxygen in metal oxide minerals (Brown et al. 1986), Al in aluminosilicate minerals and glasses and aluminum-(oxyhydr)oxide minerals (e.g., Brown et al. 1983; McKeown et al. 1985a; Ildefonse et al. 1994, 1995, 1998; Li et al. 1995b,f; Chester et al. 1999; Doyle et al. 1999a; Bugaev et al. 2000a), Si and Al in diatom frustules (Ghelen et al. 2002), Si and P in amorphous and poorly crystalline silicates (Di Cicco et al. 1990; Lagarde et al. 1993; Ildefonse et al. 1995; Li et al. 1993, 1994a,d, 1995a,e, 1996; Fleet et al. 1997), and Na in silicate materials (e.g., Greaves et al. 1981; McKeown et al. 1985b; Mazara et al. 2000). In addition, bulk-sensitive L-edge XAFS measurements have been made on first-row transition elements (Ti, Cr, V, Mn, Fe) in minerals of interest in low temperature geochemistry and environmental science by Schofield et al. (1995), Charnock et al. (1996), Kendelewicz (2000a), and Brown et al. (2001a). Soft-X-ray XAFS spectroscopy has also been used extensively to study the local coordination environment of sulfur in a variety of earth and environmental materials (e.g., Sainctavit et al. 1986; Kasrai et al. 1988; Huffman et al. 1989b,c, 1990, 1991, 1993, 1995; Huggins et al. 1991; Taghiei et al. 1992; Vairavamurthy et al. 1993; Eglinton et al. 1994; Li et al. 1994b,c,d, 1995d; Zhu et al. 1995; Kasrai et al. 1996; Morra et al. 1997; Vairavamurthy 1998; Grossman et al. 1999; Sarret et al. 1999b,c; Myneni 2000, and this volume; Myneni et al. 2000; Fararell et al. 2001; Olivella et al. 2002). The speciation of nitrogen has been investigated in coals and asphaltenes by Mitra-Kirtley et al. (1993a,b) and Mullins et al. (1993) and in "geomacromolecules" by Vairavamurthy and Wang (2002) using N K-edge XANES. Surface-sensitive SEXAFS and NEXAFS methods have been used extensively to study the speciation of hydrocarbons chemisorbed on metal surfaces (Stöhr 1988, 1992; Stohr and Anders 2000) and in polymer films (e.g., Kikuma and Tonner 1996; Smith et al. 2001).

Accuracy of XAFS-derived structural parameters. Since its introduction as a SR method in the 1970's (Sayers et al. 1971, 1972, 1975; Lytle et al. 1975; Stern et al. 1975), the accuracy of structural parameters derived from XAFS spectroscopy has increased significantly. This is due primarily to the development of curved-wave scattering theory that can be used to calculate the back-scattering amplitude and phase-shift functions for essentially any atomic environment required to fit EXAFS data to a structural model, as well as multiple-scattering contributions to XAFS spectra (e.g., Natoli and Benfatto, 1986; Rehr et al. 1991; Zabinsky et al. 1995; Ankudinov et al. 1998; Rehr and Albers 2000; Rehr and Ankudinov 2001a,b). Various computer codes are available to calculate EXAFS spectra for a specific arrangement of atoms around an absorbing ion, including GNXAS (Natoli and Benfatto 1986; Filiponi et al. 1991; Westri et al. 1995), EXCURV92 (Lee and Pendry 1975; Gurman et al. 1984; Binsted 1998), and FEFF (Rehr and Albers 1990; Rehr et al. 1991, 1992; Zabinsky et al. 1995; Ankudinov et al. 1998). The computer code FEFF (*http://leonardo.phys.washington.edu/feff/*), in particular, has helped revolutionize the application of XAFS spectroscopy to complex materials because it is user friendly and effectively eliminates the need for XAFS spectra from well-characterized model compounds with known composition and structures which have structural environments of the element of interest that are very similar to those in the unknown material. In the early days of XAFS spectroscopy, back-scattering amplitude and phase-shift functions had to be extracted from the fits of EXAFS spectra of carefully selected model compounds, then used in fitting the EXAFS spectra of an unknown compound to a structural model. This is still a reliable and widely used method in XAFS analysis, and web-based libraries of EXAFS spectra of standard model compounds are now being established for general use (e.g., Newville et al. 1999b).

Use of appropriate data collection, data analysis, and error analysis protocols, which are described in the Standards and Criteria Reports of the International XAFS Society (*http://ixs.iit.edu/*), can result in accuracies of EXAFS-derived first-neighbor and second-neighbor interatomic distances of ±0.01 Å and ±0.03 Å, respectively, or better, and accuracies of first- and second-neighbor coordination numbers of ±0.5 atoms and ±1 atoms, respectively, or better, in complex structures (e.g., Teo 1986; O'Day et al. 1994a) (see also Newville et al. 1999). Although these accuracies are not as good as modern single crystal X-ray diffraction structural determinations and refinements, which can, in practice, be as low as ±0.005 Å or better for first-neighbor interatomic distances, XAFS spectroscopy can provide accuracies similar to those for crystalline materials for a wide range of element environments, including cation adsorption complexes on mineral surfaces, in silicate glasses and melts, and in aqueous or hydrothermal solutions. Conventional X-ray scattering experiments on these types of materials are not element specific and thus cannot easily determine interatomic distances for cations like Na or K in silicate glasses, for example.

New XAFS data analysis protocols, including wavelet analysis, principle component analysis, and explicit treatment of multi-electronic excitations, multiple scattering interferences, and anharmonic effects are improving the accuracy of EXAFS analysis of environmental, biological, and geochemical samples (Wasserman 1997; Wasserman et al. 1999; Ressler et al. 2000; Beauchemin et al. 2002; Farges 2002; Farges et al. 2002; Frenkel et al. 2002; Isaure et al. 2002; Munoz et al. 2002).

Theoretical prediction of XANES spectra. Curved-wave multiple scattering theory can also be used to predict XANES spectra of specific elements that match experimental spectra reasonably well (e.g., Cabaret et al. 1996a,b,c, 1998, 1999, 2001; Farges et al. 1997, 2001; Ildefonse et al. 1998; Doyle et al. 1999a; Levelut et al. 2001). This relatively recent development is extending the usefulness of XANES spectroscopy from a method for determining the valence states and coordination numbers of cations and anions in complex earth and environmental materials (e.g., Waychunas et al. 1983; Galoisy et al. 2001; Wilke et al. 2001) to one that can provide unique information on the average medium-range structure (out to a radial distance of ≈ 5Å) around a cation or anion in a poorly crystalline or amorphous material (e.g., Farges et al. 1996a,b; Cabaret et al. 2001). XANES spectroscopy, which is dominated by multiple scattering in crystalline materials, is inherently more sensitive to medium-range order around an element relative to EXAFS spectroscopy because of the much longer mean free path of photoelectrons in the XANES energy region (up to ≈ 100 eV above an absorption edge). In more locally disordered materials, such as silicate glasses and melts, however, single scattering from first neighbors dominates the XANES spectrum, making it possible to determine interatomic distances from Fourier transformation of XANES spectra (Bugaev et al. 2000b, 2002).

Web-based guides to EXAFS data analysis and data analysis programs. A guide to experimental XAFS procedures can be found at the following website: *http://www-ssrl.slac.stanford.edu/mes/XAFS/INDEX.HTML*. Another can be found at the GSECARS website. A number of excellent computer codes are available for analyzing XAFS data at the following website: *http://cars9.uchicago.edu/IXS-cgi/XAFS_Programs*. The first author and his research group have used the following XAFS data analysis software in a variety of applications and recommend these data analysis packages:

EXAFSPAK: George and Pickering (1992) (*http://ssrl.slac.stanford.edu/exafspak.html*)

FEFFIT: Newville et al. (1995) (*http://cars9.uchicago.edu/ifeffit/index.html*)

WinXAS: Ressler (1997): (*http://ixs.csrri.iit.edu/catalog/XAFS_Programs/winxas*)

Examples of low temperature geochemical and environmental science applications of XAFS spectroscopy. Recent applications of XAFS spectroscopy to environmental and low temperature geochemical samples are reviewed by Brown et al. (1999c), Fendorf (1999), O'Day (1999), Brown and Parks (2001), Manceau et al. (this volume), Myneni (this volume), and Waychunas (this volume). Examples of XAFS studies on materials of relevance to low temperature geochemistry and environmental science include the following:

(1) sorption of aqueous metal ions on high surface area minerals (Hayes et al. 1987; see Table 1 in the Appendix for many additional references)

(2) interaction of aqueous metal ions with natural organic matter and organic acids (Allen et al. 1996a; Hesterberg et al. 1997b, 2001; Sarret et al. 1997; Xia et al. 1997a,b, 1999; Denecke et al. 1998a,b, 2002; Rossberg et al. 2000; Alcacio et al. 2001; Liu et al. 2001; Pompe et al. 2001; Schmeide et al. 2001) (see Table 2 in the Appendix)

(3) interaction of aqueous metal ions with microbial and fungal cell walls (Sarret et al. 1998a, 1999a, 2001; Templeton et al. 1999, 2002a; Soderholm et al. 2000; Hennig et al. 2001; Kelly et al. 2001, 2002; Webb et al. 2001; Panak et al. 2002)

(4) determination of the structures and composition of metal ion complexes in aqueous solutions (Fontaine et al. 1978; Apted et al. 1985; Mosselmans et al. 1999; Frenkel et al. 2001; Sandstrom et al. 2001), hydrothermal solutions (Fulton et al. 1996; Seward et al. 1996; Mayanovic et al. 1999), and natural fluid inclusions (Anderson et al. 1995, 1998a; Mayanovic et al. 1996, 1997) (see section on water and metal ions in aqueous solutions for additional examples)

(5) structural studies of iron oxides and hydroxides (Waychunas et al. 1983, 1986; Combes et al. 1986 1989a,b, 1990; Manceau and Combes 1988; Manceau et al. 1990b; Hazemann et al. 1992a,b; Manceau and Drits 1993; Wilke et al. 2001)

(6) structural studies of manganese oxides and hydroxides (Manceau and Combes 1988; Manceau et al. 1992a,b,c; Drits et al. 1997a; Manceau and Gates 1997; Silvester et al. 1997; Drits et al. 1998; Lanson et al. 2000)

(7) characterization of Mn oxidation reaction products at sediment/water interfaces (Wehrli et al. 1995)

(8) selenium transformations in soils and sediments (Tokunaga et al. 1994; Pickering et al. 1995; Tokunaga et al. 1996, 1997)

(9) characterization of "neoformed" phyllosilicates (Dähn et al. 2002a)

(10) XAFS studies of the crystal chemistry of kaolinites and low temperature iron and manganese hydroxides (Muller et al. 1995)

(11) XAFS studies of the crystal chemistry of iron silicate scale deposits in the Salton Sea geothermal field (Manceau et al. 1995)

(12) XAFS studies of trace element environments in natural and synthetic goethites (Manceau et al. 2000b)

(13) nucleation and growth of iron oxides in the presence of phosphate (Rose et al. 1996)

(14) formation of ferric hydroxide-nitrate oligomers in aqueous solutions (Rose et al. 1997)

(15) mechanism of formation of iron oxyhydroxide-chloride polymers in solution (Bottero et al. 1994)

(16) oxidation-reduction mechanisms for iron in smectites (Drits and Manceau 2000; Manceau et al. 2000c,d; Gates et al. 2002)

(17) mechanism of Cr(III) and As(III) oxidation by hydrous manganese oxides (Fendorf 1995; Silvester et al. 1995; Tournassat et al. 2002)

(18) cation distributions in phyllosilicates (Decarrau et al. 1987; Manceau and Calas 1986,1987; Manceau 1990; Manceau et al. 1990a; Drits et al. 1997b; Muller et al. 1997; Manceau et al. 1998; Gates et al. 2002)

(19) XAFS studies of actinides in minerals, glasses, melts, contaminated soils, aqueous solutions, cementitious materials, and humic materials (e.g., Petit-Maire et al. 1986; Chisholm-Brause et al. 1992a,b; Dent et al. 1992; Farges et al. 1992, 1993b; Allen et al. 1994, 1996a,b; Chisholm-Brause et al. 1994; Thompson et al. 1995, 1997; Morris et al. 1996; Allen et al. 1997a,b; Biwer et al. 1997; Clark et al. 1997; Duff et al. 1997; Farges and Brown 1997; Giaquinta et al. 1997a,b; Conradson et al. 1998; Denecke et al. 1998a,b; Clark et al. 1999; Moyes et al. 2000; Rossberg et al. 2000; Sylwester et al. 2000b,c; Antonio et al. 2001; Duff 2001; Duff et al. 2001; Williams et al. 2001; Brown et al. 2002; Denecke et al. 2002; Farges et al. 2002; Pompe et al. 2002; Schmeide et al. 2002)

(20) technetium and uranium speciation in cement waste forms (Shuh et al. 1994; Allen et al. 1997c; Sylwester et al. 2000a)

(21) pertechnetate radiolysis products in highly alkaline solutions (Lukens et al. 2001, 2002)

(22) XAFS studies of the cycling of Mn in eutrophic lakes (Friedl et al. 1997a,b)

(23) removal of selenocyanate from water (Manceau and Gallup 1997)

(24) XAFS studies of metal sulfides and sulfosalts (Sainctavit et al. 1986; Li et al. 1994b,c,d; Mosselmans et al. 1995a,b; Pattrick et al. 1997)

(25) XAFS characterization of trace elements in apatites (Sery et al. 1996; Gaillard et al. 2002; Rakovan et al. 2002)

(26) XAFS characterization of trace elements in carbonates and corals (Pingitore et al. 1992; Reeder et al. 1994; Lamble et al. 1995, 1997; Greegor et al. 1997; Parkman et al. 1998; Sturchio et al. 1998; Reeder et al. 1999, 2000, 2001; Pingitore et al. 2002a,b)

(27) XAFS characterization of the hydration layer of obsidian (Lytle et al. 2002)

(28) XAFS characterization of trace elements in coals, kerogens, and oil fly ash (Silk et al. 1989; Huffman et al. 1989a,b,c, 1990, 1991, 1992, 1993, 1994, 1995, 2000; Huggins and Huffman 1991, 1995, 1996, 1999; Huggins et al. 1988a,b, 1991, 1993, 1995, 1996, 1997, 1998, 1999, 2000a,b,c; Taghiei et al. 1992; Zhu et al. 1995; Wasserman et al. 1996; Galbreath et al. 1998, 2000; Kolker et al. 2000; Goodarzi and Huggins 2001; Sarret et al. 2002)

(29) microbial desulfurization of crude oil (Grossman et al. 1999)

(30) formation of sphalerite in natural sulfate-reducing biofilms (Labrenz et al. 2000)

(31) reduction of Cr(VI) to Cr(III) by magnetite (Peterson et al. 1996, 1997a,b; Kendelewicz et al. 2000a; Brown et al. 2001a), aquatic and wetland plants (Hunter et al. 1997; Lytle et al. 1998), thiols in organosulfur and humic substances (Taylor et al. 2000; Szulczewski et al. 2001), and dissimilatory iron-reducing bacteria (Wielinga et al. 2001)

(32) accumulation and transformation of heavy metals and metalloids in plants (Salt et al. 1995, 1997; Lytle et al. 1996; Hunter et al. 1997; Pickering et al. 2000a,b)

(33) arsenic mobilization by dissimilatory iron-reducing bacteria (Cummings et al. 1999)

(34) mechanism of lichen resistance to metal pollution (Sarret et al. 1998b)

(35) Mn oxidation by marine bacteria (Bargar et al. 2000b)

(36) XAFS studies of sulfur in metal sulfide minerals (Sainctavit et al. 1986; Kasrai et al. 1988; Li et al. 1994b,c,d, 1995d; Farrell and Fleet 2001), natural organic matter (Kasrai et al. 1996; Morra et al. 1997; Xia et al. 1998; Sarret et al. 1999b,c; Beauchemin et al. 2002), and aqueous solutions (Myneni 2000; Myneni et al. 2000)

(37) mechanisms of trace element removal from seawater (Helz et al. 1996; Brown and Parks 2001)

(38) XAFS studies of the effectiveness of reactive barriers in removing U(VI) from contaminated groundwater (Fuller et al. 2002a,b) and sorption of U(VI) on soil particles (Bostick et al. 2002)

(39) molecular-level speciation of chlorine in decaying plant litter (Myneni 2002)

(40) studies of metal-sulfur association in metalliferous peats (Martinez et al. 2002)

This partial list of XAFS-based studies illustrates the range of topics impacted by this versatile SR method. Many other XAFS-based studies relevant to low temperature geochemical and environmental samples and processes are cited throughout this chapter and in other chapters in this volume (Manceau et al., Myneni, Sutton et al., and Waychunas).

SR-based methods related to XAFS spectroscopy. In addition to XAFS methods, there is also the possibility of measuring differences in the real and imaginary portions of anomalous scattering of an element above and below its X-ray absorption edge in X-ray scattering experiments referred to as differential anomalous scattering (DAS). The data sets are normalized and subtracted yielding scattering data with greatly enhanced contributions from the chosen element. Fourier transformation of these data results in a radial distribution function that includes only pair correlations involving the specific atom. This information is similar to that from EXAFS but differs in several distinct ways. Because DAS data are not limited in the low k-range as are EXAFS data, DAS pair correlation information extends to much larger distances than that from EXAFS and is not adversely affected by large degrees of static disorder (e.g., Kortright et al. 1983; Ludwig et al. 1986). A variant of DAS is referred to as differential anomalous fine structure (DAFS) (Stragier et al. 1992; Sorensen et al. 1994). The longer range distance information from DAS or DAFS is necessary to probe intermediate-range (5-10 Å) structure in solids and liquids, which is not accurately obtained from EXAFS and is not easily retrievable from conventional X-ray scattering methods. Although the DAFS method has not been used much on materials relevant to environmental science and low temperature geochemistry, there have been several DAFS studies that determined the distribution of iron in the magnetite structure by conducting XAFS-like energy scans for particular X-ray reflections that are sensitive to cations in particular sites in a crystalline solid (Kobayashi et al. 1998; Frenkel et al. 1999).

X-ray microprobe, spectromicroscopy, and microtomography methods

When a focused beam of X-rays is used to probe environmental and low temperature geochemical samples, spatially resolved elemental mapping can be accomplished by moving the sample under the fixed incident X-ray beam which has been tuned to a particular absorption edge energy. Spatial resolution of ≤ 1 μm resolution is now possible using Kirkpatrick-Baez focusing mirrors or tapered capillaries, and ≈ 150 nm resolution can be achieved using Fresnel zone plates at hard X-ray energies (> 5 keV: Kemner et al. 1999); 50-100 nm resolution is possible using zone plates at soft X-ray/VUV energies

(< 1500 eV) (e.g., Sutton and Rivers 1999; Tonner et al. 1999; Bertsch and Hunter 2001; Sutton et al., this volume). Chemical speciation can also be mapped at these same spatial resolutions if sufficient differences in XANES spectra exist for different chemical species of the same element.

Since the first hard X-ray fluorescence synchrotron microprobe was developed at the NSLS in the 1970's (Sparks 1980), there have been a number of significant applications of this method to environmental and low temperature geochemical samples (e.g., Thompson et al. 1988; Bajt et al. 1993; Sutton et al. 1993, 1994, 1995; Bertsch et al. 1994; Kaplan et al. 1994; Tokunaga et al. 1994; Smith and Rivers 1995; Tokunaga et al. 1998; Sutton and Rivers 1999; Kemner et al. 2000; Lai et al. 2000; Gilbert et al. 2001). The review by Sutton et al. (this volume) provides details. Fluorescence X-ray microtomography has been developed and applied to environmental samples at the APS (ID-13) (Hansel et al. 2001) and the ESRF (ID22) (Chukaline et al. 2002). There have also been a number of recent applications of X-ray microscopy (scanning transmission X-ray microscopy (STXM) and transmission X-ray microscopy (TXM)) (e.g., Myneni et al. 1998, 1999; Rothe et al. 1999; Pecher et al. 2000; de Stasio et al. 2001) and X-ray photoemission electron microscopy (XPEEM) (e.g., de Stasio et al. 2000) to environmental samples. STXM and TXM are soft X-ray/VUV methods that utilize zone plates to focus the X-ray beam down to < 100 nm diameters. XPEEM uses electromagnetic lenses like transmission electron microscopes to accomplish focusing of the image (Scholl et al. 2002). The development of X-ray waveguide nanostructures has recently opened the door to coherent X-ray point sources having nanometer-scale beam cross sections (Pfeiffer et al. 2002).

A particularly useful function of elemental mapping is to determine the association of a given trace element with specific phases, which can indicate the site of sorption reactions (see chapters in this volume by Sutton et al. and Manceau et al.). A good example of this type of X-ray mapping is provided by the studies of Duff et al. (1999a,b) who determined the distribution and speciation of plutonium on a natural Topopah Spring tuff sample taken from 450 m depth at the proposed high-level radioactive waste repository in Yucca Mountain, NV. This material is thought to serve as a natural reactive barrier that will affect the migration of radionuclides that might escape from the deep repository. The core sample was exposed to an aqueous solution containing $^{239}Pu^{5+}$, and the resulting distribution and speciation of Pu was determined by X-ray fluorescence micro-XANES mapping. Figure 8 shows that Pu is associated primarily with Mn-oxide coatings on smectites. This study also showed that Pu(V) is oxidized to Pu(VI) in some areas of the manganese oxide coating, possibly reflecting a heterogeneous distribution of Mn(IV).

Another example of the utility of mapping the spatial distribution of trace elements and their association with other elements or phases comes from a recent X-ray fluorescence microtomography study of the spatial distribution of Pb, Fe, Mn, Zn, and As on and within the roots of *Phalaris arundinacea*, a common aquatic plant from a mine waste impacted wetland in the Coeur d'Alene Basin of northern Idaho (Hansel et al. 2001). As shown in Figure 9, Fe and Pb are closely associated in a rind on the root surface which consists of ≈ 63% ferrihydrite, ≈ 32% goethite, and ≈ 5% siderite, as determined by later XAFS analysis. More detailed XAFS analysis of these samples carried out at SSRL showed that Pb is complexed with organic functional groups, most likely those of bacterial biofilms. Arsenic, which also appears to be associated in part with iron, is present as a combination of two sorbed As species (≈ 82% As(V) and ≈ 18% As(III)). Mn and Zn occur as isolated nodules of mixed-metal carbonate (rhodochrosite/hydrozincite) on the root surface.

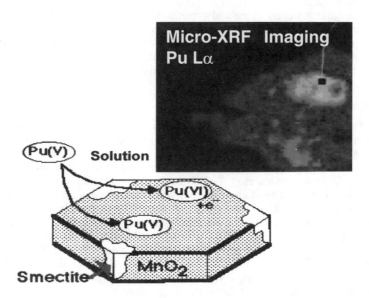

Figure 8. Micro-XRF image of Pu sorbed onto manganese oxide coatings on smectite grains in Yucca Mountain tuff reacted with a Pu(V)-containing aqueous solution. Image was collected on the insertion device beam line 13-ID-C at GSECARS (Advanced Photon Source). (after Duff et al. 1999a,b)

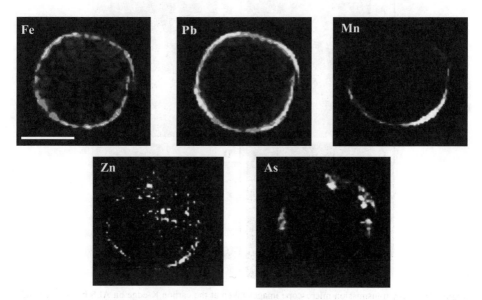

Figure 9. Metal distribution on and within a cross-section of grass root (*Phalaris arundinacea*) as determined by x-ray fluorescence microtomography at the Advanced Photon Source (GSECARS, beam line 13-ID-C). The lightest colored areas in the tomographs represent the highest concentrations of each element. The scale bar in the upper left panel is 300 μm, and is representative of the scale in the other panels. (after Hansel et al. 2001)

X-ray spectromicroscopy under *in situ* conditions is beginning to change our view of the macromolecular structure of natural organic matter, and how it changes as solution conditions are changed. An *in situ* transmission X-ray spectromicroscopy study at the ALS by Myneni and others has provided some of the first information of this type (Myneni et al. 1999). As shown in Figure 10, the macromolecular structure of fulvic acid in solution changes dramatically depending on the solution pH, ionic strength, and presence of other cations.

X-ray scattering

One of the most commonly applied SR-based scattering methods is powder X-ray diffraction, usually accompanied by Rietveld profile fitting of the diffraction pattern (e.g., Hazemann et al. 1991; Parise 1999; Lee et al. 2001). Special *in situ* reaction cells have been designed to study crystal growth and phase transformations in aqueous solutions at ambient to moderate temperatures ($\approx 200°C$) using X-ray diffraction (Cahill et al. 1998). Such cells have been used to study pyrite growth from aqueous solutions (Cahill et al. 2000). SR-based micro-XRD (MacDowell et al. 2001) is also becoming useful for identification of μm scale particles in complex environmental samples (e.g., Tamura et al. 2002; Manceau et al., this volume). Single crystal diffraction studies of the structure of very small crystals (50-160 $μm^3$ in volume) are also possible using SR sources (e.g., Pluth et al. 1997; Broach et al. 1999; Neder et al. 1999; Burns et al. 2000). Diffraction studies of such small crystals using conventional sealed or rotating anode X-ray tubes are not possible because of their very small diffracting volumes.

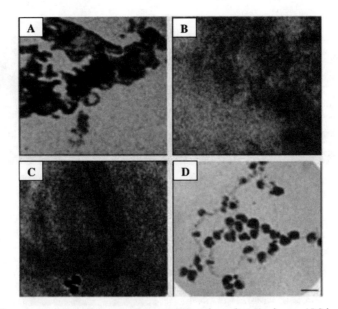

Figure 10. X-ray transmission microscope images taken at the carbon K-edge on ALS beam line 6.1.2 showing the macromolecular structure of Suwannee River fulvic acid (from the International Humic Substance Society) in aqueous solution as a function of pH and ionic strength (**A:** pH = 3.0; **B:** pH = 9.0; **C:** pH = 4.0 and $[Ca^{2+}]$ = 0.018 M; **D:** pH = 4.0 and $[Fe^{3+}]$ = 0.001 M). The scale bar ion the lower right panel is 800 nm and is representative of the scales in the other panels. [Reprinted with permission from Myneni et al. (1999). Copyright 1999 American Association for the Advancement of Science.]

Other frequently used X-ray scattering methods include small-angle X-ray scattering (SAXS) and X-ray reflectivity. Small angle X-ray scattering measures distance correlations on the order of 5 to 5000 Å and is useful in defining particle sizes and average particle shapes in the nm size range. SAXS can be used to study nucleation processes in geochemical systems as well as the size distribution of colloidal particles in aqueous solutions. Although not used extensively on earth materials to date, there have been several recent synchrotron-based SAXS studies of nucleation processes of low temperature calcium silicate minerals (Shaw et al. 2000a,b, 2002) and of ferric oxyhydroxides from solution in the presence of silicate ligands (Masion et al. 2001). The nucleation and growth of clay minerals (Carrado et al. 2000) and soot (Hessler et al. 2001) have also been investigated using SR-based SAXS measurements. Anomalous small-angle X-ray scattering (ASAXS) is a variation of SAXS in which SAXS data are taken above and below the X-ray absorption edge of an element, which can enhance the sensitivity of the technique to that element.

X-ray reflectivity at mineral/water interfaces has a range of applications in low temperature geochemistry and environmental science. Low-resolution X-ray reflectivity is useful as an *in situ* probe for determining surface roughness on single crystal samples (e.g., Chiarello et al. 1993; Teng et al. 2001) and the thickness and structure of thin films on mineral surfaces (Chiarello and Sturchio, 1994; Chiarello et al. 1997; Fenter and Sturchio 1999), whereas high-resolution X-ray reflectivity (e.g., crystal truncation rod [CTR] measurements in specular or non-specular modes) is a powerful tool for determining the structure of surfaces and interfaces (e.g., Robinson and Tweet, 1992; Chiarello and Sturchio 1995; Sturchio et al. 1997; Renaud 1999; Eng et al. 2000; Fenter et al. 2000a,b, 2001b; Cheng et al. 2001a; Trainor et al. 2002b). The theory and practical aspects of X-ray reflectivity measurements at mineral/water interfaces are reviewed, with examples of recent applications, by Fenter (this volume).

High-resolution *in situ* X-ray reflectivity studies of mineral/water interfaces have been published, to date, for calcite, barite, corundum, orthoclase, mica, and quartz. These studies have led to general insights on the extent of relaxation and reconstruction at mineral/water interfaces as well as the structure of interfacial water and other adsorbates (discussed in a subsequent portion of this chapter and in the chapter by Fenter in this volume).

X-ray standing wave methods

X-ray standing waves are generated by the interference between incident and reflected X-rays, typically when the reflected X-ray beam is large in magnitude ($R \sim 1$). The most widely used variation of XSW employs Bragg reflection from a perfect crystal, producing short-period standing waves that can be used to probe the locations of ions within crystals, adions on mineral surfaces, as well as ions within the electrical double layer at a mineral/solution interface (Cowan et al. 1980; Zegenhagen 1993; Bedzyk and Cheng, this volume). Long-period standing waves can be generated at grazing incidence (e.g., Trainor et al. 2002c) or by using synthetic multilayer materials in which layer thickness and density have been manipulated to provide a desired range of SW periods (Bedzyk 1988; Bedzyk et al. 1988; Bedzyk and Cheng, this volume). Among the earliest geochemically relevant applications of the long-period XSW method were the studies by Bedzyk and co-workers of the structure of the diffuse portion of the electrical double layer (EDL) in systems consisting of multilayer substrates made by depositing alternating crystalline layers of tungsten and silicon, capped by an amorphous silica layer overlain by a Langmuir-Blodgett film in contact with an aqueous solution containing Zn(II) and Cl(-I) ions (Bedzyk et al. 1990; Bedzyk 1990). XSW were generated above the W-Si mirror surface by total external reflection. Although these studies and a similar study by Abruna et al. (1990) are not directly relevant to mineral/aqueous solution interfaces, they were the first studies to show that the Gouy-Chapman-Grahame-Stern model of the EDL at an organic film/electrolyte solution interface is consistent with experimental observation.

The short-period XSW method has been used to probe the positions of sorbed ions (Mn(II), Co(II), Ni(II), Zn(II), As(III), Se(IV), Sr(II), Pb(II), U(VI)) within calcite and/or at the calcite surface (Qian et al. 1994; Cheng et al. 1997, 1998, 1999, 2000; Sturchio et al. 1997; Bedzyk and Cheng, this volume) and to determine the distribution of ions in the EDL at the rutile/water interface (e.g., Fenter et al. 2000c). A variant of the short-period XSW method, which is measured in the back-reflection mode, has been used at soft X-ray and VUV energies (50-1500 eV) to determine the structure of the CaO (100) (Kendelewicz et al. 1995) and galena (PbS) (100) surfaces (Kendelewicz et al. 1998a), the position of Na adatoms on the PbS (100) surface (Kendelewicz et al. 1998b), and the structure of the hydrated MgO (100) surface (Liu et al. 1998e). Recent applications of the long-period XSW method have included determination of the distribution of Pb(II) and Se(VI) ions in the EDL at metal oxide/aqueous solution interfaces (Trainor et al. 2002c) and determination of the position of Pb(II), Se(IV), and Se(VI) ions at mineral/biofilm interfaces (Templeton et al. 2001, 2002b), which will be discussed in a later section of this chapter.

X-ray photoemission spectroscopy, X-ray photoelectron diffraction, and X-ray emission spectroscopy

Laboratory-based X-ray photoelectron spectroscopy (XPS) has been used extensively to probe the surface composition and electronic structure of environmental and low temperature geochemical samples. Hochella (1988) provides a useful review of some of these studies, including the early XPS studies of mineral dissolution mechanisms (e.g., Petrovic and Berner, 1976; Holdren and Berner, 1979; Berner and Holdren, 1979), and Fadley (1978) reviews the basic concepts of XPS. Synchrotron radiation-based X-ray photoemission spectroscopy (SR-XPS), which measures total or partial electron yield or Auger electron yield (for maximum surface sensitivity) extends the applicability of laboratory-based XPS because it is not restricted to the relatively limited X-ray energies available to a typical laboratory XPS instrument (e.g., Mg Kα or Al Kα). In addition, the higher flux of synchrotron X-ray sources increases the signal-to-noise of spectra and makes the technique applicable to lower element concentrations than is possible with a laboratory X-ray photoelectron spectrometer. Because of the tunability of SR, one can select an incident X-ray energy 50 to 100 eV above the binding energy of the core-level electron of an element, which results in escape depths of the photoelectron of several Å, thus high surface sensitivity (Lindau and Spicer 1980).

There have been a number of applications of SR-based XPS to materials of relevance to environmental science and low temperature geochemistry including studies of the interaction of water vapor with metal oxide surfaces (e.g., Liu et al. 1998a,b,c,d; Kendelewicz et al. 1999; 2000b), reaction of CO_2 gas with metal oxide surfaces (Carrier et al. 1999; Doyle et al. 1999b), and oxidation of metal sulfide surfaces (e.g., Schaufuss et al. 1998; Nesbitt et al. 2000, 2002).

X-ray photoelectron diffraction (XPD) is a UHV method that exploits the masking of photoelectron scattering paths by atoms in surface layers as photoelectrons escape from a solid surface to determine surface structure (Williams et al. 1979; Holland et al. 1980; Fadley, 1992; Fadley et al. 1994, 1997). In one of the few applications of XPD to low temperature geochemical or environmental samples, Thevuthasan et al. (1999) determined the structure of the clean surface of hematite (0001).

X-ray emission spectroscopy is familiar to many geoscientists who have used an electron microprobe. For example, White and Gibbs (1967, 1968) used an electron microprobe to measure the position of the silicon Kβ and aluminum Kβ emission lines for a number of silicates and aluminosilicates, relating chemical shift of the emission lines to average Si-O and Al-O distances. Dodd and Glen (1968, 1969, 1970) and Glen and Dodd (1968) used Si Kβ X-ray emission spectra to study chemical bonding in crystalline and amorphous silicates; Albee and Chodos (1970), O'Nions and Smith (1971), and Smith and O'Nions (1971) used Fe Lβ/Lα intensity ratios to provide semi-quantiative estimates of the Fe^{2+}/Fe^{3+} content of iron-containing minerals; and deJong et al. (1981) used Si Kβ fine structure to infer the types of Q species in silicate glasses. At least in the earth sciences, this method has been in a dormant state since the early 1980's. Recently, however, X-ray emission spectroscopy has been reinvigorated using synchrotron radiation sources (Nilsson et al. 1997a,b). It is now possible to dissect, atom-by-atom and molecular orbital-by-molecular orbital, a small molecule on an otherwise clean metal surface in exquisite detail, providing unique information on the bonding within the molecule and on its bonds to the surface. Figure 11 shows angle-resolved X-ray emission spectra for glycine on Cu(110) for the C, O, and N atoms. Although this method has only been used to date on relatively simple systems under UHV conditions [e.g., glycine, formate, or acetate on Cu(110) (Hasselstrom et al. 1998); benzene on Ni(100) and Cu(110) (Weinelt et al. 1998); ammonia on Cu(110) (Hasselstrom et al. 1999); carboxylic acids on Cu(110) (Karis et al. 2000)], it has the potential to become an important technique for examining the bonding of simple molecules to clean mineral surfaces as well as "wet" surfaces in the future.

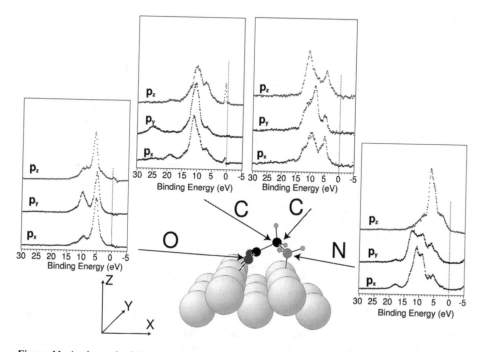

Figure 11. Angle-resolved X-ray emission spectra of glycine sorbed on Cu(110) taken at the ALS on beam line 8.0 showing the X-ray emission spectra in different polarizations [i.e., with the X-ray beam polarized along the x, y, and z axes of Cu(110)] for oxygen, the two carbon atoms, and nitrogen in the sorbed glycine molecule. (after Hasselstrom et al. 1998)

SR-based infrared spectroscopy

Applications of SR-based infrared absorption microspectroscopy to low temperature geochemistry and environmental science have only just begun to be explored (see chapter by Hirschmugl in this volume). The high brightness of infrared radiation emitted by synchrotron storage rings allows the chemical sensitivity of infrared spectroscopy to be applied to chemical imaging at diffraction limited spatial resolution (1 to 5 μm) and identification of chemical species at interfaces. Because IR spectroscopy is especially sensitive to vibrations of O-H, C-H, C-O, N-H, and C-N bonds, it has enormous potential for applications to organic compounds and the chemistry of biological materials such as living cells or bones, including phosphates, proteins, lipids, and carbohydrates. Other applications include analysis of fluid inclusion compositions and phase identification, as well as the identification of adsorbed organic species at mineral/water interfaces.

Comments on sample damage from high intensity synchrotron radiation

A typical synchrotron radiation experiment at a third-generation source using an insertion device beamline results in the deposition of a significant amount of power in a sample. Fluxes on the order of 10^{11} to 10^{13} photons sec^{-1} in a spot size ranging from several mm^2 to several μm^2 (for μXRF or μXANES experiments) are typical and may result in sample damage. Another potential source of sample damage are secondary electrons emitted from the sample. Such secondaries have an emission spectrum that is dependent upon the incident X-ray energy as well as other factors such as the nature of the solid (insulator vs. metal), crystal perfection, composition, etc. (Cazaux 2001). For low energy X-rays typical of those used to excite 1s electrons from oxygen atoms in photoemission experiments on oxide materials (600 eV), the secondary electron emission spectrum begins at about 5 eV and peaks at less than 50 eV before falling off in flux (Orlando et al. 1999). These secondaries have sufficient energy to break some chemical bonds, particularly relatively weak bonds such as O-H bonds in water or C-O bonds in organic molecules, depending upon the cross section of such bonds for the secondaries. This can lead to beam-induced hydrolysis of hydrated metal ions in aqueous solutions (Orlando et al. 1999). It can also lead to radiolysis of water in samples containing aqueous solutions, resulting in hydrated electrons that can cause reduction of redox-sensitive elements. At higher incident X-ray energies, secondary electron emission can extend to considerably higher energies (Cazaux 2001), but little is known about the extent to which these high-energy X-ray-induced secondaries cause sample damage. The extent of sample damage is usually difficult to assess unless redox reactions are observed during data collection. One obvious manifestation of X-ray beam-induced damage is a color change in a sample, which usually indicates the formation of color centers. Color changes have been observed during X-ray reflectivity studies of orthoclase and quartz at the APS (Fenter et al. 2000b; Schlegel et al. 2002), although there were no corresponding changes detected in surface reflectivity. Another manifestation of sample damage is a change in oxidation state of redox-sensitive elements in a sample. Documented examples of the latter come from recent XAFS studies of $CuCl_2$ and $AuCl_3$ in aqueous solutions by Jayanetti et al. (2001) and Farges et al. (2002), respectively. The former study found that Cu^{2+} at a concentration of 55 ppm was reduced to Cu metal clusters in solution, which they attributed to reaction of Cu ions with hydrated electron produced by the radiolysis of water in the high-flux X-ray beam of the APS. The latter study found that aqueous Au^{3+} is reduced to Au^0. XAFS studies of any samples containing water and redox-sensitive elements in a high oxidation state are potentially subject to this type of radiolysis-induced reduction. The oxidation state of the element of concern should be carefully monitored by frequent absorption edge measurements during the course of an EXAFS experiment.

A related type of reduction of redox-sensitive elements (e.g., Cr(VI), Se(VI)) has also been observed to occur in dry environmental samples by the first author and his research group in XAFS experiments at both second- and third-generation synchrotron X-ray sources. For example, a recent XAFS study of Cr speciation in bore-hole samples from the Hanford vadose zone (Catalano et al. 2001; Zachara et al. 2002) found that Cr(VI) is reduced to Cr(III) in Cr-contaminated sediments at a rate of 2.2 wt. % Cr(VI) per hr on beam line 11-2 at SSRL, which has a measured flux at 6 keV of $\approx 4 \times 10^{12}$ photons/sec using a cryogenically-cooled Si(220) double crystal monochromator with the 26-pole wiggler operating at 2 T and SPEAR-2 operating at 3 GeV and 80-100 mA (Bargar et al. 2002a). The total time the sample was exposed was about 9 hours. The initial EXAFS scans were done on a sample that contained 25.7% Cr(VI), based on the height of the pre-edge feature of the Cr K-edge XANES of this sample. Following the first EXAFS scan, which took about 40 min., the amount of Cr(VI) was 22.7%, and following the ninth EXAFS scan about 7.5 hours later, the amount of Cr(VI) was 6.3% of the total. This example shows that care should be taken to avoid such beam-induced reduction of redox-sensitive elements, particularly on insertion device beamlines at third-generation hard X-ray synchrotron sources when long data scans are required due to low element concentration. Cooling of redox-sensitive samples, where possible, has been found to retard this reduction process. Manceau et al. (this volume) discuss other examples of X-ray beam-induced changes in samples during hard X-ray EXAFS studies.

Hard X-ray synchrotron radiation experiments on samples containing microbial organisms will almost certainly kill the bacteria and stop whatever metabolic reactions are occurring. Bacteria are more tolerant of soft X-ray energies and may not be killed. Synchrotron infrared radiation do not typically cause damage to bacteria (e.g., Hirschmugl, this volume).

WATER AND METAL IONS IN AQUEOUS SOLUTIONS

Water in liquid or gaseous form is ubiquitous in most natural and laboratory systems of relevance to low temperature geochemistry and environmental science. Because of its importance, the structure and properties of liquid water under ambient conditions have been studied for decades using a variety of methods. The series of books edited by Felix Franks (1972-1982) and the short monograph by Franks (1983) provide good summaries of the older literature. Early X-ray and neutron diffraction studies of the structure of water at 25°C were published by Narten and Levy (1972) and Narten et al. (1982), respectively, and a comparison of theoretical simulations of water structure with experimental results is given by Stillinger (1980). A summary of recent molecular simulations of water structure under ambient and supercritical conditions is provided by Chialvo and Cummings (1999).

The general picture of average local water structure at 25°C that emerges from these experimental and theoretical studies is one in which each water molecule has two short OH bonds [d(O-H) ≈ 0.96 Å, bond energy ≈ 4.8 eV] and two lone-pairs (LP) of electrons, with an average H-O-H angle of 104.45° and an average LP-O-LP angle of about 114°. The dipole moment of the water molecule is about 1.83×10^{-18} esu cm, and the v_1 symmetric stretching frequency in liquid water is 3628 cm^{-1} (Eisenberg and Kauzmann 1969). Each water molecule is hydrogen bonded to four other water molecules as shown in Figure 12 (SYM), with equilibrium O···H distances lying between 2.6 and 3 Å (bond energy ≈ 0.3 eV) (Franks 1984). Water molecules undergo two types of motion in the liquid state: rapid oscillations about temporary equilibrium positions, which average about 10^{-13} s each, and slower displacements of the equilibrium positions, which happen roughly every 10^{-11} s (Eisenberg and Kauzmann 1969).

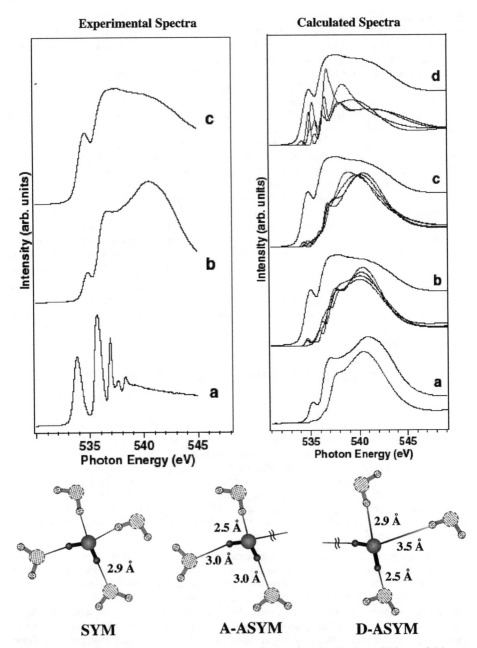

Figure 12. (upper left panel) Oxygen K-edge XANES spectra of (c) liquid water, (b) ice, and (a) water vapor. (upper right panel) Calculated oxygen K-edge XANES spectra of (a) ice and (b-d) three different models of liquid water using DFT, Including (b) SYM, (c) A-ASYM, and (d) D-ASYM. (lower panel) Three structural models of water considered. The D-assymmetric model provides the best fit of the X K-edge XANES data for liquid water among the structural models considered. The O K-XANES data were taken on beam line 8.0 at the ALS. (from Myneni et al. 2002, with permission of the Journal of Physics Condensed Matter)

One of the most recent experimental studies of liquid water structure at 25°C is that of Myneni et al. (2002) who measured oxygen K-edge XANES spectra of water and carried out density functional theory (DFT) calculations of the O K-XANES spectrum of water for different structural models. The experimental spectra of liquid water, water vapor, and ice are shown in Figure 12, together with the results of DFT calculations and several of the structural models considered for liquid water at 25°C. Because of the very short time frame sampled during the X-ray absorption process (less than a femtosecond, 10^{-15} sec) and the longer time frame of water motion, as discussed above, the resulting XANES spectrum represents the sum of a series of snapshots of frozen local geometries of water molecules on a sub-femtosecond time scale. The most notable difference between the liquid water and water ice oxygen K-XANES spectra is the large pre-edge feature in the former. The only structural model considered which reproduced this feature in the DFT calculations is one in which water is hydrogen bonded to three other water molecules in an asymmetric fashion (D-ASYM in Fig. 12). Myneni et al. (2002) suggest that this unsaturated H-bonding environment exists for a significant fraction of water molecules.

In another recent XAFS study of water structure, Wilson et al. (2000) reported the first use of EXAFS spectroscopy to determine the covalent O-H distance in water, which is 0.95 ± 0.03 Å. Hydrogen atoms are conventionally considered "invisible" to X-ray and electron diffraction methods. However, this study shows that high quality EXAFS data, coupled with accurate phase-shift and backscattering amplitude functions for H and O from curved-wave multiple scattering theory (FEFF: Ankudinov et al. 1998) and careful analysis of the atomic XAFS (i.e., the scattering of phoelectrons from electrons around the atom from which the photoelectron is ejected and from electrons in chemical bonds between atoms: Holland et al. 1978; Rehr et al. 1994), can result in enhanced sensitivity of EXAFS spectroscopy to H atoms in water and other materials. Although not directly comparable because the experiments were done on ice rather than liquid water, Issacs et al. (1999, 2000) used Compton X-ray scattering on BL ID15B at ESRF to determine the O∘∘∘H distances in ice Ih (1.72 Å and 2.85 Å). The Compton profile anisotropy was found to be consistent with a covalent rather than an electrostatic model of the hydrogen bond.

Quantitative information on the structure and dynamics of hydrated ions in aqueous solutions as a function of solution composition is essential for understanding their properties, including reactivity. The bulk of our information on the stoichiometry (and by inference, the structure) of metal ion complexes in aqueous solutions is based in large part on fits of potentiometric titration data as a function of metal ion concentration and pH (e.g., Baes and Mesmer, 1986). Structural data on hydrated ions in aqueous solutions, including hydrated radii, interatomic distances, and coordination numbers, from various types of direct and indirect experimental studies are summarized in reviews by Conway (1981), Marcus (1985, 1988), Magini et al. (1988), and Ohtaki and Radnai (1993). The results of X-ray and neutron scattering studies of ions in aqueous solutions are summarized in Enderby and Neilson (1979) and Neilson and Enderby (1989, 1996). There have also been a few combined X-ray scattering, Raman, and IR investigations of the coordination of hydrated ions in aqueous solutions (e.g., Sandstrom 1977; Sandstrom et al. 1978). In addition, molecular simulations of the structure of hydrated metal ions in aqueous solutions using different water and metal-oxygen potential functions are available for some ions (e.g., Dreisner and Cummings 1999; Chialvo and Cummings 1999). These results generally show that metal ions in aqueous solutions cause an ordering of water molecules in first and second hydration spheres around the ion, relative to bulk water (Martell and Hancock 1996). They also show that small ions with a high charge density tend to have smaller, more regular hydration spheres, whereas large ions with low charge density have larger, more irregular coordination spheres (Magini et al.

1988; Marcus 1988; Sandstrom et al. 2001), although there are some exceptions to these generalizations for specific ions, such as Hg(II) and U(VI), which form coordination complexes in solution that have two short axial bonds to oxygens and four or more longer "equatorial" bonds to water molecules, and Cu(II) which has four short equatorial bonds and two longer axial bonds to water molecules due to Jahn-Teller distortion.

SR-based EXAFS spectroscopy has contributed to the growing data base on ion hydration in aqueous solutions over the past 25 years: e.g., Ni^{2+} (Sandstrom et al. 1977); Cu^{2+} (Fontaine et al. 1978); Ni^{2+}, Cu^{2+}, Zn^{2+} (Lagarde et al. 1980); Zn^{2+} (Parkhurst et al. 1984); Ag^+ (Yamaguchi et al. 1984); Fe^{3+} (Apted et al. 1985); Cd^{2+} (Sadoc et al. 1995); Zn^{2+} (Dreier and Rabe 1986); Zn^{2+}, Cu^{2+}, Ni^{2+}, Rb^+ (Ludwig et al. 1987); Cu^{2+} (Garcia et al. 1989); In^{3+} (Marques and Lagarde 1990); Pd^{2+}, Pt^{2+} (Hellquist et al. 1991); Cu^+ (Persson et al. 1991); Au^{3+} (Farges et al. 1993a); Cu^{2+} (Filipponi et al. 1994); uranyl carbonate complexes (Allen et al. 1995b); Zn^{2+} (Anderson et al. 1995); Tl^{3+} (Blixt et al. 1995); As^{3+} (Helz et al. 1995); Cr^{3+}, Zn^{2+} (Munoz-Paez et al. 1995); Sr^{2+}, Ba^{2+} (Persson et al. 1995); Rb^+ (Fulton et al. 1996); Cu^{2+} (Mosselmans et al. 1996); Ag^+ (Seward et al. 1996); Np^{7+} (Clark et al. 1997); Ga^{3+} (Munoz-Paez et al. 1997); uranyl, neptunyl, Np^{4+}, Pu^{3+} (Allen et al. 1997b); U^{6+} (Thompson et al. 1997); Pu aquo ions (Conradson et al. 1998); Cr^{3+}, Ga^{3+}, In^{3+} (Lindqvist-Reis et al. 1998); U^{6+} (Docrat et al. 1999); trivalent lanthanide and actinide ions (Allen et al. 2000); Pu aquo ions (Conradson et al. 2000); Y^{3+} (Lindqvist-Reis et al. 2000); Sb^{5+} (Mosselmans et al. 2000); La^{3+}, Bi^{3+} (Nashlund et al. 2000a,b); Al^{3+}, Ca^{2+} (Spndberg et al. 2000); Np^{3+}, Np^{4+}, Np^{5+}, Np^{6+} (Antonio et al. 2001); Ca^{2+} (Jalilehvand et al. 2001); Ca^{2+}, Sc^{3+}, Y^{3+}, La^{3+}, U^{4+}, Th^{4+} (Sandstrom et al. 2001); Np^{7+} (Williams et al. 2001); Ga^{3+} (Clausen et al. 2002); Cu^{2+} (Persson et al. 2002).

As mentioned previously, the X-ray absorption process, which involves ejection of a photoelectron from a deep core level of an atom, occurs very rapidly (on the order of 10^{-16} sec). This process is six to ten orders of magnitude more rapid than ligand exchange between the hydration spheres of aqueous metal ions and bulk aqueous solutions (Hewkin and Prince 1970; Margerum et al. 1978). Thus XAFS methods are ideally suited for studies of metal ion complexes in aqueous solutions because ligand exchange processes occur on longer time scales than X-ray absorption, resulting in summations of billions of essentially instantaneous "snapshots" of solution complexes for a particular metal ion during the time required for accumulating a single EXAFS spectrum by conventional transmission or fluorescence-yield methods (20-40 min). Fluorescent-yield EXAFS measurements can provide this type of information on metal ion concentrations as low as 10^{-5}M (Parkhurst et al. 1984) or lower using more modern X-ray detectors and high flux beamlines.

The structure and bonding of metal ion complexes in hydrothermal and supercritical solutions (T ≤ 400°C) are poorly understood, but are of great importance in some low temperature geochemical processes, ore deposition, and environmental processes where groundwater temperatures are elevated due to natural (e.g., hot springs and geothermal fields) or anthropogenic heating (e.g., hot leachates from radioactive waste storage tanks). XAFS spectroscopy has become the method of choice for studies of such complexes at temperatures up 500°C and pressures up to 300 MPa. Such measurements are now possible because of the development of the modified diamond anvil cell (Bassett et al. 2000a,b; Fulton et al. 2001) and other types of cells (e.g., Murata et al. 1995; Wallen et al. 1996) that can hold aqueous solutions at high temperatures and high pressures, simultaneously. EXAFS studies of the following metal ions in hydrothermal solutions have been carried out: Rb^+ (Fulton et al. 1996); Co^{2+}, Mo^{6+}, Cd^{2+} (Mosselmans et al. 1996); Ag^+ (Seward et al. 1996); Zn^{2+} (Anderson et al. 1998a); Ni^{2+} (Wallen et al. 1998); Zn^{2+} (Mayanovic et al. 1999); Sr^{2+} (Seward et al. 1999); Cu^+ and Cu^{2+} (Fulton et al. 2000); In^{3+} (Seward et al. 2000); Cr^{6+} (Hoffman et al. 2001); Zn^{2+} (Mayanovic et al. 2001); La^{3+} (Anderson et al. 2002); Yb^{3+} (Mayanovic et al. 2002). One unexpected finding from most of these studies is the observation that first-shell interatomic distances and coordination numbers around metal ions in hydrothermal solutions decrease with increasing temperature, rather than increase as normally expected because of thermal expansion effects. Anharmonic effects, which could explain these shorter than expected distances and smaller than expected coordination numbers derived from EXAFS analysis, were explicitly accounted for in these studies and are not sufficient to explain these structural changes. This interesting observation has been interpreted in the case of Zn-chloride solutions, where Zn^{2+} is tetrahedrally coordinated by Cl ions, as being due to progressive reduction of the dielectric constant of water with increasing temperature, which leads to a reduction in hydrogen bonding, increased thermal energy of the water molecules, enhanced interactions of Zn 3d (t$_2$) and Cl 3p orbitals, and greater repulsion between t$_2$ and e bonding vs. antibonding orbitals (Anderson et al. 1998a, Mayanovic et al. 1999). Seward et al. (1996, 1999) carried out EXAFS studies of Ag$^+$ and Sr^{2+} as a function of temperature and explained the Ag-O and Sr-O bond length contractions and decreased first-shell coordination numbers with increasing temperature as being due to the changing nature of the ion-solvent interaction. In a high temperature EXAFS study of Rb$^+$ in aqueous solutions, Fulton et al. (1996) found a similar contraction of the Rb-O(first-shell) distance with increasing temperature, and attributed this effect to increased binding of Rb$^+$ to nearest-neighbor water molecules as the less tightly bound higher-shell waters increase in distance from Rb$^+$ with increasing temperature. Similar average bond length contractions with increasing temperature were also found for Yb^{3+} (Mayanovic et al. 2002) and La^{3+} (Anderson et al. 2002) in aqueous solutions.

The EXAFS study of Ni^{2+} in hydrothermal bromide-containing solutions by Wallen et al. (1998) shows a slight Ni-O bond length expansion with increasing temperature, which contrasts with the EXAFS results for the other ions examined in hydrothermal solutions to date. The situation for NiBr solutions is complicated, however, by the formation of bromidonickel(II) complexes and increased hydrolysis with increasing temperature, so it is difficult to separate these effects from ion-solvent interactions on changes in Ni-O distance and first-shell coordination number.

Fluid inclusions in minerals represent trapped growth media and natural brines, and are seldom larger than 1 mm in diameter; the most abundant inclusions are typically in the 1-10 μm size range (Roedder 1984). Such inclusions are typically water solutions containing less than 10 wt. % solutes, including major amounts of Na$^+$, K$^+$, Ca^{2+}, Mg^{2+}, Cl$^-$, and SO$_4^{2-}$, and lesser amounts of other ions. Although fluid inclusions in minerals are most commonly used to determine the temperature and pressure of a rock or mineral at the time of fluid trapping (Roedder 1984) and of the tendency of an element like Cu to preferentially partition into the fluid phase during magmatic differentiation and crystallization (e.g., Lowenstern et al. 1991), they also serve as useful sample vessels for high T-P XAFS studies of selected metal ions in solution using μXAFS spectroscopy. A few such studies have been performed over the past five years, including μEXAFS studies of Zn^{2+} in hypersaline fluid inclusions (Anderson et al. 1995, 1998a,b; Mayanovic et al. 1996, 1997). The results of these studies are similar to those on ZnCl$_2$ solutions carried out in a modified diamond anvil cell discussed above, showing Zn-ligand bond contraction and a reduction in Zn coordination number with increasing temperature.

CHEMISTRY AT MINERAL/AQUEOUS SOLUTION INTERFACES

The interfaces between solids and aqueous solutions and between solids and gases play an enormously important role in the geochemistry of the earth's near-surface environment. Indeed, the key mechanisms and rates by which toxins move from their source through the environment and ultimately into the biosphere are determined largely by physical and chemical processes at these interfaces. Such interfaces can be as conceptually "simple" as water or water vapor in contact with an atomically flat, defect-free surface of a model oxide like MgO or as complex as a biofilm or the irregular, defective surface of a soil particle in contact with groundwater. Phenomena occurring at interfaces control or affect such diverse environmental issues as groundwater quality and contaminant transport, plant nutrient uptake, the toxicity of individual molecules for living cells, and atmospheric chemistry and air quality. For example, the breakdown of CFC's in the atmosphere is facilitated on the surfaces of dust and ice particles in polar stratosphere clouds. Furthermore, sea salt provides a reactive surface on which tropospheric NO_x is sorbed, resulting in the formation of HNO_3 or HCl. Reactions between atmospheric gases and solid surfaces are also important in the water-unsaturated zone of the earth's crust, particularly in arid regions where the groundwater table is very deep. This general class of heterogeneous gas-solid reactions has only recently been recognized as playing a significant role in mediating the composition of the biosphere.

The most important class of interfacial reactions affecting the environment involves aqueous solutions reacting with mineral surfaces in the earth's outer crust (Stumm et al. 1987). Such reactions are largely responsible for the compositions of the oceans and the world's fresh water supplies and provide critical controls in the biogeochemical cycling of chemical elements and compounds. Furthermore, the release of inorganic nutrients to soils from such processes is essential for the growth of plants and the production of much of the world's food supply. These solid surfaces also play an important role in filtering potentially harmful microbes, viruses, and chemical contaminants from groundwaters because they can chemically bind these substances to mineral surface sites, thus retarding their movement in groundwater. Although this "filtering" capacity of soils is great, there are limits, which depend on the types and amounts of contaminants entering the soil system and the nature and properties of the soil.

Below, we discuss examples of chemical reactions at environmental interfaces studied by SR methods. We start with a brief discussion of SR studies of relatively simple model systems, such as $MgO-H_2O$ and $Al_2O_3-H_2O$, continue with a brief discussion of the structure and properties of water at solid/water interfaces, and progress to more complex interfacial systems, including those with microbial biofilms present.

The initial reaction of water with clean metal oxide surfaces

Even though there have been hundreds of studies of the interaction of water with clean metal and metal oxide surfaces over the past 30 years using a variety of surface science methods (Thiel and Madey 1987; Henderson 2002) there is little fundamental understanding of how liquid water reacts with mineral surfaces except in the simplest cases (Brown 2001). For example, the interaction of water vapor with the clean MgO (100) surface has been studied using SR-photoemission spectroscopy as a function of $p(H_2O)$ at 298 K (Liu et al. 1998a,b). This surface was prepared by cleaving a MgO crystal in a UHV preparation chamber (base pressure of 5×10^{-11} torr), exposing it sequentially to different $p(H_2O)$, transferring the sample to a UHV analysis chamber, and collecting oxygen 1s and valence band spectra between exposures to water vapor. As shown in Figure 13 (middle panel), the O 1s photoemission spectra indicate that water vapor reacts primarily with defect sites on the Mg(100) surface at $p(H_2O) \leq 10^{-4}$ torr and with terrace sites at $p(H_2O) > 10^{-4}$ torr. Evidence for this conclusion is the appearance of a low intensity feature at a kinetic energy of ≈ 80.6 eV in the O 1s spectra and at ≈ 60 eV in the O 2s + valence band spectra, which are caused by the dissociation of water and the formation of OH groups on the MgO (100) surface. The main O 1s feature centered at ≈ 83 eV is due to lattice

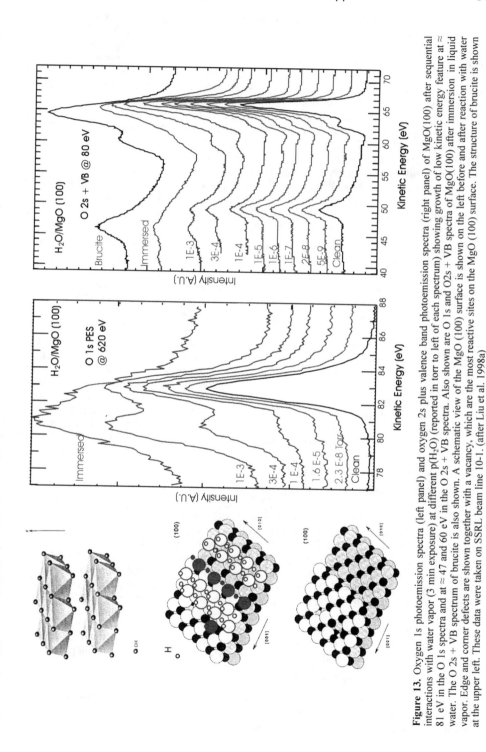

Figure 13. Oxygen 1s photoemission spectra (left panel) and oxygen 2s plus valence band photoemission spectra (right panel) of MgO(100) after sequential interactions with water vapor (3 min exposure) at different p(H₂O) (reported in torr to left of each spectrum) showing growth of low kinetic energy feature at ≈ 81 eV in the O 1s spectra and at ≈ 47 and 60 eV in the O 2s + VB spectra. Also shown are O 1s and O2s + VB spectra of MgO(100) after immersion in liquid water. The O 2s + VB spectrum of brucite is also shown. A schematic view of the MgO (100) surface is shown on the left before and after reaction with water vapor. Edge and corner defects are shown together with a vacancy, which are the most reactive sites on the MgO (100) surface. The structure of brucite is shown at the upper left. These data were taken on SSRL beam line 10-1. (after Liu et al. 1998a)

oxygens. The nature of these interactions is shown schematically in the structural drawings, which depict two MgO (100) surfaces, one showing terrace sites, a step defect, and a vacancy defect prior to exposure to water vapor, and the second showing hydroxyl groups on the surface following exposure at $p(H_2O) > 3\times10^{-4}$ torr. At exposure pressures less than this "threshold pressure", water chemisorbs dissociatively on these types of defects, which are thought to be more energetic and reactive sites than terrace sites on MgO(100) (Langel and Parinello 1994; Scamehorn et al. 1994). This low kinetic energy shoulder does not grow appreciably in intensity as $p(H_2O)$ is increased from 10^{-9} to 10^{-4} torr, even after prolonged exposures (several hours) of the surface to water vapor (up to 1.8×10^4 Langmuirs (L), where 1 L corresponds to 10^{-6} torr-sec). However, when the defect density of these surfaces was increased (from $\approx 5\%$ to 35%), this low kinetic energy shoulder increased significantly in intensity at $p(H_2O) < 10^{-3}$ torr (Liu et al. 1998b). In addition, when $p(H_2O)$ was raised to 3×10^{-4} torr with an exposure time of 3 min (corresponding to an exposure of 5.4×10^4 L), the dissociation reaction was rapid as indicated by the rapid increase in intensity of the low kinetic energy feature, which continued to grow in intensity with exposure of the surface to higher water pressures (Liu et al. 1998a).

These photoemission results were initially at variance with high-level quantum chemical calculations, which were based on the interaction of one water molecule per surface unit cell and indicated that the interaction of water with terrace sites is endothermic (Langel and Parinello 1994; Scamehorn et al. 1994). However, more recent density functional calculations (Giordano et al. 1998; Odelius 1999) in which a more realistic 3×2 ordered array of water molecules was placed on the MgO (100) surface predict that water should dissociate on terrace sites, in agreement with the photoemission results. An important interaction revealed by these calculations is hydrogen bonding between adjacent water molecules on the surface, which appears to be required for dissociation of water on MgO(100). Apparently, when a critical density of water molecules is reached ($\approx 50\%$ monolayer coverage) on these surfaces, this hydrogen bonding interaction, coupled with the surface interaction, destabilizes the water molecule and it dissociates. Also shown in Figure 13 are O 1s and O 2s + VB spectra of MgO(100) immersed in liquid water and of brucite ($Mg(OH)_2$). The spectral features indicative of hydroxyl groups on the MgO(100) surfaces are amplified in these spectra. Prolonged exposure of MgO to water will result in the formation of brucite, whose structure is shown in the upper left panel of Figure 13.

Another example where theory and experiment appear to agree involves the interaction of water vapor with α-Al_2O_3(0001). Oxygen 1s photoemission results indicate that significant dissociative chemisorption of water molecules does not occur below ≈ 1 torr $p(H_2O)$ (Liu et al. 1998d). However, following exposure of the alumina surface to water vapor above this "threshold $p(H_2O)$", a low kinetic energy feature in the O 1s spectrum grows quickly, indicating increasing levels of dissociative chemisorption of water on terrace sites. The results of recent density functional calculations of the interaction of water on α-Al_2O_3 (0001) as a function of water coverage (Hass et al. 1998; Wang et al. 2000) are in substantial agreement with the photoemission results of Liu et al. (1998d). Similar photoemission experiments on α-Fe_2O_3 (0001) indicate a "threshold $p(H_2O)$" of 10^{-4} torr, above which water dissociates on terrace sites (Liu et al. 1998d). These observations raise the interesting question as to why the "threshold $p(H_2O)$" values of isostructural corundum and hematite differ by about five orders of magnitude.

The structure of hydrated mineral surfaces

Following the reaction of water or water vapor with a metal oxide surface, as discussed above, one might expect the surface structure to be different than a simple termination of the bulk structure for normally anhydrous minerals. This is in fact what has been found in a few cases studied to date using high-resolution X-ray reflectivity at the Advanced Photon Source. For example, a recent CTR study of the hydrated α-Al_2O_3 (0001) surface in contact with humid air (Eng et al. 2000) showed that the surface structure undergoes significant relaxation, relative to the vacuum-terminated clean surface (Guenard et al. 1997), has an oxygen termination rather than the aluminum termination of the clean surface, and has a partially ordered layer of physisorbed water molecules (Fig. 14). The surface oxygens are coordinated by two octahedrally coordinated Al atoms. In contrast, the more reactive α-Al_2O_3 ($1\bar{1}02$) surface, when hydrated, is terminated by about equal proportions of one-, two-, and three-coordinated oxygens, indicating that both relaxation and reconstruction have occurred relative to the bulk termination along this plane.

Preliminary CTR scattering data on the hydrated α-Fe_2O_3 (0001) surface suggest two distinct types of structure, one that is oxygen terminated and one that is iron terminated (Trainor et al. 2002d). Similar conclusions have been reached by Shaikhutdinov and

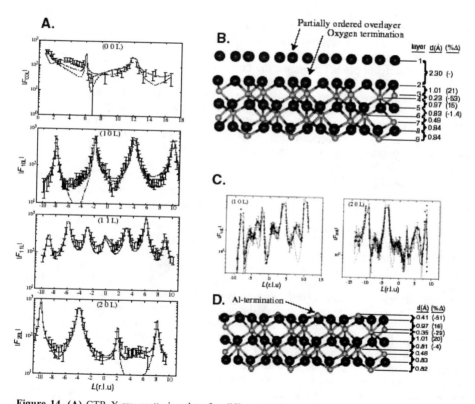

Figure 14. (**A**) CTR X-ray scattering data for different HKL rods for the hydrated α-Al_2O_3 (0001) surface, showing fits of the data; (**B**) best-fit structural model of the hydrated α-Al_2O_3 (0001) surface, showing a partially ordered layer of water molecules at 2.30 Å above the surface; (**C**) CTR X-ray scattering data for two rods for the clean α-Al_2O_3 (0001) surface (as determined by Guenard et al. 1997); (**D**) best fit model structural model of the Al-terminated clean (0001) surface of α-Al_2O_3 (from Guenard et al. 1997). To the right of each structural model are the distances of each layer from the surface oxygen layer, and in parentheses, the changes in distance (%) relative to the corundum structure. Data were taken on APS Beamline 13-ID-C. (from Trainor 2001, with permission)

Weiss (1999) and Weiss and Ranke (2002) based on STM and LEED observations as a function of oxygen pressure, by Wang et al. (1998) based on density functional calculations as a function of oxygen pressure, and by Stack et al. (2001a) based on *in situ* STM observations of the α-Fe_2O_3 (0001) surface under water. The observed differences between the α-Fe_2O_3 (0001) and α-Al_2O_3 (0001) surfaces may explain the major differences in "threshold p(H_2O)" discussed in the previous section.

In situ X-ray reflectivity studies of water structure at mineral/water interfaces

Once a mineral surface has been immersed in water, the interaction of water with the mineral surface plays a dominant role in controlling further reactivity of the surface with other dissolved ions (see earlier discussion of the initial reaction of water with clean mineral surfaces). Mineral surfaces in water are generally covered with surface hydroxyl groups, and pH-dependent protonation reactions on these hydroxyls control the speciation of the surface and influence interactions of other dissolved ions with the surface (Stumm 1992). These interactions include adsorption, dissolution, and precipitation.

In situ high-resolution X-ray reflectivity studies of mineral/water interfaces are consistent with hydroxylation and water adsorption at mineral surfaces. The first such measurements were performed at the calcite/water interface at pH values ranging from 6.8 to 12.1 (Fenter et al. 2000a). The reflectivity data indicate that the structure of the calcite (104) cleavage surface does not change significantly over this range of conditions. The best-fit model at pH 8.3 proposes the presence of 1.0 ± 0.4 monolayers of hydroxyl or water molecules at 2.50 ± 0.12 Å above the surface Ca ions and slight rotations of the carbonate groups toward the (104) plane. Important implications of these results are (1) there is no evidence for a disordered, hydrated calcite layer at the interface, such as that postulated by Davis et al. (1987) to explain sorption kinetics of Cd on calcite and (2) high-resolution X-ray reflectivity can be used *in situ* to measure the coverage and structure of surface complexes at mineral/water interfaces. These results confirmed a major assumption of the surface complexation model for calcite that was proposed by Van Cappellen et al. (1993), i. e., the presence of the surface site >Ca(OH)$_2$; however, no evidence for the surface complex >CO_3Ca^+ was found under conditions where the Van Cappellen et al. (1993) model predicted that it should be present.

Other recent applications of high-resolution X-ray reflectivity to investigations of water at mineral surfaces include barite (Fenter et al. 2001b), orthoclase (Fenter et al. 2000b, 2002), quartz (Schlegel et al. 2002), and muscovite (Cheng et al. 2001a). The studies of barite, orthoclase, and muscovite are reviewed thoroughly by Fenter (this volume). The study of the muscovite(001)/water interface by Cheng et al. (2001a) is especially noteworthy in its identification of two types of adsorbed water: (1) a coverage of 1.0 ± 0.2 water molecules per surface ditrigonal cavity (site having negative structural charge), at a distance of 1.3 ± 0.2 Å above the mean position of the relaxed surface oxygen plane (base of tetrahedral sheet), presumably in the form of hydronium replacing K^+ ions lost to solution, and (2) several layers of water ordered in the surface-normal direction, with the first layer at a distance of 2.5 ± 0.2 Å above the surface and subsequent layers approximately equally spaced out to about 10 Å. The second type of water is interpreted to be a H-bonded network, and can be thought of as being analogous to the hydration shell about a metal ion in solution. This is the first clear indication of interfacial water structure from X-ray reflectivity at an ambient mineral/water interface, and the resulting structure is consistent with independent molecular dynamics models (Park and Sposito 2002).

Comments on the structure and properties of water at solid/aqueous solution interfaces

Water very near mineral/solution interfaces is affected by electrostatic and hydrogen bonding interactions with the mineral surface, resulting in changes in the normal hydrogen bonding network and different structure and dynamics relative to bulk water. These differences are manifested as an order of magnitude difference in the dielectric constant (ε) of interfacial water (≈ 6) relative to that of bulk water (≈ 78) (Bockris and Reddy 1973; Bockris and Jeng 1990, Bockris and Kahn 1993). A more recent estimate of ε for interfacial water near a mica (001) surface comes from an atomic force microscopy (AFM) study by Teschke et al. (2000). Based on fits of these data to a dielectric exchange force model, this study concluded that ε is close to 4 for water at the mica surface (L = 0 Å), increases to ≈ 30 at L = 100 Å, and to ≈ 78 at L > 250 Å for pure water. Based on similar AFM force measurements made on solutions containing 10^{-3} M LiCl, KCl, NaCl, and $MgCl_2$, in separate experiments, Teschke et al. also concluded that the presence of cations at the negatively charged mica surface causes increases in ε, presumably due to restructuring of water molecules around these ions near the mica (001) surface. These AFM results, which suggest that water does not become bulk-like until quite a distance from the interface, are at variance with the models of interfacial water developed by Israelachvili and Pashley (1983), Cheng et al. (2001a), and Park and Sposito (2002), which indicate that water has a bulk-like structure at a distance of ≤ 10 Å from the interface. This disagreement may reflect artifacts of some of the methods used to infer the structure and properties of water near interfaces.

A sum frequency generation (SFG) study of mica (001) surfaces in contact with a controlled atmosphere with relative humidity values ranging from 20% to 90% (pH estimated to be 5.7, which is the value of rainwater buffered by the CO_2 content of the atmosphere) (Miranda et al. 1998) concluded that water vapor at the mica(001)/water interface has an "icelike" structure at room temperature. This conclusion is consistent with the results of a molecular dynamics simulation of water vapor on the mica (001) surface (Odelius et al. 1997). In contrast, the results of an X-ray reflectivity study of the mica (001)/bulk water interface (pH 5.7) (Cheng et al. 2001a), which was discussed in the previous section, and those from a recent MD-simulation of water structure at the muscovite/water interface (Park and Sposito 2002) are not consistent with the presence of an ice-like layer of water. Instead, these more recent studies propose a model in which interfacial water is more disordered and more labile. The mica surface is negatively charged, so one might expect the hydrogen dipole of the water molecules to be oriented toward the surface, as indicated by the MD-simulations of Park and Sposito (2002).

Although not directly comparable to the mineral/water interface studies discussed above, a crystal truncation rod X-ray scattering study of the Ag(111)/water interface (Toney et al. 1994) derived a somewhat similar view of interfacial water, concluding that the first monolayer of water molecules adjacent to the Ag(111) surface has an oxygen up-proton down orientation for a negatively charged surface (which can be inverted by reversing the charge on the surface), and, by implication, that the hydrogen bonding network is disrupted in this layer. The study of Toney et al., which concluded that the first water layer adjacent to Ag(111) has a significantly higher density than bulk water, differs, however, from the suggestion of Etzler and Fagundas (1987) that water at the quartz/solution interface is less dense than bulk water. The results of a recent high-resolution X-ray reflectivity study of quartz/water interfaces (Schlegel et al. 2002) support the suggestion by Etzler and Fagundas. Differences in the behavior of water at metal vs. insulator interfaces, however, are not clear.

Du et al. (1994) carried out a SFG vibrational spectroscopy study of quartz/aqueous solution interfaces at room temperature. They found a strong SFG signal at high pH (12), where the surface should be strongly negatively charged because of the low pH_{pzc} of quartz (2-3) (Parks 1965). The strong negative surface charge should cause water molecules near the surface (estimated to be at least three monolayers) to be oriented with their positively charged hydrogen dipoles pointing toward the surface, leading to a high degree of order in these layers relative to bulk water. At pH 2, where the surface charge of quartz is neutral or slightly positive, a considerably smaller but still relatively strong SHG signal was measured by Du et al., indicating a relatively high degree of order of interfacial water molecules and leading to the suggestion that the oxygen dipoles of these water molecules point toward the quartz surface. At pH values between 2 and 12, the SFG signal was found to be very weak, indicating that water molecules are disordered in the vicinity of the quartz surface over this pH range. This conclusion is consistent with the model for water at the quartz/water interface developed by Schlegel et al. (2002).

There have also been SFG (Yeganeh et al. 1999) and combined SHG and AFM (Stack et al. 2001b) studies of the corundum (001)/water interface as a function of pH, with interesting and conflicting results. Yeganeh et al. found that the SFG signal decreases as a function of pH, reaches a minimum at pH 8, then rises at higher pH values, which was interpreted as indicating a positively charged surface below pH 8, a negatively charged surface above pH 8, and different orientations of the water dipole above and below pH 8, which they assumed to be the isoelectric point of corundum (001). The SHG/AFM study of Stack et al. (2001b) found that the pH_{pzc} of corundum is between 5 and 6, which is significantly lower than values (8 to 9.4) reported in other studies (Parks, 1965; Sverjensky, 1994; Sverjensky and Sahai, 1996; Yeganeh et al. 1999). The reason for this difference in measured pH_{pzc} between these two studies is not clear.

High-resolution X-ray reflectivity studies of mineral/water interfaces are consistent with adsorbed water density being equivalent to the site densities of exposed metal cations at the surfaces of calcite (104) (Fenter et al. 2000a) and barite (001) and (210) (Fenter at al. 2001b), which are lower than that of bulk water. The fully hydrated quartz (100) and (101) surfaces have adsorbed water densities consistent with the surface site density of silanols (Schlegel et al. 2002), and the orthoclase (001) surface (Fenter et al. 2000b) has adsorbed water density consistent with the surface site density of silanols plus aluminols. In all of these cases, there is little or no observable perturbation of the water above the surface, and relaxations, which may affect the outermost several unit cells, are limited to no more than a few tenths of an Å at most. At the orthoclase (100) surface, K^+ ions are absent, presumably exchanged by hydronium ions from the water (Fenter et al. 2000b). As discussed above, the muscovite(001)/water interface is unique in having two adsorbed water layers forming a dense water layer, plus additional water structuring as far as 1 nm from the surface (Cheng et al. 2001a).

Molecular dynamics simulations of the EDL at the TiO_2/aqueous solution interface have been carried out by Chialvo et al. (2002) and Predota et al. (2002). These simulations indicate that water is structured at the interface, that Sr^{2+} ions form dominantly inner-sphere complexes at the TiO_2 (110) surface, and that Rb^+ ions form dominantly outer-sphere complexes at the interface. These results agree with those from a small-period X-ray standing wave study of Sr^{2+} and Rb^+ ions at the TiO_2(110)/solution interface (Fenter et al. 2000c) as well as with predictions from the triple layer capacitance model of Sverjensky (2001).

All of these studies point to more structured water in the layer adjacent to the mineral surface and, by inference, differences in hydrogen bonding and electrostatic interactions in the surface water layer relative to bulk water. The first monolayer of water at the quartz/, calcite/, barite/, and orthoclase/aqueous solution interfaces was found to be less dense than bulk water, whereas water at the muscovite/solution interface was found to be about the same density as bulk water. Based on these results, there does not appear to be a simple correlation of the density of the first monolayer of water with the hydrophobicity or hydrophylicity of the mineral surfaces as one might expect. These differences between interfacial and bulk water undoubtedly affect the interactions of metal ions with hydrated mineral surfaces. It is likely that the dielectric properties of the mineral affect these interactions as well, as suggested by James and Healy (1972), with high dielectric constant mineral surfaces affecting the hydration sphere of cations within the β-plane more than low dielectric constant mineral surfaces. Thus high dielectric phases like TiO_2 and MnO_2 are expected to favor loss of waters of hydration around cations in the β-plane of the EDL, whereas lower dielectric phases like alumina and quartz are expected to favor fully hydrated cations.

A number of things are missing from this rather crude picture, including detailed information on the average structure of water in the interfacial region of a solid in contact with bulk water from direct experimental studies. Additional studies are needed to determine if the D-ASYM structure proposed for bulk water by Myneni et al. (2002) (Fig. 11) is also present in the interfacial region and how abundant this structure is vs. the SYM structure for water. Very little is known from direct measurements about the properties of water in the interfacial region, such as density, dielectric constant, dielectric relaxation, ligand exchange rates, and viscosity. Also needed is experimental confirmation of the structure of water around cations and anions in the EDL predicted in MD simulations such as those of Chialvo et al. (2002), Park and Sposito (2002), and Predota et al. (2002). One of the more poorly understood details of the EDL at mineral/solution interfaces is the hydrolysis of cations, which appears to be more significant in the interface region at least for some cations (e.g., Pb(II): Bargar et al. 1997a) than in bulk solution. It is not clear if such hydrolysis occurs as the result of a ligand exchange reaction when the cation complex bonds to the mineral surface, or in the diffuse layer of the EDL as the cation approaches the charged mineral surface. Such information is of critical importance in surface complexation modeling, which requires as input chemical reactions with accurate descriptions of the stoichiometry of reactants and products. It is not clear at this point what experimental method, if any, will yield this level of detail for metal oxide/water and silicate mineral/water interfaces.

XSW and X-ray reflectivity studies of adions and the EDL at mineral/water interfaces

The ability to infer aspects of water structure directly at the mineral/water interface using high-resolution X-ray reflectivity is an important complement to vibrational spectroscopy methods. X-ray reflectivity and other techniques, notably X-ray standing waves, can be used to derive a well-constrained molecular-scale model of the structure and distribution of sorbed ions at or near the mineral/water interface (including physisorbed ions and those in the diffuse portion of the EDL).

The first study to apply XSW methods to determine the location of a sorbed ion at a mineral surface (as well as a trace impurity in a bulk mineral) was that of Qian et al. (1994), who used short-period XSW from the calcite (104) reflection to determine the location of Pb sorbed to calcite from aqueous solution. It was shown that Pb at the calcite surface was well ordered and occurred in the same position as Ca. In addition, it was determined that Mn in the bulk substituted for Ca. This first study was done ex situ (the crystal was examined under He after removal from solution).

Further study of Pb sorbed to calcite (104) was performed using the first combination of *in situ* high-resolution X-ray reflectivity, XSW, and AFM (Sturchio et al.

1997). This study used XSW to locate Pb atoms in three dimensions, employing the (104), (006), and (024) reflections of calcite; Pb was found in the Ca site, and the *in situ* results confirmed the earlier ex-situ results of Qian et al. (1994). In addition, X-ray reflectivity data showed that Pb occurred mainly (> 70%) in the surface layer of calcite as a dilute solid solution of Pb in calcite. Thus Pb had been incorporated in the calcite surface, most likely by dissolution-reprecipitation of calcite in the presence of the initially calcite-undersaturated Pb-bearing aqueous solution. A number of additional studies of ion incorporation at the calcite surface and in bulk calcite have been performed using XSW (combined with surface XAFS in some cases) by Cheng et al. (1997, 1998, 1999, 2000, 2001b). These studies are summarized in the chapter by Bedzyk and Cheng (this volume). Complementary studies of the interaction of cations with the calcite surface using X-ray microprobe, XAFS, and other methods have been performed by R.J. Reeder and co-workers (Reeder 1996; Reeder et al. 1999, 2000, 2001).

Organic anions adsorbed on minerals can form well-ordered thin films, as shown by Fenter and Sturchio (1999) in an *in situ* X-ray reflectivity study of stearate on calcite. This paper demonstrated the use of X-ray reflectivity to characterize such thin films as well as to measure a sorption isotherm. There is great potential for using *in situ* X-ray reflectivity to investigate the interactions of organic ions and molecules with mineral surfaces. Further work in this direction has been pursued by Nagy et al. (2001) in a study of the adsorption of humic acids on muscovite, and by Fenter et al. (2001a) in a study of HEDP adsorption on barite.

Mineral surfaces tend to have an electrostatic charge, and the distribution of ions at the mineral/water interface is intimately tied to this surface charge. The conventional model of the EDL has surface excesses and deficits of aqueous solute ions, distributed between the so-called condensed (or Stern) and diffuse (or Guoy-Chapman) layers, to balance the fixed charge of the mineral surface (see Hunter 1987 and Brown and Parks 2001 for detailed discussions of the Gouy-Chapman-Grahame-Stern EDL model). Surface charge can develop through chemical reactions at the surface (e.g., ionization of surface functional groups or adsorption of charged ions) or through isomorphous replacements in the lattice (e.g., substitution of Al^{3+} for Si^{4+} or Mg^{2+}) that cause structural charge. Despite the large number of studies of the EDL during the past century, there is little direct knowledge of its structure at the molecular scale. This situation is beginning to change with the advent of SR methods. For example, an *in situ* XSW study of the EDL structure at the rutile(110)/water interface showed that short-period XSW generated by Bragg diffraction can measure ion locations within the condensed layer and the partitioning of ions between the condensed and diffuse layers (Fenter et al. 2000c). Further investigations of the EDL at the rutile/water interface have shown that the location of condensed layer ions does not vary with solution ionic strength (Zhang et al. 2001). Ongoing work using Bragg XSW methods to obtain three-dimensional "holographic" images of ion locations (Zhang et al. unpublished) indicates that the ions Rb^+, Sr^{2+}, Zn^{2+}, and Y^{3+} all have different positions at the rutile (110)/water interface, and synthetic multilayer mirrors are being used with TiO_2 overgrowths to investigate ion distributions in the diffuse layer of the rutile/water interface.

Additional information on the structure of the mineral/water interface has come from neutron diffraction studies. For example, a high-resolution time-of-flight neutron diffraction study of the EDL at the vermiculite (001)/water interface (Williams et al. 1998) found evidence from inverse Monte Carlo modeling of the neutron scattering data that isotopically labeled propylammonium counterions, $C_3H_7NH_3^+$ and $C_3D_7NH_3^+$, are separated from the vermiculate surface by two layers of partially ordered water molecules. Williams et al. (1998) also found that the density of the counterions reached a maximum at the center of the interlayer region, which they claim is at odds with the predictions of the classical Gouy-Chapman-Stern EDL model because it places the maximum in counterion density on the surface in the Stern layer. However, because the potentials extending into the interlayer solution from the two clay layers should be relatively symmetrical, unlike the classical EDL model which considers only one surface, perhaps it is not surprising that the counterions have a maximum concentration at the center of this region. Furthermore, if we accept the idea (Bockris and Reddy 1973) that counterions are non-bonded and remain fully hydrated, leaving approximately two water molecules between the closest approaching ions and the surface, then the counterions will be crowded toward the center of an interphase region narrowed by the thickness of four layers of ions-free water. A similar neutron diffraction study of hydrated Li^+ ions in the hydrated interlayer region of vermiculite also found that the Li counterions are located midway between the clay platelets, forming octahedral hydration complexes with six water molecules (Skipper et al. 1995). This behavior contrasts with that of the larger alkali metal ions Na, K, and Cs, which prefer to bind directly to vermiculite surfaces rather than fully solvate. These results, together with the X-ray standing wave results of Zhang et al. (2001), raise important questions about the applicability of EDL models in confined spaces. Additional molecular-scale experimental studies should help to determine what interfacial model is most appropriate for confined spaces.

Sorption reactions of aqueous metal ions with mineral surfaces – the XAFS perspective

Among the most important applications of XAFS spectroscopy is the study of sorption reaction products at mineral/water interfaces under *in situ* conditions. As shown in Tables 1 and 2 (see Appendix), there have been over 200 XAFS studies of various heavy metals sorbed on the surfaces of minerals, amorphous solids, natural organic matter, and microbial organisms and fungi. Such measurements have yielded a wealth of information on the structure, stoichiometry, attachment geometry (inner vs. outer-sphere, monodentate vs. bidentate or tridentate), the presence of multinuclear complexes, and the presence of ternary surface complexes when complexing ligands are present in solution. Surface complexation modeling alone (e.g., Davis and Kent 1990; Dzombak and Morel 1990; Criscensti and Sverjensky 1999) cannot yield this level of detail reliably because such models are based on macroscopic measurements (Sposito 1986; Brown 1990). However, in combination with the results of XAFS spectroscopy, such modeling can yield accurate descriptions of surface complexation reaction products (e.g., Katz and Hayes 1995a,b; Hayes and Katz 1996; Brown et al. 1999c).

Because several recent papers have reviewed the applications of XAFS spectroscopy to sorption complexes at mineral/solution interfaces (e.g., Brown et al. 1999c; Brown and Parks 2001), here we list many of the sorption systems that have been studied over the past 15 years using XAFS spectroscopy methods (Appendix — Tables 1 and 2) without detailed discussion of results. The interested reader is directed to the individual papers listed in Tables 1 and 2 (see Appendix) for experimental details and results and to Brown and Parks (2001) for a detailed discussion of many sorption systems of relevance to low temperature geochemistry and environmental science.

What are some of the most important findings from these XAFS studies of metal ion sorption processes? One is the discovery that metal ion complexes at mineral/water interfaces are often different from those in bulk aqueous solutions. These differences include higher degrees of hydrolysis and different first-shell coordination environments (e.g., Pb(II) surface complexes: Bargar et al. 1997a), and a higher proportion of multinuclear complexes (e.g., Co(II) and Cr(III) surface complexes on alumina: Chisholm-Brause et al. 1990a; Fitts et al. 2000) for surface complexes vs. solution complexes. These differences are likely related to differences in the properties of water at interfaces vs. in bulk solutions (see earlier section on "comments on the structure and properties of water at solid/aqueous solution interfaces"). Early XAFS studies of metal ion sorption on metal oxides (e.g., Co(II) sorption at the γ-Al$_2$O$_3$/solution interface: Chisholm-Brause et al. 1989a, 1990a, 1991) also showed that divalent transition metal ions tend to form isolated mononuclear, inner-sphere surface complexes at the lowest metal ion sorption densities (< 10% of an effective monolayer), multinuclear complexes with increasing sorption density, and precipitates at sorption densities approaching or above effective monolayer coverages. As shown by TEM examination of the high sorption density samples, these precipitates may or may not be associated with the mineral surface. Another discovery that merits attention is the finding that mineral dissolution in aquatic systems may be followed by precipitation of a new phase that incorporates ions derived from both the dissolution process and the aqueous solution (e.g., Co(II), Ni(II), and Zn(II) ions sorbed on alumina or phyllosilicates: d'Espinose de la Caillerie et al. 1995a,b; 1997; Scheidegger et al. 1997; Towle et al. 1997; Thompson et al. 1999a,b, 2000; Trainor et al. 1999). This process results in the formation of Co, Ni, and Zn layered double hydroxides, which have a much larger capacity for sequestering contaminant ions than a static mineral surface with a limited number of reactive sites per unit surface area. Other EXAFS studies of sorption reaction products found that some contaminant or trace metal ions are incorporated in the structure of the solid phases on which they adsorb by a "neoformation" process. Such studies of divalent transition metal ion sorption on clay minerals found that the adions attach to edge sites of the clay minerals (e.g., Co(II) and Zn(II) neoformation on hectorite: Schlegel et al. 1999a,b, 2001a,b). Another important recent finding is that some of the so-called "indifferent electrolytes" that are used in sorption studies, including nitrate and perchlorate, are not necessarily indifferent. EXAFS evidence suggests that they form ternary complexes with cations such as U(VI) sorbed on hematite (Bargar et al. 2002c) and Pb(II) sorbed on ferrihydrite (Trevedi et al. 2002).

Each of these findings is beginning to affect surface complexation modeling of macroscopic uptake of cations and anions from aqueous solutions onto mineral surfaces, where chemical reactions describing surface sites and sorption products must be assumed in the absence of molecular-level information on reactants and products. It is now known that different chemical reactions with different sorption products may fit macroscopic uptake data equally well (e.g., Hayes and Katz 1996; Brown et al. 1999c), confirming the suggestion of Sposito (1986) that microscopic information cannot be extracted from macroscopic observations. EXAFS spectroscopy is also being applied to the interaction of cations and anions with microbial cell walls, as described in the next section, and is showing that assumptions made about reactive functional groups in bacterial cell walls are not always consistent with direct EXAFS spectroscopic observation.

EXAFS studies of metal ion coordination sites in bacterial and fungal cell walls

Synchrotron-based experimental approaches are well suited for studying the interaction of cations and anions with bacterial cells, although only a few systems have been studied to date. EXAFS spectroscopy can be used in some cases to determine which functional groups (e.g., carboxylic, phosphoryl, phenolic hydroxyl, thiol) within a biofilm or bacterial or fungal cell wall are reactive to trace metals (e.g., Sarret et al. 1998a, 1999a; Kelly et al. 2001, 2002; Boyanov et al. 2002; Webb et al. 2001). For example, Sarret et al. (1999a) used EXAFS spectroscopy and solution chemistry experiments to show that Pb(II) reacts with carboxyl and phosphoryl sites in the cell wall of the fungus *Penicillium chrysogenum*, with weak binding of Pb(II) to carboxyl groups accounting for 95% of Pb(II) uptake and strong binding of Pb(II) to less abundant phosphoryl groups accounting for 5% of the uptake. In EXAFS studies of the sorption products of Cd(II) and U(VI) on *Bacillus subtilis*, Kelly et al. (2002) and Boyanov et al. (2002) found that U(VI) and Cd(II), respectively, bind to phosphoryl groups at low pH and that U(VI) and Cd(II) bind increasingly to carboxyl groups at pH > 3.2 and pH > 4.4, respectively. There is also evidence that Cd(II) binds to an additional site (tentatively thought to be deprotonated phosphoryl ligands) at pH 7.8. These EXAFS results for Cd(II) refine earlier models of Cd(II) sorption on the cell walls of *B. subtilis* derived from acid-base titration data and surface complexation modeling (Fein et al. 1997). These macroscopic modeling results suggest that Cd(II) binds primarily to carboxyl groups at pH < 5 (Fein et al. 1997). Hennig et al. (2001) have also used EXAFS spectroscopy to show that U(VI) binds primarily to phosphoryl groups in the cell walls of *Bacillus cereus* and *Bacillus sphaericus*. In addition, Pu(VI) was found to bind preferentially to phosphoryl groups in the cell walls of *Bacillus sphareicus* (Panak et al. 2002). The interaction of Zn(II) with five microbial species cultured from a lake sediment was studied by Webb et al. (2001) using EXAFS spectroscopy. This study showed that the coordination environment of Zn(II) differs significantly among the five bacteria, ranging from first-shell sulfur ligands to six first-shell nitrogen/oxygen ligands. In two of the bacterial species, evidence for mixed coordination environments was found. Although these findings are preliminary, this study suggests that at least for Zn(II), it is doubtful that a single adsorption edge would be consistent with the uptake of Zn(II) on the cell walls of a variety of bacterial species (*cf.*, Yee and Fein, 2001).

Heavy metal speciation and catalysis at biofilm/mineral interfaces

Microbial biofilms are widespread in soils and form microenvironments in which aqueous chemical conditions differ from those of the host ground water. Reactive functional groups on bacterial surfaces and in exopolysaccharides provide a large array of binding sites for metals, posing the question of whether or not bacterial biomass plays a dominant role in controlling metal ion migration in soils and aquifers. Such processes could have major implications for ground water quality, as the migration and toxicity of heavy metal contaminants in the environment are controlled by interactions between the metal solutes, aqueous solutions, and soil materials. Another important issue concerns the extent to which biofilm coatings on mineral surfaces change the intrinsic reactivity of the coated mineral surface. Recently, Templeton et al. (2001) used X-ray standing wave-fluorescence yield measurements, coupled with XAFS, to probe the distributions of Pb(II), a common toxic soil contaminant, within mineral-biofilm-water systems (Fig. 15). The main purpose of this study was to determine how biofilm coatings change the reactivity of metal oxide surfaces. *In situ* XSW and GI-XAFS measurements were performed on *Berkholderia cepacia* biofilms grown on single-crystal α-Al_2O_3 (0001) and ($1\bar{1}02$) and α-Fe_2O_3 (0001) surfaces, subsequently reacted for 2 hr with aqueous solutions containing various Pb concentrations (ranging from 1 μM to 150 μM) at pH 6 and an ionic strength of 0.005 M. At a Pb concentration of 1 μM, the FY intensity for the α-Al_2O_3 ($1\bar{1}02$) and α-FeO_3 (0001) surfaces peaks at the critical angles of the two substrates ($\approx$ 160 mdeg and 185 mdeg, respectively, at 14 keV), indicating that Pb(II) is located primarily at the corundum or hematite surfaces at this concentration (Fig. 16). With increasing Pb concentration, there are two FY peaks, one occurring at the critical angle of each substrate and one occurring at $\approx$ 60 mdeg for the α-Al_2O_3 ($1\bar{1}02$)-coated surface and at about 85 mdeg for the α-Fe_2O_3 (0001)-coated surface. The growing intensity at the lower incidence angles with increasing [Pb] indicates that Pb(II) is also binding to sites in the *B. cepacia* biofilm coating. At all Pb concentrations studied, the FY-data indicate that Pb(II) binds primarily to functional groups in the biofilm on the α-Al_2O_3 (0001) sample. The results of this study show that Pb(II) binds initially to reactive sites on the α-Al_2O_3 ($1\bar{1}02$) and α-Fe_2O_3 (0001) surfaces even with a biofilm coating that covers essentially the entire mineral surface, as shown by SEM and confocal microscopy studies. The order of reactivity of these biofilm-coated surfaces for Pb(II) [α-Fe_2O_3 (0001) > α-Al_2O_3 ($1\bar{1}02$) >> α-Al_2O_3 (0001)] is the same as that observed in uptake and EXAFS studies of Pb(II) sorption on biofilm-free alumina and hematite surfaces (see Bargar et al. 1996, 1997a,b,c).

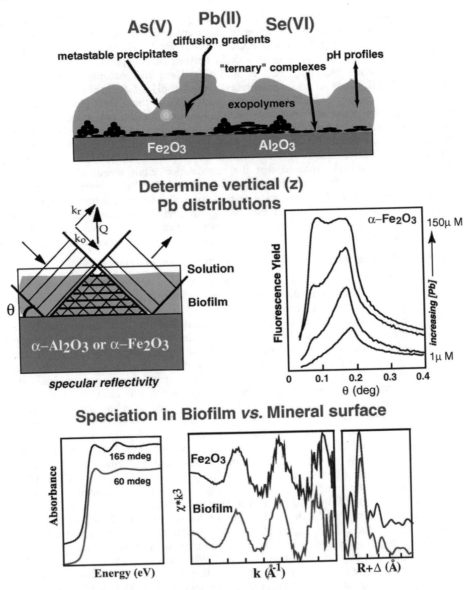

Figure 15. X-ray standing wave and grazing-incidence XAFS studies of the interaction of heavy metal(loid)s with a biofilm-coated metal oxide surface. A model of the biofilm-coated metal oxide surface is shown schematically in upper panel. Also shown are (1) how XSW's are generated (left middle panel); (2) fluorescence-yield XSW data for Pb at different solution concentrations on *B. cepacia*-coated α-Fe$_2$O$_3$ (0001) as a function of incidence angle of the X-ray beam (right middle panel); (3) Pb L$_{III}$-XANES data from a *B. cepacia*-coated α-Fe$_2$O$_3$ (0001) surface after reaction with 1 μM Pb(II)(aq) at two different incident X-ray angles, one sensitive to Pb(II) at the biofilm/mineral interface (165 mdeg) and one sensitive to Pb(II) in the biofilm (60 mdeg) (left lower panel); and (4) Pb L$_{III}$-GI-XAFS spectra (and Fourier transforms) of the Pb(II)/*B. cepacia*/α-Fe$_2$O$_3$ (0001) sample at the two different X-ray incidence angles (the upper at 165 mdeg and the lower at 60 mdeg) (lower right panel). Data were taken on SSRL beam line 6-2.

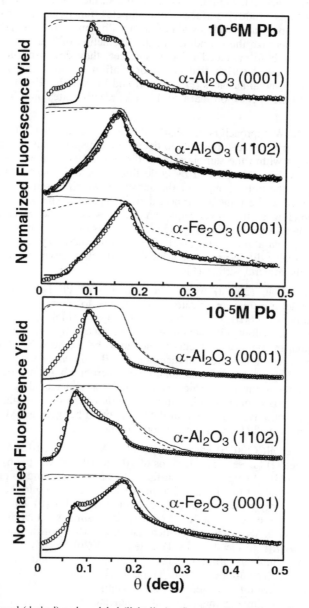

Figure 16. Measured (dashed) and modeled (light line) reflectivity (Log I_o/I_1) profiles and Pb Lα FY profiles (circles) with model fits (heavy line) for the α-Al_2O_3 (0001), α-Al_2O_3 ($1\overline{1}02$), and α-Fe_2O_3 (0001) at 10^{-6} M and 10^{-5} M [Pb]. Data were taken on SSRL beam line 6-2. (from Templeton et al. 2001, with permission)

The study by Templeton et al. (2001) also shows that the *B. cepacia* biofilm does not block all reactive sites on the alumina and hematite surfaces and that sites on the α-Al_2O_3 ($1\bar{1}02$) and α-Fe_2O_3 (0001) surfaces "outcompete" functional groups in the biofim (including the exopolysaccharide exudate) at low Pb concentrations. Although it is not appropriate to generalize these findings to the interaction of aqueous heavy metal ions with all NOM- or biofilm-coated mineral surfaces, they do raise questions about the generalization discussed earlier that NOM or biofilm coatings change the adsorption characteristics of mineral surfaces (*cf.*, Neihoff and Loeb 1972, 1974), and they are also inconsistent with the suggestion that NOM blocks reactive sites on mineral surfaces (*cf.*, Davis 1984).

When the XSW approach is coupled with grazing-incidence (GI)-XAFS and micro-XAFS spectroscopy, it is also possible to determine the speciation of each metal(loid) at separate locations within the sample (i.e., at the mineral surface and within the biofilm matrix). This can be accomplished by collecting GI-XAFS data at an incidence angle corresponding to the critical angle of the substrate, which is sensitive primarily to the metal(loid) sorbed at the mineral surface, and, in separate experiments, at an incidence angle corresponding to the lower angle FY peak, which is sensitive primarily to the metal(loid) sorbed in the biofilm (Fig. 15). The results of these types of experiments for the Pb(II)-*B. cepacia*-alumina and Pb(II)-*B. cepacia*-hematite systems (Templeton et al. 1999) indicate major differences in the local coordination environment of Pb(II) in the biofilm vs. on the mineral surface (Fig. 15).

Templeton et al. (2002a) used a combination of Pb L_{III}-XAFS and μXANES spectroscopy and transmission electron microscopy to show that *B. cepacia* causes biomineralization of Pb(II) in the form of highly insoluble pyromorphite at Pb(II) concentrations well below supersaturation with respect to pyromorphite. The phosphate in these minimal medium experiments is though to be provided by *B. cepacia*, and the pyromorphite forms on the outer cell membrane of *B. cepacia*. These types of studies are beginning to provide unique information on how microbial biofilms affect metal sorption processes at mineral surfaces, which is essential for understanding the transport and bioavailability of toxic metal ions in natural systems where such biofilms exist. They are also allowing quantitative evaluation of the competition between NOM (or biofilms) and the mineral substrates they coat for metal ion binding.

Templeton et al. (2002b) carried out similar XSW and GI-EXAFS studies of the interaction of Se(VI) (at Se concentrations of 10^{-5} to 10^{-3} M and pH 6) with *B. cepacia*-coated α-Al_2O_3 ($1\bar{1}02$) single crystal surfaces. Selenite was found to preferentially bind to the alumina surface at low [Se], and to increasingly partition into the biofilm at higher [Se]. Se(VI) is rapidly reduced to Se(IV) and red elemental Se(0) when the *B. cepacia* is metabolically active, based on Se K-XANES spectra. Similar reduction to Se(0) was also found in an earlier XANES study by Buchanan et al. (1995) of the interaction of Se(VI) with *Bacillus subtilis*. In the study by Templeton (2002b), Se(IV) was found to partition strongly to the alumina/biofilm interface, whereas Se(VI) and Se(0) were found to partition preferentially into the biofilm. These results indicate that the intrinsic reactivity of the alumina surface for selenite, like Pb(II) (Templeton et al. 2001), is not appreciably affected by the biofilm coating. They also indicate that biofilm coatings on mineral surfaces capable of reducing Se(VI) to Se(IV) and Se(0) can lead to sequestration of Se and reduction of its potential bioavailability.

These results emphasize the importance of mineral surfaces as sinks for heavy metals in natural environments, even when coated by biofilms, and raise questions about conventional assumptions that biofilm and organic coatings on mineral surfaces can change the way in which aqueous heavy metals interact with these surfaces. They also show that biofilms can transform redox-sensitive elements like Se into less toxic and less mobile forms, and in concert with the mineral substrates on which they occur, can lead to enhanced sequestration of reduced forms of these elements. Biomineralization was also found to result in the formation of a highly insoluble form of Pb on cell membranes in *B. cepacia* biofilms.

Examples of SR-based studies of complex environmental samples

As discussed in the Introduction to this chapter, SR-based methods are beginning to have a major impact on studies of complex environmental samples such as heavy metal contaminated or polluted soils and sediments, from both anthropogenic and natural sources. XAFS and μXAFS methods in particular have provided information on contaminant/pollutant element speciation, spatial distribution at μm scales, and phase association for the following elements (listed in order of increasing atomic number): **Chromium** (Kemner et al. 1997b; Peterson et al. 1997b; Szulczewski et al. 1997; Jardine et al. 2001; Zachara et al. 2002); **Copper** (Parkman et al. 1996; Sayers et al. 1997; Zhou et al. 1999); **Zinc** (Parkman et al. 1996; Hesterberg et al. 1997a; O'Day et al. 1998, 2000a; Manceau et al. 2000a; Isaure et al. 2002; Juillot et al. 2002; Roberts et al. 2002); **Arsenic** (Foster et al. 1997, 1998; Rochette et al. 1998; La Force et al. 2000; Savage et al. 2000; Kneebone et al. 2002; Moll et al. 2002; Morin et al. 2002); **Selenium** (Tokunaga et al. 1994; Pickering et al. 1995; Tokunaga et al. 1996, 1997, 1998); **Cadmium** (O'Day et al. 1998); **Mercury** (Kim et al. 1999, 2000, 2002b,c; Shaw et al. 2002b); **Lead** (Cotter-Howells et al. 1994; Manceau et al. 1996; Hesterberg et al. 1997a; O'Day et al. 1998; Cotter-Howells et al. 1999; Morin et al. 1999, 2001; Ostergren et al. 1999; Ryan et al. 2001; Mangold and Calmano 2002); **Uranium** (Allen et al. 1994; Bertsch et al. 1994, 1997; Schulze et al. 1995; Morris et al. 1996; Duff et al. 1997, 2000; Hunter and Bertsch 1998; Thompson et al. 1998; Allard et al. 1999; Bertsch and Hunter 2001; Bostick et al. 2002); **Plutonium** (Duff et al. 1999b; Neu et al. 1999a). There have also been XAFS studies of the speciation of natural chlorinated organic matter in decaying plant litter (Myneni 2002), the speciation and transformation of pollutant species in hyperaccumulating plants (Salt et al. 1995, 1997; Lytle et al. 1996, 1998; Hunter et al. 1997; Pickering et al. 2000a,b), and the speciation of trace elements in coal and fly ash (see references by Huffman, Huggins, and co-workers). In addition, soft X-ray spectromicroscopy methods, such as transmission X-ray microscopy, are revealing the macromolecular details of natural organic matter such as fulvic acid in aqueous solutions for the first time under *in situ* experimental conditions (Myneni et al. 1999).

Selected examples of the use of EXAFS spectroscopy to help determine the speciation of heavy metal and metalloid pollutants in soils, sediments, and mine tailings are reviewed below. Other examples are given by Manceau et al. (this volume) and Sutton et al. (this volume).

Chromium speciation in the vadose zone beneath the Hanford S-SX nuclear waste tank farm. Chromium is present in relatively high concentrations in the high-level nuclear waste tanks at Hanford, WA and has been found at concentrations as high as 37,000 ppm in contaminated sediments in the vadose zone beneath the tanks (Zachara et al. 2002). At the high pH values (12-14) of the tank supernatant, aqueous chromium is expected to be dominantly in the chromate (CrO_4^{2-}) form (Baes and Mesmer, 1986). This form of chromium is highly mobile in aquatic systems and highly toxic. In order to assess the environmental impact of chromium in tank leachates that have made their way into the vadose zone beneath the Hanford Tank Farm (Fig. 17), a quantitative knowledge of chromium speciation is essential. More specifically, it must be determined to what extent Cr(VI) species have been reduced, via electron transfer reactions, to Cr(III) species in the sediments in the vadose zone. In contrast to Cr(VI), Cr(III) species should be relatively

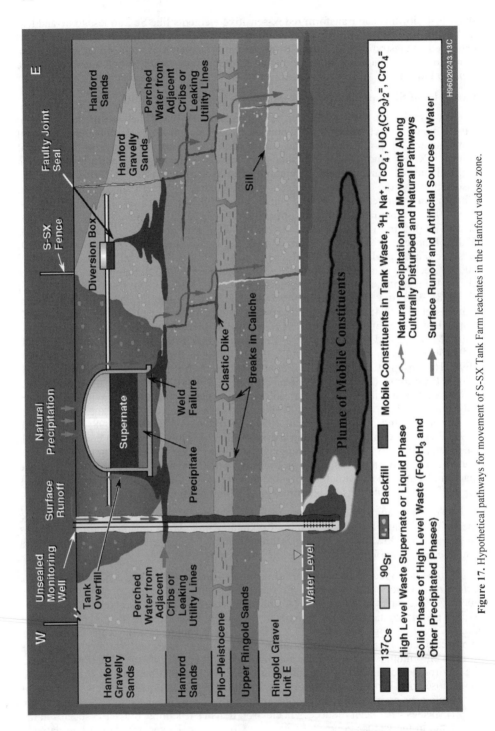

Figure 17. Hypothetical pathways for movement of S-SX Tank Farm leachates in the Hanford vadose zone.

immobile in the unsaturated and saturated zones beneath the Hanford Tank Farm and of low toxicity to biological organisms. As discussed previously, the pre-edge feature of Cr K-edge XANES spectra is very sensitive to differences in Cr oxidation state. The 1s→3d pre-edge height can be calibrated using mechanical mixtures of appropriate Cr(III)- and Cr(VI)-containing model compounds (Fig. 18), and this calibration can be used to determine the ratio of Cr(VI) to Cr(III) in Cr-contaminated soil or sediment samples for which Cr K-XANES data are collected in the same manner on the same beamline on which the calibration measurements were performed.

Cr K-edge XANES studies of four "hot" bore-hole samples from the Hanford S-SX Tank Farm were conducted at SSRL using the new Molecular Environmental Science Beamline (11-2), which is equipped for radioactive sample handling. The main sources of radioactivity in these samples were ^{90}Sr and ^{137}Cs, so special procedures had to be used in packaging these samples at Pacific Northwest National Laboratory for XAFS analysis, and special sample shipping and handling procedures were employed. The XANES spectra of these samples are shown in Figure 19 (Catalano et al. 2002; Zachara et al. 2002). Three of the four samples from the Hanford S-SX Tank Farm bore-holes (41-09-39-6AB, 41-09-7ABC, and SX-108-8A) contain 58-69% Cr(VI), whereas the fourth sample (SX-108-7AB) contains 27% Cr(VI). These data suggest that while significant reduction of chromium has occurred, most likely via abiotic pathways due to the lack of significant numbers of microorganisms in the Hanford vadose zone, it has not been complete, assuming that the bulk of Cr introduced into the Hanford vadose zone was Cr(VI). Possible reasons for this incomplete reduction include (1) insufficient reductants in the vadose zone, (2) the presence of non-conductive carbonate coatings on Fe(II)-bearing minerals, (3) the build-up

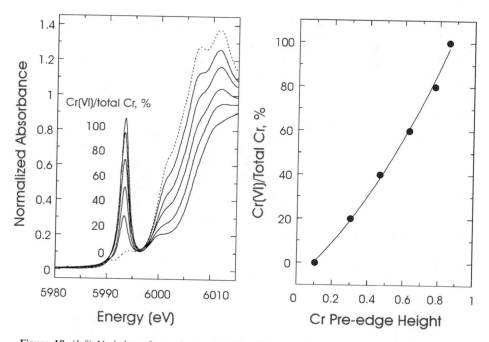

Figure 18. (*left*) Variation of pre-edge peak height of Cr K-edge XANES spectra as a function of Cr(VI)/total Cr. (*right*) Plot of Cr(VI)/total Cr *vs.* Cr pre-edge peak height. (after Peterson et al. 1996, based on XANES data taken at SSRL on beam line 4-3)

Brown & Sturchio

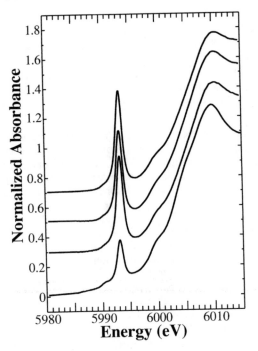

Figure 19. Normalized Cr K-edge XANES spectra of the Hanford S-SX Tank Farm contaminated sediment samples taken at SSRL. From top to bottom: 41-09-39-7ABC, 41-09-39-6AB, SX-108-8A, SX-108-7A. Data were taken on SSRL BL 11-2. (after Catalano et al. 2001; Zachara et al. 2002)

of non-conducting Cr(III)-hydroxide coatings on reductant surfaces, which can quickly stop the electron transfer process necessary for the Cr(VI) to Cr(III) reduction reaction, (4) the lack of sufficient water in the vadose zone to dissolve Fe(II)-bearing minerals and provide aqueous Fe(II) for reduction of Cr(VI), or (5) some other process that has inhibited the transfer of electrons from reductants to Cr(VI).

Speciation of zinc in mining wastes and contaminated soils. The molecular-level speciation of Zn(II) has been studied using EXAFS spectroscopy and other methods in contaminated stream sediments from the U.S. Tri-State Mining District (O'Day et al. 1998), in contaminated estuarine sediments in northern California (O'Day et al. 2000a), in smelter-contaminated soils in northern France (Manceau et al. 2000a; Juillot et al. 2002; Manceau et al., this volume), and in a Zn-contaminated dredged sediment (Isaure et al. 2002). In the case of the Zn-contaminated stream sediments (O'Day et al. 1998), the primary Zn ore mineral is sphalerite, which has dissolved in part to release Zn into solution. This study suggests that the released Zn is predominantly present in high-iron sediments as sorbed species on ferrihydrite and goethite, whereas in low-iron sediments, it is predominantly present as sorbed species on ferrihydrite or as precipitated Zn-hydroxide. For the estuarine sediments, O'Day et al. (2000a) suggested that Zn is present as a mixture of sphalerite and Zn associated with clay minerals (smectite or chlorite) substituted into the clay's octahedral layer or sorbed onto the edges of clay minerals.

In the case of smelter-contaminated soils, the study by Manceau et al. (2000a) used EXAFS (both powder and polarized) and powder XRD data to show that Zn is present in the high-temperature phases franklinite ($ZnFeO_4$), willemite (Zn_2SiO_4), hemimorphite ($Zn_4Si_2O_7(OH)_2 \cdot H_2O$), and Zn-containing magnetite-franklinite solid solution. Zinc released during the weathering of these phases is thought to be present in the clay soil fraction as "neoformed" Zn(II)-phyllosilicate, Zn(II) incorporated in Fe-(oxyhydr)oxides, and Zn(II) sorbed onto birnessite. The study by Juillot et al. (2002) of Zn-contaminated soil samples from similar localities used a combination of EXAFS spectroscopy, μSXRF, XRD, and selective chemical extractions to show that Zn is present as sphalerite (and wurtzite) and Zn(II)-bearing magnetite in the dense soil fraction, whereas it is associated with soil organic matter and present as Zn(II)-Al(III)-hydrotalcite, Zn(II)-bearing clay, and Zn(II)-bearing magnetite-franklinite solid solution in the clay fraction.

The differences in Zn(II)-containing phase association determined by Juillot et al. (2002) vs. Manceau et al. (2000a) in smelter-contaminated soils show that in complex phase assemblages, EXAFS data alone may not be sufficient to uniquely identify the types of phases with which a contaminant is associated. In these cases, other complementary methods, such as selective chemical extraction and TEM, are needed, and even then it may not be possible to uniquely identify all the phases that contain the contaminant element or to resolve differences in interpretation.

The XAFS study of Zn in a contaminated dredged sediment by Isaure et al. (2002) used a combination of methods (EXAFS, μXAFS, μPIXE, powder XRD, electron microprobe analysis, ICP-AES analysis, and selective chemical extractions) to identify three primary Zn-containing minerals (sphalerite, willemite, zincite) plus zinc associated with Fe-oxyhydroxide and phyllosilicates that was released by chemical weathering.

Speciation of As in mining wastes, soils, and sediments. The word *arsenic* has become synonymous with the word *poison* because of the use of As-containing compounds as poisons throughout much of the history of western civilization (Azcue and Nriagu 1994). This element has received a great deal of public attention because of its links to certain types of cancers and its high levels in some drinking water supplies (Nordstrom 2002). The recent lowering of the maximum contaminant level (MCL) of As from 50 ppb to 10 ppb in the U.S. and other parts of the world is a reflection of the concern about As as a public health hazard. As-contaminated drinking water (e.g., Ganges delta, West Bengal, India: Chatterjee et al. 1995; and Bangladesh: Bhattacharya et al. 1997; Vietnam: Berg et al. 2001) and As-contaminated coal (e.g., Guizhou Province, southwestern China: Finkelman et al. 1999) in a number of localities are taking a major toll in terms of human health, and emphasize the need for closer coupling of geochemical studies of the natural sources and mitigation of pollutants like As with efforts to educate the public about the dangers of long-term exposure to such pollutants. Although not as well publicized, there are also many examples of As-contaminated soils and sediments in other areas that impact humans. One such area is the Mother Lode Gold District of the Sierra Nevada foothills, California. Another is a natural As geochemical anomaly in Central France. XAFS spectroscopy has been used to determine the speciation of As in both of these cases, as well as in naturally polluted aquifer sediments in the Ganges delta, Bangladesh. These three examples are discussed below.

Arsenic is a significant pollutant from acid mine drainage associated with gold mines in the Sierra Nevada Mother Lode District in California, USA, due mainly to its release from arsenious pyrites (Foster et al. 1997, 1998a; Savage et al. 2000). XAFS spectroscopy, combined with electron microprobe analysis and Rietveld analysis of powder diffraction data, has been used to determine the relative proportions of different As species in three California mine wastes: a fully oxidized tailings (Ruth Mine), a partially oxidized tailings (Argonaut Mine), and a roasted sulfide ore (Spenceville Mine) (Foster et al. 1998a). Arsenic K-edge XANES analysis showed that As(V) is the predominant oxidation state in the mine samples, but mixed oxidation states were observed in the Argonaut mine-waste. Linear combination fitting of component (model compound) spectra to the mine tailings samples resulted in identification of the following As(V) species: scorodite ($FeAsO_4 \cdot 2H_2O$) and As(V) adsorbed on goethite (α-FeOOH) and/or gibbsite (γ-Al(OH)$_3$). Non-linear, least squares fits of mine waste EXAFS spectra indicated variable As speciation in each of the three mine wastes. Foster et al. (1998a) concluded that ferric oxyhydroxides and aluminosilicates (probably clay) bind roughly equal portions of As(V) in the Ruth Mine sample. The Argonaut Mine tailings were found to contain $\approx$ 20% reduced As bound in arsenopyrite (FeAsS) and arsenical pyrite (FeS$_{2-x}$As$_x$) and $\approx$ 80% As(V) in a ferric arsenate precipitate such as scorodite. Roasted sulfide ore of the Spenceville Mine was found to contain As(V) substituted for sulfate in the crystal structure of jarosite (KFe$_3$(SO$_4$)$_2$(OH)$_6$), and sorbed to the surfaces of hematite grains, which were identified in the Rietveld XRD analysis. One of the most important findings of this study was the presence of As sorbed to mineral surfaces, which is a form that is potentially more bioavailable than As bound in insoluble crystalline or X-ray amorphous precipitates.

Evidence for As(V) sorbed on Fe-oxides has also been found in a soil that developed above a regional As geochemical anomaly (up to 900 ppm As in soils overlying saprolites containing up to 5200 ppm As) at Echassiéres, Allier, France by Morin et al. (2002). The underlying bedrock was mica schist that had been mineralized by later hydrothermal solutions, resulting in the deposition of arsenopyrite and löllingite. XRD and µXRD, SEM, and electron microprobe analyses were used to show that As released by the alteration of these two minerals is incorporated in the iron arsenate pharmacosiderite. EXAFS spectroscopy was used to show that As(V) is sorbed on iron oxides, and that the proportion of sorbed As(V) to pharmacosiderite increases from the saprolite to the topsoil above this As-mineralized mica schist.

A much-publicized example of As pollution that is affecting millions of people is the arsenic poisoning of alluvial aquifers underlying the Ganges delta of Bangladesh (Bhattacharya et al. 1997; Nickson et al. 1998; Acharyya et al. 1999; Chowdhury et al. 1999; McArthur 1999; Foster et al. 2000; Nickson et al. 2000; Nordstrom 2002). Some of the drinking water obtained from shallow wells in this area contain as much as 1000 ppb As, with most containing in excess of 50 ppb. Although there is some controversy about the source of the As (i.e., is it released by the breakdown of arsenious pyrite or by the reductive dissolution of Fe-oxyhydroxide to which As is sorbed: Acharyya et al. 1999; Chowdhury et al. 1999; McArthur 1999; Nickson et al. 2000), As K-XAFS spectroscopy (Foster et al. 2000) has clearly shown that in the soil above the aquifer As(III) is associated with waterlogged, reduced sediments and that As(V) is predominantly associated with clay sediments containing Fe-oxyhydroxides. In the aquifer sediment, As(III) is dominantly found in reduced micaceous, fine-grained sandy sediment at depths of 6 to 25 meters ("reduced sediment"), and As(V) is dominantly found in medium- to coarse-grained Fe-oxyhydroxide-coated quartz sand and weathered mica at depths of 26-48 meters ("oxidized sediment"). The results of EXAFS analysis are consistent with As(III) sorbed to Al-hydroxide, clay minerals, or altered micas in the "reduced sediment" zone and with As(V) sorbed to Fe-oxyhydroxide coatings in the "oxidized sediment" zone. The study by Foster et al. (2000) supports previous suggestions (e.g., Nickson et al. 1998, 2000) that Fe-oxyhydroxide is a major repository of As in the aquifer sediments. However, the reason for its release into groundwater is still under study.

Selenium pollution in the Central Valley of California. Selenium occurs naturally in sediments and soils in many parts of the Western U.S. as well as other parts of the world (e.g., China). When such soils are irrigated for agricultural purposes, this element becomes soluble and is transported in agricultural drainage waters to ponds and reservoirs where it becomes concentrated in water-borne plants and animals (up to 3000 ppm) (Frankenberger and Benson 1994). One result of this concentration process was discovered in the early 1980's by government scientists at the Kesterson National Wildlife Refuge in Merced

County, California. Wildlife, particularly waterfowl, died or were born deformed from consumption of high levels of selenium (Ohlendorf and Santolo 1994). Similar problems have since been documented at nine sites in eight Western states comprising some 1.5 million acres of farmland. This problem has major financial and health implications in the San Joaquin Valley of California, a vast area that produces a significant portion of vegetables, fruits, and other crops in the U.S.. If farmers were prevented from draining irrigation waters in this region, the rapid build up of salts in the soil would quickly make production of these crops difficult or impossible. About 500,000 acres of farmland in the San Joaquin Valley—one quarter of the Valley's agricultural acreage—are at stake, with an annual crop production worth about $500 million.

One of the keys to understanding and potentially solving this problem is knowledge of the speciation of selenium in the soils and groundwaters. To provide molecular-level information on the chemical forms of selenium present in Se-contaminated soils from the Kesterson Reservoir area, XAFS spectroscopy studies were carried out which showed that selenate and selenite are present in the top few cm of soil adjacent to the drainage ponds but are reduced to elemental selenium at lower soil levels (Pickering et al. 1995; Tokunaga et al. 1997) (Fig. 20). As shown in Figure 20, Se K-XANES spectra are sensitive to Se in various chemical forms, including those in which Se is bound to amino acids. Following controlled laboratory studies of soil columns to which selenate-containing solutions were

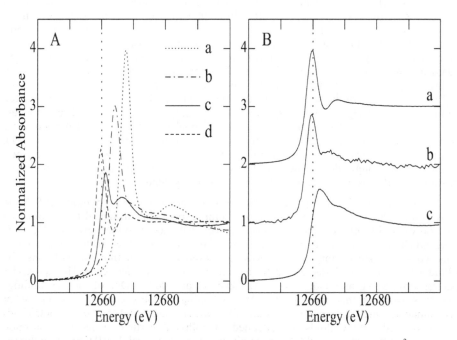

Figure 20. Se K-edge XANES spectra of **(A)** Se-containing model compounds: (a) SeO_4^{2-} (aq), (b) SeO_3^{2-} (aq), (c) selenomethionine (aq), (d) elemental (red) selenium (monoclinic); **(B)** (a) Se in Kesterson soil at depth of 0 to 0.05 m (340 ppm Se), (b) Se in Kesterson Reservoir soil at depth of 0.05 to 0.15 m (40 ppm Se), (c) Se in mushroom (*Agaricus bernardii*) collected adjacent to the reservoir (500 ppm Se). Linear combination fits of the Se K-XANES of the three Kesterson samples resulted in the following components: (a) 97% elemental Se + 3% aqueous selenite; (b) 86% elemental selenium + 14% aqueous selenite; (c) 71% selenomethionine + 11% aqueous selenite + 18% selenocystine. Data were taken on SSRL beam line 4-3. (after Pickering et al. 1995)

added, Tokunaga et al. (1996, 1998) used μXRF and μXANES to show that selenate is rapidly converted into elemental selenium in these columns. The reduction can occur via both biotic (Oremland et al. 1989; Macy 1994; Losi and Frankenberger 1998) and abiotic (Myneni et al. 1997a) pathways (see Fig. 7). However, when reoxidized to the selenate form during irrigation, selenium becomes highly mobile and is transported as an aqueous complex in drainage waters. Various solutions to this problem have been proposed, including bacterial reduction, immobilization, and removal of selenium from drainage waters or the use of drainage waters to irrigate land on which salt-resistant plants such as cotton, Eucalyptus trees, and atriplex are grown. None have been adopted and some skeptics doubt that a viable solution, which satisfies environmental, financial, and political constraints, will be found. An eventual solution to this problem will require a detailed knowledge of the redox chemistry of selenium, its speciation in soils and groundwaters, and the effect of microbial organisms and inorganic reductants on its speciation.

Mercury pollution in the California Coast Range. Mining of mercury ores has created a legacy of Hg-contaminated mine wastes in Hg-mineralized areas such as the California Coast Range. In addition, the use of elemental Hg in gold amalgamation processes during historic gold mining has resulted in gold-mining wastes containing elevated levels of Hg. These polluted sites pose significant environmental hazards locally (Yamauchi and Fowler 1994) and contribute to the Hg pollution of water bodies within the drainage areas of the mine sites, including drinking water reservoirs and San Francisco Bay (EPA 1997). The geologic variety of the primary Hg ores and the various ore processing methods used has resulted in chemically complex, heterogeneous mine wastes in which the distribution and composition of Hg-bearing species are poorly understood. Furthermore, weathering and dispersal of mine wastes redistribute Hg and in some cases change its chemical form, which may increase its bioavailability to organisms. Kim and co-workers (Kim et al. 1999, 2000, 2002b,c) have used XAFS spectroscopy, in combination with transmission and analytical electron microscopy, XRD, electron microprobe analyses, and sequential chemical extractions to determine the dominant types of Hg-containing species in Hg-mining wastes in the California Coast Range, including calcined ore in tailings piles. Cinnabar (HgS, hex) and metacinnabar (HgS, cub) are the dominant mercury species in most of the samples examined (Fig. 21). Mercury in these forms does not pose much of an environmental hazard because of their extremely low solubilities. However, highly soluble mercuric chloride species were identified in calcines generated from hot-spring mercury deposits.

In conjunction with the speciation studies described above, Lowry et al. (2002) and Shaw et al. (2002b) conducted laboratory column experiments designed to examine the transport of Hg by colloids. Starting materials were calcines from the New Idria Hg mine in central California as well as raw mine wastes from other localities, including the Sulfur Bank Mine in Lake County, CA. Following colloid generation from the columns, the colloidal fraction was characterized by a combination of powder XRD, Hg L_{III}-XAFS spectroscopy, SEM, and analytical TEM (Shaw et al. 2002b). One of the key findings from this work is that most of the Hg-containing colloids consist of cinnabar or metacinnabar, which is surprising in light of the widely held belief that heavy metal(loid) ions associated with colloids are attached to colloidal particle surfaces as sorption complexes. While there could be a small fraction of sorbed Hg(II) on particles of hematite and alunite/jarosites, which are common non-Hg-bearing constituents of the colloidal fraction in many Hg mine wastes, the bulk of the mercury is present as the crystalline mercuric sulfides cinnabar or metacinnabar.

Speciation of lead in mining wastes and contaminated soils and sediments. Lead is the most common heavy metal pollutant at the earth's surface and is associated with a

XAFS spectroscopy

Sample Locale	Composition	Residual
Turkey Run Mine $[Hg]_T$=1060 ppm	58% Cinnabar, HgS (hex) 42% Metacinnabar, HgS (cub)	0.036
Oat Hill Mine $[Hg]_T$=940 ppm	58% Cinnabar, HgS (hex) 19% Mercuric Chloride, $HgCl_2$ 13% Corderoite, $Hg_3S_2Cl_2$ 10% Terlinguite, Hg_2OCl	0.281
Corona Mine $[Hg]_T$=550 ppm	50% Cinnabar, HgS (hex) 39% Metacinnabar, HgS (cub) 11% Schuetteite, $HgSO_4$	0.052
Sulfur Bank Mine $[Hg]_T$=250 ppm	43% Metacinnabar, HgS (cub) 30% Corderoite, $Hg_3S_2Cl_2$ 20% Cinnabar, HgS (hex) 7% Mercurous Chloride, HgCl	0.185
Gambonini Mine $[Hg]_T$=230 ppm	84% Metacinnabar, HgS (cub) 16% Cinnabar, HgS (hex)	0.326

TEM micrograph and EDX spectra

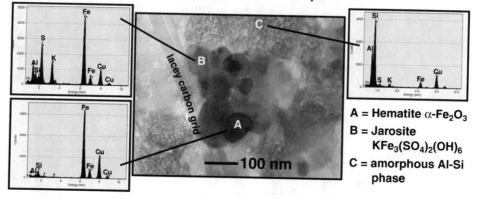

A = Hematite α-Fe_2O_3

B = Jarosite $KFe_3(SO_4)_2(OH)_6$

C = amorphous Al-Si phase

Figure 21. Mercury speciation in Hg-mine wastes from the California Coast Range as determined by Hg L_{III}-EXAFS spectroscopy. Also shown are TEM images and EDX analyses of colloidal phases generated from these mine wastes. XAFS data were taken on SSRL beamline 11-2 (after Kim et al. 2000; Shaw et al. 2002b)

number of human disorders, including learning disabilities in the young. There have been a number of XAFS studies of Pb(II) in contaminated soils and mine wastes which provide insights to its speciation in such media, the relative importance of adsorbed Pb(II) and the phases with which it preferentially associates, and the effects of aging time on the speciation of Pb(II) (e.g., Cotter-Howells et al. 1994; Manceau et al. 1996; Hesterberg et al.

1997a; O'Day et al. 1998, 2000a; Ostergren et al. 1999; Morin et al. 1999, 2001). For example, the XAFS studies by Ostergren et al. (1999) of Pb-contaminated mine tailings from Leadville, CO, USA, and by Morin et al. (1999) of Pb-contaminated soils near a major Pb-Zn smelter in northern France, both found evidence for significant Pb(II) sorption on mineral phases, including iron and manganese hydroxides, and on humic materials.

The approach used in the study by Ostergren et al. (1999) is shown in part in Figure 22. The samples examined in this study came from two types of tailings piles from Leadville, Colorado, one dominated by metal sulfides with classic acid mine drainage (Apache Tailings), and one dominated by carbonate country rock, which buffers the pH to near-neutral values (Hamms Tailings). Electron microprobe analysis identified

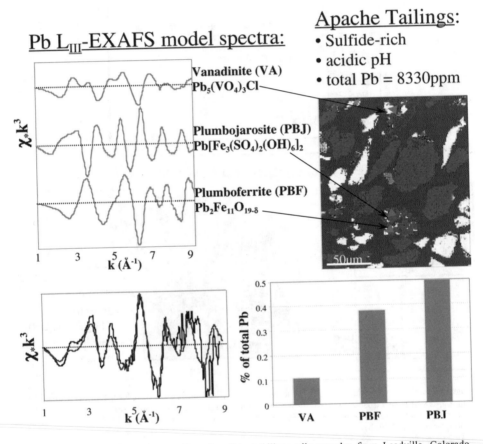

Figure 22. Composite of the XAFS study of two tailings pile samples from Leadville, Colorado. Backscattered electron image of Apache tailings sample from Leadville is shown in the upper right panel. Pb L$_{III}$-EXAFS spectra of Pb-containing model compounds are shown in upper left panel. Pb L$_{III}$-EXAFS spectra of the Apache tailings sample, together with the best-fit model derived from linear combinations of the Pb-containing model compound spectra, are shown in the lower left panel. Bar graph showing the percentage of components in the tailings samples is shown in the lower right panel. Data were taken on SSRL beam line 4-3. (after Ostergren et al. 1999; Brown et al. 1999b) (*Figure continued on next page*)

vanadinite, plumbojarrosite, and plumboferrite in the Apache Tailings and pyromorphite and hydrocerrusite in the Hamms tailings. Evidence for these Pb-containing phases was also observed in powder X-ray diffraction patterns. Sequential chemical extractions using a MgCl₂ step resulted in no removal of Pb (suggesting that no Pb is weakly bound to particles as outer-sphere complexes). This step was followed by an EDTA step, which resulted in removal of ≈ 45% of the Pb from the Hamms tailings samples and ≈ 32% of

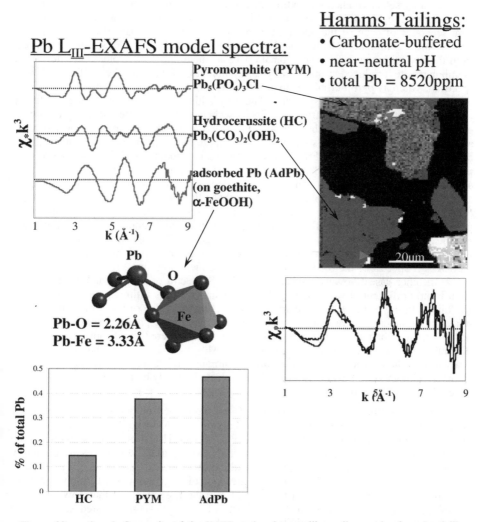

Figure 22 continued. Composite of the XAFS study of two tailings pile samples from Leadville, Colorado. Backscattered electron image of Hamms tailings sample from Leadville is shown in the upper right panel. Pb L_III-EXAFS spectra of Pb-containing model compounds are shown in upper left panel. Pb L_III-EXAFS spectra of the Hamms tailings sample, together with the best-fit model derived from linear combinations of the Pb-containing model compound spectra, are shown at the bottom right. Bar graph showing the percentage of components in each of the tailings samples is shown in the lower left panel. A model of the Pb(II) sorption complex on an Fe(O,OH)₆ octahedron of goethite is shown above the bar graph. Data were taken on SSRL beam line 4-3. (after Ostergren et al. 1999; Brown et al. 1999b)

that from the Apache Tailings samples. This latter extraction step should remove more strongly sorbed Pb from particle surfaces. X-ray photoelectron spectroscopy was also performed on polished sections of both samples, but was not definitive because of the similarity of Pb $4f_{7/2}$ chemical shifts of the two tailings samples. However, the Pb $4f_{7/2}$ XPS data (as well as XAFS spectra) did indicate reduction of surface-bound Pb following the EDTA treatment step and some alteration and redistribution of surface-bound Pb species following the $MgCl_2$ step (see also Calmano et al. 2001 for a discussion of the artifacts introduced by selective chemical extractions of Pb from environmental samples that were also examined by XAFS spectroscopy). Pb L_{III}-edge EXAFS spectra were collected on pure separates of each of the above Pb-containing crystalline phases (Fig. 22) and were used in linear combination least-squares fits of the Pb L_{III}-edge EXAFS spectra of samples from the two tailings piles. The combination 12% vanadinite + 38% plumboferrite + 50% plumbojarrosite produced a statistically acceptable fit of the Apache Tailings EXAFS spectrum, whereas pyromorphite + hydrocerrusite alone produced a relatively poor fit of the Hamms Tailings EXAFS spectrum. The only additional component needed to fit the Hamms Tailings EXAFS spectrum was Pb(II) sorbed in inner-sphere bidentate fashion on goethite – a crystalline phase that was also identified in the electron microprobe and XRD analysis of the Hamms Tailings sample. This component was the most significant in the fit of the Hamms Tailings EXAFS spectrum, comprising about 50% of the total Pb, with pyromorphite comprising $\approx$ 38% and hydrocerrusite comprising $\approx$ 12%. The presence of adsorbed Pb(II) in the Hamms Tailings is consistent with its expected sorption behavior as a function of pH. In low pH solutions, such as those flowing through the Apache Tailings, Pb(II) should not sorb appreciably on any of the phases present. However, in near-neutral pH solutions, such as those flowing through the buffered Hamms Tailings, significant Pb(II) sorption on phases such as goethite is expected (e.g., Hayes and Leckie, 1987). The linear combination fitting approach works in this case because of significant differences in the EXAFS spectra of the model compounds used, including Pb(II) sorbed on goethite at pH 7. In other cases, however, where the EXAFS spectra of model compounds are not significantly different, it would not work. The accuracy of this fitting procedure was tested by "blind" fitting of the XAFS spectra of samples that consisted of known mixtures of the model compounds and was found to be within ± 10% of the phases present (i.e., phases comprising less than about 10% of a fit should not be considered significant). This example illustrates the unique information provided by XAFS analysis, in combination with other analytical methods. The presence of significant amounts of Pb adsorbed on goethite in the Hamms Tailings sample would not have been detected directly by any of the methods used except for XAFS spectroscopy.

The XAFS study of a Pb-contaminated soil developed on a geochemical anomaly arising from a Pb-Zn stratabound deposit in Largentière (Ardèche, France) (Morin et al., 2001) showed that plumbogummite [$(PbAl_3(PO_4)_2(OH)_5 \bullet H_2O)$] is the most abundant Pb phase in the soil profile, and that Pb(II)-Mn-(hydr)oxide surface complexes are gradually replaced by Pb(II)-surface complexes with other phases, possibly Pb(II)-organic complexes, upward in the soil profile. The presence of large amounts of Pb-phosphate in this soil after many thousands of years of weathering suggests that low-solubility phosphates may be important long-term hosts of Pb in Pb-contaminated soils having sufficiently high phosphorous activities to cause formation of these phases.

Uranium pollution in soils from the Fernald, Ohio Uranium Processing Plant. Uranium pollution in soils at the Fernald, Ohio Uranium Processing Plant has been studied using powder XAFS methods on BL 4-2 at SSRL (Allen et al. 1994) and μXANES mapping methods at the NSLS on BL X-26A (Hunter and Bertsch 1998). Attempts to remediate this soil by carbonate washing methods, which remove sorbed uranyl ions from mineral surfaces due to the formation of highly stable uranyl carbonato solution complexes above pH 7, resulted in incomplete removal of the uranium from the polluted soil. The two XAFS studies cited above identified an insoluble U^{4+}-containing phase in addition to the soluble uranyl species, which explained the ineffectiveness of the remediation methods that were used. These molecular-level studies illustrate the major benefit of understanding the species of a pollutant present at a polluted site prior to choosing a particular remediation technology for its removal. Although these polluted soils were eventually dug up and transported to an EPA-approved disposal site, modification of the soil washing methods could have resulted in effective "*in situ*" removal of the uranium in this case.

Remediation of uranium-contaminated groundwater. The study by Fuller et al. (2002a,b) presents an interesting example of the use of XAFS spectroscopy and SR-powder diffraction to evaluate a remediation strategy for removing U(VI) from contaminated groundwater at Fry Canyon, Utah, USA. The remediation plan involved the placement of a reactive barrier consisting of bone-char pellets in a trench that was dug perpendicular to the direction of flow of the U-containing groundwater plume. The hypothesis tested was that an insoluble U(VI)-phosphate, such as autunite or chernikovite, would precipitate following reaction of the U-contaminated groundwater with inorganic phosphate dissolved from the apatite in the bone-char. However, neither XAFS results nor XRD results showed any evidence for U-phosphate precipitates even though U(VI) in the groundwater was attenuated by the reactive barrier. EXAFS analysis of the reacted bone-char showed that the mechanism of U(VI) attenuation involved sorption of U(VI) on bone-char surfaces, without precipitation of a detectable U(VI)- phosphate.

Plutonium pollution in soils at the Rocky Flats, Colorado Environmental Technology Site. During the decades-long operation of Rocky Flats, Colorado as a Pu processing plant, significant pollution of the underlying soils occurred due to spills of processing liquids. The risk associated with trace Pu contamination in soils is controlled to a large extent by its speciation, as discussed earlier. If incorporated in an insoluble phase, Pu poses less environmental risk because its rate of release will be extremely slow. In contrast, Pu in readily leachable phases or sorbed weakly on mineral surfaces poses a significant short-term environmental risk. In order to assess the speciation of Pu in Rocky Flats soils at the 903 drum storage site, Neu et al. (1999a) carried out Pu L_{II}-edge EXAFS studies on soil samples from this site at SSRL and found that the Pu is predominantly present as $PuO_2(s)$, which is highly insoluble under soil water conditions. This work substantiates earlier assessments that solute transport models are not applicable to Pu migration in Rocky Flats soils. Rather, particulate transport appears to be the dominant transport mechanism for Pu migration at the site. These findings translate directly into substantial cost savings because they allow the remediation contractors to focus efforts on erosion modeling and construction of dams and barriers to control particulate transport rather than removing huge quantities of soils from the site or implementing very costly soil washing procedures, which would likely be ineffective due to the insoluble nature of $PuO_2(s)$.

Development of tank waste Np and Pu separations technology. More than 300 million gallons of fission-product-bearing tank waste exist within the DOE nuclear weapons complex. The requirement to clean up these facilities is driving development of technologies that can separate high-level radionuclides, to be vitrified for disposal, from benign tank waste components such as Al-oxide. One key problem with the proposed alkaline (2.5N NaOH) sludge-washing process, designed to remove amphoteric Al from the tank sludges, is co-removal of Pu and Np (Tait et al. 2001). EXAFS measurements performed at SSRL have defined the molecular structures and compositions of dissolved Np and Pu species in sludge wash solutions, namely $[AnO_2(OH)_4]^{2-}$ and $[AnO_2(OH)_5]^{3-}$

(An = Np, Pu) (Clark et al. 1997; Neu et al. 1999b). Although U does not form these amphoteric complexes in NaOH solution, this work shows that Np and Pu form highly soluble anionic species under these tank waste conditions (Fig. 23). The discovery and characterization of these anionic species are requisite fundamental steps leading to modeling and design of enhanced chemical separations processes. Proper design of such processes will save U.S. taxpayers money and natural resources.

FUTURE SR DEVELOPMENTS AND THEIR POTENTIAL IMPACT ON RESEARCH IN LOW TEMPERATURE GEOCHEMISTRY AND ENVIRONMENTAL SCIENCE

A number of new high-brightness third-generation synchrotron X-ray sources will come on line over the next few years, including SPEAR3 at SSRL (SLAC) in the U.S., Diamond in the U.K., Soleil in France, the Canadian Light Source in Canada, and an upgraded National Synchrotron Light Source at Brookhaven National Laboratory in the U.S.. These powerful facilities will add to the capabilities of existing third-generation SR sources (the APS and ALS in the U.S., ESRF in France, BESSY-II in Germany, and SPring-8 in Japan) and will result in enhanced access to third-generation SR sources for researchers in all fields, including low temperature geochemistry and environmental science.

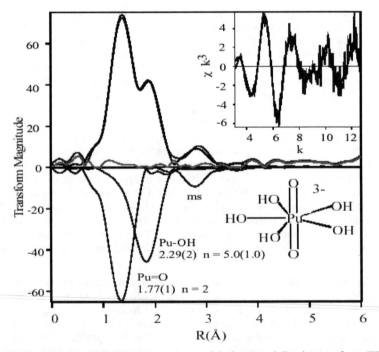

Figure 23. Plutonium L_{III}-EXAFS spectrum (upper right inset) and Fourier transform (FT) (main figure) of PuO_2^{2+} in 3.5 M TMAOH solution. The FT has been fit with two main features (shown pointing downward) corresponding to two Pu-O pair correlations at 1.77 Å and five Pu-OH pair correlations at 2.29 Å. The structure of the PuO_2^{2+} aqueous species is shown in the lower right inset drawing. The EXAFS data were taken on SSRL beam line 4-2. (after Neu et al. 1999b)

What capabilities do third-generation sources provide to earth and environmental science users? Because of their increased brightness, third-generation sources are more optimized for micro-spectroscopy and X-ray microscopy experiments than second-generation sources. This attribute has been taken advantage of at the ALS, which has the largest number of X-ray microscopy beam lines of any synchrotron radiation facility in the world. In addition, micro-focussing is a major emphasis at the APS and ESRF, where high-brightness hard X-rays are used for many types of micro-beam experiments. We foresee a major increase in studies of environmental and low temperature geochemical samples that will make use of these X-ray microscopes on soft X-ray/VUV beamlines (such as those at the ALS) and the micro-focussing spectroscopy and scattering beamlines at hard X-ray facilities (such as the APS and ESRF).

One research area that is certain to benefit from use of these facilities is the study of molecular-scale mechanisms of bioremediation and phytoremediation, where knowledge of the spatial distribution of contaminant species at the cellular level is critical for understanding reaction mechanisms and locations within or external to cells. Microfluorescence tomography is already beginning to yield this information in three dimensions at spatial scales of a few microns. Another growth area that will exploit X-ray microscopes is the characterization of natural organic matter and its interaction with mineral surfaces under *in situ* conditions. An X-ray microscopy study of fulvic acid in aqueous solutions at the ALS has already provided the first direct images of the macromolecular conformation of fulvic acid under *in situ* conditions (Myneni et al. 1999), which show that assumed structures published in textbooks are not correct. A third area is the study of colloidal particles and their interaction with aqueous solutes and microbial organisms. The 20-100 nm resolution possible with modern X-ray microscopes will make it possible to examine these interactions under *in situ* conditions at a level of detail that is far beyond what was possible only a few years ago. Construction of the new Molecular Environmental Science beamline at the ALS has been completed and the new facility should be ready for general users in 2003. It will be optimized for STXM, PEEM, X-ray emission, and XPS studies of environmental samples under *in situ* conditions (i.e., wet samples). This new beam line will take advantage of photon in-photon out methods and newly developed differentially pumped UHV sample chambers (e.g., Ogletree et al. 2000) that will allow soft X-ray XAFS and VUV photoemission spectroscopy studies of environmental samples such as natural organic matter and microbial organisms under live conditions. New types of X-ray imaging may evolve to take advantage of improvements in two-dimensional X-ray detector technology.

Another area that is certain to grow is SR-based infrared spectroscopy. As discussed in the RiMG chapter by Hirschmugl (this volume), the significantly higher brightness of SR-based IR sources relative to conventional IR spectrometers, which use globar or tunable laser IR sources, will make studies of dilute levels of organic molecules in environmental samples more feasible. In addition, the ability to conduct micro-IR studies on environmental samples with a diffraction-limited spatial resolution of about 10 μm on IR beamlines at the ALS and at the soft X-ray/VUV ring at the NSLS will make it possible to study the spatial distribution and phase association of natural organic matter as well as xenobiotic organic contaminants in sediments. Very little is known about how such organics interact with minerals and other phases, such as soot particles. These facilities will also likely have an impact in the field of aerosol chemistry where very little is currently known about the interaction of inorganic and organic molecules at a molecular/mechanistic level with aerosol particles under *in situ* conditions (i.e., with realistic relative humidities).

The hard X-ray facilities at the APS and ALS that are used in research on earth and environmental samples are reviewed by Sham and Rivers (this volume) and include a number of insertion device beamlines optimized for micro-focussing experiments using K-B mirrors and Fresnel zone plates, including μXRF, μXANES, μXRD, and microtomography. Some of these beamlines are also used for grazing-incidence XAFS and surface scattering (X-ray reflectivity) experiments on minerals that could not be done easily at second-generation SR sources. These high-brightness sources make it possible to conduct such experiments on small samples (< 1 mm^2), thus avoiding heterogeneities such as twin domains and phase intergrowths in natural crystals that can complicate X-ray reflectivity studies of interfaces. Another advantage of high-brightness third-generation sources is the ability to conduct XAFS, XSW, X-ray reflectivity, and X-ray emission experiments on surfaces with adsorbate coverages of less than a few percent. Such experiments will make it possible to begin studying the effects of surface defects (which are typically at levels of a few percent of the total surface sites) on the chemistry of sorption reactions. Such defects are typically the highest energy and most reactive sites on surfaces and are thus where mineral/water interface reactions are typically initiated.

μXRD studies of μm-size particles in complex natural samples are also benefiting from third-generation hard X-ray SR sources. For example, the combined μXRD and μXANES studies of heavy metal pollutant speciation in soils by Manceau and co-workers is providing detailed information on the types of solid phases present, their spatial distribution, and element and species zonation in individual mineral grains (Manceau et al., this volume). We anticipate that μXRD will become a widely used characterization method in studies of natural samples, including those in which biomineralization has has produced μm-size particles. It may also help in gaining a better understanding of metastable low-temperature phase transformations through *in situ* experimental studies.

These high-brightness hard X-ray beam lines will also make it possible to characterize contaminant and pollutant species of a variety of elements at lower concentration levels. One can now conduct XAFS experiments on heavy metal contaminants in environmental samples at concentration levels of 10-50 ppm under optimal conditions using second-generation SR sources. With the higher brightness and flux of third-generation sources, this level is being pushed into the high ppb range, and could go even lower with improvements in X-ray detectors and X-ray sources. This improvement in sensitivity also translates into faster data collection on samples with dilute contaminant or pollutant levels that now takes hours, thus improving sample throughput and making it possible to examine a statistically meaningful number of samples from a particular site.

Another area that will benefit from improvements in current SR sources is kinetic studies of chemical reactions at mineral and microbial surfaces and in the rhizosphere of plant roots in soils, including real-time studies of redox transformations of environmental pollutants such as chromium, arsenic, selenium, uranium, and plutonium. Higher X-ray fluxes, coupled with improvements in experimental methods such as dispersive XAFS, will translate into faster collection of XAFS (and X-ray scattering) data, thus making it possible to study chemical reactions with rates on the millisecond to microsecond time scale. Unfortunately, higher fluxes also translate into greater X-ray beam-induced damage to certain types of samples, including many of interest to geochemists and environmental scientists.

Fourth-generation SR sources, currently under development in the U.S. and Germany, will result in a highly coherent X-ray beam (an X-ray free electron laser) with pulses that are as short as 50-100 fsec containing 10^{12} photons (at 1.5 Å wavelength) to 10^{13} photons (at 15 Å wavelength) and peak brightnesses that are 4 orders of magnitude greater than those of undulator magnet SR sources on third-generation storage rings (Fig. 24). One design for such sources will make use of a very long undulator magnet (110 m) at the end of a linear electron accelerator, such as the LINAC at SLAC in California, USA. The U.S. Department of Energy has approved the conceptual design plan of the Linac Coherent Light Source (LCLS) at SLAC (LCLS 1998), which will be the first fourth-generation SR facility in the world and is scheduled for completion in 2008. A second fourth-generation SR source (TESLA) is currently scheduled for completion at DESY in Germany in 2011.

A partial preview of the LCLS will be provided by the Sub-Picosecond Photon Source (SPPS), which is currently under construction at SSRL/SLAC. The SPPS will utilize a 12-meter chicane compressor at the 1/3 point of the SLAC LINAC to generate bunches of 80 fsec length. The peak brightness of this pulse will be 20 times greater than that at ESRF, which has a bunch length of 8×10^4 fsec, and will have 2×10^7 photons at 0.1% bandwidth (vs. 3×10^5 photons per bunch at ESRF) with a repetition rate of 30 Hz, compared with 900 Hz at the ESRF. The SPPS facility will be operational in 2003.

What types of experiments are fourth-generation SR sources optimized for, and will any be of relevance to low temperature geochemistry or environmental science? The Scientific Advisory Committee of the LCLS has selected six experiments for the early phase of the project (LCLS 2000). Those of potential relevance to geochemical and

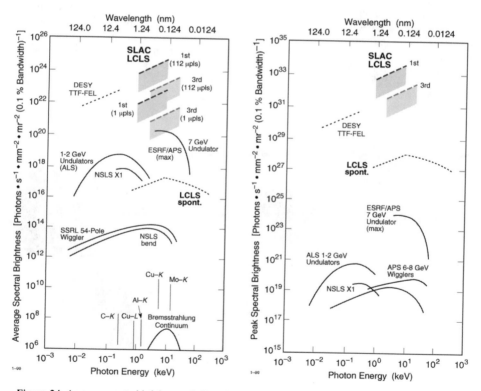

Figure 24. Average spectral brightness (left) and peak spectral brightness (right) vs. photon energy for a number of X-ray sources compared with the LCLS. (from the LCLS Design Study Report (LCLS 1998))

environmental science samples include (1) X-ray scattering from single molecules, (2) XAFS spectroscopy and X-ray scattering on very short-lived intermediates in chemical reactions that are stimulated by laser pump-probe techniques (femtochemistry), (3) X-ray scattering and X-ray spectroscopy studies of nanophase melting and nucleation dynamics and vibrational modes and capillary waves in nanoparticles, and (4) X-ray scattering studies of collective mode dynamics in amorphous materials. One problem in all of these experiments is that the sample volume irradiated will be vaporized shortly after the X-ray pulse strikes it. However, high-level MD simulations of the interaction of a 2-50 fsec pulse of 12 keV X-rays containing 3×10^{12} photons with the T4 lysozyme molecule (Fig. 25) indicate that this molecule will survive long enough for X-ray scattering data to be collected with sufficient S/N for analysis (Neutze et al. 2000). In those cases where additional scattering data are needed to fully characterize a molecule, it will be possible to conduct a series of individual scattering (or spectroscopic) experiments on a number of different molecules of the same material and sum the data sets prior to analysis.

Fourth-generation SR sources will permit the collection of an entire X-ray diffraction pattern or, potentially, an X-ray absorption spectrum in a very short time frame ($\approx$ 50-80 fsec), which will allow studies of the mechanisms of chemical reactions at an unprecedented level of detail. Most chemical bonds break and form in a time frame of 10-100 fsec, so one can imagine "capturing" the structure of a molecule during a reaction process that is stimulated by pump-probe laser excitation in an X-ray scattering experiment. This is a very exciting prospect, not only for chemists and biologists, but also for geochemists, environmental chemists, and soil chemists who are interested in the mechanisms and kinetics of reactions such as those that result in the breakdown of ozone on atmospheric ice particles, the sequestration of natural organic matter or plutonium in groundwater on mineral surfaces, the growth of crystals from aqueous solutions, and the incorporation of trace elements into growth surfaces of minerals.

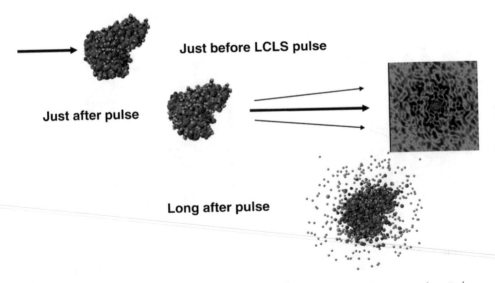

Figure 25. Simulation of a 10 fsec pulse containing 3×10^{12} photons (12 keV X-ray energy) scattering from a T4 lysozyme molecule showing the scattering pattern and molecule before interaction, just after the pulse interacts with it, and long after the interaction. (after Neutze et al. 2000)

The full answer to the questions posed above will have to await the availability of fourth-generation SR sources and will be best addressed by the next generation of synchrotron scientists who will build their careers around such facilities. Just as it was not possible to fully anticipate or predict the many scientific applications and new experimental methods made possible by current SR sources before they were developed, it is not possible to fully anticipate the exciting new experimental methods and scientific discoveries that will result from fourth-generation SR sources.

ACKNOWLEDGMENTS

We wish to thank Farrel Lytle (The EXAFS Co.) and Herman Winick (SSRL) for helpful discussions on the history of synchrotron radiation sources and the first applications of synchrotron radiation. We also thank John Bargar (SSRL) for providing Figure 3, Paul Bertsch (Savanah River Ecology Laboratory) for providing Figure 6, Colleen Hansel (Stanford University) for providing the data used to prepare Figure 9, Anders Nilsson (SSRL) for providing Figure 11, Satish Myneni (Princeton University) for providing Figure 12, Alexis Templeton (Scripps Institution of Oceanography) for preparing Figure 15, Ingrid Pickering (SSRL) for preparing Figure 20, Chris Kim (Stanford University) for preparing Figure 21, and David Clark (LANL) for supplying Figure 23. Francois Farges (Universite Marne-la-Vallee, France) is thanked for helpful discussions on EXAFS data analysis protocols, Paul Fenter (Argonne National Laboratory) is thanked for a detailed review of an earlier version of this manuscript, and Jeff Catalano (Stanford University) is thanked for help with the information on uranium in Table 1 as well as for help in reformatting several of the figures. The SR-based studies by Brown and co-workers discussed in this review were generously supported by the Department of Energy (BES (FG03-93ER14347-A009) and EMSP (FG07-99ER15024 and FG07-99ER15022)), the National Science Foundation (NSF-EGB (EAR-9905755) and NSF-CHE-CRAEMS (CHE-0089215)), and the EPA-STAR Program (EPA-R827634). The SR-based studies of Sturchio and co-workers reported herein were supported by the Department of Energy (BES) though contract W-31-109-Eng-38 and grant DE-FG03-99ER14979.

On behalf of the geochemists, environmental scientists, soil scientists, and others who have used or will use SR sources in the U.S. to address low temperature geochemical and environmental science problems, we thank the Department of Energy (Office of Basic Energy Sciences – Materials Research and Chemical Sciences Programs) for providing the funding to build and maintain these sources. Thanks also go to the DOE-Geosciences and Chemical Sciences Programs and the NSF-Earth Sciences Facilities and Instrumentation Program for providing funds to construct and operate beam lines devoted in large part to geochemical and environmental science applications, a number of which are discussed in this chapter and RiMG volume. Dan Weill and David Lambert (NSF), Bob Marianelli and Bill Millman (DOE-Chemical Sciences), Bill Luth and Nick Woodward (DOE-Geosciences), Iran Thomas (Materials Research Program), and Patricia Dehmer (DOE-BES) played critical roles in providing this funding.

REFERENCES

Abbasi SA, Soni R (1984) Teratogenic effects of chromium(VI) in the environment as evidenced by the impact of larvae of amphibian *Rand tigrina*: Implications in the environmental management of chromium. Int J Environ Studies 23:131-137

Abruna HD, Bommarito GM, Acevedo D (1990) The study of solid/liquid interfaces with X-ray standing waves. Science 250:69-74

Acharyya SK, Chakraborty P, Lahiri S, Raymahashay BC, Guha S, Bhowmik A (1999) Arsenic poisoning in the Ganges delta. Nature 401:545

Albee AL, Chodos AA (1970) Semiquantitative electron microprobe determination of Fe^{2+}/Fe^{3+} and Mn^{2+}/Mn^{3+} in oxides and silicates and its application to petrologic problems. Am Mineral 55:491-501

Alcacio TE, Hesterberg D, Chou JW, Martin JD, Beauchemin S, Sayers DE (2001) Molecular scale characteristics of Cu(II) bonding in goethite-humate complexes. Geochim Cosmochim Acta 65:1355-1366

Allard T, Ildefonse Ph, Beaucaire K, Calas G (1999) Structural chemistry of uranium associated with Si, Al, Fe gels in a granitic uranium mine. Chem Geol 158:81-103

Allen PG, Berg JM, Chisholm-Brause CJ, Conradson SD, Donohoe RJ, Morris DE, Musgrave JA, Tait CD (1994) Determining uranium speciation in contaminated soils by molecular spectroscopic methods: Examples from the uranium in soils integrated demonstration. Technology Programs: Radioactive Waste Management and Environmental Restoration 3:2063-2068

Allen PG, Bucher JJ, Clark DL, Edelstein NM, Ekberg SA, Gohdes JW, Hudson EA, Kaltsoyannis N, Lukens WW et al. (1995a) Multinuclear NMR, Raman, EXAFS, and X-ray diffraction studies of uranyl carbonate complexes in near-neutral aqueous solution. X-ray structure of $[C(NH_2)_3]_6[(UO_2)_3(CO_3)_6]\cdot6.5H_2O$. Inorg Chem 34:4797-4807

Allen PG, Bucher JJ, Shuh DK, Edelstein NM, Craig I (2000) Coordination chemistry of trivalent lanthanide and actinide ions in dilute and concentrated chloride solutions. Inorg Chem 39:595-601

Allen PG, Bucher JJ, Shuh DK, Edelstein NM, Reich T (1997b) Investigation of aquo and chloro complexes of UO_2^{2+}, NpO_2^+, Np^{4+}, and Pu^{3+} by X-ray absorption fine structure spectroscopy. Inorg Chem 36:4676-4683

Allen PG, Conradson SD, Wilson MS, Gottesfeld S, Raistrick ID (1993) Real time structural electrochemistry of platinum clusters using dispersive XAFS. Mat Res Soc Symp Proc 307 (Applications of Synchrotron Radiation Techniques to Materials Science):51-56

Allen PG, Conradson SD, Wilson MS, Gottesfeld S, Raistrick ID, Valerio J, Lovato M (1995b) Direct observation of surface oxide formation and reduction on platinum clusters by time-resolved X-ray absorption spectroscopy. J. Electroanal Chem 384:99-103

Allen PG, Shuh DK, Bucher JJ, Edelstein NM, Palmer CEA, Silva RJ, Nguyen SN, Marquez LN, Hudson EA (1996b) Determinations of uranium structures by EXAFS: schoepite and other U(VI) oxide precipitates. Radiochim Acta 75:47-53

Allen PG, Shuh DK, Bucher JJ, Edelstein NM, Reich T, Denecke MA, Nitsche H (1996a) EXAFS determinations of uranium structures: The uranyl ion complexed with tartaric, citric, and malic acids. Inorg Chem 35:784-787

Allen PG, Shuh DK, Bucher JJ, Edelstein NM, Reich T, Denecke MA, Nitsche H (1997a) Chemical speciation studies of radionuclides by XAFS. J de Physique IV 7 (Colloque C2, X-Ray Absorption Fine Structure, Vol. 2):789-792

Allen PG, Siemering GS, Shuh DK, Bucher JJ, Edelstein NM, Langton CA, Clark SB, Reich T, Denecke MA (1997c) Technetium speciation in cement waste forms determined by X-ray absorption fine structure spectroscopy. Radiochim Acta 76:77-86

Allen PG, Sylwester ER, Zhao PH, Kersting AB, Zavarin M (2001) Surface interactions and reactivity of actinide ions with mineral colloids. Abstracts of Papers, 222nd ACS National Meeting, Chicago, IL, United States, August 26-30, 2001, GEOC-042

Alloway BJ, Ayres DC (1993) Chemical Principles of Environmental Pollution. Blackie Academic & Professional, London

Altarelli M, Schlacter F, Cross J (1998) Making ultrabright X-rays. Scientific American 268:66-73

Anderson AJ, Jayanetti S, Mayanovic RA, Bassett WA, Chou, I-M (2002) X-ray spectroscopic investigations of fluids in the hydrothermal diamond anvil cell: the hydration structure of aqueous La^{3+} up to 300°C and 1600 bars. Am Mineral 87:262-268

Anderson AJ, Mayanovic RA, Bajt S (1995) Determination of the local structure and speciation of zinc in individual hypersaline fluid inclusions by micro-XAFS. Canadian Mineral 33:499-508

Anderson AJ, Mayanovic RA, Bajt S (1998a) A microbeam XAFS study of aqueous chlorozinc complexing to 430°C in fluid inclusions from the Knaumuehle granitic pegmatite, Saxonian Granulite Massif, Germany. Canadian Mineral 36:511-524

Anderson AJ, Mayanovic RA, Chou I-M (1998b) Micro-beam XAFS measurements of zinc complexes in highly saline fluid inclusions to 400°C before and after experimental re-equilibration at high hydrogen pressure. Mineral Mag 62A:55-56

Ankudinov AL, Ravel B, Rehr JJ, Conradson SD (1998) Real-space multiple-scattering calculation and interpretation of X-ray-absorption near-edge structure. Phys Rev B 58:7565-7576

Antonio MR, Soderholm L, Song I (1997) Design of spectroelectrochemical cell for *in situ* X-ray absorption fine structure measurements of bulk solution species. J Appl Electrochem 27:784-792

Antonio MR, Soderholm L, Williams CW, Blaudeau J-P, Bursten BE (2001) Neptunium redox speciation. Radiochim Acta 89:17-25

Apted MJ, Waychunas GA, Brown GE Jr (1985) Structure and speciation of iron complexes in aqueous solutions determined by X-ray absorption spectroscopy. Geochim Cosmochim Acta 492081-2089

Arai Y, Elzinga EJ, Sparks DL (2001) X-ray absorption spectroscopic investigation of arsenite and arsenate adsorption at the aluminum oxide-water interface. J Colloid Interface Sci 235:80-88.

Au K-K, Penisson AC, Yang S, O'Melia CR (1999) Natural organic matter at oxide/water interfaces: Complexation and conformation. Geochim Cosmochim Acta 63:2903-2917

Axe L, Bunker GB, Anderson PR, Tyson TA (1998) An XAFS analysis of strontium at the hydrous ferric oxide surface. J Colloid Interface Sci 199:44-52

Axe L, Tyson T, Trivedi P, Morrison T (2000) Local structure analysis of strontium to hydrous manganese oxide. J Colloid Interface Sci 224:408-416

Azcue JM, Nriagu JO (1994) Arsenic: historical perspectives. In: Nriagu JO (ed) Arsenic in the Environment. Part 1. Cycling and Characterization. John Wiley & Sons Inc, New York, p 1-15

Babich H, Stotzky G (1983) Influence of chemical speciation on the toxicity of heavy metals to the microbiota. Adv Environ Sci Technol 13:1-46

Baes CF Jr, Mesmer RE (1986) The Hydrolysis of Cations, R.F. Krieger Pub Co, Malabar, Florida

Baldwin GC (1975) Physics Today 28:9

Bajt S, Clark SB, Sutton SR, Rivers ML, Smith JV (1993) Synchrotron X-ray microprobe determination of chromate content using X-ray absorption near-edge structure. Anal Chem 65:1800-1804

Balistrieri L, Brewer PG, Murray JW (1981) Scavenging residence times of trace metals and surface chemistry of sinking particles in the deep ocean. Deep-Sea Res 28A:101-121

Banfield JF, Barker WW, Welch SA, Taunton A (1999) Biological impact on mineral dissolution: application of the lichen model to understanding mineral weathering in the rhizosphere. Proc Nat Acad Sci USA 96:3404-3411

Banfield JF, Hamers RJ (1997) Processes at minerals and surfaces with relevance to microorganisms and prebiotic synthesis. Rev Mineral 35:82-122

Banfield JF, Navrotsky A (eds) (2001) Nanoparticles and the Environment. Rev Mineral Geochem 44, Mineral Soc Am, Washington, DC

Banfield JF, Zhang H (2001) Nanoparticles in the environment. Rev Mineral Geochem 44:1-58

Bargar JR, Brown GE Jr, Evans I, Rabedeau T, Rowen M, Rogers J (2002a) A new hard X-ray XAFS spectroscopy facility for environmental samples, including actinides, at the Stanford Synchrotron Radiation Laboratory. Proceedings of the Euroconference and NEA Workshop on Speciation, Techniques, and Facilities for Radioactive Materials at Synchrotron Light Sources, Grenoble, France, Sept. 10-12, 2000, Nuclear Energy Agency/Organization for Economic Co-operation and Development, AEN/NEA 2002, Paris, p 169-176

Bargar JR, Brown GE Jr, Parks GA (1995) XAFS study of Pb(II) sorption at the α-Al$_2$O$_3$/water interface. Physica B 208 & 209:455-456

Bargar JR, Brown GE Jr, Parks GA (1997a) Surface complexation of Pb(II) at oxide/water interfaces: I. XAFS and bond-valence determination of mononuclear and polynuclear Pb(II) sorption products on aluminum oxides. Geochim Cosmochim Acta 61:2617-2637

Bargar JR, Brown GE Jr, Parks GA (1997b) Surface complexation of Pb(II) at oxide/water interfaces: II. XAFS and bond-valence determination of mononuclear Pb(II) sorption products and surface functional groups on iron oxides. Geochim Cosmochim Acta 61:2639-2652

Bargar JR, Brown GE Jr, Parks GA (1998) Surface complexation of Pb(II) at oxide/water interfaces. III: XAFS determination of Pb(II) and Pb(II)-chloro adsorption complexes on goethite and alumina. Geochim Cosmochim Acta 62:193-207

Bargar JR, Fuller CC, Davis JA (2002c) U(VI)-anion ternary complex formation on hematite. Abstracts of Papers, 223rd ACS National Meeting, Orlando, FL, United States, April 7-11, 2002, GEOC-120

Bargar JR, Persson P, Brown GE Jr (1997d) XAFS study of Pb(II)-Chloro- and Hg(II)-Chloro-ternary complexes on goethite. J de Physique IV 7 (Colloque C2, X-Ray Absorption Fine Structure, Vol. 2):825-826

Bargar JR, Persson P, Brown GE Jr (1999b) Outer-sphere adsorption of Pb(II)EDTA on goethite. Geochim Cosmochim Acta 63:2957-2969

Bargar JR, Reitmeyer R, Davis JA (1999a) Spectroscopic confirmation of uranium(VI)-carbonato adsorption complexes on hematite. Environ Sci Technol 33:2481-2484

Bargar JR, Reitmeyer R, Lenhart JJ, Davis JA (2000a) Characterization of U(VI)-carbonato ternary complexes on hematite: EXAFS and electrophoretic mobility measurements. Geochim Cosmochim Acta 64, 2737-2749

Bargar JR, Tebo BM, Villinski JE (2000b) *In situ* characterization of Mn(II) oxidation by spores of the marine Bacillus sp. strain SG-1. Geochim Cosmochim Acta 64:2775-2778

Bargar JR, Towle SN, Brown GE Jr, Parks GA (1996) Outer-sphere lead(II) adsorbed at specific surface sites on single crystal α-alumina. Geochim Cosmochim Acta 60:3541-3547

Bargar JR, Towle SN, Brown GE Jr, Parks GA (1997c) Structure, composition, and reactivity of Pb(II) and Co(II) sorption products and surface functional groups on single-crystal α-Al₂O₃. J Colloid Interface Sci 85:473-493

Bargar JR, Trainor TP, Fitts JP, Chambers SA, Brown GE Jr (2002b) *In situ* grazing incidence EXAFS study of Pb(II) chemisorption on hematite (0001) and (1̄102). Langmuir (submitted).

Bassett WA, Anderson AJ, Mayanovic RA, Chou I-M (2000a) Hydrothermal diamond anvil cell for XAFS studies of first-row transition elements in aqueous solution up to supercritical conditions. Chem Geol 167:3-10

Bassett WA, Anderson AJ, Mayanovic RA, Chou I-M (2000b) Modified hydrothermal diamond anvil cells for XAFS analyses of elements with low energy absorption edges in aqueous solutions at sub- and supercritical conditions. Zeitschrift für Kristallographie 215:711-717

Beauchemin S, Hesterberg D, Beauchemin M (2002) Principal component analysis approach for modeling sulfur K-XANES spectra of humic acids. Soil Sci Soc Am J 66:83-91

Bedzyk MJ (1988) New trends in X-ray standing waves. Nucl Instrum Meth Phys Res A 266:679-683

Bedzyk MJ (1990) Measuring the diffuse-double layer at an electrochemical interface with long period X-ray standing waves. Synchrotron Radiation News 3:25-29

Bedzyk MJ, Bommarito GM, Caffrey M, Penner TL (1990) Diffuse double layer at a membrane-aqueous interface measured with X-ray standing waves. Science 248:52-56

Bedzyk MJ, Hiderback DH, Bomarito GM, Caffrey M, Schildkraut JS (1988) X-ray standing waves: A molecular yardstick for biological membranes. Science 241:1788-1791

Berg M, Tran HC, Nguyen TC, Pham HV, Schertenleib R, Giger W (2001) Arsenic contamination in groundwater and drinking water in Vietnam: A human health threat. Environ Sci Technol 35:2621-2626

Berner RA, Holdren GR Jr (1979) Mechanism of feldspar weathering. II. Observations of feldspars from soils. Geochim Cosmochim Acta 43:1173-1186

Berrodier I, Farges F, Benedetti M, Brown GE Jr (1999) Adsorption of Au on iron oxy-hydroxides using Au L₃-edge XAFS spectroscopy. J Synchrotron Rad 6:651-652

Berrodier I, Farges F, Benedetti M, Brown GE Jr, Deveughéle (2002) Adsorption mechanisms of gold on iron and aluminum oxyhydroxides. Part II: Spectroscopic studies of gold sorbed on ferrihydrite, goethite and bohemite. Geochim Cosmochim Acta (submitted)

Bertsch PM, Hunter DB (2001) Applications of synchrotron-based X-ray microprobes. Chem Rev 101:1809-1842

Bertsch PM, Hunter DB, Nuessle PR, Clark SB (1997) Molecular characterization of contaminants in soils by spatially resolved XRF and XANES spectroscopy. J de Physique IV 7 (Colloque C2, X-Ray Absorption Fine Structure, Vol. 2):817-818

Bertsch PM, Hunter DB, Sutton SR, Bajt S, Rivers ML (1994) *In situ* chemical speciation of uranium in soils and sediments by micro X-ray absorption spectroscopy. Environ Sci Technol 28:980-984

Bertsch PM and Seaman JC (1999) Characterization of complex mineral assemblages: Implications for contaminant transport and environmental remediation. Proc Nat Acad Sci USA 96:3350-3357

Beyersmann D, Koester A, Buttner B, Flessel P (1984) Model reactions of chromium compounds with mammalian and bacterial cells. Toxicol Environ Chem 8:279-286

Bhattacharya P, Chatterjee D, Jacks G (1997) Occurrence of arsenic contaminated groundwater in alluvial aquifers from Delta Plains, Eastern India: Options for safe drinking water supply. Int Jour Water Resources Management 13:79-92

Bidoglio G, Gibson PN, O'Gorman M, Robert KJ (1993) X-ray absorption spectroscopy investigation of surface redox transformations of thallium and chromium on colloidal mineral oxides. Geochim Cosmochim Acta 57:2389-2394

Binstead N (1998) EXCURV98 Program. CLRC Daresbury Laboratory, Warrington, UK

Biwer BM, Soderholm L, Greegor RB, Lytle FW (1997) Uranium speciation in glass corrosion layers: an XAFS study. Mat Res Soc Symp Proc 465(Scientific Basis for Nuclear Waste Management XX):229-236

Blewitt JP (1946) Radiation losses in the induction electron accelerator. Phys Rev 69, 87-95

Blixt J, Glaser J, Mink J, Persson I, Persson P, Sandstroem M (1995) Structure of thallium(III) chloride, bromide, and cyanide complexes in aqueous solution. J Am Chem Soc 117:5089-5104

Blonski S, Garofalini SH (1993) Molecular dynamics simulation of α-alumina and γ-alumina surfaces. Surf Sci 295:263-274

Bochatay L, Persson P (2000) Metal ion coordination at the water-manganite (γ-MnOOH) interface. J Colloid Interface Sci 229:593-599

Bochatay L, Persson P, Lövgren L, Brown GE Jr (1997) XAFS study of Cu(II) at the water-goethite (α-FeOOH) interface. J de Physique IV 7 (Colloque C2, X-Ray Absorption Fine Structure, Vol. 2):819-820

Bochatay L, Persson P, Sjoberg, S (2000) Metal ion coordination at the water-manganite (γ-MnOOH) interface. J Colloid Interface Sci 229:584-592

Bockris JO, Jeng KT (1990) Water structure at interfaces: the present situation. Adv Colloid Interface Sci 33:1-54

Bockris J O'M, Reddy AKN (1973) Modern Electrochemistry, Vol. 2: An Introduction to an Interdisciplinary Area.

Bockris J O'M, Khan SUM (1993) Surface Electrochemistry. A Molecular Level Approach. Plenum Press, New York

Bonneviot L, Clause O, Che M, Manceau A, Decarreau A, Villain F, Bazin D, Dexpert H (1989a) Investigation by EXAFS of the effect of pH on the structure of nickel(2+) ions impregnated on silica. Physica B 158:43-44

Bonneviot L, Clause O, Che M, Manceau A, Dexpert H (1989b) EXAFS characterization of the adsorption sites of nickel ammine and ethyldiamine complexes on a silica surface. Catalysis Today 6:39-46

Bornebusch H, Clausen BS, Steffensen G, Lützenkirchen-Hecht D, Frahm R (1999) A new approach for QEXAFS data acquisition. J Synchrotron Rad 6:209-211

Bostick BC, Fendorf S, Barnett MO, Jardine PM, Brooks SC (2002) Uranyl surface complexes formed on subsurface media from DOE facilities. Soil Sci Soc Am J (2002) 66:99-108

Bottero JY, Manceau A, Villieras F, Tchoubar D (1994) Structure and mechanisms of formation of iron oxide hydroxide (chloride) polymers. Langmuir 10:316-319

Bourg ACM, Joss S, Schindler PW, (1979) Ternary surface complexes: 2. Complex formation in the system silica-Cu(II)-2,2' bipyridyl". Chimia 33:19-21

Boyanov MI, Kelly SD, Kemner KM, Bunker BA, Fein JB, Fowle DA (2002) Adsorption of cadmium to *B. subtilis* bacterial cell walls – a pH-dependent XAFS spectroscopy study. Geochim Cosmochim Acta: (in press)

Brenner SG, Hansel CM, Wielinga BW, Barber TM, Fendorf S (2002) Reductive dissolution and biomineralization of iron hydroxide under dynamic flow conditions. Environ Sci Technol 36:1705-1711

Broach RW, Bedard RL, Song SG, Pluth JJ, Bram A, Riekel C, Weber H-P (1999) Synthesis and characterization of ZP-4 ($KznPO_4 \cdot 0.8H_2O$), a new zincophosphate microporous material: Structure solution from a $2.5 \times 2.5 \times 8$ μm single crystal using a third generation synchrotron X-ray source. Chem Materials 11:2076-2080

Brown GE Jr (1990) Spectroscopic studies of chemisorption reaction mechanisms at oxide/water interfaces. Rev Mineral 23:309-363

Brown GE Jr (2001) How minerals react with water. Science 294:67-69

Brown GE Jr, Calas G, Waychunas GA, Petiau J (1988) X-ray absorption spectroscopy and its applications in mineralogy and geochemistry. Rev Mineral 18:431-512

Brown GE Jr, Chambers SA, Amonette JE, Rustad JR, Kendelewicz T, Liu P, Doyle CS, Grolimund D, Foster-Mills NS, Joyce SA, Thevuthasan S (2001) Interaction of water and aqueous chromium ions with iron oxide surfaces. *In:* Eller PG, Heineman WR (eds) Nuclear Site Remediation - First Accomplishments of the Environmental Management Science Program, Am Chem Soc Symp Ser 778, American Chemical Society, Columbus, OH, p 212-246

Brown GE Jr, Dikmen FD, Waychunas GA (1983) Total electron yield K-XANES and EXAFS investigation of aluminum in amorphous and crystalline aluminosilicates. Stanford Synchrotron Radiation Laboratory Report 83/01:148-149

Brown GE Jr, Farges F, Bargar JR, Berbeco HT (2002) Actinides in silicate glasses and melts and on mineral surfaces: Information on local coordination environments from XAFS spectroscopy and bond valence theory. Proceedings of the Euroconference and NEA Workshop on Speciation, Techniques, and Facilities for Radioactive Materials at Synchrotron Light Sources, Grenoble, France, Sept. 10-12, 2000 Nuclear Energy Agency/Organization for Economic Co-operation and Development, AEN/NEA 2002, Paris, p 15-31

Brown GE Jr, Farges F, Calas G (1995a) X-ray scattering and X-ray spectroscopy studies of silicate melts. Rev Mineral 32:317-410

Brown GE Jr, Foster AL, Ostergren JD (1999b) Mineral surfaces and bioavailability of heavy metals: a molecular-scale perspective. Proc Nat Acad Sci USA 96:3388-3395

Brown GE Jr, Henrich VE, Casey WH, Clark DL, Eggleston C, Felmy A, Goodman DW, Grätzel M, Maciel G, McCarthy MI, Nealson K, Sverjensky DA, Toney MF, Zachara JM (1999a) Metal oxide surfaces and their interactions with aqueous solutions and microbial organisms. Chem Rev 99:77-174

Brown GE Jr, Parks GA (2001) Sorption of trace elements from aqueous media: Modern perspectives from spectroscopic studies and comments on adsorption in the marine environment. Int Geol Rev 43:963-1073

Brown GE Jr, Parks GA, Bargar JR, Towle SN (1999c) Use of X-ray absorption spectroscopy to study reaction mechanisms at metal oxide-water interfaces. *In*: Sparks DL, Grundl T (eds) Kinetics and Mechanisms of Reactions at the Mineral/Water Interface, Amer Chem Soc Symp Series 715. Amer Chem Soc, Columbus Ohio, p 14-37

Brown GE Jr, Parks GA, O'Day PA (1995b) Sorption at mineral-water interfaces: macroscopic and microscopic perspectives. *In:* Vaughan DJ, Pattrick RAD (eds) Mineral Surfaces, Mineral Soc Ser 5, Chapman & Hall, London, p 129-183

Brown GE Jr, Waychunas GA, Stöhr J, Sette F (1986) Near-edge structure of oxygen in inorganic oxides: effects of local geometry and cation type. J de Physique 47 (Colloque C8):685-689

Buchanan BB, Bucher JJ, Carlson DE, Edelstein NM, Hudson EA, Kaltsoyannis N, Leighton T, Lukens W, Shuh DK, Nitsche H, Reich T, Roberts K, Torretto P, Woicik J, Yang W-S, Yee A, Yee BC (1995) A XANES and EXAFS investigation of the speciation of selenite following bacterial metabolization. Inorg Chem 34:1617-1619

Buffle J, Wilkinson KJ, Stoll S, Filella M, Zhang J (1998) A generalized description of aquatic colloid interactions: The three-colloidal component approach. Environ Sci Technol 32:2887-2899

Bugaev LA, Ildefonse Ph, Flank AM, Sokolenko AP, Dmitrienko HV (2000a) Aluminum K-XANES spectra in minerals as the source of information on their local atomic structure. J Phys Condensed Matter 10:5463-5473

Bugaev LA, Ildefonse Ph, Flank AM, Sokolenko AP, Dmitrienko HV (2000b) Determination of interatomic distances and coordination numbers by K-XANES in crystalline minerals with distorted local structure. J Phys Condensed Matter 12:1119-1131

Bugaev LA, Sokolenko AP, Dmitrienko HV, Flank A-M (2002) Fourier filtration of XANES as a source of quantitative information of interatomic distances and coordination numbers in crystalline minerals and amorphous compounds. Phys Rev B 65:024105/1-024105/8

Bunce N (1991) Environmental Chemistry. Wuerz Publishing Ltd, Winnipeg, Canada

Bunker DJ, Jones MJ, Charnock JM, Livens FR, Pattrick RAD, Collinson D (2002) EXAFS studies of co-precipitation and adsorption reactions of Tc. Proceedings of the Euroconference and NEA Workshop on Speciation, Techniques, and Facilities for Radioactive Materials at Synchrotron Light Sources, Grenoble, France, Sept. 10-12, 2000 Nuclear Energy Agency/Organization for Economic Co-operation and Development, AEN/NEA 2002, Paris, p 207-213

Burns PC, Pluth JJ, Smith JV, Eng P, Steele I, Housley RM (2000) Quetzalcoatlite: a new octahedral-tetrahedral structure from a $2 \times 2 \times 40 \ \mu m^3$ crystal at the Advanced Photon Source-GSE-CARS facility. Am Mineral 85:604-607

Cabaret D, Joly Y, Renevier H, Natoli CR (1999) Pre-edge structure analysis of Ti K-edge polarized X-ray absorption spectra in TiO_2 by full-potential XANES calculations. J Synchrotron Rad 6:258-260

Cabaret D, Le Grand M, Ramos A, Flank A-M, Rossano S, Galoisy L, Calas G, Ghaleb D (2001) Medium range structure of borosilicate glasses from Si K-edge XANES: a combined approach based on multiple scattering and molecular dynamics calculations. J Non-Crystal Solids 289:1-8

Cabaret D, Sainctavit P, Ildefonse Ph, Calas G, Flank A-M (1996a) Full multiple scattering calculations at Al K-edge on alumino-silicates and Al-oxides. Phys Chem Minerals 23:226

Cabaret D, Sainctavit P, Ildefonse Ph, Flank A-M (1996b) Full multiple-scattering calculations on silicates and oxides at the Al K edge. J Phys Condensed Matter 8:3691-3704

Cabaret D, Sainctavit P, Ildefonse Ph, Flank A-M (1996c) Full multiple scattering calculations on silicates and oxides at Al K-edge. J Electron Spectrosc Related Phenomena 79:21-24

Cabaret D, Sainctavit P, Ildefonse Ph, Flank A-M (1998) Full multiple scattering calculations of the X-ray absorption near edge structure at the magnesium K-edge in pyroxene. Am Mineral 83:300-304

Cadle RD (1966) Particles in the Atmosphere and Space. Reinhold, New York

Cahill CL, Benning LG, Barnes HL, Parise JB (2000) *In situ* time-resolved X-ray diffraction of iron sulfides during hydrothermal pyrite growth. Chem Geol 167:53-63

Cahill CL, Benning LG, Norby P, Clark SM, Schoonen MAA, Parise JB (1998) *In situ* X-ray diffraction apparatus and its application to hydrothermal reactions of iron sulfide growth and phase transformations. Mineral Mag 62A:267-268

Calas G, Brown GE Jr, Waychunas GA, Petiau J (1987) X-ray absorption spectroscopic studies of silicate glasses and minerals. Phys Chem Minerals 15:19-29

Calmano W, Mangold S, Welter E (2001) An XAFS investigation of the artifacts caused by sequential extraction analyses of Pb-contaminated soils. Fresenius J Anal Chem 371:823-830

Carrado KA, Xu L, Gregory D, Song K, Siefert S, Botto RE (2000) Crystallization of a layered silicate clay as monitored by small-angle X-ray scattering and NMR. Chem Mat 12:3052-3059

Carrier X, Doyle CS, Kendelewicz T, Brown GE Jr (1999) Reaction of CO_2 with MgO(100) surfaces. Surf Rev Lett 6:1237-1245

Catalano JG, Warner JA, Chen C-C, Yamakawa I, Newville M, Sutton SR, Ainsworth CC, Zachara JM, Traina SJ, Brown GE Jr (2001) Speciation of chromium in Hanford Tank Farm SX-108 and 41-09-39 core samples determined by X-ray absorption spectroscopy. *In:* Groundwater/Vadose Zone Integration Project Report, S-SX FIR Appendix E, Digest of S&T Evaluations, p E-145-E-162

Cazaux J (2001) Electron- and X-ray-induced electron emissions from insulators. Polymer International 50:748-755

Charnock JM, Henderson CMB, Mosselmans JFW, Pattrick RAD (1996) 3d transitional metal L-edge X-ray absorption studies of the dichalcogenides of Fe, Co and Ni. Phys Chem Minerals 23:403-408

Charlet L, Manceau A (1992a) *In situ* characterization of heavy metal surface reactions: the chromium case. Int J Environ Anal Chem 46:97-108

Charlet L, Manceau A (1992b) X-ray absorption spectroscopic study of the sorption of Cr(III) at the oxide-water interface. II. Adsorption, coprecipitation, and surface precipitation on hydrous ferric oxide. J Colloid Interface Sci 148:443-458

Charlet L, Manceau A (1993) Structure, formation, and reactivity of hydrous oxide particles: Insights from X-ray absorption spectroscopy. *In:* Buffle J, van Leeuwen HP (eds) Environmental Particles, Vol. 2, Lewis Publishers, Boca Raton, FL, p 117-164

Charlet L, Manceau A (1994) Evidence for the neoformation of clays upon sorption of Co(II) and Ni(II) on silicates. Geochim Cosmochim Acta 58:2577-2582

Chatterjee A, Das D, Mandal BK, Chowdhury TR, Samanta G, Chakraborti D (1995) Arsenic in groundwater in six districts of West Bengal, India: The biggest arsenic calamity in the world. Part I. Arsenic species in drinking water and urine of affected people. Analyst 120:643-650

Cheah S-F, Brown GE Jr, Parks GA (1997) The effect of substrate type and 2,2'-bipyridine on the sorption of copper(II) on silica and alumina. *In:* Voigt JA, Bunker BC, Casey W, Wood TE, Crossey LJ (eds) Aqueous Chemistry and Geochemistry of Oxides, Oxyhydroxides, and Related Materials, Mat Res Soc Symp Proc 432:231-236

Cheah S-F, Brown GE Jr, Parks GA (1998) XAFS spectroscopic study of Cu(II) sorption on amorphous SiO_2 and γ-Al_2O_3: Effect of substrate and time on sorption complexes. J Colloid Interface Sci 208:110-128

Cheah S-F, Brown GE Jr, Parks GA (1999) Structure and composition of copper(II)-2,2'-bipyridine sorption complexes on amorphous SiO_2. Geochim Cosmochim Acta 63:3229-3246

Cheah S-F, Brown GE Jr, Parks GA (2000) XAFS study of copper model compounds and copper(II) sorption on amorphous SiO_2, γ-Al_2O_3, and anatase. Am Mineral 85:118-132

Chen C-C, Hayes KF (1999) X-ray absorption spectroscopy investigation of aqueous Co(II) and Sr(II) sorption at clay-water interfaces. Geochim Cosmochim Acta 63:3205-3215

Chen C-C, Papelis C, Hayes KF (1998) Extended X-ray absorption fine structure (EXAFS) analysis of aqueous Sr(II) ion sorption at clay-water interfaces. *In:* Jenne EA (ed) Adsorption of Metals by Geomedia, Academic Press, New York, p 333-348

Chen LX, Liu T, Thurnauer MC, Csencsits R, Rajh T (2002) Fe_2O_3 nanoparticle structures investigated by X-ray absorption near-edge structure, surface modifications, and model calculations. J Phys Chem B 106:8539-8546

Chen LX, Rajh T, Jager W, Nedeljkovic J, Thurnauer MC (1999) X-ray absorption reveals surface structure of titanium dioxide nanoparticles. J Synchrotron Rad 6:445-447

Chen LX, Rajh T, Micic O, Wang Z, Tiede DM, Thurnauer M (1997a) Photocatalytic reduction of heavy metal ions on derivatized titanium dioxide nano-particle surface studied by XAFS. Nucl Instrum Meth Phys Res Sec B: Beam Interactions with Materials and Atoms 133:8-14

Chen LX, Rajh T, Wang Z, Thurnauer MC (1997b) XAFS studies of surface structures of TiO_2 nanoparticles and photocatalytic reduction of metal ions. J Phys Chem B 101: 10688-10697

Cheng L, Fenter P, Nagy K, Schlegel M, and Sturchio NC (2001a) Molecular scale density oscillations in water adjacent to a mica surface. Phys Rev Lett 87:156103:1-4

Cheng L, Fenter P, Sturchio NC, Zhang Z, Bedzyk MJ (1999) X-ray standing wave study of arsenite incorporation at the calcite surface. Geochim Cosmochim Acta 63:3153-3157

Cheng L, Lyman PF, Sturchio NC, Bedzyk MJ (1997) X-ray standing wave investigation of the surface structure of selenite anions adsorbed on calcite. Surf Sci 382:L690-L695

Cheng L, Sturchio NC, Bedzyk MJ (2000) Local structure of Co^{2+} incorporated at the calcite surface. An X-ray standing wave and SEXAFS study. Phys Rev B 61:4877-4883

Cheng L, Sturchio NC, Bedzyk MJ (2001b) Impurity structure in a molecular ionic crystal: Atomic-scale X-ray study of $CaCO_3$:Mn^{2+}. Phys Rev B 63:144104:1-7

Cheng L, Sturchio NC, Woicik JC, Kemner KM, Lyman PF, Bedzyk MJ (1998) High-resolution structural study of zinc ion incorporation at the calcite cleavage surface. Surf Sci 415:L976-L982

Chester AW, Absil RPA, Kennedy GJ, Lagarde P, Flank AM (1999) X-ray absorption spectroscopy of transition aluminas. J Synchrotron Rad 6:448-450

Chialvo AA, Cummings PT (1999) Molecular-based modeling of water and aqueous solutions at supercritical conditions. Advances in Chem Phys 109:115-205

Chialvo AA, Predota M, Cummings PT (2002) Molecular-based study of the electric double layer in hydrothermal systems. Abstracts of Papers, 223rd Am Chem Soc National Meeting, Orlando, FL, United States, April 7-11, 2002, GEOC-076

Chiarello RP, Sturchio NC (1995) The calcite ($10\bar{1}4$) cleavage surface in water: early results of a crystal truncation rod study. Geochim Cosmochim Acta 59:4557-4561

Chiarello RP, Sturchio NC, Grace JD, Geissbuhler P, Sorensen LB, Cheng L, Xu S (1997) Otavite-calcite solid-solution formation at the calcite-water interface studied *in situ* by synchrotron X-ray scattering. Geochim Cosmochim Acta 61:1467-1474

Chiarello RP, Wogelius RA, Sturchio NC (1993) *In situ* synchrotron X-ray reflectivity measurements at the calcite-water interface. Geochim Cosmochim Acta 57:4103-4110

Chisholm-Brause CJ, Brown GE Jr, Parks GA (1989a) EXAFS investigation of aqueous Co(II) adsorbed on oxide surfaces *in situ*. Physica B 158:646-648

Chisholm-Brause CJ, Brown GE Jr, Parks GA (1991) *In situ* EXAFS study of changes in Co(II) sorption complexes on α-Al$_2$O$_3$ with increasing sorption densities. *In:* Hasnain S (ed) XAFS VI, Sixth Internat Conf on X-ray Absorption Fine Structure, Ellis Horwood Ltd, Publishers, Chichester, UK, p 263-265

Chisholm-Brause CJ, Conradson SD, Buscher CT, Eller PG, Morris DE (1994) Speciation of uranyl sorbed at multiple binding sites on montmorillonite. Geochim Cosmochim Acta 58:3625-3631

Chisholm-Brause, C.J., Conradson SD, Eller PG, Morris DE (1992a) Changes in uranium(VI) speciation upon sorption onto montmorillonite from aqueous and organic solutions. Mat Res Soc Symp Proc 257 (Scientific Basis for Nuclear Waste Management XV):315-322

Chisholm-Brause CJ, Hayes KF, Roe AL, Brown GE Jr, Parks GA, Leckie JO (1990b) Spectroscopic investigation of Pb(II) complexes at the γ-Al$_2$O$_3$/water interface. Geochim Cosmochim Acta 54:1897-1909

Chisholm-Brause, CJ, Morris DE, Richard RE (1992b) Speciation of uranium(VI) sorption complexes on montmorillonite. Water-Rock Interactions, Proc 7th Int Symp 1:137-140

Chisholm-Brause CJ, O'Day PA, Brown GE Jr, Parks GA (1990a) Evidence for multinuclear metal-ion complexes at solid/water interfaces from X-ray absorption spectroscopy. Nature 348:528-531

Chisholm-Brause CJ, Roe AL, Hayes KF, Brown GE Jr, Parks GA, Leckie JO (1989b) XANES and EXAFS study of aqueous Pb(II) adsorbed on oxide surfaces. Physica B 158,674-676

Chowdhury TR, Basu GK, Mandal BK, Biswas BK, Samanta G, Chowdhury UK, Chanda CR, Lodh D, Roy SL, Saha KC, Roy S, Kabir S, Quamruzzaman Q, Chakraborti D (1999) Arsenic poisoning in the Ganges delta. Nature 401:545-546

Chukaline M, Simionovici A, Snigirev A. (2002) X-ray fluorescence microtomography with synchrotron radiation: image reconstruction. Proceedings of the Euroconference and NEA Workshop on Speciation, Techniques, and Facilities for Radioactive Materials at Synchrotron Light Sources, Grenoble, France, Sept. 10-12, 2000 Nuclear Energy Agency/Organization for Economic Co-operation and Development, AEN/NEA 2002, Paris, p 215-221

Clark DL, Conradson SD, Donohoe RJ, Keogh DW, Morris DE, Palmer PD, Rogers RD Tait CD (1999) Chemical speciation of the uranyl ion under highly alkaline conditions. Synthesis, structures, and oxo ligand exchange dynamics. Inorg Chem 38:1456-1466

Clark DL, Conradson SD, Ekberg SA, Hess NJ, Janecky DR, Neu MP, Palmer PD, Tait CD (1996) A multi-method approach to actinide speciation applied to pentavalent neptunium carbonate complexation. New J Chem 20:211-220

Clark DL, Conradson SD, Neu MP, Palmer PD, Runde W, Tait CD (1997) XAFS structural determination of Np(VII). Evidence for a trans dioxo cation under alkaline solution conditions. J Am Chem Soc 119:5259-5260

Clausen M, Oehman L-O, Kubicki JD, Persson P (2002) Characterisation of gallium(III)-acetate complexes in aqueous solution: A potentiometric, EXAFS, IR and molecular orbital modelling study. J Chem Soc Dalton Trans 12:2559-2564

Collins CR, Ragnarsdottir KV, Sherman DM (1999a) Effect of inorganic and organic ligands on the mechanism of cadmium sorption to goethite. Geochim Cosmochim Acta 63:2989-3002

Collins CR, Sherman DM, Ragnarsdottir KV (1998) The adsorption mechanism of Sr^{2+} on the surface of goethite. Radiochim Acta 81:201-206

Collins CR, Sherman DM, Ragnarsdottir KV (1999b) Surface complexation of Hg^{2+} on goethite: Mechanism from EXAFS spectroscopy and density functional calculations. J Colloid Interface Sci 219:345-350

Combes J-M, Chisholm-Brause CJ, Brown GE Jr, Parks GA, Conradson SD, Eller PG, Triay I, Meier A (1992) EXAFS spectroscopic study of neptunium (V) sorbed at the α-FeOOH/water interface. Environ Sci Technol 26:376-382

Combes J-M, Manceau A, Calas G (1986) Study of the local structure in poorly ordered precursors of iron oxyhydroxides. J de Physique (Colloque C8, Vol. 2):697-C8/701

Combes J-M, Manceau A, Calas G (1989a) XAS study of the evolution of local order around iron(III) in the solution to gel to iron oxide (α-Fe_2O_3) transformation. Physica B (Amsterdam) 158:419-420

Combes J-M, Manceau A, Calas G (1990) Formation of ferric oxides from aqueous solutions: A polyhedral approach by X-ray absorption spectroscopy: II. Hematite formation from ferric gels. Geochim Cosmochim Acta 54:1083-1091

Combes J-M, Manceau A, Calas G, Bottero JY (1989b) Formation of ferric oxides from aqueous solutions: A polyhedral approach by X-ray absorption spectroscopy: I. Hydrolysis and formation of ferric gels. Geochim Cosmochim Acta 53:583-594

Conradson SD (1998) Application of X-ray absorption fine structure spectroscopy to materials and environmental science. Appl Spectrosc 52:252A-279A

Conradson SD, Clark DL, Neu MP, Runde W, Tait CD (2000) Characterizing the plutonium aquo ions by XAFS spectroscopy. Los Alamos Science 26 (Vol. 2):418-421

Conradson SD, Mahamid IA, Clark DL, Hess NJ, Hudson EA, Neu MP, Palmer PD, Runde WH, Tait CD (1998) Oxidation state determination of plutonium aquo ions using X-ray absorption spectroscopy. Polyhedron 17:599-602

Conway BE (1981) Ionic hydration in chemistry and biophysics. *In:* Studies in Physical and Theoretical Chemistry 12, Elsevier Scientific Publishing Co., Amsterdam, p 59-74

Cotter-Howells JD, Champness PE, Charnock JM (1999) Mineralogy of lead-phosphorous grains in the roots of *Agrostis capillaries* L. by ATEM and EXAFS. Mineral Mag 63:777-789

Cotter-Howells JD, Champness PE, Charnock JM, Pattrick RAD (1994) Identification of pyromorphite in mine-waste contaminated soils by ATEM and EXAFS. J Soil Sci 45: 393-402

Cowan PL, Golovchenko JA, Robins MF (1980) X-ray standing wave at crystal surfaces. Phys Rev Lett 44:1680-1683

Cramer SP, Scott RA (1981) New fluorescence detection system for X-ray absorption spectroscopy. Rev Sci Instrum 52:395-399

Criscenti L, Sverjensky DA (1999) The role of electrolyte anions (ClO_4^-, NO_3^-, and Cl^-) in divalent metal (M^{2+}) adsorption on oxide and hydroxide surfaces in salt solutions. Am J Sci 299:828-899

Cummings DE, Caccavo F Jr, Fendorf S, Rosenzweig RF (1999) Arsenic mobilization by the dissimilatory Fe(III)-reducing bacterium *Shewanella alga* BrY. Environ Sci Technol 33:723-729

Dähn R, Scheidegger AM, Manceau A, Baeyens B, Bradbury MH (2002b) Local structure of Th complexes on montmorillonite clay minerals determined by extended X-ray absorption fine structure (EXAFS) spectroscopy. Proceedings of the Euroconference and NEA Workshop on Speciation, Techniques, and Facilities for Radioactive Materials at Synchrotron Light Sources, Grenoble, France, Sept. 10-12, 2000 Nuclear Energy Agency/Organization for Economic Co-operation and Development, AEN/NEA 2002, Paris, p 75-81

Dähn R, Scheidegger AM, Manceau A, Schlegel ML, Baeyens B, Bradbury MH, (2001) Ni clay neoformation on montmorillonite surface. J Synchrotron Rad 8:533-535

Dähn R, Scheidegger AM, Manceau A, Schlegel ML, Baeyens B, Bradbury MH, Morales M (2002a) Neoformation of Ni phyllosilicate upon Ni uptake on montmorillonite: A kinetics study by powder and polarized extended X-ray absorption fine structure spectroscopy. Geochim Cosmochim Acta 66:2335-2347

Dardenne K, Schäfer MA, Denecke MA, Rothe J. (2002) XAFS investigation of lanthanide sorption onto ferrihydrite and transformation products by tempering at 75°C. Proceedings of the Euroconference and NEA Workshop on Speciation, Techniques, and Facilities for Radioactive Materials at Synchrotron Light Sources, Grenoble, France, Sept. 10-12, 2000 Nuclear Energy Agency/Organization for Economic Co-operation and Development, AEN/NEA 2002, Paris, p 223-228

Davis JA (1984) Complexation of trace metals by adsorbed natural organic matter. Geochim Cosmochim Acta 48:679-691

Davis JA, Fuller CC, Cook AD (1987) A model for trace metal sorption processes at the calcite surface: Adsorption of cadmium ion(2+) and subsequent solid solution formation. Geochim Cosmochim Acta 51:1477-1490

Davis JA, Gloor R (1981) Adsorption of dissolved organics in lake water by aluminum oxide. Effect of molecular weight. Environ Sci Technol 15:1223-1229

Davis JA, Kent DB (1990) Surface complexation modeling in aqueous geochemistry. Rev Mineral 23:177-260

Decarreau A, Colin F, Herbillon A, Manceau A, Nahon D, Paquet H, Trauth-Badaud D, Trescases JJ (1987) Domain segregation in nickel-iron-magnesium smectites. Clays Clay Minerals 35:1-10

de Jong BHWS, Keefer KD, Brown GE Jr, Taylor CM (1981) Polymerization of silicate and aluminate tetrahedra in glasses, melts, and aqueous solutions. III. Local silicon environments and internal nucleation in silicate glasses. Geochim Cosmochim Acta 45:1291-308

Denecke MA, Dardenne K, Marquardt M, Rothe J, Jensen MP (2002) Speciation of the actinide humate complexation by XAFS. Proceedings of the Euroconference and NEA Workshop on Speciation, Techniques, and Facilities for Radioactive Materials at Synchrotron Light Sources, Grenoble, France, Sept. 10-12, 2000 Nuclear Energy Agency/Organization for Economic Co-operation and Development, AEN/NEA 2002, Paris, p 93-104

Denecke MA, Reich T, Bubner M, Pompe S, Heise KH, Nitsche H, Allen PG, Bucher JJ, Edelstein NM, Shuh DK (1998a) Determination of structural parameters of uranyl ions complexed with organic acids using EXAFS. J Alloys and Compounds 271-273:123-127.

Denecke MA, Reich T, Pompe S, Bubner M, Heise KH, Nitsche H, Allen PG Bucher JJ, Edelstein NM, Shuh DK, Czerwinski KR (1998b) EXAFS investigations of the interaction of humic acids and model compounds with uranyl cations in solid complexes. Radiochim Acta 82:103-108

Dent AJ, Ramsay JDF, Swanton SW (1992) An EXAFS study of uranyl ions in solutions and sorbed onto silica and montmorillonite clay colloids. J Colloid Interface Sci 150: 45-60

d'Espinose de la Caillerie J-B, Kermarec M, Clause O (1995a) Impregnation of γ-alumina with Co(II) and Ni(II) ions at neutral pH: Hydrotalcite-type coprecipitate formation and characterization. J Am Chem Soc 117:11471-11481

d'Espinose de la Caillerie J-B, Bobin C, Rebours B, Clause O (1995b) Alumina/water interfacial phenomena during impregnation. In: Preparation of Catalysts VI – Scientific Basis for the Preparation of Heterogeneous Catalysts, Poincelet G et al. (eds), Elsevier Science B.V., New York, p 169-184

de Stasio G, Gilbert B, Frazer BH, Nealson KH, Conrad PG, Livi V, Labrenz M, Banfield JF (2001) The multi-disciplinarity of spectro-microscopy: from geomicrobiology to archeology. J Electron Spectros Relat Phenom 114-116:997-1003

de Stasio G, Gilbert B, Nelson T, Hansen R, Wallace J, Mercanti D, Capozi M, Baudat PA, Perfetti P, Margaritondo G, Tonner BP (2000) Feasibility tests of transmission X-ray photoelectron emission microscopy of wet samples. Rev Sci Instrum 71:11-14

Di Cicco A, Bianconi A, Coluzza C, Rudolf P, Lagarde P, Flank AM, Marcelli A (1990) XANES study of structural disorder in amorphous silicon. J Non-Crystal Solids 116:27-32

Docrat TI, Mosselmans JFW, Charnock JM, Whiteley MW, Collison D, Livens FR, Jones C, Edmiston MJ (1999) X-ray absorption spectroscopy of tricarbonato-dioxouranate(V), $[UO_2(CO_3)_3]^{5-}$, in aqueous solution. Inorg Chem 38:1879-1882

Dodd CG, Glen GL (1968) Chemical bonding studies of silicates and oxides by X-ray K-emissionspectroscopy. J. Appl Phys 39:5377-5384

Dodd CG, Glen GL (1969) Survey of chemical bonding in silicate minerals by X-ray emission spectroscopy. Am Mineral 54:1297-1309

Dodd CG, Glen GL (1970) Studies of chemical bonding in glasses by X-ray emission spectroscopy. J Am Ceram Soc 53:322-325

Dodge CJ, Francis AJ, Gillow JB, Halada GP, Eng C, Clayton CR (2002) Association of urnium with iron oxides typically formed on corroding steel surfaces. Environ Sci Technol 36:3504-3511

Doelsch E, Rose J, Masion A, Bottero JY, Nahon D, Bertsch PM (2000) Speciation and crystal chemistry of iron(III) chloride hydrolyzed in the presence of SiO_4 ligands. 1. An Fe K-edge EXAFS study. Langmuir 16:4726-4731

Doelsch E, Rose J, Masion A, Bottero JY, Nahon D, Bertsch PM (2002) Hydrolysis of iron(II) chloride under anoxic conditions and influence of SiO_4 ligands. Langmuir 18:4292-4299

Doyle CS, Carrier X, Kendelewicz T, Brown GE Jr (1999b) Reaction of CO_2 with CaO(100) surfaces. Surf Rev Lett 6:1247-1254

Doyle CS, Traina SJ, Ruppert H, Kendelewicz T, Rehr JJ, Brown GE Jr (1999a) Al-XANES studies of aluminum-rich surface phases in the soil environment. J Synchrotron Rad 6:621-623

Dreier P, Rabe P (1986) EXAFS study of the Zn^{2+} coordination in aqueous solutions. J de Physique 47 (Colloque C8, suppl 12):809-812

Driesner T, Cummings PT (1999) Molecular simulation of the temperature- and density-dependence of ionic hydration in aqueous $SrCl_2$ solutions using rigid and flexible water models. J Chem Phys 111:5141-5149

Drits VA, Dainyak LG, Muller F, Besson G, Manceau A (1997b) Isomorphous cation distribution in celadonites, glauconites and Fe-illites determined by infrared, Mossbauer and EXAFS spectroscopies. Clay Minerals 32:153-179

Drits VA, Lanson B, Gorshkov AI, Manceau A (1998) Substructure and superstructure of four-layer Ca-exchanged birnessite. Am Mineral 83:97-118

Drits VA, Manceau A (2000) A model for the mechanism of Fe^{3+} to Fe^{2+} reduction in dioctahedral smectites. Clays Clay Minerals 48:185-195

Drits VA, Silvester E, Gorshkov AI, Manceau A (1997a) Structure of synthetic monoclinic Na-rich birnessite and hexagonal birnessite: I. Results from X-ray diffraction and selected-area electron diffraction. Am Mineral 82:946-961

Du Q, Freysz E, Shen YR (1994) Vibrational spectra of water molecules at quartz/water interfaces. Phys Rev Lett 72:238-241

Duff MC (2001) Speciation and transformations of sorbed Pu on geologic materials: Wet chemical and spectroscopic observations. Radioactivity in the Environment 1 (Plutonium in the Environment):139-157

Duff MC, Amrhein C, Bertsch PM, Hunter DB (1997) The chemistry of uranium in evaporation pond sediment in the San Joaquin Valley, California, USA, using X-ray fluorescence and XANES techniques. Geochim Cosmochim Acta 61:73-81

Duff MC, Hunter DB, Triay IR, Bertsch PM, Kitten J, Vaniman DT (2001) Comparison of two micro-analytical methods for detecting the spatial distribution of sorbed Pu on geologic materials. J Contaminant Hydrology 47:211-218

Duff MC, Hunter DB, Triay IR, Bertsch PM, Reed DT, Sutton SR, Shea-McCarthy G, Kitten J, Eng P, Chipera SJ, Vaniman DT (1999a) Mineral associations and average oxidation states of sorbed Pu on tuff. Environ Sci Technol 33:2163-2169

Duff MC, Morris DE, Hunter DB, Bertsch PM (2000) Spectroscopic characterization of uranium in evaporation basin sediments. Geochim Cosmochim Acta 64:1535-1550

Duff MC, Newville M, Hunter DB, Bertsch PM, Sutton SR, Triay IR, Vaniman DT, Eng P, Rivers ML (1999b) Micro-XAS studies with sorbed plutonium on tuff. J Synchrotron Rad 6:350-352

Dzombak DA, Morel FMM (1990) Surface Complexation Modeling. John Wiley & Sons, New York

Eglinton TI, Irvine JE, Vairavamurthy A, Zhou W, Manowitz, B (1994) Formation and diagenesis of macromolecular organic sulfur in Peru margin sediments. Org Geochem 22:781-799

Eick MJ, Fendorf SE (1998) Reaction sequence of nickel(II) with kaolinite: mineral dissolution and surface complexation and precipitation. Soil Sci Soc Am J 62:1257-1267

Eisenberg D, Kauzmann W (1969) The Structure and Properties of Water. Oxford University Press, Oxford, UK

Elder FR, Gurewitsch AM, Langmuir RV, Pollack HC (1947) Radiation from electrons in a synchrotron. Phys Rev 71:829-830

Elzinga E, Peak D, Sparks DL (2001) Spectroscopic studies of Pb(II)-sulfate interactions at the goethite-water interface. Geochim Cosmochim Acta 65:2219-2230

Elzinga E, Sparks DL (1999) Nickel sorption mechanisms in a pyrophyllite-montmorillonite mixture. J Colloid Interface Sci 213:506-512

Enderby JE, Neilson GW (1979) X-ray and neutron scattering by aqueous solutions of electrolytes. *In:* Franks F (ed) Water: A Comprehensive Treatise 6:1-46, 411-436

Eng PJ, Trainor TP, Brown GE Jr, Waychunas GA, Newville M, Sutton SR, Rivers ML (2000) Structure of the hydrated α-Al_2O_3 (0001) surface. Science 288:1029-1033

England KER, Charnock JM, Pattrick RAD, Vaughan DJ (1999a) Surface oxidation studies of chalcopyrite and pyrite by glancing-angle X-ray absorption spectroscopy (REFLEXAFS). Mineral Mag 63:559-566

England KER, Pattrick RAD, Charnock JM, Mosselmans JFW (1999b) Zinc and lead sorption on the surface of $CuFeS_2$ during flotation: a fluorescence REFLEXAFS study. Internat J Mineral Processing 57:59-71

EPA (1997) Mercury Study Report to Congress. U.S. Environmental Protection Agency, Office of Air Quality Planning and Standards and Office of Research and Development

Etzler FM, Fagundas DM (1987) The extent of vicinal water. Implications for the density of water in silica pores. J Colloid Interface Sci 115:513-519

Fadley CS (1978) Basic concepts of X-ray photoelectron spectroscopy. Electron Spectrosc Theory Tech Appl 2:1-156

Fadley CS (1992) The study of surface structures by photoelectron diffraction and Auger electron diffraction. Synchrotron Rad Res 1:421-518

Fadley CS, Thevuthasan S, Kaduwela AP, Westphal C, Kim YJ, Ynzunza R, Len P, Tober E, Zhang F (1994) Photoelectron diffraction and holography: Present status and future prospects. J Electron Spectrosc Relat Phenom 68:19-47

Fadley CS, Chen Y, Couch RE, Daimon H, Denecke R, Galloway H, Hussain Z, Kaduwela AP, Kim YJ, Len PM, Liesegang J, Menchero J, Morais J, Palomares J, Ruebush SD, Ryce S, Salmeron MB, Schattke W, Thevuthasan S, Tober ED, Van Hove MA, Wang Z, Ynzunza RX, Zaninovich JJ (1997) Surface, interface, and nanostructure characterization with photoelectron diffraction and photoelectron and X-ray holography. J Surf Anal 3:334-364

Farges F (2002) Improving data reduction methods in X-ray absorption fine structure spectroscopy. Part 1: pre-edge and XANES, resolution, convolution and component analyses. Phys Chem Minerals (submitted)

Farges F, Berrodier I, Benedetti M, Brown GE Jr, Deveughele M. (2002) Absorption mechanisms of gold on iron- and aluminum-(oxy)hydroxides. Part I. Theoretical aspects of Au L_{III}-edge XAFS spectra. Geochim Cosmochim Acta (submitted)

Farges F, Brown GE Jr (1996) An empirical model for the anharmonic analysis of high-temperature XAFS spectra of oxide compounds with applications to the coordination environment of Ni in NiO, γ-Ni_2SiO_4 and Ni-bearing Na-disilicate glass and melt. Chem Geol 128:93-106

Farges F, Brown GE Jr (1997) Coordination of actinides in silicate melts. J de Physique IV 7 (Colloque C2, X-Ray Absorption Fine Structure, Vol. 2):1009-1010

Farges F, Brown GE Jr, Navrotsky A, Gan H, Rehr JJ (1996b) Coordination chemistry of titanium(IV) in silicate glasses and melts. Part II. Glasses at ambient temperature and pressure. Geochim Cosmochim Acta 60:3039-3053

Farges F, Brown GE Jr, Navrotsky A, Gan H, Rehr JJ (1996a) Coordination chemistry of titanium(IV) in silicate glasses and melts. Part III. Glasses and melts from ambient to high temperatures. Geochim Cosmochim Acta 60:3055-3065

Farges F, Brown GE Jr, Petit P-E, Munoz M (2001) Transition elements in water-bearing silicate glasses/melts. Part I. A high resolution and anharmonic EXAFS analysis of Ni coordination environments in crystals, glasses, and melts. Geochim Cosmochim Acta 65:1665-1678

Farges F, Brown GE Jr, Rehr JJ (1997) Ti K-edge XANES studies of Ti coordination and disorder in oxide compounds: Comparison between theory and experiment. Phys Rev B 56:1809-1819

Farges F, Ewing RC, Brown GE Jr (1993b) The structure of aperiodic, metamict (Ca, Th)ZrTi$_2$O$_7$: an EXAFS study of the Zr, Th and U sites. J Materials Res 8:1983-1995

Farges F, Harfouche M, Petit P-E, Brown GE Jr (2002) Actinides in earth materials: The importance of natural analogues. Proceedings of the Euroconference and NEA Workshop on Speciation, Techniques, and Facilities for Radioactive Materials at Synchrotron Light Sources, Grenoble, France, Sept. 10-12, 2000, Nuclear Energy Agency/Organization for Economic Co-operation and Development, AEN/NEA 2002, Paris, p 63-74

Farges F, Ponader CW, Brown GE Jr (1991) Structural environments of incompatible elements in silicate glass/melt systems: I. Zr at trace levels. Geochim Cosmochim Acta 55:1563-1574

Farges F, Ponader CW, Calas G, Brown GE Jr (1992) Structural environments of incompatible elements in silicate glass/melt systems: II. U$^{(IV)}$, U$^{(V)}$, and U$^{(VI)}$. Geochim Cosmochim Acta 56:4205-4220

Farges F, Sharps JA, Brown GE Jr (1993a) Local environment around gold(III) in aqueous chloride solutions: an EXAFS spectroscopy study. Geochim Cosmochim Acta 57:1243-1252

Farquhar ML, Charnock JM, England KER, Vaughan DJ (1996) Adsorption of Cu^{2+} on the (0001) plane of mica: A REFLEXAFS and XPS study. J Colloid Interface Sci 177:561-567

Farquhar ML, Vaughan DL, Hughes CR, Charnock JM, England KER (1997) Experimental studies of the interaction of aqueous metal cations with mineral substrates: lead, cadmium, and copper with perthitic feldspar, muscovite, and biotite. Geochim Cosmochim Acta 61:3051-3064

Farrell SP, Fleet ME (2001) Sulfur K-edge XANES study of local electronic structure in ternary monosulfide solid solution [(Fe, Co, Ni)$_{0.923}$S]. Phys Chem Minerals 28:17-27

Fein JB, Daughney CJ, Yee N, Davis TA (1997) A chemical equilibrium model for metal adsorption onto bacterial surfaces. Geochim Cosmochim Acta 61:3319-3328

Fendorf SE (1995) Surface reactions of chromium in soils and waters. Geoderma 67:55-71

Fendorf SE (1999) Fundamental aspects and applications of X-ray absorption spectroscopy in clay and soil science. *In:* Schulze DG, Stucki JW, Bertsch PM (eds) Synchrotron X-ray Methods in Clay Science, Clay Minerals Society Workshop Lectures 9. The Clay Minerals Society, Boulder, Colorado, p 19-67

Fendorf SE, Eick MJ, Grossl P, Sparks DL (1997) Arsenate and chromate retention mechanism on goethite. I. Surface structure. Environ Sci Technol 31:315-320

Fendorf SE, Jardine PM, Patterson RR, Taylor DL, Brooks SC (1999) Pyrolusite surface transformations measured in real-time during the reactive transport of Co(II)EDTA^{2-}. Geochim Cosmochim Acta 63:3049-3057

Fendorf SE, Jardine PM, Taylor DL, Brooks SC, Rochette EA (1998) Auto-inhibition of oxide mineral reductive capacity toward Co(II)EDTA. ACS Symp Ser 715 (Mineral-Water Interfacial Reactions), American Chemical Society, Washington, DC, p 358-371

Fendorf SE, Lamble GM, Stapleton MG, Kelley MJ, Sparks DL (1994) Mechanisms of chromium(III) sorption on silica. 1. Cr(III) surface structure derived by extended X-ray absorption fine structure spectroscopy. Environ Sci Technol 28:284-289

Fendorf S, Wielinga BW, Hansel CM (2000) Chromium transformations in natural environments: The role of biological and abiological processes in chromium(VI) reduction. Int Geol Rev 42:691-701

Fenter P, Cheng L, Rihs S, Machesky M, Bedzyk MJ, Sturchio NC (2000c) Electrical double-layer structure at the rutile-water interface as observed *in situ* with small-period X-ray standing waves. J Colloid Interface Sci 225:154-165

Fenter P, Geissbuhler P, DiMasi E, Srajer E, Sorensen LB, Sturchio NC (2000a) Surface speciation of calcite observed *in situ* by high-resolution X-ray reflectivity. Geochim Cosmochim Acta 64:1221-1228

Fenter P, McBride MT, Srajer G, Sturchio NC, Bosbach D (2001b) Structure of barite (001) – and (210) – water interfaces. J Phys Chem B 105:8112-8119

Fenter P, McBride MT, Srajer G, Sturchio NC, Bosbach D (2001a) Structure of HEDP adsorbed at the barite-water interface. (Abstract) 222nd American Chemical Society Annual Meeting, Chicago, IL, Aug. 26-30, 2001

Fenter P, Park C, Cheng L, Zhang Z, Krekeler M, Sturchio NC (2002) Orthoclase dissolution kinetics probed by *in situ* X-ray reflectivity: Effects of temperature, pH and crystal orientation. Geochim Cosmochim Acta 66:(in press)

Fenter P, Sturchio NC (1999) Structure and growth of stearate monolayers on calcite: first results of an *in situ* X-ray reflectivity study. Geochim Cosmochim Acta 63:3145-3152

Fenter P, Teng H, Geissbuhler P, Hanchar JM, Nagy KL, Sturchio NC (2000b) Atomic-scale structure of the orthoclase (001)-water interface measured with high-resolution X-ray reflectivity. Geochim Cosmochim Acta 64:3663-3673

Filipponi A, D'Angelo P, Pavel NV, Di Cicco A (1994) Triplet correlations in the hydration shell of aquaions. Chem Phys Lett 225:150-155

Filipponi A, Di Cicco A, Tyson TA, Natoli CR (1991) "Ab-initio" modeling of X-ray absorption spectra. Solid State Commun 78:265-268

Finkelman RB, Belkin HE, Zheng B (1999) Health impacts of domestic coal use in China. Proc Natl Acad Sci USA 96:3427-3431

Fitts JP, Brown GE Jr, Parks GA (2000) Structural evolution of Cr(III) polymeric species at the γ-Al$_2$O$_3$/water interface. Environ Sci Technol 34:5122-5128

Fitts JP, Persson P, Brown GE Jr, Parks GA (1999b) Structure and bonding of Cu(II)-glutamate complexes at the α-Al$_2$O$_3$-water interface. J Colloid Interface Sci 220:133-147

Fitts JP, Trainor TP, Grolimund D, Bargar JR, Parks GA, Brown GE Jr (1999a) Grazing-incidence XAFS investigations of Cu(II) sorption products at the oxide-water interface. J Synchrotron Rad 6:627-629

Flank A-M, Fontaine A, Jucha A, Lemonnier M, Williams C (1982) Extended X-ray absorption fine structure in dispersive mode. J Phys Lett 43:315-319

Fleet ME, Muthupari S, Kasrai M, Prabakar S (1997) Sixfold coordinated Si in alkali and alkali-CaO silicophosphate glasses by Si K-edge XANES spectroscopy. J Non-Crystal Solids 220:85-92

Fontaine A, Baudelet F, Dartyge E, Guay D, Itie JP, Polian A, Tolentino H, Tourillon G (1992) Two time-dependent, focus-dependent experiments using the energy-dispersive spectrometer at LURE. Rev Sci Instrum 63 (1, Pt. 2B):960-965

Fontaine A, Lagarde P, Raoux D, Fontana MP, Maisano G, Migliardo P, Wanderlingh F (1978) Extended X-ray-absorption fine-structure studies of local ordering in highly concentrated aqueous solutions of copper(II) bromide. Phys Rev Lett 41:504-507

Ford FG, Kemner KM, Bertsch PM (1999a) Influence of sorbate-sorbent interactions on the crystallization kinetics of nickel- and lead-ferrihydrite coprecipitates. Geochim Cosmochim Acta 63:39-48

Ford FG, Scheinost AC, Scheckel KG, Sparks DL (1999b) The link between clay mineral weathering and the stabilization of nickel surface precipitates. Environ Sci Technol 33:3140-3144

Ford RG, Sparks DL (2000) The nature of Zn precipitates formed in the presence of pyrophyllite. Environ Sci Technol 34:2479-2483

Foster AL, Breit GN, Welch AH, Whitney JW, Yount JC, Alam MM, Islam MK, Islam MN, Islam MS (2000) *In situ* identification of arsenic species in soil and aquifer sediment from Ramrail, Brahmanbaria, Bangladesh. EOS Trans Am Geophys Union 81, Fall Meeting Suppl, Abstract No H21D-01

Foster AL, Brown GE Jr, Parks GA (1998b) XAFS study of photocatalyzed, hetero-geneous As(III) oxidation on kaolin and anatase. Environ Sci Technol 32:1444-1452

Foster AL, Brown GE Jr, Parks GA (2002) XAFS study of As(V) and Se(IV) sorption complexes on hydrous Mn oxides. Geochim Cosmochim Acta (in press)

Foster AL, Brown GE Jr, Parks GA, Tingle TN, Voigt D, Brantley SL (1997) XAFS determination of As(V) associated with Fe(III) oxyhydroxides in weathered mine tailings and contaminated soil from California, USA. J de Physique IV 7 (Colloque C2, X-Ray Absorption Fine Structure, Vol. 2):815-816

Foster AL, Brown GE Jr, Tingle TN, Parks GA (1998a) Quantitative arsenic speciation in mine tailings using X-ray absorption spectroscopy. Am Mineral 83:553-568

Frahm R, Wong J, Holt JB, Larson EM, Rupp B, Waide PA (1993) Real-time probe of reaction centers in solid combustions by QEXAFS on the sub-second time scale. Japan J Appl Phys Part 1 32 (Suppl. 32-2, XAFS VII):185-187

Frankenberger WT Jr, Benson SL (eds) (1994) Selenium in the Environment. Marcel Dekker Inc, New York

Franks F (1972-1982) Water: A Comprehensive Treatise, Vols. 1-6, Plenum Press, New York

Franks F (1984) Water. The Royal Society of Chemistry, London

Frenkel AI, Cross JO, Fanning DM, Robinson IK (1999) DAFS analysis of magnetite. J Synchrotron Rad 6:332-334

Frenkel AI, Kleifeld O, Wasserman SR, Sagi I (2002) Phase speciation by extended X-ray absorption fine structure spectroscopy. J Chem Phys 116:9449-9456

Frenkel AI, Vairavamurthy A, Newville M (2001) A study of the coordination environment in aqueous cadmium-thiol complexes by EXAFS spectroscopy: experimental vs theoretical standards. J Synchrotron Rad 8:669-671

Friedl G, Wehrli B, Manceau A (1997a) Solid phases in the cycling of manganese in eutrophic lakes: new insights from EXAFS spectroscopy. Geochim Cosmochim Acta 61:275-290

Friedl G, Wehrli B, Manceau A (1997b) Solid phases in the cycling of manganese in eutrophic lakes: New insights from EXAFS spectroscopy. Geochim Cosmochim Acta 61:3277

Fuhrmann M, Bajt S, Schoonen MAA (1998) Sorption of iodine on minerals investigated by X-ray absorption near edge structure (XANES) and ^{125}I tracer sorption experiments. Appl Geochem 13:127-141

Fuhrmann M, Lasat MM, Ebbs SD, Kochian LV, Cornish J (2002) Uptake of cesium-137 and strontium-90 from contaminated soil by three plant species; application to phytoremediation. J Environ Qual 31:904-909

Fukushima Y, Okamoto T (1987) Extended X-ray absorption fine structure of cobalt-exchanged sepiolite. *In:* Proc Int Clay Conf Denver 1985, Schultz LG, van Olphen H, Mumpton FA (eds), The Clay Mineral Soc, Bloomington, IN, p 9-16

Fuller CC, Bargar, JR, Davis JA, Piana MJ (2002a) Mechanisms of uranium interactions with hydroxyapatite: Implications for groundwater remediation. Environ Sci Technol 36:158-165

Fuller CC, Bargar JR, Piana MJ, and Davis JA (2002b) Molecular-scale characterization of uranium sorption by apatite materials from a permeable reactive barrier demonstration. Environ Sci Technol (submitted)

Fulton JL, Darab JG, Hoffmann MM (2001) X-ray absorption spectroscopy and imaging of heterogeneous hydrothermal mixtures using a diamond microreactor cell. Rev Sci Instrum 72:2117-2122

Fulton JL, Hoffmann MM, Darab JG, Palmer BJ, Stern EA (2000) Copper(I) and copper(II) coordination structure under hydrothermal conditions at 325°C: An X-ray absorption fine structrure and molecular dynamics study. J Phys Chem A 104:11651-11663

Fulton JL, Pfund DM, Wallen SL, Newville M, Stern EA, Ma Y (1996) Rubidium ion hydration in ambient and supercritical water. J Chem Phys 105:2161-2166

Gaillard C, Den AC, Conradson SD (2002) An X-ray absorption near edge spectroscopy study of trace amount technetium implanted in apatite. Phys Chem Chem Phys 4:2499-2500

Gaillard JF, Webb SM, Quintana JPG (2001) Quick X-ray absorption spectroscopy for determining metal speciation in environmental samples. J Synchrotron Rad 8:928-930

Galbreath KC, Toman DL, Zygarlicke CJ, Huggins FE, Huffman GP, Wong JL (2000) Nickel speciation of residual oil fly ash and ambient particulate matter using X-ray absorption spectroscopy. J Air Waste Manage Assoc 50:1876-1886

Galbreath KC, Zygarlicke CJ, Huggins FE, Huffman GP, Wong JL (1998) Chemical speciation of nickel in residual oil ash. Energy Fuels 12:818-822

Galoisy L, Calas G, Arrio MA (2001) High resolution XANES spectra of iron in minerals and glasses: structural information from the pre-edge region. Chem Geol 174:307-319

Garcia J, Benfatto M, Natoli CR, Bianconi A, Fontaine A, Tolentino H (1989) The quantitative Jahn-Teller distortion of the copper (2+) site in aqueous solution by XANES spectroscopy. Chem Phys 132:295-307

Gates WP, Slade PG, Manceau A, Lanson B (2002) Site occupancies by iron in nontronites. Clays Clay Minerals 50:223-239

Gauglhofer J, Bianchi V (1992) Chromium. *In*: Merian E (ed) Metals and Their Compounds in the Environment. VCH, Weinheim, Germany, p 853-878

Gauthier C, Solé VA, Signorato R, Goulon J, Moguiline E (1999) The ESRF Beamline ID26: X-ray absorption on ultradilute samples. J Synchrotron Rad 6:164-166

George GN, Pickering IJ (1992) EXAFSPAK: A Suite of Computer Programs for Analysis of X-Ray Absorption Spectra. Stanford Synchrotron Radiation Laboratory, Stanford, CA (*http://ssrl.slac.stanford.edu/exafspak.html*)

Ghelen M, Beck L, Calas G, Flank A-M, Van Bennekom AJ, Van Beusekom JEE (2002) Unraveling the atomic structure of biogenic silica: Evidence of the structural association of Al and Si in diatom frustules. Geochim Cosmochim Acta 66:1601-1609

Giaquinta DM, Soderholm L, Yuchs SE, Wasserman SR (1997a) The speciation of uranium in a smectite clay. Evidence for catalyzed uranyl reduction. Radiochim Acta 76:113-121

Giaquinta DM, Soderholm L, Yuchs SE, Wasserman SR (1997b) Hydrolysis of uranium and thorium in surface-modified bentonite under hydrothermal conditions. J Alloys and Compounds 249:142-145

Gilbert B, Margaritondo G, Douglas S, Nealson KH, Egerton RF, Rempfer GF, de Stasio G (2001) XANES microspectroscopy of biominerals with photoconductive charge compensation. J Electron Spectrosc Relat Phenom 114-116:1005-1011

Giordano L, Goniakowski J, Suzanne J (1998) Partial dissociation of water molecules in the (3×2) water monolayer deposited on the MgO (100) surface. Phys Rev Lett 81: 1271-1273

Gleiter H (1992) Nanostructured materials. Adv Materials (Weinheim, Germany) 4:474-481.

Glen GL, Dodd CG (1968) Use of molecular orbital theory to interpret X-ray K-absorption spectral data. J Appl Phys 39:5372-5377

Goodarzi F, Huggins FE (2001) Monitoring the species of arsenic, chromium and nickel in milled coal, bottom ash and fly ash from a pulverized coal-fired power plant in western Canada. J Environ Monitoring 3:1-6

Gratz AJ, Hillner PE, Hansma PK (1993) Step dynamics and spiral growth on calcite. Geochim Cosmochim Acta 57:491-495

Greaves GN, Fontaine A, Lagarde P, Raoux D, Gurman SJ (1981) Local structure of silicate glasses. Nature 293:611-616

Greegor RB, Pingitore NE Jr, Lytle FW (1997) Strontianite in coral skeletal aragonite. Science 275:1452-1454

Grolimund D, Borkovec M, Barmettler K, Sticher H (1996) Colloid-facilitated transport of strongly sorbing contaminants in natural porous media: A laboratory column study. Environ Sci Technol 30:3118-3123

Grolimund D, Kendelewicz T, Trainor TP, Liu P, Fitts JP, Chambers SA, Brown GE Jr (1999) Identification of Cr species at the solution-hematite interface after Cr(VI)-Cr(III) reduction using GI-XAFS and Cr L-edge NEXAFS. J Synchrotron Rad 6:612-614

Grolimund D, Warner JA, Carrier X, Brown GE Jr (2000) Chemical behavior of strontium at the solid-liquid interface of amorphous manganese oxides: A molecular-level study using EXAFS. (Abstract) J Conference Abstracts 52:461

Grossman MJ, Lee MK, Prince RC, Garrett KK, George GN, Pickering IJ (1999) Microbial desulfurization of a crude oil middle-distillate fraction: analysis of the extent of sulfur removal and the effect of removal on remaining sulfur. Appl Environ Microbiol 65:181-188

Guenard P, Renaud G, Barbier A, Gautier-Soyer M (1997) Determination of the α-$Al_2O_3(0001)$ surface relaxation and termination by measurements of crystal truncation rods. Surf Rev Lett 5:321-324

Guilhaumou N, Dumas P, Carr GL, Williams GP (1998) Synchrotron infrared microspectrometry applied to petrography in micrometer-scale ranges: fluid chemical analysis and mapping. Appl Spectrosc 52:1029-1034

Gurman SJ, Binsted N, Ross I (1984) A rapid, exact curved-wave theory for EXAFS calculations. J de Physique C 17:143-151

Hansel CM, Fendorf S, Sutton S, Newville M (2001) Characterization of Fe plaque and associated metals on the roots of mine-waste impacted aquatic plants. Environ Sci Technol 35:3863-3868

Hasnain SS, Hodgson KO (1999) Structure of metal centers in proteins at subatomic resolution. J Synchrotron Rad 6:852-864

Hass KC, Schneider WF, Curioni A, Andreoni W (1998) The chemistry of water on alumina surfaces: Reaction dynamics from first principles. Science 282:265-268

Hasselstrom J, Fohlisch A, Karis O, Wassdahl N, Weinelt M, Nilsson A, Nyberg M, Pettersson LGM, Stohr J (1999) Ammonia adsorbed on Cu(110): An angle resolved X-ray spectroscopic and ab initio study. J Chem Phys 110:4880-4890

Hasselstrom J, Karis O, Weinelt M, Wassdahl N, Nilsson A, Nyberg M, Pettersson LGM, Samant MG, Stohr J (1998) The adsorption structure of glycine adsorbed on Cu(110); comparison with formate and acetate/Cu(110). Surf Sci 407:221-236

Hayes KF, Katz LE (1996) Application of X-ray absorption spectroscopy for surface complexation modeling of metal ion sorption. *In:* Brady PV (ed) Physics and Chemistry of Mineral Surfaces, CRC Press, Boca Raton, FL, p 147-223

Hayes KF, Leckie JO (1987) Modeling ionic strength effects on cation sorption at hydrous oxide/solution interfaces. J Colloid Interface Sci 115:564-572

Hayes KF, Roe AL, Brown GE Jr, Hodgson KO, Leckie JO, Parks GA (1987) *In situ* X-ray absorption study of surface complexes at oxide/water interfaces: selenium oxyanions on α-FeOOH. Science 238:783-786

Hazemann J-L, Berar JF, Manceau A (1991) Rietveld studies of the aluminum-iron substitution in synthetic goethite. Mater Sci Forum 79-82

Hazemann J-L, Manceau A, Sainctavit P, Malgrange C (1992) Structure of the α-Fe$_x$Al$_{1x}$OOH solid solution. I. Evidence by polarized EXAFS for an epitaxial growth of hematite-like clusters in Fe-diaspore. Phys Chem Minerals 19:25-38

Heald SM, Chen H, Tranquada JM (1988) Glancing-angle extended X-ray absorption fine structure and reflectivity studies of interfacial regions. Phys Rev B 38:1016-1026

Hellquist B, Bengtsson LA, Holmberg B, Hedman B, Persson I, Elding LI (1991) Structures of solvated cations of palladium(II) and platinum(II) in dimethyl sulfoxide, acetonitrile, and aqueous solution studied by EXAFS and LAXS. Acta Chem Scand 45:449-455

Helz GR, Miller CV, Charnock JM, Mosselmans JFW, Pattrick RAD, Garner CD, Vaughan DJ (1996) Mechanism of molybdenum removal from the sea and its concentration in black shales: EXAFS evidence. Geochim Cosmochim Acta 60:3631-3642

Helz GR, Tossell JA, Charnock JM, Pattrick RAD, Vaughan DJ, Garner CD (1995) Oligomerization in As(III) sulfide solutions: Theoretical constraints and spectroscopic evidence. Geochim Cosmochim Acta 59:4591-4604

Henderson MA (2002) The interaction of water with solid surfaces: fundamental aspects revisited. Surf Sci Rep 285:1-308

Hennig C, Panak PJ, Reich T, Rossberg A, Raff J, Selenska-Pobell S, Matz W, Bucher JJ, Bernhard G, Nitsche H (2001) EXAFS investigation of uranium(VI) complexes formed at *Bacillus cereus* and *Bacillus sphaericus* surfaces. Radiochim Acta 89:625-631

Hessler JP, Seifert S, Winans RE, Fletcher TH (2001) Small-angle X-ray studies of soot inception and growth. Faraday Discuss 119:395-407

Hesterberg D, Chou JW, Hutchison KJ, and Sayers DE (2001) Bonding of Hg(II) to reduced organic sulfur in humic acid as affected by S/Hg ratio. Environ Sci Technol 35:2741-2745

Hesterberg D, Sayers DE, Zhou W, Plummer GM, Robarge WP (1997a) X-ray absorption spectroscopy of lead and zinc speciation in a contaminated groundwater aquifer. Environ Sci Technol 31:2840-2846

Hesterberg D, Sayers DE, Zhou W, Robarge WP, Plummer GM (1997b) XAFS characterization of copper in model aqueous systems of humic acid and illite. J Phys IV 7 (Colloque C2, X-Ray Absorption Fine Structure, Vol. 2): 833-834

Hesterberg D, Zhou W, Hutchison KJ, Beauchemin S, Sayers DE (1999) XAFS study of adsorbed and mineral forms of phosphate. J Synchrotron Rad 6:636-638

Hewkin DJ, Prince RH (1970) The mechanism of octahedral complex formation by labile metal ions. Coord Chem Rev 5:45-73

Hirschmugl CJ (2002) Frontiers in infrared spectroscopy at surfaces and interfaces. Surf Sci 500:577-604

Hochella MF Jr (1988) Auger electron and photoelectron spectroscopies. Rev Mineral 18:573-637

Hochella MF Jr (1995) Mineral surfaces: Their characterization and their chemical, physical, and reactive properties. *In:* Vaughan DJ, Pattrick RAD (eds) Mineral Surfaces, Mineral Soc Ser 5, Chapman & Hall, London, p 17-60

Hoffmann MM, Darab JG, Fulton JL (2001) An infrared and X-ray absorption study of the structure and equilibrium of chromate, bichromate, and dichromate in high-temperature aqueous solutions. J Phys Chem A 105:6876-6885

Holdren GR Jr, Berner RA (1979) Mechanism of feldspar weathering. I. Experimental studies. Geochim Cosmochim Acta 43:1161-1171

Holland BW, Pendry JB, Pettifer RF, Bordas J (1978) Atomic origin of structure in EXAFS experiments. J Phys C 11:633-642

Holland BW, Woolfson MS, Woodruff DP, Johnson PD, Norman D, Farrell HH, Traum MM, Smith NV (1980) Structural sensitivity of photoelectron diffraction azimuthal patterns. Solid State Commun 35:225-227

Hudson EA, Allen PG, Terminello LJ, Denecke MA, Reich T (1996) Polarized X-ray-absorption spectroscopy of the uranyl ion: comparison of experiment and theory. Phys Rev B 54:156-165

Hudson EA, Terminello LJ, Viann BE, Denecke M, Reich T, Allen PG, Bucher JJ,. Shuh DK, Edelstein NM (1999) The structure of uranium(VI) sorption complexes on vermiculite and hydrobiotite. Clays Clay Minerals 47:439-457

Huffman GP, Huggins FE, Francis HE, Mitra S, Shah N (1990) Structural characterization of sulfur in bioprocessed coal. Coal Sci Technol 16 (Processes Utilizing High-Sulfur Coals 3):21-32

Huffman GP, Huggins FE, Levasseur AA, Durant JF, Lytle FW, Greegor RB, Mehta A (1989a) Investigation of atomic structures of calcium in ash and deposits produced during the combustion of lignite and bituminous coal. Fuel 68:238-242

Huffman GP, Huggins FE, Mitra S, Shah N, Lytle FW Greegor RB (1989b) Forms of occurrence of sulfur and chlorine in coal. Physica B (Amsterdam) 158:225-226

Huffman GP, Huggins FE, Mitra S, Shah N, Pugmire RJ, Davis B, Lytle FW, Greegor RB (1989c) Investigation of the molecular structure of organic sulfur in coal by XAFS spectroscopy. Energy Fuels 3:200-205

Huffman GP, Huggins FE, Shah N (1992) XAFS spectroscopy studies of critical elements in coal and coal derivatives. Adv Coal Spectrosc 29-47

Huffman GP, Huggins FE, Shah N, Huggins R, Linak WP, Miller CA, Pugmire RJ, Meuzelaar HLC, Seehra MS, Manivannan A (2000) Characterization of fine particulate matter produced by combustion of residual fuel oil. J Air Waste Manage Assoc 50:1106-1114

Huffman GP, Huggins FE, Shah N, Zhao J (1994) Speciation of arsenic and chromium in coal and combustion ash by XAFS spectroscopy. Fuel Process Technol 39:47-62

Huffman GP, Mitra S, Huggins FE, Shah N, Vaidya S, Lu F (1991) Quantitative analysis of all major forms of sulfur in coal by X-ray absorption fine structure spectroscopy. Energy Fuels 5:574-581

Huffman GP, Shah N, Huggins FE, Lu F, Zhao J (1993) Further sulfur speciation studies by sulfur K-edge XANES spectroscopy. Coal Sci Technol 21 (Processes Utilizing High-Sulfur Coals V):1-13

Huffman GP, Shah N, Huggins FE, Stock LM, Chatterjee K, Kilbane JJ II, Chou I-M, Buchanan DH (1995) Sulfur speciation of desulfurized coals by XANES spectroscopy. Fuel 74:549-555

Huggins FE, Goodarzi F, Lafferty CJ (1996) Mode of occurrence of arsenic in subbituminous coals. Energy Fuels 10:1001-1004

Huggins FE, Huffman GP (1991) An XAFS investigation of the form-of-occurrence of chlorine in U.S. coals. Coal Sci. Technol. 17 (Chlorine Coal):43-61

Huggins FE, Huffman GP (1995) Chlorine in coal: an XAFS spectroscopic investigation. Fuel 74:556-569

Huggins FE, Huffman GP (1996) Modes of occurrence of trace elements in coal from XAFS spectroscopy. Int J Coal Geol 32:31-53

Huggins FE, Huffman GP (1999) Use of XAFS spectroscopy for element speciation in environmentally important energy-related materials. Int J Soc Mater Eng Resources 7:230-241

Huggins FE, Huffman GP, Dunham GE, Senior CL (1999) XAFS examination of mercury sorption on three activated carbons. Energy Fuels 13:114-121

Huggins FE, Huffman GP, Robertson JD (2000a) Speciation of elements in NIST particulate matter SRMs 1648 and 1650. J. Hazard. Mater 74:1-23

Huggins FE, Huffman GP, N. Shah, R.G. Jenkins, F.W. Lytle, and R.B. Greegor (1988a) Further EXAFS examination of the state of calcium in pyrolyzed char. Fuel 67:938-941

Huggins FE, Mitra S, Vaidya S, Taghiei MM, Lu F, Shah N, Huffman GP (1991) The quantitative determination of all major inorganic and organic sulfur forms in coal from XAFS spectroscopy: method and applications. Coal Sci Technol 18 (Processes Utilizing High-Sulfur Coals 4):13-42

Huggins FE, Najih M, Huffman GP (1998) Direct speciation of chromium in coal combustion byproducts by X-ray absorption fine-structure spectroscopy. Fuel 78: 233-242

Huggins FE, Parekh BK, Robertson JD, Huffman GP (1995) Modes of occurrence of trace elements in coal: Geochemical constraints from XAFS and PIXE spectroscopic analysis of advanced coal cleaning tests. Coal Sci Technol 24:175-178

Huggins FE, Shah N, Huffman GP, Lytle FW, Greegor RB, Jenkins RG (1988b) *In situ* XAFS investigation of calcium and potassium catalytic species during pyrolysis and gasification of lignite chars. Fuel 67:1662-1667

Huggins FE, Shah N, Huffman GP, Kolker A, Crowley S, Palmer CA, Finkelman RB (2000b) Mode of occurrence of chromium in four US coals. Fuel Process Technol 63:79-92

Huggins FE, Shah N, Huffman GP, Robertson JD (2000c) XAFS spectroscopic characterization of elements in combustion ash and fine particulate matter. Fuel Process Technol 65-66:203-218

Huggins FE, Shah N, Zhao J, Lu F, Huffman GP (1993) Nondestructive determination of trace element speciation in coal and coal ash by XAFS spectroscopy. Energy Fuels 7:482-489

Huggins FE, Srikantapura BK, Parekh L, Blanchard, and J.D. Robertson (1997) XANES spectroscopic characterization of selected elements in deep-cleaned fractions of Kentucky no. 9 coal. Energy Fuels 11:691-701

Huggins FE, Zhao J, Huffman GP, Kuo C-H, Tarrer AR (1996) Investigation of zinc additives in coliquefaction of waste lubricating oil and a bituminous coal. J Environ Sci Health, Part A Environ Sci Eng Toxic Hazard Subst Control A31:1755-1766

Hunter DB, Bertsch PM (1998) *In situ* examination of uranium contaminated soil particles by micro-X-ray absorption and micro-fluorescence spectroscopies. J Radioanal Nucl Chem 234:237-242

Hunter DB, Bertsch PM, Kemner KM, Clark SB (1997) Distribution and chemical speciation of metals and metalloids in biota collected from contaminated environments by spatially resolved XRF, XANES, and EXAFS. J Phys IV 7 (Colloque C2, X-Ray Absorption Fine Structure, Vol. 2):767-771

Hunter KA (1980) Microelectrophoretic properties of natural surface-active organic matter in coastal seawater. Limnology and Oceanography 25:807-822

Hunter KA, Liss PS (1979) The surface charge of suspended particles in estuarine and coastal waters. Nature 282:823-825

Hunter RJ (1987) Foundations of Colloid Science, Vol. 1. Oxford University Press, Oxford

Hunter RJ (1993) Introduction to Modern Colloid Science. Oxford University Press, Oxford

Ildefonse Ph, Cabaret D, Sainctavit P, Calas G, Flank A-M, Lagarde P (1998) Aluminum X-ray absorption near edge structure in model compounds and Earth surface minerals. Phys Chem Minerals 25:112-121

Ildefonse Ph, Calas G, Flank A-M, Lagarde P (1995) Low Z elements (Mg, Al, and Si) K-edge X-ray absorption spectroscopy in minerals and disordered systems. Nucl Instrum Methods Phys Res Sect B 97:172-175

Ildefonse Ph, Kirkpatrick RJ, Montez B, Calas G, Flank A-M, Lagarde P (1994) ^{27}Al MAS NMR and aluminum X-ray absorption near edge structure study of imogolite and allophanes. Clays Clay Minerals 42:276-287

Israelachvili JN, Pashley RM (1983) Molecular layering of water at surfaces and origin of repulsive hydration forces. Nature 306:249-250

Isaure M-P, Laboudigue A, Manceau A, Sarret G, Tiffreau C, Trocellier P, Lamble G, Hazemann J-L, Chateigner D (2002) Quantitative Zn speciation in a contaminated dredged sediment by µPIXE, µSXRF, EXAFS spectroscopy and principal component analysis. Geochim Cosmochim Acta 66:1549-1567

Isaacs ED, Shukla A, Platzman PM, Hamann DR, Barbiellini B, Tulk CA (1999) Covalency of the hydrogen bond in ice: A direct X-ray measurement. Phys Rev Lett 82:600-603

Isaacs ED, Shukla A, Platzman PM, Hamann DR, Barbiellini B, Tulk CA (2000) Compton scattering evidence for covalency of the hydrogen bond in ice. J Phys Chem Solids 61:403-406

Itie JP, Polian A, Calas G, Petiau J, Fontaine A, Tolentino H (1989) Pressure-induced coordination changes in crystalline and vitreous germanium dioxide. Phys Rev Lett 63:398-401

Jalilehvand F, Spangberg D, Lindqvist-Reis P, Hermansson K, Persson I, Sandstrom M (2001) Hydration of the calcium ion. An EXAFS, large-angle X-ray scattering, and molecular dynamics simulation study. J Am Chem Soc 123:431-441

James RO, Healy TW (1972) The adsorption of hydrolyzable metal ions at the oxide-water interface. III. A thermodynamic model of adsorption. J Colloid Interface Sci 40:65-81

Jardine PM, Fendorf SE, Mayes MA, Larsen IL, Brooks SC, Bailey WB (1999) Fate and transport of hexavalent chromium in undisturbed heterogeneous soil. Environ Sci Technol 33:2939-2944

Jayanetti S, Mayanovic RA, Anderson AJ, Bassett WA, Chou I-M (2001) Analysis of radiation-induced small Cu particle cluster formation in aqueous CuCl$_2$. J Chem Phys 115:954-962.

Juillot F, Morin G, Ildefonse Ph, Trainor TP, Benedetti M, Galoisy L, Calas G, Brown GE Jr (2002) Occurrence of Zn/Al hydrotalcite in smelter-impacted soils from Northern France: Evidence from EXAFS spectroscopy and chemical extractions. Am Mineral (in press)

Kaplan DI, Bertsch PM, Adriano DD, Orlandini KA (1994) Actinide association with groundwater colloids in a coastal plain aquifer. Radiochim Acta 66/67:181-187

Kaplan DI, Hunter DB, Bertsch PM, Bajt S, Adriano DC (1994) Application of synchrotron X-ray fluorescence spectroscopy and energy dispersive X-ray analysis to identify contaminant metals on groundwater colloids. Environ Sci Technol 28:1186-1189

Karis O, Hasselstrom J, Wassdahl N, Weinelt M, Nilsson A, Nyberg M, Pettersson LGM, Stohr J, Samant MG (2000) The bonding of simple carboxylic acids on Cu(110). J Chem Phys 112:8146-8155

Kasrai M, Brown JR, Bancroft GM, Yin Z, Tan KH (1996) Sulfur characterization in coal from X-ray absorption near edge spectroscopy. Int J Coal Geol 32:107-135

Kasrai M, Fleet ME, Sham TK, Bancroft GM, Tan KH, Brown JR (1988) A XANES study of the S L-edge in sulfide minerals: application to interatomic distance determination. Solid State Commun 68:507-511

Katz LE, Hayes KF (1995a) Surface complexation modeling: I. Strategy for modeling monomer complex formation at moderate surface coverage. J Colloid Interface Sci 170:477-490

Katz LE, Hayes KF (1995b) Surface complexation modeling: II. Strategy for modeling polymer and precipitation reactions at high surface coverage. J Colloid Interface Sci 170:491-501

Kelly SD, Boyanov MI, Bunker BA, Fein JB, Fowle DA, Yee N, Kemner KM (2001) XAFS determination of the bacterial cell wall functional groups responsible for complexation of Cd and U as a function of pH. J Synchrotron Rad 8:946-948

Kelly SD, Kemner KM, Fein JB, Fowle DA, Boyanov MI, Bunker BA, Yee N (2002) X-ray-absorption fine-structure determination of pH-dependent U-bacterial cell wall interactions. Geochim Cosmochim Acta 66:(in press)

Kemner KM, Hunter DB, Bertsch PM, Kirkland JP, Elam WT (1997a) Determination of site specific binding environments of surface sorbed cesium on clay minerals by Cs-EXAFS. J de Physique IV 7(C2, X-Ray Absorption Fine Structure, Vol. 2):777-779

Kemner KM, Hunter DB, Gall EJ, Bertsch PM, Kirkland JP, Elam WT (1997b) Molecular characterization of Cr phases in contaminated soils by Cr and Fe EXAFS: a tool for evaluating chemical remediation strategies. J Phys IV 7 (C2, X-Ray Absorption Fine Structure, Vol. 2): 811-812

Kemner KM, Lai B, Maser J, Schneegurt MA, Cai Z, Ilinski P, Kulpa CF, Legnini DG, Nealson KH, Pratt ST, Rodrigues W, Lee-Tischler M, Yun W (2000) Use of the high-energy X-ray microprobe at the Advanced Photon Source to investigate the interactions between metals and bacteria. AIP Conf Proc 507 (X-ray Microscopy): 319-322

Kemner KM, Yun W, Cai Z, Lai B, Lee H-R, Maser J, Legnini DG, Rodriques W, Jastrow JD, Miller RM, Pratt ST, Schneegurt MA, Kulpa CF Jr (1999) Using zone plates for X-ray microimaging and microspectroscopy in environmental samples. J Synchrotron Rad 6:639-641

Kendelewicz T, Doyle CS, Carrier X, Brown GE Jr (1999) Reaction of water with clean surfaces of MnO(100). Surf Rev Lett 6:1255-1263

Kendelewicz T, Liu P, Brown GE Jr, Nelson EJ (1998a) Atomic geometry of the PbS(100) surface. Surf Sci 395:229-238

Kendelewicz T, Liu P, Brown GE Jr, Nelson EJ (1998b) Interaction of sodium overlayers with cleaved surfaces of galena (PbS(100)): Evidence for exchange reactions. Surf Sci 411:10-21

Kendelewicz T, Liu P, Doyle CS, Brown GE Jr (2000a) Spectroscopic study of the reaction of $Cr(VI)_{aq}$ with Fe_3O_4 (111) surfaces. Surf Sci 469:144-163

Kendelewicz T, Liu P, Doyle CS, Brown GE Jr, Nelson EJ, Chambers SA (2000b) Reaction of water with the (100) and (111) surfaces of Fe_3O_4. Surf Sci 453:32-46

Kendelewicz T, Liu P, Labiosa WB, Brown GE Jr (1995) Surface EXAFS and X-ray standing wave study of the cleaved CaO(100) surface. Physica B 208 & 209:441-442

Kersting AB, Efurd DW, Finnegan DL, Rokop DJ, Smith DK, Thompson JL (1999) Migration of plutonium in ground water at the Nevada Test Site. Nature 397:56-59

Kieffer F (1992) Metals as essential trace elements for plants, animals, and humans. *In*: Merian E (ed) Metals and Their Compounds in the Environment. VCH, Weinheim, Germany, p 481-489

Kikuma J, Tonner BP (1996) XANES spectra of a variety of widely used organic polymers at the C K-edge. J Electron Spectrosc Relat Phenom 82:53-60

Kim CS, Bloom NS, Rytuba JJ, Brown GE Jr (2002a) Mercury speciation by extended X-ray absorption fine structure (EXAFS) spectroscopy and sequential chemical extractions: An intercomparison of speciation methods. Environ Sci Technol (submitted)

Kim CS, Brown GE Jr, Rytuba JJ (2000) Characterization and speciation of mercury-bearing mine wastes using X-ray absorption spectroscopy (XAS). Science of the Total Environment 261:157-168

Kim CS, Rytuba JJ, Brown GE Jr (1999) Utility of EXAFS in characterization and speciation of mercury-bearing mine wastes. J Synchrotron Rad 6:648-650

Kim CS, Rytuba JJ, Brown GE Jr (2002b) EXAFS study of mercury(II) sorption on Fe- and Al-(hydr)oxides: II. Effects of chloride and sulfate. J Colloid Interface Sci (submitted)

Kim CS, Rytuba JJ, Brown GE Jr (2002c) EXAFS study of mercury(II) sorption on Fe- and Al-(hydr)oxides: I. Effects of pH. J Colloid Interface Sci (submitted)

Kim CS, Rytuba JJ, Brown GE Jr (2002d) Geological and anthropogenic factors influencing mercury speciation in mine wastes. Appl Geochem (submitted)

Kneebone PE, O'Day PA, Jones N, Hering JG (2002) Deposition and fate of arsenic in iron- and arsenic-enriched reservoir sediments. Environ Sci Technol 36:381-386

Kobayashi K, Kawata H, Mori K (1998) Site specification on normal and magnetic XANES of ferrimagnetic Fe_3O_4 by means of resonant magnetic Bragg scattering. J Synchrotron Rad 5:972-975

Kolker A, Huggins FE, Palmer CA, Shah N, Crowley SS, Huffman GP, Finkelman RB (2000) Mode of occurrence of arsenic in four US coals. Fuel Process Technol 63:167-178

Koningsberger D, Mojet B, Miller J, Ramaker D (1999) XAFS spectroscopy in catalysis research: AXAFS and shape resonances. J Synchrotron Rad 6:135-141

Kortright J, Warburton W, Bienenstock A (1983) Anomalous X-ray scattering and its relationship to EXAFS. Springer Ser Chem Phys, Vol. 27, Springer-Verlag, Berlin, p 362-372

Labrenz M, Druschel GK, Thomsen-Ebert T, Gilbert B, Welch SA, Kemner KM, Logan GA, Summons RE, de Stasio G, Bond PL, Lai B, Kelly SD, Banfield JF (2000) Formation of sphalerite (ZnS) deposits in natural biofilms of sulfate-reducing bacteria. Science 290:1744-1745

La Force MJ, Hansel CM, Fendorf S (2000) Arsenic speciation, seasonal transformations, and co-distribution with iron in a mine waste-influenced palustrine emergent wetland. Environ Sci Technol 34:3937-3943

Lagarde P, Flank A-M, Itie JP (1993) Polarized XANES spectra of quartz: application to the structure of densified silica. Japan J Appl Phys Part 1 32 (Suppl. 32-2, XAFS VII):613-615

Lagarde P, Fontaine A, Raoux D, Sadoc A, Migliardo P (1980) EXAFS studies of strong electrolytic solutions. J Chem Phys 72:3061-3069

Lai B, Kemner KM, Maser J, Schneegurt MA, Cai Z, Ilinski P, Kulpa CF, Legnini DG, Nealson KH, Pratt ST, Rodrigues W, Tischler ML, Yun W (2000) High-resolution X-ray imaging for microbiology at the Advanced Photon Source. AIP Conf Proc 506 (X-ray and Inner-Shell Processes):585-589

Lamble GM, Lee JF, Staudt WJ, Reeder RJ (1995) Structural studies of selenate incorporation into calcite crystals. Physica B (Amsterdam) 208 & 209:589-590

Lamble GM, Reeder RJ, Northrup PA (1997) Characterization of heavy metal incorporation in calcite by XAFS spectroscopy. J de Physique IV 7 (Colloque C2, X-ray Absorption Fine Structure, Vol. 2):793-797

Langel W, Parinello M (1994) Hydrolysis at stepped MgO surfaces. Phys Rev Lett 73: 504-507

Lanson B, Drits VA, Silvester E, Manceau A (2000) Structure of H-exchanged hexagonal birnessite and its mechanism of formation from Na-rich monoclinic buserite at low pH. Am Mineral 85:826-838

LCLS (1998) LCLS Design Study Report. SLAC–R-521, Revised December 1998, UC-414, Stanford Linear Accelerator Center, Stanford, CA

LCLS (2000) LCLS: The First Experiments. Stanford Synchrotron Radiation Laboratory, Stanford, CA, September 2000

Lee PA, Pendry JB (1975) Theory of the Extended X-ray Absorption Fine Structure. Phys Rev B 11:2795-2811

Lee Y, Hriljac JA, Vogt T, Parise JB, Artioli G (2001) First structural investigation of a super-hydrated zeolite. J Am Chem Soc 123:12732-12733

Lenhart JJ, Bargar JR, Davis JA (2001) Spectroscopic evidence for ternary surface complexes in the lead(II)-malonic acid-hematite system. J Colloid Interface Sci 234:448-452

Levelut C, Cabaret D, Benoit M, Jund P, Flank A-M (2001) Multiple scattering calculations of the XANES Si K-edge in amorphous silica. J Non-Crystal Solids 293-295:100-104

Lewis RJ Sr (ed) (1991) Carcinogenically Active Chemicals. Van Nostrand Reinhold, New York

Li D, Bancroft GM, Fleet ME (1996) Coordination of Si in $Na_2O-SiO_2-P_2O_5$ glasses using Si K- and L-edge XANES. Am Mineral 81:111-118

Li D, Bancroft GM, Fleet ME, Feng XH (1995a) Silicon K-edge XANES spectra of silicate minerals. Phys Chem Minerals 22:115-122

Li D, Bancroft GM, Fleet ME, Feng XH, Pan Y (1995b) Al K-edge XANES spectra of aluminosilicate minerals. Am Mineral 80:432-440

Li D, Bancroft GM, Kasrai M, Fleet ME, Feng X, Tan K (1995c) Polarized X-ray absorption spectra and electronic structure of molybdenite ($2H-MoS_2$). Phys Chem Minerals 22:123-128

Li D, Bancroft GM, Kasrai M, Fleet ME, Feng X, Tan K (1995d) S K- and L-edge X-ray absorption spectroscopy of metal sulfides and sulfates: applications in mineralogy and geochemistry. Canadian Mineral 33:949-960

Li D, Bancroft GM, Kasrai M, Fleet ME, Feng XH, Tan KH (1994a) High-resolution Si and P K- and L-edge XANES spectra of crystalline SiP_2O_7 and amorphous $SiO_2-P_2O_5$. Am Mineral 79:785-788

Li D, Bancroft GM, Kasrai M, Fleet ME, Feng XH, Tan KH, Yang BX (1993) High-resolution Si K- and $L_{2,3}$-edge XANES of α-quartz and stishovite. Solid State Commun 87:613-617

Li D, Bancroft GM, Kasrai M, Fleet ME, Feng XH, Tan KH, Yang BX (1994b) Sulfur K- and L-edge XANES and electronic structure of zinc, cadmium and mercury monosulfides: a comparative study. J Phys Chem Solids 55:535-543

Li D, Bancroft GM, Kasrai M, Fleet ME, Feng XH, Yang BX, Tan KH (1994c) S K- and L-edge XANES and electronic structure of some copper sulfide minerals. Phys Chem Minerals 21:317-324

Li D, Bancroft GM, Kasrai M, Fleet ME, Secco RA, Feng XH, Tan KH, Yang BX (1994d) X-ray absorption spectroscopy of silicon dioxide (SiO_2) polymorphs: the structural characterization of opal. Am Mineral 79:622-632

Li D, Bancroft GM, Kasrai M, Fleet ME, Yang BX, Feng XH, Tan K, Peng M (1994e) Sulfur K- and L-edge X-ray absorption spectroscopy of sphalerite, chalcopyrite and stannite. Phys Chem Mineral 20:489-499

Li D, Fleet ME, Bancroft GM, Kasrai M, Pan Y (1995e) Local structure of Si and P in SiO_2-P_2O_5 and Na_2O-SiO_2-P_2O_5 glasses: a XANES study. J Non-Crystal Solids 188:181-189

Li D, Secco RA, Bancroft GM, Fleet ME (1995f) Pressure induced coordination change of Al in silicate melts from Al K-edge XANES of high pressure $NaAlSi_2O_6$-$NaAlSi_3O_8$ glasses. Geophys Res Lett 22:3111-3114

Lindau I, Spicer WE (1980) Photoemission as a tool to study solids and surfaces. *In:* Winick H, Doniach S (eds) Synchrotron Radiation Research, Plenum Press, New York, p 159-221

Lindqvist-Reis P, Lamble K, Pattanaik S, Persson I, Sandstroem, M (2000) Hydration of the yttrium(III) ion in aqueous solution. An X-ray diffraction and XAFS structural study. J Phys Chem B 104:402-408

Lindqvist-Reis P, Munoz-Paez A, Diaz-Moreno S, Pattanaik S, Persson I, Sandstroem M (1998) The structure of the hydrated gallium(III), indium(III), and chromium(III) ions in aqueous solution. A large angle X-ray scattering and EXAFS study. Inorg Chem 37:6675-6683

Liu C, Frenkel AI, Vairavamurthy A, Huang PM (2001) Sorption of cadmium on humic acid: Mechanistic and kinetic studies with atomic force microscopy and X-ray absorption fine structure spectroscopy. Canadian J Soil Sci 81 (3, Spec. Issue):337-348

Liu P, Kendelewicz T, Brown GE Jr, Parks GA (1998a) Reaction of water with MgO (100) surfaces: I. Synchrotron X-ray photoemission spectroscopy studies of low defect surfaces. Surf Sci 412/413:287-314

Liu P, Kendelewicz T, Brown GE Jr (1998b) Reaction of water with MgO (100) surfaces: II. Synchrotron X-ray photoemission spectroscopy studies of defective surfaces. Surf Sci 412/413:315-332

Liu P, Kendelewicz T, Brown GE Jr, Parks GA, Pianetta P (1998c) Reaction of water with vacuum-cleaved CaO(100) surfaces: An X-ray photoemission spectroscopy study. Surf Sci 416:326-340

Liu P, Kendelewicz T, Brown GE Jr, Nelson EJ, Chambers SA (1998d) Reaction of water with α-Al_2O_3 and α-Fe_2O_3 (0001) surfaces: synchrotron X-ray photoemission studies and thermodynamic calculations. Surf Sci 417:53-65

Liu P, Kendelewicz T, Nelson EJ, Brown GE Jr (1998e) Reaction of water with MgO(100) surfaces: III. X-ray standing wave studies. Surf Sci 415:156-169

Losi ME, Frankenberger WT Jr (1998) Microbial oxidation and solubilization of precipitated elemental selenium in soil. J Environ Qual 27:836-843

Lowenstern JB, Mahood GA, Rivers ML, Sutton SR (1991) Evidence for extreme partitioning of copper into a magmatic vapor phase. Science 252:1405-1409

Lowry GV, Shaw S, Kim CS, Rytuba JJ, Brown GE Jr (2002) Particle-facilitated mercury transport from New Idria and Sulphur Bank mercury mine tailings. 1. Column experiments and macroscopic analysis. Environ. Sci. Technol. (submitted)

Lu R, Goncharov A, Mao H-K, Hemley RJ (1999) Synchrotron infrared microspec-troscopy: Applications to hydrous minerals. *In*: Schulze DG, Stucki JW, Bertsch PM (eds) Synchrotron X-ray Methods in Clay Science, Clay Minerals Society Workshop Lectures 9. The Clay Minerals Society, Boulder, Colorado, p 164-182

Ludwig KF Jr, Warburton WK, Fontaine A (1987) X-ray studies of concentrated aqueous solutions. J Chem Phys 87:620-629

Lukens WW Jr, Bucher JJ Edelstein NM, Shuh DK (2001) Radiolysis of TcO_4^- in alkaline, nitrate solutions: Reduction by NO_3^{2-}. J Phys Chem A 105:9611-9615

Lukens WW Jr, Bucher JJ Edelstein NM, Shuh DK (2002) Products of pertechnetate radiolysis in highly alkaline solution: Structure of $TcO_2 \cdot xH_2O$. Environ Sci Technol 36:1124-1129

Lytle CM, Lytle FW, Smith BN (1996) Use of XAS to determine the chemical speciation of bioaccumulated manganese in *Potamogeton pectinatus*. J Environ Qual 25:311-316

Lytle CM, Lytle FW, Yang N, Qian J-H, Hansen D, Zayed A, Terry N (1998) Reduction of Cr(VI) to Cr(III) by wetland plants: Potential for *in situ* heavy metal detoxification. Environ Sci Technol 32:3087-3093

Lytle FW (1966) Determination of interatomic distances from X-ray absorption fine structure. Advances in X-ray Analysis 9:398-409

Lytle FW (1989) Experimental X-ray absorption spectroscopy. *In*: Winick H, Xian D, Ye M, Huang T (eds) Applications of Synchrotron Radiation. Gordon and Breach, New York, p 135-224

Lytle FW (1999) The EXAFS family tree: a personal history of the development of extended X-ray absorption fine structure. J Synchrotron Rad 6:123-134

Lytle FW, Greegor RB, Sandstrom DR, Marques DR, Wong J, Spiro CL, Huffman GP, Huggins FE (1984) Measurement of soft X-ray absorption with a fluorescence ion chamber detector. Nucl Instrumen Methods 226:542-548

Lytle FW, Pingitore NE Jr (2002) Iron valence in the hydration layer of obsidian: characterization by X-ray absorption spectroscopy. Microchemical J 71:185-191

Lytle FW, Sayers DE, Stern EA (1975) Extended X-ray-absorption fine-structure technique. II. Experimental practice and selected results. Phys Rev B 11:4825-4835

Lytle FW, Sayers DE, Stern EA (1982) The history and modern practice of EXAFS spectroscopy. *In*: Bonnelle C, Mandé (eds) Advances in X-ray Spectroscopy. Pergamon Press, New York, p 267-286

Lytle FW, Via GH, Sinfelt JH (1980) X-ray absorption spectroscopy: catalyst applications. *In*: Winick H, Doniach S (eds) Synchrotron Radiation Research. Plenum Press, New York, p 401-424

MacDowell AA, Celestre RS, Tamura N, Spolenak R, Valek BC, Brown WL, Bravman JC, Padmore HA, Batterman BW, Patel JR (2001) Submicron X-ray diffraction. Nucl Instrum Methods A 468:936-943

Macy JM (1994) Biochemistry of selenium metabolism by *Thauera selenatis* gen. nov. sp. nov. and use of the organism for bioremediation of selenium oxyanions in San Joaquin Valley drainage water. *In:* Frankenberger WT Jr, Benson S (eds) Selenium in the Environment. Marcel Decker, Inc., New York, p 421-444

Madden RP, Codling K (1963) New autoionizing atomic energy levels in He, Ne, and Ar. Phys Rev Lett 10:516-518

Magini M, Licheri G, Paschina G, Piccaluga G, Pinna G (1988) X-ray Diffraction of Ions in Aqueous Solutions: Hydration and Complex Formation. CRC Press, Boca Raton, Florida

Maire G, Garin F, Bernhardt P, Girard P, Schmitt JL, Dartyge E, Dexpert H, Fontaine A, Jucha A, Lagarde P (1986) Note on a dynamic study of a complete catalytic process by X-ray absorption spectroscopy in dispersive mode: application to iridium-copper/alumina catalysts. Appl Catal 26:305-312

Masion A, Doelsch E, Rose J, Moustier S, Bottero JY, Bertsch PM (2001) Speciation and crystal chemistry of iron(III) chloride hydrolyzed in the presence of SiO_4 ligands. 3. Semilocal scale structure of the aggregates. Langmuir 17:4753-4757

Manceau A (1990) Distribution of cations among the octahedra of phyllosilicates: Insight from EXAFS. Canadian Mineral 28:321-328

Manceau A (1995) The mechanism of anion adsorption on iron oxides: evidence for the bonding of arsenate tetrahedra on free $Fe(O,OH)_6$ edges. Geochim Cosmochim Acta 59:3647-3653

Manceau A, Boisset MC, Sarret G, Hazemann J-L, Mench M, Cambier P, Prost R (1996) Direct determination of lead speciation in contaminated soils by EXAFS spectroscopy. Environ Sci Technol 30:1540-1552

Manceau A, Bonnin D, Kaiser P, Fretigny C (1988) Polarized EXAFS of biotite and chlorite. Phys Chem Minerals 16:180-185

Manceau A, Bonnin D, Stone WEE, Sanz J (1990a) Distribution of iron in the octahedral sheet of trioctahedral micas by polarized EXAFS. Comparison with NMR results. Phys Chem Minerals 17:363-370

Manceau A, Calas G (1986) Nickel-bearing clay minerals: II. Intracrystalline distribution of nickel: An X-ray absorption study. Clay Minerals 21:341-360

Manceau A, Calas G (1987) Absence of evidence for Ni/Si substitution in phyllosilicates. Clay Minerals 22:357-362

Manceau, A, Charlet L (1992) X-ray absorption spectroscopic study of the sorption of Cr(III) at the oxide-water interface. I. Molecular mechanisms of Cr(III) oxidation on Mn oxides. J Colloid Interface Sci 148:425-442

Manceau A, Charlet L (1994) The mechanism of selenate adsorption on goethite and hydrous ferric oxide. J Colloid Interface Sci 168:87-94

Manceau A, Charlet L, Boisset MC, Didier B, Spadini L (1992a) Sorption and speciation of heavy metals on hydrous Fe and Mn oxides. From microscopic to macroscopic. Applied Clay Sci 7:201-223

Manceau A, Chateigner D, Gates WP (1998) Polarized EXAFS, distance-valence least-squares modeling (DVLS), and quantitative texture analysis approaches to the structural refinement of Garfield nontronite. Phys Chem Minerals 25:347-365

Manceau A, Combes J-M (1988) Structure of manganese and iron oxides and oxyhydroxides: A topological approach by EXAFS. Phys Chem Minerals 15:283-295

Manceau A, Combes J-M, Calas G (1990b) New data and a revised structural model for ferrihydrite: Comment. Clays Clay Minerals 38:331-334

Manceau A, Decarreau A (1988) Extended X-ray absorption fine-structure study of cobalt-exchanged sepiolite: Comment on a paper by Y. Fukushima and T. Okamoto. Clays Clay Minerals 36:382-383

Manceau A, Drits VA (1993) Local structure of ferrihydrite and feroxyhite [feroxyhyte] by EXAFS spectroscopy. Clay Minerals 28(2):165-184

Manceau A, Drits VA, Lanson B, Chateigner D, Wu J, Huo D, Gates WP, Stucki JW (2000c) Oxidation-reduction mechanism of iron in dioctahedral smectites: II. Crystal chemistry of reduced Garfield nontronite. Am Mineral 85:153-172

Manceau A, Drits VA, Silvester E, Bartoli C, Lanson B (1997) Structural mechanism of Co^{2+} oxidation by the phyllomanganate buserite. Am Mineral 82:1150-1175

Manceau A, Gallup DL (1997) Removal of selenocyanate in water by precipitation: Characterization of copper-selenium precipitate by X-ray diffraction, infrared, and X-ray absorption spectroscopy. Environ Sci Technol 31:968-976

Manceau A, Gates WP (1997) Surface structural model for ferrihydrite. Clays Clay Minerals 45:448-460

Manceau A, Gorshkov AI, Drits VA (1992b) Structural chemistry of manganese, iron, cobalt, and nickel in manganese hydrous oxides: Part I. Information from XANES spectroscopy. Am Mineral 77:1133-1143

Manceau A, Gorshkov AI, Drits VA (1992c) Structural chemistry of manganese, iron, cobalt, and nickel in manganese hydrous oxides: Part II. Information from EXAFS spectroscopy and electron and X-ray diffraction. Am Mineral 77:1144-1157

Manceau A, Ildefonse Ph, Hazemann J-L, Fland A-M, Gallup D (1995) Crystal chemistry of hydrous iron silicate scale deposits at the Salton Sea geothermal field. Clays Clay Minerals 43:304-317

Manceau A, Lanson B, Drits VA (2002b) Structure of heavy metal sorbed birnessite. Part III: Results from powder and polarized extended X-ray absorption fine structure spectroscopy. Geochim Cosmochim Acta 66:2639-2663

Manceau A, Lanson B, Drits VA, Chateigner D, Gates WP, Wu J, Huo D, Stucki JW (2000d) Oxidation-reduction mechanism of iron in dioctahedral smectites: I. Crystal chemistry of oxidized reference nontronites. Am Mineral 85:133-152

Manceau A, Lanson B, Schlegel ML, Harge JC, Musso M, Eybert-Berard L, Hazemann JL, Chataignier D, Lamble GM (2000a) Quantitative Zn speciation in smelter-contaminated soils by EXAFS spectroscopy. Am J Sci 300:289-343

Manceau A, Nagy KL, Spadini L, Ragnarsdottir KV (2000e) Influence of anionic layer structure of Fe-oxyhydroxides on the structure of Cd surface complexes. J Colloid Interface Sci 228:306-316

Manceau A, Schlegel M, Chateigner D, Lanson B, Bartoli C, Gates W (1999a) Application of polarized EXAFS to fine-grained layered minerals. *In:* Schulze DG, Stucki JW, Bertsch PM (eds) Synchrotron X-ray Methods in Clay Science, Clay Minerals Society Workshop Lectures 9. The Clay Minerals Society, Boulder, Colorado, p 69-114

Manceau A, Schlegel ML, Musso M, Sole VA, Gauthier C, Petit PE, Trolard F (2000b) Crystal chemistry of trace elements in natural and synthetic goethite. Geochim Cosmochim Acta 64:3643-3661

Manceau A, Schelegel ML, Nagy KL, Charlet L (1999b) Evidence for the formation of trioctahedral clay upon sorption of Co^{2+} on quartz. J Colloid Interface Sci 220:181-197

Mangold S, Calmano W (2002) Heavy-metal speciation of contaminated soils by sequential extraction and X-ray absorption fine structure spectroscopy (XAFS). Proceedings of the Euroconference and NEA Workshop on Speciation, Techniques, and Facilities for Radioactive Materials at Synchrotron Light Sources, Grenoble, France, Sept. 10-12, 2000 Nuclear Energy Agency/Organization for Economic Co-operation and Development, AEN/NEA 2002, Paris, p 253-259

Manning BA, Fendorf SE, Bostick B, Suarez DL (2002) Arsenic(III) oxidation and arsenic(V) adsorption reactions on synthetic birnessite. Environ Sci Technol (2002) 36:976-981

Manning BA, Fendorf SE, Goldberg S (1998) Surface structure and stability of arsenic(III) on goethite: spectroscopi evidence for inner-sphere complexes. Environ Sci Technol 32:2383-2388

Mansour AN, Melendres CA, Wong J (1998) *In situ* X-ray absorption spectroscopic study of electrodeposited nickel oxide films during redox reactions. J Electrochem Soc 145:1121-1125

Marcus Y (1985) Ion Solvation. John Wiley & Sons, New York

Marcus Y (1988) Ionic radii in aqueous solutions. Chem Rev 88:1475-1498

Margerum DW, Cayley GR, Weatherburn DC, Pagenkopf GK (1978) Kinetics and mechanisms of complex formation and ligand exchange. *In:* Martell AE (ed) Coordination Chemistry, Vol. 2, ACS Monograph 174, American Chemical Society, Washington DC, p 1-220

Marques MI de Barros, Lagarde P (1990) EXAFS studies of local order in indium tribromide aqueous solutions. J Phys Condensed Matter 2:231-238

Martell AE, Hancock RD (1996) Metal Complexes in Aqueous Solutions. Plenum Press, New York

Martinez CE, McBride MB, Kandianis MT, Duxbury JM, Yoon S-J, Bleam WF (2002) Zinc-sulfur and cadmium-sulfur association in metalliferous peats: Evidence from spectroscopy, distribution coefficients, and phytoavailability. Environ Sci Technol 36:3683-3689

Matocha CJ, Elzinga EJ, Sparks DL (2001) Reactivity of Pb(II) at the Mn(III,IV) (oxyhydr)oxide-water interface. Environ Sci Technol 35:2967-2972

Mayer LM (1994a) Relationships between mineral surfaces and organic carbon concentrations in soils and sediments. Chem Geol 114:347-363

Mayer LM (1994b) Surface area control of organic carbon accumulation in continental shelf sediments. Geochim Cosmochim Acta 58:1271-1284

Mayer LM (1999) Extent of coverage of mineral surfaces by organic matter in marine sediments. Geochim Cosmochim Acta 63:207-215

Mayer LM, Xing B (2001) Organic matter-surface area relationships in acid soils. Soil Sci Soc Am J 65:250-258

Mayanovic RA, Anderson AJ, Bajt S (1996) Microbeam XAFS investigations on fluid inclusions. Materials Research Society Symposium Proceedings 437 (Applications of Synchrotron Radiation Techniques to Materials Science III):201-206

Mayanovic RA, Anderson AJ, Bajt S (1997) Microbeam XAFS studies on fluid inclusions at high temperatures. J de Physique IV 7 (Colloque C2, X-ray Absorption Fine Structure, Vol. 2):1029-1030

Mayanovic RA, Anderson AJ, Bassett WA, Chou, I-M (1999) XAFS measurements on zinc chloride aqueous solutions from ambient to supercritical conditions using the diamond anvil cell. J Synchrotron Rad 6:195-197

Mayanovic RA, Anderson AJ, Bassett WA, Chou, I-M (2001) Hydrogen bond breaking in aqueous solutions near the critical point. Chem Phys Lett 336:212-218.

Mayanovic RA, Jayanetti S, Anderson AJ, Bassett WA, Chou, I-M (2002) The structure of Yb^{3+} aquo ion and chloro complexes in aqueous solutions at up to 500°C and 270 MPa. J Phys Chem A 106:6591-6599

Mazzara C, Jupille J, Flank A-M, Lagarde P (2000) Stereochemical order around sodium in amorphous silica. J Phys Chem B 104:3438-3445

McArthur JM (1999) Arsenic poinoning in the Ganges delta. Nature 401:546-547

McHale JM, Auroux A, Perrotta AJ, Navrotsky A (1997) Surface energies and thermodynamic phase stability in nanocrystalline aluminas. Science 277:788-791

McKeown DA, Waychunas GA, Brown GE Jr (1985a) EXAFS study of the coordination environment of aluminum in a series of silica-rich glasses and selected minerals within the sodium aluminosilicate system. J Non-Crystal Solids 74:349-371

McKeown DA, Waychunas GA, Brown GE Jr (1985b) EXAFS and XANES study of the local coordination environment of sodium in a series of silica-rich glasses and selected minerals within the sodium aluminosilicate system. J Non-Crystal Solids 74:325-348

Miranda PB, Xu L, Shen YR, Salmeron M (1998) Icelike water monolayers adsorbed on mica at room temperature. Phys Rev Lett 81:5876-5879

Mitra-Kirtley S, Mullins OC, Branthaver JF, Cramer SP (1993a) Nitrogen chemistry of kerogens and bitumens from X-ray absorption near-edge structure spectroscopy. Energy Fuels 7:1128-1134

Mitra-Kirtley S, Mullins OC, van Elp J, Cramer SP (1993b) Nitrogen chemical structure in petroleum asphaltene and coal by X-ray absorption spectroscopy. Fuel 72:133-135

Moll H, Zänker H, Richter W, Brendler V, Reich T, Hennig C, Roßberg A, Funke H, Kluge A (2002) XAS study of acid rock drainage samples from an abandoned Zn-Pb-Ag mine at Frieberg, Germany. Proceedings of the Euroconference and NEA Workshop on Speciation, Techniques, and Facilities for Radioactive Materials at Synchrotron Light Sources, Grenoble, France, Sept. 10-12, 2000 Nuclear Energy Agency/Organization for Economic Co-operation and Development, AEN/NEA 2002, Paris, p 263-269

Morin G, Juillot F, Ildefonse P, Calas G, Samama J-C, Chevallier P, Brown GE Jr (2001) Mineralogy of lead in a soil developed on a Pb-mineralized sandstone (Largentiére, France). Am Mineral 86:92-104

Morin G, Juillot F, Ostergren JD, Ildefonse P, Calas G, Brown GE Jr (1999) XAFS determination of the chemical form of lead in smelter-contaminated soils and mine tailings: Importance of adsorption processes. Am Mineral 84:420-434

Morin G, Lecocq D, Juillot F, Calas G, Ildefonse Ph, Belin S, Brios V, Dillman P, Chevallier P, Gautier C, Sole A, Petit P-E, Borensztajn S (2002) EXAFS evidence of sorbed arsenic(V) and pharmacosiderite in a soil overlying the Echassiéres geochemical anomaly, Allier, France. Bull Soc geol. France 173:281-291

Morra MJ, Fendorf SE, Brown PD (1997) Speciation of sulfur in humic and fulvic acids using X-ray absorption near-edge structure (XANES) spectroscopy. Geochim Cosmochim Acta 61:683-688

Morris DE, Allen PG, Berg JM, Chisholm-Brause CJ, Conradson SD, Donohoe RJ, Hess NJ, Musgrave JA, Tait CD (1996) Speciation of uranium in Fernald soils by molecular spectroscopic methods: characterization of untreated soils. Environ Sci Technol 30:2322-2331

Morton JD, Semrau JD, Hayes KF (2001) An X-ray absorption spectroscopy study of the structure and reversibility of copper adsorbed to montmorillonite clay. Geochim Cosmochim Acta 65:2709-2722

Mosselmans JFW, Charnock JM, Garner CD, Pattrick RAD, Vaughan DJ (1995a) A XAS study of the structural changes undergone by amorphous copper sulfides when precipitated from solution. Physica B (Amsterdam) 208 & 209:609-610

Mosselmans JFW, Helz GR, Pattrick RAD, Charnock JM, Vaughan DJ (2000) A study of speciation of Sb in bisulfide solutions by X-ray absorption spectroscopy. Applied Geochem 15:879-889

Mosselmans JFW, Pattrick RAD, Charnock JM, Sole VA (1999) EXAFS of copper in hydrosulfide solutions at very low concentrations: implications for the speciation of copper in natural waters. Mineral Mag 63:769-772

Mosselmans JFW, Pattick RAD, van der Laan G, Charnock JM, Vaughan DJ, Henderson CMB, Garner CD (1995b) X-ray absorption near-edge spectra of transition metal disulfides FeS_2 (pyrite and marcasite), CoS_2, NiS_2 and CuS_2, and their isomorphs FeAsS and CoAsS. Phys Chem Minerals 22:311-317

Mosselmans JFW, Schofield PF, Charnock JM, Garner CD, Pattrick RAD, Vaughan DJ (1996) X-ray absorption studies of metal complexes in aqueous solution at elevated temperatures. Chem Geol 127:339-350

Moyes LN, Jones MJ, Reed WA, Livens FR, Charnock JM, Mosselmans JFW, Hennig C, Vaughan DJ, Pattrick RAD (2002) An X-ray absorption spectroscopy study of neptunium(V) reactions with mackinawite (FeS). Environ Sci Technol 36:179-183

Moyes LN, Parkman RH, Charnock JM, Vaughan DJ, Livens FR, Hughes CR, Braithwaite A (2000) Uranium uptake from aqueous solution by interaction with goethite, lepidocrocite, muscovite, and mackinawaite: An X-ray absorption spectroscopy study. Environ Sci Technol 34:1062-1068

Muller F, Besson G, Manceau A, Drits V-A (1997) Distribution of isomorphous cations within octahedral sheets in montmorillonite from Camp-Bertaux. Phys Chem Minerals 24:159-166

Muller J-P, Manceau A, Calas G, Allard T, Ildefonse P, Hazemann, J-L (1995) Crystal chemistry of kaolinite and Fe-Mn oxides: relation with formation conditions of low temperature systems. Am J Sci 295:1115-1155

Mullins OC, Mitra-Kirtley S, van Elp J, Cramer SP (1993) Molecular structure of nitrogen in coal from XANES spectroscopy. Appl Spectrosc 47:1268-1275

Munoz M, Argoul P, Farges F (2002) Improving data reduction methods in X-ray absorption fine structure spectroscopy. Part 2: Continuous Cauchy wavelet transform analysis of EXAFS spectra. Phys Chem Minerals (submitted)

Munoz-Paez A, Diaz-Moreno S, Sanchez Macos E, Martinez JM, Pappalardo RR, Persson I, Sandstrom M, Pattanaik S, Lindquist-Reis P (1997) EXAFS study of the hydration structure of Ga^{3+} aqueous solution. Comparison of data from two laboratories. J de Physique IV 7 (Colloque C2, X-ray Absorption Fine Structure, Vol. 1):647-648

Munoz-Paez A, Pappalardo RR, Marcos ES (1995) Determination of the second hydration shell of Cr^{3+} and Zn^{2+} in aqueous solutions by extended X-ray absorption fine structure. J Am Chem Soc 117:11710-1172

Murata T, Nakagawa K, Kimura A, Otoda N, Shimoyama I (1995) Design of the cell and gas-handling system for the X-ray absorption study of supercritical fluid material. Rev Sci Instrum 66:1437-1439

Myneni SCB (2000) X-ray and vibrational spectroscopy of sulfate in earth materials. Rev Mineral Geochem 40:113-172

Myneni SCB (2002) Formation of stable chlorinated hydrocarbons in weathering plant material. Science 295:1039-1041

Myneni SCB, Brown JT, Martinez GA, Meyer-Ilse W (1999) Imaging of humic substance macromolecular structures in water and soils. Science 286:1335-1337

Myneni SCB, Luo Y, Naslund LA, Cavalleri M, Ojamae L, Ogasawara H, Pelmenschikov A, Wernet Ph, Vaterlein P, Heske C, Hussain Z, Pettersson LGM, Nilsson A (2002) Spectroscopic probing of local hydrogen-bonding structures in liquid water. J Phys Condensed Matter 14:L213-L219

Myneni SCB, Tokunaga TK, Brown GE Jr (1997a) Abiotic selenium redox transformations in the presence of Fe(II,III) hydroxides. Science 278:1106-1109

Myneni SCB, Traina SJ, Logan TJ, Waychunas GA (1997b) Oxyanion behavior in alkaline environments: Sorption and desorption of arsenate in ettringite. Environ Sci Technol 31:1761-1768

Myneni SCB, Warwick TA, Martinez GA, Meigs G (1998) C-functional group chemistry of humic substances and their spatial variation in soils. ALS Website: (*http://alspubs.lbl.gov/AbstractManager/uploads/myneni2.pdf*)

Myneni SCB, Waychunas GA, Traina SJ, Brown GE Jr (2000) Molecular investigation of sulfate complexation on Fe-oxide-surfaces. Div Environ Chem Preprints of Extended Abstracts 20, 220th American Chemical Society Meeting, Washington, DC, p 531-532

Nagy KL, Schlegel ML, Fenter P, Cheng L, Sturchio NC (2001) Structure of natural organic matter sorbed on muscovite as determined by surface X-ray reflectivity. Eleventh Annual V M Goldschmidt Conference, Hot Springs, VA, May 20-24

Narten AH, Levy HA (1972) Liquid water. Scattering of X-rays. *In:* Water: A Comprehensive Treatise Franks F (ed) 1:311-332

Narten AH, Thiessen WE, Blum L (1982) Atom pair distribution functions of liquid water at 25°C from neutron diffraction. Science 217:1033-1034

Naslund J, Lindqvist-Reis P, Persson I, Sandstrom M (2000a) Steric effects control the structure of the solvated lanthanum(III) ion in aqueous, dimethyl sulfoxide, and N,N'-dimethylpropyleneurea solution. An EXAFS and large-angle X-ray scattering study. Inorg Chem 39:4006-4011

Naslund J, Persson I, Sandstrom M (2000b) Solvation of the bismuth(III) ion by water, dimethyl sulfoxide, N,N'-dimethylpropyleneurea, and N,N-dimethylthioformamide. An EXAFS, large-angle X-ray scattering, and crystallographic structural study. Inorg Chem 39:4012-4021

NAS/NRC (2002) Dietary Reference Intakes for Vitamin A, Vitamin K, Arsenic, Boron, Chromium, Copper, Iodine, Iron, Manganese, Molybdenum, Nickel, Silicon, Vanadium, and Zinc. National Academy Press, Washington, DC

Natoli CR, Benfatto M (1986) A unifying scheme of interpretation of X-ray absorption spectra based on the multiple scattering theory. J de Physique (Colloque C8, Vol. 1):C8/11-C8/23

Neder RB, Burghammer M, Grasl T, Schulz H, Bram A, Fiedler S (1999) Refinement of the kwolinite structure from single-crystal synchrotron data. Clays Clay Minerals 47:487-494

Neihof RA, Loeb GI (1972) The surface charge of particulate matter in seawater. Limnology and Oceanography 17:7-16

Neihof RA, Loeb GI (1974) Dissolved organic matter in seawater and the electrical charge of immersed surfaces. J Marine Res 32:5-12

Neilson GW, Enderby JE (1989) The coordination of metal aqua ions. Adv Inorg Chem 34:195-218

Neilson GW, Enderby JE (1996) Aqueous solutions and neutron scattering. J Phys Chem 100:1317-1322

Nesbitt HW, Scaini M, Hochst H, Bancroft GM, Schaufuss AG, Szargan R (2000) Synchrotron XPS evidence for Fe^{2+}-S and Fe^{3+}-S surface species on pyrite fracture-surfaces, and their 3D electronic states. Am Mineral 85:850-857

Nesbitt HW, Uhlig I, Szargan R (2002) Surface reconstruction and As-polymerization at fractured loellingite ($FeAs_2$) surfaces. Am Mineral 87:1000-1004

Neu MP, Clark DL, Conradson SD, Donohoe RJ, Gordon JC, Keogh DW, Morris DE, Rogers RD, Scott BL, Tait CD (1999b) Structure and stability of actinides (U, Np, Pu) under strongly alkaline radioactive waste tank conditions. Book of Abstracts, 218th ACS National Meeting, New Orleans, LA, Aug. 22-26, 1999

Neu MP, Runde WH, Clark DL, Conradson SD, Efurd DW, Janecky DR, Kaszuba JP, Tait CD, Haire RG (1999a) Plutonium speciation and its effects on environmental migration. Book of Abstracts, 218th ACS National Meeting, New Orleans, LA, Aug. 22-26, 1999

Neutze R, Wouts R, van der Spoel D, Weckert E, Hajdu J (2000) Potential for biomolecular imaging with femtosecond X-ray pulses. Nature 406:752-757

Newville M, Boyanov BI, Sayers DE (1999a) Estimation of uncertainties in XAFS data. J Synchrotron Rad 6:264-265

Newville M, Carroll SA, O'Day PA, Waychunas GA, Ebert M (1999b) A web-based library of XAFS data on model compounds. J Synchrotron Rad 6:276-277

Newville M, Ravel B, Haskel D, Rehr JJ, Stern EA, Yacoby Y (1995) Analysis of multiple-scattering XAFS data using theoretical standards. Physica B 208&209:154-155 (*http://cars9.uchicago.edu/ifeffit/index.html*)

Nickson R, McArthur J, Burgess W, Ahmed KM, Ravenscroft P, Rahman M (1998) Arsenic poisoning of Bangladesh groundwater. Nature 395:338

Nickson RT, McArthur JM, Ravenscroft P, Burgess WG, Ahmed KM (2000) Mechanism of arsenic release to groundwater, Bangladesh and West Bengal. Appl Geochem 15:403-413

Nilsson A, Wassdahl N, Weinelt M, Karis O, Wiell T, Bennich P, Hasselstrom J, Fohlisch A, Stohr J, Samant M (1997a) Local probing of the surface chemical bond using X-ray emission spectroscopy. Appl Physics A: Materials Science & Processing A 65:147-154

Nilsson A, Weinelt M, Wiell T, Bennich P, Karis O, Wassdahl N, Stohr J, Samant MG (1997b) An atom-specific look at the surface chemical bond. Phys Rev Lett 78:2847-2850

Nordstrom DK (2002) Worldwide occurrences of arsenic in ground water. Science 296:2143-2145

O'Day PA (1999) Molecular environmental geochemistry. Rev Geophys 37:249-274

O'Day PA, Brown GE Jr, Parks GA (1994b) X-ray absorption spectroscopy of cobalt(II) multinuclear surface complexes and surface precipitates on kaolinite. J Colloid Interface Sci 165:269-289

O'Day PA, Carroll SA, Randall A, Martinelli RE, Anderson SL, Jelinski J, Knezovich JP (2000a) Metal speciation and bioavailability in contaminated estuary sediments, Alameda Naval Air Station, California. Environ Sci Technol 34:3665-3673

O'Day PA, Carroll SA, Waychunas GA (1998) Rock-water interactions controlling zinc, cadmium, and lead concentrations in surface waters and sediments, U.S. Tri-State Mining District. I. Molecular identification using X-ray absorption spectroscopy. Environ Sci Technol 32:943-955

O'Day PA, Chisholm-Brause CJ, Towle SN, Parks GA, Brown GE Jr (1996) X-ray absorption spectroscopy of Co(II) sorption complexes on quartz (α-SiO$_2$) and rutile (TiO$_2$). Geochim Cosmochim Acta 60:2515-2532

O'Day PA, Newville M, Neuhoff PS, Sahai N, Carroll SA (2000b) X-ray absorption spectroscopy of strontium(II) coordination I. Static and thermal disorder in crystalline, hydrated, and precipitated solids and in aqueous solution. J Colloid Interface Sci 222:184-197

O'Day PA, Parks GA, Brown GE Jr (1994c) Molecular structure and binding sites of cobalt (II) surface complexes on kaolinite from X-ray absorption spectroscopy. Clays Clay Minerals 42:337-355

O'Day PA, Rehr JJ, Zabinsky SI, Brown GE Jr (1994a) Extended X-ray Absorption Fine Structure (EXAFS) analysis of disorder and multiple-scattering in complex crystalline solids. J Am Chem Soc 116:2938-2949

Odelius M (1999) Mixed molecular and dissociative water adsorption on MgO(100). Phys Rev Lett 82:3919-3922

Odelius M, Bernasconi M, Parrinello M (1997) Two dimensional ice adsorbed on mica surface. Phys Rev Lett 78:2855-2858

Ogletree DFB, Fadley C, Hussain Z, Lebedev G, Salmeron M (2000) Photoelectron spectroscopy at ten torr. 8th international Conference on Electronic Spectroscopy and Structure, Lawrence Berkeley National Laboratory, 408

Ohlendorf HM, Santolo GM (1994) Kesterson Reservoir – past, present, and future: An ecological risk assessment. *In:* Frankenberger WT Jr, Benson S (eds) Selenium in the Environment. Marcel Dekker Inc, New York, p 69-117

Ohtaki T, Radnai T (1993) Structure and dynamics of hydrated ions. Chem Rev 93:1157-1204

Olivella MA, Palacios JM, Vairavamurthy A, del Rio JC, de las Heras FXC (2002) A study of sulfur functionalities in fossil fuels using destructive- (ASTM and Py-GC-MS) and non-destructive- (SEM-EDX, XANES and XPS) techniques. Fuel 81:405-411

O'Nions RK, Smith DGW (1971) Investigations of the L$_{II,III}$ X-ray emission spectra of Fe by electron microprobe. Part 2. The Fe L$_{II,III}$ spectra of Fe and Fe-Ti oxides. Am Mineral 56:1452-1463

Oremland RS, Hollibaugh JT, Maest AS, Presser TS, Miller LG, Culbertson CW (1989) Selenate reduction to elemental selenium by anaerobic bacteria in sediments and culture: biogeochemical significance of a novel, sulfate-independent respiration. Appl Environ Microbiol 55:2333-2343

Orlando TM, Kimmel GA, Simpson WC (1999) Quantum-resolved electron stimulated interface reactions: D$_2$ formation from D$_2$O films. Nucl Instrum Meth Phys Res Sect B: Beam Interactions with Materials and Atoms 157:183-190

Ostergren JD, Brown GE Jr, Parks GA, Persson P (2000a) Inorganic ligand effects on Pb(II) sorption to goethite (α-FeOOH): II. Sulfate. J Colloid Interface Sci 225:483-493

Ostergren JD, Brown GE Jr, Parks GA, Tingle TN (1999) Quantitative lead speciation in selected mine tailings from Leadville, CO. Environ Sci Technol 33:1627-1636

Ostergren JD, Trainor TP, Bargar JR, Brown GE Jr, Parks GA (2000b) Inorganic ligand effects on Pb(II) sorption to goethite (α-FeOOH): I. Carbonate. J Colloid Interface Sci 225:466-482

Östhols E, Manceau A, Farges F, Charlet L (1997) Adsorption of thorium on amorphous silica: an EXAFS study. J Colloid Interface Sci 194:10-21

Panak PJ, Booth CH, Caulder DL, Bucher JJ, Shuh DK, Nitsche H (2002) X-ray absorption fine structure spectroscopy of plutonium complexes with bacillus sphaericus. Radiochim Acta 90:315-321

Papelis C, Brown GE Jr, Parks GA, Leckie JO (1995) X-ray absorption spectroscopic studies of cadmium and selenite adsorption on aluminum oxides. Langmuir 11:2041-2048

Papelis C, Hayes KF (1996) Distinguishing between interlayer and external sorption sites of clay minerals using X-ray absorption spectroscopy. Colloids and Surfaces A 107:89-96

Parise J (1999) New opportunities for microcrystalline and powder diffractometry at synchrotron sources. *In:* Schulze DG, Stucki JW, Bertsch PM (eds) Synchrotron X-ray Methods in Clay Science, Clay Minerals Society Workshop Lectures 9. The Clay Minerals Society, Boulder, Colorado, p 115-145

Park SH, Sposito G (2002) Structure of water adsorbed at a mica surface. Phys Rev Lett 89:085501-1–085501-3

Parkhurst DA, Brown GE Jr, Parks GA, Waychunas GA (1984) Structural study of zinc complexes in aqueous chloride solutions by fluorescence EXAFS spectroscopy. Abstracts with Program Geol Soc Am Ann Mtg 16:618

Parkman RH, Charnock JM, Livens FR, Vaughan DJ (1998) A study of the interaction of strontium ions in aqueous solution with the surfaces of calcite and kaolinite. Geochim Cosmochim Acta 62:1481-1492

Parkman RH, Curtis CD, Vaughan DJ, Charnock JM (1996) Metal fixation and mobilisation in the sediments of Afon Goch estuary – Duals Bay, Anglesey. Appl Geochem 11:203-210

Parks GA (1965) The isoelectric points of solid oxides, solid hydroxides, and aqueous hydroxo complex systems. Chem Rev 65:177-198

Pascarelli S, Neisius T, De Panfilis S, Bonfim M, Pizzini S, Mackay K, David S, Fontaine A, San Miguel A, Itie JP, Gauthier M, Polian A (1999) Dispersive XAS at third-generation sources: strengths and limitations. J Synchrotron Rad 6:146-148

Paschin YV, Kozachenko VL, Sal'nikova LE (1983) Differential mutagenic response at the HGPRT locus in V-79 and CHO cells after treatment with chromate. Mutat Res 122:362-365

Patterson RR, Fendorf S, Fendorf M (1997) Reduction of hexavalent chromium by amorphous iron sulfide. Environ Sci Technol 31:2039-2044

Pattrick RAD, Charnock JM, England KER, Mosselmans JFW, Wright K (1998) Lead sorption on the surface of ZnS with relevance to flotation: a fluorescence Reflexafs study. Minerals Eng 11:1025-1033

Pattrick RAD, England KER, Charnock JM, Mosselmans JFW (1999) Copper activation of sphalerite and its reaction with xanthate in relation to flotation: an X-ray absorption spectroscopy (reflection extended X-ray absorption fine structure) investigation. Internat J Mineral Processing 55:247-265

Pattrick RAD, Mosselmans JFW, Charnock JM, England KER, Helz GR, Garner CD, Vaughan DJ (1997) The structure of amorphous copper sulfide precipitates: an X-ray absorption study. Geochim Cosmochim Acta 61:2023-2036

Pecher K, Kneedler E, Rothe J, Meigs G, Warwick T, Nealson K, Tonner B (2000) Charge state mapping of mixed-valent iron and manganese mineral particles using scanning transmission X-ray microscopy (STXM). AIP Conf Proc 507 (X-ray Microscopy):291-300

Persson I, Persson P, Sandstrom M, Ullstroem A-S (2002) Structure of Jahn-Teller distorted solvated copper(II) ions in solution, and in solids with apparently regular octahedral coordination geometry. J Chem Soc Dalton Trans 7:1256-1265

Persson I, Sandstroem M, Steel AT, Zapatero MJ, Aakesson R (1991) A large-angle X-ray scattering, XAFS, and vibrational spectroscopic study of copper(I) halide complexes in dimethyl sulfoxide, acetonitrile, pyridine, and aqueous solutions. Inorg Chem 30:4075-4081

Persson I, Sandstroem M, Yokoyama H, Chaudhry M (1995) Structure of the solvated strontium and barium ions in aqueous, dimethyl sulfoxide and pyridine solution, and crystal structure of strontium and barium hydroxide octahydrate. Zeitschrift fuer Naturforschung, A Phys Sci 50:21-37

Persson P, Parks GA, Brown GE Jr (1995) Adsorption and local environment of Co(II) at the zinc oxide- and zinc sulfide-aqueous interfaces. Langmuir 11:3782-3794

Peterson ML, Brown GE Jr, Parks GA (1996) Direct XAFS evidence for heterogeneous redox at the aqueous chromium/magnetite interface. Colloids and Surfaces 107:77-88

Peterson ML, Brown GE Jr, Parks GA, Stein CL (1997b) Differential redox and sorption of Cr(III/VI) on natural silicate and oxide minerals: EXAFS and XANES results. Geochim Cosmochim Acta 61:3399-3412

Peterson ML, White AF, Brown GE Jr, Parks GA (1997a) Surface passivation of magnetite (Fe_3O_4) by reaction with aqueous Cr(VI): XAFS and TEM results. Environ Sci Technol 31:1573-1576

Petit-Maire D, Petiau J, Calas G, Jacquet-Francillon N (1986) Local structure around actinides in borosilicate glasses. J de Physique 47 (Colloque 8):849-852

Petrovic R, Berner RA, Goldhaber MB (1976) Rate control in dissolution of alkali feldspars. I. Study of residual feldspar grains by X-ray photoelectron spectroscopy. Geochim Cosmochim Acta 40:537-548

Pfalzer P, Urbach JP, Klemm M, Horn S, Den Boer ML, Frenkel AI, Kirkland JP (1999) Elimination of self-absorption in fluorescence hard X-ray absorption spectra. Phys Rev B 60:9335-9339

Pfeiffer F, Davd C, Burghammer M, Riekl C, Salditt T (2002) Two-dimensional X-ray waveguides and point sources. Science 297:230-234

Pickering IJ, Brown GE Jr, Tokunaga T (1995) X-ray absorption spectroscopy of selenium transformations in Kesterson Reservoir soils. Environ Sci Technol 29:2456-2459

Pickering IJ, Prince RC, George MJ, Smith RD, George GN, Salt DE (2000a) Reduction and coordination of arsenic in Indian mustard. Plant Physiol 122:1171-1177

Pickering IJ, Prince RC, Salt DE, George GN (2000b) Quantitative, chemically specific imaging of selenium transformation in plants. Proc Nat Acad Sci USA 97:10717-10722

Pingitore NE Jr, Iglesias A, Bruce A, Lytle F, Wellington GM (2002a) Valences of iron and copper in coral skeleton: X-ray absorption spectroscopy analysis. Microchemical J 71:205-210

Pingitore NE Jr, Iglesias A, Lytle F, Wellington GM (2002b) X-ray absorption spectroscopy of uranium at low ppm levels in coral skeletal aragonite. Microchemical J 71:261-266

Pingitore NE Jr, Lytle FW, Davies BM, Eastman MP, Eller PG, Larson EM (1992) Mode of incorporation of strontium ion (2+) in calcite: determination by X-ray absorption spectroscopy. Geochim Cosmochim Acta 56:1531-1538

Pompe S, Schmeide K, Reich T, Hennig C, Funke H, Roßberg A, Geipei G, Brendler V, Heise KH, Bernhard G (2002) Neptunium(V) complexation by various humic acids in solution studied by EXAFS and NIR spectroscopy. Proceedings of the Euroconference and NEA Workshop on Speciation, Techniques, and Facilities for Radioactive Materials at Synchrotron Light Sources, Grenoble, France, Sept. 10-12, 2000 Nuclear Energy Agency/Organization for Economic Co-operation and Development, AEN/NEA 2002, Paris, p 277-284

Predota M, Chialvo AA, Cummings PT (2002) Molecular simulation of aqueous systems at TiO_2 surfaces. Abstracts of Papers, 223rd Am Chem Soc National Meeting, Orlando, FL, United States, April 7-11, 2002, GEOC-075

Pluth JJ, Smith JV, Pushcharovsky DY, Semenov EI, Bram A, Riekel C, Weber H-P, Broach RW (1997) Third-generation synchrotron X-ray diffraction of 6-μm crystal of raite, $\approx Na_3Mn_3Ti_{0.25}Si_8O_{20}(OH)_2 \cdot 10H_2O$, opens up new chemistry and physics of low-temperature minerals. Proc Nat Acad Sci USA 94:12263-12267

Qian Y, Sturchio NC, Chiarello RP, Lyman PF Lee T-L, Bedzyk MJ (1994) Lattice location of trace elements within minerals and at their surfaces with X-ray standing waves. Science 265:1555-1557

Rajh T, Nedeljkovic JM, Chen LX, Poluektov O, Thurnauer MC (1999) Improving optical and charge separation properties of nanocrystalline TiO_2 by surface modification with vitamin C. J Phys Chem B 103:3515-3519

Rakovan J, Reeder RJ, Elzinga EJ, Cherniak DJ, Tait CD, Morris DE (2002) Structural characterization of U(VI) in apatite by X-ray absorption spectroscopy. Environ Sci Technol 36:3114-3117

Randall SR, Sherman DM, Ragnarsdottir KV (1998) An extended X-ray absorption fine structure spectroscopy investigation of cadmium sorption on cryptomelane ($KMn_{18}O_{16}$). Chem Geol 151:95-106

Randall SR, Sherman DM, Ragnarsdottir KV, Collins CR (1999) The mechanism of cadmium surface complexation on iron oxyhydroxide minerals. Geochim Cosmochim Acta 63:2971-2987

Ransom B, Bennett RH, Baerwald R, Shea K (1997) TEM study of *in situ* organic matter on continental margins: occurrence and the "Monolayer" hypothesis. Marine Geol 138:1-9

Redden G, Bargar JR, Bencheikh-Latmani R (2001) Citrate-enhanced uranyl adsorption on goethite: An EXAFS analysis. J Colloid Interface Sci 244:211-219

Redden GD, Li J, Leckie JO (1998) Adsorption of U(VI) and citric acid on goethite, gibbsite, and kaolinite: Comparing results for binary and ternary systems. *In:* Jenne EA (ed) Adsorption of Metals by Geomedia: Variables, Mechanisms, and Model Applications, Academic Press, p 291-315

Reeder RJ (1996) Interaction of divalent cobalt, zinc, cadmium, and barium with the calcite surface during layer growth. Geochim Cosmochim Acta 60:1543-1552

Reeder RJ, Lamble GM, Lee J-F, Staudt WJ (1994) Mechanism of SeO_4^{2-} substitution calcite: an XAFS study. Geochim Cosmochim Acta 58:5639-5646

Reeder RJ, Lamble GM, Northrup PA (1999) XAFS study of the coordination and local relaxation around Co^{2+}, Zn^{2+}, Pb^{2+}, and Ba^{2+} trace elements in calcite. Am Mineral 84:1049-1060

Reeder RJ, Nugent M, Lamble GM, Tait CD, Morris DE (2000) Uranyl incorporation into calcite and aragonite: XAFS and luminescence studies. Environ Sci Technol 34:638-644.

Reeder RJ, Nugent M, Tait CD, Morris DE, Heald SM, Beck KM, Hess WP, Lanzirotti A (2001) Coprecipitation of uranium(VI) with calcite: XAFS, micro-XAS, and luminescence characterization. Geochim Cosmochim Acta 65:3491-3503

Rehr JJ, Albers RC (1990) Scattering matrix formation of curved wave multiple scattering theory – Application to X-ray absorption fine structure. Phys Rev B 41:8139-8149

Rehr JJ, Albers RC (2000) Theoretical approaches to X-ray absorption fine structure. Rev Mod Phys 72:621-654

Rehr JJ, Ankudinov AL (2001a) New developments in the theory of X-ray absorption and core photoemission. J Electron Spectrosc Relat Phenom 114-116:1115-1121

Rehr JJ, Ankudinov AL (2001b) Progress and challenges in the theory and interpretation of X-ray spectra. J Synchrotron Rad 8:61-65

Rehr JJ, Booth CH, Bridges F, Zabinsky SI (1994) X-ray-absorption fine structure in embedded atoms. Phys Rev B 49:12347-12350

Rehr JJ, Mustre de Leon J, Zabinsky SI, Albers RC (1991) Theoretical X-ray absorption fine structure standards. J Amer Chem Soc 113:5135-5140

Rehr JJ, Zabinsky SI, Albers RC (1992) High-order multiple scattering calculations of X-ray absorption fine structure. Phys Rev Lett 69:3397-4000

Reich T, Moll H, Arnold T, Denecke MA, Hennig C, Geipel G, Bernhard G, Nitsche H, Allen PG, Bucher JJ, Edelstein NM, Shuh DK (1998) An EXAFS study of uranium(VI) sorption onto silica gel and ferrihydrite. J Electron Spectrosc Relat Phenom 96:237-243

Renaud G (1998) Oxide surfaces and metal/oxide interfaces studied by grazing incidence X-ray scattering. Surf Sci Rept 32:1-90

Ressler T (1997) WinXAS: A new software package not only for the analysis of energy-dispersive XAS data. J de Physique IV 7 (Colloque C2, X-Ray Absorption Fine Structure, Vol. 1):269-270 (*http://ixs.csrri.iit.edu/catalog/XAFS_Programs/winxas*)

Ressler T, Wong J, Roos J, Smith IL (2000) Quantitative speciation of Mn-bearing particulates emitted from autos burning (methylcyclopentadienyl)manganese tricarbonyl-added gasolines using XANES spectroscopy. Environ Sci Technol 34:950-958

Roberts DR, Scheidegger AM, Sparks DL (1999) Kinetics of mixed Ni-Al precipitate formation on a soil clay fraction. Environ Sci Technol 33:3749-3754

Roberts DR, Scheinost AC, Sparks DL (2002) Zinc speciation is a smelter-contaminated soil profile using bulk and microscopic techniques. Environ Sci Technol 36:1742-1750

Robinson IK, Tweet DJ (1992) Surface X-ray diffraction. Rep Prog Phys 55:599-651

Rochette EA, Li GC, Fendorf SE (1998) Stability of arsenate minerals in soil under biotically-generated reducing conditions. Soil Sci Soc Am J 62:1530-1537

Roe AL, Hayes KF, Chisholm-Brause CJ, Brown GE Jr, Hodgson KO, Parks GA, Leckie JO (1991) X-ray absorption study of lead complexes at α-FeOOH/water interfaces. Langmuir 7:367-373

Roedder E (1984) Fluid Inclusions. Rev Mineral Vol. 12, Mineralogical Society of America, Washington D.C.

Rose J, Manceau A, Bottero J-Y, Masion A, Garcia F (1996) Nucleation and growth mechanisms of Fe oxyhydroxide in the presence of PO_4 ions. 1. Fe K-edge EXAFS study. Langmuir 12:6701-6707

Rose J, Manceau A, Masion A, Bottero J-Y (1997) Structure and mechanisms of formation of $FeOOH(NO_3)$ oligomers in the early stages of hydrolysis. Langmuir 13:3240-3246

Rose J, Moulin I, Masion A, Bertsch PM, Wiesner MR, Bottero J-Y, Mosnier F, Haehnel C (2001) X-ray absorption spectroscopy study of immobilization processes for heavy metals in calcium silicate hydrates. 2. Zinc. Langmuir 17:3658-3665

Rossberg A, Baraniak L, Reich T, Hennig C, Bernhard G, Nitsche H (2000) EXAFS structural analysis of aqueous uranium(VI) complexes with lignin degradation products. Radiochim Acta 88:593-597

Rothe J, Kneedler EM, Pecher K, Tonner BP, Nealson KH, Grundi T, Meyer-Ilse W, Warwick T (1999) Spectromicroscopy of Mn distributions in micronodules produced by biomineralization. J Synchrotron Rad 6:359-361

Ryan JA, Zhang P, Hesterberg DA, Chou J, Sayers DE (2001) Formation of chloropyromorphite in lead-contaminated soil amended with hydroxyapatite. Environ Sci Technol 35:3798-3803

Sadoc A, Lagarde P, Vlaic G (1985) EXAFS evidence for local order in aqueous solutions of cadmium bromide. J Phys C 18:23-31

Sahai N, Carroll SA, Roberts S, O'Day PA (2000) X-ray absorption spectroscopy of strontium(II) coordination. II. Sorption and precipitation at kaolinite, amorphous silica and goethite surfaces. J Colloid Interface Sci 222:198-212

Sainctavit Ph, Calas G, Petiau J, Karnatak J, Esteva JM, Brown GE Jr (1986) Electronic structure from X-ray K-edges in ZnS:Fe and $CuFeS_2$. J de Physique 47 (Colloque C8):C8411-C8414

Salt DE, Pickering IJ, Prince RC, Gleba D, Dushenkov S, Smith RD, Raskin I (1997) Metal accumulation by aquacultured seedlings of indian mustard. Environ Sci Technol 31:1636-1644.

Salt DE, Prince RC, Pickering IJ, Raskin I (1995) Mechanisms of cadmium mobility and accumulation in Indian mustard. Plant Physiol 109:1427-1433

Sandstrom DR, Dodgen HW, Lytle FW (1977) Study of nickel(II) coordination in aqueous solution by EXAFS analysis. J Chem Phys 67:473-476

Sandstrom M (1977) An X-ray diffraction and Raman study of mercury(II) chloride complexes in aqueous solution. Evidence for the formation of polynuclear complexes. Acta Chem Scand Ser A 31:141-150

Sandstrom M, Persson I, Ahrland S (1978) On the coordination around mercury(II), cadmium(II) and zinc(II) in dimethyl sulfoxide and aqueous solutions. An X-ray diffraction, Raman and infrared investigation. Acta Chem Scand Ser A 32:607-625

Sandstrom M, Persson I, Jalilehvand F, Lindquist-Reis P, Spangberg D, Hermansson K (2001) Hydration of some large and highly charged metal ions. J Synchrotron Rad 8:657-659

Sarret G, Connan J, Kasrai M, Bancroft GM, Charrie-Duhaut A, Lemoine S, Adam P, Albrecht P, Eybert-Berard L (1999b) Chemical forms of sulfur in geological and archeological asphaltenes from Middle East, France, and Spain determined by sulfur K- and L-edge X-ray absorption near-edge structure spectroscopy. Geochim Cosmochim Acta 63:3767-3779

Sarret G, Connan J, Kasrai M, Eybert-Berard L, Bancroft GM (1999c) Characterization of sulfur in asphaltenes by sulfur K- and L-edge XANES spectroscopy. J Synchrotron Rad 6:670-672

Sarret G, Manceau A, Cuny D, Van Haluwyn C, Deruelle S, Hazemann J-L, Soldo Y, Eybert-Berard L, Menthonnex J-J (1998b) Mechanisms of lichen resistance to metal pollution. Environ Sci Technol 32:3325-3330

Sarret G, Manceau A, Hazemann J-L, Gomez A, Mench M (1997) EXAFS study of the nature of zinc complexation sites in humic substances as a function of Zn concentration. J Phys IV 7:799-802

Sarret G, Manceau A, Spadini L, Roux J-C, Hazemann J-L, Soldo Y, Eybert-Berard L, Menthonnex J-J (1998a) Structural determination of Zn and Pb binding sites in *Penicillium chrysogenum* cell wall by EXAFS spectroscopy. Environ Sci Technol 32:1648-1655

Sarret G, Manceau A, Spadini L, Roux J-C, Hazemann J-L, Soldo Y, Eybert-Berard L, Menthonnex J-J (1999a) Structural determination of Pb binding sites in *Penicillium chruysogenum* cell walls by EXAFS spectroscopy and solution chemistry. J Synchrotron Rad 6:414-416

Sarret G, Mongenot T, Connan J, Derenne S, Kasrai M, Bancroft GM, Largeau C (2002) Sulfur speciation in kerogens of the Orbagnoux deposit (Upper Kimmeridgian, Jura) by XANES spectroscopy and pyrolysis. Organic Geochem 33:877-895

Sarret G, Vangronsveld J, Manceau A, Musso M, D'Haen J, Menthonnex J-J, Hazemann J-L (2001) Accumulation forms of Zn and Pb in *Phaseolus vulgaris* in the presence and absence of EDTA. Environ Sci Technol 35:2854-2859

Savage KS, Tingle TN, O'Day PA, Waychunas GA, Bird DK (2000) Arsenic speciation in pyrite and secondary weathering phases, Mother Lode Gold District, Tuolumne County, California. Appl Geochem 15:1219-1244

Sayers DE, Hesterberg D, Zhou W, Robarge WP, Plummer GM (1997) XAFS characterization of copper contamination in the unsaturated and saturated zones of a soil profile. J de Physique IV 7 (Colloque C2, X-Ray Absorption Fine Structure, Vol. 2):831-832

Sayers DE, Lytle FW, Stern EA (1972) Structure determination of amorphous germanium, germanium oxide, and germanium selenide by Fourier analysis of extended X-ray absorption fine structure (EXAFS). J Non-Crystal Solids 8-10:401-407

Sayers DE, Stern EA, Lytle FW (1971) New technique for investigating non-crystalline structures. Fourier analysis of the extended X-ray-absorption fine structure. Phys Rev Lett 27:1204-1207

Sayers DE, Stern EA, Lytle FW (1975) New method to measure structural disorder. Application to germanium dioxide glass. Phys Rev Lett 35:584-587

Scamehorn CA, Harrison NM, McCarthy MI (1994) Water chemistry on surface defect sites: Chemidissociation versus physisorption on MgO(100). J Chem Phys 101:1547-1554.

Schaufuss AG, Nesbitt HW, Kartio I, Laajalehto K, Bancroft GM, Szargan R (1998) Reactivity of surface chemical states on fractured pyrite. Surf Sci 411:321-328

Scheckel KG, Scheinost AC, Ford RG, Sparks DL (2000) Stability of layered Ni hydroxide surface precipitates-a dissolution kinetics study. Geochim Cosmochim Acta 64:2727-2735

Scheckel KG, Sparks DL (2000) Kinetics of the formation and dissolution of Ni precipitates in a gibbsite/amorphous silica mixture. J Colloid Interface Sci 229:222-229

Scheidegger AM, Lamble GM, Sparks DL (1996) Investigation of Ni sorption on pyrophyllite: an XAFS study. Environ Sci Technol 30:548-554

Scheidegger AM, Lamble GM, Sparks DL (1997) Spectroscopic evidence for the formation of mixed-cation hydroxide phases upon metal sorption on clays and aluminum oxides. J Colloid Interface Sci 186:118-128

Scheidegger AM, Strawn DG, Lamble GM, Sparks DL (1998) The kinetics of mixed Ni-Al hydroxide formation on clay and aluminum oxide minerals: A time-resolved XAFS study. Geochim Cosmochim Acta 62:2233-2245

Scheinost AC, Sparks DL (2000) Formation of layered single- and double-metal hydroxide precipitates at the mineral/water interface. A multiple-scattering XAFS analysis. J Colloid Interface Sci 223:167-178

Scheinost AC, Abend S, Pandya KI, Sparks DL (2001) Kinetic controls on Cu and Pb sorption by ferrihydrite. Environ Sci Technol 35:1090-1096

Schlegel ML, Charlet L, Manceau A (1998) Adsorption mechanism of Co(II) on hectorite and its consequences on the dissolution process: Insight from polarized EXAFS and kinetic chemical studies. Mineral Mag 62A:1337-1338

Schlegel ML, Charlet L, Manceau A (1999a) Sorption of metal ions on clay minerals II. Mechanism of Co sorption on hectorite at high and low ionic strength and impact on the sorbent stability. J Colloid Interface Sci 220:392-405

Schlegel ML, Manceau A, Charlet L, Chateigner D, Hazemann J-L (2001a) Sorption of metal ions on clay minerals. III. Nucleation and epitaxial growth of Zn phyllosilicate on the edges of hectorite. Geochim Cosmochim Acta 65:4155-4170

Schlegel ML, Manceau A, Charlet L, Hazemann J-L (2001b) Adsorption mechanisms of Zn on hectorite as a function of time, pH, and ionic strength. Am J Sci 301:798-830

Schlegel ML, Manceau A, Chateigner D, Charlet L (1999b) Sorption of metal ions on clay minerals, I. Polarized EXAFS evidence for the adsorption of Co on the edges of hectorite particles. J Colloid Interface Sci 215:140-158

Schlegel ML, Nagy KL, Fenter P, Sturchio NC (2002) Structures of quartz (010)- and (110)-water interfaces determined by X-ray reflectivity and atomic force microscopy of natural growth surfaces. Geochim Cosmochim Acta 66:3037-3054

Schmeide K, Pompe S, Reich T, Hennig C, Funke H, Robberg A, Geipel G, Brendler V, Heise KH, Bernhard G (2002) Interaction of neptumium(IV) with humic substances studied by XAFS spectroscopy. Proceedings of the Euroconference and NEA Workshop on Speciation, Techniques, and Facilities for Radioactive Materials at Synchrotron Light Sources, Grenoble, France, Sept. 10-12, 2000 Nuclear Energy Agency/Organization for Economic Co-operation and Development, AEN/NEA 2002, Paris, p 287-294

Schofield PF, Henderson CMB, Cressey G, van der Laan G (1995) 2p X-ray absorption spectroscopy in the earth sciences. J Synchrotron Rad 2:93-98

Scholl A, Ohldag H, Nolting F, Stohr J, Padmore HA (2002) X-ray photoemission electron microscopy, a tool for the investigation of complex magnetic structures. Rev Sci Instrum 73 (3, Pt. 2):1362-1366

Schulze DG, Bertsch PM (1995) Synchrotron X-ray techniques in soil, plant, and environmental research. Adv Agron 55:1-66

Schulze DG, Stucki JW, Bertsch PM (eds) (1999) Synchrotron X-ray Methods in Clay Science, Clay Minerals Society Workshop Lectures 9. The Clay Minerals Society, Boulder, Colorado

Schulze DG, Sutton SR, Bajt S (1995) Determining manganese oxidation state in soils using X-ray absorption near-edge structure (XANES) spectroscopy. Soil Sci Soc Am J 59:1540-1548

Sery A, Manceau A, Greaves GN (1996) Chemical state of Cd in apatite phosphate ores as determined by EXAFS spectroscopy. Am Mineral 81:864-873

Seward TM, Henderson CMB, Charnock JM (2000) Indium(III) chloride complexing and solvation in hydrothermal solutions to 350°C: an EXAFS study. Chem Geol 167:117-127

Seward TM, Henderson CMB, Charnock JM, Dobson BR (1996) An X-ray absorption (EXAFS) spectroscopic study of aquated Ag^+ in hydrothermal solutions to 350°C. Geochim Cosmochim Acta 60:2273-2282

Seward TM, Henderson CMB, Charnock JM, Driesner T (1999) An EXAFS study of solvation and ion pairing in aqueous strontium solutions to 300°C. Geochim Cosmochim Acta 63:2409-2418

Shaikhutdinov Sh K, Weiss W (1999) Oxygen pressure dependence of the α-Fe_2O_3(0001) surface structure. Surf Sci 432:L627-L634

Shaw S, Clark SM, Henderson CMB (2000a) Hydrothermal formation of the calcium silicate hydrates, tobermorite ($Ca_5Si_6O_{16}(OH)_2 \cdot 4H_2O$) and xonotlite ($Ca_6Si_6O_{17}$ $(OH)_2$): an *in situ* synchrotron study. Chem Geol 167:129-140

Shaw S, Henderson CMB, Clark SM (2002a) *In situ* synchrotron study of the kinetics, thermodynamics, and reaction mechanisms of the hydrothermal crystallization of gyrolite, $Ca_{16}Si_{24}O_{60}(OH)_8 \cdot 14H_2O$. Am Mineral 87:533-541

Shaw S, Henderson CMB, Komanschek BU (2000b) Dehydration/recrystallization mechanisms, energetics, and kinetics of hydrated calcium silicate minerals: an *in situ* TGA/DSC and synchrotron radiation SAXS/WAXS study. Chem Geol 167:141-159

Shaw S, Lowry GV, Kim CS, Rytuba JJ, Brown GE Jr (2002b) Particle-facilitated mercury transport from New Idria and Sulphur Bank mercury mine tailings. 2. Microscopic and spectroscopic analysis. Environ Sci Technol (submitted)

Shuh DK, Kaltsoyannis N, Bucher JJ, Edelstein NM, Clark SB, Nitsche H, Almahamid I, Torretto P, Lukens W et al. (1994) Environmental applications of XANES: speciation of Tc in cement after chemical treatment and Se after bacterial uptake. Materials Research Society Symposium Proceedings 344 (Materials and Processes for Environmental Protection):323-328

Siegel RW (1996) Creating nanophase materials. Scientific American 275:74-79

Silk JE, Hansen LD.; Eatough DJ, Hill MW, Mangelson NF, Lytle FW, Greegor RB (1989) Chemical characterization of vanadium, nickel, and arsenic in oil fly-ash samples using EXAFS and XANES spectroscopy. Physica B (Amsterdam) 158:247-248

Silvester E, Charlet L, Manceau A (1995) Mechanism of chromium(III) oxidation by Na-buserite. J Phys Chem 99:16662-16669

Silvester E, Manceau A, Drits VA (1997) Structure of synthetic monoclinic Na-rich birnessite and hexagonal birnessite: II. Results from chemical studies and EXAFS spectroscopy. Am Mineral 82:962-978

Sinfelt JH, Via GH, Lytle FW (1984) Application of EXAFS in catalysis. Structure of bimetallic cluster catalysts. Catal Rev - Sci Eng 26:81-140

Skipper NT, Smalley MV, Williams GD, Soper AK, Thompson CH (1995) Direct measurement of the electric double-layer structure in hydrated lithium vermiculite clays by neutron diffraction. J Phys Chem 99:14201-14204

Smith AP, Urquhart SG, Winesett DA, Mitchell G, Ade H (2001) Use of near edge X-ray absorption fine structure spectromicroscopy to characterize multicomponent polymeric systems. Appl Spectrosc 55:1676-1681

Smith DGW, O'Nions RK (1971) Investigations of the $L_{II,III}$ X-ray emission spectra of Fe by electron microprobe. Part 1. Some aspects of the Fe $L_{II,III}$ X-ray emission spectra from metallic iron and hematite. Brit J Appl Phys 4:147-159

Smith JV, Rivers ML (1995) Synchrotron X-ray microanalysis. Mineral Soc Ser 6 (Microprobe Techniques in the Earth Sciences), p 163-233

Snow ET, Xu LS (1989) Effects of chromium(III) on DNA replication in vitro. Biol Trace Element Research 21:61-72

Soderholm L, Antonio MR, Williams C, Wasserman SR (1999) XANES spectroelectrochemistry: A new method for determining formal potentials. Anal Chem 71:4622-4628

Soderholm L, Liu GK, Antonio MR, Lytle FW (1998) X-ray excited optical luminescence (XEOL) detection of X-ray absorption fine structure (XAFS). J Chem Phys 109:6745-6752.

Soderholm L, Williams CW, Antonio MR, Tischler ML, Markos M (2000) The influence of *Desulfovibrio desulfuricans* on neptunium chemistry. Mat Res Soc Symp Proc 590 (Applications of Synchrotron Radiation Techniques to Materials Science V):27-32

Solé VA, Gauthier C, Goulon J, Natali F (1999) Undulator QEXAFS at the ESRF beamline ID26. J Synchrotron Rad 6:174-175

Sorensen LB, Cross JO, Newville M, Ravel B, Rehr JJ, Stragier H (1994) Diffraction anomalous fine structure: Unifying X-ray diffraction and X-ray absorption with DAFS *In:* Materlik G, Sparks CJ, Fischer K (eds) Resonant Anomalous X-ray Scattering: Theory and Applications, North-Holland, p 389-420

Spadini L, Manceau A, Schindler PW, Charlet L (1994) Structure and stability of Cd^{2+} surface complexes on ferric oxides. J Colloid Interface Sci 168:73-86

Sparks CJ Jr (1980) X-ray fluorescence microprobe for chemical analysis. *In:* Winick H, Doniach S (eds) Synchrotron Radiation Research, Plenum Press, New York, p 459-512

Spngberg D, Hermansson K, Lindqvist-Reis P, Jalilehvand F, Sandstroem M, Persson I (2000) Model extended X-ray absorption fine structure (EXAFS) spectra from molecular dynamics data for Ca^{2+} and Al^{3+} aqueous solutions. J Phys Chem B 104:10467-10472

Sposito G. (1986) Distinguishing adsorption from surface precipitation. *In:* Geochemical Processes at Mineral Surfaces, Davis JA, Hayes KF (eds) ACS Symposium Series 323, The American Chemical Society, Washington, DC, p 217-228

Stack AG, Eggleston CM, Higgins SR, Pribyl R, Nichols J (2001a) Oxygen versus iron termination of hematite (001) surfaces: STM imaging in air and in aqueous solutions (abstract). Eleventh Annual VM Goldschmidt Conference, May 20-24, 2001, Hot Springs, VA, 3824.pdf

Stack AG, Higgins SR, Eggleston CM (2001b) Point of zero charge of a corundum-water interface probed with optical Second Harmonic Generation (SHG) and Atomic Force Microscopy (AFM): New approaches to oxide surface charge. Geochim Cosmochim Acta 65:3055-3063

Stern EA, Sayers DE, Lytle FW (1975) Extended X-ray-absorption fine-structure technique. III. Determination of physical parameters. Phys Rev B 11:4836-4846

Stern RH (1987) Role of speciation in metal toxicity. Proc Second Nordic Symp on Trace Elements in Human Health and Disease. Environ Health 26, World Health Org, Copenhagen, p 3-7

Stillinger FH (1980) Water revisited. Science 209:451-457

Stöhr J (1988) SEXAFS: Everything you always wanted to know about SEXAFS but were afraid to ask. In: Konigsberger DC, Prins R (eds) X-ray Absorption: Principles, Applications Techniques of EXAFS, SEXAFS and XANES, John Wiley &Sons, New York, p 443-571

Stöhr J (1992) NEXAFS Spectroscopy. Springer-Verlag, Heidelberg, Germany

Stöhr J, Anders S (2000) X-ray spectromicroscopy of complex materials and surfaces. IBM Journal of Research and Development 44:535-551

Stone AT, Morgan JJ (1987) Reductive dissolution of metal oxides. *In:* Stumm W (ed) Aquatic Surface Chemistry, Wiley-Interscience, New York, p 221-254

Stragier H, Cross JO, Rehr JJ, Sorensen LB, Bouldin CE, Woicik JC (1992) Diffraction anomalous fine structure: A new X-ray structural technique. Phys Rev Lett 69:3064-3067

Strawn DG, Scheidegger AM, Sparks DL (1998) Kinetics and mechanisms of Pb(II) sorption and desorption at the aluminum oxide-water interface. Environ Sci Technol 32:2596-2601

Strawn DG, Sparks DL (1999) The use of XAFS to distinguish between inner- and outer-sphere lead adsorption complexes on montmorillonite. J Colloid Interface Sci 216: 257-269

Strawn DG, Sparks DL (2000) Effects of soil organic matter on the kinetics and mechanisms of Pb(II) sorption and desorption in soil. Soil Sci Soc Am J 64:144-156

Stumm W (1992) Chemistry of the Solid-Water Interface. John Wiley & Sons, Inc, New York

Stumm W and Furrer G (1987) The dissolution of oxides and aluminum silicates: Examples of surface-coordination-controlled kinetics. *In:* Aquatic Surface Chemistry, Stumm W (ed) Wiley-Interscience, New York, p 197-219

Stumm W, Werhli B, Wieland E (1987) Surface complexation and its impact on geochemical kinetics. Croatica Chemica Acta 60:429-456

Sturchio NC, Antonio MR, Soderholm L, Sutton SR, Brannon JC (1998) Tetravalent uranium in calcite. Science 281:971-973

Sturchio NC, Chiarello RP, Cheng L, Lyman PF, Bedzyk MJ, Qian Y, You H, Yee D, Geissbuhler P, Sorensen LB, Liang Y, Baer DR (1997) Lead adsorption at the calcite-water interface: synchrotron X-ray standing wave and X-ray reflectivity studies. Geochim Cosmochim Acta 61:251-263

Sutton SR, Bajt S, Delaney J, Schulze D, Tokunaga TK (1995) Synchrotron X-ray fluorescence microprobe: Quantification and mapping of mixed valence state samples using micro-XANES. Rev Sci Instrum 66:1464-1467

Sutton S, Rivers M (1999) Hard X-ray synchrotron microprobe techniques and applications. *In*: Schulze DG, Stucki JW, Bertsch PM (eds) Synchrotron X-ray Methods in Clay Science, Clay Minerals Society Workshop Lectures 9. The Clay Minerals Society, Boulder, Colorado, p 146-163

Sutton SR, Rivers ML, Bajt S, Jones K (1993) Synchrotron X-ray fluorescence microprobe analysis with bending magnets and insertion devices. Nucl Instrum Methods Phys Res Sect B 75:553-558

Sutton SR, Rivers ML, Bajt S, Jones K, Smith JV (1994) Synchrotron X-ray fluorescence microprobe: a microanalytical instrument for trace element studies in geochemistry, cosmochemistry, and the soil and environmental sciences. Nucl Instrum Methods Phys Res Sect A 347:412-416

Sverjensky DA (1994) Zero-point-of-charge prediction from crystal chemistry and solvation theory. Geochim Cosmochim Acta 58:3123-3129

Sverjensky DA (2001) Interpretation and prediction of triple-layer model capacitances and the structure of the oxide-electrolyte-water interface. Geochim Cosmochim Acta 65:3643-3655

Sverjensky DA, Sahai N (1996) Theoretical prediction of single-site surface protonation equilibrium constants for oxides and silicates in water. Geochim Cosmochim Acta 60:3773-3797

Sylwester ER, Allen PG, Zhao P, Viani BE (2000a) Interactions of uranium and neptunium with cementitious materials studied by XAFS. Materials Research Society Symposium Proceedings 608 (Scientific Basis for Nuclear Waste Management XXIII):307-312

Sylwester ER, Hudson EA, Allen PG (2000c) Surface interactions of actinide ions with geologic materials studied by XAFS. Mat Res Soc Symp Proc 590:9-16

Sylwester ER, Hudson EA, Allen PG (2000b) The structure of uranium (VI) sorption complexes on silica, alumina, and montmorillonite. Geochim Cosmochim Acta 64: 2431-2438

Szulczewski MD, Helmke PA, Bleam WF (1997) Comparison of XANES analyses and extractions to determine chromium speciation in contaminated soils. Environ Sci Technol 31:2954-2959

Szulczewski MD, Helmke PA, Bleam WF (2001) XANES spectroscopy studies of Cr(VI) reduction by thiols in organosulfur compounds and humic substances. Environ Sci Technol 35:1134-1141

Taghiei MM, Huggins FE, Shah N, Huffman GP (1992) *In situ* X-ray absorption fine structure spectroscopy investigation of sulfur functional groups in coal during pyrolysis and oxidation. Energy Fuels 6:293-300

Tait CD, Clark DL, Conradson SD, Donohoe RJ, Gordon JC, Gordon PL, Keogh DW, Konze WV, Morris DE (2001) Effect of aluminate on the speciation of the actinides under tank waste conditions. Abstracts of Papers, 222nd ACS National Meeting, Chicago, IL, United States, August 26-30, 2001

Tamura N, Celestre RS, MacDowell AA, Padmore HA, Spolenak R, Valek BC, Meier Chang N, Manceau A, Patel JR (2002) Submicron X-ray diffraction and its applications to problems in materials and environmental science. Rev Sci Instrum 73 (3, Pt. 2):1369-1372

Taylor RW, Shen S, Bleam WF, Tu S-I (2000) Chromate removal by dithionite-reduced clays: evidence from direct X-ray adsorption near edge spectroscopy (XANES) of chromate reduction at clay surfaces. Clays Clay Minerals 48:648-654

Templeton AS, Ostergren JD, Trainor TP, Foster AL, Traina SJ, Spormann AM, Brown GE Jr (1999) XAFS and XSW studies of the distribution and chemical speciation of Pb sorbed to biofilms on α-Al_2O_3 and α-FeOOH surfaces. J Synchrotron Rad 6:642-644

Templeton AS, Spormann AM, Brown GE Jr (2002c) Speciation of Pb sorbed by *Burkholderia cepacia*/goethite composites. Environ Sci Technol (submitted)

Templeton AS, Trainor TP, Traina SJ, Spormann AM, Brown GE Jr (2001) Pb(II) distribution at biofilm-metal oxide interfaces. Proc Nat Acad Sci USA 98:11897-11902

Templeton AS, Trainor TP, Spormann AM, Brown GE Jr (2002b) Selenium speciation and partitioning within *B. cepacia* biofilms formed on metal oxide surfaces. Geochim Cosmochim Acta (submitted)

Templeton AS, Trainor TP, Spormann AM, Newville M, Sutton S, Dohnalkova A, Gorby Y, Brown GE Jr (2002a) Sorption *vs.* biomineralization of Pb(II) within *Burkholderia cepacia* biofilms on alumina. Environ Sci Technol (submitted).

Teng HH, Fenter P, Cheng L, Sturchio NC (2001) Resolving orthoclase dissolution processes with atomic force microscopy and X-ray reflectivity. Geochim Cosmochim Acta 65:3459-3474

Teo BK (1986) EXAFS: Basic Principles and Data Analysis. Springert-Verlag, Berlin

Teschke O, Ceotte G, de Souza EF (2000) Interfacial aqueous solutions dielectric constant measurements using atomic force microscopy. Chem Phys Lett 326:328-334

Thevuthasan S, Kim YJ, Yi SI, Chambers SA, Morais J, Denecke R, Fadley CS, Liu P, Kendelewicz T, Brown GE Jr (1999) Surface structure of MBE-grown α-Fe_2O_3(0001) by intermediate-energy X-ray photoelectron diffraction. Surf Sci 425:276-286

Thiel PA, Madey TF (1987) The interaction of water with solid surfaces: fundamental aspects. Surf Sci Rep 7:211-385

Thompson AC, Underwood JH, Wu Y, Giauque RD, Jones KW, Rivers ML (1988) Elemental measurements with an X-ray microprobe of biological and geological samples with femtogram sensitivity. Nucl Instrum Meth Phys Res A 266:318-323

Thompson HA, Brown GE Jr, Parks GA (1995) Low and ambient temperature XAFS study of U(VI) in solids and aqueous solutions. Physica B 208 & 209:167-168

Thompson HA, Brown GE Jr, Parks GA (1997) XAFS spectroscopic study of uranyl coordination in solids and aqueous solution. Am Mineral 82:483-496

Thompson HA, Parks GA, Brown GE Jr (1998) Structure and composition of uranium(VI) sorption complexes at the kaolinite-water interface. *In:* Adsorption of Metals by Geomedia: Variables, Mechanisms, and Model Applications, Jenne EA (ed) Academic Press, p 349-370

Thompson HA, Parks GA, Brown GE Jr (1999a) Ambient-temperature synthesis, evolution, and characterization of cobalt-aluminum hydrotalcite-like solids. Clays Clay Minerals 47:425-438

Thompson HA, Parks GA, Brown GE Jr (1999b) Dynamic interactions of dissolution, surface adsorption, and precipitation in an aging cobalt(II)-clay-water system. Geochim Cosmochim Acta 63:1767-1779

Thompson HA, Parks GA, Brown GE Jr (2000) Formation and release of cobalt(II) sorption and precipitation products in aging kaolinite-water slurries. J Colloid Interface Sci 222:241-253

Tipping E (1981) The adsorption of aquatic humic substances by iron oxides. Geochim Cosmochim Acta 45:191-199

Tipping E, Cooke D (1982) The effects of adsorbed humic substances on the surface charge of goethite (α-FeOOH) in freshwaters. Geochim Cosmochim Acta 49:75-80

Tokunaga TK, Brown GE Jr, Pickering IJ, Sutton SR, Bajt S (1997) Selenium transport between ponded waters and sediments. Environ Sci Technol 31:1419-1425

Tokunaga TK, Pickering IJ, Brown GE Jr (1996) X-ray absorption spectroscopy studies of selenium transformations in ponded sediments. Soil Sci Soc Am J 60:781-790

Tokunaga TK, Sutton SR, Bajt S (1994) Mapping of selenium concentrations in soil aggregates with synchrotron X-ray fluorescence microprobe. Soil Sci 158:421-434

Tokunaga TK, Sutton SR, Bajt S, Nuessle P, Shea-McCarthy G (1998) Selenium diffusion and reduction at the water-sediment boundary: Micro-XANES spectroscopy of reactive transport. Environ Sci Technol 32:1092-1098

Tomboulian DH, Hartman PL (1956) Spectral and angular distribution of ultraviolet radiation from the 300-m.e.v. Cornell synchrotron. Phys Rev 102:1423-1447

Toney MF, Howard JN, Richer J, Borges GL, Gordon JG, Melroy OM, Wiesler DG, Yee D, Sorensen LB (1994) Voltage-dependent ordering of water molecules at an electrode-electrolyte interface. Nature 368:444-446

Tonner BP, Droubay T, Denlinger J, Meyer-Ilse W, Warwick T, Rothe J, Kneedler E, Pecher K, Nealson K, Grundl T (1999) Soft X-ray spectroscopy and imaging of interfacial chemistry in environmental specimens. Surf Interface Anal 27:247-258

Tournassat C, Charlet L, Bosbach D, Manceau A (2002) Arsenic(III) oxidation by birnessite and precipitation of manganese(II) arsenate. Environmen Sci Technol 36:493-500

Towle SN, Bargar JR, Brown GE Jr, Parks GA (1997) Surface precipitation of Co(II)(aq) on Al_2O_3. J Colloid Interface Sci 187:62-82

Towle SN, Bargar JR, Brown GE Jr, Parks GA, Barbee TW Jr (1995a) Effect of surface structure on the adsorption of Co(II) on α-Al_2O_3: a glancing angle XAFS study. Materials Research Society Symposium Proceedings 357 (Structure and Properties of Interfaces in Ceramics):23-28

Towle SN, Bargar JR, Persson P, Brown GE Jr, Parks GA (1995b) XAFS study of Co(II) sorption at the α-Al_2O_3-water interface. Physica B 208 & 209:439-440

Towle SN, Brown GE Jr, Parks GA (1999a) Sorption of Co(II) on metal oxide surfaces: I. Identification of specific binding sites of Co(II) on (110) and (001) surfaces of TiO_2 (rutile) by grazing-incidence XAFS spectroscopy. J Colloid Interface Sci 217:299-311

Towle SN, Bargar JR, Brown GE Jr, Parks GA (1999b) Sorption of Co(II) on metal oxide surfaces: II. Identification of Co(II)(aq) adsorption sites on (1-102) and (0001) surfaces of α-Al_2O_3 by grazing-incidence XAFS spectroscopy. J Colloid Interface Sci 217:312-321

Trainor TP (2001) X-ray Scattering and X-ray Absorption Spectroscopy Studies of the Structure and Reactivity of Aluminum Oxide Surfaces. Ph.D. Dissertation. Department of Geological & Environmental Sciences, Stanford University, Stanford, CA, USA

Trainor TP, Brown GE Jr, Parks GA (2000) Adsorption and precipitation of aqueous Zn(II) on alumina powders. J Colloid. Interface Sci 231:359-372

Trainor TP, Eng P, Brown GE Jr, Robinson IK, De Santis M (2002b) Crystal truncation rod diffraction study of the clean and hydrated α-Al_2O_3 (1-102) surface. Surf Sci 496:238-250

Trainor TP, Eng P, Brown GE Jr, Wayuchunas GA, Newville M, Sutton S, Rivers M (2002d) Crystal truncation rod diffraction study of the hydrated α-Fe_2O_3 (0001) surface. 2001 Activity Report, Advanced Photon Source

Trainor TP, Fitts JP, Grolimund D, Bargar JR, Brown GE Jr (1999) Grazing-incidence XAFS studies of aqueous Zn(II) on sapphire single crystals. J Synchrotron Rad 6:618-620

Trainor TP, Fitts JP, Templeton AS, Grolimund D, Brown GE Jr (2002a) Grazing-incidence XAFS study of aqueous Zn(II) sorption on α-Al_2O_3 single crystals. J Colloid Interface Sci 244:239-244

Trainor TP, Templeton AS, Brown GE Jr, Parks GA (2002c) Application of the long-period X-ray standing wave technique to the analysis of surface reactivity: Pb(II) sorption at α-Al_2O_3/aqueous solution interfaces in the presence and absence of Se(VI). Langmuir 18:5782-5791

Trivedi P, Axe L, Tyson TA (2001a) An analysis of Zn sorption to amorphous versus crystalline iron oxides using XAS. J Colloid Interface Sci 244:230-238

Trivedi P, Axe L, Tyson TA (2001b) XAS Studies of Ni and Zn sorbed to hydrous manganese oxide. Environmen Sci Technol 35:4515-4521

Trivedi PI, Sparks DL, Dyer JA, Pandya K (2002) Spectroscopic evaluation of the role of background anions on metal sorption to ferrihydrite. Abstracts of Papers, 223rd ACS National Meeting, Orlando, FL, United States, April 7-11, 2002, GEOC-131

Tröger L, Arvanitis D, Baberschke K, Michaelis H, Grtimm U, Zschech E (1992) Ful correction of the self-absorption in soft-fluorescence extended X-ray absorption fine structure. Phys Rev B 46:3283-3289

Um W, Papelis C (2001) Sorption mechanisms of Sr(II) and Pb(II) on zeolitized tuffs from the Nevada test site. Abstracts of Papers, 222nd ACS National Meeting, Chicago, IL, United States, August 26-30, 2001

Vairavamurthy A (1998) Using X-ray absorption to probe sulfur oxidation states in complex molecules. Spectrochim Acta, Part A: Molecular and Biomolecular Spectroscopy 54A:2009-2017

Vairavamurthy A, Manowitz B, Luther GW III, Jeon Y (1993) Oxidation state of sulfur in thiosulfate and implications for anaerobic energy metabolism. Geochim Cosmochim Acta 57:1619-1623

Vairavamurthy A, Wang S (2002) Organic nitrogen in geomacromolecules: Insights on speciation and transformation with K-edge XANES spectroscopy. Environ Sci Technol 36:3050-3056

Vairavamurthy A, Zhou W, Eglinton T, Manowitz B (1994) Sulfonates: a novel class of organic sulfur compounds in marine sediments. Geochim Cosmochim Acta 58:4681-4687

Van Cappellen P, Charlet L, Stumm W, Wersin P (1993) A surface complexation model of the carbonate mineral-aqueous solution interface. Geochim Cosmochim Acta 57:3505-3518

Van der Lelie D, Schwitzguébel JP, Galss DJ, Vangronsveld J, Baker A (2001) Assessing phytoremediation's progress in the United States and Europe. Environ Sci Technol 35:446-452

Waite TD, Davis JA, Payne TE, Waychunas GA, Xu N (1994) Uranium(IV) adsorption to ferrihydrite: application of a surface complexation model. Geochim Cosmochim Acta 58:5465-78

Wallen SL, Palmer BJ, Fulton JL (1998) The ion pairing and hydration structure of Ni^{2+} in supercritical water at 425°C determined by X-ray absorption fine structure and molecular dynamics studies. J Chem Phys 108:4039-4046

Wallen SL, Pfund DM, Fulton JL, Yonker CR, Newville M, Ma Y (1996) High-pressure, capillary X-ray absorption fine structure cell for studies of liquid and supercritical fluid solutions. Rev Sci Instrum 67:1-3

Wang X-G, Weiss W, Shaikhutdinov Sh K, Ritter M, Petersen M, Wagner F, Schlögl R, Scheffler M (1998) The hematite (α-Fe_2O_3) (0001) surface: evidence for domains of distinct chemistry. Phys Rev Lett 81:1038-1041

Wang X-G, Chaka A, Scheffler M (2000) Effect of the environment on the α-Al_2O_3 (0001) surface structures. Phys Rev Lett 84:3650-3653

Wasserman SR (1997) The analysis of mixtures: Application of principal component analysis to XAS spectra. J de Physique IV **7** (Colloque C2, X-Ray Absorption Fine Structure, Vol. 1):203-205

Wasserman SR, Allen PG, Shuh DK, Bucher JJ, Edelstein NM (1999) EXAFS and principal component analysis: a new shell game. J Synchrotron Rad 6:284-286

Wasserman SR, Winans RE, MacBeth R (1996) Iron species in Argonne premium coal samples. An investigation using X-ray absorption spectroscopy. Energy Fuels 10:392-400

Waychunas GA, Apted MJ, Brown GE Jr (1983) X-ray K-edge absorption spectra of Fe minerals and model compounds: I. Near edge structure. Phys Chem Minerals 10:1-9

Waychunas GA, Brown GE Jr (1990) Polarized X-ray absorption spectroscopy of metal ions in minerals: identification of near edge electronic transitions and scattering resonances and application to site geometry determinations. Phys Chem Minerals 17:420-430

Waychunas GA, Brown GE Jr (1994) Fluorescence yield XANES and EXAFS experiments: application to highly dilute and surface samples. Advances in X-ray Analysis 37:607-617

Waychunas GA, Brown GE Jr, Apted MJ (1986) X-ray K-edge absorption spectra of Fe minerals and model compounds: II. EXAFS. Phys Chem Minerals 13:31-47

Waychunas GA, Davis JA, Fuller CC (1995) Geometry of sorbed arsenate on ferrihydrite and crystalline FeOOH: re-evaluation of EXAFS results and topological factors in predicting sorbate geometry, and evidence for monodentate complexes. Geochim Cosmochim Acta 59:3655-61

Waychunas GA, Davis JA, Reitmeyer R (1999) GIXAFS study of Fe^{3+} sorption and precipitation on natural quartz surfaces. J Synchrotron Rad 6:615-617

Waychunas GA, Fuller CC, Davis JA (2002a) Surface complexation and precipitate geometry for aqueous Zn(II) sorption on ferrihydrite I: X-ray absorption extended fine structure spectroscopy analysis. Geochim Cosmochim Acta 66:1119-1137

Waychunas GA, Fuller CC, Davis JA, Rehr JJ (2002b) Surface complexation and precipitate geometry for aqueous Zn(II) sorption on ferrihydrite: II. XANES analysis. Geochim Cosmochim Acta (in press).

Waychunas GA, Fuller CC, Rea BA, Davis JA (1996) Wide angle X-ray scattering (WAXS) study of "two-line" ferrihydrite structure: effect of arsenate sorption and counterion variation and comparison with EXAFS results. Geochim Cosmochim Acta 60:1765-1781

Waychunas GA, Rea BA, Fuller CC, Davis JA (1993) Surface chemistry of ferrihydrite: Part 1. EXAFS studies of the geometry of coprecipitated and adsorbed arsenate. Geochim Cosmochim Acta 57:2251-69

Webb SM, Gaillard J-F, Jackson BE, Stahl DA (2001) An EXAFS study of zinc coordination in microbial cells. J Synchrotron Rad 8:943-945

Webb SM, Leppard GG, Gaillard J-F (2000) Zinc speciation in a contaminated aquatic environment: Characterization of environmental particles by analytical electron microscopy. Environ Sci Technol 34:1926-1933

Weesner FJ, Bleam WF (1997) X-ray absorption and EPR spectroscopic characterization of adsorbed copper(II) complexes at the boehmite (AlOOH) surface. J Colloid Interface Sci 196:79-86

Weesner FJ, Bleam WF (1998) Binding characteristics of Pb^{2+} on anion-modified and pristine hydrous oxide surfaces studied by electrophoretic mobility and X-ray absorption spectroscopy. J Colloid Interface Sci 205:380-389

Wehrli B, Friedl G, Manceau A (1995) Reaction rates and products of manganese oxidation at the sediment-water interface. Advances in Chemistry Series 244 (Aquatic Chemistry):111-134

Weinelt M, Wassdahl N, Wiell T, Karis O, Hasselstrom J, Bennich P, Nilsson A, Stöhr J, Samant M (1998) Electronic structure of benzene on Ni(100) and Cu(110): An X-ray-spectroscopy study. Phys Rev B 58:7351-7360

Weiss W, Ranke W (2002) Surface chemistry and catalysis on well-defined epitaxial iron-oxide layers. Prog Surf Sci 70:1-151

Welter E, Calmano W, Mangold S, Troger L (1999) Chemical speciation of heavy metals in soils by use of XAFS spectroscopy and electron microscopical techniques. Fresenius J Anal Chem 364:238-244

Westre TE, Di Cicco A, Filipponi A, Natoli CR, Hedman B, Solomon EI, Hodgson KO (1995) GNXAS, a multiple-scattering approach to EXAFS analysis: Methodology and applications to iron complexes. J Am Chem Soc 117:1566-1583

Wharton MJ, Atkins B, Charnock JM, Livens FR, Pattrick RAD, Collison D (1999) An X-ray absorption spectroscopy study of the coprecipitation of Tc and Re with mackinawite (FeS). Appl Geochem 15:347-354

White EW, Gibbs GV (1967) Structural and chemical effects on the silicon K_β X-ray line for silicates. Am Mineral 52:985-993

White EW, Gibbs GV (1968) Structural and chemical effects on the aluminum K_β X-ray emission band among aluminum-containing silicates and aluminum oxides. Am Mineral 54:931-936

Wielinga B, Mizuba MM, Hansel CM, Fendorf S (2001) Iron promoted reduction of chromate by dissimilatory iron-reducing bacteria. Environ Sci Technol 35:522-527

Wilke M, Farges F, Petit P-E, Brown GE Jr, Martin F (2001) Oxidation state and coordination of Fe in minerals: an Fe K-XANES study. Am Mineral 86:714-730

Williams GD, Soper AK, Skipper NT, Smalley MV (1998) High resolution structural study of an electrical double layer by neutron diffraction. J Phys Chem B 102:8945-8949

Wilson KR, Tobin JG, Ankudinov AL, Rehr JJ, Saykally RJ (2000) Extended X-ray absorption fine structure from hydrogen atoms in water. Phys Rev Lett 85:4289-4292

Williams CW, Blaudeau J-P, Sullivan JC, Antonio MR, Bursten B, Soderholm L (2001) The coordination geometry of Np(VII) in alkaline solution. J Am Chem Soc 123:4346-4347

Williams GP, Cerrina F, McGovern IT, Lapeyre GJ (1979) Surface structure effects via photoexcited core electron diffraction for sodium on nickel. Solid State Commun 31:15-18

Winick H (1987) Synchrotron radiation. Scientific American 257(5):88-99

Wong J, Froeba M, Frahm R (1995) A critical assessment of the QEXAFS method. Rev Sci Instrum 66:1517-1520

Xia K, Bleam WF, Helmke PA (1997a) Studies of the nature of Cu^{2+} and Pb^{2+} binding sites in soil humic substances using X-ray absorption spectroscopy. Geochim Cosmochim Acta 61:2211-2221.

Xia K, Bleam WF, Helmke PA (1997b) Studies of the nature of binding sites of first row transition elements bound to aquatic and soil humic substances using X-ray absorption spectroscopy. Geochim Cosmochim Acta 61:2223-2235

Xia K, Mehadi A, Taylor RW, Bleam WF (1997c) X-ray absorption and electron paramagnetic resonance studies of Cu(II) sorbed to silica: surface-induced precipitation at low surface coverages. J Colloid Interface Sci 185:252-257

Xia K, Skyllberg UL, Bleam WF, Bloom PR, Nater EA, Helmke PA (1999) X-ray absorption spectroscopic evidence for the complexation of Hg(II) by reduced sulfur in soil humic substances. Environ Sci Technol 33:257-261

Xia K, Weesner F, Bleam WF, Bloom PR, Skyllberg UL, Helmke PA (1998) XANES studies of oxidation states of sulfur in aquatic and soil humic substances. Soil Sci Soc Am J 62:1240-1246

Yamaguchi T, Lindquist O, Boyce JB, Claeson T (1984) Determination of the hydration structure of silver ions in aqueous silver perchlorate and nitrate solutions from EXAFS using synchrotron radiation. Acta Chem Scand A 38:423-428

Yamauchi H, Fowler BA (1994) Toxicity and metabolism of inorganic and methylated arsenicals. *In:* Arsenic in the Environment. Part 2. Human Health and Ecosystem Effects. Nriagu JO (ed) John Wiley & Sons, New York, p 35-53

Yeganeh MS, Dougal SM, Pink HS (1999) Vibrational spectroscopy of water at liquid/solid interfaces: Crossing the isoelectric point of a solid surface. Phys Rev Lett 83:1179-1182

Zabinsky SI, Rehr JJ, Ankudinov A, Albers RC, Eller MJ (1995) Multiple-scattering calculations of X-ray-absorption spectra. Phys Rev B 52:2995-3009

Zachara JM, Ainsworth CC, Brown GE Jr, Catalano JG, McKinley JP, Oafoku O, Smith SC, Szecsody JE, Traina SJ, Warner JA (2002) Chromium speciation and mobility in a high level nuclear waste vadose zone plume. Geochim Cosmochim Acta (submitted)

Zegenhagen J (1993) Surface structure determination with X-ray standing waves. Surf Sci Reports 18:199-271

Zhang Z, Cheng L, Fenter P, Sturchio NC, Bedzyk M, Machesky ML, Wesolowski DJ (2001) Ionic strength dependence of Zn^{2+} and Sr^{2+} ion adsorption at the rutile/aqueous interface using X-ray standing waves. Preprints and Extended Abstr ACS Nat Meet, Am Chem Soc, Div Environ Chem 41:335-336

Zhao P, Allen PG, Sylwester ER, Viani BE (2000) The sorption of uranium(VI) and neptunium(V) onto hydrothermally altered concrete. Radiochim Acta 88:729-736

Zhou W, Hesterberg D, Hansen PD, Hutchison KJ, Sayers DE (1999) Stability of copper sulfide in a contaminated soil. J Synchrotron Rad 6: 630-632

Zhu Z, Gu Y, George SC, Wilson MA, Huggins FE, Huffman GP (1995) A study of sulfur and nitrogen in two high sulfur coals. Coal Sci Technol 24:1681-1684

On the following pages....

APPENDIX — TABLES 1 AND 2

Table 1. Summary of XAFS spectroscopy studies of metal(loid) ion sorption complexes at mineral/solution interfaces in model systems in the absence of complexing ligands[1] (arranged in increasing order of atomic number of the sorbate ion). Individual references should be consulted for pH, surface coverages (Γ), metal concentration, ionic strength values, background electrolyte type, and other experimental variables. The dominant sorbate geometry or phase listed below should not be generalized beyond the range of variables considered in each study.

Sorbate	Sorbent	Dominant Sorbate Geometry or Phase	Reference
P(V)	goethite	phosphate tetrahedron	Hesterberg et al. (1999)
Cr(VI)	magnetite	Cr(III)OOH precipitate	Peterson et al. (1996)
	magnetite	Cr(III)OOH precipitate	Peterson et al. (1997)
	magnetite (100)*	am-(Cr,Fe)OOH precipitate	Kendelewicz et al. (2000)
	goethite	N.D.**	Bidoglio et al. (1993)
	goethite	IS-mono and bidentate	Fendorf et al. (1997)
	hematite (0001)*	IS-multinuclear or precipitate	Grolimund et al. (1999)
	am-FeS	(Cr(III),Fe)(OH)$_3$ precipitate	Patterson et al. (1997)
Cr(III)	am-SiO$_2$	IS-monodentate	Fendorf et al. (1994)
	γ-Al$_2$O$_3$	IS-multinuclear	Fitts et al. (2000)
	α-Al$_2$O$_3$ (0001)*	IS-bidentate	Grolimund et al. (1999)
	HFO	(Cr,Fe)OOH precipitate	Charlet and Manceau (1992a)
	goethite	(Cr,Fe)OOH precipitate	Charlet and Manceau (1992b)
	hematite (0001)*	IS-multinuclear	Grolimund et al. (1999)
	HMO	IS-multidentate (initial) OS (CrO$_4^{2-}$) (seconds later)	Manceau and Charlet (1992)
	Na-buserite	OS (initial), then CrO$_4^{2-}$)	Silvester et al. (1995)
Fe(III)	quartz ($10\bar{1}1$)*	IS-bidentate	Waychunas et al. (1999)
Co(II)	α-SiO$_2$	IS-multinuclear	O'Day et al. (1996)
	α-SiO$_2$	Co(II)-phyllosilicate	Manceau et al. (1999b)
	γ-Al$_2$O$_3$	IS-multinuclear	Chisholm-Brause et al. (1989a,1990a)
	γ-Al$_2$O$_3$	IS-multinuclear	Chisholm-Brause et al. (1991)
	γ-Al$_2$O$_3$	Co(II)-Al LDH precipitate	d'Espinose de la Caillerie et al. (1995a)
	α-Al$_2$O$_3$	IS-multinuclear	Hayes and Katz (1996)
	α-Al$_2$O$_3$ (0001)*	IS-multidentate	Towle et al. (1995a)
	α-Al$_2$O$_3$	Co(II)-Al LDH precipitate	Towle et al. (1995b,1997)
	α-Al$_2$O$_3$ (0001)*	IS-tridentate	Towle et al. (1999b)
	α-Al$_2$O$_3$ (1-102)*	IS-tetradentate	Towle et al. (1999b)
	TiO$_2$ (rutile)	IS-multinuclear	Chisholm-Brause et al. (1990a)
	TiO$_2$ (rutile)	IS-multinuclear	O'Day et al. (1996)
	TiO$_2$ (rutile) (110)*	IS-mono- & bidentate	Towle et al. (1999a)
	TiO$_2$ (rutile) (001)*	IS-bi- & tridentate	Towle et al. (1999a)
	buserite	Co(III)-sub for Mn(III)	Manceau et al. (1997)
	ZnO	IS-bidentate([VII]Co+[IV]Co)	Persson et al. (1995)
	ZnS	IS-bidentate([IV]Co to S+O) Co(OH)$_2$-like precipitate at high Γ	Persson et al. (1995)
	calcite	lattice substitution	Lamble et al. (1997)
	calcite	lattice substitution	Reeder et al. (1999)
	kaolinite	IS (edge sites)	Charlet and Manceau (1994)

Sorbate	Sorbent	Dominant Sorbate Geometry or Phase	Reference
Co(II)	kaolinite	IS-multinuclear	O'Day et al. (1994b)
	kaolinite	IS-bidentate	O'Day et al. (1994c)
	kaolinite	OS (charge sites-low pH) Co(II)-Al LDH precipitate (high pH)	Chen and Hayes (1999)
	kaolinite	Co(II)-Al LDH precipitate	Thompson et al. (1999a,b, 2000)
	kaolinite	Co(II)-Al LDH precipitate	Thompson et al. (2000)
	illite	OS (charge sites-low pH) Co(II)-Al LDH precipitate (high pH)	Chen and Hayes (1999)
	hectorite	IS (edge sites)	Schlegel et al. (1998, 1999a)
	hectorite	OS (charge sites-low pH) Co(II)-Al LDH precipitate (high pH)	Chen and Hayes (1999)
	hectorite	IS (edge sites)	Schlegel et al. (1998,1999a,b)
	smectite	OS (charge sites) (low pH) IS-polynuclear (higher pH and I.S.)	Papelis and Hayes (1996)
	smectite	OS (charge sites-low pH) Co(II)-Al LDH precipitate (high pH)	Chen and Hayes (1999)
	sepiolite	IS (exchange site)	Fukushima and Okamoto (1987)
	sepiolite	IS (edge site)	Manceau and Decarreau (1988) Charlet and Manceau (1994)
	humic materials	IS (1 C 2nd-neighbor)	Xia et al. (1997b)
Ni(II)	am-SiO_2	Ni-hydrosilicate precipitate	Bonneviot et al. (1989a,b)
	am-SiO_2	α-Ni(OH)$_2$ precipitate	Scheinost and Sparks (2000)
	γ-Al$_2$O$_3$	Ni(II)-Al LDH precipitate	d'Espinose de la Caillerie et al. (1995a)
	Al-oxides	Ni(II)-Al LDH precipitate	Scheidegger et al. (1996,1997,1998)
	gibbsite	Ni(II)-Al LDH precipitate	Scheckel et al. (2000)
	gibbsite	Ni(II)-Al LDH precipitate	Scheinost and Sparks (2000)
	gibbsite/ am-SiO_2 mixture	α-Ni(OH)$_2$ (low Γ) Ni-phyllosilicate (high Γ)	Scheckel and Sparks (2000)
	ferrihydrite	Ni-coprecipitate	Ford et al. (1999a)
	HMO	OS-mononuclear	Trivedi et al. (2001b)
	kaolinite	Ni(II)-Al LDH precipitate	Eick and Fendorf (1998)
	pyrophyllite	Ni(II)-Al LDH precipitate	Scheidegger et al. (1996,1997,1998)
	pyrophyllite	Ni(II)-Al LDH precipitate	Elzinga and Sparks (1999)
	pyrophyllite	Ni(II)-Al LDH precipitate	Ford et al. (1999b)
	pyrophyllite	Ni(II)-Al LDH precipitate	Scheinost and Sparks (2000)
	pyrophyllite	Ni(II)-Al LDH precipitate	Scheckel et al. (2000)
	smectite	Ni(II)-Al LDH precipitate	Scheidegger et al. (1996,1997,1998)
	smectite	IS (edge sites)	Elzinga and Sparks (1999)
	smectite	IS (edge sites)	Dähn et al. (2001,2002a)
	pyrophyllite/smectite mixture	Ni(II)-Al LDH precipitate + IS (edge sites)	Elzinga and Sparks (1999)
	talc	α-Ni(OH)$_2$ precipitate	Scheinost and Sparks (2000)
	talc	α-Ni(OH)$_2$ precipitate	Scheckel et al. (2000)
	soil clay fraction	Ni(II)-Al LDH precipitate	Roberts et al. (1999)
	humic materials	IS (2 C 2nd-neighbors)	Xia et al. (1997b)

Sorbate	Sorbent	Dominant Sorbate Geometry or Phase	Reference
Cu(II)	am-SiO$_2$	IS-monodentate (mono + dimers) + Cu(OH)$_2$ precipitate	Cheah et al. (1997, 1998)
	silica	Cu(OH)$_2$ clusters	Xia et al. (1997c)
	γ-Al$_2$O$_3$	IS-mono or bidentate	Cheah et al. (1997, 1998)
	α-Al$_2$O$_3$ (0001)*	IS-bidentate	Fitts et al. (1999a)
	boehmite	IS (low Γ) OS (high Γ)	Weesner and Bleam (1997)
	goethite	multidentate polymer (pH 8)	Bochatay et al. (1997)
	ferrihydrite	IS-bidentate (monomers + dimers)	Scheinost et al. (2001)
	TiO$_2$ (anatase)	IS-dimers	Cheah et al. (2000)
	birnessite	IS-tridentate, interlayer	Manceau et al. (2002b)
	muscovite (001)*	IS (at defects)	Farquhar et al. (1996, 1997)
	biotite (001)*	OS, IS, and ion exchange	Farquhar et al. (1997)
	feldspar (001) or (010)*	OS, IS, and ion exchange	Farquhar et al. (1997)
	montmorillonite	OS + edge sites(mono+di)	Morton et al. (2001)
	ZnSe (111)*	IS or covellite-like w/xanthate	Pattrick et al. (1999)
	humic materials	IS (4 C 2nd-neighbors)	Xia et al. (1997a, 1997b)
Zn(II)	γ-Al$_2$O$_3$	Zn(II)-Al LDH precipitate	d'Espinose de la Caillerie et al. (1995b)
	γ-Al$_2$O$_3$	IS-bidentate [IV]Zn (low Γ) Zn(II)-Al precipitate (high Γ)	Trainor et al. (2000)
	α-Al$_2$O$_3$(0001)*	IS-bidentate	Trainor et al. (1999, 2002a)
	ferrihydrite	IS-bidentate([IV]Zn) (pH 6)	Waychunas et al. (1993, 1996, 2002a,b)
	HFO	OS-[VII]Zn	Trivedi et al. (2001a)
	goethite	IS-[IV]Zn	Trivedi et al. (2001a)
	goethite	IS-bidentate [VII]Zn	Juillot et al. (2002)
	HFO (2L)	IS-bidentate [IV]Zn	Juillot et al. (2002)
	birnessite	IS-[VII]Zn (high Γ)	Silvester et al. (1997)
	birnessite	IS-[IV]Zn (low Γ) tridentate, interlayer	Manceau et al. (2002b)
	γ-MnOOH	IS-[IV]Zn (pH 6.2) IS-[VII]Zn (pH 9.9)	Bochatay and Persson (2000)
	HMO	OS-mononuclear	Trivedi et al. (2001b)
	pyrophyllite	Zn(II)-Al LDH precipitate	Ford and Sparks (2000)
	hectorite	IS (edge sites)	Manceau et al. (2000)
	hectorite	IS (edge sites)	Schlegel et al. (2001a)
	hectorite	OS (exchange sites) (initial) IS (edge sites) (slow uptake)	Schlegel et al. (2001b)
	calcite	lattice substitution	Lamble et al. (1997)
	calcite	lattice substitution	Reeder et al. (1999)
	Ca-silicate hydrates	IS-[IV]Zn linked to silicate chain	Rose et al. (2001)
	CuFeS$_2$	IS (to Fe sites)	England et al. (1999b)
	peat humic acid	IS-[VII]Zn	Juillot et al. (2002)
	humic substances	IS-[VII]Zn+[IV]Zn (low [Zn]) IS-[VII]Zn (intermediate [Zn]) OS (high [Zn])	Sarret et al. (1997)
	soil humic materials	IS (2 S,4 O 1st-neighbors) (1-2 C 2nd-neighbors)	Xia et al. (1997b)
	Phaseolus vulgaris	Zn-phosphate-dihydrate	Sarret et al. (2001)

Sorbate	Sorbent	Dominant Sorbate Geometry or Phase	Reference
Zn(II)	*Penicillium chrysogenum*	IS (bonded to 4 PO_4 groups) + bonded to carboxyl groups at site saturation	Sarret et al. (1998a)
	Obligate anaerobe X	IS (bonded to 4S)	Webb et al. (2000)
	Obligate anaerobe Y	IS (bonded to 3S+1O/N)	Webb et al. (2000)
	Obligate anaerobe Z	IS (bonded to 2S+2O/N)	Webb et al. (2000)
	Obligate anaerobe Mt	IS (bonded to 4 O/N)	Webb et al. (2000)
	Obligate anaerobe C1	IS (bonded to 5 O/N)	Webb et al. (2000)
As(V)	gibbsite	IS-bidentate	Foster et al. (1998a)
	γ-Al_2O_3	IS-bidentate	Arai et al. (2001)
	ferrihydrite	IS-bidentate	Waychunas et al. (1993,1995)
	HFO	IS-bidentate	Manceau (1995)
	goethite	IS-bidentate	Waychunas et al. (1993,1995)
	goethite	IS-monodentate (low Γ) IS-bidentate (higher Γ)	Fendorf et al. (1997)
	goethite	IS-bidentate	Foster et al. (1998a)
	birnessite	IS-bidentate	Foster et al. (2002)
	birnessite	IS-bidentate	Manning et al. (2002)
	ettringite	OS (in channels)	Myneni et al. (1997b)
As(III)	γ-Al_2O_3	IS-bidentate	Arai et al. (2001)
	goethite	IS-bidentate	Manning et al. (1998)
	anatase	oxidized to As(V)	Foster et al. (1998b)
	birnessite	$MnHAs(V)O_4{\cdot}H_2O$ precip.	Tournassat et al. (2002)
	kaolinite	oxidized to As(V)	Foster et al. (1998b)
Se(VI)	δ-Al_2O_3	IS	Papelis et al. (1995)
	goethite	OS	Hayes et al. (1987)
	HFO	IS-bidentate	Manceau and Charlet (1994)
	calcite	lattice substitution	Reeder et al. (1994)
	calcite	lattice substitution	Lamble et al. (1995)
	Burkholderia cepacia	reduced to Se(IV) + Se(0)	Templeton et al. (2002b)
Se(IV)	goethite	IS-bidentate	Hayes et al. (1987)
	HFO	IS-bidentate	Manceau and Charlet (1994)
	HMO	IS-bidentate	Foster et al. (2002)
	Bacillus subtilis	reduced to Se(0)	Buchanan et al. (1995)
Sr(II)	am-SiO_2	OS	Sahai et al. (2000), O'Day et al. (2000b)
	goethite	IS-bidentate	Collins et al. (1998)
	goethite	OS	Sahai et al. (2000), O'Day et al. (2000b)
	HFO	OS	Axe et al. (1998)
	HMO	IS-bidentate	Grolimund et al. (2000)
	HMO	OS	Axe et al. (2000)
	kaolinite	OS-mononuclear	Chen et al. (1998)
	kaolinite	IS-edge sites (low Γ) OS (higher Γ)	Parkman et al. (1998)
	kaolinite	OS	Chen and Hayes (1999)
	kaolinite	OS	Sahai et al. (2000), O'Day et al. (2000b)
	illite	OS-mononuclear	Chen et al. (1998)

Sorbate	Sorbent	Dominant Sorbate Geometry or Phase	Reference
Sr(II)	illite	OS	Chen and Hayes (1999)
	hectorite	OS-mononuclear	Chen et al. (1998)
	hectorite	OS	Chen and Hayes (1999)
	montmorillonite	OS-mononuclear	Chen et al. (1998)
	smectite	OS	Chen and Hayes (1999)
	calcite	lattice substitution (low Γ) strontianite precipitate (high Γ)	Parkman et al. (1998)
	zeolitized tuff	OS-mononuclear	Um and Papelis (2001)
Tc(VII)	FeS (mackinawite)	Tc(IV)S$_2$-like precipitate	Wharton et al. (1999)
	FeS (mackinawite)	Tc(IV)S$_2$-like precipitate	Bunker et al. (2002)
	green rust	Tc(IV)O$_2$-like precipitate	Bunker et al. (2002)
Tc(IV)	FeS (mackinawite)	Tc(IV)S$_2$-like precipitate	Wharton et al. (1999)
Cd(II)	δ-Al$_2$O$_3$	Cd-precipitate	Papelis et al. (1995)
	HFO	IS (edge sites)	Spadini et al. (1994)
	goethite	lattice substitution	Spadini et al. (1994)
	goethite	IS-bidentate	Randall et al. (1999)
	goethite	IS-mononuclear	Manceau et al. (2000e)
	cryptomelane	IS (in channels)	Randall et al. (1998)
	lepidocrocite	IS-bi- & tridentate	Randall et al. (1999)
	lepidocrocite	IS-bidentate	Manceau et al. (2000e)
	akaganeite	IS	Randall et al. (1999)
	schwertmannite	IS	Randall et al. (1999)
	γ-MnOOH	IS-bidentate (mononuclear)	Bochatay et al. (2000)
	feldspar(001) or (010)*	OS	Farquhar et al. (1997)
	muscovite(001)*	OS	Farquhar et al. (1997)
	biotite(001)*	OS, IS, and exchange sites	Farquhar et al. (1997)
	humic acid	IS	Liu et al. (2001)
	Bacillus subtilis	IS (bonded to carboxyl groups)	Kelly et al. (2001)
	Bacillus subtilis	IS (bonded to phosphoryl groups at pH < 4.4 and to carboxyl groups at higher pH)	Boyanov et al. (2002)
I(-I)	magnetite	IS (I$^-$ sorbed)	Fuhrmann et al. (1998)
	biotite	IS (IO$_3^-$ sorbed)	Fuhrmann et al. (1998)
	pyrite	IS (IO$_3^-$ reduced to I$_2$)	Fuhrmann et al. (1998)
Cs(I)	smectite	OS/IS	Kemner et al. (1997b)
	phlogopite	OS/IS	Kemner et al. (1997b)
Ba(II)	calcite	lattice substitution	Reeder et al. (1999)
Re(VII)	FeS (mackinawite)	Re-S-Fe precipitate	Wharton et al. (1999)
Re(IV)	FeS (mackinawite)	Re(IV)O$_2$-like precipitate	Wharton et al. (1999)
Au(III)	goethite	IS-bidentate	Berrodier et al. (1999)
	goethite	IS-bidentate	Berrodier et al. (2002)
Hg(II)	gibbsite	IS-bidentate	Kim et al. (2002a)
	goethite	IS-bidentate	Collins et al. (1999b)
	goethite	IS-bidentate	Kim et al. (2002a)
	humic acid	IS (bonded to S)	Hesterberg et al. (2001)
	humic material	IS (bonded to S)	Xia et al. (1999)

Sorbate	Sorbent	Dominant Sorbate Geometry or Phase	Reference
Tl(I)	δ-MnO$_2$	Tl$_2$O$_3$ precipitate	Bidoglio et al. (1993)
Pb(II)	γ-Al$_2$O$_3$	IS-monodentate	Chisholm-Brause (1989b, 1990b)
	α-Al$_2$O$_3$	IS-bidentate, multinuclear	Bargar et al. (1995)
	α-Al$_2$O$_3$	IS-bidentate+ dimers	Bargar et al. (1997a)
	γ-Al$_2$O$_3$	IS-bidentate	Strawn et al. (1998)
	α-Al$_2$O$_3$ ($1\bar{1}02$)*	IS-bidentate	Bargar et al. (1996, 1997c)
	α-Al$_2$O$_3$ (0001)*	OS (specifically adsorbed)	Bargar (1996, 1997c)
	boehmite	IS-multinuclear	Weesner and Bleam (1998)
	goethite	IS-mononuclear (low Γ)	Roe et al. (1991)
	goethite	IS-bidentate, mononuclear	Bargar et al. (1997b)
	goethite	IS-mononuclear	Weesner and Bleam (1998)
	goethite	IS-bidentate	Ostergren et al. (1999)
	goethite	IS-bidentate	Morin et al. (1999)
	HFO	IS-bidentate IS-multinuclear (high Γ)	Manceau et al. (1992a)
	ferrihydrite	IS-bidentate	Ford et al. (1999a)
	ferrihydrite	IS-bidentate	Scheinost et al. (2001)
	hematite	IS-bidentate, mononuclear	Bargar et al. (1997b)
	α-Fe$_2$O$_3$ (0001)*	IS-multinuclear	Bargar et al. (2002b)
	α-Fe$_2$O$_3$ ($1\bar{1}02$)*	IS-multinuclear	Bargar et al. (2002b)
	birnessite	IS-bidentate-multinuclear	Manceau et al. (1992a)
	birnessite	IS-monodentate	Morin et al. (1999)
	birnessite	IS (1 Mn 2nd neighbor	Matocha et al. (2001)
	birnessite	IS-tridentate, interlayer	Manceau et al. (2002b)
	γ-MnOOH	IS (2 Mn 2nd neighbors)	Matocha et al. (2001)
	montmorillonite	OS (low I.S., pH 4.5-6.4) OS/IS (high I.S., pH 6.8)	Strawn and Sparks (1999)
	calcite	lattice substitution	Reeder et al. (1999)
	Ca-silicate hydrates	IS-Pb linked to silicate chain	Rose et al. (2000)
	zeolitized tuff	IS-polynuclear	Um and Papelis (2001)
	ZnS	IS/OS-? (bonded to O)	Pattrick et al. (1998)
	CuFeS$_2$	IS (bonded to S sites)	England et al. (1999b)
	humic materials	IS-$^{[IV]}$Pb (2 C neighbors)	Xia et al. (1997a)
	humate	N.D.**	Morin et al. (1999)
	Penicillium chrysogenum	IS (95% bonded to carboxyl groups) IS (5% bonded to phosphoryl groups)	Sarret et al. (1998a,1999a)
	Phaseolus vulgaris	cerrusite precipitate	Sarret et al. (2001)
	Burkholderia cepacia	pyromorphite precipitate	Templeton et al. (1999,2002c)
Lu(III)	goethite	lattice substitution	Dardenne et al. (2002)
	ferrihydrite	lattice substitution	Dardenne et al. (2002)
	hematite	lattice substitution	Dardenne et al. (2002)
Th(IV)	am-SiO$_2$	IS-bidentate	Östhols et al. (1997)
	montmorillonite	IS-bidentate (low Γ) Th(OH)$_2$-like precip. (high Γ)	Dähn et al. (2002b)

Sorbate	Sorbent	Dominant Sorbate Geometry or Phase	Reference
U(VI)	SiO$_2$	N.D.**	Dent et al. (1992)
	silica gel	IS-mononuclear	Reich et al. (1998)
	silica	IS-bidentate	Sylwester et al. (2000b,c)
	alumina	IS-bidentate	Sylwester et al. (2000b,c)
	HFO	IS-bidentate	Manceau et al. (1992a)
	ferrihydrite	IS-mononuclear	Reich et al. (1998)
	ferrihydrite	IS-bidentate	Waite et al. (1994)
	ferrihydrite	IS-bidentate	Dodge et al. (2002)
	goethite	IS-bidentate	Moyes et al. (2000)
	lepidocrocite	IS-bidentate	Moyes et al. (2000)
	lepidocrocite	IS-bidentate	Dodge et al. (2002)
	kaolinite	IS (mixed sites)	Thompson et al. (1998)
	smectite	OS (exchange sites)	Dent et al. (1992)
	smectite	OS (multiple sites)	Chisholm-Brause et al. (1992a,1994)
	montmorillonite	OS (exchange, low pH) IS (neutral pH)	Sylwester et al. (2000b,c)
	hydrobiotite	OS (exchange sites) IS to precipitate (w/drying)	Hudson et al. (1999)
	vermiculite	OS (exchange sites) IS to precipitate (w/drying)	Hudson et al. (1999)
	muscovite	U(VI) hydroxide precipitate	Moyes et al. (2000)
	calcite	lattice substitution	Reeder et al. (2001)
	aragonite	lattice substitution	Reeder et al. (2001)
	apatite	IS-bidentate	Fuller et al. (2002a)
	mackinawaite	IS (low Γ) reduction to U(IV) precipitate (higher Γ)	Moyes et al. (2000)
	cementitious materials	N.D.**	Sylwester et al. (2000a)
	altered concrete	N.D.**	Zhao et al. (2000)
	Bacillus cereus	IS (bonded to phosphate groups)	Hennig et al. (2001)
	Bacillus sphaericus	IS (bonded to phosphate groups)	Hennig et al. (2001)
	Bacillus subtilis	IS (bonded to phosphate groups)	Kelly et al. (2001,2002)
	Bacillus subtilis	IS (bonded to phosphate groups at pH < 3.2 and increasingly to carboxyl groups at pH 3.2 to 4.8)	Kelly et al. (2002)
Np(V)	goethite	IS suggested	Combes et al. (1992)
	mackinawite	IS-tridentate(bonded to surface S)	Moyes et al. (2002)
	cementitious materials	N.D.**	Sylwester et al. (2000a)
	altered concrete	N.D.**	Zhao et al. (2000)
Pu(VI)	*Bacillus sphaericus*	IS (bonded to phosphate groups)	Panak et al. (2002)

[1] Many of the experiments leading to the information reported above were done in controlled atmospheres. However, some were done in ambient air, and may be affected by CO_2 and other atmospheric gases. Therefore, complexing ligands such as HCO_3^- or CO_3^{2-} or may be present in solution. In addition, it has recently been suggested that "indifferent" electrolyte species such as nitrate and perchlorate form ternary surface complexes with cations such as U(VI) (Bargar et al. 2002) and Pb(II) (Trivedi et al. 2002), on the basis of EXAFS analysis. Thus, these "indifferent" electrolyte species may also act as complexing ligands for sorbing cations, although they are normally assumed not to do so.

[*] Indicates single crystal substrates and GI-EXAFS or UHV XAFS experiments. All other samples are high surface area powdered solids or bacterial, fungal, or organic substances.

[**]Not determined or not determinable from data

Table 2. XAFS spectroscopy studies of selected metal cations on metal-(oxyhydr)oxides in the presence of selected inorganic and organic ligands in model systems.

Ternary System	*References*
Cobalt	
Co(II)-EDTA-pyrolusite	Fendorf et al. (1998, 1999)
Nickel	
Ni(II)-EDTA-am-SiO$_2$	Bonneviot et al. (1989a,b)
Copper	
Cu(II)-humate-goethite	Alcacio et al. (2001)
Cu(II)-bipyridine-γ-Al$_2$O$_3$	Cheah et al. (1997)
Cu(II)-bipyridine-am-SiO$_2$	Cheah et al. (1997, 1999)
Cu(II)-glutamate-α-Al$_2$O$_3$	Fitts et al. (1999b)
Cu(II)-humate-illite	Hesterberg et al. (1997b)
Cu(II)-xanthate-ZnS	Pattrick et al. (1999)
Cadmium	
Cd(II)-sulfate-goethite	Collins et al. (1999a)
Cd(II)-phosphate-goethite	Collins et al. (1999a)
Mercury	
Hg(II)-chloride-goethite	Bargar et al. (1997d)
Hg(II)-chloride-goethite	Kim et al. (2002a)
Hg(II)-sulfate-goethite	Kim et al. (2002a)
Lead	
Pb(II)-sulfate-goethite	Weesner and Bleam (1998)
Pb(II)-sulfate-boehmite	Weesner and Bleam (1998)
Pb(II)-phosphate-boehmite	Weesner and Bleam (1998)
Pb(II)-phosphate-goethite	Weesner and Bleam (1998)
Pb(II)-chloride-goethite	Bargar et al. (1997d, 1998)
Pb(II)-EDTA-goethite	Bargar et al. (1999b)
Pb(II)-carbonate-goethite	Ostergren et al. (2000b)
Pb(II)-sulfate-goethite	Ostergren et al. (2000a)
Pb(II)-sulfate-goethite	Elzinga et al. (2001)
Pb(II)-malonate-hematite	Lenhart et al. (2001)
Pb(II)-organic matter-soil particles	Strawn and Sparks (2000)
Pb(II)-EDTA-*Phaseolus vulgaris*	Sarret et al. (2001)
Uranium	
U(VI)-carbonate-hematite	Bargar et al. (1999a, 2000a)
U(VI)-citrate-goethite	Redden et al. (2001)
U(VI)-phosphate-apatite	Fuller et al. (2002a)
U(VI)-carbonate-apatite	Fuller et al. (2002a)

2   A Brief Overview of Synchrotron Radiation

T. K. Sham[1] and Mark L. Rivers[2]

[1]*Department of Chemistry*
Chemistry Building
University of Western Ontario
London, Ontario, N6A 5B7, Canada

[2]*Consortium for Advanced Radiation Sources*
Building 434A
Argonne National Laboratory
9700 S. Cass Avenue
Argonne, Illinois, 60439, U.S.A.

ABSTRACT

The generation of synchrotron radiation from a storage ring is described with emphasis placed on aspects relevant to synchrotron radiation users. Topics include the history of synchrotron radiation, the machinery that makes a storage ring work, the dipole and the insertion device sources, the importance of storage ring energy, current and emittance, the properties of synchrotron radiation and their relevance to the study of the interaction of light with matter.

INTRODUCTION

When high-energy charged particles, such as electrons or positrons, are traveling in a circular orbit at relativistic energies (close to the speed of light), they emit electromagnetic radiation tangential to the orbit. The radiation thus emitted is called synchrotron radiation or synchrotron light (Winick 1994). Electron accelerators that confine these relativistic electrons in the orbit are known as storage rings and synchrotron light sources.

Synchrotron radiation began to be applied to problems in geochemistry beginning in the early 1980's, introducing techniques that are completely impractical on laboratory X-ray sources. These include spectroscopy, microprobe and surface scattering, methods that reveal chemical, atomic and electronic structure of materials.

Synchrotron radiation technology has evolved from the first observation in the laboratory in 1947 (Elder et al. 1947) to the third generation storage rings of today. Although fourth generation sources such as free electron lasers and energy recovery linear accelerators are being developed, the third generation storage ring is arguably the most versatile synchrotron technology for a broad range of experimental techniques important in geochemistry and environmental science. To take full advantage of what the synchrotron light source can deliver, it is desirable that users understand some of the fundamental features of synchrotron radiation, and some important properties of the storage ring and radiation sources. This introduction is intended to present an overview of synchrotron radiation to new and potential synchrotron users. Emphasis is placed on the essential elements of synchrotron radiation technology from the users' perspective. More specific applications of synchrotron radiation and details of each technique are covered in other chapters in this volume.

With this objective in mind, this introduction will be organized as follows. First, a brief overview of the generation of synchrotron radiation is given. This is followed by a

1529-6466/00/0049-0002$05.00

description of synchrotron storage rings and their history. We then present the characteristics of various sources, including bending magnets, wigglers and undulators. This is followed by the description of the important elements of an electron storage ring and their relevance to the quality of the photon beam, and a description of the beamlines that deliver the photons to the sample. We conclude with a brief discussion of the interactions between light and matter. The present status of synchrotron radiation technology worldwide is summarized in a table that lists synchrotron radiation facilities in the world.

How does an electron emit light?

When an electron is accelerated, it produces electromagnetic radiation. The power radiated by an accelerated charge, P, whose speed is significantly less than the speed of light, is described in classical electrodynamics by the Larmor formula. In a synchrotron, the electrons are moving at a speed close to the speed of light (relativistic electrons). The power radiated by relativistic electrons is given by

$$P = \frac{2}{3} \frac{e^2 \gamma^2}{m_o^2 c^3} \left| \frac{dp}{dt} \right|^2 \tag{1}$$

where m_o is the rest mass of the particle, e is the electron charge, c is the speed of light, $p = m_o v$, is the momentum and $(1/m_o) \cdot (dp/dt)$ is the acceleration, γ is the ratio of the mass m of the relativistic electron to its rest mass, m_o, and it is expressed as

$$\gamma = \frac{E}{m_o c^2} = \frac{1}{\sqrt{1-\beta^2}}, \quad \beta = \frac{v}{c} \tag{2}$$

For non-relativistic electrons, $\beta \sim 0$, $\gamma = 1$, for relativistic electrons, $\beta \sim 1$ and with convenient units,

$$\gamma = 1957 \, E \, [\text{GeV}] \tag{3}$$

where E is the energy of the electron in GeV. For example, the Advanced Light Source (ALS) at the Lawrence Berkeley National Laboratory operates at 1.9 GeV and $\gamma = 3718$, while the Advanced Photon Source (APS) at Argonne National Laboratory operates at 7 GeV and $\gamma = 13699$.

A comparison of the radiation pattern from a non-relativistic and a relativistic electron is shown in Figure 1. The opening angle ψ (the half-angle of the divergent synchrotron light above or below the orbit plane) is $\sim 1/\gamma$. Thus higher electron energy implies smaller opening angle of the emitted radiation and brighter synchrotron light (Krinsky et al. 1983; Wille 2000).

Let us consider the circular motion of an electron in an orbit through an angle $d\alpha$, as shown in Figure 2. We have $dp = p \, d\alpha$, thus

$$\frac{dp}{dt} = p \frac{d\alpha}{dt} = p \frac{v}{\rho} \tag{4}$$

where p is the momentum, v the velocity, ρ the radius of curvature.

For relativistic electrons, $v \sim c$, we can substitute dp/dt above into Equation (1), and replace γ with $E/m_o c^2$ and pc with E. We then obtain the radiated power of an electron per turn when it is bent by a dipole magnet with a radius of curvature ρ.

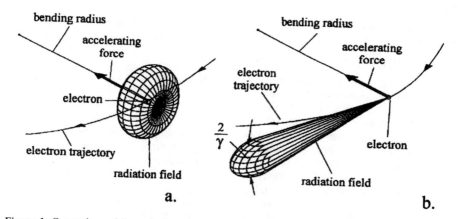

Figure 1. Comparison of the radiation pattern from a.) non-relativistic and b.) relativistic electrons. [Used by permission of the editor, Oxford University Press, from Wille (2000), Fig. 2.3, p. 35]

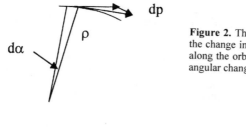

Figure 2. The electron orbit showing the change in classical momentum dp along the orbit and the corresponding angular change, $d\alpha$.

$$P = \frac{2}{3} \frac{e^2 c}{(m_o c^2)^4} \frac{E^4}{\rho^2} \tag{5}$$

If the radius is the same everywhere in the orbit, the energy loss, ΔE, per electron per turn to synchrotron radiation can be obtained with

$$\Delta E = \oint_{orbit} P dt = P \frac{2\pi\rho}{c} \tag{6}$$

Substituting Equation (5) into Equation (6), and using convenient units, we can obtain a simple formula

$$\Delta E[\text{keV}] = 88.5 \frac{E^4[\text{GeV}^4]}{\rho[\text{m}]} \tag{7}$$

Equation (7) shows that the energy loss goes as the fourth power of the electron energy. Synchrotron radiation becomes noticeable for electrons with energy of a few tens of MeV and significant at GeV. For example, for a radius of 12.2 m, at $E = 100$ MeV, the radiated energy per turn is of the order of eV, but at 3.5 GeV ΔE becomes 10^6 eV. Thus, more synchrotron radiation is produced at high-energy rings. Most of the storage rings built today are at energies of the order of 2-8 GeV.

Figures of merit

It is important to quantify the figures of merit for comparing X-ray sources, and for evaluating the suitability of a source for a particular technique or experiment.

Three figures of merit are typically used.

- The *flux* is the number of photons/second/horizontal angle (θ)/bandwidth, integrated over the entire vertical angle (ψ). The bandwidth is typically specified as 0.1%, e.g., at 10 keV photon energy, a 0.1% bandwidth includes all photons within an energy range of 9095 to 1005 eV. The flux is relevant figure of merit for an experiment with a large sample intercepting the entire beam in the vertical direction. It is also the figure of merit that is important for an experiment that uses X-ray optics to collect a large horizontal and/or vertical fan of radiation and focus it on the sample, but where the focal spot size is not particularly important. On a typical synchrotron bending magnet source the available horizontal angle is typically 2-20 mrad, or 0.1 to 1 degree, so angular units of mrad are typically used. The flux is conserved along a beamline as long as photons are not lost by any optical elements.

- The *brightness*, is the flux / vertical angle (ψ), and is thus the number of photons per solid angle. It is the appropriate figure of merit for an experiment which uses a collimator or pinhole to select a small beam. If a beamline includes focusing optics then the brightness is not conserved, it can be changed by the optical elements.

- The *brilliance* is the brightness/source area. It is the relevant figure of merit for a beamline with focusing optics where the focal spot size is important for the experiment, for example an X-ray microprobe or microscope. Thus the smaller the spatial cross-section of the bunches of electron/positrons (i.e., from a storage ring of lower emittance, defined later), the better the brilliance. In order to be able to image the entire source with small aberrations, the design of the optical imaging system is much easier if the beam divergence is small. Hence, the third-generation sources with low emittance and low divergence yield much higher brilliance than the earlier sources. Brilliance is conserved along the beamline even if there are focusing optics. Such optics can change the size and/or divergence of the X-ray beam, but not the product of the size and divergence.

Note that the above definitions of brightness and brilliance are used at the APS. Others use the term brightness to refer to what we call brilliance here.

SYNCHROTRON STORAGE RINGS

A typical storage ring comprises symmetric straight and curved sections of stainless steel (or aluminum) tubes under ultra-high vacuum. Magnets which bend (dipoles) and focus (quadrupoles and sextapoles) the particle beam are installed at various places around the ring to confine the electrons in a prescribed orbit and maintain a small beam size. One or more radio frequency (r.f.) cavities are installed in the straight sections to replenish the energy lost to synchrotron radiation. Electrons are usually first accelerated with a linear accelerator then with a booster synchrotron to the desired energy of the storage ring. The pre-accelerated electrons are then injected into the storage ring. In low energy rings, the pre-acceleration is sometimes done with a device called microtron. The injection instrumentation typically includes pulsed magnets, namely, an inflector to first bend the incoming electron beam into the ring and then kickers in the ring to briefly defect stored beam next to the inflector to accept the incoming beam. More details can be found in the literature (Winick 1994; Wille 2000). The circumference of a storage ring is typically 50-1000 m. The electrons, which travel at almost the velocity of light, thus take about 0.15 to 3 μs for the circuit.

Third generation storage rings are designed for insertion devices known as wigglers and undulators in the straight sections of the storage ring, between the dipole magnets. They are usually arranged in such a way that the magnetic field is normal and sinusoidal to the electron orbit. The magnetic field produces an oscillation in the electron beam trajectory that is also sinusoidal and lies in the horizontal plane. These insertion devices can produce extremely bright beams, and are discussed later.

The layout of the Aladdin storage ring at the Synchrotron Radiation Center, University of Wisconsin-Madison is shown in Figure 3. Figure 4 shows a layout of bending magnet and insertion device. Figure 5 shows some of these elements in the VUV ring of the National Synchrotron Light Source (NSLS).

As the electron beam is circulating in an orbit, it is steered and focused by the

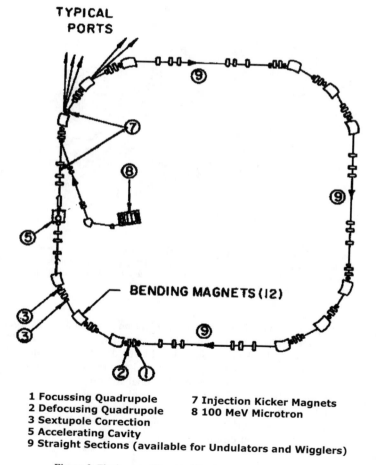

TYPICAL PORTS

BENDING MAGNETS (12)

1 Focussing Quadrupole
2 Defocusing Quadrupole
3 Sextupole Correction
5 Accelerating Cavity
7 Injection Kicker Magnets
8 100 MeV Microtron
9 Straight Sections (available for Undulators and Wigglers)

Figure 3. The layout of the Aladdin ring (courtesy of SRC).

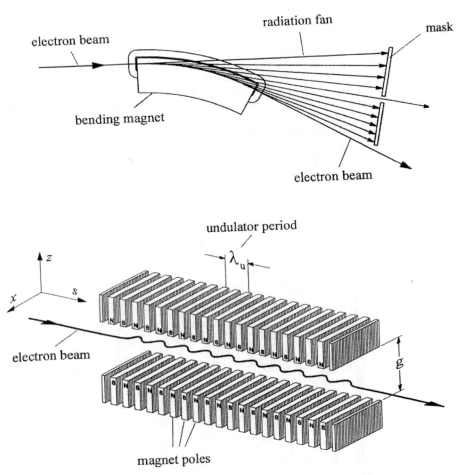

Figure 4. Schematic of a bending magnet and an insertion device, showing the magnet arrangement, the radiation fan, and the exaggerated trajectory of the electron through the insertion device. The maximum angle of deflection in the undulator is K/γ (Eqn. 8). [Used by permission of the editor, Oxford University Press, from Wille (2000), Figs. 2.6 and 2.7, p. 40-41]

magnets described above. The dimension of the electron beam and its divergence vary at different points of the orbit. Synchrotron light is emitted whenever the electron beam is accelerated (i.e., deflected by a magnetic field). Let us denote the standard deviation of the dimension of the electron beam in the horizontal and the vertical position as σ_x and σ_y, respectively. Then the source area is proportional to $\sigma_x\sigma_y$. Similarly, we denote the divergence of the beam in the horizontal and vertical plane as σ_x' and σ_y', respectively and the solid angle of emission is related to $\sigma_x'\sigma_y'$. The deviation of the beam position and angle relative to the ideal orbit are correlated. This correlation is depicted in phase space plots in terms of electron emittance, ε_x and ε_y, where the horizontal emittance is defined as,

$$\varepsilon_x = \sigma_x\sigma_x' \text{ [mm-mrad or nm-rad]} \tag{8}$$

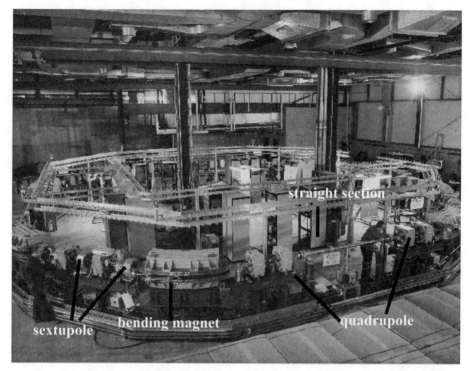

Figure 5. VUV ring at the National Synchrotron Light Source (NSLS) (Courtesy of J. Godel).

and the vertical emittance is

$$\varepsilon_y = \sigma_y \sigma_y' \quad [\text{mm-mrad or nm-rad}] \tag{9}$$

The emittance ε_x and ε_y are conserved (Liouville's theorem); that is that although the deviation in beam dimension and angle may change at different points of the ring, their product is invariant.

History of synchrotron radiation

Figure 6 is a plot of brightness (brilliance in our terminology) versus time (Winick 1994) since the discovery of X-rays, from the conventional hot-cathode tube in the 1940's to the early bending-magnet sources (1970's), to an undulator on a third-generation synchrotron source (1990's), to the next-generation free-electron lasers (FELs). It can be seen from Figure 6 that unprecedented brilliance is being achieved today and even more brilliant machines are expected in the future. Table 1 summarizes the history of synchrotron radiation research and sources.

The development of synchrotron radiation can be conveniently followed in terms of generations. Each generation witnessed some distinct progress in synchrotron technology and usage. Table 2 summarizes the sources that are currently in operation around the world.

First-generation rings. First-generation hard X-ray sources were parasitic on accelerators used for high-energy physics, and provision of time for materials scientists

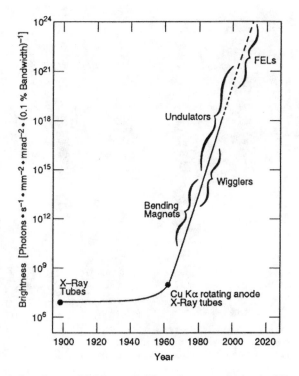

Figure 6. The evolution of spectral brightness (brilliance in our terminology) of X-ray sources through time. [Used by permission of the editor, World Scientific, from Winick (1994), Fig. 1.4, p. 6]

was erratic. As an accelerator became obsolete for high-energy physics, more time was released for X-ray research so long as the accelerator was not closed down. Several are still functioning well, and expanding operations after important upgrades (e.g., the Cornell High Energy Synchrotron Source (CHESS) on the 5.6 GeV Cornell Electron Storage Ring (CESR) at Cornell University, New York, and the Stanford Synchrotron Research Laboratory (SSRL) on the 3 GeV SPEAR ring at Stanford University, California, now turned over for dedicated X-ray generation). The bending magnets for X-ray production were fitted into parts of an existing ring, and the electron bunches tended to be wide and laterally unstable. Before upgrades to tighten the beam, these sources were best suited for experiments on large samples.

Second-generation rings. Second-generation synchrotrons were designed for fully dedicated operation using an array of bending magnets spaced around the entire ring. Typical examples are the 2 GeV Synchrotron Radiation Source (SRS) at Daresbury, England; the 0.7 and 2.5 GeV rings at the NSLS, and the 2.5 GeV Photon Factory at Tsukuba, Japan. The 2-2.5 GeV energy was chosen because of the efficient generation of 1-25 keV X-rays from the ~1-Tesla bending magnets. The NSLS rings were designed to get the small beam diameter needed for high brilliance, and the SRS ring was retrofitted to improve the brilliance.

Third generation rings. Third generation synchrotrons were designed to optimize the number of straight sections to take advantage of the increased brilliance provided by

Table 1. Brief history of synchrotron radiation research.

Period	Relevant Development
Pre-synchrotron era *Before 1947*	Light emission by non-relativitic electrons (Larmor, 1897) Discovery of X-rays (Röntgen, 1895)
Birth and Infancy *1947*	Principle of synchrotron (McMillian, Veksler, 1945) First observation of synchrotron radiation (GE Research Laboratory in Schenectady, NY; April 1947)
Zero Generation *50's - 60's*	A few scattered experimenters used electron accelerators to perform SR experiments, SR remained a nuisance for high energy physics, no users community
First Generation *60's – 70's*	Regular use of SR in research began (e.g., HASYLAB in Germany, Daresbury in UK and NBS in US). In late 60's, small synchrotrons were built (e.g., Tantalus at Stoughton, DCI at Orsay, INS-SOR in Tokyo, SURF II at NBS). In early 70's, parasitic operation began and quickly expanded in high-energy facilities (e.g., SSRL at Stanford). User community emerged.
Second Generation *Late 70's – early 90's*	SR research took off worldwide. Fully-dedicated VUV and X-ray rings were funded and built (NSLS, Photon Factory, etc.). Most came on-line in mid 80's. Bending magnet (dipole) source was emphasized. User community expended to span a large number of disciplines. Demand of high quality SR (from IR to hard X-rays) increased dramatically.
Third Generation *Early 90's – present*	Insertion devices with periodically alternated magnetic fields in the straight section (wigglers and undulators) quickly led to the implementation of insertion-device based rings of the 90's (e.g., APS, ALS, ESRF, SRRC, Spring8). These are arguably the most versatile rings for a wide variety of applications. They are still being commissioned, e.g., the Canadian Light Source.
Fourth Generation *Under development*	Special applications: e.g., Free Electron Laser and Energy Recovery Linear Accelerators and others.

undulators, and to a lesser extent by wigglers. There are three hard X-ray third-generation sources in the world: the European Synchron Radiation Facility (ESRF) in Grenoble, France (6 GeV), the APS at Argonne (7 GeV), and Super Photon ring-8 (Spring-8) in Japan (8 GeV). Third third-generation soft X-ray sources include the ALS at Berkeley and ELLETRA in Italy.

Fourth generation sources. There are two potential future light sources with the potential for significantly increased peak brilliance compared to third-generation undulator sources. The first is the Energy Recovery Linac (ERL). This device would use a linear accelerator with energy recovered from the electrons after a single pass (or a few passes) around a ring with undulator beamlines. The horizontal emittance would be significantly less than a storage ring, and so the brilliance could be more than 10 times that of a third-generation source. A more radically new source is the Free Electron Laser (FEL), which would increase the brilliance by 4-5 orders of magnitude compared to an undulator. If hard X-ray FEL's are successfully built they will probably be limited to specialized experiments where problems of low repetition rates and sample damage are not important.

Table 2. Storage ring synchrotron radiation sources planned*, under construction**, and operating (November, 2001).

Location		Ring (Inst.)	Energy (GeV)
AUSTRALIA		Boomerang*	3
BRAZIL	Compinas	LNLS-1	1.35
		LNLS-2	2
CANADA	Saskatoon	CLS** (Canadian Light Source)	2.5-2.9
CHINA (PRC)	Beijing	BEPC (Inst. High En. Phys.)	1.5-2.8
		BLS (Inst. High En. Phys.)	2.2-2.5
	Hefei	NSRL (Univ. Sci. Tech. Of China)	0.8
	Shanghai	SSRF *(Inst. Nucl. Res.)	3.5
DENMARK	Aarhus	ASTRID (ISA)	0.6
		ASTRID II (ISA)	1.4
ENGLAND	Daresbury	SRS (Daresbury)	2
		DIAMOND*(Daresbury/Appleton)	3.0
		SINBAD*(Daresbury)	0.6
FRANCE	Grenoble	ESRF	6
	Orsay	DCI (LURE)	1.8
		SuperACO (LURE)	0.8
		SINBAD*	2.5-2.75
GERMANY	Berlin	BESSY I **(*to become SESAME of the middle East*)	0.8
		BESSY II	1.7-1.9
	Bonn	ELSA (Bonn Univ.)	1.5-3.5
	Dortmund	DELTA (Dortmund Univ.)	1.5
	Hamburg	DORIS III (HASYLAB/DESY)	4.5-5.3
		PETRA II (HASYLAB/DESY)	7-14
	Karlsruhe	ANKA (FZK)	2.5
INDIA	Indore	INDUS-I **(Ctr. Adv. Tech.)	0.45
		INDUS-II **(Ctr. Adv. Tech.)	2.5
ITALY	Frascati	DAΦNE	0.51
		ELETTRA (Synch. Trieste)	1.5-2
JAPAN	Hiroshima	HISOR (Hiroshima Univ.)	0.7
	Ichihara	Nano-hana (Japan SOR Inc.)	1.5-2
	Kashiwa	VSX (Univ. of Tokyo-ISSP)	2-2.5
	Kusatsu	AURORA (Ritsumaiken Univ.)	0.6
	Kyoto	KSR (Kyoto Univ.)	0.3
	Nishi Harima	Spring-8 (JASRI)	8
		Subaru (Himeji Inst. Tech.)	1-1.5
	Okazaki	UVSOR (Inst. Mol. Science)	0.75
		UVSOR-II*(Inst. Mol. Science)	1.0
	Sendai	TLS*(Tohoku Univ.)	1.5

Location		Ring (Inst.)	Energy (GeV)
JAPAN	Tsukuba	TERAS (ElectroTech. Lab.)	0.8
		NIJI II (ElectoTech. Lab.)	0.6
		NIJI IV (ElectoTech. Lab.)	0.5
		Photon Factory (KEK)	2.5-3
		Accumulator Ring (KEK)	6
KOREA	Pohang	Pohang Light Source	2
	Seoul	CESS (Seoul Nat. Univ.)	0.1
MIDDLE EAST		SESAME	1
RUSSIA	Moscow	Siberia I (Kurchatov Inst.)	0.45
		Siberia II (Kurchatov Inst.)	2.5
	Dubna	DELSY (JINR)	0.6-1.2
	Novosibirsk	VEPP-2M (BINP)	0.7
		VEPP-3 (BINP)	2.2
		VEPP-4 (BINP)	5-7
		Siberia-SM (BINP)	0.8
	Zelenograd	TNK (F.V. Lukin Inst.)	1.2-1.6
SINGAPORE		Helios2 (Univ. of Singapore)	0.7
SPAIN	Barcelona	Catalonia SR Lab	2.5-3
SWEDEN	Lund	MAX I (Univ. of Lund)	0.55
		MAX II (Univ. of Lund)	1.5
		New Ring*(Univ. of Lund)	0.7
SWITZERLAND	Villigen	SLS (Paul Scherrer Inst.)	2.4
TAIWAN (ROC)	Hsinchu	SRRC (Synch. Rad Res. Ctr.)	1.3-1.5
THAILAND	Nakhon Ratchasima	SIAM *(Suranaree Univ. of Tech.)	1.0
UKRAINE	Kharkov	Pulse Strecher/Synch. Rad.	0.75-2
	Kiev	ISI-800 (UNSC)	0.7-1.0
USA	Argonne, IL	APS (Argonne Nat. Lab.)	7
	Baton Rouge, LA	CAMD (Louisiana State Univ.)	1.4
	Berkeley, CA	ALS (Lawrence Berkeley Lab.)	1.5-1.9
	Durham, NC	FELL (Duke University)	1-1.3
	Gaithersburg, MD	SURF III (NIST)	0.4
	Ithaca, NY	CESR (CHESS/Cornell Univ.)	5.5
	Stanford, CA	SPEAR2 (SSRL/SLAC)	3
		SPEAR3 **(SSRL/SLAC)	3
	Stoughton, WI	Aladdin (Synch. Rad. Center)	0.8-1
	Upton, NY	NSLS I (Brookhaven Nat. Lab)	0.80
		NSLS II (Brookhaven Nat. Lab)	2.5-2.8

Note: *The websites of these facilities can be found through the Links to other facilities at the CLS website (www.cls.usask.ca/media/links.shtml).*

Bending magnet radiation

When electrons or positrons (used by some storage rings) are travelling in the magnetic field of a bending magnet, they are subjected to a Lorentz force that accelerates the electron centrifugally and bends the electrons into a circular orbit. The bending results in the emission of synchrotron radiation tangent to the orbit. The instantaneous jet of X-rays sweeps out an arc as the electron bunch passes through a bending magnet. Thus the horizontal divergence is increased by the arc subtended by the magnet at the center of curvature (Fig. 4), and if the ring were a continuous circle the radiation fan would be a thin disk centered on the orbit plane. The light thus produced is called synchrotron radiation or synchrotron light from a bending magnet (Krinsky et al. 1983; Turner 1990). It exhibits a continuum of radiation characterized by a critical photon energy ε_c (keV) or critical wavelength λ_c (Å). By definition half of the radiated power is above ε_c and half below. ε_c is determined by the electron energy: the higher the electron energy, the higher the critical energy, with the relationship given by

$$\varepsilon_c = \frac{2.218E^3}{\rho} = 0.665BE^2 \tag{10}$$

where ρ is the radius in meters of the electron orbit at the bending magnet, E is the particle beam energy in GeV, and B is the magnetic field in Tesla. To reach the ultra-violet/soft X-ray region, the electrons in a circular ring with conventional 1-Tesla bending magnets need to be accelerated up to about 1 GeV. For hard X-rays (> 5 keV), more than 2 GeV is needed. Thus, the NSLS X-ray ring with $E = 2.8$ GeV, $\rho = 6.875$ m, $B = 1.36$ Tesla has $\varepsilon_c = 7.1$ keV, and the APS with $E = 7$ GeV, $\rho = 38.96$ m, $B = 0.6$ Tesla has $\varepsilon_c = 19.5$ keV.

For bending magnets, the key parameter for experiments using a collimator is the brightness, or flux per vertical angle:

$$B_{BM} = 1.33 \times 10^{13} E^2 I H_2(\varepsilon/\varepsilon_c) \tag{11}$$

where E is again the storage ring energy in GeV, I is the ring current in Ampere, and $H_2(\varepsilon/\varepsilon_c)$ is a function plotted in Figure 7. Note that B_{BM} at a fixed fraction of the critical energy increases linearly with the ring current and by the square of the ring energy.

Figure 8 shows the brilliance of several bending-magnet sources, including the NSLS, the APS and the new superconducting bending magnets at the ALS at Berkeley. The parameters of these sources are summarized in Table 3. Note how the high-energy rings provide similar brilliance below 5 keV to the low-energy rings. Above 10 keV, the brilliance of the high-energy rings drops off much less than for the low-energy rings.

The vertical opening angle, ψ, above and below the plane of the orbit defines a cone of synchrotron radiation from a bending magnet source that is a function of ($\varepsilon/\varepsilon_c$). ψ (the RMS half opening angle) can be approximately expressed as (Krinsky et al. 1983; Green 1975; Kim 1986)

$$\psi \text{ [mrad]} \sim \frac{0.32(\varepsilon/\varepsilon_c)^{-0.55}}{E} \tag{12}$$

for energies from 1-10 times the critical energy. Hence higher energy electrons will produce a more collimated photon beam at the critical energy. Recalling that $\gamma = 1957 \cdot E$ (Eqn. 1), then at the critical energy we get

$$\psi \text{ [mrad]} \sim 630/\gamma \tag{13}$$

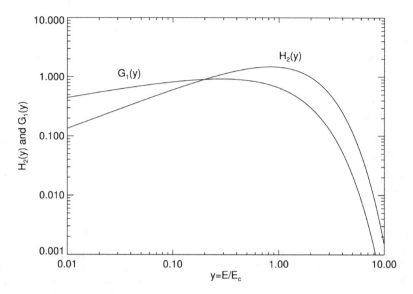

Figure 7. The brightness function ($H_2(y)$) and the flux function ($G_1(y)$) for a bending magnet source. Both functions fall slowly for energies less than the critical energy, and fall rapidly for energies above the critical energy.

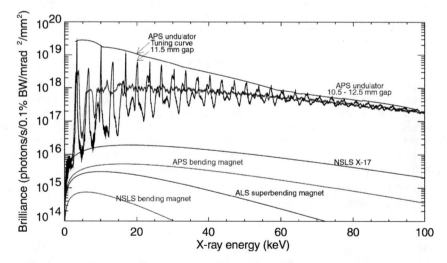

Figure 8. Brilliance of the APS undulator in untapered mode (11.5 mm gap) and tuning curve for available gap range, APS undulator in tapered mode (10.5-12.5 mm gap), APS bending magnet, NSLS superconducting wiggler X-17, NSLS bending magnet, and ALS superconducting bending magnet. For XAFS (requiring ~ 1 keV scannable range), the undulator is scanned in unison with the monochromator. For energy-dispersive diffraction the undulator is run in tapered mode, where the spectrum is quite smooth above 20 keV.

Table 3. Parameters of synchrotron sources plotted in Figure 8.

Source	E (GeV)	Magnetic field (T)	# poles	Source size (σ_x, σ_y; mm)
NSLS bending magnet	2.8	1.4	1	0.393, 0.025
NSLS X-17 superconducting wiggler	2.8	4.2	5	0.353, 0.006
ALS superconducting bending magnet	1.9	5	1	0.098, 0.015
APS bending magnet	7	6	1	0.109, 0.027
APS undulator A	7	0.8	144	0.252, 0.012

Converting to radians, and considering the full, rather than the half angle, we often approximate

$$2\psi \; [\text{rad}] \sim 1/\gamma \tag{14}$$

$1/\gamma$ decreases from 510 micro-radian (µrad) at $E = 1$ GeV to 73 µrad at 7 GeV.

For experiments that utilize the entire vertical fan, either with large samples or with focusing optics, the important parameter is the flux integrated over all vertical angles, which is the brightness above. The equation for the flux (photons/sec/0.1% bandwidth /horizontal mrad) is:

$$F_{BM} = 2.457 \times 10^{13} \, EIG_1 \left(\varepsilon / \varepsilon_c \right) \tag{15}$$

where E and I are as in Equation (11) above, and G_1 is another function plotted in Figure 7. We see that the flux varies only linearly with ring energy, and falls off rapidly above the critical energy (almost 1000 times at $\varepsilon = 10\varepsilon_c$). The flux falls off much more slowly below the critical energy (less than a factor of 10 at $\varepsilon = 0.001\varepsilon_c$)

INSERTION DEVICES

The next factor important for gain of brightness and brilliance was the invention of insertion devices. An insertion device consists of an array of N pairs of magnets inserted into a straight section between two bending magnets (Fig. 4). The magnet pairs alternate in polarity causing the electrons to follow a sinusoidal path. The devices are carefully designed so that when the electrons exit the device their position and direction is the same as if the device were not present. The design of the third-generation sources has been driven by emphasis on these insertion devices in the straight sections.

Every insertion device can be characterized by the deflection parameter, K_{id}, related to the peak magnetic field B_0 (Tesla) by

$$K_{id} = 0.934 \lambda_{id} B_0 \tag{16}$$

where λ_{id} is the magnetic period of the insertion device in cm. The angular deflection of the orbit is equal to K_{id}/γ, and so K_{id} is the ratio of the angular deflection of the beam to the natural opening angle of the radiation, $1/\gamma$. The two principal types of insertion devices, wigglers and undulators, differ fundamentally in their values of K_{id}, which controls the properties of the radiation emitted.

Wiggler radiation

A wiggler is an insertion device with $K_{id} \gg 1$, typically in the range 10 to 60. Thus a wiggler uses relatively long magnetic periods with large magnetic fields. Many new wigglers have N = 20-50, with magnetic periods of 5 to 15 cm. Wigglers give the electrons a sizable angular deflection (large compared with the natural emission angle) as they traverse the device. A wiggler with N periods produces a continuum of radiation that approximates the superposition of the radiation from 2N bending magnets of the same magnetic field. If the magnetic field of the wiggler is greater than that of the bending magnets on the storage ring, as is often the case, then it shifts the critical energy ε_C to a higher value. Thus, wigglers are often used to produce harder X-rays with higher flux from a medium energy storage ring. For example, the bending magnets at the X-ray ring at the NSLS (2.8 GeV) have a magnetic field of 1.36 T, and the critical energy is thus 7.1 keV. The X-17 superconducting wiggler typically operates at 4.2 Tesla when the ring runs at 2.8 GeV, and thus a critical energy of 21.9 keV. The wiggler thus has a much higher flux at high energies. This wiggler has 5 poles, while the wigglers on the X-21 and X-25 beamlines at the NSLS have 27 poles, and the flux and brightness is increased by these factors above the bending magnets at their respective critical energies.

The vertical divergence of a wiggler is the same as for a BM with the same field strength ($\sim 1/\gamma$ at the critical energy). Because the beams from the left and right wiggles are not collinear, the horizontal divergence of a wiggler is greater than the vertical divergence by a factor K_{id}.

A simple wiggler produces linearly polarized radiation on-axis, because the electron/positron beam is essentially traveling in coplanar arcs, just as for a bending magnet. The polarization from a wiggler can be modified deliberately by geometrical changes of the magnet arrays. For example, two arrays at 90° can produce in-plane circularly-polarized X-rays.

Undulator radiation

The second type of insertion device, the undulator, uses weaker magnetic fields, and/or shorter periods, so that $K_{id} \approx 1$. Each pulse generates radiation out of phase with the next pulse by the time taken to reach the next magnet. Only those photons with wavelength equal to this difference, or a subharmonic (n) thereof, are reinforced, and the white spectrum from a bending magnet and a wiggler is replaced by peaks at the harmonic wavelengths (Figs. 8 and 9). The spectrum has lower mean energy than for a bending magnet or a wiggler, but with harmonic peaks of greatly enhanced brightness and brilliance. At these harmonic energies the X-ray beam is highly collimated in both the horizontal and vertical directions.

The on-axis energy of the nth harmonic in keV is

$$\varepsilon_n = 0.95 n E^2 \Big/ \left[\lambda_{id} \left(1 + K_{id}{}^2 / 2 \right) \right] \tag{17}$$

Note that reduction of the vertical spacing between each magnet pair causes B_0 to increase, and the consequent increase in K_{id} drives ε_n to lower energy. It is important to emphasize this inverse relation between increasing magnetic field and decreasing photon energy of each harmonic.

As K_{id} increases, the higher harmonics become more important and more closely spaced in energy, until the undulator effectively becomes a wiggler for $K_{id} > 10$. Note that the effective critical energy does increase as K_{id} increases, because of the increasing contributions of the higher harmonics.

The on-axis brightness of the nth harmonic is

$$B_N = 1.74 \times 10^{14} N^2 E^2 I F_n(K_{id}) \tag{18}$$

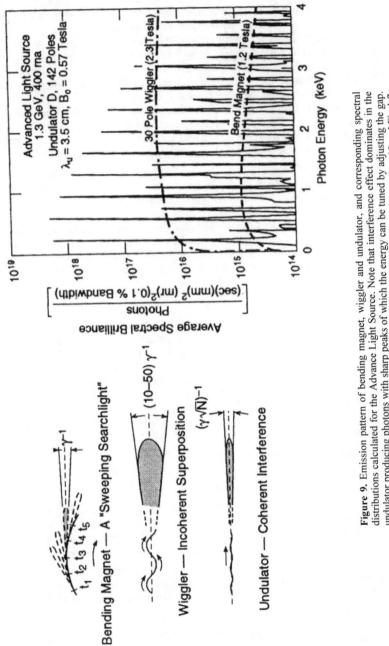

Figure 9. Emission pattern of bending magnet, wiggler and undulator, and corresponding spectral distributions calculated for the Advance Light Source. Note that interference effect dominates in the undulator producing photons with sharp peaks of which the energy can be tuned by adjusting the gap. [Used by permission of the editor, World Scientific, from Winick (1994), Fig. 1.6, p. 10 and Fig.1.7, p.11)]

where $F_n(K_{id})$ is plotted in Figure 10. Observe how the first harmonic loses its dominance over the third harmonic as K_{id} reaches 1.5, and how the fifth (and higher) harmonics take over for K_{id} above 2.

The characteristic opening angle in both the horizontal and vertical directions of the nth harmonic for an ideal ring with zero emittance is

$$\sigma_{r'} = \sqrt{(\lambda_n / 2L)} \qquad (19)$$

where λ_n is the photon wavelength and L is the length of the undulator. For APS Undulator "A," with a harmonic at, for example, 12.4 keV (= 1 Å = 0.1 nm) and $L = 72$ periods × 3.3 cm spacing, $\sigma_{r'}$ becomes 4.6 μrad. This is 17 times smaller than the intrinsic opening angle for the instantaneous acceleration in a bending magnet at the APS. In a real storage ring with non-zero emittance the actual divergence of the undulator beam will be the quadrature sum of the divergence from Equation (19), $\sigma_{r'}$ and the electron beam divergence in the x or y directions, $\sqrt{(\sigma_{r'}^2 + \sigma_{(x,y)'}^2)}$. For the APS, the vertical electron beam divergence, $\sigma_{y'}$ with the new low-emittance lattice is 3 μrad, so the divergence of the undulator radiation is increased from 4.6 μrad to 5.5 μrad. In the horizontal direction $\sigma_{x'} = 15.6$, so the undulator opening angle is increased from 4.6 to 16.3 μrad, an increase of a factor of 3.5. The broadening of the opening angle reduces the peak brightness by $\sigma_{x'}\sigma_{y'}/\sigma_{r'}^2$. For the above parameters, the reduction is about a factor of 4.

Odd harmonics ($n = 1, 3$) are ideally on-axis, and even ones are off-axis. For a real ring with non-zero emittance, however, the phase envelopes of the harmonics overlap, and all are accessible on-axis. The formal bandwidth of each harmonic wavelength is $1/nN$, but deviation of the magnetic field from ideal causes broadening of the peaks. In general, the odd harmonics have linear polarization on-axis, while the polarization of the even harmonics is complex.

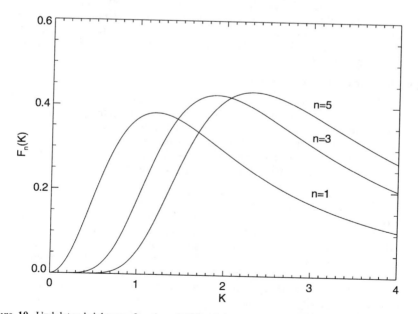

Figure 10. Undulator brightness function, $F_n(K)$, of the odd harmonics of an undulator versus the deflection parameter K_{id}.

The wavelength (and energy) of each harmonic can be changed simply by varying the magnetic field. The usual way this is accomplished is to mount the upper and lower magnets on racks that are moved symmetrically about the vacuum tube containing the electron beam. As the gap is changed, so is the magnetic field and the energy of each harmonic. For example, at the APS the gap of the standard 3.3 cm period undulator can be changed from 11 mm to 40 mm, over which range K varies from 2.6 to 0.16, and the energy of the first harmonic varies from 3.2 keV to 13.9 keV. Over this range in gap the third harmonic energy changes from 9.6 keV to 41.7 keV. The range of the second harmonic (6.4 to 28 keV) overlaps those of the first and third harmonics. This is illustrated in Figure 11, which shows the calculated spectrum for the APS Undulator "A" at a variety of gaps and K values. The observed spectra match these calculated ones very closely. The first harmonic is represented by a slightly ragged peak near 13 keV. The peak splays out asymmetrically to 1% of peak brightness at about 10 and 13.5 keV. The second harmonic is nearly flat-topped from 20 to 26 keV, and the third harmonic is rather irregular.

The harmonics can be broadened by either tapering the gap along the length of the undulator, or by sliding the magnets sideways in opposite directions. This broadening can be used for XAS, which requires monochromator tuning through a bandwidth of about 10%. Figure 8 shows the calculated spectrum of a 3.3 cm period undulator at the APS with a large taper, the gap varying from 10.5 mm at one end to 12.5 mm at the other.

Very important for the success of brilliance-driven experiments is development of optical elements that do not warp under a high heat load. The power from an APS undulator at closed gap is more than 6 kW, and the peak power density is more than 160 W/mm^2. These values are enough to melt any uncooled material, and are particularly challenging for the first crystal of the beamline monochromator, which must not distort more than a few microradians.

High harmonics (> 9) are usable, so these undulators are capable of producing high energy X-rays to beyond 50 keV. Specialized undulators allow changing the orientation of the magnetic field, which allows control of the polarization. A schematic of an undulator that produces desired polarization is shown in Figure 12. These devices are ideal for polarization sensitive measurements such as X-ray magnetic circular dichroism (XMCD) of magnetic materials.

Important storage ring parameters

The electrons are slowly lost by scattering from each other and from atoms and molecules of the residual vacuum, and it is customary to dump the beam when about half the electrons are lost. New bunches are accelerated in the booster ring and injected into the synchrotron storage ring. The electrons in a bunch repel each other, and this plus other factors limits the electron current attainable in a ring. The lifetime before dumping and re-injection is strongly governed by the quality of the vacuum. .

The early synchrotron pioneers suffered extreme frustration from short lifetime of the beam (few hours at most) and spatial instability of the beam (up to a millimeter). Current rings are delivering beam with a reliability well over 95%, with a lifetime before dumping near one day, and with lateral and vertical stability, controlled by sensors and fast feedback circuits, in the micrometer range. Some days are reserved for special experiments and instrumental upgrades leaving somewhere between 200 and 250 days per year for standard experiments. The APS now operates in a "top-up" mode, in which electrons are injected frequently, every two minutes. This produces a near constant current in the storage ring (less than 1% variation), which leads to stable power on the beamline optics, more integrated flux on the sample, and fewer problems with detector non-linearities and intensity normalization.

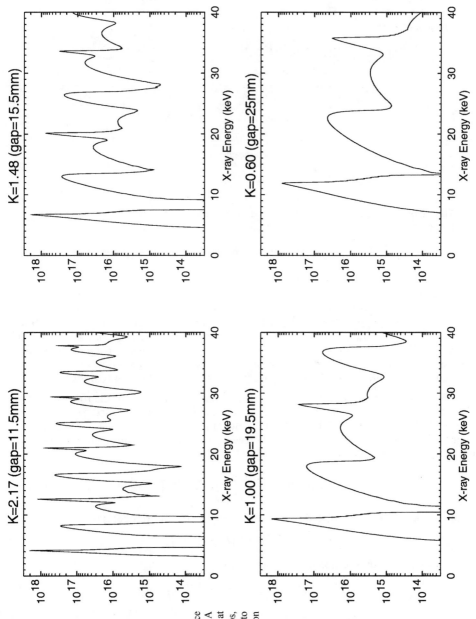

Figure 11. Brilliance of APS undulator A (3.3 cm period) at four different gaps, corresponding to different deflection parameters K_u.

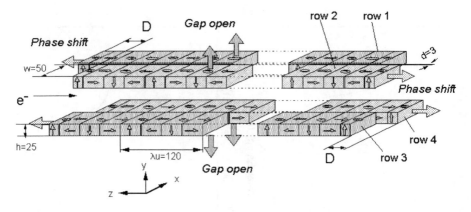

The magnetic structure of the APPLE II consists of two pairs of arrays of permanent magnets.

Figure 12. Schematic of an undulator with adjustable polarization (after Agui 2001).

The most relevant parameters of a storage ring are its energy, E, operating current, I, emittance, ε and radio frequency $\nu_{r.f.}$. To synchrotron users, they translate into practical photon energy range, intensity (flux), brilliance (beam size and angle) and time structure, respectively. For dipole radiation from bending magnets, storage rings operating with an energy of ~ 1 GeV, or below are VUV rings (practical photon energy ranging from IR to several keV), and rings with ~ 2 GeV or above are X-ray rings (practical photon energy from a few keV to tens of keV) while those in between are the soft X-ray rings.

The choice of the bending radius has some bearing on the size of the ring, hence the size of the conventional facility. In practice, the field strength is limited to ~ 1.5 T for conventional magnets and 5 T for super-conducting magnets. Thus, the X-ray rings are larger than VUV rings. We can express the energy loss of an electron per turn in terms of the energy and the magnetic field.

$$\Delta E[\text{keV}]/turn = 88.5\frac{E^4[\text{GeV}]}{\rho[\text{m}]} = 26.6E^3[\text{GeV}]B[\text{T}] \qquad (20)$$

It follows that the power radiated (kW) for a stored current of I (ampere) is

$$P[\text{kW}] = 26.6I[\text{A}]E^3[\text{GeV}]B[\text{T}] \qquad (21)$$

The energy loss in an insertion device is given by

$$\Delta E[\text{keV}] = 0.633E^2[\text{GeV}]\langle B^2[\text{T}]\rangle L[\text{m}] \qquad (22)$$

where $<B^2>$ is the average value of the square of the magnetic field over the length of the device L (Winick 1994).

The electrons circulate in the storage ring are bunched by the r.f. system which replenishes the energy loss of the electron to synchrotron radiation, as described by Equations (20)-(22). This is done by a time varying electric field. It should be noted that the ring circumference L_c (ideal orbit) must be an integral multiple, h (harmonic number), of the r.f. wavelength, $c\nu_{r.f.}$ resulting in the circulating period $T_0 = L_c/c = h/\nu_{r.f.}$. For example, an electron travelling in a 120 m circumference ring has a period of 4×10^{-7} sec or a frequency or 2.5 MHz. Thus synchrotron radiation is a pulse. The bunch length and the maximum number of bunches is determined by the frequency of the r.f. system. The dimension and angle of emission of the bunch determines the brilliance. The length of the bunch determines the pulse width and the spacing between bunches ($1/v_{r.f.}$ in time and $c/v_{r.f.}$ in distance) determines the repetition rate. Thus, single bunch mode with a long repetition rate is sometimes desirable for experiments that utilize delayed photon emission, such as nuclear resonance scattering. Typical bunch lengths (for a 500 MHz r.f. system) are much smaller than 1 ns, and bunch spacing is about 2 ns. The total number of the electrons in all bunches in the storage ring per circulated time amounts to the current. Since the light intensity scales linearly with the current, higher current means higher flux.

The size and the angular divergence of the electron bunch is a very important parameter since it is related to the brilliance of the photon beam. As noted earlier, the vertical size, σ_y and angular spread σ_y' of the electrons in the circular orbit in the direction of motion is associated with a storage ring parameter called emittance. We recall that although the vertical size and angular divergence of an electron beam vary in various locations of the ring, their product, the emittance, is a constant. The emittance is often discussed in terms of a phase space ellipse as shown in Figure 13 (Gudat and Kunz 1979). At locations where the β function (see below) is at a minimum (typically in the

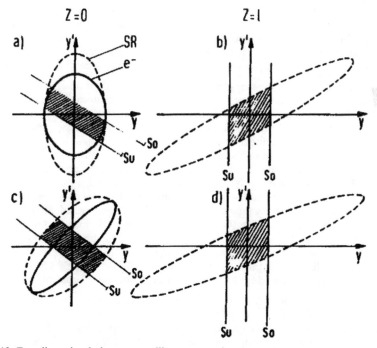

Figure 13. Two-dimensional phase space ellipse representing the spatial and angular deviation of the electron beam in the y (vertical) and y' direction. Note that the shape of the ellipse is upright at z = 0 (a) but changes at different points of the ring say at z = 1 (b) but the area is conserved. If a slit (S_o and S_u) is placed at the z = 1 position, it can be projected back to the origin. (c) and (d) show the same transformation. Note that the area intersected by the slit boundaries is also conserved. The synchrotron light ellipse (SR) is also shown. [Used by permission of the editor, Springer-Verlag, from Gudat (1979), Fig. 3.7, p. 68]

center of straight sections) the phase space ellipse is upright and the relationship $\varepsilon_y = \sigma_y \sigma_y'$ holds. Figure 13 also shows the corresponding ellipse for the photon beam denoted SR and the effect of a slit intersecting the beam. At different points of the orbit, the vertical and horizontal plane rotate (Fig. 13 c-d). The invariant parameters are the area of the ellipse and the intersections with the y and y' axis. If we define x as the horizontal co-ordinate and y the vertical co-ordinate along the orbit s, then the emittance is defined by a four-dimensional phase space $(x, x'; y, y')$. A detailed phase space analysis of a storage ring can be found in the literature (e.g., Krinsky et al. 1983).

The emittance of a storage ring (characterized by the area of the ellipse) is invariant through out the ring. The emittance, together with other storage ring parameters such as the β function (describes the transverse oscillation of electrons in the ring, β oscillation) and the dispersion function (η) determines the beam size (Wille 2000). η is a property of the magnetic lattice. The electron emittance together with the opening angle of the synchrotron radiation determines the brilliance of the photon beam at a given wavelength. In general, smaller emittance results in smaller angular divergence and a brighter beam. The brilliance of the photon beam at a given wavelength increases as the size and divergence of the electron beam become smaller until it reaches the diffraction limit. For a given wavelength, the diffraction limit is the lowest limit at a given wavelength below which the brilliance can no longer be improved (Winick 1994). This is about one tenth of the wavelength ($\lambda / 4\pi$). X-rays from many third generation sources have reached the intrinsic angular opening of the beam in the vertical direction (Eqn. 19), meaning that further reductions in the vertical electron beam divergence do not improve the brilliance. However, they have not reached the intrinsic divergence in the horizontal direction, and have not reached the intrinsic source size in either direction. The vertical electron beam size has become so small, however, that it is very difficult to produce X-ray optics with slope errors that are small enough to preserve the vertical brilliance of the radiation.

All third generation storage rings have low emittance as their design objective. Since emittance (hence brilliance) is a parameter inherent to the storage ring, it cannot be improved by beamline optics. Low emittance (brighter photon beam) is critical for microscopy and other applications that require a small spot size on the sample. It should also be noted that the key element in a storage ring design is its lattice. A lattice of a storage ring is the arrangement of the magnetic system that confines the electrons in the storage ring. A lattice contains units called cells. The unit cells are the basic building blocks of the lattice. Combined with straight sections, they define a ring's symmetry. The most common cell is the Double-Bend Achromat (DBA). It contains two bending magnets separated by focussing magnets (Fig. 3). The lattice based on DBA is often known as Chasman-Green Lattice (Green 1976). The NSLS rings at the Brookhaven National Laboratory are based on this design, as is the Canadian Light Source. Another common Lattice is the Triple-Bend Achromat (TBA). In TBA, a third bending magnet is inserted symmetrically between the outer two. It was adopted in the ALS design among others. It tends to help smaller rings to achieve lower emittance (Cornacchia 1994).

Stability (spatial, temporal, flux/brilliance and polarization) is arguably the most important gauge of the performance of the ring. Users should be concerned with stability since it defines the quality of the photon beam. For example, at the APS the undulator beam has a vertical beam divergence of about 5.4 µrad. If the centroid position of this beam on the sample or optics is to be stable to within 10% of this angle, then the angular stability of the electron beam in the accelerator must be better than 0.54 µrad over time scales from hours to fractions of a second. This corresponds to a stability of 0.2 µm over the 2.5 m length of the undulator! The actual stability at the APS vertical beam position has been measured to be less than 1 µm over periods from seconds to a day, and the variation in vertical angle is less than 0.2 μrad over these time scales. The quality of the photon beam that finally enters the users' experiment is dependent upon not only the performance of the ring, but also the quality and stability of the beamline optics.

In short, the journey of an electron to emitting synchrotron radiation can be summarized. Electrons are first generated and accelerated in a linear accelerator. Further acceleration is usually carried out in a booster synchrotron to the desired energy. Electrons with the desired energy are subsequently injected into a storage ring (at a straight section) and circulate around the orbit. The r.f. cavity (at a straight section) bunches the electrons and replenishes the energy loss to synchrotron radiation (electron travels in a storage ring in time ranging from tens of nanoseconds to microseconds, corresponding to a frequency of MHz and hundreds of MHz, the r.f. band). Again, whenever the electron is passing through either a bending magnet or an insertion device it will be accelerated and emit synchrotron radiation. The rest of the system helps to confine the electrons in the design orbit and replenishes the lost energy.

BEAMLINE AND EXPERIMENTAL STATIONS

Beamline design

Briefly, a beamline comprises three components: the front-end, beamline optics and experimental station downstream. Experimental stations at the end of the beamline are built to take advantage of the photons the source and the beamline optics can provide. Thus, synchrotron light instrumentation development and its optimum usage is an integrated process that evolves through the interaction between the machine group and the users who often start out by wanting to do a specific type of experiment that demands photons with certain characteristics. Examples of specific beamline design and applications will be dealt with elsewhere in this volume.

Synchrotron beamlines: general

This section concentrates on general design features. It ignores many technical details critically important to the success of a beamline. These details are generally unnoticed, and indeed should be, by an analyst visiting a user-friendly general-service facility.

Nearly all beamlines, including those for XRF, XAS and XRD, contain a similar set of components. The X-rays from a BM or an ID are transported to the experimental station by a very sophisticated set of components designed to protect the ring from catastrophic vacuum failure, and the scientists from X-rays.

The front-end, which is typically behind the main concrete shielding wall, contains vacuum isolation valves and/or Be windows to protect the storage ring from vacuum mishaps on the beamline, and water cooled apertures to limit the horizontal extent of the beam. Most important, it contains a thick metal safety shutter which, when closed, prevents X-rays from passing further down the beamline. The safety shutter cannot be opened unless the beamline and the experimental stations are interlocked.

The beam transport typically consists of one or more Be windows, to isolate the beamline vacuum from that of the storage ring, and evacuated beam pipes ~ 10-20 cm in diameter to conduct the beam into the experimental stations. Along the beam transport, between the storage ring and the experimental stations, there may be one or more optics enclosures, which contain X-ray optical devices such a monochromator and a focusing mirror.

Most synchrotron beamlines are equipped with monochromators which consist of two Si or Ge crystals which diffract a single wavelength. The output beam is parallel to, but slightly offset vertically from, the incoming white beam. All beamlines for XAS require a monochromator. XRF experiments, on the other hand, can use either the original white synchrotron beam or a monochromatized beam.

Many synchrotron beamlines are equipped with a focusing mirror to increase the photon flux on the sample. Such a mirror works by the principle of total external reflection at grazing incidence angle (3-8 mrad). At such a shallow angle, a mirror 500-1000 mm long is required to collect the full vertical fan of radiation from the synchrotron, which is typically 3-5 mm tall at the position of the mirror. A mirror placed halfway between the source point in the storage ring and the experimental station is called a 1:1 mirror. In principle it produces a focused beam of X-rays of the same size as that of the electron or positron beam in the storage ring. At the NSLS, for example, the size would be about 0.025 mm (vertical) by 0.4 mm (horizontal). To produce the desired spot size, the mirror must be fabricated with slope error less than a few μradians. The mirror also acts to remove higher harmonics that are passed by the beamline monochromator, since the mirror will typically not reflect the high-energy X-rays.

To produce micrometer sized focused beams, one employs highly demagnifying optics to image the source onto the sample. Such optics can include Kirkpatrick-Baez mirrors, Fresnel zone plates, tapered capillaries, and compound refractive lenses, all of which have been used to produce submicron focal spots at third generation storage rings.

The experimental station for hard X-ray experiments is typically a steel or steel/lead enclosure. The experimenter has access to the station to change samples, and align equipment so long as the shutter is closed. When ready for data collection, the station is interlocked to prevent access and the safety shutter is opened.

The experimental station contains manipulators to positions the sample, optics to view the sample, and detectors to measure the fluorescent, scattered or diffracted X-rays. The details of these apparatus depend entirely on the type of experiment to be performed. For example, an XRF microprobe requires a precision X-Y-Z sample stage, high quality microscope and multi-element fluorescence detector. A surface scattering experiment, on the other hand requires a 4-circle goniometer and a low-noise photon counting detector.

OTHER CHARACTERISTICS OF SYNCHROTRON RADIATION

Synchrotron radiation is an extremely versatile light source for a large number of experimental investigations because of its following properties: 1.) brightness (highly collimated in the forward direction), 2.) tunability (broad energy coverage from Infra-red to gamma rays that can be continuous with bending magnets and wigglers, or tunable with undulators), 3.) polarization (linear polarized in the orbit plane, elliptically polarized above and below the plane, tunable polarization with specially designed undulators, Fig. 12), 4.) Time- structure (nanosecond to sub-nanosecond pulses with microsecond to nanosecond repetition rates) and 5.) coherence (laser like properties from undulators). We now examine some of these properties.

Polarization

Polarization is a very important property of synchrotron radiation. It provides the capability for the study of magnetic and optical circular and linear dichroism, for polarization dependent EXAFS, and a variety of other experiments. On-axis, (in the case of a single electron in the orbit plane, i.e., zero emittance) the photon is 100% polarized with the E vector parallel to the plane. Above and below the plane the radiation is elliptically polarized and the degree of linear polarization is defined as:

$$p = \left(I_{\parallel} - I_{\perp}\right) / \left(I_{\parallel} + I_{\perp}\right) \tag{23}$$

where $I_\parallel$ and $I_\perp$ are intensities of the parallel and perpendicular components. The degree of polarization is illustrated in Figure 14. We see from Figure 14 that the degree of polarization can be controlled with an aperture. In fact, this was exactly what was done in the early XMCD experiments in which partially circular-polarized light was selected by blocking photons either above or below the plane.

The polarization in the plane of the orbit, together with the small natural vertical opening angle of the radiation from bending magnets and wigglers, is the reason that most beamline monochromators diffract in the vertical direction, rather than the horizontal direction.

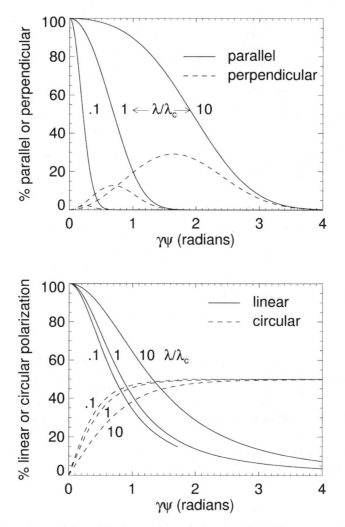

Figure 14. Dependence of the polarization on the vertical angle ψ (top). Three curves with different (λ/λ_c) values are plotted. The x-axis is in $\gamma\psi$ units, making it universal. The bottom curve shows the percent linear and circular polarization as a function of $\gamma\psi$ for the same three values of (λ/λ_c).

Polarization is also very important for X-ray fluorescence experiments because of its effect on the scattered X-ray background. In the case of 100% linear polarization there can be no elastic or inelastic scatter at 90° in the plane of polarization. By placing the fluorescence detector at 90° in the polarization (horizontal) plane, the scattered background is greatly reduced, compared to an unpolarized laboratory source, or to a detector placed in a non-optimal position.

As shown in Figure 12, polarization tunable or semi-tunable insertion devices are now being implemented to provide photons with desired polarization. Some of these devices are known as CPU (cicular polarized undulator) and EPU (elliptically polarized undulator).

Time structure

As noted above, electrons injected into a storage ring are bunched by the r.f cavity that uses an oscillating voltage to replenish the energy loss by the electrons to synchrotron radiation. The r.f. cavity operates at a frequency which falls into the radio frequency bandwidth. It produces regions of stability for the electrons coming through called buckets that need or need not be filled. Thus the number of bunches varies from single bunch to as many as hundreds, depending on the frequency of the cavity and the circumference of the ring. The bunch length determines the pulse width of the synchrotron light. In general, high frequency and high voltage are required to produce short bunches. Figure 15 illustrates the time structure in a typical second and third generation source. It is interesting to note that the NSLS ring (r.f. = 50MHz) has a pulse width of ~ 2 ns with a repetition rate of ~ 20 ns and some of the buckets are not filled while ALS has a considerably shorter pulse width of 35 ps with a much faster repetition rate of 2 ns.

THE INTERACTION OF LIGHT WITH MATTER

The versatile synchrotron light source provides a large number of opportunities for experiments. Let's now outline some of the ways light and matter interacts. In the context of this discussion, we use the dual particle and wave properties of light interchangeably. It is convenient to consider light as an electromagnetic wave to begin. Figure 16 shows the electromagnetic spectrum of light and the corresponding size of the object it can gauge. It is also useful to note that E (eV) = $h\nu$ = hc/λ and in convenient units

$$E \text{ (eV)} = 12398.5/\lambda \text{ (Å)} \qquad (24)$$

Scattering

Upon interacting with matter, light can scatter, i.e., change direction (momentum) with or without energy loss. If it suffers no loss of energy, the process is called elastic scattering. If it loses some of the energy, it is called inelastic scattering. The scattering of light can occur in a random manner when light encounters an irregular object (incoherent scattering) or in a coherent manner when it encounters an ordered array of objects (diffraction). Figure 17 illustrates the ramifications of the scattering of light.

Elastic scattering using synchrotron radiation X-rays provides tools for small crystal crystallography, protein crystallography, powder diffraction and small angle scattering. Inelastic scattering provides even more tools to investigate a dynamic range of physical phenomena, from large energy loss events such as Compton scattering to small energy loss events such as phonon and magnetic scattering or even inelastic nuclear resonance events.

An example of X-ray energy loss spectrum of C_6H_6 with excitation energy of 7270 eV (Hayashi et al. 2002) is shown in Figure 18.

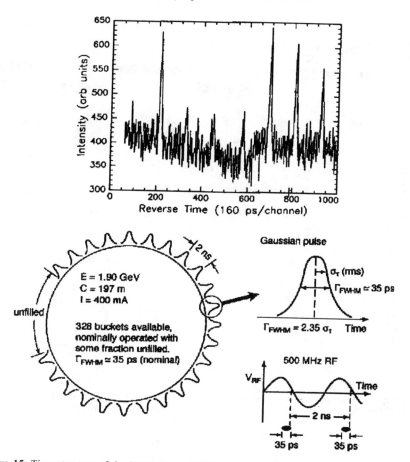

Figure 15. Time structure of the VUV ring at NSLS (top, experimental data) and at ALS (bottom, schematic). [Used by permission of the editor, Cambridge University Press, from Attwood (2000), Fig. 5.25, p. 169]

Absorption

Another behaviour of light-matter interaction is photoabsorption in which the photon is annihilated and its energy is used to excite the system. Synchrotron X-ray absorption provides a number of useful techniques for materials analysis and fabrication.

The tunability of X-rays greatly facilitates techniques such as X-ray absorption fine structures (XAFS), an element sensitive tool for local structure and bonding investigation, and photoelectron spectroscopy (PES) and related phenomena such as Auger and X-ray fluorescence and emission spectroscopy. The photoelectron technique is surface sensitive and with the added advantage of tunability, it can be used widely in surface analysis.

The tunable photon can also be used as a scalpel to fragment molecules in a site - specific manner. Photo-induced chemical processes can be devised for lithography applications.

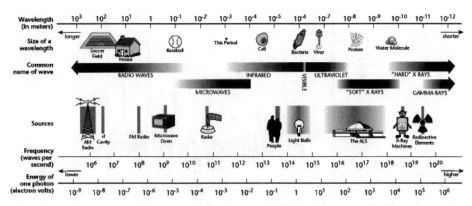

Figure 16. The electromagnetic spectrum. Also shown are objects with comparable dimension as the wavelength (from ALS website).

(a) Isotropic scattering from a point object

(b) Non-isotropic scattering from a partially ordered system

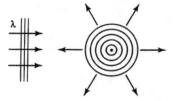

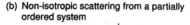

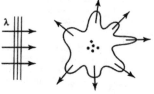

(c) Diffraction by an ordered array of atoms, as in a crystal

(d) Diffraction from a well-defined geometric structure, such as a pinhole

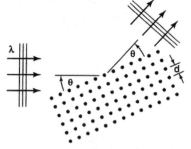

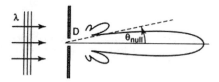

(e) Refraction at an interface

(f) Total external reflection

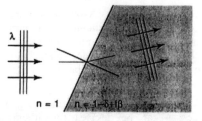

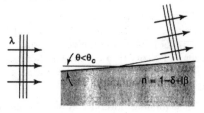

Figure 17. Scattering, diffraction and refraction of light. [Used by permission of the editor, Cambridge University Press, from Attwood (2000), Fig. 1.13, p. 19]

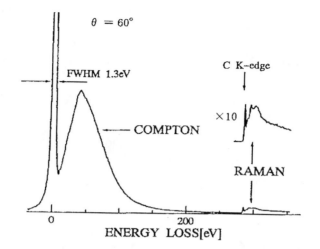

Figure 18. X-ray energy loss spectrum of C_6H_6 with an excitation energy of 7270 eV. [Used by permission of the editor, World Scientific, from Hayashi (2002), Fig. 7, p. 865]

Another application of the absorption process is to use a tunable X-ray microbeam for imaging, microscopy or even radiation-therapy. Tuning the photon to below and above an appropriate edge results in enhanced image contrast, very desirable for tomographic imaging. Furthermore, with the development of sub-micron beams, synchrotron radiation has become a powerful tool for the structural characterization of fine-grained, heterogeneous materials, such as rocks and soils.

The overall picture

Although we have discussed the scattering and absorption of light with matter separately, they are competing processes involving interaction primarily with bound electrons in matter. Both processes are taking place simultaneously. Under certain conditions, one process dominates the other. With a tunable light source, users can therefore select the right photon energy to do the right experiment. The refractive index, n, deals with how light turns going from one medium to the other, and its formulation contains both processes.

$$n = 1 - \delta + i\beta = 1 - \left(n_a r_e \lambda^2 / 2\pi\right)\left(f_1^0 - i f_2^0\right) \qquad (25)$$

Where δ contains information about scattering (f_1^0) and β is related to absorption (f_2^0). Both terms are strongly photon energy dependent. The scattering factor and absorption coefficient of Cu is shown in Figure 19 (Henke et al. 1993). We can clearly see the abrupt change of the scattering factors and the concomitant change of the absorption coefficient at the absorption edge. There are also new research opportunities in resonance spectroscopy and anomalous scattering at the absorption edge.

It is interesting to note that light can see an object if its wavelength is smaller or comparable to the size of the object (diffraction limit; see Fig. 16). For example, optical microscope cannot reveal images smaller than the wavelength of visible light (300-700 nm). Techniques using shorter wavelength are being developed to image smaller objects.

Finally, in light-matter interaction, one may conveniently consider the incident light (photon) as an oscillating hammer (electric field, or A vector). This hammer either gets

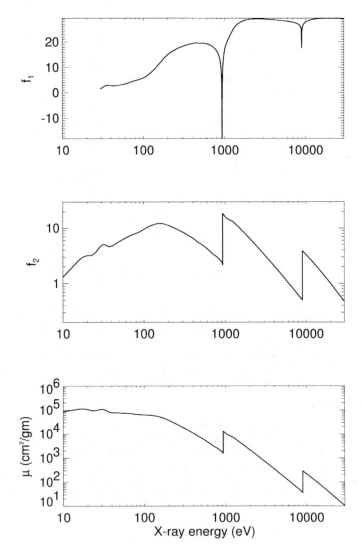

Figure 19. Atomic scattering factors and absorption coefficient for Cu (data from Henke et al. 1993)

bounced by the electrons or knocks the electrons out of their orbital into an excited state that can be bound, quasi bound, or in the continuum.

Tunable VUV and soft X-rays from synchrotron light sources are ideally suited to study surface, near surface (thin film) and interface phenomena. These photons can access many useful core levels of many elements (site specific), they usually do not penetrate very deep and they produce electrons with very short escape depths (surface sensitivity). On the other hand, hard X-rays with great penetration power are ideally suited for all kinds of imaging, such as tomography of rocks and soils.

ACKNOWLEDGMENTS

Useful comments from Dr. Mike A. Green of the Synchrotron Radiation Center, University of Wisconsin-Madison on an earlier version of the manuscript are acknowledged. Research at the University of Western Ontario was supported by the Natural Science and Engineering Research Council (NSERC) of Canada.

REFERENCES

Agui A, Yoshigoe A, Nakatani T, Matsushita T, Saitoh Y, Yokoya A, Tanaka H, Miyahara Y, Shimada T, Takeuchi M, Bizen T, Sasaki S, Takao M, Aoyagi H, Kudo TP, Satoh K, Wu S, Hiramatsu Y, Ohkuma H (2001) First operation of circular dichroism measurements with periodic photon-helicity switching by a variable polarizing undulator at BL23SU at Spring-8. Rev Sci Instru 72:3191-3197

Attwood D (2000) Soft X-ray and Extreme Ultraviolet Radiation, Principles and Applications. Cambridge University Press, Cambridge, U.K.

Conacchia M (1994) Lattices. *In:* Synchrotron Radiation Sources a Primer. Winick H (ed) World Scientific, Singapore, p 30-58

Elder FR, Gurewitsch AM, Langmuir AM, Pollock HC (1947) Radiation from electrons in a synchrotron. Phys Rev 71:829-830

Green GK (1976) Spectra and optics of synchrotron radiation. Brookhaven National Laboratory Report BNL 50522 (National Technical Information Service, Springfield, VA, USA).

Gudat W, Kunz C (1979) Instrumentation for spectroscopy and other applications. *In:* Synchrotron Radiation, Techniques and Applications. Kunz C (ed) Springer Verlag, Berlin, p 55

Hayashi H, Udagawa Y, Gillet JM, Caliebe WA, Kao CC (2002) Chemical application of inelastic X-ray scattering. *In:* Sham TK (ed) Chemical Application of Synchrotron Radiation. Vol. II. World Scientific, Singapore, p 850- 908

Henke BL, Gullikson EM, Davis JC(1993) X-ray Interactions: photoabsorption, scattering, transmission, and reflection at E =50 –30,000 eV, Z =1-92. Atomic Data and Nucl Data Tables 54:181-342

Hofmann A (1990) Characteristics of synchrotron radiation. *In:* Synchrotron Radiation and Free Electron Lasers. Turner. S (ed) CERN Accelerator School Proceedings, CERN 90-03, Geneva, Switzerland. p 115-137

Jackson JD (1998) Classical Electrodynamics. 3rd edition. Wiley, New York

Kim KJ (1985) Characteristics of Synchrotron Radiation. *In:* X-ray Data Booklet. Vaughn D (ed). Lawrence Berkeley Laboratory PUB-490

Krinsky S, Perlman ML, Watson RE (1983) Characteristics of Synchrotron Radiation and Its Sources. *In:* Handbook on Synchrotron Radiation Vol. 1A. Koch E.E (ed) North-Holland, Amsterdam, The Netherlands

McMillan EM (1945) The synchrotron. A proposed high energy particle accelerator. Phys Rev 68:143-144

Schwinger J (1949) On the classical radiation of accelerated electrons. Phys Rev 75:1912-1925

Turner S (ed) (1990) Synchrotron Radiation and Free Electron Lasers. CERN Accelerator School Proceedings, CERN 90-03, Geneva, Switzerland

Veksler V (1945) A new method of acceleration of relativistic particles. J Phys (USSR) 9:153-157

Wille K (2000) The Physics of Particle Accelerators, an Introduction. Oxford University Press, Oxford, U.K.

Winick H (ed) (1994) Synchrotron Radiation Sources, a Primer. World Scientific, Singapore

Winick H (1980) Properties of synchrotron radiation. *In:* Synchrotron Radiation Research. Winick H, Doniach, S. (eds) Plenum, New York p 11-15

3

X-ray Reflectivity
as a Probe of Mineral-Fluid Interfaces:
A User Guide

Paul A. Fenter

Environmental Research Division
Argonne National Laboratory, ER-203
Argonne, Illinois, 60439-4843, U.S.A.

INTRODUCTION

The many tools with which one can probe the atomic-scale structures of surfaces include electron-, ion-, and X-ray based techniques (e.g., low energy electron diffraction, Rutherford ion backscattering, X-ray diffraction, photoelectron diffraction), as well as scanning probe microscopies (e.g., scanning tunneling microscopy and atomic force microscopy) (Somorjai 1981; Woodruff and Delchar 1986; Zangwill 1988; van Hove 1999). These tools have been extremely valuable for revealing surfaces structures and processes at ultra-high vacuum conditions. However, most of these surface-sensitive techniques suffer from the substantial shortcoming, from the perspective of mineral-fluid interface studies, that they cannot be applied to surfaces in contact with water. It is preferable to measure mineral-fluid interface structures *in situ* for direct insight into the geochemical phenomena of interest because there is no reason to assume that a mineral surface can be removed from an aqueous solution without substantially modifying the surface structure or properties.

X-rays are an ideal probe of mineral-water interfaces. X-rays readily penetrate macroscopic amounts of water and can therefore investigate the mineral-water interface directly, *in situ*. X-rays can measure atomic scale structures, such as the separation of individual atoms or molecules, because X-ray wavelengths are comparable to atomic sizes (Warren 1990). In fact, the interaction of X-rays with matter is known at a very fundamental level, and X-ray based techniques can provide truly quantitative data concerning the arrangements of atoms through a variety of approaches, such as crystallography and X-ray absorption spectroscopy (Als-Nielsen and McMorrow 2001). These characteristics can also be used to study the structure of the mineral-fluid interface (e.g., atomic locations, bond lengths) with sub-Ångstrom precision.

Of the many X-ray based techniques available, a very powerful approach for probing interfacial structures is based on the measurement of X-ray reflectivity. The X-ray reflectivity is simply defined as the ratio of the reflected and incident X-ray fluxes. In the simple case of the mirror-like reflection of X-rays from a surface or interface, i.e., specular reflectivity, the structure is measured along the surface normal direction. Lateral structures are probed by non-specular reflectivity. The measurement and interpretation of X-ray reflectivity data (i.e., the angular distribution of X-rays scattered elastically from a surface or interface) (Als-Nielsen 1987; Feidenhans'l 1989; Robinson 1991; Robinson and Tweet 1992) are derived from the same theoretical foundation as X-ray crystallography, a technique used widely to study the structure of bulk (three-dimensional or 3D) materials (Warren 1990; Als-Nielsen and McMorrow 2001). The immense power of the crystallographic techniques developed over the past century can therefore be applied to determine nearly all aspects of interfacial structure. An important characteristic of X-ray reflectivity data is that they are not only sensitive to, but also specifically derived from interfacial structures.

1529-6466/00/0049-0003$10.00

As will be shown below, X-ray data for surfaces and bulk materials appear in different forms (rods vs. points) due to the different dimensionality of these systems (2D vs. 3D). The magnitudes of the signals in surface and bulk X-ray scattering measurements are also quite different, as they are proportional to the number of scattering atoms. The substantially fewer atoms at a surface ($\sim 10^{14}/cm^2$) than in bulk material ($\sim 10^{23}/cm^3$) necessitates the use of powerful X-ray sources. Thus, surface crystallography measurements are almost always performed at synchrotron sources that routinely produce X-ray beams having extremely high flux and low divergence.

It is straightforward to predict the X-ray reflectivity for any given surface or interfacial model because the interaction of X-rays with matter is understood at a fundamental level. Such a calculation can be compared directly with experimental data to test the validity of any particular model. Models can therefore be optimized through direct, quantitative comparison of calculated and measured scattering intensities. More generally, mathematical techniques have been developed that should allow the structure to be derived directly from the data (instead of by optimization of a particular model) (Yacoby et al 2000; Saldin et al. 2001; Takahashi et al. 2001). In the end, it is the fundamental knowledge of the X-ray scattering process that gives X-ray scattering techniques their great power.

A number of excellent articles and books describe the general features of X-ray scattering (Warren 1990; Als-Nielsen and McMorrow 2001), and its use as a probe of surfaces and interfaces (Als-Nielsen 1987; Feidenhans'l 1989; Robinson 1991; Robinson and Tweet 1992). This chapter contains a comprehensive description of X-ray reflectivity as a probe of mineral-water interfaces, including brief discussion of those aspects discussed previously in terms of the vacuum-solid interface. We use simple calculations to provide insight into the utility and sensitivity of X-ray reflectivity for probing interfacial structures. Beyond this introductory material, we emphasize aspects that are unique to measurements at the mineral-fluid interface (sample cell geometry, signal-to-background issues, etc.) and include a description of how a measurement is performed. We end with a few examples from recent work showing how these measurements reveal fundamental properties and processes at the mineral-fluid interface including (1) structure and termination of crystals, (2) adsorption structure and conformation of large molecules, (3) behavior of water near mineral-water interfaces, (4) defect structures of surfaces, and (5) real-time measurements of surface reactions. Additional examples of applications of X-ray reflectivity to mineral-fluid interfaces are presented by Sturchio and Brown (this volume).

Review of relevant literature

Although X-ray reflectivity was first described as a surface-sensitive technique by Parratt (1954), it was not until the late 1970's and early 1980's that it was demonstrated to be a viable technique for probing the atomic-scale structure of buried interfaces and free surfaces (Marra et al. 1979; Eisenberger and Marra 1981; Robinson 1983). X-ray scattering was applied soon thereafter to the solid-liquid interface, although primarily to study the structure of electrified aqueous-metal interfaces. (See papers by Toney 1994, Ocko 1995, and You and Nagy 1999 and references therein.) The application of X-ray scattering to probe the structure of the mineral-water interface was pioneered by Chiarello, Sturchio, and colleagues in the early 1990's with measurements of small angle X-ray reflectivity at the calcite-water interface (Chiarello et al. 1993), heteroepitaxial film growth on calcite (Chiarello and Sturchio 1994), and high-resolution crystal truncation rod measurements (Chiarello and Sturchio 1995). Studies continuing to the present have applied these techniques to solid solution formation at the calcite-water interface (Chiarello et al. 1997); ion adsorption (Sturchio et al. 1997); large molecular adsorption

(Fenter and Sturchio 1999); molecular-scale structure and speciation measurements of calcite, orthoclase and barite interfaces (Fenter et al. 2000a,b, 2001); measurements of real-time dissolution processes (Teng et al. 2001) and dissolution kinetics (Fenter et al. 2002); the three-dimensional structure of the calcite-water interface (Geissbuhler 2000); and the structure of natural growth surfaces (Schlegel et al. 2002). These techniques have been used to probe crystal growth phenomena, including studies of the crystal-fluid interface (De Vries et al. 1998, 1999; Arsic et al. 2001), the structure of water at electrified oxide-water interfaces (Chu et al. 2001), and the oligoclase-water interface (Farquhar et al. 1999). There have been a limited number studies of oxide and mineral surfaces in the absence of a macroscopically thick solution layer, such as studies of alumina surfaces in humid environments (Eng et al. 2000; Trainor et al. 2002) and rutile in ultra-high vacuum (Charlton et al. 1997). (See Renaud (1998) for a complete review the use of X-ray scattering to probe oxide surfaces). X-ray reflectivity and scattering techniques have also been used to probe such diverse phenomena as the structuring of fluids near the solid-liquid interface (Yu et al. 1999, 2000a; Huisman et al. 1997; Cheng et al. 2001) and the association of ions with organic monolayers, either the sub-phase of monolayers at the air-water interface (Leveiller et al. 1991; Kmetko 2001) or monolayers supported by solid substrates (Li et al. 1995). A much larger number of measurements have been performed by using these techniques to study metal and semiconductor surfaces in ultra-high vacuum.

Two observations can be derived from this brief review of previous studies. First, X-ray reflectivity techniques have been applied to a diverse range of phenomena, providing a powerful and robust probe of interfacial properties. Second, the use of these techniques in studying the mineral-fluid interface has been limited to a few research groups. This suggests that application of these techniques to mineral-fluid interface structures and processes is still in its infancy. This review uses examples of recent work carried out at Argonne National Laboratory that were chosen to represent the diverse range of mineral-fluid interface phenomena that can be studied with X-ray reflectivity techniques.

ELEMENTARY SCATTERING THEORY: INTERFERENCE AND BRAGG'S LAW

Interference and reciprocal space

The basis for understanding X-ray scattering phenomena is *interference*. That is, waves that are in phase add constructively, and waves that are out of phase add destructively. This phenomenon is observed for all waves (e.g., photons and sound) and even particles as a results of their de Broglie wavelength (e.g., electrons, neutrons, atoms), and can be demonstrated by the interference of waves on the surface of water. X-rays scatter primarily from electrons, and the angular distribution of X-rays scattered by a single electron is isotropic (except for a slowly varying polarization factor that will be described below) (Warren 1990). The angular variation of scattering intensities observed in experiments, including the location and shape of Bragg peaks, is due to interference phenomena. This is highlighted in Figure 1, in which the angular variation of scattering intensity is shown for a few different objects. (Scattering intensities refer to the count rate of elastically scattered photons in the detector. In contrast, the reflectivity is the ratio of the scattered and incident photon intensities.) The objects in Figure 1 include an electron, an atom, a molecule, and a crystal. These calculations show that the total scattering intensity increases as an object contains more electrons, and the scattering intensity varies more rapidly as a function of angle as the object becomes larger.

The concept of *phase* is used to quantify interference phenomena. As in any optical phenomenon, the relative phase of two photons is determined by the difference in their

Fenter

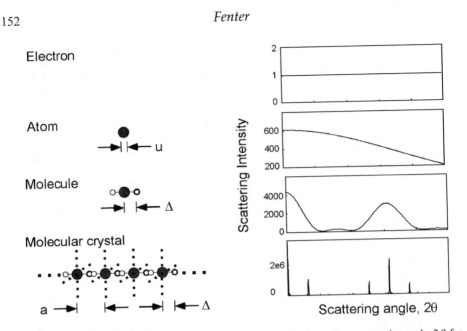

Figure 1. The hierarchy of scattering, illustrated with the scattering intensity vs. scattering angle, 2θ, for a series of objects: an electron, an atom, a molecule, and a crystal. Scattering intensity increases rapidly with the number of electrons in the sample, and the width of the intensity variation as a function of 2θ decreases with the increasing spatial extent of the sample. Note that the bulk Bragg peak intensities are modulated by the molecular form factor for a single molecule.

optical path lengths as they travel from a single source (far from the sample), scatter from two points in a sample, then travel to a detector (again, far from the sample). The path length for each photon is the product of the geometric path length and the index of refraction of the materials through which the photon passes. At X-ray wavelengths, the index of refraction, n, is close to unity for most materials and is defined by the relation, $n = 1 - \delta$, where $\delta \ll 1$. For example, $\delta = 0$ for vacuum and $\delta = 2 \times 10^{-6}$ and 8×10^{-7} for calcite and water, respectively (for a photon energy of 17 keV) (Feidenhans'l 1989). Under most conditions the path length difference for two photons (i.e., the phase difference) scattering from two scattering centers (e.g., atoms) can be determined by using only the geometric path length difference. This situation is idealized in Figure 2A, where X-rays are shown to scatter from two parallel layers separated by a spacing, d, with the angle of the incident and reflected X-ray beams with respect to these planes equal to θ. As for all wave phenomena, the amplitude of the sum of two or more waves depends upon the coherent sum of their amplitudes, ε_i, where the amplitude is a complex number. (See Appendix 1 for a brief summary of complex numbers.) Two waves interfere *constructively* when they are in phase, so that the net amplitude is the *sum* of the two individual amplitudes ($|\varepsilon_{tot}| = |\varepsilon_1| + |\varepsilon_2|$), and *destructively* when they are out of phase, so that the net amplitude is the *difference* between the two individual amplitudes ($|\varepsilon_{tot}| = |\varepsilon_1| - |\varepsilon_2|$).

The condition for constructive interference is simply that the path length difference for the two photons is an integer multiple of wavelengths (as shown in Fig. 2A), or

$$n\lambda = 2d \sin(\theta) \qquad (1)$$

Equation (1) is Bragg's law, which describes the angles at which Bragg reflections occur

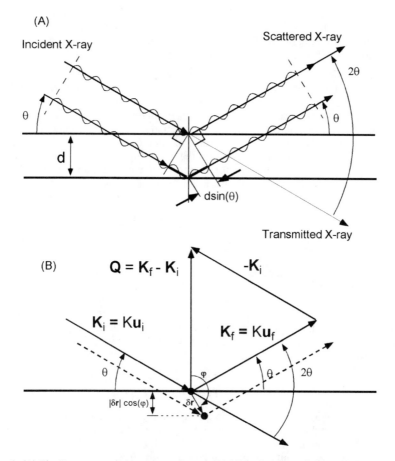

Figure 2. (A) The Bragg scattering geometry, shown to highlight the change in X-ray path length for X-rays scattering from two parallel layers. The two paths differ in length by two wavelengths. (B) The momentum transfer, **Q**, is related to the incident and reflected photon wave vector through the relation, **Q** = **K**$_f$ - **K**$_i$. For X-rays that scatter from two points differing in position by δr, the difference in phase depends only on δr cos(φ), corresponding to the projection of δr along the direction defined by **Q**.

for crystals as a function of the wavelength, λ, the Bragg plane spacing, d, and the diffraction order, n (n has integer values corresponding to the Bragg conditions). (See Appendix 2 for a list of terminology, symbols and definitions.) Figure 2A is drawn for the condition $n = 2$, so that the phase shift between neighboring planes is exactly two X-ray wavelengths. Equation (1) highlights a general difficulty in the interpretation of diffraction patterns: the *spacings* between atoms are derived from the measurements of *angles* (e.g., at which the X-rays scatter most strongly), or more generally from the angular variation of scattered intensities. Thus, it is necessary to learn how to "read" diffraction patterns. Also shown in Figure 2A is the scattering angle, 2θ, which is the angle between the incident and scattered photon directions. The geometry as drawn in Fig. 2A is general for all Bragg diffraction conditions, although the physical surface plane need not coincide with a particular Bragg plane. Consequently it is necessary to distinguish between the Bragg angle, θ, which is measured with respect to the Bragg plane, from the incident angle, α_i, which is measured with respect to the surface plane. For specular (i.e., mirror-like) reflectivity, these two planes coincide and the incident and reflected angles are equal to half the scattering angle. More generally, the detector angle and the angle of the beam with respect to the sample are controlled independently.

Instead of describing interference conditions in terms of angles, it is useful to describe the scattering process in "reciprocal space" in terms of momentum transfer, **Q**, which is a vector having both a magnitude and a direction. (Here, vectors are expressed in bold type, and scalar quantities are written in plain type.) The magnitude of the momentum transfer is related to the scattering angle but is more general, because the momentum transfer at a particular scattering condition (i.e., a Bragg peak) is independent of the choice of X-ray wavelength, whereas the scattering angle of a Bragg peak changes with the X-ray wavelength. The vector nature of the momentum transfer, Figure 2B, allows for an elegant and compact description of single-crystal diffraction phenomena. The momentum transfer is related to the scattering angle, 2θ, by the relation,

$$\mathbf{Q} = \mathbf{K}_f - \mathbf{K}_i = K(\mathbf{u}_f - \mathbf{u}_i) \tag{2a}$$

$$Q = |\mathbf{Q}| = \frac{4\pi}{\lambda}\sin(2\theta/2) \tag{2b}$$

where $\mathbf{u}_f$ ($\mathbf{u}_i$) is the unit vector specifying the reflected (incident) beam direction, having a wave vector $K = 2\pi/\lambda$. Q has units of inverse Ångstroms (Å^{-1}). Here, we explicitly write the expression in terms of the scattering angle, 2θ, instead of the angle of incidence, θ, to highlight that $|\mathbf{Q}|$ is controlled solely by the detector angle, 2θ. That is, **Q** refers to the scattering condition that is being probed by a particular scattering angle and is controlled separately from the orientation of the sample that determines if a diffraction peak is observed. To see this, we rewrite Bragg's law, Equation (1), in reciprocal space,

$$\mathbf{Q}_{HKL} = [H(2\pi/a), K(2\pi/b), L(2\pi/c)] \tag{3a}$$

$$Q_{HKL} = |\mathbf{Q}_{HKL}| = n2\pi/d_{HKL} \tag{3b}$$

where HKL are the Miller indices of the crystal having lattice constants a, b, and c in the three principle crystallographic directions (Fig. 3A), n is the diffraction order and d_{HKL} is the Bragg plane spacing, and $d_{HKL} = 1/|H/a, K/b, L/c|$. The X-ray wavelength does not appear in this expression and Bragg peaks of a given structure have unique positions in reciprocal space, independent of the wavelength.

The real space structure of a simple atomic lattice shown in Figure 3A results in the reciprocal space structure in Figure 3B having regularly spaced Bragg points that appear as discrete spots. Equation (3a) demonstrates that satisfying the Bragg condition for a particular (HKL) diffraction condition for a single crystal requires that two separate conditions be satisfied. First, it is necessary to choose the magnitude of the momentum transfer so that Equation (3b) is satisfied; experimentally, this fixes the scattering angle, 2θ, through Equation (2b). Next it is necessary to orient the momentum transfer to coincide with the points in reciprocal space corresponding to the HKL Bragg condition, as shown in Figure 3B for three diffraction conditions. Figure 3B is drawn in a sample coordinate system (in which the Bragg peaks are fixed), and the momentum transfer rotates to coincide with the Bragg conditions. Experimentally, it is usually the sample that is rotated to allow the Bragg points defined by the crystal lattice to coincide with the momentum transfer vector, **Q**.

The scattered intensity measured far from the sample is calculated by summing the scattering amplitudes, ε_j, for each atom, j, in the sample through the following relation:

$$I \propto \left|\sum \varepsilon_j\right|^2 = |F|^2 \tag{4}$$

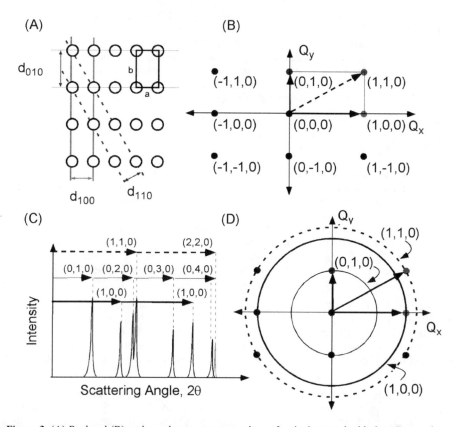

Figure 3. (A) Real and (B) reciprocal space representations of a single crystal with three Bragg planes noted in (A) and their corresponding Bragg conditions shown in (B). Note that the momentum transfer for each Bragg peak in (B) has an orientation that is perpendicular to the Bragg planes noted in (A). A schematic powder diffraction pattern of the structure shown in (A) is shown in (C) with each reflection identified. A reciprocal space description of a powdered sample is shown in (D), in which each Bragg spot is broadened into Bragg rings.

where the magnitude of a complex number, ε, is written $|\varepsilon|^2 = \varepsilon\varepsilon^*$. The scattering amplitude, ε_j, is proportional to the product of its scattering factor, f_j, and phase factor, $e^{i\varphi}$. The sum of all scattering amplitudes within the sample is called the structure factor, F. Consequently, the scattering amplitude, ε, is a complex number whose *amplitude* reflects the *scattering strength*, and whose phase factor determines the constructive and destructive *interference conditions* for any particular value of **Q**. The diffraction intensities are proportional to the magnitude of the scattering amplitudes. Consequently the intensities do not depend upon an absolute phase, but instead on the relative phases for each scattering event.

The phase factor, $e^{i\phi}$, determines the phase of any particular atom for any arbitrary scattering condition, **Q**. The phase for scattering from an object at a location, **r**, with respect to an arbitrarily chosen origin is

$$\phi = \mathbf{Q} \cdot \mathbf{r} = |\mathbf{Q}|\,|\mathbf{r}|\cos(\varphi) \tag{5}$$

where "·" denotes the dot product of two vectors, and φ is the angle separating the vectors **Q** and **r**, as shown in Figure 2B. This expression reveals that the phase difference, $\Delta\phi = \phi_i - \phi_j$, in scattering from two objects separated by δr for a particular value of **Q** depends only on $|\delta r| \cos(\varphi)$, or the separation of atoms along the direction of the vector defined by **Q**. For the case of specular reflection (i.e., with **Q** along the surface-normal direction), this means that the reflected intensity depends solely on the distribution of atomic positions along the surface normal direction and is independent of an atom's lateral location.

Powder diffraction vs. single crystal diffraction. Having reviewed the concepts of Bragg diffraction and reciprocal space, we now make a connection between powder diffraction and single-crystal diffraction. In powder diffraction, the sample consists of many randomly oriented crystallites. The sample may be held fixed, and the scattering intensity is measured as a function of scattering angle, 2θ. A schematic powder diffraction pattern is shown in Figure 3C. All diffraction peaks from a given crystalline phase in the sample are observed in this geometry, because a continuous distribution of crystallite orientations ensures that a subset of crystals is always oriented to Bragg diffract. It is only necessary to orient the detector properly to observe the diffraction signal; that is, the condition that the momentum transfer is coincident with a Bragg point, as shown in Figure 3B, is always satisfied because a subset of the crystallites in the powder sample will always have the correct angular orientation. In reciprocal space, this results in a smearing out of the Bragg points into Bragg "rings" (Fig. 3D).

In powder diffraction, the "background" between the Bragg peaks is of little use, because it derives from a number of different sources (e.g., elastic and inelastic scattering). As we will see below, however, it is precisely in this region that surface reflectivity signals appear. Nearly all surface scattering measurements to date have been performed with high-quality single crystal surfaces, with few exceptions (e.g., Horn et al. 1978).

The "N-slit" diffraction pattern. We start with a simple example of Bragg diffraction from a small perfect crystal to demonstrate how X-ray reflectivity can be used to probe interfaces. Consider the structure factor for a lattice containing N layers having a regular separation, c, with each layer having a scattering strength, f_0. (Here, f_0 is approximately the atomic number, Z; a more accurate expression is described below.) In this case, each scattering center represents a crystal plane. We assume here for simplicity that the momentum transfer is oriented normal to these planes; in this case the phase , from Equation (4), can be calculated for scattering from any particular plane, j, as $\phi_j = \mathbf{Q} \cdot \mathbf{r}_j = |Q| |r_j| = Qz_j$, where $z_j = -(j-1)c$. (Here we choose the origin as the surface layer at $z = 0$). In summing over the contributions from the N scattering centers while keeping track of their different phase factors, we obtain the "structure factor":

$$F = f_0\left\{1 + \exp(-iQc) + \exp(-i2Qc) + \ldots\exp\left[-i(N-1)Qc\right]\right\} \qquad (6a)$$

In this chapter, we follow the convention that the surface is at $z = 0$, and distances below the crystal surface are negative. Here, the structure factor, F, is a complex number. This sum can easily be transformed into a closed form expression because this is a simple geometric series: $1 + x + x^2 + \ldots x^{N-1} = (x^N - 1)/(x - 1)$. Using this relationship, we find that

$$F = f_0 \frac{\exp(-iNQc) - 1}{\exp(-iQc) - 1}$$

$$= f_0 \exp\left(\frac{-i(N-1)Qc}{2}\right)\left[\frac{\sin(NQc/2)}{\sin(Qc/2)}\right] \qquad (6b)$$

The scattering intensity is then:

$$I \propto |F|^2 = |f_0|^2 \left[\frac{\sin^2(NQc/2)}{\sin^2(Qc/2)} \right] \tag{6c}$$

The calculated intensities for a series of structures with different N are shown in Figure 4. The scattering intensity is plotted over a wide Q range (Fig. 4A) showing that Bragg peaks appear at regular intervals, and in this simple case, with identical shapes and intensities. These peaks appear wherever the denominator in Equation (6c) approaches 0 or, equivalently, when $Qc/2 = n\pi$ (where n is any integer). This expression is identical to Bragg's law when written in reciprocal space (Eqn. 3). In Figure 4B the scattering intensity is plotted near the first Bragg condition for selected numbers of layers, N. Little scattering is observed on this linear scale for $N < 4$, but the intensity grows rapidly and develops into sharp Bragg peaks with increasing N. The peak intensity varies as $|f_0|^2 N^2$, and the width varies approximately as $\Delta Q \sim 2\pi/(Nc)$, resulting in an integrated intensity that varies as $\sim |f_0|^2 N$. Therefore, as N increases, the Bragg peaks become sharper and more intense. For an infinite crystal, the diffraction peaks become extremely sharp, and the Bragg condition is satisfied only at the "points" in reciprocal space that exactly satisfy Bragg's law.

So far the X-ray scattering intensity has been discussed in terms of the Bragg diffraction from a bulk *crystal* having a finite number of layers, N. These calculations also contain a small but significant scattering from the *surfaces* of this N-layer crystal. This surface contribution is difficult to see when the data are presented on a linear scale, because the surface scattering intensity is weaker by orders of magnitude than the intensity from the bulk crystal. In fact, a small but significant scattering intensity is observed at essentially all values of Q for these structures, as can be seen in a plot with a

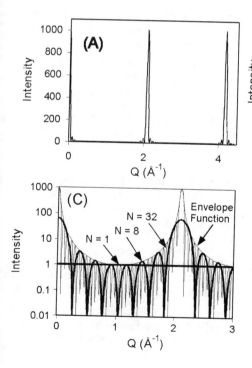

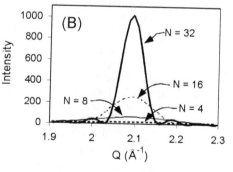

Figure 4. The intensity variation of an N-layer crystal (often referred to as an "N-slit" diffraction pattern). (A) The intensity, plotted as $|F|^2/f_0^2$, is shown over a broad Q range for N = 32. (B) The first-order diffraction peak is shown in detail for N = 4, 8, 16, 32. In these units, the peak intensity varies as N^2. (C) The same functional form, plotted on a logarithmic scale, shows both the bulk Bragg peak near Q = 2.1 and the continuously modulated intensity found between the bulk Bragg peaks.

logarithmic intensity scale (Fig. 4C). The surface scattering intensity is observed as a continuous but strongly modulated "rod" of intensity, in contrast to the bulk Bragg peaks that appear as "points." (That this is a "rod" of intensity, that is sharp laterally and diffuse along the Q_z direction, is shown explicitly below.) For $N = 1$, the intensity is perfectly flat, indicating a lack of interference (because a single layer has nothing with which to interfere). For $N > 1$, the intensity increases near the Bragg peak position (as shown in Fig. 4C), and Bragg peaks develop in both intensity and sharpness as N becomes large.

The intensity away from the Bragg peaks shows a strong modulation with a period (in Q- or angle-space) that decreases as the crystal becomes thicker. An examination of these calculations shows that the oscillation maxima occur at $Q = n2\pi/(Nc)$ or, equivalently, that the oscillation has a period, $\Delta Q = 2\pi/(Nc)$. To understand the source of this modulation, recall that X-rays scattering from two arbitrary layers having a separation d will scatter constructively at angles specified by Bragg's law, $Q = n2\pi/d$ (Eqn. 3). Correspondence between these two expressions requires that $d = Nc$, implying that the modulation of the scattering intensity is due to interference of X-rays that reflect from the top and bottom interfaces of the crystal. The intensity maxima are observed when these X-rays scatter constructively (i.e., where Bragg's law is satisfied with $d = Nc$). These intensity oscillations are often referred to as Kissig fringes, particularly when they are observed as weak oscillations just above and below the Bragg peak, as seen in Figure 4A, for $N = 32$.

This same conclusion can be reached through an analysis of these intensity calculations in angle space. In this case, the angular period of the oscillations, $\Delta(2\theta)$, is related to the plane separation by Bragg's Law, by $n\lambda = 2d\sin(2\theta/2)$. Or, the angular separation of these constructive interference conditions, $\Delta(2\theta)$, is given by the relation:

$$\lambda = 2d\Delta\left[\sin(2\theta/2)\right] \approx d\cos(2\theta/2)\Delta(2\theta) \approx d\Delta(2\theta), \text{ or } \Delta(2\theta) \approx \lambda/d \qquad (7)$$

Here we use the approximation $\cos(\theta) \approx 1$ which is valid if d/λ is large. If the data were plotted as a function of 2θ, then the conversion of the oscillation period to a distance would correspond to a spacing of $d = Nc$, as derived above.

The intensity maxima of these oscillations are defined by an "envelope function" giving an overall U-shape to the intensity between the Bragg conditions when plotted on a logarithmic intensity scale, with a functional form of $1/\sin^2(Qc/2)$ independent of the value of N. The magnitude of this envelope function near the halfway point between the Bragg peaks (i.e., $Q = \pi/c$) is exactly equal to the intensity for $N = 1$; that is, the scattering intensity near the "mid-zone" (at successive maxima) for a crystal of finite thickness is independent of the number of layers in the crystal, and the scattering intensity is equal in magnitude to that from a single isolated layer.

These observations concerning the scattering intensity and the period of the intensity oscillation reveal that the scattering intensity between bulk Bragg peaks is primarily due to X-rays that scatter from the upper and lower *surfaces* of the thin crystal instead of the interior "bulk" of the crystal. This phenomenon is analogous to the observation of color fringes in soap bubbles, where the interference is between photons that are reflected from the top and bottom interfaces of the soap film. In the present case, the use of monochromatic X-rays results in modulation of the reflected intensity; if we were to use a broad-band ("white") X-ray beam, we would see the "color" (i.e., the wavelength) of the reflected X-rays change continuously between the Bragg peaks, just as with soap bubbles.

The crystal truncation rod (CTR). We now calculate the scattering intensity for a semi-infinite lattice, i.e., which has only one reflecting interface. This sum is nearly identical to that of the N-slit diffraction pattern, except for an additional factor that is needed to account for linear attenuation of X-rays as they pass through the lattice which must be included as the number of layers, N, approaches infinity. The linear attenuation of X-rays over a distance, x, follows the form $\exp(-x/\Lambda)$, where Λ is the linear attenuation length. Even though X-ray attenuation is weak in most solids over atomic dimensions, it is nonetheless finite in size, and it must be taken into account explicitly when the crystal thickness becomes comparable to the X-ray penetration depth. For example, calcite has a penetration length of $\Lambda \sim 50\ \mu$m for an X-ray wavelength of 1.5 Å. This corresponds to a typical per-layer attenuation factor of $\eta = \exp(-c/\Lambda) = 0.99994$ (where $c = 3.035$ Å, corresponds to the plane spacing of the (104) cleavage plane). Including this small but significant attenuation factor in the structure factor calculation gives the following expression:

$$F_{ctr} = f_0\left[1 + \eta\exp(-iQc) + \eta^2\exp(-i2Qc) + \ldots + \eta^{N-1}\exp(-i(N-1)Qc)\right] \quad (8a)$$

$$= f_0\frac{\eta^N\exp(-iNQc) - 1}{\eta\exp(-iQc) - 1} \quad (8b)$$

$$F_{ctr} = f_0\frac{1}{\left(1 - e^{-iQc}\right)} \quad (8c)$$

$$I_{ctr} \propto |F|^2 = \frac{|f_0|^2}{4\sin^2(Qc/2)} \quad (8d)$$

Equations (8a and 8b) are written for a finite crystal thickness with the assumption that the structure factor is evaluated in the limit where N becomes very large. Therefore, we use the fact that $\eta^N = 0$ as N becomes very large (since $\eta < 1$); because typical values of η differ from 1 by only a very small amount, we approximate $\eta = 1$ in the denominator of Equation (8b). The resulting expression (Eqn. 8d) is plotted (Fig. 5A) with the N-slit diffraction pattern for $N = 1$ and $N = 32$. This expression has the strong enhancement of the scattering intensity near the Bragg condition and the general U-shape that we observed for the finite crystal, but it does not show high frequency oscillations between the Bragg peaks. This is because X-rays that scatter from the top interface have nothing with which to interfere. (There is no "bottom" interface.) Meanwhile, the intensity just above or below a bulk Bragg peak is insensitive to whether the crystal has a finite number of layers or is semi-infinite. (The intensity exactly at the Bragg peak is still very sensitive to the exact thickness of the layer, as seen in Figs. 4B and 4C.) The structure factor (Eqn. 8c) for a semi-infinite lattice is known as a "crystal truncation rod" or CTR (Robinson 1986); that is, it represents the structure factor of a truncated (semi-infinite) lattice. This is a very general result applicable to any two-dimensional interface. This further confirms that the U-shaped variation of the overall intensity between Bragg peaks for an N-layer crystal is caused by scattering from the crystal surfaces.

Note that the CTR for a semi-infinite crystal has a scattering intensity (in units of $|F|^2/f_0^2$) of 0.25 at the first anti-Bragg condition (i.e., $Q = \pi/a$), as compared to the value of 1 for the N-slit diffraction pattern (at successive maxima). To understand this observation, note that the CTR structure factor at this "mid-zone" position, $Q = \pi/c$, is $F_{ctr} = f_0[1/(1 - e^{-iQc})] = f_0[1/(1 - e^{-i\pi})] = f_0/2$, or half the scattering strength of a single layer! Consequently, the scattering intensity for a single interface is given by $|F_{ctr}|^2 = f_0^2/4$. This appears counterintuitive, given the observation above that the scattering intensity midway between Bragg reflections from an N-layer crystal has an intensity of f_0^2, which is the same as the intensity from a single layer of atoms. However, since the scattering factor of a single interface is $f_0/2$, the net scattering intensity for two interfaces is $|f_0/2 + f_0/2|^2 = f_0^2$

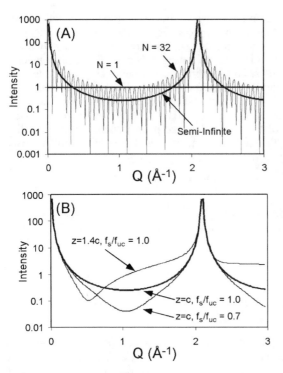

Figure 5. (A) The crystal truncation rod (CTR) structure factor, shown along with the N-layer structure factor for $N = 1$ and 32 (again plotted as $|F|^2/f_{uc}^2$). (B) The CTR structure factor for an ideally terminated surface, with two surfaces where the outermost layer occupation f_s/f_{uc} or position z is modified. These relatively small changes in have a substantial (~10-fold), highly Q-dependent effect on the reflected intensity that is measured.

when the two interfaces interfere constructively and $|f_0/2 - f_0/2|^2 = 0$ when the two interfaces interfere destructively as seen in Figure 5A. The fact that the structure factor of an interface, at $Q = \pi/c$, is one-half the structure factor of a single molecular layer is what makes the scattering intensity near the anti-Bragg condition the same for a single layer and a crystal with N layers (the latter observed at successive maxima). The rapid oscillations between the Bragg peaks for the N-layer crystal can therefore be attributed to the oscillation of the phase factor between 1 and −1, as expected from the application of Bragg's law with $d = Nc$.

In deriving Equation (8), we have implicitly assumed that the incident beam intensity is not significantly depleted as the X-rays reflect from subsequent layers of the crystal at the Bragg condition. (We have only accounted for linear attenuation of the X-ray beam.) This is known as the kinematic approximation, and it is valid when the fraction of the incident beam that is reflected from the crystal (i.e., the reflectivity) is small (<<1). However, we know from Figures 4B and 4C that the scattering intensity becomes very large near a Bragg peak; Equation (8) is not valid very close to the Bragg peak positions when N is large. It is possible to derive a complete expression for the scattering intensity that is also correct at the bulk Bragg position, but such expressions are substantially more complicated and are referred to as dynamical diffraction theory (Bedzyk and Cheng, this volume). This theory does not substantially improve our ability to determine the *surface* structure, although it is necessary to explain quantitatively the scattering intensity from the bulk of a perfect crystal. Therefore, we avoid measurements very close to the bulk Bragg positions and safely ignore dynamical diffraction effects in our analysis of X-ray reflectivity data.

The CTR shape is sensitive to the termination of the crystal surface. Calculations show (Fig. 5B) that CTR data are sensitive not only to the presence of the crystal termination but also to the detailed termination of the lattice. Here, we compare the scattering intensity for a semi-infinite lattice in which the outermost surface layer is ideally terminated with a bulk-like termination, or has been modified, either by its position, d_s, or its scattering strength, f_s. The total structure factor of the crystal with a modified surface is just the sum of individual structure factors for all atoms in the crystal (a table of commonly used structure factors is given in Appendix 3). Conceptually, this quantity can be broken into two parts, consisting of contributions from the modified surface layer, F_{surf}, and from the semi-infinite substrate, F_{sub}:

$$F = F_{sub} + F_{surf}$$
$$= \frac{F_{uc}}{1 - e^{-iQc}} + F_s \exp(iQd_s) \tag{9a}$$

where F_{uc} is the unit cell structure factor of the bulk crystal (the outermost layer of the unperturbed crystal is at $z = 0$), F_s is the structure factor of the surface layer (calculated assuming that it is located at $z = 0$), and $\exp(iQd_s)$ is an extra phase factor to translate the surface layer to the height $z = d_s$. If we make the simplifying assumption that F_{uc} and F_s are real, the intensity is then:

$$I \propto FF^* = \left\{ \frac{F_{uc}}{\left(1 - e^{-iQc}\right)} + F_s \exp(iQd_s) \right\} \left\{ \frac{F_{uc}}{\left(1 - e^{iQc}\right)} + F_s \exp(-iQd_s) \right\}$$
$$= \frac{|F_{uc}|^2}{4\sin^2(Qc/2)} + F_s^2 + \frac{2F_{uc}F_s \sin[Q(c/2 - d_s)]}{2\sin(Qc/2)} \tag{9b}$$

This result is plotted in Figure 5B for the case in which either the scattering factor or the location of the outermost layer is changed with respect to the ideally-truncated bulk structure, representing the phenomena of adsorption and surface relaxation, respectively. As expected, if $d_s = c$ and $F_s = F_{uc}$, then the result is identical to the calculation for $F_{surf} = 0$, because the addition of an otherwise identical layer to a semi-infinite crystal should have no effect on the scattering intensity. Changes in the position of the outermost layer by a few tenths of a d spacing, or the scattering amplitude by a modest amount, result in order-of-magnitude changes in the reflectivity that are highly Q-dependent. These calculations show that the CTR is extremely sensitive to the surface termination. An interesting phenomenon is observed when the surface layer is identical to the substrate layers but is only half-occupied (i.e., $F_s = F_{uc}/2$, and $d_s = c$). In this case the reflected intensity is exactly zero at $Q = \pi/c$! This result is independent of the crystal structure, as represented by F_{uc}. [This can bee seen from Eqn. 9a: $F = F_{uc}/(1 - (-1)) + (F_{uc}/2)(-1) = F_{uc}/2 - F_{uc}/2 = 0$; $I = |F|^2 = 0$.] This scattering condition is known at the first "anti-Bragg" condition (as compared with the Bragg condition at $Q = 2\pi/c$). A half-filled layer acts as a perfect anti-reflective coating at this scattering condition. We describe below how this phenomenon can be used to probe dissolution processes and kinetics through real-time measurements of the reflectivity at the anti-Bragg condition.

This calculation also illustrates a general result in diffraction theory: the scattering intensity for any compound object, $A + B$, can be written as:

$$I \propto |A + B|^2 = AA^* + BB^* + AB^* + A^*B = \left(|A|^2 + |B|^2\right) + \left(AB^* + A^*B\right) \tag{10}$$

In other words, the scattering intensity from a compound object, $A + B$, is the *incoherent* sum of the scattering intensities from the individual objects (e.g., $|A|^2 + |B|^2$), plus a *cross term* representing the interference of X-rays scattered by the two objects ($AB^* + A^*B$). As seen explicitly in Equation (9b) above, the height of the surface layer with respect to the substrate, d_s, appears only in the cross term.

Experimentally, the "signal" that establishes the separation can be thought of as a Q-dependent structure factor, $F_{uc}/[2\sin(Qc/2)]$ (this is previously known since it depends only on the bulk crystal structure through F_{uc}), modulated by an additional multiplicative factor, $F_s\sin[Q(c/2 - d_s)]$, that contains the unknown structural information, specifically the scattering factor (approximately its atomic number) and the location of the unknown scatterer, F_s and d_s, respectively. To determine uniquely the cross term in Equation (9) (i.e., the term proportional to $F_{uc}F_s$) requires observing the sinusoidal variation of this cross term over a broad range of reciprocal space through a high-resolution X-ray reflectivity measurement.

SCATTERING INTENSITY AND REFLECTIVITY

The largely qualitative description of X-ray scattering presented above is meant to provide insight into the overall features that would be observed in a surface diffraction experiment. The main value of X-ray scattering as a structural tool, however, is that the technique is truly quantitative, so that calculated intensities can be compared directly to experimentally measured scattering intensities. To take advantage of the full power of the X-ray scattering technique requires all relevant aspects of the X-ray scattering process to be included in the calculation of the scattering intensities. Because this subject has been covered previously in much detail (Feidenhans'l 1989; Robinson 1991; Robinson and Tweet 1992), an overall description of the numerous factors is given here for completeness, with reference to more detailed discussions where available.

In surface crystallographic measurements we are concerned primarily with the *integrated* scattering intensity, measured by a rocking scan (i.e., where the scattering intensity is measured as a function of a spectrometer angle, typically θ, and the integrated intensity is the integrated weight of a sharp peak corresponding to the reflectivity, after subtracting any background signals; this is described in more detail below). As originally derived by Robinson (1991), the integrated scattering intensity, I_{int}, of any two-dimensional structure can be written with no adjustable parameters as

$$I_{int} = I_0 \left(\frac{e^2}{mc^2}\right)^2 T_{cell} \left[\frac{\left(A\lambda^3 Pol / a_{uc}^2\right)}{\Omega}\right]\left[\frac{1}{\sin(2\theta)}\right]|F|^2\left(\frac{\Delta Q_z}{2\pi}\right) \tag{11}$$

where I_0 is the incident photon intensity (in units of photons per area per second), $r_e = e^2/mc^2$ is the classical electron radius, T_{cell} is the transmission factor of X-rays through the cell, A is the "active area" of the surface (i.e., the area of the surface that participates in the diffraction measurement), λ is the X-ray wavelength, Pol is a polarization factor, a_{uc} is the area of the surface unit cell, Ω is the angular velocity in the rocking scan, $1/\sin(2\theta)$ is the Lorentz factor, $|F|^2$ is the structure factor of the sample, and ΔQ_z is the length of the surface rod that is integrated in the rocking scan (e.g., due to a finite detector resolution). I_{int} therefore has units of photons per second.

Many of these factors depend on the magnitude and/or orientation of the momentum transfer, $\mathbf{Q}$, and in some cases depend upon the choice of X-ray spectrometer and scattering geometry. (See, for example, Appendix 4 for description of the relationship between spectrometer angles and reciprocal space coordinates for a four-circle spectrometer.) It is therefore not practical to provide a general expression for all possible experimental configurations. Instead, simple examples are used to highlight the roles of each factor, with an emphasis on the application of this expression to specular reflectivity measurements. Because the structural information of interest is found in the structure factor, $|F|^2$, the experiment should be designed so that all other factors are robust and can be quantified to a known level of precision. Some of the factors in Equation (11) require only minimal explanation. The other factors are explained in more detail below.

The classical radius of an electron is the effective size of an electron as seen by X-rays, $r_e = e^2/mc^2 = 2.818 \times 10^{-5}$ Å. This small size is responsible for the fact that X-rays interact weakly with matter. For instance, we can estimate the fraction of X-rays that interact with electrons in passing through a single layer of calcite. With a Thompson cross section of $\sim 8 \times 10^{-10}$ Å^2 per electron, and 50 electrons per $CaCO_3$, and 5×10^{14} $CaCO_3$/cm^2, we find that only ≈ 1 out of 10^9 photons interact with calcite as they pass through a single calcite layer! The weakness of this interaction is an important factor that allows us to calculate X-ray scattering intensities quantitatively, but it also requires very strong X-ray sources for routine measurements of samples having few atoms (such as surfaces and interfaces).

The factor Ω denotes the angular scan rate (e.g., radians/sec) in a rocking scan. Although measurements can be performed with a constant scan rate, most measurements are done by counting for a particular amount of time, τ, at regularly spaced intervals, $\delta\theta$, through the rocking scan with an average scan rate of $\Omega = \delta\theta/\tau$. The integrated intensity is then determined by least-squares fitting of the rocking curve data (as described below). The raw data are usually normalized to the signal from a beam monitor to eliminate systematic variations of the incident beam intensity that are typical of synchrotron sources. The beam monitor is normally either an ion chamber that measures photoelectric production in a gas, or a scintillator detector that monitors photons scattered by air in the beam path. In either case, the monitor signal, M (measured in counts per second, or cps), is designed to be proportional to the incident beam flux, I_0A_0 (in photons per second), or $M = \alpha_{mon}I_0A_0$, where α_{mon} is a prefactor that can be calibrated independently or calculated on the basis of the monitor design and the known X-ray absorption and scattering cross sections, and A_0 is the beam cross sectional area. The monitor signal for each point in a rocking curve will vary as $M = \alpha_{mon}I_0A_0/\Omega$. The quantity that is usually measured is the scattering intensity normalized to the beam monitor:

$$\frac{I_{int}}{M} = \left(\frac{e^2}{mc^2}\right)^2 \left(\frac{T_{cell}}{\alpha_{mon}}\right)\left[\frac{(A/A_0)\lambda^3 Pol}{a_{uc}^2}\right]\left[\frac{1}{\sin(2\theta)}\right]|F|^2\left(\frac{\Delta Q_z}{2\pi}\right) \tag{12a}$$

To present the data in terms of an absolute reflectivity (i.e., the ratio of the reflected to incident flux) we normalize all integrated intensities by the integrated signal from a scan through the "straight-through beam," i.e., at $2\theta = 0°$ with the sample retracted (a "2θ arm-zero" scan). The detector slits are normally larger than the beam cross section, so the 2θ arm-zero scan is resolution limited with a width defined by the angular detector size, $\Delta(2\theta)$. In this case, the arm-zero scan normalized to the monitor signal is:

$$\frac{I_{beam}}{M} = \left(I_0A_0/\Omega\right)\frac{\Delta(2\theta)}{\alpha_{mon}I_0A_0/\Omega} = \frac{\Delta(2\theta)}{\alpha_{mon}} \tag{12b}$$

The absolute reflectivity can be written as follows:

$$R = \frac{(I_{int}/M)}{(I_{beam}/2M)}$$

$$= \left(\frac{e^2}{mc^2}\right)^2 T_{cell} \left[\frac{(A/A_0)\lambda^3 Pol}{a_{uc}^2}\right]\left[\frac{1}{\sin(2\theta)}\right]|F|^2\left[\frac{\Delta Q_z}{\pi\Delta(2\theta)}\right] \qquad (13)$$

Under most circumstances, it is not convenient to measure the incident and reflected beams directly with the same detector because scintillator counters have a substantially smaller dynamic range than the $\sim 10^5$- to 10^{10}-fold difference between the direct beam flux and reflected beam flux. Instead, direct knowledge of the detector resolution, $\Delta(2\theta)$, and the conversion factor between the monitor signal and the incident beam flux, α_{mon}, can be used to estimate the absolute reflectivity. Furthermore, the absolute reflectivity is well constrained by measurements close to bulk Bragg features or at the total external reflection condition near $2\theta \sim 0°$. These intensities are dominated by bulk properties of the substrate and provide an independent calibration on the absolute reflectivity scale.

X-ray reflectivity results are sometimes reported in terms of "structure factor." In this case, the data are presented as the square root of the measured integrated intensities. Suitable normalization enables the conversion of this scale to a plot of scattering amplitude in units of electron scattering.

The rest of the factors, described below, are separated into "intrinsic" and "sample- and cell-dependent" correction factors. Each of these requires some discussion.

Intrinsic factors

Structure factor. The structure factor for a collection of atoms is

$$F = \sum_j f_j(Q)\exp(i\mathbf{Q}\cdot\mathbf{R}_j)\exp\left[-\tfrac{1}{2}(Qu_j)^2\right] \qquad (14)$$

where f_j, $\mathbf{R}_j$, and u_j are the atomic scattering factor, position, and vibrational amplitude of atom j. This sum is over all atoms in the sample. The atomic scattering factor, $f(Q)$, is a slowly varying function that takes into account the finite spatial breadth of the core and valence electron distributions (Fig. 1). These scattering factors are parameterized in the form of a sum of Gaussians, where appropriate parameters are tabulated for all atoms and for different chemical states of each atom (Hubbell et al. 1975). The scattering factor of an atom at $Q \sim 0$ is insensitive to the spatial distribution of electrons in the atomic orbitals. Consequently $f_j(Q = 0) = Z$, where Z is the atomic number of that element. The Debye-Waller factor, $\exp[-\frac{1}{2}(Qu)^2]$, accounts for the finite distribution of atomic locations, for instance due to thermal vibrations. A typical value for the vibrational amplitude of an atom in a crystal at ambient conditions is $u < 0.1$ Å. For more complex distributions, e.g., layering in fluid water, the same functional form can be used to describe the time-averaged distribution of a particular atom.

We re-derive the CTR structure factor for a 3D semi-infinite crystal to show its rod-like character explicitly. We assume an orthogonal bulk crystal structure with lattice parameters (a, b, c) where the surface is defined by the a-b plane (Fig. 6A). The lattice vectors are oriented so that $\mathbf{Q} = [Q_x, Q_y, Q_z] = [(2\pi/a)H, (2\pi/b)K, (2\pi/c)L]$, where H, K, and L are the surface Miller indices. If the structure of the crystal is assumed to be identical for every unit cell in the crystal (including all surface layers), then the sum in Equation (14) can be rewritten so that the structure factor of the whole crystal is expressed as the unit cell structure factor, F_{uc}, multiplied by the phase factors that translate the unit cell structure factor to each and every unit cell in the crystal, both laterally within the surface plane, and vertically into the crystal. In this case,

$$F = \left\{ \sum_{uc} f_j(Q) \exp(i\mathbf{Q} \cdot \mathbf{R}_j) \exp\left[-\tfrac{1}{2}(Qu_j)^2 \right] \right\}$$
$$\times \left[\sum \exp(inQ_x a) \right] \left[\sum \exp(imQ_y b) \right] \left[\sum \exp(-ipQ_z c) \right]$$

$$(15)$$

$$|F|^2 = |F_{uc}|^2 \left[\frac{\sin(NQ_x a/2)}{\sin(Q_x a/2)} \right]^2 \left[\frac{\sin(MQ_y b/2)}{\sin(Q_y b/2)} \right]^2 \left[\frac{1}{\sin(Q_z c/2)} \right]^2$$

$$= |F_{uc}|^2 \frac{\delta(Q_x - H2\pi/a)\delta(Q_y - K2\pi/b)}{\sin^2(Q_z c/2)}$$

Here the sums are for $n = 0, \pm 1, \pm 2 \dots$; $m = 0, \pm 1, \pm 2 \dots$; and $p = 0, 1, 2 \dots$, etc. In Equation (15), we have taken the limit where N and M approach infinity. This shows that the scattering from a terminated crystal lattice is in the form of rods that satisfy a 2D Bragg condition, $\mathbf{Q}_{//} = (Q_x, Q_y) = (H2\pi/a, K2\pi/b)$, with H and K having integer values.

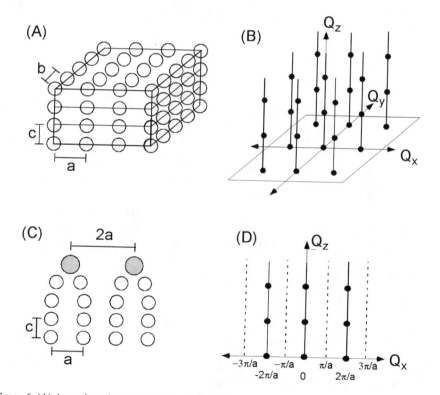

Figure 6. (A) A terminated crystal having surface lattice spacings, a and b, and layer spacing c along the surface normal direction. (B) The reciprocal-space structure for the structure in (A); note that every Bragg peak is intersected by a CTR. (C) Side view of the surface in (A) whose surface symmetry has been modified by adsorption resulting in a doubling of the unit cell dimension, 2a, and by a subsequent lateral surface relaxation in the outermost substrate layer. (D). A reciprocal space schematic of the surface in (C). The doubling of the surface unit cell results in new surface rods at $Q_x = \pm \pi/a$, $\pm 3\pi/a \dots$(shown as vertical dashed lines) that do not intersect any bulk Bragg peaks.

The scattering intensity is, in principle, non-zero at any point along the CTRs, and the bulk Bragg peaks are located along each surface rod, notated by surface Bragg indices H and K, at Q_z values dictated by the bulk crystal structure. The CTR structure is shown schematically in Figure 6B, where the thin vertical lines are the CTRs, and the spots on these rods correspond to bulk Bragg conditions.

The structure factor in Equation (14) can be rewritten in a form that allows for substantial conceptual simplification. Any mineral-fluid interface consists of three parts, the semi-infinite bulk crystal structure that is known in principle a priori, the distorted surface region (whose depth is not known a priori) and the fluid layer above the surface. The total structure factor, F_{tot}, can be written as follows:

$$F_{tot} = F_{uc} F_{ctr} + F_{surf} + F_{water.} \tag{16a}$$

Each of these terms is identical to the structure factor as described in Equation (14), except that the summation is over a subset of atoms. F_{uc} is summed only over atoms within a single bulk unit cell, F_{surf} is summed over all near-surface atoms that might be displaced from their ideal bulk lattice positions (typically 2-3 layers deep into the crystal) plus any adsorbed layers attached to the surface, and F_{water} describes the fluid structure above the interface, including any structuring of the fluid near the mineral surface. Equation (16a) can be rewritten to express the scattering intensity,

$$I_{tot} \propto \left| F_{uc} \right|^2 \left| F_{ctr} \right|^2 + \left[F_{uc} F_{ctr} F_{surf}^* + F_{uc}^* F_{ctr}^* F_{surf} + F_{uc} F_{ctr} F_{water}^* + F_{uc}^* F_{ctr}^* F_{water} \right]$$
$$+ \left| F_{surf} \right|^2 + \left[F_{water} F_{surf}^* + F_{water}^* F_{surf} \right] + \left| F_{water} \right|^2 \tag{16b}$$

where these terms are ordered in terms of decreasing contributions to the reflectivity for typical systems. Equation (16b) highlights an important subtlety of the CTR approach. The ability to calculate the *surface* reflectivity quantitatively depends on precise knowledge of the bulk crystal structure. This can be easily seen in Equation (16b) where certain terms contain information specific to the bulk crystal structure (e.g., $|F_{uc}|^2|F_{ctr}|^2$), but also terms that can be strongly modulated by the contributions from the bulk crystal structure, as indicated by terms that contain F_{uc} (e.g., $F_{uc}F_{ctr}F_{surf}^*$). Additional terms are derived solely from interfacial structure (e.g., $|F_{surf}|^2$).

Two points, however, should be taken into account. First, natural crystals can show significant variability that depends upon the growth conditions and locality (e.g., solid solutions and incorporation of impurities). It is necessary to measure the bulk crystal structure of such samples before it is possible to determine the surface structure using the CTR approach for such samples. Second, the CTR intensities can depend on the type of form factors (e.g., neutral or ionic form factors) used in the bulk structure analysis. At minimum, the calculated bulk Bragg reflectivities must reproduce the observed values precisely; internal consistency requires that we use the same atomic form factors that were used in the determination of the bulk crystal structure. Similarly, the bulk vibrational amplitudes derived from the original bulk crystal structure analysis must be used. In many cases, vibrational amplitudes are anisotropic and are therefore described by a tensor. The appropriate projection of the vibrations for each scattering condition, $\mathbf{Q}$, needs to be included in the expression for F_{uc}.

Crystal truncation rods vs. surface rods. All bulk Bragg points in Figure 6B are intersected by a crystal truncation rod. This is a fundamental property of a terminated lattice (Robinson 1986). The truncation rods are always oriented perpendicular to the physical surface plane, even when the surface plane and the crystallographic plane do not coincide (e.g., a miscut crystal). However, additional "surface rods" that do not intersect bulk Bragg points are often observed. Such rods would be associated with specific structural arrangements of the surface (e.g., a 2D adsorbate lattice or a surface reconstruction) where the surface lattice has a larger unit cell than that found for an ideally terminated crystal (Fig. 6C). These additional rods appear in the diffraction pattern between the CTRs (Fig. 6D) where the substrate lattice makes no contribution to the scattering intensity. The intensity of these rods will therefore not show the typical CTR shape described above; instead the surface rods can be calculated as:

$$F_{surf\text{-}rod} = F_{surf} + F_{water} \qquad (17)$$

These surface rods will typically be much flatter as a function of Q_z because surface structures are typically confined to a depth of a few Ångstroms. As in the case of the N-layer diffraction pattern, the width of a diffraction feature varies approximately as, $\Delta Q \sim 2\pi/R_{max}$, where R_{max} is the size of the object in the direction being probed. Consequently one can expect that the intensity along the surface rods to modulate slowly as a function of Q_z. More generally, the shape of the surface rod in the Q_z direction provides direct information concerning the vertical structure of this surface phase with no contribution from the bulk crystal structure. This avoids some of the complications that were discussed above concerning uncertainties in the bulk crystal structure. But, as is implicit in Equation (17), the surface rods will have contributions from any substrate layers that are displaced with respect to the bulk crystal structure. So if a surface rod appears due to the adsorption of an ion with a larger unit cell, the substrate will contribute to the surface rod if there are any significant displacements of the substrate layers in the symmetry of the adsorbate layer (as shown in the outermost substrate layer in Fig. 6C), as is common for chemisorbed layers.

Sample- and cell-dependent factors

Surface roughness factor, $|B|^2$. All surfaces exhibit steps and other defects that must be accounted for in any quantitative comparison of data and calculations. As shown in Figure 5, the CTR is sensitive to the presence of partially occupied layers, even if those layers are exactly in the bulk-like location. Steps modify the reflectivity because X-rays that scatter from terraces at different heights interfere with each other. The reflectivity therefore is sensitive to the distribution of terrace heights. A net increase in roughness results in a decrease in reflectivity near the anti-Bragg condition. However, this decrease is not uniform as a function of Q. To calculate the effect of roughness, we must first determine the lateral length scale over which the roughness occurs. Two extreme examples are useful. If the typical lateral step spacing is small with respect to the lateral coherence length of the measurement, the presence of steps needs to be averaged coherently in the structure factor, as

$$I_{rough} \propto \left| \sum_j H_j F_{ideal} e^{i\phi(j)} \right|^2 \qquad (18a)$$

Here $F_{ideal}(Q)$ is the structure factor for a smooth surface (as described in Eqn. 16), H_j is the terrace height distribution (i.e., the probability that the crystal is terminated at a particular terrace height; $\Sigma_j H_j = 1$), and $\phi(j)$ is the phase corresponding to each terrace height. At the other extreme, if the lateral step-spacing is large with respect to the lateral coherence length of the X-ray beam, the steps should be included by averaging the intensities incoherently as,

$$I_{rough} \propto \sum_j H_j |F_{ideal}|^2 \sim |F_{ideal}|^2 \qquad (18b)$$

In this case, the X-ray beam typically sees a perfect surface over a typical coherence length, and the reflectivity becomes insensitive to the presence and distribution of the remaining steps.

In practice, steps are always present to some degree and we account for them in a structure factor calculation by guessing an appropriate height distribution function, $H(n)$, of the surface terraces (or equivalently with "occupation factors" for each layer). For example, a surface that is perfectly flat, except that 20% of the surface area is a single unit cell higher, has $H(0) = 0.8$ and $H(1) = 0.2$, with $H = 0$ for all other n. (The same surface is described in terms of occupation factors with $occ(n) = 1$ for $n \leq 0$, $occ(1) = 0.2$, and $occ(n) = 0$ for $n \geq 2$.)

If the structure factor for an ideally smooth surface, F_{ideal}, is written by summing laterally over a single surface unit cell and vertically over all z (including the semi-infinite sum over the bulk crystal, the sum over the surface layer and any fluid layer above the surface) as written in Equation (16), the structure factor for a rough surface can be written:

$$F_{rough} = \sum_n F_{ideal} H(n) \exp(iQz_n) = F_{ideal} \sum_n H(n) \exp(iQz_n) = F_{ideal} B(Q) \qquad (19)$$

Here, surface roughness modifies the reflectivity through a single multiplicative factor, $B(Q)$. This simplifies the inclusion of steps into structure factor calculations, and the main challenge is to determine an appropriate form for $H(n)$.

To gain insight into the effect of surface roughness, we reproduce a calculation by Robinson (1986). We assume that the layers of a crystal are fully occupied up to a certain height (with complete layers at $z = 0$, -c, $-2c$...). Above this height the laterally averaged occupation factors decrease according to the relation $occ_n = \beta^n$ (with $\beta < 1$) for layers at $z = c$, $2c$, $3c$..., corresponding to $n = 1, 2, 3$..., etc. In terms of a terrace height distribution, this model implies that $H(n) = (1 - \beta)\beta^n$, where $n \geq 0$, and $H(n) = 0$ for $n < 0$. Here we calculate the roughness factor as

$$B(Q) = (1 - \beta) + (1 - \beta)\beta e^{iQc} + (1 - \beta)\beta^2 e^{i2Qc} ...$$
$$= (1 - \beta)/(1 - \beta e^{iQc}) \qquad (20a)$$

The effect on the reflectivity will have the following functional form:

$$|B(Q)|^2 = \frac{(1 - \beta)^2}{1 + \beta^2 - 2\beta\cos(Qc)} \qquad (20b)$$

This factor results in a strong decrease in reflectivity between the bulk Bragg peaks that depends strongly on the value of β. At the midpoint between Bragg reflections ($Q_z = \pi/c$, $3\pi/c$...), $|B|^2 = [(1 - \beta)/(1 + \beta)]^2$. However, near the Bragg conditions ($Q_z = 2\pi/c$, $4\pi/c$...), $|B|^2 \approx [(1 - \beta)/(1 - \beta)]^2 = 1$, and there is no change in the reflectivity close to the Bragg peaks. This latter result is expected, because terraces that differ in height by single unit cells scatter coherently near the bulk Bragg condition.

The exact functional form of $B(Q)$ depends on the variation in occupation factors from layer to layer, although the same qualitative trends are found for terrace height distributions having a single preferred height. The particular assumptions made to derive Equation (20) are not applicable to all systems. The sensitivity to different terrace height distribution functions is shown in Figure 7, where the roughness factor for the Robinson model is compared with results for a model that assumes a Gaussian distribution of terrace heights. To compare the functional forms of $B(Q)$, we have arbitrarily chosen a root-mean-square (rms) width for the Gaussian distribution that provides the same value of $B(Q)$ at $Q = \pi/c$ for two particular values of β in the Robinson model. These two models have essentially identical functional forms when the surface is relatively smooth (e.g., $\beta = 0.02$, corresponding to an rms roughness of 0.14 lattice spacings). However, the

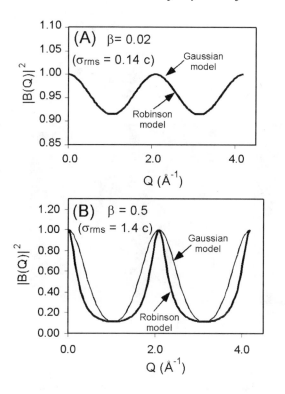

Figure 7. Calculated roughness factors, $|B(Q)|^2$, for (A) modestly and (B) substantially roughened surfaces for the Robinson model described by Equation (20), and for a Gaussian distribution of terrace heights. The exact form of $|B|^2$ depends on the terrace height distribution for the highly roughened surface in (B).

exact form of $B(Q)$ becomes highly model dependent when the surface is rough (e.g., $\beta = 0.5$, corresponding to an rms roughness of 1.4 lattice spacings). In this case, discrepancies between these two simple models exceed a factor of 2. An inappropriate choice of terrace height distribution for a rough surface can therefore have a substantial impact on the structural analysis of a rough surface. In contrast, if the surface is relatively smooth (e.g., $\beta \ll 1$), any reasonable assumption concerning the form of $H(n)$ will give essentially the same answer.

X-ray transmission through a thin-film cell. The most common cell geometry for studying the liquid-solid interface is the "thin-film" cell, in which a plastic film holds a thin layer of water in contact with the surface (Fig. 8A). The advantages of this cell are its simplicity and the minimal background signals that result from diffuse scattering from the solution layer and plastic film, having a total thickness, D_w. These properties make this cell the best choice for high-resolution structural studies. Linear attenuation of an X-ray beam through a material varies as $\exp(-PL/\Lambda)$, where PL and Λ are the path length of X-rays and the linear attenuation length in the medium, respectively. For a thin-film cell, the path length through the layer varies as $PL = 2D_w/\sin(\theta)$. The angular variation of the attenuation factor can be significant, but it depends on the values of the experimental parameters. If we assume a symmetric scattering geometry (i.e., where the angle of the incident beam with respect to the surface plane, α, is equal to the exit angle, the transmission through the cell can be written as follows:

$$T_{cell} = \exp\left[\frac{-2D_w}{\Lambda \sin(\alpha)}\right] = \exp\left[\frac{-8\pi D_w}{Q_z \lambda \Lambda}\right] \tag{21}$$

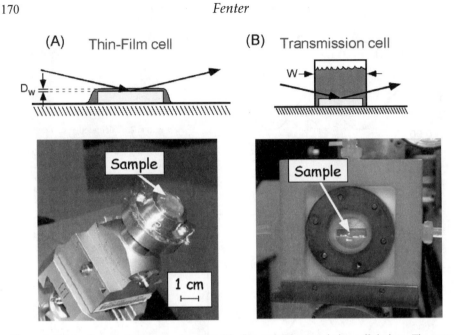

Figure 8. Schematic and photograph of the (A) thin-film and (B) transmission cell designs. The water film thickness, D_w, and cell width, W, for the two cell designs are shown.

Depending on the momentum transfer, Q_z; photon wavelength, λ; and water thickness, this correction can vary substantially. For instance, the attenuation correction as a function of Q_z is shown in Figure 9A for a photon energy of 20 keV and film thicknesses of $D_w = 10$ μm and 200 μm. (These thicknesses include an 8 μm-thick kapton film whose attenuation factor is similar to that of water.) For reference, the attenuation of a 200 μm-thick layer at 20 keV is equivalent to that of a 35 μm-thick water film at 8 keV. The magnitude of the correction varies strongly with Q for $D_w = 200$ μm and weakly for $D_w = 10$ μm. Clearly these attenuation corrections must be known accurately if reliable surface structures are to be determined. But how accurately?

To make this correction robust for realistic situation (i.e., where there might be a variation in water film thickness over the macroscopic area defined by the X-ray beam foot print), quantitative measurements are best done where the exponential term can be approximated by a linear expansion such as $e^{-x} \sim 1 - x$, so that the unavoidable variations in film thickness over the macroscopic areas under the beam footprint can be described quantitatively by a single average film thickness. If we specify a precision of 2% in the expansion, the exponent in the attenuation factor must have a value < 0.2. This implies that the smallest acceptable Q would correspond to $Q_{min}(\text{Å}^{-1}) = 125\ D_w(\mu m)/[\lambda(\text{Å})\ \Lambda(\mu m)]$, where the units are shown explicitly for each variable. For values of $D_w = 10$ μm, E = 20 keV ($\lambda = 0.62$ Å), and $\Lambda = 10,000$ μm, we find $Q_{min} = 0.2$ Å^{-1}, or 10% of the reciprocal lattice cell spacing (for a nominal lattice spacing of 3Å, corresponding to a calcite lattice). This condition is highlighted as a dashed horizontal line in Figure 9A, and it is satisfied over nearly the entire Q range for $D_w = 10$ μm. However, when this same condition is applied for the conditions E = 20 keV and $D_w = 200$ μm [equivalent to E = 8 keV ($\lambda = 1.55$ Å) and a $D_w = 35$ μm] it is satisfied only for $Q_{min} = 3.8$ Å^{-1}, or ≈2 reciprocal lattice spacings for a calcite CTR. In other words, the range of reciprocal space that is inaccessible to quantitative measurements varies strongly with the experimental conditions.

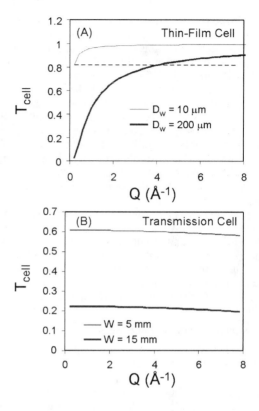

Figure 9. Q-dependent transmission factors at photon energy of 20 keV for the (A) thin-film and (B) transmission cells. Transmission factors, T_{cell}, below the dashed line in (A) correspond to values that cannot be calculated quantitatively in the presence of a non-uniform film thickness within the X-ray beam footprint.

This constraint is actually somewhat stricter than necessary if the distribution of film thicknesses is symmetric about a mean value. In this case, the average value of the attenuation correction, $T = e^{-x}$, can be written as

$$\langle T \rangle = \langle e^{-x} \rangle = 1 - \langle x \rangle + \frac{1}{2}\langle x^2 \rangle = 1 - x_0 + \frac{1}{2}x_0^2 + \frac{1}{2}\sigma_x^2 \tag{22}$$

if x is assumed to have a Gaussian distribution with an average value of x_0 and a width of σ_x. For values of $D_w = 200$ μm, $\sigma(D_w) = 100$ μm, and $E \sim 20$ keV, we find $Q_{min} \sim 2.5$ Å^{-1}.

The choice of photon energy and solution thickness is therefore critical if the reflectivity is to be measured quantitatively over a broad range of reciprocal space, especially at small values of Q_z. When the variation of film thickness within the X-ray footprint is sufficiently large, the angular variation of the attenuation correction can differ substantially from the simple form derived in Equation (21), and it cannot be modeled by a single effective water thickness. Substantial systematic error is therefore associated with measurements performed with an excessive film thickness.

Although the thin-film cell has many advantages, it also has some disadvantages. The composition of the thin water film can be modified readily if the surface reacts with the solution. For example, dissolving a single layer of a crystal having $\sim10^{14}$ atoms/cm^2 in a 2 μm-thick solution layer corresponds to a change of solution composition by ~0.002 M. The thin plastic film may also be semi-permeable to gases, so the solution composition might be modified if care is not taken to control the atmosphere above the plastic film with an inert gas (e.g., N_2, He).

X-ray transmission through a transmission cell. The transmission cell (Fig. 8B) is a geometry that has two main advantages over the thin-film cell. First, as shown in Figure 9B, the cell maintains a macroscopically thick solution above the surface whose properties (temperature, pH, ionic strength…) can be controlled precisely by flowing a solution with well-defined composition over the sample during the measurement, even under reactive conditions. Second, the variation of the attenuation correction as a function of incident angle is very small, and it can be quantified easily. In this geometry with a photon energy of 20 keV, a typical cell width is $W = 5$ mm (Fig. 8). If the cell width is equal to the sample length, a beam having a 0.1 mm vertical beam size can reflect from the surface without spilling off the edges of the crystal surface at the first mid-zone condition, $Q_z = \pi/c$. In this geometry, the path length, PL, varies as $PL = W/\cos(\alpha)$, and the attenuation factor is $T_{cell}(\alpha) = \exp\{-W/[\Lambda\cos(\alpha)]\}$. A plot of the Q-dependent attenuation for this cell geometry is shown in Figure 9B. For typical values of $E = 20$ keV, $\Lambda = 10$ mm, and $W = 5$ mm, we find $T_{cell}(\alpha) = 0.61$ and 0.59 at $Q = 0$ and 6 Å^{-1}, respectively. In other words, the attenuation changes by only 3% over the full range of reciprocal space measured in a typical high-resolution reflectivity measurement. The attenuation factor is relatively large, but it varies minimally with angle (or Q_z). Moreover, there are no adjustable parameters, because the cell width, W, can be calibrated by measuring the beam attenuation of the straight-through beam: $W = \Lambda \ln[T_{cell}(\alpha = 0)]$. The main disadvantage of the transmission cell is the strong attenuation of the beam through the cell, which results in a substantial background signal that makes high-resolution reflectivity measurements difficult or impossible.

Spectrometer-dependent factors

Standard spectrometer configurations include the "four circle" (Busing and Levy 1967; Mochrie 1988), "five circle" (Vlieg et al. 1987), "z-axis" (Bloch 1985), "six circle" (Lohmeier and Vlieg 1993), "Kappa" (Robinson 1995), and "4S+2D" (You 1999) spectrometers. These spectrometers are designed to control the sample and detector angles precisely to reach particular scattering conditions. The number of spectrometer degrees of freedom for each spectrometer exceeds the three degrees of freedom necessary to reach a particular scattering condition (e.g., defined by Q_x, Q_y, and Q_z). This feature allows for a choice of distinct "modes" that use the extra degrees of freedom to choose how to reach a particular scattering condition so that the scattering signal can be optimized. In the case of surface scattering this usually means choosing how the incident and reflected X-ray beams are oriented with respect to the surface plane. Typical choices include fixed angle of incidence, fixed exit angle, and a symmetric geometry in which the angle of incidence is kept equal to the exit angle. The spectrometer-dependent normalization factors, described below, are dependent upon the type of spectrometer used, the mode in which the spectrometer is used, and the type of scan chosen for the rocking scan. The general features of these normalization factors are described here for completeness, but the interested reader can find more details elsewhere (Feidenhans'l 1989; Robinson 1991).

Detector resolution, ΔQ_z. Under normal experimental conditions where the incident beam divergence is small, the scattering intensities depend on the angular resolution of the detector (e.g., the detector slit aperture). A critical aspect of the reflectivity measurement is knowledge of the instrumental resolution at every value of **Q**. Because the surface reflectivity is in the form of Bragg "rods," the measured intensity depends on how much of the rod is accepted into the detector slit along the Q_z direction, or $\Delta Q_z(\mathbf{Q})$, where $\mathbf{Q} = (Q_x, Q_y, Q_z)$. For this discussion, Q_x and Q_y are assumed to lie within the surface plane, and Q_z is along the surface-normal direction. For this discussion we define the concept of the "scattering plane" as the plane containing the incident and reflected wave vectors. Perpendicular to this plane is a direction that is transverse to this scattering plane.

To illustrate the variation of $\Delta Q_z(\mathbf{Q})$, we must discuss the Q resolution of the scattering measurement in reciprocal space. Consider the experimental resolution in the idealized case where the incident beam is monochromatic (with typical energy resolution of $\Delta E/E \sim 10^{-4}$) and non-divergent (with a typical beam divergence of $0.005°$). The range of reciprocal space probed by a given sample and detector orientation is determined by integrating over all values of Q that are accessible by changing the reflected wavevector, $\mathbf{K}_f$, within the detector slit dimensions. This range of reciprocal space that is probed at any given Q is referred to as the "resolution function". The detector slit size is often different within the scattering plane (i.e., the plane containing both the incident and reflected wave vectors, $\mathbf{K}_i$ and $\mathbf{K}_f$, which corresponds to the plane of Fig. 10A) and transverse to the scattering plane (i.e., perpendicular to the plane of Fig. 10A). Since we have assumed that the incident beam is non-divergent and monochromatic, the range of exit angles accepted by the resolution-defining detector slits is represented in reciprocal

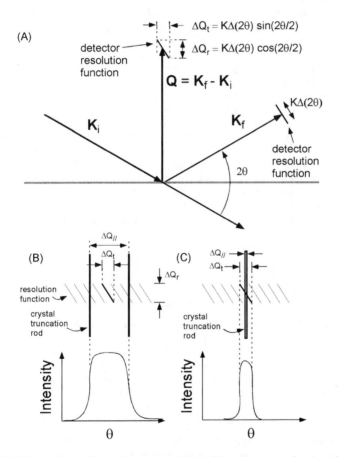

Figure 10. (A) The resolution of an arbitrary scattering condition, shown as a function of the detector slit size, $K\Delta(2\theta)$. (B) and (C) A rocking scan is measured by scanning resolution function across a specular CTR (B) The surface rod is broader than the transverse resolution, ΔQ_t so that the rocking curve width is determined by the intrinsic width of the CTR. (C) The surface rod shape is more narrow than the transverse resolution, ΔQ_t, and the shape of the measured rocking curve is "resolution limited" (i.e., determined by the slit size instead of the intrinsic line width).

space by a plane that is perpendicular to the exit beam direction ($\mathbf{K}_f$) as shown in Figure 10A. Since $\mathbf{Q} = \mathbf{K}_f - \mathbf{K}_i$, the plane that defines the detector resolution function in reciprocal space is then tilted, with respect to $\mathbf{Q}$, by an angle that is half the size of the total scattering angle. The dimensions of the resolution function both within and transverse to the scattering plane are determined by the angular detector size. In reciprocal space, the Q range corresponding to the angular size of the slit can be written as $\Delta Q = K\Delta(2\theta)$, where $\Delta(2\theta) = \Delta_{v\_det}/R_{det}$ is the angular size of the detector within the scattering plane for detector aperture Δ_{v_det} at distance R_{det} from the sample ($K = 2\pi/\lambda$). Transverse to the scattering plane the angular size of the detector is $\Delta\gamma = \Delta_{h\_det}/R_{det}$, where Δ_{h_det} is the horizontal detector slit size, resulting in a Q range transverse to the scattering plane of $\Delta Q_\perp = K\Delta\gamma$.

This leads to the important concept of "resolution limited" measurements. Any X-ray scattering measurement is a convolution of the intrinsic reciprocal space structure associated with the sample and the instrumental resolution. When the structure factor varies slowly with respect to the defining resolution element, that feature is said to be resolved, and the measured shape and intensities are intrinsic to the sample. In contrast, when the structure factor varies rapidly over the Q-range defined by the resolution function, the measured shape of that feature can be substantially modified from the intrinsic shape; this is known as "resolution limited". Since the shape of diffraction peaks contains important information (typically concerning the degree of order in a sample, it is important to compare measured peak widths with the calculated resolution width to know if a measurement reflects the instrumental resolution or the sample behavior; only widths that are substantially broader than the instrumental width are representative of the intrinsic sample properties. Figures 10B and 10C illustrate this for a rocking scan of a crystal truncation rod as a given resolution element is scanned across two surface truncation rods having different lateral widths (e.g., due to different surface domain sizes or mosaic widths). The measured rocking curve width is determined by the width of the surface rod when the rod is broader than the transverse resolution, $\Delta Q_t = K\Delta(2\theta)\sin[(2\theta)/2]$ (Fig. 10B), and is determined by the transverse detector resolution when the rocking curve is more narrow than the lateral resolution (Fig. 10C). Hence the rocking scan in Figure 10C is resolution limited. In between these two extremes, the measured width, ΔQ_{meas}, can be approximated as the sum of the intrinsic and instrumental widths added in quadrature: $\Delta Q_{meas} = [\Delta Q_{//}^2 + \Delta Q_t^2]^{1/2}$, where $\Delta Q_{//}$ is the intrinsic lateral width of the surface rod. While this example specifically addressed the effect of resolution on the shape of a rocking scan, the general concept is true for any X-ray scattering measurement. A more complete description of this phenomenon is given by Toney et al. (1993).

The projection of the detector resolution along the direction defined by Q (i.e., the radial direction) is $\Delta Q_r = K\Delta(2\theta)\cos[(2\theta)/2]$. For specular reflectivity, this corresponds to the length of the specular rod that is integrated in a rocking scan, as shown in Figures 10B and 10C (whether the lateral width of the specular rod is narrow or broad). For typical values of $E = 20$ keV, a slit size of 0.25 mm, and a detector distance of 700 mm, $\Delta Q_r \sim K\Delta(2\theta) \approx 0.0036$ Å^{-1}, or 0.2% of the reciprocal lattice spacing (assuming $d \sim 3$ Å, corresponding to the calcite (104) reflection). Therefore, as many as ~500 independent measurements can be made of the CTR intensity between each bulk Bragg peak.

The transverse resolution normal to the scattering plane must also be considered. The resolution transverse to the scattering plane is $\Delta Q_\perp = K\Delta\gamma$. In any measurement, if the radial resolution width is the defining resolution element, the transverse resolution width must integrate fully across the Bragg rod in both transverse directions to obtain a complete integration of the intensity along Q_x and Q_y, so that the integrated intensity is independent of any variation in the rocking curve peak shape as a function of Q_z. Only when this condition is met can the intensities be compared quantitatively to structure factor calculations. Operationally, by measuring the reflectivity with a rocking scan (as shown in Figs. 10B and 10C) we ensure that the intensity is properly integrated along ΔQ_t. One must separately verify that the resolution width transverse to the scattering plane, $\Delta Q_\perp$, is broader than the intrinsic shape of the reflectivity rod at all values of Q_z.

For non-specular reflectivity measurements, the length of the rod that is integrated during a rocking scan depends on many details of the experimental configuration, including the spectrometer mode, the choice of rocking scan, etc. These issues have been described previously (Feidenhans'l 1989; Lohmeier and Vlieg 1993).

Lorentz and polarization factors. The Lorentz factor, $1/\sin(2\theta)$, is associated with the fact that the calculations are performed in rectangular coordinates while the measurements are all performed with spectrometers that are controlled in an angular coordinate system. For most experimental configurations, the Lorentz factor has the familiar form $L = 1/\sin(2\theta)$ when the rocking curves are measured by scanning the θ motor. However, other functional forms are found for different spectrometers (Feidenhans'l 1989).

The polarization factor, *Pol*, takes into account the variation in scattering intensity caused by the beam polarization, as can be anticipated because elastic scattering is forbidden along the polarization direction of the incident beam. Synchrotron sources typically produce X-ray beams that are linearly polarized in the horizontal plane. Most synchrotron-based scattering measurements are performed with the scattering angle measured in the vertical plane, so that $P = 1$. For horizontal scattering spectrometers with a horizontally-polarized beam, the polarization factor is $P = \cos(2\theta)$. A number of spectrometers scan the detector in both the vertical and horizontal planes.

Active area corrections (specular reflectivity). Active area corrections are needed because the diffraction signal is proportional to the number of surface atoms within the beam. This factor is proportional to the surface area that reflects the incident beam and is simultaneously seen by the detector (i.e., the overlap of the incident beam and detector "footprints"). As the sample angle changes, so does the amount of diffracting surface (Fig. 11A). These corrections are simple for specular reflectivity in which the incident and exit angles (α_i and α_f, respectively) are equal. In this case, the incident beam has a cross section of Δ_t by Δ_v in the transverse and vertical directions, respectively, and the beam footprint has dimensions Δ_t by $\Delta_v/\sin(\alpha_i)$. The projection of the detector slits onto the surface results in a similar detector footprint [Δ_{t_det} by $\Delta_{v\_det}/\sin(\alpha_f)$]. For specular reflectivity, the detector slits should be set to accept the full incident beam so that the detector footprint is always larger than the incident beam footprint. Therefore, the full active area correction is determined solely by the variation of the incident beam footprint and varies as $A = \Delta_t\Delta_v/\sin(\alpha_i)$. This active area correction is general to all spectrometers for specular reflectivity measurements, as long as the incident beam does not spill-off the edges of the sample at small incident angles.

Active area corrections (non-specular reflectivity). The active area correction for non-specular reflectivity becomes more complicated because it relies on determining the overlap of the incident and detector footprints, which might not have a simple analytical form. In Figures 11B-D, we show representative examples. For a four-circle spectrometer in the symmetric scattering mode (i.e., $\alpha_i = \alpha_f$), with an incident beam shape of $\Delta_t = 0.5$ mm by $\Delta_v = 0.1$ mm and a square 10 mm by 10 mm sample, we show the incident and detector footprints at selected values of Q_z for a non-specular CTR with $Q_{//} = 1.2$ Å^{-1} and Q_z varying from 1.8 Å^{-1} to 7.2 Å^{-1}. This calculation shows that the beam footprint becomes highly skewed as Q_z increases at constant $Q_{//}$, and the overlap of the incident and

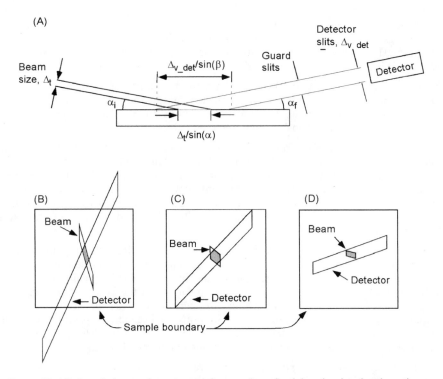

Figure 11. (A) A typical scattering geometry for specular reflectivity, showing that the active area is determined by the incident beam cross section and the incident angle, α_i. (B-D) An image of the active area for non-specular reflectivity measurements, with the overlap of the incident and detector footprints shown in three regimes at Q_z values of (B)1.8 Å^{-1}, (C) 3.6 Å^{-1}, and (D) 7.2 Å^{-1}, each calculated for $Q_{//}$ = 1.2 Å^{-1}. In each case the large square indicates the 10 mm by 10 mm sample boundaries, and the sample is viewed from above the surface plane. (B) and (D) show regimes where the reflectivity can be measured with minimal systematic error because the active area will be insensitive to imperfections in the lineup that may shift the positions of the incident beam and detector footprints from their expected locations.

detector footprints changes dramatically as a function of Q_z. At Q_z = 1.8 Å^{-1} (Fig. 11B) the active area is limited by the sides of the beam and detector footprints, providing a simple geometric correction for active area vs. Q_z. At Q_z = 7.2 Å^{-1} (Fig. 11D) the incident beam footprint is fully contained by the detector footprint, and the active area is determined solely by the incident beam footprint. Between these two conditions is a regime where the active area depends on the way the edges of the two footprints overlap, Q_z = 3.6 Å^{-1} (Fig. 11C). It is preferable to have a simple, straightforward correction (as in Figs. 11B and 11D where the intersection of these footprints forms a simple parallelogram) that does not depend on details that might be poorly established because of imperfections in the spectrometer line-up (as in Fig. 11C).

Absolute specular reflectivity

We can now derive a general expression for the absolute specular reflectivity, based upon Equation (13), where the intensity is measured by a rocking scan at each point. Using the following expressions A/A_0 = 1/sin($2\theta/2$) and ΔQ_z = $K\Delta(2\theta)$ cos[($2\theta)/2$] = $(2\pi/\lambda)\,\Delta(2\theta)$ cos[($2\theta)/2$],

$$R = \frac{r_e^2 \lambda^2 T_{cell} |F|^2}{a_{uc}^2 \sin^2(2\theta/2)}$$

$$= \left[\frac{4\pi r_e}{Q a_{uc}}\right]^2 T_{cell} |B(Q)|^2 |F_{uc} F_{ctr} + F_{surf} + F_{water}|^2 \qquad (23)$$

This expression is rather simple. In this case, the only unknown quantities are the structure factor of the near-surface region, $F_{surf} + F_{water}$, and the roughness factor, $|B(Q)|^2$. Although the detector resolution, ΔQ_z, appears in the expression for the scattering intensity (Eqn. 11), it does not appear explicitly in the expression for the reflectivity (Eqn. 23). The absolute specular reflectivity is therefore a well-defined quantity that is independent of the experimental details (e.g., detector slit sizes). This makes specular reflectivity relatively straightforward to measure, and to correlate these measured reflectivities to the structure factor. An expression for non-specular reflectivity measurements can be similarly derived, although the active area corrections and detector resolution factor may not have a simple analytical form and may exhibit a strong dependence upon the scattering condition, **Q**, as well as extrinsic factors such as spectrometer mode and detector resolution. Non-specular reflectivity measurements therefore require much greater care than specular reflectivity measurements to insure that the numerous factors in Equation (13) (especially the active area and detector resolution factors) are established with a known precision, so that the structural information (e.g., in the structure factor) can be determined precisely.

PRACTICAL ISSUES

While the reflectivity signal changes by more than 10 orders of magnitude between bulk Bragg reflections along a CTR, the background signal may be flat, or it may be largest when the reflectivity is smallest. The greatest practical limitation in high-resolution CTR measurements of the mineral-fluid interface structure is often distinguishing small reflectivity signals against the background signals. Although reflectivities of $< 10^{-10}$ are not uncommon for high-resolution CTR measurements, it is not typically the small reflectivity that is the limiting factor for these measurements, because the incident flux can be as high as $\sim 10^{11}$ to 10^{12} photons per second, resulting in a detector count rate of 10-100 photons per second, which is in itself straightforward to measure. Instead, the relatively large background signals from the sample cell, the water, and the bulk crystal can dominate this modest reflectivity signal. A significant challenge is therefore to minimize the background signals to a level where the small reflectivity signals can be measured reliably.

Measuring reflectivity signals

The reflectivity signal can be distinguished from any background because it is in the form of a narrow rod oriented precisely along the surface-normal direction. The background signal therefore can be distinguished from the reflectivity signal by measuring the reflectivity both at and away from the specular reflection condition. As described above, it is the integrated intensity that is of most interest for structural measurements. The magnitude of the reflectivity should be independent of the method by which it is determined, but its uncertainty can depend strongly on many factors including how the measurement is performed, the counting statistics, and whether the measurement is well-optimized

Rocking scans. The reflectivity signal is determined from these data in three general ways. The most quantitative and straightforward method is to measure the reflectivity as a function of incident angle (i.e., a rocking scan) so that the reflectivity and background signals are well resolved (Fig. 12A). The reflectivity is then determined by fitting these data to a line-shape that includes a linear background and a function corresponding to the rocking curve shape (e.g., Gaussian, Lorentzian, etc.). The peak width and peak height are determined from least-squares fitting (Bevington 1969), and the integrated reflectivity, I_{int}, is calculated from these quantities. The optimization relies on a quality of fit that assesses the degree of fit between the calculated and measured data. This is the χ^2 factor,

$$\chi^2 = \frac{1}{(N - N_p)} \sum_j \left[\left(I_j - I_{calc} \right) / \sigma_j \right]^2 \tag{24}$$

Here, I_j and σ_j are the intensity and uncertainty of the j'th data point in the rocking curve (the uncertainty is assumed to be due to counting statistics, whereby the measurement of N photons is uncertain by $\sqrt{N}$); I_{calc} is the calculated intensity for a particular set of parameters (peak position, width and intensity, etc.); the sum is performed over all data

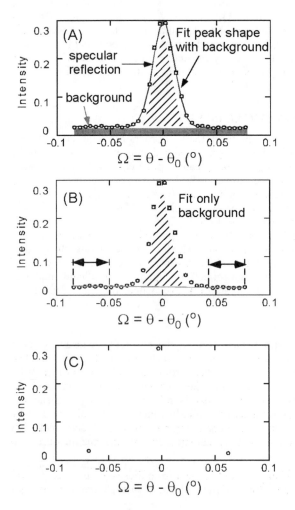

Figure 12. Background subtraction techniques. (A) The area is derived by fitting the rocking curve shape with a Gaussian and a linear background. The reflectivity is proportional to the integrated areas (hatched). (B) The background is fitted and subtracted numerically from the data which are then integrated numerically. (C) The reflectivity is determined by measuring the reflectivity at the peak of the rocking curve, and the background is determined by measurements at two points away from the Bragg peak.

points in the rocking scan; N is the total number of data points; and N_p is the number of parameters used to fit the data. The best-fit structure is one in which χ^2 is minimized. Since each term in Equation (24) is normalized by the statistical uncertainty of that data point, an optimal fit corresponds to a situation in which the typical deviation between the data and calculation is comparable to the statistical error for that point. $\chi^2 \approx N/(N-N_p) \approx 1$ corresponds to an optimal fit. (N_p is typically much smaller than N.) Any reduction of χ^2 below this level corresponds to fitting the statistical noise in the data. χ^2 has typical values of ≈ 1 to 5 for best-fit structures of high resolution X-ray reflectivity data. Values above this level might indicate systematic error in the measurement, an inappropriate structural model, or other sources of error.

Since the uncertainties in each data point in a rocking scan are determined primarily by counting statistics, the uncertainty of the integrated intensity can also be derived from the uncertainties in the peak width and peak height (Bevington 1969). These uncertainties will be used to determine the uncertainties in the derived interfacial structure. Typically, there is a residual uncertainty in the reflectivity that is difficult to avoid. This uncertainty may be derived, for example, from the uncertainties in any of the prefactors in Equation (11), detector saturation at high counting rates, or the finite resolution of the detector. Given these error sources, we typically enforce a minimal fractional error in the integrated intensity (typically ≈ 1 to 2%) for each data point depending upon the sample system. This prevents the fitting process from being dominated by a few data points that have systematic error in excess of their statistical error.

A second approach to determining the integrated reflectivity is to numerically integrate the rocking scan. This is done by fitting a linear background to the data points away from the specular reflection condition, typically by using 5-10 data points on each side of the peak, as shown in Figure 12B. This fitting can be done with a closed-form numerical calculation (Bevington 1969). The fitted background is then calculated at each data point, $b(\theta)$, and subtracted from each data point, $I(\theta)$. The integrated intensity of the peak is then calculated, $I_{int} = \Sigma [I(q) - b(q)] \, \delta\theta$, where $\delta\theta$ is the angular spacing of the individual data points. Although this approach can result in larger error bars than when the data are fitted with an appropriate peak shape, it has a number of uses. First, the numerically integrated intensity is valuable as a check for comparison with the reflectivity determined by least-squares fitting the rocking scan to a particular line shape. The numerical approach is generally the most accurate means of determining the reflectivity when the rocking curve is not well described by a simple functional form (e.g., when the reflectivity has multiple peaks or an asymmetric "tail" due to sample mosaic). Under these conditions, significant systematic error is associated with "force-fitting" the peak to an inappropriate line shape. The major source of systematic error in this approach is when a true background level is not obtained in the rocking scan. Care must be taken especially with rocking scans that have a non-Gaussian (e.g., Lorentzian) line shape, whose tails might converge slowly to the true background level.

A third option is to measure the reflectivity by a three-point measurement (one measurement on the peak, plus one on the background on each side of the peak) as shown in Figure 12C. The main advantage of this method is the (~10-fold) faster scan time, which is useful when real-time measurements are needed. The drawback is that one must know a priori that the rocking scan shape does not change either as a function of incident angle (e.g., during a CTR measurement) or as a function of time (e.g., during a real-time measurement) because this approach only measures the peak reflectivity which is proportional to integrated reflectivity only if the peak width and line shape do not change. The location of the background points must be chosen carefully to avoid any contribution from the true reflectivity (especially true when the rocking scan has a Lorentzian line shape, with an intensity variation that decreases slowly away from the specular condition). Unfortunately, this method is also subject to systematic error. In addition to changes in peak shape, the sample orientation may drift with time (due to thermal stresses) resulting in a measured reflectivity that is systematically reduced with respect to the true integrated reflectivity.

Counting statistics. Beyond the question of how to determine the reflectivity from a set of data, we must also consider the statistical limitations and the associated question of how long to count photons at each point in a typical scan. For instance, for a peak reflectivity signal of $R = 100$ cps, the error due to counting statistics scales as the square root of the total number of counts (Bevington 1969). This measurement will have 10% error for a counting time of 1 sec, and 1% error for a counting time of 100 sec. However, in the presence of a finite background counting rate, b, the time, t, to obtain a specified level of error will always be larger than the time needed in the absence of a background. In the presence of a background signal, the peak count *rate*, $P = R + b$ (in counts per second), must be differentiated from the background count *rate*, b. The error in the difference of two signals will scale as sqrt$[t(R + 2b)]$, where t is the counting time in seconds, while the actual reflectivity signal will vary as $[tP - tb] = [t(R+b) - tb] = tR$, where R and b represent the reflectivity and background count rates, and t represents the counting time at each point. This results in a fractional error in the reflectivity signal of $\sigma_R/R =$ sqrt$[(R + 2b)/t]/R$. The influence of the background signal on the error of the reflectivity signal is shown in Figure 13, where the necessary counting time is plotted vs. background count rate for a fixed reflectivity count rate and desired statistical error, with a reflectivity signal (20 cps). This calculation shows that the counting time necessary to maintain a given precision increases substantially as either the desired error decreases or the background signal increases. Once the background signal is small compared with the reflectivity (i.e., high signal-to-background ratio), there is little change in the necessary counting time as the background decreases. The time to obtain a set of data depends critically on both the reflectivity and background signal levels. For any given measurement, the best way to minimize counting time is to minimize the background signal, especially when the reflectivity is small.

Background signals from water and X-ray window. Measurements of the mineral-fluid interface require that X-rays propagate through water and an X-ray window. The

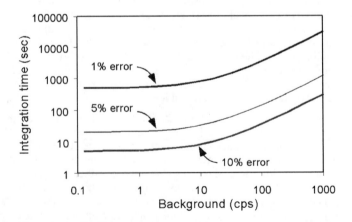

Figure 13. The integration time as a function of the background signal level for a fixed signal count rate of 20 counts per second. Each curve corresponds to a different fractional uncertainty. Note the strong increase in counting time when the background count rate is larger than the signal count rate.

attenuation and scattering of the X-ray beam through the water and kapton films contributes to the background signals in the raw data, and it is useful to know how these contributions depend upon the scattering condition. The scattering intensity as a function of Q is shown for kapton and water in Figure 14. These measurements were performed in a transmission geometry (Fig. 14A) with a photon energy of 15.8 keV. The raw data for a ~50 μm-thick kapton film and a ~2.6 mm-thick water layer confined between two 50 μm -thick kapton windows are shown in Figure 14B. To emphasize the relative strength of the kapton and water contributions, we show, in Figure 14C, the same data after we subtracted out the kapton contribution from the composite data and then scaled the data so that both curves represent the scattering intensity associated with a 50 μm-thick layer. These data show that the intrinsic contribution of kapton to the background signal is larger than that of water for equivalent thicknesses. The kapton film also exhibits Bragg peaks in the small-Q regime. So, given that the kapton film in a typical thin-film

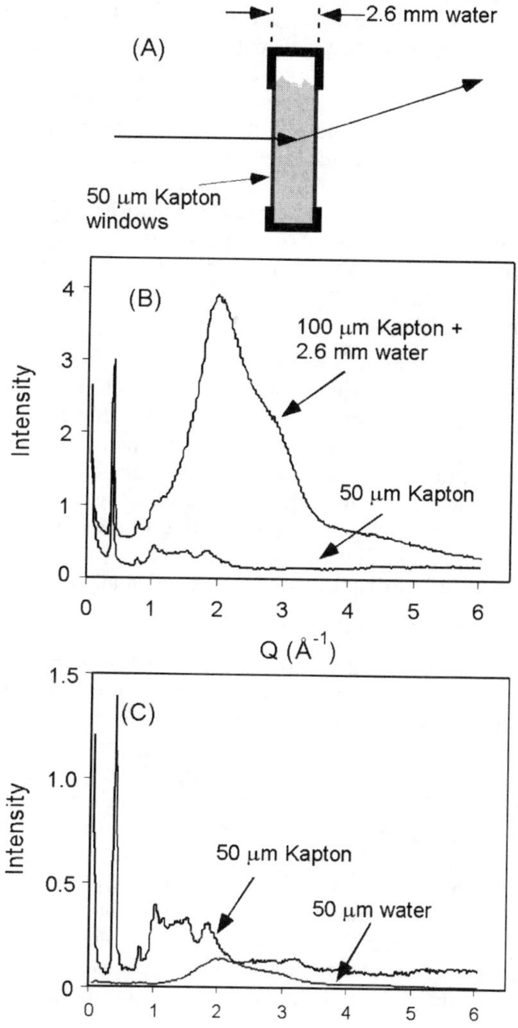

Figure 14. The scattering intensity from a single 50 μm-thick kapton layer and two 50 μm-thick kapton layers confining a 2.6 mm-thick water layer, as a function of Q. The scattering geometry is shown in (A), and the data are shown in (B). The same data are shown in (C) after we subtracted the kapton contribution from the kapton-water data and scaled the measured intensity to reveal the individual contributions from kapton and water for equivalent layer thicknesses.

measurement has a thickness of ≈8 μm, while the water thickness can be as small as ≈2 μm, the main contribution to background signal will be from the kapton. Aside from the two strong kapton Bragg peaks, the kapton contributes a broad lobe of intensity between 1 and 2 Å^{-1}, while the water contributes in a similar way between 1.5 and 3 Å^{-1}. Because the kapton and water signals shown in Figure 14C reflect transmission through a fixed water film thickness, measurements in a thin-film cell will vary with an additional multiplicative factor of $\sim 1/Q$ associated with the changing path length through the film as a function of incident angle in a grazing incidence geometry.

Optimizing reflectivity signals. The reflectivity signal is measured most easily, for a given background count rate, when the peak intensity of a rocking scan is maximized. Since the integrated intensity of a rocking scan is related to the interfacial structure, the only way to increase the peak intensity (for a given beam intensity) is to reduce the angular width of the peak in the rocking scan; that is, the rocking scan width should be as narrow as possible. The rocking scan width may depend on a number of different issues. If the crystal is imperfect due to angular (i.e., mosaic) disorder, all reflections will be broadened by the mosaic width. Even if the bulk crystal quality is perfect, a surface with small average terrace size will exhibit broadened rods in reciprocal space, with a lateral width of $\Delta Q_{//} = 2\pi/L$ (Fig. 10B), where L is the typical step spacing in that direction. In each of these two cases, a substantial improvement of the data quality is best obtained by improving the sample quality. Finally, the rocking curve width may be limited by the transverse Q range accepted by the detector (i.e., the width is resolution limited). In the case of a resolution-limited rocking curve scan for specular reflectivity, as shown in Figure 10C, the rocking scan width will be $\Delta Q_t = K\Delta(2\theta) \sin[(2\theta)/2]$ when measured in reciprocal space and $\Delta\theta = \Delta(2\theta)/2$ (i.e., independent of Q_z) when measured in angle space. The best signal to background ratio will be obtained when the detector slit size is chosen so that the transverse detector resolution, ΔQ_t, matches the lateral width of the surface rod, $\Delta Q_{//}$.

In the case of non-specular reflectivity, the overlap of the detector resolution with the non-specular rod can have a significant effect on the rocking scan width. This effect can be controlled by choice of spectrometer mode. Because the tilt of the plane that defines the detector resolution function is defined by the exit angle of the scattered photon, one can minimize this source of broadening by measuring the reflectivity in a "fixed exit angle mode" in which the exit angle is kept at a small value just above the surface plane. In this case, the plane corresponding to the detector resolution function is almost parallel to the non-specular rod, and the rocking curve shape can be much sharper, resulting in better signal-to-background ratios and smaller statistical error.

Surface inhomogeneities

An implicit assumption in this chapter has been that, except for steps, the surface structure is homogeneous. This may not generally be the case, especially for poorly characterized surfaces. As in the above discussion of the modification of the structure factor by steps, there are two limiting cases to consider. If the surface is heterogeneous over distances that are large compared with the lateral coherence length, $L = 2\pi/\Delta Q_{//}$, then the reflectivity from regions that have different structures add incoherently. However, if the heterogeneities exist over distances that are within the lateral resolution, then the reflectivities add coherently. In the case of specular reflectivity, these domains will all add together, and the derived structure will reflect this laterally averaged termination. In the case of an incoherently averaged heterogeneous surface, the incoherently averaged reflectivity data might look very different from the individual reflectivity spectra. In particular, because the reflectivity at a specific scattering condition might differ by orders of magnitude for different terminations of a given surface, even a minor incoherent heterogeneity can in some cases dominate the signal, especially if the dominant structure has a very low reflectivity. In this case, much information concerning the structure is lost in a specular reflectivity experiment. In non-specular reflectivity measurements, the contribution from each co-existing phase can be measured separately if the structures have different lateral symmetries (e.g., unit cell size) because the reciprocal lattice rods for each phase will be arranged differently in reciprocal space.

The spatial resolution of a CTR, or how many data are needed?

The spatial resolution of a scattering measurement can be estimated as $L_{res} = 2\pi/Q_{max}$, where Q_{max} is the range of the scattering data. The term resolution is used here in the optical sense as the ability to resolve two objects. Minerals may have complex structures, and the ability to make definitive statements about the chemistry of an interface ultimately resides in the ability to resolve different atoms. It is then useful to have a way to visualize a structure at a given finite resolution to understand what features of the model are uniquely defined. What follows is an approximate analysis of the experimental resolution (Fenter et al. 2001) derived from the two equivalent definitions of the Patterson function (Warren 1990) which is defined below.

We first assume that, at any specified resolution determined by Q_{max}, the effective electron density of each particular atom, $\rho_{eff}(z)$, can be written as a Gaussian, in which the position z_0 is determined by the structure, the width u_{eff} is determined by either the intrinsic electron distribution or the experimental resolution, and the integral weight corresponds to the atomic number of that element, Z, or

$$\rho_{eff}(z) = \frac{Z}{(2\pi)^{1/2} u_{eff}} \exp\left[-\frac{1}{2}\left(\frac{z - z_0}{u_{eff}}\right)^2\right] \tag{25}$$

For generality, the net effective width $u_{eff}(j)$, for each atom j, is determined by adding the resolution width u_{res}, and the true vibrational amplitude σ_j, in quadrature, or $u_{eff}(j) = (u_{res}^2 + \sigma_j^2)^{1/2}$, where u_{res} is assigned the value $u_{res} = 1.1/Q_{max}$, and Q_{max} is the maximum momentum transfer in the measurement.

To motivate this choice of u_{res}, we note that the Patterson function, $P(z)$, is defined as the density-density correlation function. It can be derived separately from the electron density, $\rho(z)$, and the experimental data, $I(Q)$. It can be written as (Warren 1990):

$$P(z) = \int \rho_{eff}(z)\rho_{eff}(z + \zeta)d\zeta \tag{26a}$$

$$P(z) = \int I(Q)\cos(Qz)dQ \tag{26b}$$

where we assign $\rho_{eff}(x)$ to be the *effective* electron density of the material at the finite resolution of the measurement defined by Q_{max}, probed in the reflectivity measurement. In Equation (26a) the integral is over all space ($-\infty \leq \zeta \leq \infty$). In Equation (26b), the integral is over the range $0 < Q < Q_{max}$.

If ρ_{eff} is written as a series of Gaussians to represent the electron density of each atom, the integral in Equation (26a) represents the convolution of two Gaussians corresponding to density-density correlations, exhibiting peaks at the interatomic spacings and at the origin. The convolution of two Gaussians, each having width u_{res} is a Gaussian with width $\sqrt{2}u_{res}$, and $P(z)$ can be written as a sum of Gaussians having widths of $\sqrt{2}u_{res}$. Each Gaussian term in the Patterson function will peak at the interatomic spacings, or at the origin (the latter term corresponds to the self-convolution of each atom's electron density).

To determine a proper form for u_{eff}, we use the scattering intensity for an isolated dimer with an atomic separation, Δ. In this case, $I(Q) \propto |(1 + e^{iQ\Delta})|^2 = 2 + 2\cos(Q\Delta)$. The intensity for any structure can be written as a constant term plus a series of cosine terms corresponding to interatomic spacings. The Patterson function for this case can be written as $P(x) = \sin[Q_{max}(x - \Delta)]/(x - \Delta) + \sin[Q_{max}x]/x$. When the resolution is reasonably good (i.e., if $Q_{max} > 2 \text{ Å}^{-1}$, corresponding to a resolution of < 3 Å), the numerical value of $u_{res} \approx 1.1/Q_{max}$ provides good correspondence between the two expressions, Equations (26a,b).

To illustrate the usefulness of this formalism, consider the simple example of scattering from a dimer. The scattering intensity corresponding to a dimer, $I(Q) \propto 2 + 2\cos(Q\Delta) = 4\cos^2(Q\Delta/2)$ with an atom-atom spacing of $\Delta = 1$ Å is shown in Figure 15A. Also shown is the structure factor of a single broadened Gaussian, $I(Q) \propto \exp[-(Qu)^2]$, where the value of $u = 0.55$ Å was chosen to highlight the close similarity in the structure factors between the dimer and the broadened Gaussian for $Q < 3 \text{ Å}^{-1}$. We therefore do not expect to be able to differentiate these two structures unless $Q_{max} > 3 \text{ Å}^{-1}$.

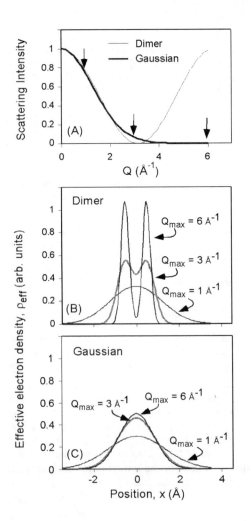

Figure 15. (A) The scattering intensity for two atoms separated by 1 Å (e.g., a dimer) as a function of Q, along with the structure factor of a broadened Gaussian having a width, $u = 0.55$ Å. Note the similarity of shapes for $Q < 2 \text{ Å}^{-1}$. The effective electron density of a (B) dimer and (C) broadened Gaussian, plotted at selected values of Q_{max}. (Q_{max} is inversely related to the resolution of a scattering measurement). The values of Q_{max} for the plots in (B and C) are shown as arrows in (A).

The effective electron density for the dimer and broadened Gaussian, from Equation (25), are shown at different values of Q_{max} in Figures 15B and 15C (with the corresponding values of Q_{max} highlighted by arrows in Fig. 15A). Not surprisingly, the effective electron density of the dimer does not resolve the individual atoms for $Q_{max} = 1$ Å^{-1} (corresponding to a resolution of ≈ 6 Å), where the effective electron density is indistinguishable from that of a single broadened Gaussian (Fig. 15C). The effective electron density *begins* to resolve these two atoms at $Q_{max} = 3$ Å^{-1} (corresponding to a resolution of ≈ 2 Å) where the two structure factors in Figure 15A begin to differ in form. However, if $Q_{max} = 6$ Å^{-1} (corresponding to a resolution of ~1 Å), the two atoms are fully resolved. This illustrates that two atoms in a structure can be resolved only when Q_{max} is large enough to reveal the oscillatory nature of the structure factor corresponding to this interatomic spacing (Fig. 15A) and therefore to distinguish this dimer structure factor from that of a single broadened Gaussian. Measurements must be made to high resolution (~1 Å) to determine atomic/molecular-scale features.

EXAMPLES

Deconstructing a CTR: the orthoclase(001)-water interface

We now "deconstruct" a CTR measurement of a freshly-cleaved orthoclase (001)-water interface to show many experimental details that are not usually presented in research articles. A typical experimental setup is shown schematically in Figure (16). The beam energy is selected with a monochromator, and the beam is reflected by a mirror that can focus the X-rays and remove the "harmonics" [e.g., $\lambda/3$ for a Si(111) monochromator] that are Bragg-reflected by the monochromator at the same scattering condition that is used to select the primary beam energy. An appropriate beam size is selected with an incident slit (Slit1 in Fig. 16), and the beam flux is monitored throughout the experiment, before and after the defining beam slits, by using monitors M0 and M1. (M1 is used to normalize out time-dependent variations in incident beam signal that are typical at synchrotron sources.) Monitors are designed to provide a precise measurement of the beam flux with minimal perturbation of the beam itself; i.e., either through photo-electron production in an ionization chamber, or by measuring the scattering of the beam as it transmits through air. Often, a beam attenuator ("filter box" in Fig. 16) is used to reduce the incident flux for measurements where the scattered flux is too large to be measured by the detector (because of the finite temporal resolution of the detector). Attenuators, typically metal sheets having different thicknesses (e.g., Al for E ~ 8 keV,

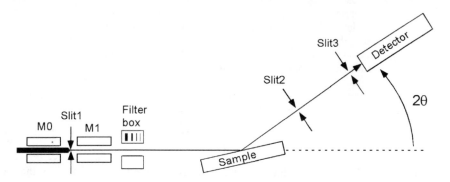

Figure 16. The geometry of a typical CTR measurement, shown schematically with beam monitors, slits, a sample, and detector.

Mo for E ~ 18 keV), enable the detector to measure signals that vary in count rate by many orders of magnitude. The attenuator factor for each filter is calibrated so that the absolute reflectivity can be determined from the detector signal. The beam is incident on the sample and scattered into the detector at an angle of 2θ with respect to the incident beam direction. The resolution defining slits (Slit3 in Fig. 16) are set to a particular resolution, and the guard slits (Slit2 in Fig. 16) are chosen to minimize the background signals without changing the angular acceptance of the detector slits with respect to the sample.

In Figure 17A, we show selected "raw" rocking curves for the orthoclase (001)-water interface over a wide range of Q_z values. In each of these scans the detector is held fixed, and the incident angle is scanned to reveal the surface reflectivity (which exhibits a narrow peak when the incident angle corresponds to one-half the scattering angle, 2θ) and a flat background due to incoherent scattering. These measurements were performed in a thin-film cell in which an 8 μm-thick kapton film was used to maintain a thin water layer in contact with the mineral surface (see Fig. 8A). In this plot, a number of rocking scans are shown, and the peak in each rocking scan is at a different incident angle, α, corresponding to half the scattering angle, 2θ. A silver attenuator foil, whose attenuation factor was calibrated separately, was used to reduce the count rate in the detector by a factor of ~38 for selected rocking scans whose peak counting rates were too high to be measured quantitatively by the detector. The first step in analyzing these data is multiplying the measured intensities in each rocking scan by the appropriate attenuation factor so that the data are all directly comparable (Fig. 17B). (The attenuation factor is 1 for scans where no attenuator is used.) These data, plotted on a logarithmic scale, show that the reflectivity signal changes dramatically along the CTR. We also find, in this case, that the background signal is actually *largest* where the surface reflectivity signal is *smallest* (see Fig. 17C).

Selected rocking scans (Fig. 18) demonstrate how the reflectivity and background signals are separated. These data show a similar resolution-limited line shape at all momentum transfers. The specular reflection is clearly observed as a narrow peak superimposed over a linear background. The presence of a significant surface reflectivity over a broad range of scattering angles, coupled with the absence of any visible "diffuse" scattering near the specular direction (either in the form of a broadened rocking curve or as "tails" near the specular direction) at any scattering angle, indicates that the orthoclase-water interface in the present study is atomically smooth with large flat terraces.

The signal-to-background ratio varies very substantially as a function of Q_z. (Note the change in vertical scale in the scans in Fig. 18.) The solid lines are fits to the data for a Gaussian peak shape and a linear background. To estimate the surface domain size, we make use of the relation between the rocking scan width, $\Delta Q_{//}$, and coherence length, L. In reciprocal space, this relation is expressed as $\Delta Q_{//} = 2\pi/L$ (the exact prefactor in this expression depends upon the correlations of defects in the sample). For these rocking curve scans, the magnitude of the momentum transfer is fixed and we change only its direction with respect to the sample surface normal direction. For specular reflectivity, the rocking curve widths in angle and reciprocal space are related by $\Delta Q_{//} = Q_z \Delta\theta$. Therefore, the average surface coherence length can be written as

$$L = \frac{2\pi}{Q_z \Delta\theta} \tag{27}$$

The rocking scan data in Figure 18 are resolution limited, as indicated by the facts that the rocking scan width in angle space does not vary as a function of Q_z, and its magnitude, $\Delta\theta \sim 0.02°$, corresponds to one-half the angular size of the detector within the

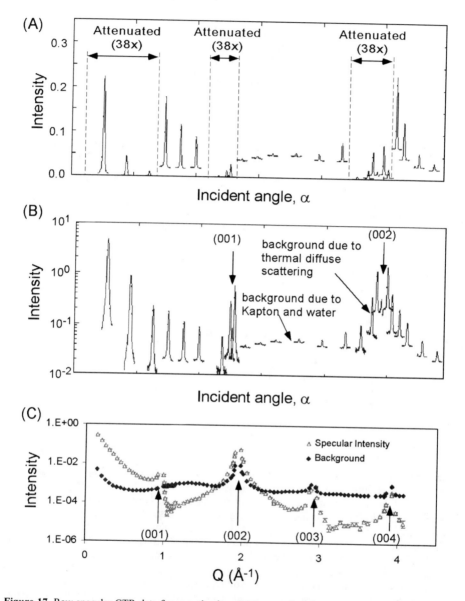

Figure 17. Raw specular CTR data for an orthoclase (001)-water interface measured in a thin-film cell at various levels of analysis. (A) Selected rocking curves are shown with abrupt changes in reflectivity due to changes in the filter setting. (Normalizing the detector signal to the monitor signal does not remove these changes because the filter is downstream from the monitor.) The peak for each rocking curve is observed at the specular condition, $\alpha = 2\theta/2$. (B) The monitor normalization has been divided out, and the data are now presented on a logarithmic scale. (C) The integrated reflectivities and background signals are plotted vs. Q_z. The reflectivity signal varies much more strongly than the background signal, and the typical measurement is made when the background is substantially larger than the reflectivity.

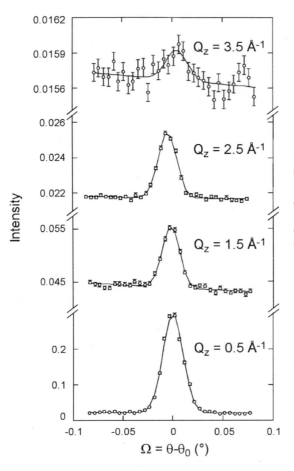

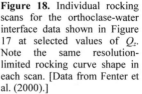

Figure 18. Individual rocking scans for the orthoclase-water interface data shown in Figure 17 at selected values of Q_z. Note the same resolution-limited rocking curve shape in each scan. [Data from Fenter et al. (2000).]

scattering plane, as is expected when the width is resolution-limited in a specular scattering geometry. This is illustrated in Figure 10C, in which the lateral intrinsic width of the CTR is more narrow than the experimental angular resolution. In this case, we find a lower limit for the surface domain size by using the rocking scan at the smallest values of Q_z. With the rocking curve at $Q_z = 0.5$ Å^{-1}, we find L > 4 μm. These data illustrate the need for high-quality surfaces in performing high resolution X-ray reflectivity measurements. Since the integrated intensity is conserved for any change in peak shape, any increase in the width of the rocking scan (either due to a surface having a typical domain size of less than 4 μm, or a substrate mosaic disorder of > 0.02°) would diminish our ability to distinguish the reflectivity from the incoherent background, especially where the reflectivity is already substantially smaller than the background signals and consequently difficult to observe.

We plot the integrated area of each of these rocking curves as a function of Q_z, as shown in Figure 17C, along with the background signal level. The background counting rates are plotted after multiplying by the rocking scan width ($\Delta\theta = 0.02$°), so that the plotted background signal is integrated over the width of the rocking curve and reflects the relative strengths of the reflectivity and background signals in the rocking scans. The specular reflectivity data follow the expected form of a CTR (Robinson 1986) with the reflectivity R largest near a strong bulk Bragg peak [notated as (001), (002), and (003), as shown in Fig. 17C] and much smaller between the bulk Bragg peaks.

A casual comparison of these data shows that the background signal is frequently much larger than the reflectivity signal. Comparison of the background signals derived from the reflectivity measurement (Fig. 17C) and the elastic scattering from kapton and water (Fig. 14) indicates that the broad peak in the background signal near 1.5 Å^{-1} can be attributed to scattering from the kapton film. The increase in the background close to the bulk Bragg peaks is caused by thermal diffuse scattering from the substrate lattice (Warren 1990). The strong increase in background at $Q < 0.5$ Å^{-1} suggests that it is from the water and kapton films, as any thin-film contribution to the background should vary as $1/\sin(\theta) \sim 1/Q$. Otherwise, we find a rather flat background signal as a function of Q, which suggests that this Q-independent contribution is from the substrate crystal (this is because the crystal is substantially thicker than the linear penetration depth of X-rays). This background from the substrate crystal depends primarily on the X-ray penetration depth, and has a strong variation with the photon energy ($\Lambda \sim E^3$) (Als-Nielsen and McMorrow 2001). However, reducing the photon energy increases the sensitivity to systematic errors associated with non-uniformities in the water and kapton film thickness (Fig. 9). It is generally necessary to choose a photon energy that minimizes both background from the substrate and systematic error associated with linear absorption through the thin film cell. We have found that energies ranging from 15 keV to 20 keV are optimal, consistent with the conclusions of other investigators (de Vries et al. 1999). Because we have used a thin-film cell that minimizes the background signals due to the water and plastic films above the surface, and because the bulk contributions to the background cannot be avoided, a small reflectivity signal in the presence of a large background must be determined by counting for sufficiently long times to distinguish the small reflectivity from the large background signal (Fig. 13).

Mineral termination. The first step in understanding the CTR data is to do model calculations of the mineral surface in which the surface is assumed to be ideally terminated, with no structural relaxations or rearrangements at the surface. The surface plane of the orthoclase (001) cleavage surface can be formed either by breaking bonds between the Si_1 and O_1 atoms (as drawn in Fig. 19A) for the α termination or by breaking bonds between Si_2 and O_5 atoms for the β termination. Because the type and number of bonds broken during cleavage of these two faces are different, the structure and composition of this surface is expected to depend strongly on which surface is exposed.

We show CTR calculations (Fenter et al. 2000b) for ideally terminated orthoclase (001) surfaces for the two terminations (Fig. 19B). The differences in calculated reflectivities are quite dramatic as a function of Q_z. In particular, there is a large difference between the α and β terminations just above and below the (001) Bragg peak. In each case there is a strong "dip" in the reflectivity, but this dip appears just above the (001) Bragg peak for the α termination and just below for the β termination. The measured reflectivity exhibits a strong asymmetry just above and below the (001) Bragg condition, with a substantially lower reflectivity *above* the (001) Bragg peak. At higher Q_z values, there is little difference between these two terminations near the (002) Bragg peak, but differences are again observed near the (003) Bragg peak that resemble those near the (001) Bragg peak. The data near the (003) Bragg peak also follow the qualitative trends associated with the α termination. This model calculation therefore indicates the α termination is exposed by cleaving the crystal since it provides a better description of the qualitative features in the CTR data than does the β termination. There are, however, substantial discrepancies between the calculated reflectivity for this ideally terminated structure and the experimental data. Therefore, important aspects of the mineral termination and structure have yet to be described properly.

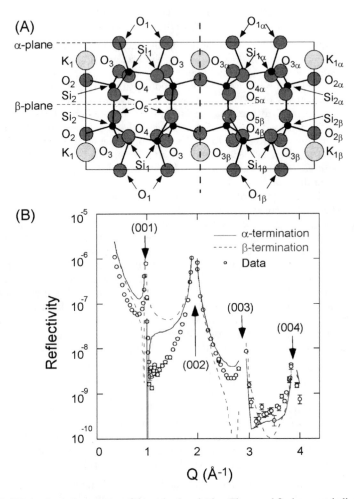

Figure 19. (A) A schematic structure of the orthoclase lattice. The α and β planes are indicated on the left. (B) The orthoclase-water interface reflectivity, along with a structure factor calculation for the α- and β-terminated lattices. Note that the same asymmetry is found in the data and the calculation for the α termination. [Figure 19B used by permission of the editor of Geochimica et Cosmochimica Acta, from Fenter et al. (2000b), Fig. 5, p. 3668.]

Designing a CTR calculation. The most widely used approach to determine atomic-scale structures is through least-squares fitting by directly comparing the data and a calculation through the quality of fit (Eqn. 24). One limitation of this approach is the possibility of finding a local minimum in the quality of fit that is distinct from the actual structure. Even for an ideal data set, care in the fitting process is often necessary to derive a structure that is reasonable with respect to the finite resolution of the data. Some aspects of the structure might not be determined uniquely. For "real" data sets, potential sources of systematic error must also be taken into account explicitly.

The most fundamental problem in fitting surface scattering data is that the depth to which the interfacial structure might be perturbed is not known. The surface is, by definition, all aspects of the interfacial structure that deviate from the known bulk structure. Displacements of near-surface atoms (i.e., structural relaxations) are a universal phenomenon of surfaces caused by the lower crystallographic coordination of the surface atoms with respect to their bulk counterparts (Zangwill 1988). A number of other structural phenomena are possible. The surface composition can be modified by chemical reaction, or the surface can be restructured with a lateral period different from that of the known bulk crystal structure; this is known as a surface reconstruction (Zangwill 1988). As Equation (16a) shows, the structure factor can be written as the sum of bulk, surface, and water structures. The first and third terms are typically known, but the structure factor for the surface is not. The main challenge is writing the surface structure factor in a sufficiently generally form so that the surface structure can be determined by direct comparison with the data.

The most straightforward way to do this is to fit the data by allowing the location of each atom in the surface layer to vary. This approach is not usually practical. First, an over-constrained fit is desirable, with the number of independent data points substantially larger than the number of parameters. For bulk crystallographic measurements, a ratio of 10 independent data points per parameter is typical (Giacovazzo 1992). For surface crystallographic measurements a ratio of at least 3-5 data points per parameter is considered sufficient, but the confidence in the results increases with the magnitude of this ratio. With surface relaxation and restructuring penetrating 2-3 layers into a crystal (with ~10 atoms per unit cell), we can expect to need as many as ≈30 parameters to describe uniquely a typical structure. An ideal CTR data set will likely have at least ≈100 to 150 independent data points. Thus, minimizing the number of fitting parameters, where possible, is always beneficial. A second problem with a fitting approach is that the data may not resolve certain features uniquely, leading to a high degree of correlation between nominally independent parameters. For example, two symmetry-distinct atoms that are found at the same height with respect to the surface plane in the ideal bulk structure will probably displace differently at the surface (e.g., one goes up, the other goes down). Yet it might be impossible to tell which atom goes which way, for example, if only the vertical structure is obtained by using specular reflectivity.

It is best to limit, wherever possible, the number of parameters used in the fitting process. It might be obvious that some structural degrees of freedom do not need to be explored during fitting. An example is calcite, an ionic crystal of calcium and carbonate ions. The carbon-oxygen bonds in the carbonate ions are unlikely to be distorted, although the orientation and position of the carbonate ion can change substantially (Fenter 2000a).

In the case of orthoclase, a more subtle approach is needed. Aside from the potassium atoms, all of the silicon, aluminum, and oxygen atoms form a covalently bonded network in which the positions and bond angles are likely to be highly correlated. For a specular reflectivity measurement, the heights of each of these atoms must be specified separately, and 13 parameters are needed to specify the vertical silicate structure for a single unit cell. If the structural relaxations penetrate two or more layers into the crystal, the number of parameters needed to specify the full surface structure grows substantially.

A reasonable approach to minimizing the number of fitting parameters for a network structure such as orthoclase is to assume that if a particular atom moves, its neighbors must move a similar distance. Thus, one can specify the position of a subset of the atoms (e.g., the tetrahedral sites) and assume that each oxygen atom will be displaced by the average of the displacements of the two tetrahedral sites to which it is bound. Through this approach, one can reduce the number of parameters needed to specify the silicate structure to 4 per unit cell from the 13 required for an unconstrained approach.

Beyond the positions of these atoms, we also need to specify their occupation factors. Feldspars are widely reported to exhibit incongruent dissolution, whereby the alkali and aluminum atoms are dissolved preferentially. Thus, we allow for removal of individual species at the surface by varying the occupation of the near-surface potassium ions. This is an additional multiplicative factor in the expression for the structure factor (Eqn. 14) for each atom in the crystal, and varies between 0 and 1.

We must also consider the fluid side of the orthoclase-water interface. First, there is the termination of the exposed tetrahedra (e.g., with hydroxyl groups). Half of the surface tetrahedral sites will be undercoordinated immediately after cleaving, and these sites are likely to react with water to hydroxylate the surface. This reaction can be accounted for by allowing the location and occupation of the outermost oxygen atoms (e.g., O1 sites) to vary in the fitting process. The fluid water might exhibit layering, or there might be some specific interaction of the water molecules with the orthoclase surface (e.g., through hydrogen bonding). Although bulk water will not contribute directly to the CTR intensities at large Q_z (because liquid water has no order along the surface-normal direction), it modifies the reflectivity at small Q_z by changing the density discontinuity between substrate and fluid layers. In addition, we must allow for any modification of the fluid water structure near the surface (e.g., in the form of layering).

There are two general choices for modeling the fluid structure. The first is to use an error-function profile whose position and width are determined while the height of the error function is constrained to correspond to the bulk electron density of the fluid layer ($\rho_{e^-} = 0.0333$ e$^-$/Å^3 for water). Any specific adsorption to the mineral surface can be included by adding into the structure factor sum additional "adsorbate" atoms whose position and occupation are determined by the fitting procedure. A second approach is to use a "layered" water structure. (See section on the mica-water interface, below.)

In fitting the structure factor data, we restrict the fitting models to different degrees of complexity to assess which aspects of the model are actually necessary to fit the data quantitatively. For example, a layered fluid structure might be part of a model that fits the data adequately, but it is only a compelling part of the result if its absence results in a substantially worse fit (when we have accounted for the different number of parameters). As in the case of fitting the individual rocking scans, the quality of fit is usually chosen to be the χ^2 function (Eqn. 24) and the uncertainties for each structural parameter are derived directly from the least-squares fitting process.

Orthoclase(001)-water interface structure. Optimization of the structural model through a direct comparison of structure factor calculations with the data results in a best-fit calculation shown with the data in Figure 20A, and the corresponding structural model is shown in Figure 20B (Fenter et al. 2001b). These results confirm that orthoclase cleaves along the α plane.

We find that the K coverage in the outermost layer ($K_{1\alpha}$) is 0 ± 0.08 monolayer (ML) (where a monolayer is defined as the 2D density of K atoms in the ideal bulk-truncated α bilayer, or 1 ML = 1 atom/55.76 Å^2), revealing that the outermost K layer was completely removed upon interaction with the aqueous solution, while K in the next and deeper layers was unchanged with respect to the expected value of 1.0 ML.

The derived coverage of the surface $O_{1\alpha}$ atom is 1.9 ± 0.25 ML, revealing that the terminal tetrahedral sites are fully hydroxylated. Although this result is not unexpected, it is notable because symmetry considerations require half of the $O_{1\alpha}$ atoms to be removed upon cleavage. The difference between the observed and the expected coverages (2.0 ML) is insignificant, given the statistical uncertainty in the measurement. The $O_{1\alpha}$ atom is 1.24 ± 0.15 Å higher than the top layer $Si_{1\alpha}$ atom, a value 0.21 Å smaller than the height

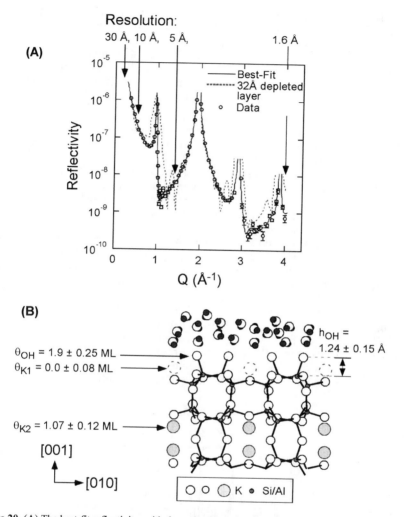

Figure 20. (A) The best-fit reflectivity, with the orthoclase-water interface data and a separate model for a 32 Å-thick K depleted layer. (B) The model corresponding to the best-fit curve in (A). [Figure 20A used by permission of the editor of Geochimica et Cosmochimica Acta, from Fenter et al. (2000b), Fig. 6a, p. 3668.]

difference of these two atoms for a bulk-terminated surface structure. This inward relaxation might indicate one or more possible structural distortion mechanisms, such as a tilting of the Si-O bond or a Si-O bond length contraction due to the reduced valence of an O-H bond with respect to that of an O-Si bond (Brown 1976). A complex relaxation profile derived from the data in Figure 20A is also included in the best-fit model. This relaxation is largest at the surface (as large as ~0.15 Å) and decays slowly into the crystal with displacements as deep as 20 Å. The rapid removal of K at the surface is consistent with experimental alkali feldspar dissolution studies (e.g., Garrels and Howard 1957; Busenberg and Clemency 1976; Chou and Wollast 1984; Blum and Lasaga 1991; Stillings et al. 1995).

Previous measurements of the oligoclase-water interface with low-resolution X-ray reflectivity (Farquhar et al. 1999) reported the rapid (< 30 min) formation of a 32 Å-thick, Na-depleted layer at a chemomechanically polished oligoclase (001) surface in contact with deionized water. A K-depleted layer as thick as that inferred by Farquhar et al. (1999) is inconsistent with our data, as can be seen for the calculation of an orthoclase surface identical to the best-fit structure described above (solid line, Fig. 20A), except for the absence of K^+ ions in the outermost 32 Å (dashed line, Fig. 20A).

Visualizing the spatial resolution of a CTR. The orthoclase-water interface results are useful to demonstrate the relationship between the range of measured data, the resulting spatial resolution, and what can be determined from the reflectivity data. Using the formalism that we derived above (Eqn. 14 and Fig. 15), we show in Figure 21 the derived effective electron densities for selected resolutions (i.e., different values of Q_{max}) for a single structure. This structure is the same as the best-fit orthoclase-water interface structure described above, except that we have arbitrarily removed the K atoms from the outermost 32 Å surface layer. As Figure 21 shows, the effective electron density profile changes dramatically as a function of the resolution with respect to the intrinsic electron density profile, as the resolution varies from 1.6 Å to 30 Å. (The corresponding Q_{max} for each resolution is noted in Fig. 20A.) The resolution of 1.6 Å corresponds to the full data set shown in Figure 20A, while that of 30 Å corresponds to a aforementioned small-angle X-ray reflectivity study of Farquhar et al. (1999). Although some features of the intrinsic structure are lost at 1.6 Å resolution, the data clearly resolve the individual silicate layers and the K layers in the bulk crystal structure. Looking carefully above and below $z = 12$ Å in Figure 21 (the location of the $KAlSi_3O_8$-$HAlSi_3O_8$ boundary), one can see the individual contributions from the missing potassium atoms. In contrast, under conditions appropriate to small-angle reflectivity with a resolution of ~30 Å, no Ångstrom-scale features associated with individual atoms within the surface layer are observed, although we see an overall change in the density of the surface layer corresponding to the depletion

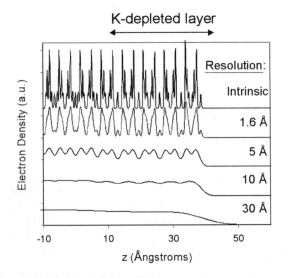

Figure 21. The effective electron density profiles as a function of the resolution (offset vertically) for an orthoclase (001) surface having a 32 Å thick potassium depleted layer. The values of Q corresponding to these resolutions are shown by vertical arrows in Figure 20A.

of K atoms. This implies that if a surface layer with a reduced density is observed, this features cannot be uniquely attributed to the depletion of any individual component of the structure if the resolution is not sufficient to resolve individual atomic contributions.

This semi-quantitative approach of displaying the electron density profile at the resolution appropriate to the data shows that the range of data acquired in a CTR measurement is closely coupled with the aspects of the structure that can be uniquely determined from the data. This approach is also a useful way to test the uniqueness of a particular set of fitting results. Different models that reproduce the reflectivity data equally well might suggest a lack of uniqueness in the CTR data, but in many cases actually correspond to identical electron density profiles. In some cases, best-fit structural models simply cannot be distinguished on the basis of the X-ray reflectivity data alone.

Probing large molecule adsorption: stearate monolayers on calcite

In the above example, the atomic-scale structure of the orthoclase-water interface was determined. Here, we describe how a similar approach can be used to probe the structure and conformation of larger, nanometer-sized molecules adsorbed to mineral surfaces. Our initial measurements in this area involved the interaction of fatty acids with the calcite surface (Fenter and Sturchio 1999). Fatty acids, hydrocarbon chains having terminal carboxyl groups, are representative of a wide range of naturally occurring organic compounds that are soluble in aqueous solutions and might interact with mineral surfaces. In the measurements described below, a transmission cell (Fig. 8B) was used, and the sample environment was controlled by flushing and filling the cell with a solution of known composition, then sealing the cell during the X-ray scattering measurements.

Figure 22A shows the X-ray reflectivity of the calcite-methanol interface where the fluid is a 5 mM stearic acid solution in methanol. The calcite surfaces are prepared by cleaving an optically-clear calcite crystal resulting in atomically smooth surfaces. The reflectivity was measured as a function of momentum transfer by performing rocking scans across the specular rod at each point and fitting each rocking scan to a Gaussian peak and a linear background. The data, plotted on a log scale, show the expected strong reflectivity near $Q = 2.067$ Å^{-1} = $2\pi/a_{104}$, corresponding to the calcite (104) Bragg peak position. Similar data for the calcite-water interface, taken in the same transmission cell, are shown in Figure 22A for comparison. Although the experimental data for the calcite-water interface show a typical smoothly varying CTR shape, the data taken in contact with the stearic acid solution clearly exhibit an additional strong periodic modulation of the reflectivity that varies by nearly two orders of magnitude.

To interpret these data, we show model calculations for X-rays reflecting from a surface coated by a thin film (Fig. 23A) calculated using a Fresnel formalism which includes the effect of total external reflection at small angle but does not include any atomic-scale structural information (Parrat 1954). Here, the internal structure of the substrate lattice and organic film do not contribute to the reflectivity (e.g., see effective electron densities in Fig. 21); instead, only the relative densities of these films, the film thickness, and the interfacial widths contribute to the reflectivity (Tidswell et al. 1990). The reflectivities of bare and film-coated surfaces are shown in Figure 23B. The bare substrate exhibits a monotonically decreasing reflectivity, corresponding to a functional form known as the Fresnel reflectivity, R_f (Als-Nielsen and Mohwald 1991) which shows total external reflection near $Q \approx 0$, and a rapid decay of the reflectivity, $R_f \sim 1/Q^4$. For incident angles above the total external reflection condition, this functional form is equivalent to the CTR shape for $Q < \pi/c$ as can be seen when Equation (23) is simplified at small values of Q [e.g., $\sin(Qc/2) \sim Qc/2$], resulting in $R = 16\pi^2 r_e^2 \rho_{e^-}^2/Q^4$ (ignoring the cell transmission and surface roughness factors, T_{cell} and $|B|^2$, respectively). The addition of a

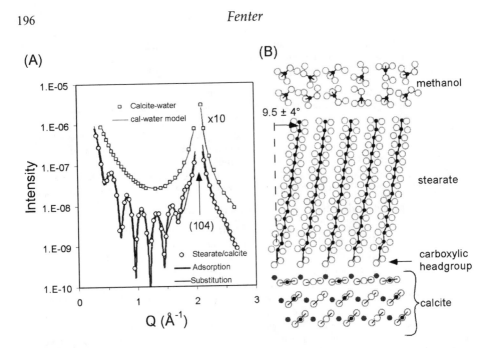

Figure 22. (A) The specular reflectivity of the calcite-water interface (squares), with similar data for a calcite surface in contact with a 5 mM solution of stearic acid in methanol. (The calcite-water data are offset vertically by a factor of 10, for clarity). The measurements were performed in a transmission cell. The solid lines are best-fit structure factor calculations for selected models. The thick solid line through the stearate data is optimized for an extended stearate molecule adsorbed on top of the calcite surface, as shown in (B). The thin solid line through the stearate data is optimized for the carboxylic head group substituting for surface carbonate ions of the calcite lattice. (B) Best-fit model for the stearate monolayer adsorbed on calcite.

film introduces a sinusoidal interference pattern. Bragg's law, as written in Equation (3), suggests that the period of oscillations can be related to the film thickness, L, by $\Delta Q = 2\pi/L$.

Comparison of these model calculations to the data (Fig. 22A) reveals that the oscillations indicate the formation of a thin film. The period of these oscillations, $\Delta Q \sim 0.24$ Å^{-1}, corresponds to a spacing of $L = 2\pi/\Delta Q = 26$ Å. This observation can be attributed to the ~24 Å length of a single stearate molecule, suggesting that the stearate molecules are "standing up" on the calcite surface in an all-trans configuration. Therefore, we tentatively conclude that the stearate film consists of a single monolayer. The depth of the oscillations in the data (in Fig. 22A) are not constant, especially near the (104) Bragg condition; in contrast, the oscillation magnitude is constant for the simple model shown in Figure 23. This additional slow modulation of the reflectivity found for stearate/calcite suggests that additional internal structure(s) of the film and substrate surface must be included before a quantitative assessment can be made concerning particular aspects of the internal film structure, such as the molecular orientation.

Through comparison of the data to representative atomistic models of the stearate-calcite interface structure, we can learn about the detailed structure of this film, the stearate coverage, and its structural relationship to the substrate lattice. In these calculations, we include the positions, scattering factors f_i, and rms widths (usually interpreted as vibrational amplitudes) of each atom in the structure. We use the results of Vand et al. (1951) for the stearic acid molecular structure near the -COOH head group,

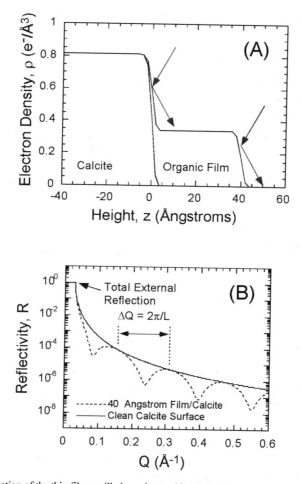

Figure 23. Explanation of the thin-film oscillations observed in Figure 22. (A) A electron density profile is shown for a 40 Å-thick organic film on a calcite surface. (Here it is assumed that the measurement is done at small angles, and consequently the internal structure of the film and substrate are not resolved.) X-rays that reflect from the organic film surface and the film-substrate interface interfere, producing intensity oscillations, as shown in (B) as a function of the momentum transfer along the surface normal direction. The period of the oscillations is directly related to the film thickness through $\Delta Q = 2\pi/L$.

and we parameterize the monolayer structure by changing the position and orientation of the molecules while we keep the internal molecular structure unchanged. The interaction of the -COOH group with the calcite surface is probably through the terminal -OH group, so we use this O atom as a reference point to describe the calcite-stearate molecule interface structure. The bold solid line in Figure 22A corresponds to the best-fit structure and provides an explanation for nearly all of the features in the experimental data. However, the quality of fit for this best-fit structure, $\chi^2 \sim 4$, is somewhat higher than expected on the basis of the statistical uncertainty of the experimental data ($\chi^2 \sim 1$) but within the range of χ^2 values found in similar studies, suggesting that the result reflects the intrinsic structure of the stearate monolayer on calcite.

The stearate molecule could conceivably interact with the calcite surface through a substitutional mechanism in which the carboxylic acid head group of the stearate molecule displaces a carbonate group in the outermost calcite layer. As Figure 22A shows (see the thin line through stearate/calcite data) a model optimized with this assumption can reproduce the period of the oscillation fringes, but it cannot reproduce the scattering intensity, especially for $1 < Q < 2$ Å^{-1}. This demonstrates the sensitivity of X-ray reflectivity to the internal organic monolayer structure, specifically to the stearate-calcite interfacial structure.

The best-fit structure, shown schematically in Figure 22B, includes a stearate coverage of 1.02 ± 0.05 ML. (1 ML = 4.95×10^{14}/cm^2 and is the density of Ca sites on the calcite surface.) The tilt of the hydrocarbon chain is determined to be $9.5 \pm 4°$ with respect to the surface normal. This confirms our earlier qualitative assessment that the stearate molecules are "standing–up." Another aspect of the structure that can be assessed is the calcite-stearate bonding. The vertical Ca-O distance is determined to be 2.05 ± 0.15 Å. This distance is smaller than the expected Ca-O bond length (~2.4 Å), but we measured only the component of the Ca-O spacing along the surface-normal direction. Thus, if we assume a Ca-O bond length of 2.4 Å, the data indicate that this bond is tilted by ~30° with respect to the surface normal. Furthermore, because the "resolution" of the experiment, $2\pi/Q_{max} \sim 2.5$ Å, was larger than the derived Ca-O spacing, this parameter might be sensitive to systematic error associated with experimental factors, limitations of our structural models, or both.

We find that a finite roughness of the calcite surface must be included to achieve quantitative agreement between the calculated and experimental reflectivities. We use the model described in Equation (20) (Robinson 1986) that assumes that the roughness is the result of individual monatomic steps within the (approximately micrometer-sized) lateral coherence length of these measurements. We found a surface roughness of 1.1 Å. Independent atomic force microscopy measurements have shown that the step density of these calcite cleavage surfaces is very small (Sturchio et al. 1997).

Other models can give insight into the structure and binding of the stearate molecules at the calcite surface. A monolayer of bound hydroxyl groups is present at the calcite-water interface (at the Ca site) (Van Cappellen et al. 1993; Fenter et al. 2000a). Bound OH$^-$ groups were probably present on our calcite surfaces before the surfaces were exposed to the methanol solutions containing stearic acid. When we included the possibility of adsorbed OH$^-$ groups in our calculations of the stearate monolayer scattering reflectivity, the data were best fitted with an OH coverage of 0.0 ± 0.3 ML. Alternatively, if we fixed the coverage of adsorbed OH$^-$ groups in the presence of the stearate molecules at 1.0 ML, the quality of fit was degraded significantly (and the resulting model had an unrealistic Ca-O vertical spacing of 1.1 Å). This result indicates that surface hydroxyl groups are displaced from the surface after adsorption of the stearate monolayer.

Water structure near the muscovite-water interface

We have stated above that reflectivity data are sensitive to any structuring at the mineral-water interface. Examples above were concerned only with what could be described as adsorbed layers, and no compelling evidence was found for any changes in the fluid structure near the mineral-water interface with respect to fluid water. Researchers have inferred for many years that substantial structuring may occur near the muscovite-water interface (Israelachvili and Pashley 1983). High-resolution neutron scattering studies and many molecular dynamics simulations have found such structuring in the interlayer region of clay minerals (Skipper et al. 1991; Karaborni et al. 1996;

Bridgeman and Skipper 1997), but no such direct structural studies were performed for an isolated mineral-water interface. The muscovite (001) cleavage surface is ideal to probe such phenomenon for many reasons. This surface is molecularly smooth, as it is composed of the basal planes of SiO_4 tetrahedra arranged in ditrigonal rings. In water under ambient conditions, the K^+ ions are expected to desorb (Gaines 1957), making this surface effectively a hard wall (defined by the densely packed basal-plane O atoms) augmented with an array of ditrigonal cavities carrying a permanent negative charge.

The X-ray reflectivity data for the muscovite-water interface are shown in Figure 24A with the best-fit curve to the CTR data (Cheng et al. 2001). The acquisition of these data was non-trivial, given that even the highest-quality muscovite often has bulk mosaic (i.e., angular) disorder that is substantially larger (by factors of as much as 10-100) than the angular resolution of our typical measurements (~0.04°). Because the integrated intensity is conserved in a rocking scan, such broad rocking curve widths reduce the peak reflectivity in a rocking scan substantially reducing our ability to measure the reflectivity, especially at the smallest reflectivities. After developing appropriate procedures to select and mount the muscovite crystals, we obtained data with negligible mosaic contribution that reveal the intrinsic reflectivity profiles for the muscovite-water interface (Fig. 24A). The optimized structural parameters from the best fit reveal that relaxations are substantial for atoms within the outermost polyhedral layer $[d_{001} = (c/2) \sin(\beta)]$, with predominantly inward displacements as large as 0.04 ± 0.02 Å.

An interesting aspect of these data is that they show clear evidence that the water structure near the interface is modulated by the presence of the mineral surface. To model the possible water layering, we follow an elegant approach that was first used by

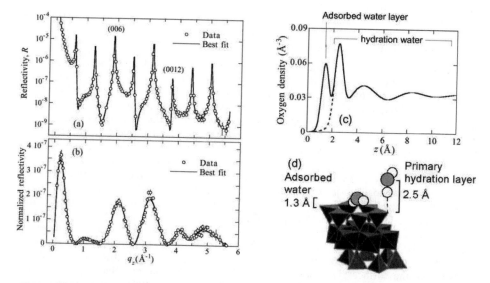

Figure 24. (A) Specular reflectivity of the muscovite-water interface, and (B) the same data normalized by the generic CTR shape, $R_{ctr} = 1/[Q \sin(Qc/2)]^2$, where c is the lattice spacing of the mica lattice. (C) The derived near-surface water structure, plotted as the electron density profile of the adsorbed and fluid water above the muscovite surface. (D) A schematic representation of the adsorbed and primary hydration-layer water molecules, corresponding to the electron density profile in (C). [Figure 24 (A-C) used by permission of the editor of Physical Review Letters, from Cheng et al. (2001), Fig. 2, p. 156103-2, and Fig. 3, p. 156103-3.]

Magnussen et al. (1995) to measure the weakly ordered fluid structures near fluid-vapor interfaces. The fluid is modeled in this approach as a series of Gaussian peaks that are separated by a fixed distance, c_w, with constant occupation factors determined by the bulk density of the fluid. However, the rms width, σ_n, of each layer, n, is constrained to follow the following form:

$$\sigma_n = [\sigma_0^2 + n\sigma_{bar}^2]^{1/2} \tag{28}$$

In this form, σ_0 is interpreted as the vibrational amplitude of the first layer of the fluid, and σ_{bar} is a parameter that specifies how quickly the vibrational amplitude increases for each successive layer into the fluid. This form can easily be included in the sum of contributions to the water structure factor, resulting in a simple closed form expression for the fluid CTR:

$$F_{water} = f_0 \exp\left[-\tfrac{1}{2}(Q\sigma_0)^2\right]$$
$$\times \left\{1 + \exp\left[-\tfrac{1}{2}(Q\sigma_{bar})^2\right]\exp(iQc_w) + \exp\left[-(Q\sigma_{bar})^2\right]\exp(2iQc_w)...\right\} \tag{29}$$

$$= f_0 \frac{\exp\left[-\tfrac{1}{2}(Q\sigma_0)^2\right]}{\left\{1 - \exp\left[-\tfrac{1}{2}(Q\sigma_{bar})^2\right]\exp(iQc_w)\right\}}$$

This expression shows that the σ_0 term acts as an overall Debye-Waller factor, as it would for any atom in the structure factor calculation, but that σ_{bar} controls the overall behavior of this structure factor. If $\sigma_{bar} = 0$, we reproduce the form of a truly crystalline CTR. We also recover the structure factor for an error function profile in the small-Q regime where this expression varies as $\sim 1/iQc_w$, as long as $Q\sigma_{bar} \ll 1$. Therefore, this expression is capable of describing structures ranging from an error function profile associated with an disordered material to a highly "layered" fluid structure.

In fitting the muscovite-water CTR data to structural models, we found that it was not possible to obtain a quantitative fit to these data with only surface relaxations and a featureless water structure modeled by an error function profile. When we included the formalism in Equation (29) for a structured water layer and allowed for separate "adsorbed" layers at the mica-water interface, we obtained quantitative agreement between the calculations and the data. The derived interfacial water structure (i.e., the water O density profile) is shown in Figure 24B. This plot also includes the broadening of the electron density due to the finite experimental resolution (~ 1.1 Å) as described by Equation (25) (Fenter et al. 2001) The resolved features are therefore expected to be uniquely determined. The adsorbed layer was located at 1.3 ± 0.2 Å above the mean position of the relaxed surface O, comparable to the 1.6-Å height of the unrelaxed K ion (Fig. 24C). The hydration structure beyond this first "adsorbed" layer has an oscillatory density profile, with a narrow first layer and substantially damped subsequent layers extending ~ 10 Å above the surface. The first peak in the hydration layer has a lateral density equivalent to 1.3 ± 0.2 water molecules per ditrigonal ring and is located at 2.5 ± 0.2 Å above the surface. This position compares well with the calculated oxygen atom heights of ~ 2.6-2.8 Å of water molecules directly hydrating the (001) face of vermiculite, montmorillonite, and talc under hydration (Skipper et al. 1991; Karaborni et al. 1996; Bridgeman and Skipper 1997).

Because water molecules in both the adsorbed and the first hydration layers directly hydrate the surface, these two layers may be considered as a single interface layer, giving a combined lateral density of $0.10 \pm 0.02/\text{Å}^2$, corresponding to the 2D density of a densely packed water layer. This is shown schematically in Figure 24C. The 1.2(3) Å height difference of water molecules within the combined first layer may contribute to the broadening of the second and subsequent layers. The peak of the second hydration layer is 2.7(2) Å (roughly the water molecular size) from the mean position of the combined first layer. The spacing between neighboring layers from the second hydration layer on, beyond which the oscillation amplitude is rapidly reduced and a constant spacing was assumed in the model, is 3.7(3) Å. This increased layer spacing at larger distances from the surface, if confirmed, suggests that additional structuring is involved as the partially ordered interfacial water is transformed into a randomly ordered structure. Variations in layer spacing, a characteristic of bond ordering of solutes, are seen in simulated interfacial water structures (Stockelmann and Hentschke 1999); they contrast with the constant layering spacing in density-ordered hard-sphere liquids (Yu et al. 1999, 2000a).

DEFECT STRUCTURES AT BARITE-WATER INTERFACES

The discussion above describes measurements of the laterally averaged structure of the mineral-water interface. In some cases, isolated defects such as steps are expected to dominate surface reactivity. X-ray reflectivity can also characterize surface defect structures directly, as demonstrated by measuring the systematic variation of the rocking scan shape as a function of Q_z in specular X-ray reflectivity data (Fenter et al. 2001) to probe the step height and step-step correlations at the barite-water interface. Representative rocking scans of the barite(001)-water interface are shown in Figure 25 at selected values of L_{rlu} for the (001) surface. (Here we plot the data versus $L_{rlu} = Q_z/(2\pi/c)$ which is the reciprocal lattice unit along the surface normal direction, where c is the unit cell spacing along the surface normal direction, to emphasize the location of the Bragg peaks that appear at integer values of L_{rlu}.). The data (Fig. 25) show that the rocking scans for the (001) surface vary in both shape and width as a function of L_{rlu}. Also shown (Figs. 26A-D) are the specular reflectivity and rocking scan widths, plotted as a function of L_{rlu}, for both the (001) and (210) surfaces. The rocking scans are narrow and have a Gaussian-like shape near $L_{rlu} = n$ (or $Q_z = 2\pi n/c_{hkl}$), where n is an integer and c_{hkl} is the unit cell dimension along the surface-normal direction ($c_{001} = 7.157$ Å and $c_{210} = 3.44$ Å). In contrast, the scans have

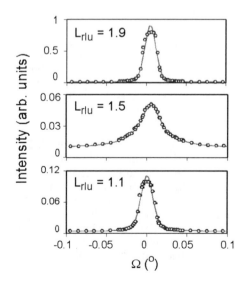

Figure 25. Rocking curve scans for the barite(001)-water interface at selected values of $L_{rlu} = Q_z/(2\pi/c_{001})$. Note that the peaks are narrow and Gaussian shaped at $L \sim 1$ and $L \sim 2$, while is the peak is broader and Lorentzian shaped near $L \sim 1.5$. [Reproduced with permission from *J. Phys. Chem. B* **2001**, *105*, 8112-8119. Copyright 2001 Am. Chem. Soc.]

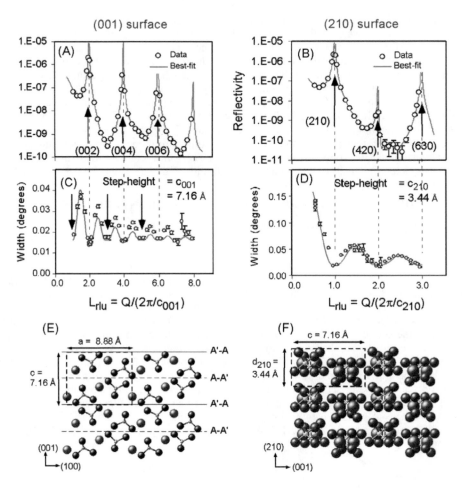

Figure 26. (A) and (B) Specular reflectivity data for the (001) and (210) surfaces as a function of $L_{rlu} = Q_z/(2\pi/c_{001})$. The variation of the rocking curve width is plotted for these two surfaces in (C) and (D), and side views of the barite lattice for each cleavage surface are shown in (E) and (F), respectively. The derived step height for the (001) surface is 7.16 Å, even though the A'-A and A-A' cleavage planes expose symmetry-equivalent surfaces. [Used by permission of the editor of Journal of Physical Chemistry B, from Fenter et al. (2001), Fig. 1 p. 8113, Fig. 4(b-c) p. 8115, Fig. 5a p. 8117 and Fig. 6a p. 8118.]

Lorentzian-like shapes and are broader near half-integer values of *n*. Their widths oscillate as a function of L_{rlu} from a maximum of ~0.04° to a minimum of <0.02°. The widths are small near the Bragg condition, because neighboring terraces interfere constructively at this point in reciprocal space, and they are larger between the Bragg peaks where neighboring terraces interfere destructively. The actual values of the widths near the subsequent minima (Figs. 26C and 26D) are a result of the finite angular resolution of the detector, so that the rocking curve shape is resolution limited as shown in Figure 10C. (This was confirmed independently by scanning the detector slit through the direct beam, revealing a measured angular detector width of $\Delta(2\theta) = 0.038°$.) Rocking scan widths broader than $\Delta\theta = \Delta(2\theta)/2 = 0.019°$ result from a broadening of the specular rod (see Fig. 10B) associated with the defect structure of the barite surface. A much stronger variation in the rocking scan width was observed for the (210) surface (Fig. 26D), with a maximum width of 0.15° at small L_{rlu} and resolution-limited widths of ~0.02° at successive minima.

A Lorentzian-like line shape indicates random defect distribution across the surface. The oscillation of the rocking curve width as a function of L_{rlu} is a direct result of constructive and destructive interference conditions that are alternately satisfied as the momentum transfer perpendicular to the surface, Q_z, increases. The vertical extent of the defects, h, is revealed through the relation $h = 2\pi/\delta Q_z = c_{hkl}/\delta L$ where δQ is the characteristic period of the oscillations. The periods of oscillations in Figures 26C and 26D are $\delta Q = 2\pi/c_{hkl}$ for both surfaces, implying that for each surface the predominant defects correspond to unit-cell-high steps, with $h = 7.2$ Å and 3.4 Å on the (001) and (210) surfaces, respectively. This conclusion is consistent with results from previous atomic force microscopy studies (Pina et al. 1998).

An estimate for the lateral domain size, $L_{//}$, can be derived from the rocking curve width, $\Delta\theta$, measured at a momentum transfer, Q, through the relation $L_{//} \sim 2\pi/(Q^*\Delta\theta)$. From this relation we estimate domain sizes of ~7,400 Å at $L_{rlu} = 1.5^*(2\pi/c_{001})$ for the (001) surface and ~2,800 Å at $L_{rlu} = 0.5^*(2\pi/c_{210})$ for the (210) surface. (This estimate ignores the dependence of the derived domain size on the rocking curve line shape; a more quantitative estimate is derived below.) In each case, the domain size can be thought of as the mean step-step spacing. Larger derived domain sizes that are >10,000 Å (corresponding to the narrowest rocking curve widths) are found at constructive interference conditions when scattering from neighboring terraces is in phase, reflecting the surface domain size when the steps are ignored. These values are expressed as lower limits, because the measurements are limited by the finite angular resolution of the detector (0.02°).

To provide a more quantitative description of the surface defect structure and distribution, we use a model for randomly distributed surface defects (Lu and Lagally 1982). With this model, the rocking curve width and its variation with Q can be calculated from a few parameters that describe the probability of encountering a defect (i.e., a step) and the phase change in encountering that defect. (For specular reflectivity, this phase change reflects the height difference across the step.) Within this model, an approximately Lorentzian line shape is reproduced.

Comparing the calculated results from this model to the measured rocking scan widths for the (210) surface (Fig. 26D), we can explain our data with single unit-cell-high steps (randomly in either the up or down direction) having a mean step-step spacing of 2,300 Å. This model properly describes the periodicity of the oscillations for the (001) surface, confirming that the unit-cell-high step is the predominant defect averaged macroscopically over the (001) surface. However, the variation of the (001) rocking scan width is not well described by this model for a defect spacing of 7,300 Å (Fig. 26C). The rate at which the rocking scan width decreases with increasing Q is overestimated, implying a smaller-than-expected surface domain size at larger incident angle (and smaller X-ray footprint). By measuring separate spots on the sample surface, we found that the domain size on this surface was inhomogeneous. The simplest explanation for these data is that the X-ray beam footprint at large values of Q (where the footprint was the smallest) was on an area with a relatively high step density, while at small values of Q the larger area of the X-ray beam footprint included areas having lower step densities.

An interesting aspect of these data is that the step height observed for the (001) surface is, in fact, a unit cell high, containing two separate $BaSO_4$ layers, as was previously observed by AFM (Pina et al. 1998). The (001) surface is formed by alternating $BaSO_4$

layers (which we refer to as A and A' layers) that are symmetry-related though a 2_1 screw axis (Fig. 26E). There are no structural or chemical differences between surfaces terminated with the A and A' layers, and there can be no thermodynamic driving force for this observation (i.e., differences in surface energies or the number of broken bonds during cleavage). The absence of half-unit cell high steps can be noted by the lack of any additional broadening near L = 1, 3, 5... in Figure 26C (shown as vertical arrows) where the sensitivity of the rocking curve width to half-unit-cell-high steps is maximized. Because there is no evidence for any broadening beyond experimental resolution at these scattering conditions, we conclude that unit-cell-high steps are not simply typical defects but instead are the predominant defects averaged over macroscopic surface regions, at least for this sample and all other samples that we have studied. The small finite lateral spatial resolution in our measurement (~ 1 μm), compared with the ~1 mm^2 to 5 mm^2 areas covered by the X-ray beam footprint, does not change this conclusion; the interference responsible for the modulation in rocking scan width is observed in each micrometer-wide region associated with the X-ray coherence length. Given the absence of evidence for steps with heights other than unit cell height, we conclude that one of the two possible surface terminations is present over most of the barite surface.

These results show that X-ray reflectivity can be a powerful probe of the defect structure at the mineral-water interface, including the defect size and distribution. In this case, the result relied primarily on variation of the rocking curve width as a function of Q_z. The rocking curve shape generally provides additional information concerning the correlations between steps (Robinson and Tweet 1992).

Observing dissolution reactions in real-time

Measurement of mineral-water interface structure during surface reactions provides direct insight into mineral reactivity and is a powerful approach for understanding complex interfacial reactions. It is currently not possible to provide a complete structural measurement with a temporal resolution of a few minutes. It is, however, possible to measure representative changes in real time in a way that provides important constraints on the dissolution process. Such measurements can also be coupled with high-resolution measurements of previously reacted surfaces to provide snapshots of the reacted surface.

Much effort has been given to understanding the low-temperature destruction of feldspar because of its importance in soil formation and various geochemical cycles; experimental studies of the rates and mechanisms of feldspar dissolution have been numerous in the past few decades (reviewed by Blum and Stillings 1995). A major obstacle to achieving a better fundamental understanding of the feldspar dissolution process, however, has been the lack of direct, *in situ* structural measurements of the reacting feldspar-solution interface at the atomic scale. X-ray reflectivity can be used to make direct measurements of Ångstrom-scale dissolution processes and kinetics at orthoclase (001) cleavage surfaces as a function of pH and temperature and to provide unambiguous constraints on the dissolution process (Teng et al. 2001).

Dissolution processes. We used X-ray reflectivity measurements to probe the changes in orthoclase surface structure and termination in real time during dissolution. Measurements of the dissolution process were performed *in situ* in flowing solutions of 0.1 M HCl and 0.1 M NaOH, having pH values (at 25 °C) of 1.1 and 12.9, respectively (Teng et al. 2001).

In these measurements, we used a simple variation of a technique previously used to probe the *growth* of semiconducting surfaces, known as "RHEED oscillations" (van der Vegt et al. 1992). In the language of these previous growth studies, the temporal variation of the reflectivity, measured at the "anti-Bragg" point, gives direct insight into the "growth mode." The anti-Bragg scattering condition is specified by the momentum transfer, Q_z, through the relation $Q_z = \pi/d_{001} = 0.48 \text{ Å}^{-1}$ (for orthoclase, $d_{001} = c \sin(\beta) = 6.459$ Å). Alternatively, one can specify the first anti-Bragg condition from Bragg's law, $n\lambda = 2d \sin(\theta)$, with $n = \frac{1}{2}$.

The physical basis for these measurements is shown in Figure 27A. X-rays that reflect from neighboring terraces at the anti-Bragg condition interfere destructively at this scattering condition. The reflectivity is maximized when there are no steps and all reflected photons are in phase. Conversely, the reflectivity is minimized when the surface is terminated by a half-occupied layer (averaged over an area corresponding to the lateral coherence length of the X-ray beam, typically 1-10 µm). This can be visualized on the basis of Figure 27A by noting that all X-rays that reflect off a particular terrace height interfere destructively with X-rays that reflect from terraces that are either higher or lower by a single unit-cell high step. More generally, the reflectivity decreases as root-mean-square (rms) width of the interface increases.

Raw X-ray reflectivity data are shown in Figure 27B for measurements during dissolution at pH = 12.9 and 73 °C. These data show that the rocking curve shape does not change during dissolution. Instead, only the peak reflectivity of the rocking curve changes, and the data show a clear non-monotonic variation as a function of time. We integrated these rocking curve data to determine the variation of the reflectivity as a function of time during dissolution. Data in Figure 27C show the time variation of the reflectivity during dissolution at acidic and alkaline pH. These data show that the reflectivity varies in an *oscillatory* fashion during dissolution. Although the data at acidic pH also show a systematic decrease of reflectivity during dissolution, the data at alkaline pH show a nearly complete recovery after a single oscillation.

To demonstrate the systematics of how an oscillatory time variation in the reflectivity is possible during dissolution, we assume that the surface is characterized by occupation factors described by an error function profile (Fig. 28A). These occupation factors can be thought of as "blocks" of orthoclase, so that occupation factors <1 represent a partially filled layer that is locally stoichiometric. The position of the error function moves continuously as the surface dissolves. For simplicity, we first assume that the interface width does not change during dissolution. The reflectivity is calculated as a function of the error function position. At the anti-Bragg condition, neighboring terraces are out of phase. The phase factor for each layer, n, varies as $\exp(inQc) = (-1)^n$, so that interfacial structure factor becomes:

$$F(Q_z = \pi/c) = occ(1) - occ(2) + occ(3) - occ(4)... = \sum(-1)^n occ(n)$$

$$= \sum_{surf}(-1)^n occ(n) + F_{ctr}(Q_z = \pi/c) \tag{30}$$

$$= \sum_{surf}(-1)^n occ(n) + \frac{1}{2}$$

Here the sum over surface sites that might have less than full occupancy is denoted as Σ_{surf}, and the structure factor of a CTR is evaluated at the anti-Bragg condition in the last line. The high sensitivity of the reflectivity to the presence of complete and half-occupied layers is not unexpected for a smooth surface, but a remarkable feature of this phenomenon is that this sensitivity is retained even in the presence of the substantial roughening shown in Figures 28C and 28D. In fact, the reflectivity approaches zero whenever a half-occupied layer is found in the surface region. If we include the possibility that the surface gets rougher as dissolution proceeds, we find that the intensity oscillations persist but become progressively weaker (Fig. 28D).

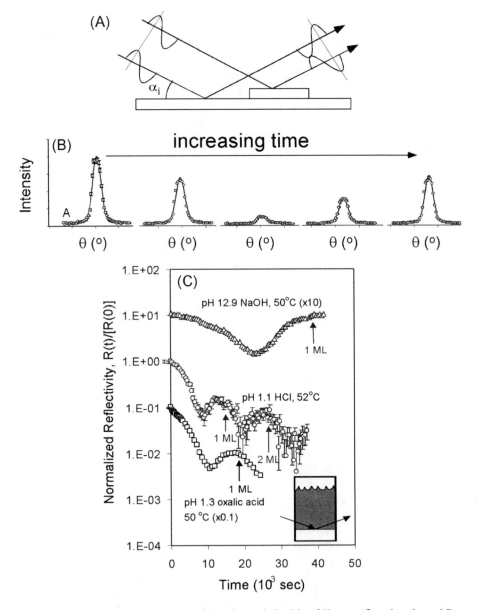

Figure 27. (A) Schematic representation of the phase relationship of X-rays reflected at the anti-Bragg condition, showing that X-rays interfere destructively when they are reflected by terraces that differ in height by a single unit cell. (B) Raw reflectivity data in the form of rocking curves, measured as a function of time during dissolution at pH 12.9 and 73 °C. (C) Integrated reflectivity data as a function of time for dissolution at acidic and alkaline conditions. Note the oscillatory variation of the reflectivity under all conditions, with a strong overall decay for only the acidic dissolution data. A schematic of the transmission cell is inset. [Figure 27c used by permission of the editor of Geochimica et Cosmochimica Acta, from Teng et al. (2001), Fig. 5, p. 3464.]

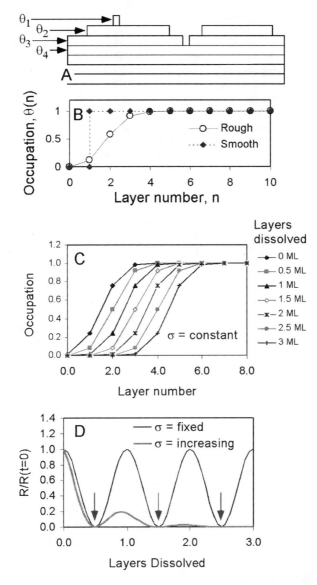

Figure 28. (A) Schematic model of the surface topography, with each layer described by an occupation factor. (B) Occupation factors for rough and smooth surfaces. (C) The time variation of the occupation factors for a model in which the occupation factors are modeled by an error function whose position changes during dissolution, but whose rms width is assumed to be unchanged by dissolution. (D) The variation of the reflectivity at the anti-Bragg condition as a function of the amount of material dissolved for the occupation factors in (C) (σ = fixed), as well as for the case where σ increases by 50% during dissolution of the three layers (σ = increasing). Note in (D) that the reflectivity goes to zero (vertical arrows) whenever the occupation profile has a half-occupied layer, as is found when the amount dissolve corresponds to 0.5 ML, 1.5 ML, and 2.5 ML. [Figure 28 (a) and (b) used by permission of the editor of Geochimica et Cosmochimica Acta, from Teng et al. (2001), Fig. 6 (a) and (b), p. 3465.]

The models in Figure 28 provide a demonstration of the high sensitivity of X-ray reflectivity to the rate of dissolution (as seen by the period of oscillations) and to the evolution of roughness during dissolution (as seen by the intensity variation). These models, however, do not provide any direct insight into the dissolution process, for example to reveal the type of reactive sites or their distribution across the surface. To gain insight into the growth process from these data, we have modified a model used to study semiconductor growth (van der Vegt et al. 1992) to describe dissolution. In this model, the occupation of a given layer, n, is given by a coverage that varies according to the following set of coupled differential equations, that relate the rate at which the coverage changes, $d\theta_n/dt$, in a particular layer, n, to the occupations found in layers n and n-1:

$$d\theta_n / dt = K_d \left(\theta_n - \theta_{n-1} \right) + K_s \left[1 - 4 \left(\theta_n - \tfrac{1}{2} \right)^2 \right] \tag{31}$$

Here θ_n is the occupation factor of the nth layer ($0 \leq \theta_n \leq 1$), and K_d and K_s are the rate constants associated with two fundamental dissolution processes. The first term (with prefactor K_d) is proportional to the number of exposed basal sites in layer n, representing a direct dissolution process that can occur at any exposed surface site. The second term (with prefactor K_s) represents the dissolution at steps that proliferate for partially dissolved layers. We solve these equations numerically. The results of these calculations are shown in Figure 29.

For $K_s/K_d = 0$ (i.e., no lateral surface processes) we get random dissolution with a monotonic decrease in the reflectivity vs. the number of dissolved layers (bold line, Fig. 29). If $K_s/K_d \gg 1$, we find a periodic variation of the reflectivity, with the reflectivity minima (maxima) corresponding to the presence of half (fully) occupied surface layers (open circles, Fig. 29B). This oscillatory pattern corresponds to layer-by-layer dissolution in which the dissolution of a given layer is essentially complete before dissolution of the subsequent layer begins. Finally, if K_s and K_d have comparable sizes (e.g., for $K_s/K_d = 2.9$), the reflectivity exhibits an oscillatory variation with an overall decrease in reflectivity due to net roughening of the surface (thin line, Fig. 29B).

These model calculations show that the experimentally observed intensity maxima in Figure 29 generally correspond to integral numbers of dissolved layers, even in the presence of surface roughening, as we also found in the more phenomenological model in Figure 28. The large "order of magnitude" changes in reflectivity (Fig. 29B) associated with dissolution of the first half-monolayer, coupled with our ability to measure the time-dependent reflectivity precisely with an uncertainty of a few percent, establishes a direct way to probe mineral dissolution rates with sub-monolayer sensitivity (picomolar sensitivity within our X-ray beam spot). The precision of these measurements results from the fact that X-ray reflectivity probes mineral dissolution kinetics in a way that is statistically averaged over the macroscopic surface area corresponding to the footprint of the X-ray beam ($\sim$8 mm^2).

Substantial discussion in the literature concerns the formation of non-stoichiometric layers during dissolution of alkali feldspars. We calculated the variation of the reflectivity for various values of K_s/K_d, assuming that H$^+$ exchange for K$^+$ was the primary reaction mechanism during our experiments, as shown in Figure 29C. The predominant difference between stoichiometric and non-stoichiometric dissolution models is the magnitude of the temporal reflectivity changes. The change in the reflectivity for nonstoichiometric dissolution models (Fig. 29C) is much smaller than that for stoichiometric dissolution models (Fig. 29B). (Note the difference in vertical scales between Figs. 29B and 29C.) This smaller change in reflectivity is a direct result of the small ($\sim$13%) change in electron density upon exchange of H$^+$ for K$^+$ in the non-stoichiometric layer.

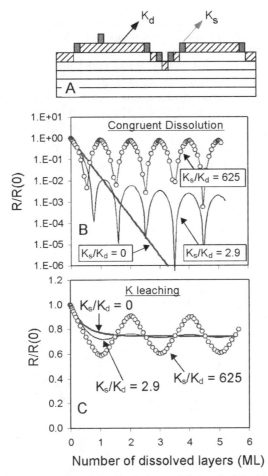

Figure 29. Schematic representation of the dissolution model described by Equation (32). The rate constants K_s and K_d describe the "step specific" and "direct" dissolution processes whose reactive sites are shown schematically. The direct process is assumed to occur at any exposed site on the surface, while the step-specific process is assumed to occur only at sites that are under-coordinated with respect to the ideal terrace sites. Calculated variation of the reflectivity at the anti-Bragg condition for (B) congruent and (C) K-leaching models described by Equation (32). The reflectivity varies by orders of magnitude for a congruent dissolution process (i.e., the removal of stoichiometric blocks of orthoclase), while a much smaller variation of the reflectivity is found if the dissolution involves the leaching of K from the lattice without the disruption of the silicate lattice. [Figure 29 (b-c) used by permission of the editor of Geochimica et Cosmochimica Acta, from Teng et al. (2001), Fig. 6 (c) and (d), p. 3465.]

From these considerations, we conclude that the data in Figure 27C reveal that dissolution at pH 12.9 is fully stoichiometric and dominated by lateral dissolution processes producing ideal layer-by-layer dissolution. In contrast, dissolution at pH 1.1 results in a strongly damped oscillatory pattern, indicative of a more random dissolution process in which the orthoclase surface is substantially disrupted and roughened. Separate CTR measurements of reacted surfaces provided a more detailed picture of the dissolving surface after 1 to 15 layers were dissolved at pH 1.1 and 12.9 (Teng et al. 2001) and fully support the conclusions reached here on the basis of the real-time dissolution data shown in Figure 27C. These CTR data reveal that the non-stoichiometric layer has a thickness of <6.5 Å after dissolution of ~15 layers at pH 1.1 at 95°C. We conclude that these data are inconsistent with a dissolution process in which exchange of H^+ for K^+ proceeds beneath the surface layer with minimal disruption of the aluminosilicate lattice.

Dissolution kinetics. A separate aspect of these data is that the dissolution *rate* can be estimated directly from the oscillation period of $R(t)/R(t=0)$. Having concluded above that dissolution at both acidic and alkaline pH proceeds by disrupting the silicate lattice, we can make use of our model predictions (Fig. 29C) that each period corresponds to the dissolution of a single layer. Our data for HCl solutions at pH 1.1 and T = 52 °C, and for NaOH solutions at pH 12.9 and T = 50 °C (Fig. 27) reveal dissolution rates of 4.0×10^{-10} and 1.5×10^{-10} mol $KAlSi_3O_8$ m^{-2} sec^{-1}, respectively (based on 4 $KAlSi_3O_8$ per surface unit cell, or 0.036 $KAlSiO_8$/Å^2).

We have extended these observations to probe the variation of dissolution rate as a function of temperature at selected pH values (Fenter et al. 2002). A series of real-time reflectivity measurements for different temperatures at pH 12.9 is shown in Figure 30A.

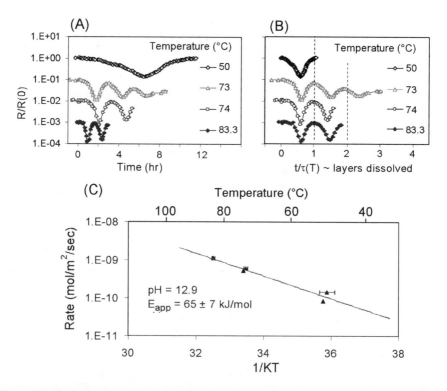

Figure 30. (A) The temporal variation of the reflectivity at selected sample temperatures at pH 12.9. The frequency of the intensity oscillations increases significantly with temperature. (B) Scaling the time axis of the data in (A) to correspond to the number of dissolved layers reveals the same intensity variation vs. the number of layers, which directly reveals a temperature-independent dissolution process under these conditions. (C) The derived dissolution rate is plotted as a function of inverse temperature, whose slope corresponds to an effective activation energy of 65 ± 7 kJ/mol.

These data have similar slightly damped oscillatory reflectivity patterns over the entire temperature range studied (50-83.3 °C), although the time between subsequent reflectivity maxima varies strongly with temperature. We determine the relative dissolution rates by normalizing the time axis by a temperature-dependent factor, $\tau(T)$, that is approximately the time to remove a single layer at temperature, T. This normalization converts the horizontal time axis into a measure of the number of layers dissolved at the (001) surface (Fig. 30B), providing a direct measure of the absolute dissolution rate. The result is a nearly identical functional form for the variation of the reflectivity as a function of t/τ. This directly indicates that although the dissolution *rate* changes substantially as a function of temperature at pH 12.9, there no significant changes occur in the dissolution *process* over this measured temperature range.

The systematic variation of the dissolution rate, $1/\tau$, as a function of inverse temperature is shown in Figure 30C for the (001) surface at pH 12.9. These data are consistent with Arrhenius behavior. Assuming that the slopes of these data are proportional to the effective activation energies for dissolution at each pH, we find a best-fit effective activation energy of 65 ± 7 kJ/mol at pH 12.9.

These measurements demonstrate that X-ray reflectivity can be used as a probe of mineral dissolution kinetics through real-time measurements of the reflectivity at the "anti-Bragg" scattering condition. Such measurements can give new insights by revealing the dissolution process at a single, well-characterized cleavage surface where the type and distribution of reactive sites can be established with confidence. Extending this approach to different crystallographic faces and different solution conditions will enable us to test our understanding of feldspar dissolution processes. This includes such fundamental issues as whether the stoichiometry, the kinetics, and even the effective activiation energy of dissolution on a particular crystal face differ from the values measured with macroscopic approaches on powder samples.

SUMMARY, CONCLUSIONS, AND PERSPECTIVES

The results described above provide a snapshot of the current (year 2002) use of X-ray reflectivity techniques to probe the mineral-fluid interface. The examples were chosen to emphasize the ability to probe a diverse range of phenomena, including mineral termination, adsorption of small and large molecules to interfaces, defect structures, and real-time measurements of mineral reactions. At present, the use of X-ray reflectivity to probe mineral-fluid interfaces is still in its infancy, especially with respect to its use to study metal and semiconductor surfaces. Nevertheless the complexity of many mineral structures, coupled with the diverse range of phenomena that are of interest to the geochemistry and environmental communities (e.g., mineral surface termination, adsorption, dissolution, growth), suggest that there is plenty of work to be done. In next few years, we anticipate substantial advances in a number of areas that should substantially improve our understanding of mineral-fluid interface phenomena. These areas are summarized below.

Non-specular reflectivity

Non-specular reflectivity measurements have been done in a few cases to probe mineral-fluid interface structure (Geissbuhler 2000; Eng et al. 2000; Trainor et al. 2002), but their application has been limited. The use of non-specular measurements directly reveals the lateral structure of these interfaces which, in many cases, is critical for distinguishing between different structural models and uniquely identifying the lateral adsorption geometry for an adsorbate. Such measurements reveal much greater detail of the interfacial structure, but they also require more time and effort to perform with respect to specular reflectivity measurements. In this sense, the non-specular measurements are probably best suited to study phenomena that are well characterized, where the scientific issues to be resolved are well defined.

Anomalous reflectivity

One of the main drawbacks of the use of X-ray scattering as a probe of interfacial phenomena (e.g., ion adsorption) is that the complete interfacial structure must be understood before any individual part of that structure (e.g., the location of an adsorbate) can be determined. More generally, X-ray reflectivity data are sensitive to the total electron density and consequently are not ideally suited to provide information concerning the distribution of a particular element, particularly if the contribution by that element to the total electron density is small (e.g., a low Z element, or a low density of high Z element). The variation in the atomic scattering factor of an element near its X-ray absorption edge (Henke et al. 1993) can be used to add an element specific sensitivity to the many powerful characteristics of X-ray reflectivity techniques described above. Such an approach, while relatively common in studying bulk structures ranging from protein crystals (Hendrickson 1999) to liquids (Saito and Waseda 2000), has only rarely been used for surface scattering measurements (Chu et al. 1999) because of the need to determine precisely changes in the scattering intensity where the reflectivity already might be quite small ($\sim 10^{-10}$). In spite of these difficulties, the anomalous scattering approach is expected to be a powerful approach for mineral-fluid interface studies.

Microbeams

Many potential benefits can be anticipated with the implementation of microbeam focussing for high-resolution reflectivity measurements. These benefits include greater photon flux on the sample, reduced active area corrections, and the ability to study samples that are not available as large, centimeter-sized single crystals. Again, relatively little has been done in this respect, although selected measurements have been performed (Eng et al. 2000). Reduction of the incident beam size from a millimeter to micrometers by using focused undulator sources at the Advanced Photon Source has already been achieved with Kirpatrick-Baez mirrors (Eng et al. 1998). Such a reduction in beam size should enable high-resolution CTR measurements that are fully analogous to those described above on the small, ~100 μm-sized crystals that are currently inaccessible to high-resolution scattering techniques. One of the main limitations to the straightforward application of such measurements will be the need to control the sample position with sub-micron precision as the sample is rotated by large angles necessary to obtain high-resolution scattering data. Beam damage might also become a significant issue because the photon intensities (photons per area per second) will be orders of magnitude higher than in an unfocused undulator beam.

Temporal resolution

Elucidation of the kinetics and mechanisms of mineral-fluid interactions requires high-resolution X-ray scattering measurements on rapid time scales. Time series analyses are desired for addressing the evolution of structure and composition at the interface, on time scales as small as milliseconds or less. The high brilliance of the third-generation synchrotron sources affords new opportunities for such time-resolved studies, because we can observe in real time the processes of adsorption/desorption and complex formation at mineral-fluid interfaces. For example, experiments using a pressure-jump relaxation techniques yield rates of adsorption and desorption of protons and hydroxide at the surface of metal oxides in the range of milliseconds to seconds (reviewed by Casey and Ludwig 1995). Our measurements of orthoclase dissolution have been limited to date by the use of point-to-point measurements during dissolution. This need can be alleviated by the use of linear and area detectors based on charge-coupled device sensors, photodiode arrays and commonly employed energy-dispersive techniques (Mills 1991; Shimura and Harada 1993). Such detectors allow simultaneous measurement of scattered X-rays across a wide solid angle on short time scales, obviating the need for point-by-point measurements, and also allow for a >10-fold increase in the range of reaction rates that can be accessed by X-ray scattering techniques (Teng et al. 2001).

Direct methods

The discussion in this chapter concerning the analysis of high-resolution reflectivity data was based on a "search and fit" procedure. This approach is straightforward but it can get trapped in "false structures" that correspond to local minima in the χ^2 space, especially for complex structures that differ substantially from the bulk truncated structure. Recent years have seen increasing applications of "direct methods" to invert experimental data directly and reveal a preliminary structure, thereby bypassing the "phase problem" (Yacoby et al. 2000; Saldin et al. 2001; Takahashi et al. 2001). In scattering experiments, only the magnitude of the structure factor, $|F|^2$, is measured, while a simple inversion of the data to reveal the electron density requires knowledge of the phase at each point through the relation $\rho(x) = \int F(Q) \, e^{iQz} \, dQ$. These methods all make use of constraints on the scattering phase that can be derived, for example, by enforcing the atomistic and positive-definite character of the electron density. Routine application of these techniques to high resolution CTR measurements is therefore likely to be a significant benefit to studies of mineral-fluid interfaces.

ACKNOWLEDGMENTS

Many people have contributed directly to the work described above. These include Neil Sturchio, who initiated the mineral-fluid interface program at Argonne National Laboratory, and many others, including Michael Bedzyk, Dirk Bosbach, Likwan Cheng, Ron Chiarello, Phillip Geissbuhler, Mary McBride, Kathy Nagy, Changyong Park, Michel Schlegel, George Srajer, H. Henry Teng and Zhan Zhang. Neil Sturchio and Kathy Nagy are also very much thanked for providing helpful and detailed reviews of this chapter, and Karen Haugen is thanked for editing the manuscript.

This work was supported by the Geosciences Research Program, Office of Basic Energy Sciences, Office of Science, U.S. Department of Energy, under contract W-31-109-ENG-38. Use of the Advanced Photon Source was supported by the Office of Basic Energy Sciences, Office of Science, U.S. Department of Energy, under contract W-31-109-ENG-38 to Argonne National Laboratory.

REFERENCES

Als-Nielsen J (1987) Solid and liquid surfaces studied by synchrotron X-ray diffraction. Top Curr Phys 43:181-222
Als-Nielsen J, Mohwald H (1991) Synchrotron X-ray scattering studies of Langmuir films. *In*: Ebashi S, Koch M, Rubenstein E (Eds) Handbook on synchrotron radiation, vol 4. North Holland, New York, p 1-53
Als-Nielsen J, McMorrow D (2001) Elements of modern X-ray physics. John Wiley and Sons, West Sussex, England
Arsic J, Reedijk MF, Sweegers AJR, Wang YS, Vlieg E (2001) Compression versus expansion on ionic crystal surfaces. Phys Rev B 64:233402(1-4)
Bevington PR (1969) Data reduction and error analysis for the physical sciences. McGraw-Hill, New York
Bloch JM (1985) Angle and index calculations for a z-axis X-ray diffractometer. J Appl Cryst 18:33-36
Blum AE, Lasaga AC (1991) The role of surface speciation in the dissolution of albite. Geochim Cosmochim Acta 55:2193-2201
Blum AE, Stillings LL (1995) Chemical weathering rates of silicate minerals. Rev Mineral 31:291–351

Bridgeman CH, Skipper NT, (1997) A Monte Carlo study of water at an uncharged clay surface. J Phys Condens Matter 9:4081-4087

Brown, ID (1976) On the geometry of the O-H···O hydrogen bond. Acta Cryst A32:24-31

Busenberg E, Clemency CV (1976) The dissolution kinetics of feldspars at 25°C and 1 atm CO_2 partial pressure. Geochim Cosmochim Acta 40:41-50

Busing WR, Levy HA (1967) Angle calculations for 3- and 4-circle X-ray and neutron diffractometers. Acta Cryst 22:457-464

Casey W, Ludwig C (1995) Silicate mineral dissolution as a ligand-exchange reaction. Rev Mineral 31:87-117

Charlton G, Hoowes PB, Nicklin CL, Steadman P, Taylor JSG, Muryn CA, Harte SP, Mercer J, McGrath R, Norman D, Turner TS, Thornton G (1997) Relaxation of $TiO_2(110)$-(1×1) using surface X-ray diffraction. Phys Rev Lett 78:495-498

Cheng L, Fenter P, Nagy KL, Schlegel ML, Sturchio NC, (2001) Molecular-scale density oscillations in water adjacent to a mica surface. Phys Rev Lett, 87:156103(1-4)

Chiarello RP, Wogelius RA, Sturchio NC (1993) *In situ* X-ray reflectivity measurements at the calcite-water interface. Geochim Cosmochim Acta 57:4103-4110

Chiarello RP, Sturchio NC (1994) Epitaxial Growth of otavite on calcite observed *in situ* by synchrotron X-ray scattering. Geochim Cosmochim Acta 58:5633-5638

Chiarello RP, Sturchio NC (1995) The calcite (10-14) cleavage surface in water: early results of a crystal truncation rod study. Geochim Cosmochim Acta 59:4557-4561

Chiarello RP, Sturchio NC, Grace JD, Geissbuhler P, Sorenson LB, Cheng L, Xu S (1997) Otavite-calcite solid solution formation at the calcite-water interface studied by *in situ* synchrotron X-ray scattering. Geochim Cosmochim Acta 61:1467-1474

Chou L, Wollast R (1984) Study of the weathering of albite at room temperature and pressure with a fluidized bed reactor. Geochim Cosmochim Acta 48:2205-2218.

Chu YS, You H, Tanzer JA, Lister TE, Nagy Z (1999) Surface resonance X-ray scattering: core-electron binding-energy shifts of Pt (111) surface atoms during electrochemical oxidation. Phys Rev Lett 83:552-555

Chu YS, Lister TE, Cullen WG, You H, Nagy Z (2001) Commensurate water monolayer at the $RuO_2(110)$/water interface. Phys Rev Lett 86:3364-3367

De Vries SA, Goedtkindt P, Bennett SL, Huisman WJ, Zwanenburg MJ, Smilgies DM, DeYoreo JJ, van Enckevort WJP, Bennema P, Vlieg E (1998) Surface atomic structure of KDP crystals in aqueous solution: an explanation of the growth shape. Phys Rev Lett 80:2229-2232

De Vries SA, Goedtkindt P, Huisman WJ, Zwanenburg MJ, Feidenhans'l R, Bennett SL, Smilgies DM, Stierle A, DeYoreo JJ, van Enckevort WJP, Bennema P, Vlieg E (1999) X-ray diffraction studies of potassium dihydrogen phosphate (KDP) crystal surfaces. J Cryst Growth 205:202-214

Eisenberger P, Marra WC (1981) X-ray diffraction study of the Ge(001) reconstruction. Phys Rev Lett 46:1081-1084.

Eng PJ, Newville M, Rivers ML, Sutton SR (1998) Dynamically figured Kirkpatrick-Baez X-ray microfocussing optics. *In*: I. McNulty (Ed.) X-ray microfocussing: Applications and Techniques SPIE Proc 3449:145

Eng PJ, Trainor TP, Brown GE Jr, Waychunas GA, Newville M, Sutton SR, Rivers ML (2000) Structure of the hydrated α-Al_2O_3 (0001) surface. Science 288:1029-1033

Farquhar ML, Wogelius RA, Tang CC (1999) *In situ* synchrotron X-ray reflectivity study of the oligoclase feldspar mineral-fluid interface. Geochim Cosmochim Acta 63:1587-1594

Feidenhans'l, R (1989) Surface structure determination by X-ray diffraction. Surf Sci Reports 10:105-188

Fenter P, Sturchio NC (1999) Structure and growth of stearate monolayers on calcite: first results of an *in situ* X-ray reflectivity study. Geochim Cosmochim Acta 63:3145-3152

Fenter P, Geissbühler P, DiMasi E, Srajer G, Sorensen LB, Sturchio NC (2000a) Surface speciation of calcite observed *in situ* by X-ray scattering. Geochim Cosmochim Acta 64:1221-1228

Fenter P, Teng H, Geissbühler P, Hanchar JM, Nagy KL, Sturchio NC (2000b) Atomic-scale structure of the orthoclase (001)-water interface measured with high-resolution X-ray reflectivity. Geochim Cosmochim Acta 64:3663-3673

Fenter P, McBride MT, Srajer G, Sturchio NC, Bosbach D (2001) Structure of barite(001) and (210)-water interfaces. J Phys Chem B 105:8112-8119

Fenter P, Park C, Cheng L, Zhang Z, Krekeler MPS, Sturchio NC (2002) Orthoclase dissolution kinetics probed by *in situ* X-ray reflectivity: effects of temperature, pH and crystal orientation. Geochim Cosmochim Acta, in press

Gaines GL (1957) The ion exchange properties of muscovite mica. J Phys Chem 61:1408-1413

Garrels RM, Howard P (1957) Reactions of feldspar and mica with water at low temperature and pressure. Proc 6th Natl Conf On Clays and Clay Minerals 68-88

Geissbuhler MP (2000)X-ray interfacial crystallography of water on calcite. PhD Dissertation, University of Washington, Seattle, WA

Giacovazzo C (1992) Crystallographic computing, *In*: Giacovazzo C (ed.) Fundamentals of Crystallography, International Union of Crystallography, Oxford University Press, p 141-228

Hendrickson WA, (1999) Maturation of MAD phasing for the determination of macromolecular structures. J Synchrotron Radiat 6:845-851

Henke BL, Gullikson EM, Davis JC (1993) At. Data Nucl. Data Tables 54:181

Horn PM, Birgeneau RJ, Heiney P, Hammonds EM, (1978) Melting of submonolayer krypton films on graphite. Phys Rev Lett 41:961-964

Hubbell JH, Veigele WJ, Briggs EA, Brown RT, Cromer DT, Howerton RJ (1975) Atomic form factors, incoherent scattering functions, and photon scattering cross sections. J Phys Chem Ref Data 4:471-538

Huisman WJ, Peters JF, Zwanenburg MJ, de Vries SA, Derry TE, Abernathy D, van der Veen JF (1997) Layering of a liquid metal in contact with a hard wall. Nature 390:379-381

Israelachvili JN, Pashley RM (1983) Molecular layering of water at surfaces and origin of repulsive hydration forces. Nature (London) 306:249-250

Karaborni S, Smit B, Heidug W, Urai J, van Oort U.E (1996). The swelling of clays: molecular simulations of the hydration of montmorillonite. Science 271:1102-1104

Kmetko J, Datta A, Evmenenko G, Dutta P (2001) The effects of divalent ions on Langmuir monolayer and subphase structure: A grazing-incidence diffraction and Bragg rod study. J Phys Chem B 105:10818-10825

Leveiller F, Jacquemain D, Lahav M, Leiserowitz L, Deutsch M, Kjaer K, Als-Nielsen J (1991) Crystallinity of the double-layer of cadmium arachidate films at the water surface. Science 252:1532:1536

Li J, Liang KS, Scoles G, Ulman A (1995) Counterion overlayers at the interface between an electrolyte and an w-functionalized monolayer self-assembled on gold. An X-ray reflectivity study. Langmuir 11:4418-4427

Lohmeier M, Vlieg E (1993) Angle calculations for a six-circle X-ray diffractometer. J Appl Cryst 26:706-716

Lu TM, Lagally MG (1982) Diffraction from surfaces with randomly distributed steps. Surf Sci 120:47-66

Magnussen OM, Ocko BM, Regan MJ, Penanen K, Pershan PS, Deutsch M (1995) X-ray reflectivity measurements of surface layering in liquid mercury. Phys Rev Lett 74:4444-4447

Marra WC, Eisenberger P, Cho AY (1979) X-ray total external reflection-Bragg diffraction: a structural study of the GaAs-Al interface. J Appl Phys 50:6927-6933

Mills D (1991) Time resolved studies. *In*: Brown GS, Moncton DE (eds) Handbook on Synchrotron Radiation, vol. 3. North Holland, New York, p 291-335.

Mochrie SGJ (1988) Four-circle angle calculations for surface diffraction. J Appl Cryst 21:1-3

Ocko BM (1995) Surface X-ray scattering and scanning tunneling microscopy studies of the Au(111) electrode. NATO ASI Ser E 288:103-119

Parratt LG (1954) Surface studies of solids by total reflection X-rays. Phys Rev 95:359-369

Pina CM, Becker U, Risthaus P, Bosbach D, Putnis A (1998) Molecular-scale mechanisms of crystsal growth in barite. Nature 395:483-486.

Renaud, G (1998) Oxide surfaces and metal/oxide interfaces studied by grazing incidence X-ray scattering, Surf Sci Rep 32:1-90

Robinson IK (1983) Direct determination of the Au(110) reconstruction by surface X-ray diffraction. Phys Rev Lett 50:1145-1148

Robinson IK (1986) Crystal truncation rods and surface roughness. Phys Rev B 33:3830-3836

Robinson IK (1991) Surface Crystallography. *In*: Brown GS, Moncton DE (eds) Handbook on Synchrotron Radiation, vol. 3, North Holland, New York, p 223-266.

Robinson IK, Tweet DJ (1992) Surface X-ray diffraction. Rep Prog Phys 55:59-651

Robinson IK, Graafsma H, Kvick A, Linderholm J (1995) First testing of the fast kappa diffractometers at National Synchrotron Light Source and European Synchrotron Radiation Facility. Rev Sci Instrum 66:1765-1767

Saito M, Waseda Y (2000). Anomalous X-ray scattering for determining the partial structural functions of binary liquids. J Synchrotron Radiat 7:152-159

Saldin DK, Harder R, Vogler H, Moritz W, Robinson IK (2001) Solving the structure completion problem in surface crystallography. Comp Phys Comm 137:12-24

Schlegel ML, Nagy KL, Fenter P, Sturchio NC (2002) Structures of prismatic and pyramidal surfaces of quartz: a combined high resolution X-ray reflectivity and atomic force microscopy study. Geochim Cosmochim Acta, in press

Shimura T, Harada J (1993). A new technique for the observation of X-ray CTR scattering by using an imaging plate detector. J Appl Cryst 26: 151-158

Skipper NT, Soper AK, McConnell JDC (1991). The structure of interlayer water in vermiculite. J Chem Phys 94:5751-5760

Somorjai G (1981) Chemistry in Two-Dimensions: Surfaces, Cornell University Press (Ithaca)

Stillings LL, Brantley SL, Machesky ML (1995) Proton adsorption at the adularia feldspar surface. Geochim Cosmochim Acta 59:1473-1482

Stockelmann E, Hentschke R (1999) A molecular-dynamics simulation study of water on NaCl(100) using a polarizable water model. J Chem Phys 110:12097-12107

Sturchio NC, Chiarello RP Cheng L, Lyman P, Bedzyk MJ, Qian Y, You H, Yee D, Geissbuhler P, Sorensen LB, Liang Y, Baer DR (1997) Lead adsorption at the calcite-water interface: synchrotron X-ray standing wave and X-ray reflectivity studies. Geochim Cosmochim Acta 61:251-263

Takahashi T, Sumitani K, Kusano S (2001) Holographic imaging of surface atoms using surface X-ray diffraction. Surf Sci 493:36-41

Teng,H, Fenter P, Cheng L, Sturchio NC (2001) Resolving orthoclase dissolution mechanisms with atomic force microscopy and x-ray reflectivity. Geochim Cosmochim Acta 65:3459-3474

Tidswell IM, Ocko BM, Pershan PS, Wasserman SR, Whitesides GM, Axe JD (1990) X-ray specular reflection studies of silicon coated by organic monolayers (alkylsiloxanes). Phys Rev B 41:1111-1128

Toney MF (1994) Studies of electrodes by *in situ* X-ray scattering. NATO ASI Ser C 432:109-25

Toney MF and Wiesler DG (1993) Instrumental effects on measurements of surface X-ray diffraction rods: resolution function and active sample area. Acta Cryst A49:624-642

Trainor TP, Eng PJ, Brown GE, Robinson IK, De Santis M (2002) Crystal truncation rod diffraction study of the α-Al$_2$O$_3$ (1102) surface. Surf Sci 496:238-250

Van Cappellen P, Charlet L, Stumm W, Wersin P (1993) A surface complexation model of the carbonate mineral-aqueous solution interface. Geochim Cosmochim Acta 57:3503-3518

van der Vegt HA, Pinxteren HM, Lohmeier M, Vlieg E, Thornton JMC (1992) Surfactant induced layer-by-layer growth of Ag on Ag(111). Phys Rev Lett 68:3335-3338

van Hove MA (1999) Atomic scale surface structure determination: comparison of techniques. Surf Interface Anal 29:36-43

Vand V, Morley WM, Lomer TR (1951) The crystal structure of lauric acid. Acta Cryst 4:324-329

Vlieg E, van der Veen JF, Macdonald JE, Miller M (1987) Angle calculations for a five-circle diffractometer used for surface X-ray diffraction. J Appl Cryst 20:330-337

Warren BE (1990) X-Ray diffraction. Dover Publications, New York

Woodruff DP, Delchar TA (1986) Modern techniques of surface science. Cambridge University Press, Cambridge

Yacoby Y, Pindak R, MacHarrie R, Pfeiffer L, Berman L, Clarke R (2000) Direct structure determination of systems with two dimensional periodicity. J Phys: Condens Matter 12:3929-2938

You H, Nagy Z (1999) Applications of synchrotron surface X-ray scattering studies of electrochemical interfaces. MRS Bull 24:36-40

You H (1999) Angle Calculations for 4S+2D six-circle diffractometer. J Appl Cryst 32:614-623

Yu CJ, Richter AG, Datta A, Durbin MK, Dutta P (1999) Observation of molecular layering in thin liquid films using X-ray reflectivity. Phys Rev Lett 82:2326-2329

Yu CJ, Richter AG, Kmetko J, Datta A, Dutta P (2000a) X-ray diffraction evidence of ordering in a normal liquid near the solid-liquid interface. Europhys Lett , 50:487-493

Yu CJ, Richter AG, Datta A, Durbin M, Dutta P (2000b) Molecular layering in a liquid on a solid substrate: An X-ray reflectivity study. Physica B (Amsterdam) 283:27-31

Zangwill A (1988) Physics at surfaces. Cambridge University Press, Cambridge

APPENDIX 1
COMPLEX NUMBERS

$e^{ix} = \cos(x) + i\sin(x)$

Real part of $e^{ix} = \text{Re}(e^{ix}) = \cos(x)$

Imaginary part of $e^{ix} = \text{Im}(e^{ix}) = \sin(x)$

$i = \sqrt{(-1)}$, or $i^2 = -1$

$e^{i\pi} = -1$

$e^{i2\pi} = 1$

For any complex number, Z:

$Z = |Z|e^{i\phi} = \text{Re}(Z) + i\,\text{Im}(Z)$

$|Z|^2 = ZZ^* = \left[\text{Re}(Z) + i\,\text{Im}(Z)\right]\left[\text{Re}(Z) - i\,\text{Im}(Z)\right] = \text{Re}(Z)^2 + \text{Im}(Z)^2\,|Z|^2$

APPENDIX 2
TERMINOLOGY, SYMBOLS AND DEFINITIONS

θ "Theta"; a spectrometer angle in a four-circle spectrometer that is typically used for rocking scans. In Bragg diffraction, this is also the angle of the incident X-ray beam with respect to the Bragg planes.

2θ "Two-theta"; the scattering angle. The angle of the scattered photon with respect to the incident beam direction.

α_i Angle of incidence of the X-ray beam with respect to the surface plane.

α_f Exit angle of scattered X-ray beam with respect to the surface plane.

A Active area of the surface (i.e., seen by both the incident beam and the detector).

a_{uc} Surface unit cell area.

β A parameter used to describe the surface roughness due to individual steps ($0 \leq \beta \leq 1$).

B Surface roughness factor.

c_w Layer spacing for the layered water model.

d Bragg plane spacing.

D_w Film thickness (including both water and plastic films) in a thin-film cell.

$\Delta(2\theta)$ Angular resolution of the detector within the scattering plane.

$\Delta\gamma$ Angular resolution of the detector transverse to the scattering plane.

ΔQ_z Length of rod that is integrated in a measurement.

ε: Scattering amplitude.

E Photon energy.

f_0 Atomic form factor.

F_{uc} Form factor for the unit cell: $F_{uc} = \sum_j f_j(Q)\exp(i\mathbf{Q}\cdot\mathbf{R}_j)\exp\left[-\frac{1}{2}(Qu_j)^2\right]$.

F_{ctr} Crystal truncation rod structure factor: $F_{ctr} = F_{uc}\left[1/\left(1-e^{-iQc}\right)\right]$.

I_0 Incident X-ray beam intensity (in units of photons per area per time).

I_{int} Integrated intensity.

HKL   Miller indices of the diffraction condition.

H_j Terrace height distribution; the probability that a surface is terminated at a particular terrace height, j.

K Photon wave vector: $K = 2\pi/\lambda$, where λ is the X-ray wavelength.

λ X-ray wavelength.

L Lorentz factor.

Λ X-ray atttenuation length.

ML   Monolayer; a measure of two-dimensional coverage, (units of atoms per area).

Ω Angular velocity of a rocking curve.

Pol Polarization factor.

P Patterson function.

PL Pathlength of X-ray beam through the sample cell.

Q Momentum transfer: $Q = (4\pi/\lambda)\sin(2\theta/2)$.

r_e classical electron radius: $r_e = 2.818\times10^{-5}$ Å.

ρ_{eff} Effective electron density at a resolution corresponding to Q_{max}.

R Reflectivity; the ratio of the reflected to incident X-ray flux.

R_{det} Distance from the sample to the detector slits.

σ_{bar} Parameter used to describe extent of water layering near a mineral surface.

σ_0 Parameter used to describe width of first water layer above mineral surface.

T_{cell} Transmission of X-rays through the sample cell.

u Vibrational amplitude of an atom.

u_{eff} Effective vibrational amplitude, including the finite resolution of the scattering data.

W Cell width for a transmission cell.

χ^2 Chi-square factor; measure of the quality of fit between data and a calculation.

z_j Height of atom *j* with respect to the surface plane.

Z Atomic number.

APPENDIX 3
STRUCTURE FACTORS

Unit cell structure factor:

$$F_{uc}(Q) = \sum_i f_i(Q)\exp[-i\mathbf{Q}\cdot\mathbf{r}]\exp[-\tfrac{1}{2}(\mathbf{Q}\cdot\mathbf{u})^2]$$

(sum is over all atoms in unit cell)

N-layer crystal:

$$z = 0, -c, -2c \ldots -(N-1)c$$

$$F_{N-slit} = F_{uc}\left(\frac{1-e^{-iNQc}}{1-e^{-iQc}}\right) = F_{uc}e^{-i(N-1)Qc/2}\left[\frac{\sin(NQc/2)}{\sin(Qc/2)}\right]$$

Crystal truncation rod (i.e., a semi-infinite lattice sum):

$$z = 0, -c, -2c\ldots$$

$$F_{ctr} = \frac{F_{uc}}{1-e^{-iQc}} = \frac{F_{uc}e^{iQc/2}}{2i\sin(Qc/2)}$$

Semi-infinite featureless lattice:

$$\rho(z) = 0 \ (z > 0); \ \rho(z) = \rho_0 \ (z \leq 0)$$

$$F_{step} = \frac{F_{uc}}{iQc}$$

Layered water:

$$z_n = (n-1)c, \ \sigma_n = (\sigma_0^2 + n\,\sigma_{bar}^2)^{1/2}:$$

$$F_{layered\text{-}water} = F_{water}\frac{\exp\left[-\tfrac{1}{2}(Q\sigma_0)^2\right]}{\left\{1-\exp\left[-\tfrac{1}{2}(Q\sigma_{bar})^2\right]\exp(iQc_w)\right\}}$$

Structure factor of a Gaussian profile:

$$\rho(z) = \frac{Z}{(2\pi\sigma^2)^{1/2}}\exp\left\{-\frac{1}{2}\left(\frac{z-z_0}{\sigma}\right)^2\right\}:$$

$$F_{gau} = Z\exp\left[-\tfrac{1}{2}(Q\sigma)^2\right]\exp(iQz_0)$$

Structure factor of a delta function profile:

$$\rho(z) = \delta(z-z_0):$$

$$F_{delta\text{-}fcn} = \exp(iQz_0)$$

APPENDIX 4
SPECTROMETER ANGLES AND RECIPROCAL SPACE

Here the relationship between four circle spectrometer angles (Busing and Levy 1967) and the value of the momentum transfer is presented for a symmetry scattering geometry (i.e., with the incident and exit angles constrained to be equal with respect to the surface plane) with the surface-normal direction aligned precisely along the ϕ axis. The four-circle spectrometer has four motors, 2θ, θ, χ, and ϕ. Q_z is oriented along the surface-normal direction, Q_x and Q_y are within the surface plane, and $Q_{//} = [Q_x^2 + Q_y^2]^{1/2}$. The relation between the motor angles and the momentum transfer, Q, can then be written:

$$\sin(2\theta/2) = Q/(2K)$$

$$\theta = 2\theta/2$$

$$\tan(\chi) = Q_z/Q_{//}$$

$$\tan(\phi) = Q_y/Q_x$$

4

X-ray Standing Wave Studies of Minerals and Mineral Surfaces: Principles and Applications

Michael J. Bedzyk[1,2] and Likwan Cheng[3]

[1]*Department of Materials Science and Engineering
Northwestern University
Evanston, Illinois, 60208, U.S.A.*

[2]*Materials Science Division*
[3]*Environmental Research Division
Argonne National Laboratory
Argonne, Illinois, 60439, U.S.A.*

INTRODUCTION

With a penetration depth ranging from microns to millimeters, Ångstrom-wavelength X-rays are an ideal probe for studying atomic-scale buried structures found in the natural environment, such as impurities in minerals and adsorbed ions at mineral-water interfaces. But this penetration depth also makes an X-ray beam inherently less useful as a spatially localized probe. Using the superposition of two coherently coupled X-ray beams, however, makes it possible to localize the X-ray intensity into interference fringes of an X-ray standing wave (XSW) field, as illustrated in Figure 1, and thereby attain a spatially localized periodic probe with a length scale equivalent to the XSW period. The XSW period is

$$D = \frac{\lambda}{2\sin\theta} = \frac{1}{Q} \qquad (1)$$

where λ is the X-ray wavelength and 2θ is the scattering angle or angle separation between the two coherently coupled wave vectors K_R and K_0. In reciprocal space, the

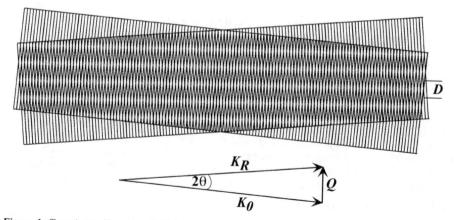

Figure 1. *Top:* A standing wave field formed from the superposition of two traveling plane waves of wavelength λ and intersection angle (scattering angle) 2θ. The standing wave period is D as defined in Equation (1). *Bottom:* The two traveling planes waves are represented in reciprocal space by wave vectors K_0 and K_R. $K_0 = K_R = 1/\lambda$. The standing wave is defined by standing-wave vector Q defined in Equation (2).

1529-6466/00/0049-0004$05.00

scattering vector, or wave vector transfer, is defined as

$$Q = K_R - K_0 \tag{2}$$

Q is in the direction perpendicular to the equal-intensity planes of the XSW and has a magnitude that is the reciprocal of D. Thus, Q is also referred to as the standing wave vector.

An X-ray standing wave can be used as an atom-specific probe via the photoelectric effect, which can be observed by photoelectron emission, fluorescence, or Auger electron emission. There are a number of mechanisms for generating an XSW. The simplest and perhaps most practical method for producing an XSW is by reflection, in which case the superposition of the reflected and incident X-ray plane waves gives rise to the standing wave. This two-beam reflection condition can be produced by (1) strong Bragg diffraction from a single crystal, (2) strong Bragg diffraction from a periodically layered synthetic microstructure, (3) total external reflection (TER) from an X-ray mirror, or (4) weak kinematical Bragg diffraction from a single-crystal thin film.

As an element-sensitive high-resolution structural probe, the XSW technique has been used in the past two decades to investigate a wide range of surface, interface, and thin film structures. These include semiconductor, metal, and oxide surfaces; electrochemical interfaces; and organic membranes. In this article, we present an introduction to the basic principles of the major types of XSW techniques. We will discuss how the XSW phase is directly linked to the substrate reflecting lattice planes or interfaces and can thereby be used to directly determine the positions of the selected elements relative to these substrate planes. We will discuss the experimental aspects of the XSW method, focusing on single-crystal diffraction XSW at a synchrotron radiation source. We will then describe advances in XSW applications in geochemistry and environmental science. These include four areas of structural investigations: impurity structures in minerals, aqueous ion adsorption and incorporation at mineral surfaces, metal distribution in organic membranes and organic matter at solid surfaces, and the electrical double-layer structure at water-solid interfaces.

X-RAY STANDING WAVES
BY BRAGG DIFFRACTION FROM A SINGLE CRYSTAL

The most commonly used means for generating an X-ray standing wave is the use of strong Bragg diffraction from a single-crystal. In 1964, using Bragg diffraction from a Ge crystal, Batterman (1964) made the first observation of the XSW effect—an angularly modulated Ge fluorescence yield across the reflection. He then used this fluorescence anomaly to locate As impurity atom sites within a Si crystal (Batterman 1969). Later, Golovchenko and coworkers realized that the XSW field generated inside the crystal extended above the crystal surface and used the XSW to determine the crystallographic registration of adsorbate atoms with respect to the underlying substrate lattice (Cowan et al. 1980; Golovchenko et al. 1982). These experiments are the early demonstrations of what has now become an established technique for atomic-resolution surface science.

The X-ray standing wave field

Following the observations of Batterman, a quantitative explanation of the fluorescence yield has been based on the dynamical diffraction theory of von Laue and Ewald (Laue 1960). The formal theory for the Bragg diffraction XSW technique has been described by Afanas'ev et al. (1978), Takahashi and Kikuta (1979), Hertel et al. (1985) and Bedzyk and Materlik (1985), as well as in a review by Zegenhagen (1993). The theory of dynamical X-ray diffraction has been reviewed by Laue (1960), Batterman and

Cole (1964) and Authier (2001).

Consider the two-beam Bragg diffraction condition described in Figure 2; the incident and the Bragg-diffracted X-ray plane waves are as follows

$$\mathcal{E}_0(\boldsymbol{r},t) = \boldsymbol{E}_0 \exp\left[-2\pi i\left(\boldsymbol{K}_0 \boldsymbol{\cdot} \boldsymbol{r} - vt\right)\right]$$
$$\mathcal{E}_H(\boldsymbol{r},t) = \boldsymbol{E}_H \exp\left[-2\pi i\left(\boldsymbol{K}_H \boldsymbol{\cdot} \boldsymbol{r} - vt\right)\right] \tag{3}$$

Here $\boldsymbol{E}_0$ and $\boldsymbol{E}_H$ are the complex amplitudes associated with the incident and diffracted X-ray plane-waves, $\boldsymbol{K}_0$ and $\boldsymbol{K}_H$ are the respective complex wave vectors inside the crystal, and v is the X-ray frequency. The two wave vectors are coupled according to the Laue condition, with

$$\boldsymbol{H} = \boldsymbol{K}_H - \boldsymbol{K}_0 \tag{4}$$

where $\boldsymbol{H} = h\boldsymbol{a}^* + k\boldsymbol{b}^* + l\boldsymbol{c}^*$ is the reciprocal lattice vector. The scalar equivalent of the Laue condition reduces to Bragg's law

$$\lambda = 2d_H \sin\theta_B \tag{5}$$

where d_H is the lattice spacing of the *hkl* crystal diffraction planes and θ_B is the geometrical Bragg angle. The interference between the incident and diffracted plane

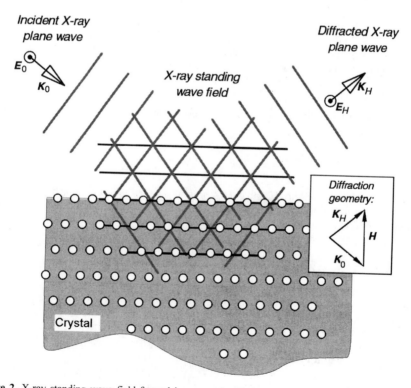

Figure 2. X-ray standing wave field formed in a crystal and above its surface by the interference of incident and Bragg-diffracted monochromatic X-ray plane waves. The inset shows the reciprocal space diagram for the Laue condition described by Equation (4).

waves results in a standing-wave field. The normalized intensity of the total E-field that gives rise to the XSW field is

$$
I(\theta,r) = \frac{\left|E_0 + E_H\right|^2}{\left|E_0\right|^2}
$$

$$
= \left[1 + R(\theta) + 2\sqrt{R(\theta)}\cos(v(\theta) - 2\pi H \cdot r)\right] \times \begin{cases} 1 & , \quad \text{above the surface} \\ e^{-\mu_z(\theta)z} & , \text{ at depth } z \text{ below surface} \end{cases}
\tag{6}
$$

where the reflectivity R is related to the E-field amplitude ratio as

$$
R = \left|\frac{E_H}{E_0}\right|^2
\tag{7}
$$

and the XSW phase, v, is identical to the relative phase between the two E-field amplitudes

$$
\frac{E_H}{E_0} = \left|\frac{E_H}{E_0}\right|\exp(iv)
\tag{8}
$$

From Equations (1) and (5), one can conclude that for Bragg diffraction the XSW periodicity is equal to the lattice d-spacing of the $H = hkl$ diffraction planes; that is, $D = d_H$.

In the following discussion, we will assume the most common case of σ-polarized symmetrical Bragg diffraction from a semi-infinite crystal with $1° < \theta_B < 89°$. Figure 2 shows the case of σ-polarization with the vector directions of the two E-fields pointing perpendicular to the scattering plane defined by the two wave vectors. The incident and exit angles of the two wave vectors with respect to the surface are equivalent for a symmetric reflection.

From dynamical diffraction theory (Batterman and Cole 1964), the E-field amplitude ratio is defined as

$$
\frac{E_H}{E_0} = -\sqrt{\frac{F_H}{F_{\bar{H}}}}\left(\eta \pm \sqrt{\eta^2 - 1}\right)
\tag{9}
$$

where F_H and $F_{\bar{H}}$ are the H and $-H$ structure factors (see chapter by Fenter, in this volume, for a definition of F_H), and η is a normalized, dimensionless complex angular parameter defined as

$$
\eta = \frac{-\Delta\theta \sin(2\theta_B) + \Gamma F_0}{\Gamma\sqrt{F_H F_{\bar{H}}}}
\tag{10}
$$

In this equation, $\Delta\theta = \theta - \theta_B$ is the relative incident angle. $\Gamma = (r_e\lambda^2)/(\pi V_c)$ is a scaling factor, where $r_e = 2.818\times10^{-5}$ Å is the classical electron radius and V_c is the volume of the unit cell (uc). (To separate the real and the imaginary parts of a complex quantity A, we use the notation $A = A' + iA''$, where A' and A'' are real quantities.) From Equations (7), (9), and (10), it can be shown that the reflectivity approaches unity over a very small arc-second angular width ω, defined as

$$
\omega = \Delta\theta_{\eta'=-1} - \Delta\theta_{\eta'=1} = \frac{2\Gamma\sqrt{F_H'F_{\bar{H}}' + F_0''^2 - F_H''F_{\bar{H}}''}}{\sin 2\theta_B}
\tag{11}
$$

This is the "Darwin width" of the reflectivity curve or "rocking curve."

From dynamical diffraction theory, the relative phase, v, of the standing wave field decreases by π radians as the incident angle is scanned from the low-angle side to the high-angle side of the rocking curve (i.e., from $\eta' = 1$ to $\eta' = -1$). According to Equation (6), this causes the standing-wave antinodal planes to move by a distance of $\frac{1}{2}d_H$ in the $-H$ direction, as illustrated in Figure 3. Also from Equation (6), if $R = 1$, the intensity at the antinode is four-times the incident intensity, $|E_0|^2$, and there is zero intensity at the node. The case of $I = 4$ at the antinode assumes that the field is being examined above the surface or at a shallow depth where $\exp(-\mu_z z) \approx 1$.

The Darwin width, ω, is dependent on both the structure factors and the wavelength (or energy E_γ) of the incident X-ray beam. For a typical low-index strong Bragg reflection from a mineral crystal ω is within the range of 5 to 100 microradians (μrad) for X-rays within the range of $E_\gamma = 5$ keV to 20 keV. The bottom portion of Figure 4 shows a calculated rocking curve $R(\theta)$ (according to Eqn. 7, 9 and 10) and the corresponding phase v (according to Eqn. 8) for the calcite $(10\bar{1}4)$ Bragg reflection at $E_\gamma = 13.8$ keV. In this case, $\omega = 20.0$ μrad $= 4.12$ arc-sec. Semi-empirically the full-width at half-maximum is FWHM $= 1.2\omega$. Notice that the center of the rocking curve is shifted slightly above the geometrical Bragg angle θ_B by ~22 μrad. This shift is the result of refraction at the crystal-air interface. In general, this shift is $\Delta\theta = \Gamma F_0' / \sin 2\theta_B$. The asymmetry in the reflectivity curve, namely the slight diminishing from $R = 1$ as the angle is increased through the strong Bragg condition, is due to the movement of the XSW. On the high-angle side the XSW antinodes align with the strong X-ray absorption planes in the crystal. Therefore absorption is higher than average on the high-angle side and weaker on the low-angle side.

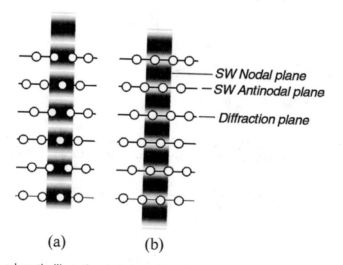

(a)   (b)

Figure 3. A schematic illustration depicting the spatial relationship between the Bragg diffraction generated XSW and the diffraction planes. The XSW antinodal planes are parallel to the diffraction planes and have the same periodicity. Advancing in angle through the strong Bragg diffraction condition (Bragg band-gap) causes the XSW to phase shift inward by $d_H / 2$. (a) On the low-angle side of the Bragg reflection, the XSW nodal planes are aligned with the diffraction planes. (b) On the high-angle side of the Bragg reflection, the antinodal planes are aligned with the diffraction planes. The position of the diffraction planes relative to the bulk lattice are defined by the maxima in the real part of the Hth Fourier component of the effective electron density (see, for example, Bedzyk and Materlik, 1985).

Bedzyk & Cheng

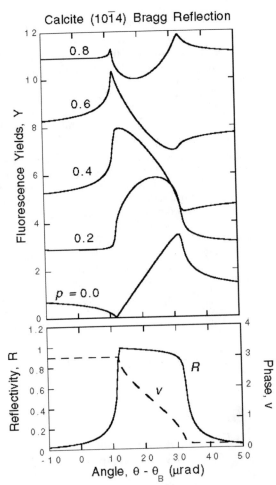

Figure 4. *Bottom*: The ideal theoretical angular dependence of the reflectivity R and XSW phase v (in radians) for the calcite $(10\bar{1}4)$ Bragg reflection at $E_\gamma = 13.8$ keV. *Top*: The corresponding theoretical normalized fluorescence yields (Eqn. 15) for coherent positions $P_{10\bar{1}4} = 0.0, 0.2, 0.4, 0.6,$ and 0.8 with coherent fraction $f_{10\bar{1}4} = 1$ and $Z(\theta) = 1$. The curves labeled 0.2 through 0.8 are given vertical offsets for purposes of clarity.

The exponential damping factor in Equation (6) accounts for attenuation effects within the crystal, in which case the effective absorption coefficient is defined as

$$\mu_z(\theta) = \frac{\mu_0}{\sin\theta_B}\left[1 + \frac{F_{\bar{H}}'}{F_0''}\left(\frac{E_H}{E_0}\right)'' + \frac{F_{\bar{H}}''}{F''}\left(\frac{E_H}{E_0}\right)'\right] \qquad (12)$$

where

$$\mu_0 = \frac{2\pi}{\lambda}\Gamma F_0''$$

is the linear absorption coefficient. The second and third terms in Equation (12) account for the extinction effect that strongly limits the X-ray penetration depth $1/\mu_z$ for a strong Bragg reflection. For example, the penetration depth for 13.8 keV X-rays at the calcite $(10\bar{1}4)$ Bragg reflection goes from 35 μm at off-Bragg conditions to 0.71 μm at the center of the Bragg rocking curve. This minimum penetration depth or extinction length is

$$\Lambda_{ext} = \text{Min}\left[\mu_e^{-1}\right] = V_c\left[4d_H r_e\left(F_0'' + \sqrt{|F_H||F_{\bar{H}}|}\right)\right]^{-1} \tag{13}$$

XSW-induced yields from photo-effect

The XSW field established inside the crystal and above the crystal surface induces photoelectron emission from atoms within the field. The excited atoms (ions), in turn, emit characteristic fluorescence X-rays and Auger electrons. In the dipole approximation, the photoelectric effect cross section is proportional to the E-field intensity at the center of the atom. (It is necessary to consider higher-order multi-pole terms in the photoelectric cross-section under special conditions, as discussed by Fischer et al. (1998) and Schreiber et al. (2001). For this review, we will assume the dipole approximation.) Therefore, with the XSW intensity from Equation (6), the normalized X-ray fluorescence yield is defined as

$$Y(\theta) = \int I(\theta, r)\rho(r)\exp\left[-\mu_f(\alpha)z\right]dr \tag{14}$$

where $\rho(r)$ is the normalized fluorescent atom distribution, and $\mu_f(\alpha)$ is the effective absorption coefficient for the emitted fluorescent X-rays which is dependent on their takeoff angle, α. Upon integration, the normalized XSW yield is given as

$$Y(\theta) = \left[1 + R(\theta) + 2\sqrt{R(\theta)}f_H\cos\left(v(\theta) - 2\pi P_H\right)\right]Z(\theta) \tag{15}$$

where the parameters f_H and P_H are the coherent fraction and coherent position, respectively. They are the amplitude and phase of the Hth-order Fourier component of the normalized distribution function

$$\mathcal{F}_H = \int_{uc}\rho(r)\exp\left(2\pi i H\cdot r\right)dr = f_H\exp\left(2\pi i P_H\right) \tag{16}$$

$Z(\theta)$ is the effective-thickness factor, which will be discussed below. $Z(\theta) = 1$ for atoms above the surface of the crystal. $Z(\theta) \sim 1$ at a depth much less than the extinction depth.

Equation (15) is the working equation for the Bragg diffraction XSW technique. In an XSW experiment, the reflectivity rocking curve $R(\theta)$ and the fluorescence yield $Y(\theta)$ are acquired simultaneously while scanning through the desired Bragg reflection of the sample crystal. Then, a rocking curve calculated according to dynamical diffraction theory is fitted to the experimentally measured $R(\theta)$ to calibrate the angle scale. The best-fit rocking curve is then used for fitting Equation (15) to the measured $Y(\theta)$. From this fit, the coherent fraction f_H and the coherent position P_H are obtained. The off-Bragg yield, which is also obtained from the fit to the measured data, is used for overall normalization of the yield, which gives the fluorescent atom concentration or surface coverage. In the top portion of Figure 4, calculated $Y(\theta)$ curves are shown for $Z(\theta) = 1$; $f_H = 1$; and $P_H = 0.0, 0.2, 0.4, 0.6$, and 0.8, respectively. These calculated curves are for the calcite (10$\bar{1}$4) reflection at $E_\gamma = 13.8$ keV. The marked change in the functional form of $Y(\theta)$ for each increment of 0.2 in d-spacing is the basis for the reliability of the Bragg diffraction XSW as a high-resolution structural probe.

The XSW structural information about the fluorescent atom is contained in f_H and P_H, which, as they are obtained from Equation (15), are model-independent quantities. If a set of f_H and P_H values is acquired up to a sufficiently high order of H, based on Equation (16), the distribution $\rho(r)$ of each fluorescent atomic species can be synthesized directly by the Fourier summation

$$\rho(r) = \sum_H f_H\exp\left[2\pi i\left(P_H - H\cdot r\right)\right] = 1 + 2\sum_{\substack{H \neq -H \\ H \neq 0}} f_H\cos\left[2\pi\left(P_H - H\cdot r\right)\right] \tag{17}$$

The above simplification to a summation of cosine terms makes use of the following: $f_0 = 1$; $f_{\bar{H}} = f_H$; and $P_{\bar{H}} = -P_H$.

Unlike conventional diffraction methods, the XSW method does not lose phase information and can therefore be used directly to map the direct-space structure from the set of Fourier coefficients collected in reciprocal space. The XSW measurement does not lose phase information, because the detector of the E-field is the fluorescent atom itself, lying within the spatial region where the fields interfere coherently with each other. In contrast, in conventional diffraction measurements the relative phase between the diffracted and incident fields is lost, because the intensities of the fields are detected far from this region of coherent spatial overlap.

Obtaining real-space $\rho(r)$ by direct Fourier inversion (Eqn. 17) has recently been experimentally demonstrated with the case of lattice and impurity atoms in muscovite (Cheng et al. 2002). This newly developed XSW direct-space imaging method, when combined with the high intensity and special optics available at the APS stations described later in this chapter, should prove to be very powerful when trying to resolve 3D multiple-site bulk atom and surface adsorbate structures. Although f_H and P_H are model independent, up until now the XSW analysis has relied on simple models to interpret the small set of f_H and P_H values. If the model is inconsistent and/or the surface structure proves to be more complicated, it is now practical to make more *hkl* measurements, apply this XSW direct-space method and skip over the simple model interpretation step. From the ~1 Å resolution 3D image a new more complicated model should emerge that when fitted to the data should give high-resolution ($\pm$ 0.02Å) adsorbate-site positions.

In the typical practice of the XSW technique, however, only a limited set of *hkl* measurements is taken, and the analysis resorts to comparing the measured f_H and P_H values to those predicted by various competing structural models. The procedures of structural analysis using f_H and P_H will be described in more detail in a later section of this chapter. It should be stressed that the Bragg XSW positional information acquired is in the same absolute coordinate system as used for describing the substrate unit cell. This unit cell and its origin were previously chosen when the structure factors F_H and $F_{\bar{H}}$ where calculated and used in Equations (9), (10), and (12). As previously derived and experimentally proven (Bedzyk and Materlik 1985), the phase of the XSW is directly linked to the phase of the structure factor. This is an essential feature of the XSW method that makes it unique; namely, it does not suffer from the well known "phase problem" of X-ray diffraction.

Extinction effect and evanescent-wave emission

The effective-thickness factor $Z(\theta)$ in Equation (15) accounts for the θ dependence of the penetration depth of the primary X-ray field (extinction effect) in conjunction with the escape depth, Λ, of the out-going secondary fluorescence X-rays. For atoms at the crystal surface (e.g., adsorbates) or at a depth much smaller than the extinction depth, the effective-thickness factor is constant at $Z(\theta) = 1$. For atoms evenly distributed throughout the semi-infinite crystal

$$Z(\theta) = \frac{\mu_0 \left(\sin \theta_B \right)^{-1} + \mu_f(\alpha)}{\mu_z(\theta) + \mu_f(\alpha)} \tag{18}$$

where $\mu_z(\theta)$ is the effective absorption coefficient of the incident X-rays (Eqn. 12) and $\mu_f(\alpha)$ is the effective absorption coefficient of the outgoing fluorescence X-rays from the crystal at takeoff angle α. To achieve surface sensitivity of substrate atoms, it is possible to reduce α to a value approaching the critical angle of the fluorescence X-rays. Under such conditions, $\mu_f(\alpha)$ dominates over $\mu_z(\theta)$ in Equation (18) and therefore the effective thickness factor is constant at $Z(\theta) = 1$. This is the evanescent-wave emission effect observed by Becker et al. (1983). The value of $\mu_f(\alpha)$ is dependent on the wavelength λ_f of the fluorescence and the index of refraction calculated at $\lambda = \lambda_f$. The index of refraction is

$$n = 1 - \delta - i\beta \tag{19}$$

where

$$\delta = \frac{1}{2\pi} N_e r_e \lambda^2 = \frac{1}{2} \Gamma F_0'$$
$$\beta = \frac{1}{4\pi} \mu_0 \lambda = \frac{1}{2} \Gamma F_0'' \tag{20}$$

and N_e is the effective electron density of the refractive medium. From the index of refraction, $\mu_f(\alpha)$ can be obtained as

$$\mu_f(\alpha) = \frac{2\sqrt{2}\pi}{\lambda_f} \left[\sqrt{(2\delta - \alpha^2)^2 + 4\beta^2} + 2\delta - \alpha^2 \right]^{\frac{1}{2}} \tag{21}$$

At takeoff angles much greater than the critical angle (i.e., $\alpha \gg \sqrt{2\delta}$), $\mu_f = \mu_0/\sin\alpha$. Below the critical angle, the escape depth $\Lambda = \mu_f^{-1}$ approaches $\frac{1}{4}(N_e r_e)^{-1/2}$.

The evanescent-wave emission effect can be used to remove the extinction effect of bulk fluorescence in XSW experiments (Lee et al. 1996). Consider the case of 3.69-keV Ca K_α fluorescence X-rays emitted from calcite. For $CaCO_3$ at this energy $\delta = 3.97 \times 10^{-5}$, $\beta = 6.84 \times 10^{-6}$, and the critical angle $\alpha_c = 0.51°$. Figure 5a shows the corresponding takeoff angle dependence of the escape depth Λ. At takeoff angles below α_c, the escape depth is ~33 Å. Figure 5b shows the angle θ dependence of the reflectivity and Ca K_α signals that were collected while scanning through the $CaCO_3$ ($10\bar{1}4$) Bragg peak. The standing wave effect in the bulk Ca K_α signal is compared to calculated yield curves (Eqn. 15) for takeoff angles 0.25°, 1.3°, 7°, and 15°, respectively. To completely eliminate the extinction effect, one would ideally acquire the fluorescence at $\alpha < \alpha_c$, but this requires high-precision apparatus. Alternatively, one may acquire data just above the sharp rise in the escape curve, for example at 1.3°, as the data were taken for this example. At much higher α, for example 15°, the extinction effect begins to dominate over the standing wave effect and one begins to lose phase (positional) sensitivity.

Structure determination using coherent fraction and coherent position

The coherent fraction f_H is a measure of the spatial distribution of the fluorescent atoms. For simplicity, it is useful to subdivide f_H into three factors:

$$f_H = C a_H D_H. \tag{22}$$

In this definition, C is the ordered fraction, a_H is the geometrical factor, and D_H is the Debye-Waller factor. All three factors range in value from 0 to 1.

To explain these quantities, consider the general case of a discrete atom distribution with N different unit cell position vectors $r_j = x_j a + y_j b + z_j c$ for the same fluorescent atom, plus an added random distribution of the same atoms. The ordered fraction C is the fraction of the atoms in the distribution that are coherently located or are crystallographically registered with the substrate crystal lattice. If the atoms' occupation fractions for the ordered positions are $c_1, c_2, ..., c_N$, respectively, the ordered fraction is

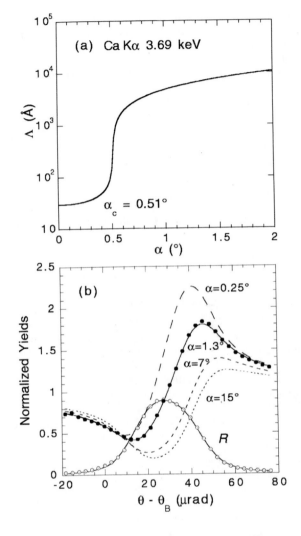

Figure 5. (a) The escape depth $\Lambda = 1/\mu_f$ (Eqn. 21) for 3.69 keV (Ca $K\alpha$ fluorescence) X-rays from calcite as a function of takeoff angle α. This illustrates the surface sensitivity of the evanescent-wave emission technique. (b) For the calcite $(10\bar{1}4)$ Bragg reflection at an incident energy of $E_\gamma = 13.5$ keV, the angular dependence of the experimental and theoretical reflectivity R and the Ca $K\alpha$ normalized fluorescence yields at various take-off angles. (See Eqn. 15 and 18.) The theory curves in (b) are convoluted with the angular emittance reflectivity curve from the upstream Si(111) monochromator and are hence smeared-out in comparison to the curves shown in Figure 4.

$$C = \sum_{j=1}^{N} c_j \qquad (23)$$

The geometrical factor a_H is the modulus of the normalized geometrical structure factor S_H for the ordered fluorescent-selected atoms:

$$S_H = \frac{1}{C} \sum_{j=1}^{N} \left[c_j \exp\left(2\pi i \mathbf{H} \cdot \mathbf{r}_j \right) \right] \qquad (24)$$

Therefore, the geometrical factor is

$$a_H = \frac{1}{C} \left| \sum_{j=1}^{N} \left[c_j \exp\left(2\pi i \mathbf{H} \cdot \mathbf{r}_j \right) \right] \right| \qquad (25)$$

The coherent position P_H is the phase of S_H:

$$P_H = \frac{1}{2\pi} \text{Arg}\left[\sum_{j=1}^{N} \left[c_j \exp\left(2\pi i H \cdot r_j\right)\right]\right] \tag{26}$$

Note that the origin for the set of r_j (and therefore the origin of P_H) in the unit cell is the same origin that was arbitrarily chosen for generating the structure factor F_H used in Equations (9) through (12). Consider the simplest case of one atom site, $N = 1$. In this case, $a_H = 1$, and P_H is the projected $H \cdot r$ fractional d-spacing position of the atom site (Fig. 6). For the case of two atom sites of equal occupancy, where $N = 2$ and $c_1 = c_2$, the geometrical factor reduces to

$$a_H = \left| \cos\left[\pi H \cdot (r_1 - r_2)\right]\right| \tag{27}$$

The coherent position in this case is the averaged fractional d-spacing position of the two sites (Fig. 6). If the two equally occupied sites have a separation of exactly one-half of a d-spacing along a particular H, then $a_H = 0$ for that particular H; this is analogous to a forbidden reflection in crystallography.

In XSW analysis, the Debye-Waller factor D_H accounts for the time-averaged spatial distribution due to thermally induced vibrations of the fluorescence-selected ordered atoms about their average lattice positions. In general, individual D_H factors can be inserted into the sum that is used to define the geometrical factor (Eqn. 24). In addition to including the dynamic (thermal vibration) distribution, one can also include the static (spatial disordering) distribution. (Kazimirov et al. 1988) Generally, D_H can be expressed in terms of the mean-square vibrational amplitude along the H direction $\langle u_H^2 \rangle$, as

$$D_H = \exp(-M) = \exp\left(-2\pi^2 \langle u_H^2 \rangle / d_H^2 \right) \tag{28}$$

The sensitivity of the Bragg diffraction XSW method for determining atomic

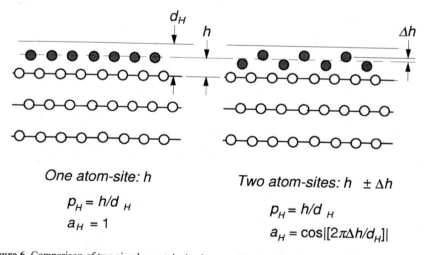

Figure 6. Comparison of two simple crystal adsorbate models: (*Left*) one adsorbate site and (*Right*) two equally occupied adsorbate sites on a crystal surface. Also shown are the corresponding calculations for the coherent positions and the geometrical factors. Both models have the same average adsorbate height relative to the bulk-like diffraction plane. The two-site model has a height difference of $2\Delta h$.

positions is illustrated in Figure 4, where the fluorescence yields are calculated for a set of coherent positions that differ by a two-tenths of the d-spacing. This particular case is for an angle scan through the calcite $(10\bar{1}4)$ Bragg reflection at $E_\gamma = 13.8$ keV. For calcite, $d_{10\bar{1}4} = 3.04$ Å, and therefore each 0.2 increment in P_H is equivalent to 0.61 Å. We have for convenience chosen to locate the origin of the calcite unit cell at the carbon atom site of the lattice. This illustration demonstrates the sensitive dependence of the functional form of $Y(\theta)$ on the value of P_H. This large change in $Y(\theta)$ is caused by the change in the standing wave phase $v(\theta)$. This phase sensitivity is the basis of the sub-Ångstrom spatial resolution of the Bragg diffraction XSW technique.

The XSW measurements are not restricted to being made along the surface-normal direction (i.e., H perpendicular to the crystal surface). They can be performed with respect to any sufficiently strong Bragg reflection of a crystal. A three-dimensional triangulation of the atom site can be obtained by combining XSW measurements by using three mutually non-collinear diffraction vectors. XSW triangulation was first demonstrated by Golovchenko et al. (1982). This is a powerful feature of the XSW technique in uniquely determining the lattice sites of a surface adsorbate or a bulk impurity. In triangulating an atom site, the point symmetry of the surface can often be used to reduce the number of required XSW measurements from three to two.

The Bragg diffraction XSW method has been used in many areas, perhaps most extensively in obtaining high-resolution adsorbate structures at semiconductor surfaces; much of this work is documented in the review by Zegenhagen (1993). In a later section of this chapter, we describe in detail the use of Bragg diffraction XSW in determining the surface structures of aqueous ions incorporated at the calcite $(10\bar{1}4)$ surface.

"Ideal crystals" vs. "real crystals" in Bragg diffraction XSW

According to Equation (11), the intrinsic Darwin width for a typical strong Bragg diffraction peak in the reflection geometry is between 5 to 100 μrad. To produce useful quantitative information from an XSW measurement, the measured reflectivity curve should reasonably match the theory. Only a few exceptionally high quality crystals qualify as ideal crystals according this criterion; among them are Si, Ge, and calcite. Most real single crystals (such as orthoclase and muscovite) contain internal imperfections that result in mosaic spreads in their Bragg rocking curves that exceed their intrinsic Darwin widths. These mosaic spreads can reduce the reflected intensity and subsequently smear out the standing wave effect. For these real crystals, the reflectivity curves cannot be fitted directly to the ideal theoretical reflectivity described by Equation (15).

To make the Bragg diffraction XSW technique adaptable to applications on real crystals, a number of theoretical and experimental methods have been developed. The most formal method is the modification of the XSW theory to include the effect of crystal imperfection on Bragg diffraction. Such theories, however, involve additional complexity beyond the theory for ideal crystals and are beyond the scope of this article. Readers are referred to an updated review on this topic by Vartanyants and Kovalchuk (2001) for details.

On the other hand, one can make XSW applications under certain conditions where the stringent requirements of the conventional theory are relaxed. One of such condition is when a Bragg reflection occurs near the back-reflection geometry, typically when $87° < \theta_B < 90°$. Under this condition, Equation (11) breaks down, and the intrinsic Darwin width is magnified to milliradians. Another condition is when the crystal is very thin; under this condition the kinematical theory can replace the dynamical theory in calculating Bragg reflection. Both of these special-cases, which will be discussed briefly below, are important alternatives when the conventional XSW technique cannot be applied.

An additional practical method for applying XSW to real crystals takes advantage of the high-brilliance of X-ray undulators at third-generation synchrotron sources. With such a source, sufficiently high X-ray intensities can be delivered with beams slitted down to as small as a 10 microns in cross section. Thus making it possible to illuminate one isolated perfect grain at the sample surface. With angular divergence matching the intrinsic Darwin width, this eliminates or significantly reduces the smearing effect of mosaic spread. Figure 7 compares the rocking curves of the muscovite (006) reflection taken at a conventional laboratory source and at an undulator synchrotron beamline. The laboratory data, taken with a larger beam size because of limited beam intensity, are not suitable for XSW analysis. The undulator data, taken with the beam dimension reduced by roughly an order of magnitude, is suitable for XSW analysis. Although the X-ray parameters are not identical for the two cases, it is clear that with reduced beam size many crystals can be suitable for XSW experiments.

The special case of back-reflection XSW

To produce XSW data that can be analyzed quantitatively by X-ray diffraction theory, it is necessary to achieve an analytically well-defined XSW field. A first test of meeting this condition is to determine if the angular dependence of the measured reflectivity curve closely matches the ideal reflectivity calculated with dynamical diffraction theory. This, of course, includes a convolution with the emittance function

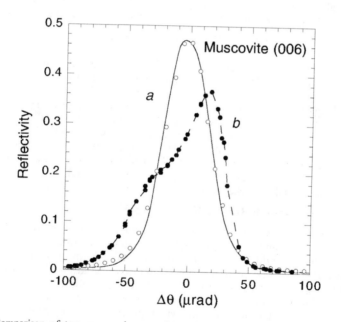

Figure 7. Comparison of two measured muscovite rocking curves for the (006) Bragg reflection acquired by using (a, open circles) an undulator synchrotron beamline with a Si(111) monochromator at $E_\gamma = 7.4$ keV, and (b, solid circles) Cu $K\alpha_1$ X-rays from a Cu fixed-anode source ($E_\gamma = 8.04$ keV) followed by a Si(111) four-bounce high-resolution monochromator and high-resolution diffractometer. The difference between the two-curves is primarily due to the size of the incident beam slit, which is 0.2 mm × 0.05 mm for (a) and 2 mm × 1 mm for (b). Only a source with the brightness of the undulator could be slitted down to such a small size as for (a) and still has sufficient flux on the sample for XSW measurements. Also shown is the best fit to the rocking curve of (a) (solid line). At 7.4 keV, the Darwin width of Si(111) is ~37 μrad and that of the muscovite (006) is ~20 μrad.

from the upstream optics. The level of difficulty in meeting this challenge can be realized by noting that the natural Darwin width of 5 to 100 μrad is typically much smaller than the mosaic spread of typical crystals. One means of circumventing this problem is to use dynamical diffraction at Bragg angles approaching 90°. In this back-reflection geometry, the angular Darwin width is measured in milliradians instead of microradians.

For $\theta_B > 88°$, the conventional dynamical diffraction theory breaks down. This is due to an approximation in the conventional theory that treats the spherical asymptotes of the dispersion surface in reciprocal space as planes. With an extended dynamical diffraction theory (Kohra and Matsushita 1972; Caticha and Caticha-Ellis 1982), the Bragg reflectivity and the standing wave E-field intensity can be properly described for $87° < \theta_B < 90°$. In this regime, the Bragg reflectivity has a much smaller energy width and a much broader angle width. The largest angle width occurs at a wavelength of

$$\lambda_b = 2d \left(1 - \frac{\Gamma \left(F_0' - F_H' \right)}{2} \right) \tag{29}$$

In this case, $\Gamma = 4 r_e d_H^2 (\pi V_c)^{-1}$. For wavelengths slightly smaller, the angular width can be expressed as

$$\omega = \Delta\theta_{y=1} - \Delta\theta_{y=-1} \tag{30}$$

where $\Delta\theta_{y=1}$ and $\Delta\theta_{y=-1}$ are the angular displacements from 90° of the low- and high-angle side of the strong Bragg diffraction condition, respectively. For a symmetric reflection,

$$\Delta\theta_{y=\pm1} = \sqrt{2(1 - \sin\theta_B) - \Gamma \left(F_0' \mp F_H' \right)} \tag{31}$$

The back-reflection XSW technique has been used mostly on metal and oxide crystals, whose rocking curves often have angular mosaicity up to ~0.1°. In application to mineral surfaces, the technique has also been used by Kendelewicz et al. (1998b) to study the exchange between Na and Pb in a Na overlayer on the PbS(100) surface. A review of the BRXSW technique was given by Woodruff (1998). Since the BRXSW method is employed at soft X-ray energies (typically from $E_\gamma = 1$ keV to 4 keV), it is not useful for *in situ* studies of the mineral-fluid interface. However, the BRXSW is a very useful probe for UHV surface science measurements of adsorbed molecules on metal single crystal surfaces (Jones et al. 2002) and for site-specific valence-band photoemission studies developed by Woicik and coworkers (Woicik et al. 2001; Kim et al. 2002).

The special case of thin-film Bragg diffraction XSW

The development of Bragg diffraction XSW from thin films is partially driven by the fact that many crystals can be grown as high quality μm-thick films but not as large-size crystals. Similar to Bragg diffraction from a bulk crystal, Kazimirov et al. (1997) showed that Bragg diffraction from a crystalline thin film also generates an X-ray standing wave field. Because the thickness of the film is much less than the extinction depth of the incident X-rays, kinematical diffraction theory can be used as a good approximation to calculate the intensity of the field. The analytical procedure is fully analogous to that for bulk-reflection XSW. However, the small thickness also results in very weak reflectivity and consequently very weak angular modulation of the fluorescence yield from an atom within or above the film. Examples of the applications of the thin-film XSW applications include that of Kazimirov et al. (2000) in investigating rare-earth element positions in superconducting films and that of Bedzyk et al. (2000) in probing the polarity of ferroelectric PbTiO$_3$ films deposited by chemical-vapor deposition on a SrTiO$_3$(001) substrate. One of the PbTiO$_3$ films was only 100 Å thick which gave the theoretically predicted peak reflectivity of only 0.02%. Consequently, the relative modulation in the XSW yield was 1%. To observe this very weak modulation and use it to quantify the polarity of the film required the collection of approximately 10^6 Pb L_α fluorescents counts at each of fifty different angle positions of the scan through the (002) PbTiO$_3$ Bragg peak. It was possible to effectively produce this measurement by using a very intense X-ray undulator beam at the APS. This technique can be applied to high purity synthetic minerals, such as, hematite α-Fe$_2$O$_3$ grown epitaxially on sapphire Al$_2$O$_3$(001).

X-RAY STANDING WAVES GENERATED BY TOTAL EXTERNAL REFLECTION

While a single-crystal XSW provides a high-resolution probe well-suited for atomic-scale structural determination, this XSW period is too small to profile larger, nanoscale structures—such as the diffuse ion distribution at the solid-liquid interface, where the Debye length is measured on a nanometer length scale. To extend the XSW technique to the nanoscale and beyond, two new, long-period XSW techniques were developed by Bedzyk and coworkers. These are XSW generated by total external reflection (TER) from a mirror surface (Bedzyk et al. 1989), which will be described in this section, and XSW generated by Bragg diffraction from a periodically layered synthetic microstructure (LSM) (Bedzyk et al. 1990), which will be described in the following section. Because the reflection condition in these two cases occurs at much smaller angles, the XSW period, D, is much longer. For TER, in particular, D varies from 1 μm to 10 nm as θ increases through the TER condition.

XSW generated by total external reflection

During total external reflection the interference between the incident and the specularly reflected X-ray plane waves produces an X-ray standing wave above the mirror surface (Fig. 8). The nodal (and antinodal) planes of this periodic E-field intensity pattern are parallel to the surface and have a variable period of D = $\lambda/(2\sin\theta)$, as defined in Equation (1). The TER condition occurs between $\theta = 0$ and $\theta = \theta_c$, the critical angle. The corresponding XSW periods are very long, ranging from $D = \infty$ to $D = D_c = \lambda/2\theta_c$, the critical period. From Equation (19) and $\theta_c = \sqrt{2\delta}$, the critical period can be written as

$$D_c = \frac{1}{2}\sqrt{\frac{\pi}{N_e r_e}} \tag{32}$$

Aside from small anomalous dispersion effects, the critical period is wavelength independent. D_c is a materials property dependent on the electron density, N_e. For a Au mirror $D_c = 80$ Å and for a Si mirror $D_c = 200$ Å.

The TER condition is the zeroth-order Bragg diffraction condition, or the condition when $H = 0$ and $d_H = \infty$. Therefore, the derivations from dynamical diffraction theory for Bragg diffraction can be applied to the case of TER by substituting F_0 for F_H. In which case, Equation (10) becomes

$$\eta = \eta' + i\eta'' = \frac{-2\theta^2 + \Gamma F_0}{\Gamma F_0} \tag{33}$$

with the substitutions

$$\eta' = 1 - 2x^2 \quad \text{and} \quad \eta'' = 2x^2 y \tag{34}$$

where $x = \theta/\theta_c$ is the normalized angle parameter and $y = \beta/\delta = F_0''/F_0' \ll 1$ is the absorption factor. From Equation (9), the complex E-field amplitude ratio becomes

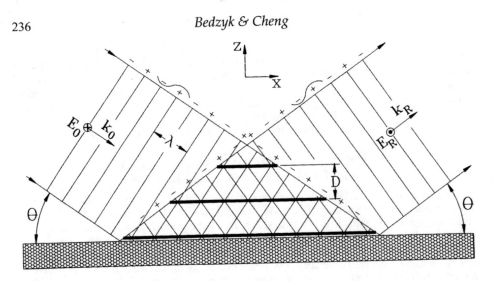

Figure 8. Schematic of XSW field formed above a mirror surface by the interference between an incident and a total-external-reflected X-ray plane wave.

$$\frac{E_R}{E_0} = -\left(\eta \pm \sqrt{\eta^2 - 1}\right) = \frac{x - \sqrt{x^2 - 1 - iy}}{x + \sqrt{x^2 - 1 - iy}} = \left|\frac{E_R}{E_0}\right| \exp(iv) \tag{35}$$

where E_R is the complex amplitude of the reflected plane wave. This amplitude ratio is identical to that derived from classical Fresnel theory (Born et al. 1999). For the simple case of no absorption, where $\beta = 0$, the reflectivity is

$$R = \left|\frac{E_R}{E_0}\right|^2 = \begin{cases} 1, & 0 \leq x \leq 1 \\ 8x^4 - 8x^3\sqrt{x^2 - 1} + 4x\sqrt{x^2 - 1} - 8x^2 + 1, & x > 1 \end{cases} \tag{36}$$

and the phase is

$$v = \begin{cases} \cos^{-1}(2x^2 - 1), & 0 \leq x \leq 1 \\ 0, & x > 1 \end{cases} \tag{37}$$

Figure 9a shows the angular dependence of the reflectivity R and phase v based on Equations (36) and (37). The phase at the mirror surface decreases from π to 0 as the mirror is tilted through the total reflection condition. Thus, at the mirror surface, where $z = 0$, the reflected plane wave is completely out of phase with the incident plane wave when $\theta = 0$. As the incident angle is increased, the phase decreases smoothly until it is completely in phase at $\theta = \theta_c$. Therefore, at $\theta = 0$, a standing wave node is at the mirror surface, and the first antinode is at infinity. As θ increases, the first antinode moves in from infinity toward the mirror surface, until it coincides with the mirror surface upon reaching $\theta = \theta_c$. At the same time, the second, third, and higher-order antinodes of the standing wave also move toward the surface, as the period D decreases based on Equation (1).

The normalized E-field intensity above the mirror surface can be expressed as

$$I(\theta, z) = \frac{|\mathcal{E}_0 + \mathcal{E}_R|^2}{|E_0|^2} = 1 + R + 2\sqrt{R}\cos(v - 2\pi Qz) \tag{38}$$

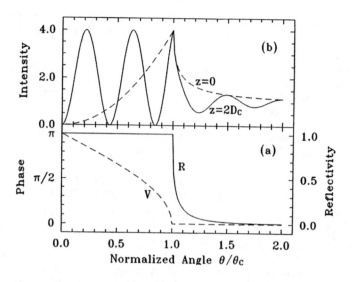

Figure 9. Under the condition of no absorption ($\beta = 0$): (a) The angular dependence of the reflectivity R and relative phase v of the reflected X-ray plane wave; (b) the angular dependence of the normalized E-field intensity at the mirror surface ($z = 0$) and at a distance $z = 2D_c$. Reprinted with permission from Bedzyk et al. 1989. Copyright 1989 by the American Physical Society.

Figure 9b shows the angular dependence of the E-field intensity at $z = 0$ (the mirror surface) and at $z = 2D_c$. The fluorescence signal for an ideally narrow single atomic plane fixed at these heights would have the same angular dependence.

The normalized fluorescence yield from an arbitrary distribution of atoms $\rho(z)$ above the mirror surface can be obtained by integrating over all values of z:

$$Y(\theta) = \int_0^\infty I(\theta, z)\, \rho(z)\, dz \tag{39}$$

With $I(\theta, z)$ calculated with Equation (38), the atom distribution profile $\rho(z)$ can be obtained by assuming a modeled distribution and fitting it to the measured yield $Y(\theta)$. For the specific cases of δ-function atom distributions in a plane at $z = xD_c$, there are $x + \frac{1}{2}$ modulations between $\theta = 0$ and $\theta = \theta_c$. The extra $\frac{1}{2}$ modulation is due to the π phase shift in v.

With TER-XSW, we measure the Fourier transform of an atom distribution over a continuous range in $Q = 1/D$, with variable period D ranging from roughly 100 Å to 1 μm. Therefore, TER-XSW is ideally suited to measure surface and interface structures of length scales in the range of 10 to 2000 Å.

The above treatment accurately describes the X-ray E-fields at and above the mirror surface for the simple case of one interface with vacuum (or air). To apply TER-XSW as a probe for studying liquid-solid interfacial structures or organic films deposited on a solid surface, it is necessary to include reflection, refraction, and absorption effects from the layers that lie between the substrate surface and the vacuum (or air). This can be accomplished by making use of the Parratt's recursion formulation (Parratt 1954) to calculate the transmitted and reflected fields at any interface. These same fields can then be calculated at any point within the slab by appropriately accounting for the X-ray absorption and refraction effects on the fields as they travel from the interface to the point.

The TER-XSW method opens the possibility of using XSW to profile nanoscale metal structures and ion distributions above solid surfaces and at fluid-solid interfaces. Applications of TER-XSW have included direct observation of the diffuse electrical double layer at the charged membrane and electrolyte interfaces (Bedzyk et al. 1990; Wang et al. 2001), structural characterization of self-assembled organic monolayers (Lin et al. 1997), and determining metal ion partitioning at oxide-biofilm interfaces (Templeton et al. 2001). These applications are discussed later in this chapter.

X-RAY STANDING WAVES FROM LAYERED SYNTHETIC MICROSTRUCTURES

For Bragg diffraction purposes, a layered synthetic microstructure (LSM) is fabricated (typically by sputter deposition) to have a depth-periodic layered structure consisting of 10 to 200 layer pairs of alternating high- and low-electron density materials, such as W and Si (Bilderback et al. 1983). Sufficient uniformity in layer thickness is obtainable in the range between 10 and 100 Å (*d*-spacing of fundamental diffraction planes from 20 Å to 200 Å). Because of the rather low number of layer pairs that affect Bragg diffractions, these optical elements have a significantly wider energy band pass and angular reflection width than do single crystals. The required quality of a LSM is that experimental reflection curves compare well with dynamical diffraction theory, and peak reflectivities are as high as 80%. Therefore, a well-defined XSW can be generated and used to probe structures deposited on an LSM surface with a periodic scale equivalent to the rather large *d*-spacing. To a good approximation, the first-order Bragg diffraction planes coincide with the centers of the high-density layers of the LSM. Above the surface of the LSM, the XSW period is again defined by Equation (1). The reflectivity can be calculated by using Parratt's recursion formulation (Parratt 1954). This same optical theory can be extended to allow the calculation of the *E*-field intensity at any position within any of the slabs over an extended angular range that includes TER. Then, Equation (39) is used to calculate the fluorescence yield. Later in this chapter, we describe the use of LSM-XSW method in determining ion distributions in organic films (Bedzyk et al. 1988). This technique was also used by for studies of electrochemical interfaces (Abruna et al. 1988; Bedzyk et al. 1986).

EXPERIMENTAL CONSIDERATIONS OF X-RAY STANDING WAVE MEASUREMENTS

The XSW technique is not exclusively a synchrotron-based technique; it can be performed with use of a conventional fixed-tube X-ray source or rotating anode. However, several practical considerations make it far more advantageous, and often essential, to perform these experiments at a synchrotron source. Some of these considerations are generally true for any X-ray experiment, while others are specific to the XSW techniques. These considerations are often of great practical importance in carrying out XSW experiments. We briefly discuss some of these factors, which include X-ray source brightness, tunability, and polarization.

The first factor has to do with the quality of the crystal for dynamical diffraction experiments. In the previous section we discussed the issue that most single crystals are not ideal but have defects that make them unsuitable for dynamical diffraction studies. To reduce the adverse smearing effect (angle averaging and therefore XSW phase averaging) of mosaicity, one may use a micrometer-size beam so that only a single grain at the surface of the entire crystal is probed. With such a highly collimated, small beam, reasonable intensities are typically achievable only at undulator X-ray sources at third-generation synchrotrons. Another frequently essential property that a synchrotron source offers is the capability of tuning the incident X-rays to a desirable energy, so that in the X-ray fluorescence spectrum the emissions from the atoms of interest are clearly pronounced, while those that could detrimentally saturate the solid-state pulse-counting fluorescence detector can be excluded. A third synchrotron advantage is its high degree of linear polarization in the orbital plane. By aligning the fluorescence detector in the polarization direction, one strongly reduces the unwanted Compton and thermal diffuse scattering components from the fluorescence spectrum.

XSW setup at the Advanced Photon Source

In view of the considerations discussed above and in earlier sections, the primary qualities of an X-ray beam suitable for XSW experiments are its brilliance, energy tunability, and polarization. The XSW experiment with a submonolayer adsorbate coverage on a less than perfect single crystal is an experiment that can take full advantage of a high-brilliance undulator source because of the requirement to produce a very intense beam on a small spot with very high vertical angular collimation.

Zegenhagen (1993) gave detailed description of a bending magnet X-ray beamline dedicated to XSW experiments at the National Synchrotron Light Source. Many of the XSW experiments described later in this chapter were performed at this beamline. Most of the optical and detection elements, and the data acquisition procedures in this setup are typical of XSW experiments in general.

Here, we describe the setup for undulator beamlines 5ID-C and 12ID-D at the Advanced Photon Source (APS). (See Fig. 10) The undulator is an insertion device composed of a linear array of magnets with alternating polarities. The electron bunches circulating inside the synchrotron storage ring undergo horizontal oscillations while passing through the undulator. The radiation given off in successive oscillations interferes. This leads to peaks approximately 100 eV wide in the radiation spectrum, which are typically 10^3 times more intense than the flux of a bending magnet beamline (Rivers, this volume). The peaks in the undulator spectrum can be tuned in energy conveniently by adjusting the undulator gap. Because of the high power density of the undulator beam, the Si(111) double-crystal beamline monochromator is cooled with liquid nitrogen. To filter out photons from higher-order harmonics (i.e., (333), (444), etc.) the angle of the second crystal of the Si(111) monochromator is typically detuned to 80% transmission. These harmonics can also be eliminated by an X-ray mirror set at an incident angle to reflect X-rays below a desired cutoff energy. At 5ID-C, for example, a pair of horizontally deflecting mirrors [just downstream of the high-heat load Si(111) monochromator] is used for harmonic rejection and for horizontal focusing. For single-crystal XSW measurements vertical focusing of the incident beam can introduce an unwanted vertical divergence, which will extend beyond the acceptance angle of the postmonochromator optics.

The postmonochromator for XSW experiments

The purpose of the postmonochromator optics for single-crystal Bragg XSW measurements is to ensure that the standing wave has a sufficiently high fringe visibility for making a high-resolution measurement of the coherent position. This is achieved by reducing angle-averaging and wavelength-averaging effects. The function of the postmonochromator is to prepare the incident beam with an angular emittance width that is 10% to 30% of the angular acceptance width of the sample, and to prevent (or dramatically

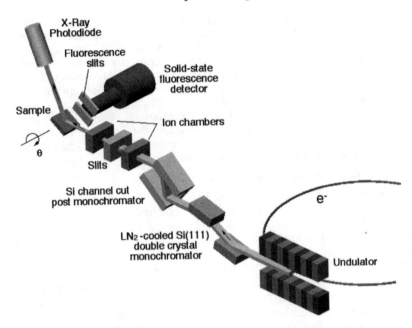

Figure 10. Schematic layout of the X-ray source, optical components, beam intensity monitors and detectors for an undulator beamline for XSW experiments at the Advanced Photon Source (from Lee 1999).

reduce) any wavelength spread from coupling into the angular acceptance of the sample. There are three typical postmonochromator solutions: (1) use of a grazing-incidence, asymmetric Bragg diffraction crystal with d-spacing matching the sample crystal reflection; (2) use of a dispersive pair of channel-cut crystals; and (3) use of a nondispersive pair of channel-cut crystals with d-spacing closely matching the sample crystal reflection.

The third option is used at 5ID-C and 12ID-D. The key components of the postmonochromator are shown in the design drawing in Figure 11. This design allows for remote controlled switching between three Si channel-cut pairs with (hhh), ($hh0$), and ($h00$) reflections. Four ion chambers are used to monitor the X-ray beam intensity as it passes down the line through each optical element. The sample is held in the center of a four-circle diffractometer. The reflected beam intensity from the sample can be monitored by a fifth ion chamber (not shown in Fig. 11) or by a pulse-counting detector (such as a photodiode detector) mounted on the 2θ arm of the diffractometer. Angular motion for each channel-cut crystal at sub-μrad precision is achieved by a piezo-driven rotary stage with a flex-pivot torsion bearing. Each channel-cut crystal is stabilized by an error-integrating feedback loop of an electronic stabilizer unit (MOSTAB) (Krolzig et al. 1984). The MOSTAB unit allows the angular detuning between the two channel-cut crystals to be fixed at a desired setting. At 100% tuning (maximum transmission), the angular emittance width from the postmonochromator will be equivalent to the Darwin width of the selected hkl reflection from Si, as defined in Equation (11). For a 25% tuning, the emittance width will be 25% as wide.

With the above setup, along with microstepping motions for the sample, it is possible to obtain reflectivity curves that are as narrow as 5 μrad. Achieving this performance

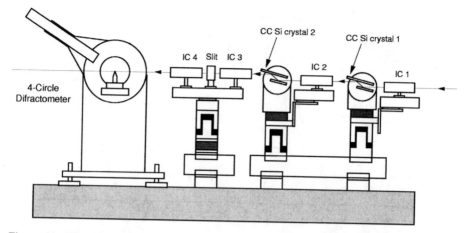

Figure 11. Side-view schematic of optical table setup used for XSW experiments at undulator beamlines 5ID-C and 12ID-D at the Advanced Photon Source (Bedzyk et al., unpublished). The postmonochromator used for single-crystal XSW experiments has two separate rotary stages for tuning the Bragg reflections of the Si channel-cut (CC) crystals, and ion chambers (IC) for monitoring the X-ray intensities.

requires highly perfect postmonochromator and sample crystals in strain-free mounts. It is also necessary to reduce mechanical vibrations and electronic noise. Usually, XSW data are acquired by accumulating repetitive scans through the rocking curve of the sample. A typical scan has 32 angle steps with a dwell time of 1 sec at each step. Long-term angular drift in the system is easily corrected within the XSW computer control program by measuring the centroid of the rocking curve after each XSW scan and adding this as an offset to the next scan. To perform the XSW scan, one can either scan the angle of the sample or scan the energy of the incident beam by scanning the angles of the postmonochromator channel cuts in unison. The latter method should be used only when the Darwin width of the sample is much smaller than that of the high-heat-load monochromator, in this case Si(111). The fluorescence slit shown in Figure 10 is an option that can be used to limit the emission takeoff angle, which can be crucial if surface sensitivity is required. The fluorescence detector is a Canberra multi-element Ultra LEGe array detector with a 25 μm Be window and a UHV interface option. An XIA DXP-4T digital X-ray processor spectroscopy system is used to pulse-height analyze the fluorescence emission. For a desired precision in the coherent position of ± 0.03 Å at a reasonably high coherent fraction, it is necessary to collect fluorescence counts until a relative counting statistical error of roughly 2% is achieved. If the background counts under the fluorescence peak are very low, this would correspond to collecting 2500 counts at each step of the scan or roughly 8×10^4 counts. As an example, for 1 ML or 5×10^{14} cm^{-2} of Zn at $E_\gamma = 10$ keV with only the Si(111) monochromator and no postmonochromator, the Zn K_α fluorescence count rate would be 10 cps at an unfocussed bend magnet and 10 kcps at the APS undulator.

As stated earlier, the postmonochromator optics should prepare the incident beam in a way that avoids angle and wavelength averaging effects that would smear (or reduce the fringe visibility of) the XSW. Figure 12a shows the λ vs. θ DuMond diagram for the APS undulator source at $E_\gamma = 12.50$ keV, the Si(111) monochromator, and the pair of Si(004) postmonochromator reflections. The slanted stripes represent the conditions where Bragg diffraction is allowed on the basis of dynamical diffraction theory. The slopes of the

Bedzyk & Cheng

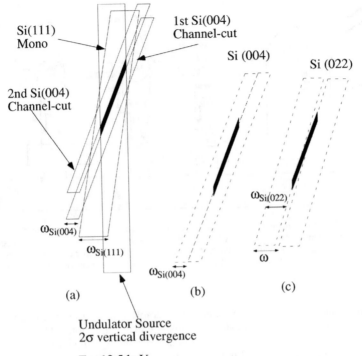

Undulator Source
2σ vertical divergence

E = 12.5 keV

Figure 12. λ vs. θ DuMond diagrams for the optics of the postmonochromator (from Rodrigues 2000).

stripes are from the derivative of Bragg's law:

$$\frac{\Delta\lambda}{\Delta\theta} = 2d_H \cos\theta_B \tag{40}$$

The angular widths of the Bragg stripes are the respective Darwin widths, ω. The APS undulator source divergence of $2\sigma = 20.$ μrad is represented by the vertical stripe. The overlap between the undulator source, the Si(111), the first Si(004), and the detuned second Si(004) is indicated by the shaded area. This shaded area represents the beam emittance profile from the postmonochromator. Figures 12b and 12c compare this postmonochromator output to the acceptance of the symmetric Si(004) and Si(022) sample reflections, respectively, as the samples are scanned from the low-angle side to the high-angle side of the rocking curve. A slight amount of dispersion exists between the Si(004) channel-cuts and the Si(022) sample reflection, but this will not cause significant smearing since the small divergence of the undulator has produced a small wavelength width. However, more intensity with the same amount of phase resolution could be produced if a pair of Si(022) channel-cuts was used for the Si(022) sample reflection.

The inclusion of the extra optical elements (e.g., the four extra bounces from the two detuned channel-cut postmonochromator crystals) in the optical path while producing the desired effect of increasing the XSW phase (or atomic positional) resolution, come at the cost of reducing the X-ray intensity incident on the sample. Using Figure 12a one can visually estimate this effect by comparing the emittance from the second Si(004) channel-cut (black area) to the emittance from the upstream Si(111) monochromator where it overlaps the undulator source divergence. Comparing the two areas gives an estimated transmission of 7% through the pair of detuned channel-cuts.

LOCATING IMPURITY LATTICE SITES IN MINERALS WITH X-RAY STANDING WAVES

Impurities in minerals

Impurities incorporated into minerals are important indicators of the geochronological history of the minerals and the environmental conditions during formation and growth (Ludwig et al. 1992; Brannon et al. 1996). Particularly important is the selective nature of atom or molecule incorporation, and how such selectivity is related to the lattice structure of the host mineral. Because the Bragg diffraction XSW technique gives the lattice position of an atom projected in the substrate unit cell, this technique is ideally suited for determining the lattice sites of bulk-incorporated impurities in a crystal. Combining this technique with polarization-dependent extended X-ray absorption fine structure spectroscopy (EXAFS), which probes the local structure surrounding the impurity, gives a complete picture of the impurity structure at atomic resolution.

Mn^{2+} lattice sites in calcite

In calcite, divalent cations are commonly found as bulk impurities. Arguments that these impurities are substitutional (with Ca^{2+}) were previously based on (1) the fact that they are isovalent to Ca^{2+} and (2) the EXAFS evidence that their near-neighbor bonding structures are identical to that of Ca^{2+} (Pingitore et al. 1988; Reeder at al 1999; Cheng et al. 2001). XSW was used to directly obtain the lattice site(s) of Mn^{2+} in calcite, giving direct evidence that Mn^{2+} is located at the Ca^{2+} site (Qian et al. 1994; Cheng et al. 2001) (see Fig. 13). Quantitative comparisons in the Mn^{2+} impurity structures with those in pure

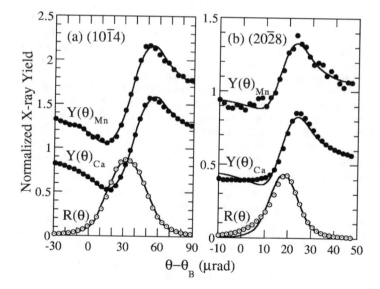

Figure 13. Reflectivity R and yields Y of Ca $K\alpha$ and Mn $K\alpha$ fluorescence from bulk calcite are shown for (a) the ($10\bar{1}4$) and (b) the ($20\bar{2}8$) Bragg reflections. Reprinted with permission from Cheng et al. 2001. Copyright 2001 by the American Physical Society.

ionic (i.e., KCl) and pure covalent (GaAs) crystals suggests that the large relaxations at the calcite impurity sites can be attributed to the flexibility in intramolecular relaxation of the first-neighbor CO_3^{2-} anions. This implies that the calcite structure possesses a high flexibility, or compliance, for accommodating impurities and it makes ion adsorption on the calcite surface an important process, because such a surface impurity could in fact be the initial stage in potential incorporation of impurities into the crystal bulk.

Phengitic ion sites in muscovite

Phengitic cations substituted at the octahedral sites of muscovite are the principal elemental variations from end-member muscovite. The major phengitic cations include Mg, Mn, Fe, and Ti. Their presence and concentrations as impurities in muscovite are considered evidence of metamorphic conditions during rock formation (Guidotti 1984). In addition to the octahedral sites, these and other ions are expected to substitute at other sites of muscovite to achieve bulk charge balance. The determination of impurity ion sites in muscovite has traditionally relied on optical spectroscopy. Although much is known, many details are not fully resolved.

By locating the lattice sites, the Bragg diffraction XSW technique offers an entirely different, structurally more direct approach for determining the lattice-sites of the impurities. In a recent experiment, a set of (00*l*) XSW measurements were made of a muscovite sample to locate the positions of Mn, Fe, and Ti substituting ions within the tetrahedral-octahedral-tetrahedral layer and the interlayer space (Cheng et al. 2002). All three impurity ions were found to be located at the position of the octahedral site.

ION ADSORPTION STRUCTURES AT MINERAL SURFACES WITH X-RAY STANDING WAVES

Ion adsorption at the calcite-water interface

The adsorption of trace aqueous ions at the calcite-water interface is one of the most intensely studied mineral-water interfacial processes. A goal of these studies is to achieve a molecular-scale understanding of the interfacial transport process and the factors that control it. Such understanding has great implications for a wide range of geochemical and environmental processes, most notably in toxic metal transport in groundwater and soils. Traditional methods for studying aqueous ion adsorption rely mostly on solution concentration measurements on batch powder samples. These methods address the issue of adsorption through its dependence on solution thermodynamics, but they are unable to probe adsorption in relation to mineral surface structures. The structural dependence of ion incorporation in calcite and other minerals is a well-accepted concept that is explained quantitatively, for example, in the case of Mn^{2+} in calcite (Cheng et al. 2001; see discussion in previous section). Surface selectivity in ion adsorption based on structural compatibility is now recognized as a key criterion determining whether adsorption will occur.

The coherently adsorbed ions—those crystallographically registered with the surface lattice—are of primary importance in an adsorption event. This is because these ions, initially incorporated as a part of a surface solid solution, have the potential through subsequent overlayer growth to become a part of a bulk solid solution. The otavite-calcite solid-solution growth on calcite (10$\overline{1}$4) observed with X-ray diffraction by Chiarello et al. (1997) is an example of such growth. This growth leads to the retention of ions as bulk impurities, with the potential for future release if the chemistry of the solution undergoes substantial change.

To address the issue of coherent adsorption at the calcite-water interface, a series of

Bragg diffraction XSW studies has been carried out on the ions' structural registration with the calcite surface lattice (Qian et al. 1994; Sturchio et al. 1997; Cheng at al. 1997 1998, 1999, 2000; Cheng 1998). To date the most systematic and extensive series of XSW technique applications to a molecular geochemical or environmental system, it covers the adsorption of the monatomic cations Co^{2+}, Ni^{2+}, Cu^{2+}, Zn^{2+}, Pb^{2+}, and Cd^{2+}; the molecular anions AsO_3^{2-}, SeO_3^{2-}, and CrO_4^{2-}; and the molecular cation UO_2^{2+}. In many cases, polarization-dependent EXAFS or surface crystal truncation rod experiments were performed to complement the XSW results. These studies are summarized in Table 1 and are described below. The structure of the calcite $(10\bar{1}4)$ surface in water is shown in the atomic force microscopy (AFM) image in Figure 14; the structure has been investigated by AFM (Ohnesorge and Binnig 1993; Liang et al. 1996) and X-ray reflectivity (Chiarello and Sturchio 1995; Fenter et al. 2000a).

Cation adsorption on calcite

Isomorphic substitution is the predominant mechanism of trace cation incorporation at the water-calcite interface. These adsorption studies therefore focus on cations that are isovalent to Ca^{2+}, beginning with cations associated with carbonates of the calcite

Table 1. Summary of XSW studies of aqueous ion adsorption at the calcite $(10\bar{1}4)$ surface: the adsorbate ions' coherent coverage, coherent fractions and coherent positions. $\Theta_c = f_{10\bar{1}4}\Theta \cdot 1$ ML corresponds to the $(10\bar{1}4)$ planar density of Ca^{2+} ions.

Adsorbate ion	Coherent coverage Θ_c [ML]	Lattice plane H	Coherent fraction f_H	Coherent position P_H	Reference
Cations					
Co^{2+}	0.09	$10\bar{1}4$	0.65(2)	0.89(1)	(1)
		$01\bar{1}8$	0.45(2)	0.88(1)	
		$2\bar{2}04$	0.36(2)	0.87(1)	
Ni^{2+}	~0.1	$10\bar{1}4$	0.41(2)	0.88(1)	(2)
Cu^{2+}	~0.09	$10\bar{1}4$	0.60(2)	0.86(1)	(2)
		$20\bar{2}8$	0.58(2)	0.71(2)	
Zn^{2+}	0.09	$10\bar{1}4$	0.77(1)	0.87(1)	(4)
		0006	0.51(2)	0.45(2)	
		$02\bar{2}4$	0.61(2)	0.85(3)	
Cd^{2+}	~0.1	$10\bar{1}4$	0.99(3)	1.00(1)	(3)
Pb^{2+}	0.05	$10\bar{1}4$	0.59(2)	1.00(1)	(5),(6)
		0006	0.64(3)	0.52(2)	
		$02\bar{2}4$	0.63(1)	0.99(1)	
UO_2^{2+}	~0.05	$10\bar{1}4$	0.14-0.65	0.84(2)	(7)
Anions					
SeO_3^{2-}	0.06	$10\bar{1}4$	0.77(5)	0.18(2)	(8)
		0006	0.74(15)	0.22(4)	
		$1\bar{1}20$	0.40(6)	0.99(3)	
AsO_3^{2-}	~0.05	$10\bar{1}4$	0.62(3)	0.15(2)	(9)
		0006	0.50(10)	0.27(6)	
CrO_4^{2-}	~0.1	$10\bar{1}4$	0.44(2)	0.03(2)	(2),(3)

References: (1) Cheng et al. 2000, (2) Cheng 1998, (3) Cheng et al., unpublished, (4) Cheng et al. 1998, (5) Qian et al. 1994, (6) Sturchio et al. 1997, (7) Sturchio et al. 2001, (8) Cheng et al. 1997, (9) Cheng et al. 1999

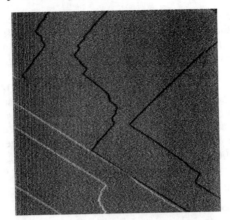

Figure 14. A 7 μm $\times$ 7 μm area of the calcite (10$\bar{1}$4) surface in water under typical conditions for XSW experiments, as imaged by AFM in an *in situ* solution cell (from archive of L. Cheng and H. Teng).

structure and extending to cations associated with carbonates of the aragonite structure.

Cobalt (II). XSW measurements on Co^{2+} adsorbed at the calcite (10$\bar{1}$4) surface were made with respect to the (10$\bar{1}$4), (2$\bar{2}$04) and (01$\bar{1}$8) Bragg reflections of calcite (Cheng et al. 2000). The XSW results are shown in Figure 15. The crystallographic ordering of the adsorbed Co^{2+} is revealed by the relatively high coherent-fraction values measured in all three (non-colinear) directions. Each of the three coherent positions gives the Co^{2+} lattice site(s) viewed in that direction as $h_H = d_H \times P_H$. For example, with respect to the (10$\bar{1}$4) lattice, the Co^{2+} ion is located at a height 3.04 Å $\times$ 0.89 = 2.70 Å above the plane. The intersection of the three positions gives the three-dimensional location of the Co^{2+} adsorbate relative to the calcite lattice. These XSW data provide an atomic-scale structural model of the Co^{2+} adsorbate structure, as shown in Figure 16. The model shows that Co^{2+} is located at the Ca^{2+} site, giving quantitative structural evidence that this adsorption is substitutional. The model further reveals that the Co^{2+} adsorbate is slightly relaxed from the ideal Ca^{2+} position. This relaxation can be explained quantitatively as being along the vertical Co-O direction. Accordingly, this relaxation is attributed to the result of surface truncation and to the smaller size of Co^{2+} in comparison to Ca^{2+}. (The Co-O distance in CoCO$_3$ is 2.11 Å, while the Ca-O distance in calcite is 2.35 Å).

Additionally, polarization-dependent surface EXAFS reveals that the Co^{2+} adsorbate maintains an octahedral coordination with neighboring oxygen atoms. This observation suggests that, in addition to the four CO$_3{}^{2-}$ in the plane and the one below, another species is bonded to the surface Co^{2+} above the plane. The Co-O nearest-neighbor distance is 2.11 Å, as in CoCO$_3$. Putting together XSW and EXAFS data, while maintaining an octahedral coordination, reduces the radius of the Co^{2+} coordination sphere compared to that of Ca^{2+}, and relaxes its center, as shown in Figure 17. The Co^{2+} ions are incorporated in the calcite (10$\bar{1}$4) surface lattice, forming a dilute two-dimensional solid solution of Ca$_{0.9}$Co$_{0.1}$CO$_3$.

Nickel(II). From similarity in the sizes of Ni^{2+} to Co^{2+}, one expects these cations to have similar adsorption structures on calcite. XSW made with respect to the (10$\bar{1}$4) planes showed that the Ni^{2+} adsorbate is located 2.7 Å above the (10$\bar{1}$4) plane. In-plane surface EXAFS measurements on the Ni^{2+} adsorbates showed a first-neighbor Ni-O distance identical to that in NiCO$_3$ (Cheng 1998).

Copper(II). The Cu^{2+} ion has a unique electronic structure that causes its nearest-neighbor coordination structure to undergo Jahn-Teller distortion, splitting the usually equidistant octahedral neighbors into four closer equatorial neighbors and two more

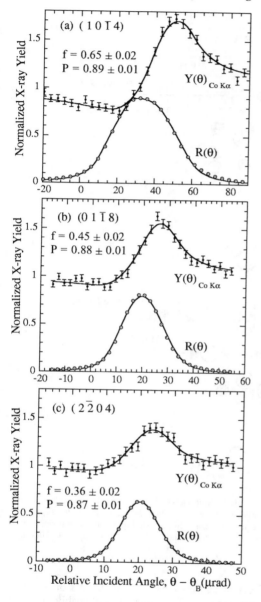

Figure 15. XSW data for Co^{2+} adsorbed on the calcite ($10\bar{1}4$) surface. Experimental data (dots) and theoretical fits (solid-line curves) of the normalized X-ray reflectivity R and Co $K\alpha$ fluorescence yield Y are shown for the (a) ($10\bar{1}4$), (b) ($01\bar{1}8$), and (c) ($2\bar{2}04$) Bragg reflections of calcite. Reprinted with permission from Cheng et al. (2000). Copyright 2000 by the American Physical Society.

distant axial neighbors (e.g., Goodenough 1998). The distinction of the Cu^{2+} structure from that of the other common divalent metals is reflected in both the amorphous phase and the bulk crystalline phase of Cu^{2+} salts. However, the effect of the distorted nearest-neighbor shell of the Cu^{2+} ion on its incorporation at the calcite surface and the ordering and local bonding structure of the resultant incorporation structure are not known. XSW measurements on Cu^{2+} adsorbed at the calcite ($10\bar{1}4$) surface were made respect to the ($10\bar{1}4$) and ($20\bar{2}8$) lattice planes (Cheng 1998). The results reveal that the Cu^{2+} adsorbate is located at a height of 2.7 Å above the ($10\bar{1}4$) lattice, an incorporation

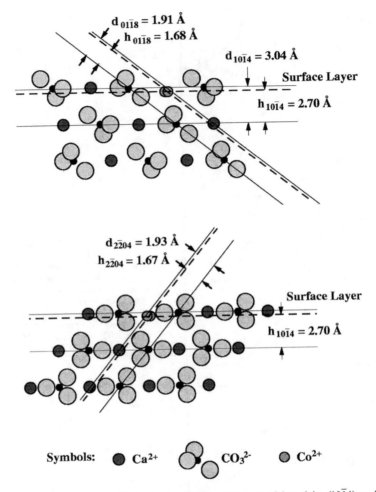

Figure 16. Structural model of Co^{2+} adsorbed on calcite. Side views of the calcite $(10\bar{1}4)$ surface show the averaged height of the Co^{2+} adsorbate relative to the each of three lattice planes with respect to which the XSW measurements are made. Note that there are two crystallographically nonequivalent Co^{2+} sites per surface unit cell. Reprinted with permission from Cheng et al. (2000). Copyright 2000 by the American Physical Society.

structure very similar to that of other divalent cations of similar size. This observation indicates that the distorted near-neighbor structure of Cu^{2+} does not noticeably affect its incorporation into the calcite surface or the site of incorporation.

Zinc (II). XSW triangulation on Zn^{2+} ion adsorption site at the calcite $(10\bar{1}4)$ surface was made with respect to the $(10\bar{1}4)$, (0006), and $(02\bar{2}4)$ Bragg reflections (Cheng et al. 1997). As for Co^{2+}, the high coherent fractions indicate that the Zn^{2+} adsorbates are well registered with the calcite surface lattice. And again, the structural model based on the coherent positions reveals that the adsorbed Zn^{2+} is located at the surface Ca^{2+} site, with its center slightly displaced from that of the ideal Ca^{2+} along the vertical Zn-O direction. The height of Zn^{2+} measured from the $(10\bar{1}4)$ lattice plane, for

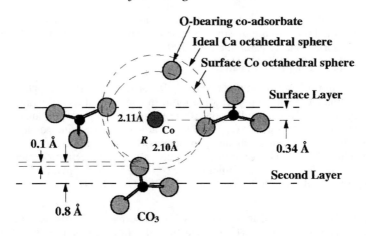

Figure 17. Side view of the structural model describing the site and the local bonding structure of Co²⁺ incorporated at the calcite (10$\bar{1}$4) surface according to XSW and polarization-dependent EXAFS results. Reprinted with permission from Cheng et al. (2000). Copyright 2000 by the American Physical Society.

example, is h = 2.64 ± 0.03 Å. In-plane EXAFS measurements made on the Zn²⁺ adsorbate showed that the Zn-O nearest-neighbor distance is 2.11 ± 0.01 Å, as in ZnCO₃, and the in-plane coordination number is 3.8 ± 0.7. Therefore, the local bonding structure of the Zn²⁺ adsorbate is also similar to that of Co²⁺.

Cadmium(II). In ionic size, Cd²⁺ is most similar to Ca²⁺ among divalent cations with carbonates of the calcite structure. XSW measurements on Cd²⁺ adsorbed at the calcite (10$\bar{1}$4) surface were made with respect to the (10$\bar{1}$4) reflection (Cheng et al., unpublished). This result reveals a highly ordered adsorbate at a position identical to that of ideal Ca²⁺. This high degree of ordering is perhaps related to the close size similarity between Cd²⁺ and Ca²⁺ (ionic radii: Cd²⁺, r = 0.97 Å; Ca²⁺, r = 0.99 Å) , which makes the incorporation structurally more compatible. This adsorbate structure is the initial stage of the three-dimensional $Cd_xCa_{1-x}CO_3$ solid solution phase observed in the X-ray diffraction studies of Chiarello et al. (1997).

Lead(II). XSW triangulation of Pb²⁺ adsorbed at the calcite (10$\bar{1}$4) surface was performed with respect to the (10$\bar{1}$4), (0006), and (02$\bar{2}$4) Bragg reflections (Qian et al. 1994; Sturchio et al. 1997). Again, high coherent fractions were observed in all three directions, indicating that the Pb²⁺, despite the fact that its carbonate is of the aragonite structure, can become incorporated at the calcite surface lattice. From the triangulation based on the coherent positions, the site of Pb²⁺ is within uncertainties that of the ideal Ca²⁺. This exact matching in the site initially appears to be surprising. However, the absence of noticeable relaxation of the Pb²⁺ position can be explained by a likely structural scenario in which a vertical expansion due to the larger size of the Pb²⁺ ion (Pb²⁺: r = 1.21 Å; Ca²⁺: r = 0.99 Å) almost exactly cancels the relaxation due to surface truncation.

Several additional measurements were made on the Pb²⁺ adsorption surface. First, comparison between *in situ* and ex situ XSW show identical results (Sturchio et al. 1997). Second, in-plane EXAFS measurements of the Pb²⁺ adsorbate showed that its nearest-neighbor Pb-O distance is 2.69 ± 0.03 Å, as in PbCO₃ (Cheng 1998). Finally, specular X-ray crystal truncation rod measurements were also made to determine independently at which surface lattice layer (or layers) the Pb^{2+} ions are incorporated. The analysis concluded that the Pb^{2+} ions are mostly incorporated within the topmost monolayer of the calcite $(10\bar{1}4)$ surface lattice (Sturchio et al. 1997).

Molecular anion adsorption on calcite

For adsorption of monatomic cations, the factor of structural compatibility can be described by a single parameter—the ionic radii. In contrast, structural compatibility for molecular anions involves consideration of the geometric shape of the ions. Oxyanions having the stoichiometric forms BO_3^{2-} and BO_4^{2-} are among the principal aqueous-phase species containing environmentally and geochemically important elements (e.g., National Research Council 1983). In order of increasing difference in molecular geometry from the CO_3^{2-} ion are the planar triangular anions (i.e., NO_3^{2-}, PO_3^{2-}), the pyramidal anions (i.e., AsO_3^{2-}, SeO_3^{2-}), and the tetrahedral anions (i.e., AsO_4^{2-}, SeO_4^{2-}, and CrO_4^{2-}) (Evans 1964).

Arsenite. The XSW measurements on AsO_3^{2-} adsorbed at the calcite $(10\bar{1}4)$ surface were made with respect to the $(10\bar{1}4)$ and (0006) reflections of calcite (Cheng et al. 1999). The data are shown in Figure 18. High coherent fractions of the As atom were observed in both directions. From the coherent positions, the projected heights of the As atom relative to the $(10\bar{1}4)$ and the (0006) planes are 0.46 ± 0.06 Å and 0.76 ± 0.17 Å, respectively. (The ideal position of C in CO_3^{2-} projects zero length in both directions.)

According to these projected heights of the As atom, a molecular structural model was constructed for AsO_3^{2-} adsorbed on the calcite surface, as shown in Figure 19. In the model, the intersecting point of the two projected positions gives the As site. Because the calcite $(10\bar{1}4)$ surface has twofold symmetry in the direction normal to the plane of view of the model, the projected height of the As atom in the direction perpendicular to the plane of view is zero. This model gives quantitative structural evidence that the adsorbed AsO_3^{2-} substitutes at the CO_3^{2-} site. Furthermore, the measured height of the As position above the CO_3^{2-} plane is roughly consistent with the projected height of the As apex above the O basal plane in the pyramidal AsO_3^{2-} anion. Therefore, it can be further inferred that the AsO_3^{2-} substitution has the specific molecular orientation that the basal O plane overlaps that of CO_3^{2-}, with the apical As pointing in the +[0001] direction of calcite. After AsO_3^{2-} incorporation, the calcite surface can be described as a dilute two-dimensional solid solution of the chemical form $Ca(AsO_3)_x(CO_3)_{1-x}$.

Selenite. The XSW measurements of the adsorption of SeO_3^{2-} on the calcite $(10\bar{1}4)$ surface were made with respect to the $(10\bar{1}4)$, (0006), and $(11\bar{2}0)$ reflections of calcite (Cheng et al. 1997). The result of SeO_3^{2-} adsorption is analogous to that of AsO_3^{2-}. Considering the geometric similarity of these two oxyanions, their similar adsorption structures are quite expected. The exact position of the Se atom located by the intersection of these three heights is displaced from the ideal C position along the +[0001] direction by 0.63 ± 0.11 Å, similar to the AsO_3^{2-} model. A similar explanation based the molecular geometry applies. For the SeO_3^{2-} anions, the distance between the apical Se and the O basal plane is typically 0.75 Å. This is within uncertainty in agreement with the measured value. Therefore, as for AsO_3^{2-}, a plausible adsorption structure of SeO_3^{2-} derived from the XSW measurements indicates that SeO_3^{2-} substitutes for CO_3^{2-}, with a preferred orientation such that the Se apex points outward from the crystal.

Chromate. The XSW measurements of CrO_4^{2-} adsorption on the calcite $(10\bar{1}4)$ surface were made with respect to the $(10\bar{1}4)$ and (0006) reflections (Cheng 1998). From the average measured coherent positions, the Cr atom is located with a displacement relative to the ideal C position of -0.12 Å normal to the $(10\bar{1}4)$ plane. This position is

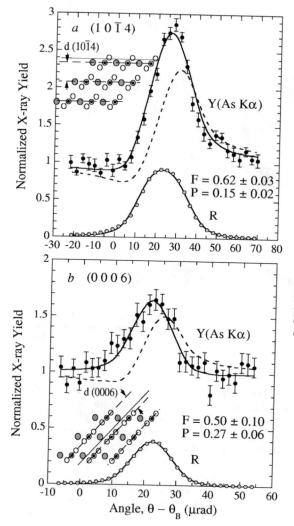

Figure 18. XSW data for AsO_3^{2-} adsorbed at the calcite $(10\bar{1}4)$ surface. Experimental data (dots) and theoretical fits (solid-line curves) of the normalized X-ray reflectivity R and As $K\alpha$ fluorescence yield Y are shown for (a) the $(10\bar{1}4)$, and (b) the (0006) Bragg reflections of calcite. Used by permission of Elsevier Science, from Cheng et al. (1999) *Geochim Cosmochim Acta*, 63, p 3153-3157.

consistent with a CrO_4^{2-} substituting at the CO_3^{2-} site. Although the CrO_4^{2-} ion is relatively large in volume, its O basal plane (side length: 2.61 Å) is only moderately larger than the CO_3^{2-} plane (side length: 2.22 Å). Therefore, the substitution structure is not surprising.

Uranyl adsorption on calcite

The UO_2^{2+} ion has a O-U-O linear geometry, with a U-O distance of 1.8 Å. Unlike the monatomic cations and molecular anions whose structural conformities with their respective counterpart in calcite lead to substitutional incorporation, the UO_2^{2+} ion does not fit in size and shape to the Ca^{2+} site. Despite this structural dissimilarity, recent EXAFS analyses suggest that UO_2^{2+} substitutes at the Ca^{2+} site (Kelly et al. 2002). These findings support the notion that calcite, as a molecular ionic crystal, possesses large structural flexibility in accommodating ions of dissimilar geometries (Cheng et al. 2001).

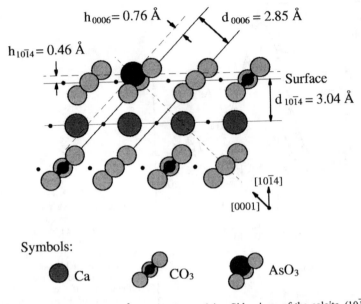

Figure 19. Structural model of AsO_3^{2-} adsorbed on calcite. Side views of the calcite $(10\bar{1}4)$ surface show the averaged height of the As position of the AsO_3^{2-} adsorbate relative to each lattice plane with respect to which the XSW measurement is made. Used by permission of Elsevier Science, from Cheng et al. (1999), *Geochim Cosmochim Acta*, 63, p 3153-3157.

Bragg diffraction XSW measurements were made of uranyl adsorbed on calcite $(10\bar{1}4)$ from a pH ~ 10.5 solution (Sturchio et al. 2001). The coherent fractions obtained range from 0.14 to 0.65. The lower values of the coherent fractions may imply adsorption at multiple sites, although the highest value indicates a high probability of adsorption at a single dominant site. The mean value of the measured coherent position in all measurements was 0.84 ± 0.02. This position was independent of dissolved U(VI) concentration, and corresponds to a distance between the U atom and the calcite $(10\bar{1}4)$ plane of 2.55 ± 0.06 Å. These results are consistent with U(VI) adsorption at the calcite surface mainly as a monodentate inner-sphere uranyl-carbonate surface complex. Steric considerations imply that the observed U(VI) surface complex may occur both at edge sites ($<\bar{4}41>_-$ and $<48\bar{1}>_-$) and on terrace areas adjacent to Ca vacancies.

PROBING METAL DISTRIBUTIONS IN ORGANIC FILMS WITH X-RAY STANDING WAVES

Metal in organic films at mineral surfaces

Natural organic matter from biosynthesis and biodegradation is ubiquitous in natural aqueous environments. Common examples of natural organic materials are extracellular polymeric matrices (biofilms) generated by bacteria, hydrocarbons from industrial effluents and fossil fuel products, and numerous humic substances. Because of their hydrophobicity, natural organic materials have low solubility in water and tend to accumulate at solid surfaces. As a result, in groundwater and soils, mineral surfaces are commonly coated with organic films.

The presence of organic films at the mineral-water interface complicates the interactions of metal ions with mineral surfaces. The organic films may in some cases inhibit metal-mineral interactions by blocking the reactive surface sites of the mineral surfaces. Certain functional groups in the organic films may form stable complexes with aqueous metals, causing the films to act as sinks for certain geochemically and environmentally critical trace elements, very much as minerals do.

In a sense, metal-mineral surface interactions and metal-organic film interaction compete at natural mineral-water interfaces. The first-order information needed to understand the role of the organic films in mineral-aqueous metal interactions is a map of the partitioning of metals at the organic-mineral interface.

The capabilities of TER-XSW and LSM-XSW in profiling metal distribution from 50 to 1000 Å above a surface offer a unique method for directly determining metal ion partitioning at mineral-organic film-aqueous interfaces. In this section, we discuss experiments using long-period XSW to locate the metal distribution in model organic films on solid surfaces, including Langmuir-Blodgett trilayers and self-assembled monolayers and multilayers (SAMs). These model film systems are analogous to organic films on mineral surfaces, and the approaches employed in these studies are directly applicable to mineral-organic film studies. We also discuss recent experiments in which TER-XSW techniques were applied to investigate the Pb^{2+} distribution at metal oxide-biofilm interfaces.

Zn^{2+} in a Langmuir-Blodgett multilayer

Bedzyk et al. (1989) used TER-XSW to locate the Zn atom layer in a Langmuir-Blodgett (LB) multilayer. The Zn atom layer was embedded in the top bilayer of the LB multilayer deposited on a gold mirror surface (Fig. 20). In this experiment, three full oscillations in the Zn Kα fluorescence yield (Fig. 20b) occurred before the gold critical angle was reached (Fig. 20a). As described above in the TER-XSW section this indicates that the Zn layer was at $z = 2.5 D_c = 200$ Å. This was more precisely determined to be 218 Å by using a χ^2 fit of the data to theoretical yields based on a layered model. From this fit, it was determined that the 2σ thickness of the Zn layer was 24 Å. After annealing to 105°C the Zn oscillations (Fig. 20c) were substantially smeared out, indicating that the Zn layer was no longer confined to such a narrow layer.

Thermally induced dynamical behavior of the Zn layer in the same membrane system was studied with long-period XSW generated by Bragg diffraction from LSM (Bedzyk et al. 1988) (see Fig. 21). In this case, the LB trilayer of Zn and Cd arachidate was deposited on an LSM consisting of 200 W/Si layer pairs with periodicity of 25 Å. Bragg-reflection and TER-XSW scans for Zn fluorescence were made at temperatures between 35°C and 140°C. Figure 22 shows the Bragg reflectivity and Zn K fluorescence modulations. The measured coherent positions for the Cd layer indicated that the thickness of the top Si layer of the LSM had been reduced from 18.75 Å to 9.8 Å by a surface-cleaning process. The Cd position in the LB film was unaffected by heating and cooling. This was, however, not the case for Zn. Initially, at 34°C, the Zn layer was located 53.4 Å above the LSM surface. Figure 22 shows the evolution of the Zn fluorescence as temperature was raised from 34°C to 100°C. Over this temperature change, the peak of the Zn fluorescence signal shifted progressively from the low-angle side to the high-angle side of the Bragg reflection. This corresponds to a movement of the Zn layer by almost one-half of the 25-Å LSM d-spacing. As the temperature was raised further, from 100°C to 110°C, the peak in Zn fluorescence shifted back toward the low-angle side, indicating that the Zn layer had moved either farther from the surface by $\sim \frac{1}{4}d$ or toward the surface by $\sim \frac{3}{4}d$. The TER-XSW measurement showed that the latter was the case.

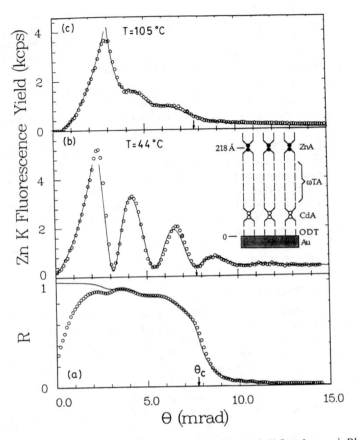

Figure 20. The angular dependence of (a) the reflectivity at $E_\gamma = 9.8$ keV for a Langmuir-Blodgett film deposited on a Au surface (see inset), (b) the Zn $K\alpha$ fluorescence yield at 44°C and (c) the Zn $K\alpha$ fluorescence yield at 105°C. The data are shown as circles, and the fits are solid lines. Reprinted with permission from Bedzyk et al. (1989). Copyright 1989 by the American Physical Society.

Br⁻ location and ordering in self-assembled multilayers

Long-period XSW has been applied by Lin et al. (1997) to locate the position and ordering of Br⁻ ions within the self-assembled organic monolayers and multilayers (SAMs). In these studies, three sets of SAM samples were examined—a (3-bromopropyl)trichlorosilane monolayer, an octachlorotrisiloxane-capped (3-bromopropyl)trichlorosilane monolayer, and a trilayer-coupled multilayer (Fig. 23). These organic structures were adsorbed on a surface of a Si/Mo LSM, and both TER-XSW and LSM-XSW measurements were made. The results are shown in Figure 24. For the monolayer sample, the LSM-XSW coherent position and fraction reveal that the Br⁻ ions were located at a mean height of 10 Å above the surface, with a Gaussian distribution width of 7.3 Å. This height is in agreement with the thickness of the monolayer (~10 Å), and the distribution width is attributed to the surface roughness (~6 Å). For the capped trilayer sample, the coherent fraction and position remain unchanged, indicating that the self-assembly produces ordered superlattice structures. For the coupled trilayer sample,

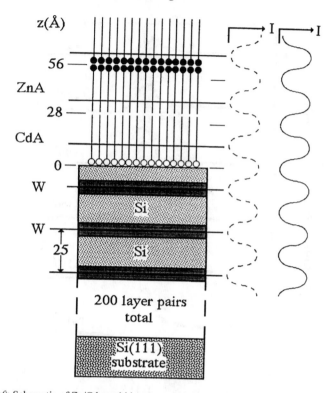

Figure 21. *Left*: Schematic of Zn/Cd arachidate Langmuir-Blodgett layers deposited on the Si surface of a W/Si layered synthetic microstructure with 25-Å period. *Right*: The intensities of the XSW on the low-angle side (dashed line) and high-angle side (solid line) of the Bragg diffraction. Reprinted with permission from Bedzyk et al. (1988). Copyright 1988 by the American Association for the Advancement of Science.

the XSW results showed that the Br^- ion in the upper layer is located at a height of 48.8 Å with a Gaussian width of 10.4 Å.

These direct structural measurements using TER-XSW and LSM-XSW provide quantification to the location and ordering of the Br^- ions in the SAMs in the surface-normal direction to a precision of 0.1 Å.

Pb^{2+} partitioning at biofilm-oxide interfaces

The TER-XSW method has recently been applied by Templeton et al. (2001) to (1) investigate the partitioning of Pb^{2+} sorbed at the biofilm-hematite (α-Fe_2O_3) and the biofilm-corundum (α-Al_2O_3) interfaces and (2) determine how surface-bound bacteria affect the adsorption of metal ions at oxide surfaces. The TER-XSW measurements were made on α-Fe_2O_3 (0001) and on α-Al_2O_3 (0001) and ($1\bar{1}02$) surfaces coated with a gram-negative chemo-organotrophic bacterium and exposed to pH 6 solutions having dilute Pb^{2+} concentrations.

The data were interpreted by assuming a simple two-component model for Pb^{2+} distribution, in which Pb^{2+} occurs at the crystal surface and within the biofilm. The fit to the XSW fluorescence yield allows differentiation between Pb^{2+} located at the crystal

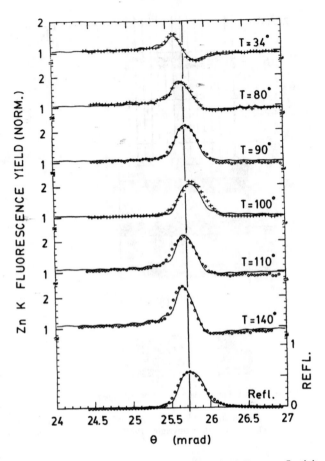

Figure 22. The experimental (dots) and theoretical (solid lines) Bragg reflectivity from the LSM described in Figure 21 at $E_\gamma = 9.8$ keV, with the normalized experimental and theoretical Zn $K\alpha$ fluorescence yields at selected temperatures. Reprinted with permission from Bedzyk et al. (1988). Copyright 1988 by the American Association for the Advancement of Science.

surface and Pb^{2+} within the overlying biofilm (Fig. 25). For the two α-Al_2O_3 surfaces, Pb^{2+} coverage on the biofilm-coated surfaces was similar to that on the biofilm-free surfaces. The inference is that the biofilms had not severely altered the reactivity of the surfaces by passivating their reactive sites (Trainer et al. 2002).

PROFILING THE ELECTRICAL DOUBLE LAYER STRUCTURE WITH X-RAY STANDING WAVES

The electrical double layer at the solid-water interface

Analytical models of double layer structures originated roughly a century ago, based on the theoretical work of Helmholtz, Gouy, Chapman, and Stern. In Figure 26, these idealized double-layer models are compared. The Helmholtz model (Fig. 26a) treats the interfacial region as equivalent to a parallel-plate capacitor, with one plate containing the

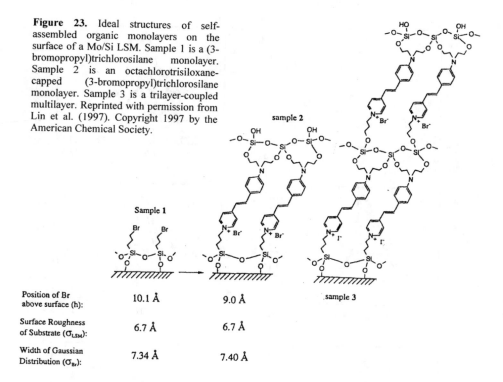

Figure 23. Ideal structures of self-assembled organic monolayers on the surface of a Mo/Si LSM. Sample 1 is a (3-bromopropyl)trichlorosilane monolayer. Sample 2 is an octachlorotrisiloxane-capped (3-bromopropyl)trichlorosilane monolayer. Sample 3 is a trilayer-coupled multilayer. Reprinted with permission from Lin et al. (1997). Copyright 1997 by the American Chemical Society.

Position of Br above surface (h):	10.1 Å	9.0 Å
Surface Roughness of Substrate (σ_{LSM}):	6.7 Å	6.7 Å
Width of Gaussian Distribution (σ_{Br}):	7.34 Å	7.40 Å

surface charge and the other plate, referred to as the outer Helmholtz plane, containing electrostatically attracted ions from the solution. This simplified rigid model of the interface neglects a number of important physical properties of the ions. The Gouy-Chapman model (Fig. 26b) takes into account the thermal motion of the ions, and thus it describes the ion distribution as a diffuse double layer. The Stern model (Fig. 26c) extends the Gouy-Chapman model by including specifically adsorbed ions at the interface. In this model the first layer of ions is constrained to lie in a Helmholtz plane with a thickness defined by the ionic radii. In general, this condensed ion layer does not completely neutralize the surface charge, and a diffuse layer exists beyond the interface electrostatically attracted to the residual net surface charge.

Lack of suitable experimental tools had in the past prevented direct verification of these models. Understanding of the double layer structure was gained mainly through analytical and simulation studies. Although certain experimental methods, such as, the surface force apparatus method (Israelachivilli 1991), are capable of probing the double-layer structure and confirming its existence, their poor lateral spatial resolution and pervasive nature prevent them from obtaining molecular-scale structural data.

The diffuse layer profiled with long-period X-ray standing waves

The development of the long-period XSW techniques led to the first direct, *in situ* observation of the electrical double layer structure by Bedzyk et al. (1990). These investigators examined a double layer structure of Zn^{2+} ions at the interface of a dilute $ZnCl_2$ aqueous solution and a charged phospholipid membrane. In this interface system, illustrated in Figure 26, the 27-nm-thick ordered membrane rests on a Si/W LSM. At the

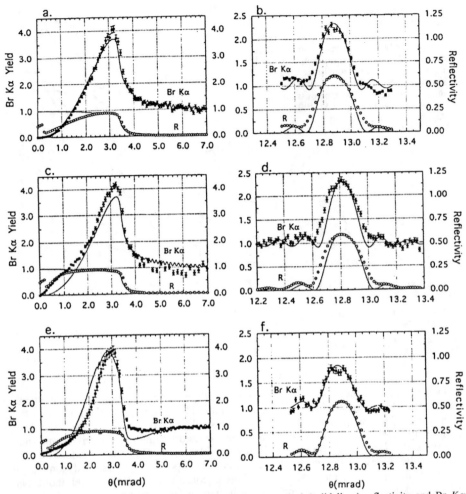

Figure 24. Experimental (open and filled circles) and theoretical (solid lines) reflectivity and Br $K\alpha$ fluorescence yields as a function of incident angle. Parts (a), (c), and (e) are the results from TER-XSW on samples 1, 2, and 3, respectively (in Fig. 23). Parts (b), (d), and (f) are the results from LSM-XSW measurements on samples 1, 2, and 3, respectively. Reprinted with permission from Lin et al. (1997). Copyright 1997 by the American Chemical Society.

top portion of the membrane, a negatively charged sheet of PO_4^{2-} ions was in contact with a 0.1 mM $ZnCl_2$ solution. At different solution pH levels, XSW measurements were then made under the TER condition. These measurements allowed derivation of the Debye length of the electrostatically attracted diffuse Zn^{2+} layer, as well as the coverage of the condensed Zn^{2+} layer as a function of solution pH.

Distribution structure and Debye length of the diffuse layer; condensed concentration. Figure 27 shows the TER-XSW results of these studies at pH 6.8, 4.4, and 2.0. In all three cases, the maximum of the Zn fluorescence yield appears at angle below the substrate critical angle. Because the first antinodal plane moves toward the

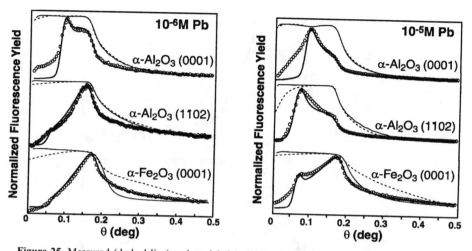

Figure 25. Measured (dashed line) and modeled (solid line) reflectivity and Pb L fluorescence yields from TER-XSW measurements of α-Al$_2$O$_3$ and α-Fe$_2$O$_3$ surfaces coated with bacteria and exposed to 10^{-6} and 10^{-5} M Pb solutions. Reprinted with permission from Templeton et al. (2001). Copyright 2001 by the National Academy of Science of the U.S.A.

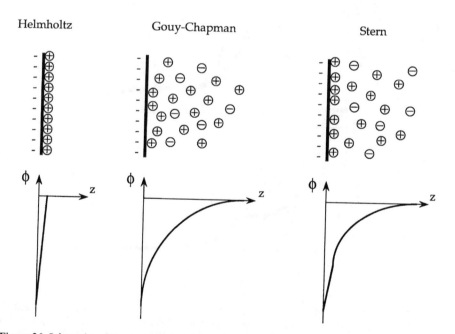

Figure 26. Schematics of the electrical double layer at a solid-liquid interface. (a) the Helmholtz model, (b) the Gouy-Chapman model, and (c) the Stern model.

Bedzyk & Cheng

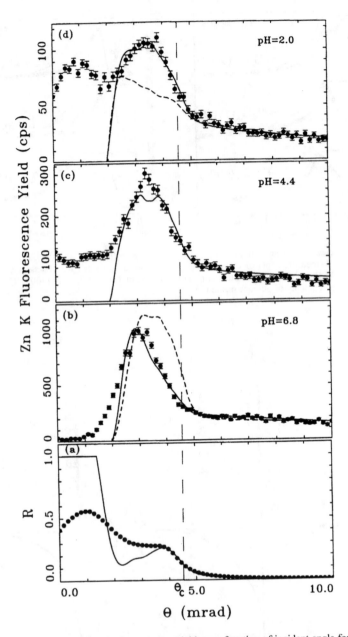

Figure 27 Reflectivity and Zn $K\alpha$ fluorescence yields as a function of incident angle for three different *p*H values from TER-XSW measurements of the diffuse double layer at the water-charged membrane surface. Reprinted with permission from Bedzyk et al. (1990). Copyright 1990 by the American Association for the Advancement of Science.

surface as the angle is increased toward the critical angle, the angular position of this maximum indicates that a concentrated layer of Zn^{2+} is located at some displaced distance above the substrate surface. For the case of pH 6.8, the results show that a calculated model based on the Helmholtz double layer structure of Zn^{2+} distribution cannot satisfactorily fit the data (Fig. 27b, dashed line). On the other hand, a model assuming an exponential decay distribution of Zn^{2+}, qualitatively based on the Gouy-Chapman-Stern model, agrees well with the data (Fig. 27b, solid line). In this model, the two model parameters are the Zn^{2+} concentration (i.e., coverage) in the condensed layer and the Debye length of the exponential decay. For the pH 6.8 solution, the Zn^{2+} surface concentration in the condensed layer is 7.7×10^{12} cm^{-2}, and the Debye length of the diffuse distribution is 58 Å. As the pH is decreased from 6.8 to 4.4 to 2.0, the Debye length decreases from 58 to 8 to 3 Å. At the same time, the condensed Zn^{2+} surface concentration changes from 7.7 to 7.7 to 4.4×10^{12} cm^{-2}. This change in the Debye length of the Zn ions represents a weakening of the electrostatic attraction of the charged interface due to protonation of the phosphate head groups of the lipid membrane.

Thermodynamic reversibility. Further investigation of the structural properties of the diffuse layer focused on its thermodynamic reversibility (Wang et al. 2001). In this study, TER-XSW was used on a similar aqueous interface system but that was supported by a Au mirror substrate. Originally set at pH 5.8, the solution (0.1 m*M* ZnCl$_2$) was titrated through a cycle of pH levels, first to pH 4.2 and then to pH 2.0, and then back to pH 4.2 and then to pH 5.8. The Zn data were fitted with the same exponential decay model. During the titrations downward in pH, the condensed layer surface concentration decreased from 2.2 to 0.67 to 0.34×10^{-12} cm^{-2}, and the Debye length decreased from 20 to 18 to 9 Å. During the subsequent titration upward in pH, both the condensed layer concentration and the Debye length tracked the corresponding values for the downward titration. These experiments proved the reversibility of the electric double layer structure.

The condensed layer measured with Bragg-diffraction X-ray standing waves

The pH dependence of the condensed Zn^{2+} concentration revealed in the TER-XSW studies is direct proof that the condensed layer is an integral part of the double layer system. However, in addition to being sensitive to solution conditions, the ions in the condensed layer, as a surface phase, must also depend on the properties of the solid surface. Thus, investigations of the condensed layer require understanding the bonding relationship between the aqueous ions and the surface atoms. Ideally, this can be gained by surface-polarized EXAFS measurements in the interface system. However, at least for oxides and many other surfaces, the first-neighbor atoms bonded to the ions in the condensed layer are often oxygen atoms, which cannot be distinguished from oxygen atoms in an aqueous hydration shell. Alternatively, one may use Bragg diffraction XSW to locate the three-dimensional surface sites of the condensed ions, as in the ion adsorption experiments on calcite described in an earlier section of this chapter. These sites give information on the possible bonding structure of the ions in the condensed layer.

Specific adsorption at electrochemical interfaces. Bragg diffraction XSW has been used to probe the structures of specifically adsorbed ions at metal-electrolyte and semiconductor-electrolyte interfaces under controlled electrochemical conditions. Among the earliest work in this category is the study of Tl adsorption from a dilute solution (mM) on Cu(111) through underpotential deposition (Materlik et al. 1987; Zegenhagen et al. 1990).

Condensed structures of Sr^{2+} and Cu^{2+} on rutile. To address condensed layer structures more directly related to mineral-aqueous interfaces, the adsorption of dilute divalent cations (~μM) on the TiO$_2$(110) surface was investigated. The studies addressed the structure of specifically adsorbed Sr^{2+} on $TiO_2(110)$ (Fenter et al. 2000b) and the structure of specifically adsorbed Cu^{2+} on $TiO_2(110)$ (Cheng et al. unpublished; Fig. 28). To prove that the condensed ions are directly bonded to surface atoms, XSW measurements in the surface-normal direction were made after the solutions had been removed (without allowing precipitation). Ordered adsorbates were found for both Sr^{2+} and Cu^{2+}, even with some surface roughness. For Sr^{2+}, the typical *ex situ* coherent fraction is ~0.3, and the coherent position is ~0.9. For Cu^{2+}, the coherent fraction is 0.36, the coherent position is 0.91. *In situ* measurements also made on the Sr^{2+} adsorption surface gave slightly higher coherent fractions than the ex situ measurements. Polarization-dependent EXAFS measurements were also taken on the Sr^{2+} interface in in-plane and surface-normal directions. A structural model based on the XSW coherent position and the EXAFS nearest-neighbor distances revealed that the condensed Sr^{2+} ion is coordinated with the surface by tetradentate bridging in octahedral coordination.

Comparison of Sr^{2+} and Zn^{2+} adsorption structures on rutile. The Sr^{2+} tetradentate adsorption structure on TiO_2 derived above is supported by recent *in situ* XSW triangulation measurement (Zhang et al., unpublished). A similar measurement of Zn^{2+} on TiO_2, however, suggests a monodendate surface site (Zhang et al., unpublished). These contrasting surface adsorption structures for the two isovalent cations can be explained by the difference in their hydration energies. For the larger Sr^{2+} ion ($r = 2.36$ Å), the hydration energy is lower, and the ion can be relatively easily dehydrated and thus coordinated at the most favorable surface site in tetradentate coordination. For the smaller Zn^{2+} ion ($r = 2.11$ Å), it is more difficult to remove hydration molecules. These results demonstrate the important competing effects between hydration structures and surface sites in aqueous ion adsorption.

CONCLUSION

While the X-ray standing wave technique has been actively used over the past two decades for investigating adsorbate structures on semiconductor surfaces, it has only seen recent use in molecular geochemistry and environmental sciences. The primary attraction of

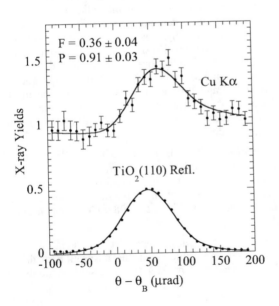

Figure 28 XSW standing wave data for Cu^{2+} adsorbed at the TiO_2 (110) surface from a dilute aqueous solution (Cheng et al. unpublished).

XSW to this area is its ability to uniquely produce high-resolution element-specific surface structural information at the mineral-fluid interface and impurity point-defect structures in minerals. The recent success in the geochemical and environmental sciences is partially due to the availability of new higher-intensity synchrotron facilities and developments of the XSW technique that allow its applicability to a larger class of crystalline substrates. It is worthwhile to emphasize the directness of the XSW method as a structural probe that comes from its elemental and phase sensitivities, and the subsequent reliability of the results. It is also important to note that no one method solves all problems, especially complex problems that deal with realistic environments. Better insights in any system are often achieved when the XSW methods are used in combination with other, complementary techniques, such as EXAFS, X-ray reflectivity and diffraction, as exemplified in a number of the example applications reviewed in this chapter. The *in situ* atomic-scale quantitative structural information from such a combined X-ray experimental approach, in correlation with scanning probe microscopy and theoretical calculations, will greatly aid in refining the knowledge base presently in place for describing the mineral-fluid interface and impurity point-defect structures in minerals.

ACKNOWLEDGMENTS

The authors thank Paul Fenter, Neil Sturchio and two anonymous reviewers for critical reviews of this chapter. This work was supported by the U. S. Department of Energy, BES under contract W-31-109-ENG-38 to Argonne National Laboratory and by the U. S. National Science Foundation under contract CHE-9810378 to the Institute of Environmental Catalysis at Northwestern University.

REFERENCES

Abruna HD, White JH, Albarelli MJ, Bommarito GM, Bedzyk MJ, McMillan M (1988) Uses of synchrotron radiation in the *in situ* study of electrochemical interfaces. J Phys Chem 92:7045-7052

Afanas'ev AM, Kohn VG (1978) External photoeffect in the diffraction of X-rays in a crystal with a perturbed layer. Sov Phys JETP 47: 154-161

Authier A (2001) Dynamical Theory of X-ray Diffraction. Oxford University Press, New York

Batterman BW (1964) Effect of dynamical diffraction in X-ray fluorescence scattering. Phys Rev 133:A759-764

Batterman BW, Cole H (1964) Dynamical diffraction of X-rays by perfect crystals. Rev Mod Phys 36:681-717

Batterman BW (1969) Detection of foreign atom sites by their X-ray fluorescence scattering. Phys Rev Lett 22:703-705

Becker RS, Golovchenko JA, Patel JR (1983) X-ray evanescent-wave absorption and emission. Phys Rev Lett 50:153-156.

Bedzyk MJ and Materlik G (1985) Two-beam dynamical diffraction solution of the phase problem: A determination with X-ray standing waves. Phys Rev B 32:6456-6463

Bedzyk MJ, Bilderback D, White J, Abruna HD, Bommarito GM (1986) Probing electrochemical interfaces with X-ray standing waves. J Phys Chem 90:4926-4928

Bedzyk MJ, Bilderback D, Bommarito GM, Caffrey M, Schildkraut JS (1988) X-ray standing waves; A molecular yardstick for biological membranes. Science 241:1788-1791

Bedzyk MJ, Bommarito GM, Schildkraut JS (1989) X-ray standing waves at a reflecting mirror surface. Phys Rev Lett 62:1376-1379

Bedzyk MJ, Bommarito GM, Caffrey M, Penner TA (1990) Diffuse-double layer at a membrane-aqueous interface measured with X-ray standing waves. Science 248:52-56

Bedzyk MJ (1992) X-ray standing wave studies of the liquid/solid interface and ultrathin organic films. *In* Springer Proceedings in Physics. Vol 61: Surface X-ray and Neutron Scattering. Zabel H, Robinson IK (eds) Springer-Verlag, Berlin, p 113-117

Bedzyk MJ, Kazimirov A, Marasco DL, Lee TL, Foster CM, Bai GR, Lyman, PF, Keane DT (2000) Probing the polarity of ferroelectric thin films with X-ray standing waves. Phys Rev B 61:R7873-R7876

Bilderback DH, Lairson BM, Barbee TW, Ice GE, Sparks CJ (1983) Design of doubly focusing, tunable 5-30 keV wide bandpass optics made from LSMS. Nucl Instrum Meth 208:251-257

Born M, Wolf E, Bhatia AB (1999) Principles of Optics: Electromagnetic Theory of Propagation, Interference and Diffraction of Light. New York: Cambridge University Press.

Brannon JC, Cole SC, Podosek FA, Ragan VM, Coveney RM Jr, Wallace MW, Bradley AJ (1996) Th-Pb and U-Pb dating of ore-stage calcite and Paleozoic fluid flow. Science 271:491-493

Caticha A, Caticha-Ellis S (1982) Dynamical theory of X-ray diffraction at Bragg angles near $\pi/2$. Phys Rev B 25:971-983

Cheng L, Lyman PF, Sturchio NC, Bedzyk MJ (1997) X-ray standing wave investigation of the surface structure of selenite anions adsorbed on calcite. Surf Sci 382:L690-L695

Cheng L, Sturchio NC, Woicik JC, Kemner KM, Lyman PF, Bedzyk MJ (1998) High-resolution structure of zinc incorporated at the calcite cleavage surface. Surf Sci 415:L976-L982

Cheng L (1998) Atomic scale study of ion incorporation at the calcite surface using synchrotron X-ray methods. PhD Dissertation, Northwestern University, Evanston, Illinois

Cheng L, Fenter P, Sturchio NC, Bedzyk MJ (1999) X-ray standing wave study of arsenite incorporation at the calcite surface. Geochim Cosmochim Acta 63:3153-3157

Cheng L, Sturchio NC, Bedzyk MJ (2000) Local structure of Co^{2+} incorporated at the calcite surface: An X-ray standing wave and SEXAFS study. Phys Rev B 61:4877-4883

Cheng L, Sturchio NC, Bedzyk MJ (2001) Impurity structure in a molecular ionic crystal: Atomic-scale X-ray study of $CaCO_3:Mn^{2+}$. Phys Rev B 63:144104(1)-144104(7)

Cheng L, Fenter P, Bedzyk MJ, Sturchio NC (2002) Direct profiling of atom distributions in mica with X-ray standing waves. Geochim Cosmochim Acta 66:A136

Chiarello RP, Sturchio NC (1995) The calcite (1014) cleavage surface in water: Early results of a crystal truncation rod study. Geochim Cosmochim Acta 59:4557-4561

Chiarello RP, Sturchio NC, Grace J, Geissbuhler P, Sorensen LB, Cheng L, Xu S (1997) Otavite-calcite solid-solution formation at the water-calcite interface studied *in situ* by synchrotron X-ray scattering. Geochim Cosmochim Acta 61:1467-1473

Cowan CE, Zachara JM, Resch CT (1990) Solution ion effects on the surface exchange of selenite on calcite. Geochim Cosmochim Acta 54:2223-2234

Cowan PL, Golovchenko JA, Robbins MF (1980) X-ray standing waves at crystal surfaces. Phys Rev Lett 57:4103-4110

Evans RC (1964) An Introduction to Crystal Chemistry. Cambridge University Press, London

Effenberger X, Mereiter HK, Zemann J (1981) Crystal structure refinements of magnesite, calcite, rhodochrosite, siderite, smithonite, and dolomite, and discussion of some aspects of the stereochemistry of calcite-type carbonates. Zeit Kristallogr 156:233-243

Fenter P, Geissbuhler P, DiMasi E, Srajer G, Sorenson LB, Sturchio NC (2000a) Surface speciation of calcie observed *in situ* by high-resolution X-ray reflectivity. Geochim Cosmochim Acta 64:1221-1228

Fenter P, Cheng L, Rihs S, Machesky M, Bedzyk MJ, Sturchio NC (2000b) Electrical double-layer structure at the rutile-water interface as observed *in situ* with small-period X-ray standing waves. J Colloid Interface Sci 225:154-165

Fischer CJ, Ithin R, Jones RG, Jackson GJ, Woodruff DP, Cowie BCC (1998) Non-dipolar photoemission effects in X-ray standing wave field determination of surface structure. J Phys Cond Matter 10:L623-L633

Golovchenko JA, Batterman BW, Brown WL (1974) Observation of internal X-ray wave fields during Bragg diffraction with an application to impurity lattice location. Phys Rev B 10:4239-4243

Golovchenko JA, Patel JR, Kaplan DR, Cowan PL, Bedzyk MJ (1982) Solution to the surface registration problem using X-ray standing waves. Phys Rev Lett 49:560-563

Goodenough JB (1998) Jahn-Teller phenomena in solids. Ann Rev Mater Sci 28:1-27

Hertel N, Materlik G, Zegenhagen J (1985) X-ray standing wave analysis of Bi implanted in Si(110). Z Phys B 58:199-204

Israelachvili JN (1991) Intermolecular and Surface Forces, 2nd ed. London: Academic Press.

Jones RG, Chan ASY, Roper MG, Skegg MP, Shuttleworth IG, Fisher CJ, Jackson GJ, Lee JJ, Woodruff DP, Singh NK, Cowie BCC (2002) X-ray standing waves at surfaces. J Phys-Condens Mat 14: 4059-4074

Kazimirov A, Kovalchuk M, Kohn V (1988) Study of the structure of individual sublattices in multicomponent $In_{0.5}Ga_{0.5}P/GaAs$ expitaxial films by X-ray standing waves. Sov Tech Phys Lett 14: 587-588

Kazimirov A, Haage T, Ortega L, Stierle A, Comin F, Zegenhagen J (1997) Excitation of a X-ray standing wave in a $SmBa_2Cu_3O_{7-\delta}$ thin film. Solid State Commun. 104: 347-350

Kazimirov A, Cao LX, Scherb G, Cheng L, Bedzyk MJ, Zegenhagen J (2000) X-ray standing wave analysis of the rear earth atomic positions in $RBa_2Cu_3O_{7-\delta}$ thin films. Solid State Comm 114:271-276

Kelly S, Newville M. Cheng L, Kemner K, Sutton S, Fenter P, Sturchio NC (2002) Uranyl incorporation in natural calcite. Environmental Science and Technology (in review)

Kendelewicz T, Liu P, Brown GE, Nelson EJ (1998a) Atomic geometry of the PbS(100) surface. Surf Sci 395:229-238

Kendelewicz T, Liu P, Brown GE, Nelson EJ (1998b) Interaction of sodium overlayers with the PbS(100) (galena) surface: Evidence for a Na-Pb exchange reaction. Surf Sci 411:10-22

Kim CY, Bedzyk MJ, Nelson EJ, Woicik JC, Berman LE (2002) Site -specific valence band photoemission study of α-Fe$_2$O$_3$, Phys Rev B 66:085115(1)-085115(4)

Kohra K, Matsushita T (1972) Some characteristics of dynamical diffraction at a Bragg angle of about $\pi/2$. Z Naturforsch A 27:484-487

Krolzig A, Materlik G, Swars M, Zegenhagen J (1984) A feedback control system for synchrotron radiation double crystal instruments. Nucl Instrum Method Phys Rev A 219:430-436

Lamble GM, Reeder RJ, Northrup PA (1997) Characterization of heavy metal incorporation in calcite by XAFS spectroscopy. J Phys IV 7:793-797

Laue M (1960) Roentgenstrahl-Interferenzen, Akademische-Verlagsgeslellschaft, Frankfurt

Lee TL, Qian Y, Lyman PF, Woicik JC, Pellegrino JG, Bedzyk MJ (1996) The use of X-ray standing waves and evanescent-wave emission to study buried strained-layer heterostructures. Physica B 221:437-440

Lee, TL (1999) High-Resolution Analysis of Adsorbate-Induced GaAs(001) Surface Structures and Strain in Buried III-V Semiconductor Heterolayers by X-ray Standing Waves. PhD Dissertation, Northwestern University, Evanston, Illinois

Liang Y, Lea AS, Baer DR, Engelhard MH (1996) Structure of the cleaved CaCO$_3$ (1014) surface in an aqueous environment. Surf Sci 351:172-182

Lin W, Lee TL, Lyman PF, Lee JJ, Bedzyk MJ, Marks TJ (1997) Atomic resolution X-ray standing wave microstructural characterization of NLO-active self-assembled chromophoric superlattices. J Am Chem Soc 119:2205-2211

Liu P, Kendelewicz T, Nelson EJ, Brown GE (1998) Reaction of water with MgO(100) surfaces: Part III. X-ray standing wave studies. Surf Sci 415:156-169

Ludwig KR, Simmons KR, Szabo BJ, Winograd IJ, Landwehr JM, Riggs AC, Hoffman RJ (1992) Mass-spectrometric thorium-230-uranium-234-uranium-238 dating of the Devils Hole calcite vein. Science 258:284-287

Materlik G, Schmaeh M, Zegenhagen J, Uelhoff W (1987) Structure determination of adsorbates on single crystal electrodes with X-ray standing waves. Ber Bunsen-Ges Phys Chem 91:292-296

Mucci A, Morse JW (1983) The incorporation of Mg^{2+} and Sr^{2+} into calcite overgrowths: Influences of growth rate and solution composition. Geochim Cosmochim Acta 47:217-233

Masel MI (1996) Principles of adsorption and reaction on solid surfaces. Wiley, New York

National Research Council Subcommittee on Selenium (1983) Selenium in Nutrition. National Academy Press, Washington

Ohnesorge F, Binnig G (1993) True atomic resolution by atomic force microscopy through repulsive and attractive forces. Science 260:1451-1456

Parratt LG (1954) Surface studies of solids by total reflection of X-rays. Phys Rev 95:359-369

Pingitore NE Jr, Eastman MP, Sandidge M, Oden K, Freiha B (1988) The coprecipitation of manganese(II) with calcite: an experimental study. Mar Chem 25:107-120

Qian Y, Sturchio NC, Chiarello RP, Lyman PF, Lee TL, Bedzyk MJ (1994) Lattice location of trace elements within minerals and at their surfaces with X-ray standing waves. Science 265:1555-1557

Reeder RJ, Lamble GM, Northrup PA (1999) XAFS study of the coordination and local relaxation around Co^{2+}, Zn^{2+}, Pb^{2+}, and Ba^{2+} trace elements in calcite. Am Min 84:1049-1060

Rodrigues WP (2000) Growth and characterization of Si/Ge heterostructures on Si(001) surface and Ge nano-dots on Si(001) surface. PhD Dissertation, Northwestern University, Evanston, Illinois

Schreiber F, Ritley KA, Vartanyants IA, Dosch H, Zegenhagen J, Cowie BCC (2001) Non-dipolar contributions in XPS detection of X-ray standing waves. Surf Sci 486:L519-L523

Stipp SL, Hochella MF (1991) Structure and bonding environments at the calcite surface as observed with X-ray photoelectron spectroscopy and low-energy electron diffraction. Geochim Cosmochim Acta 55:1723-1736

Stumm W (1992) Chemistry of the solid-water interface. Wiley, New York

Sturchio NC, Chiarello RP, Cheng L, Lyman PF, Yee D, Geissbuhler P, Bedzyk MJ, Sorensen LB, Liang Y, Baer DR, Qian Y, You H (1997) Lead sorption at the calcite-water interface: Synchrotron X-ray standing wave and reflectivity studies. Geochim Cosmochim Acta 61(2):251-263

Sturchio NC, Rihs S, Cheng L, Fenter P, Orlandini KA (2001) Uranyl adsorption on calcite observed with X-ray standing waves. Abstr Pap Am Chem S 222:23

Takahashi T, Kikuta S (1979) Variation of the yield of photoelectrons emitted from a silicon single crystal under the asymmetric diffraction condition of X-rays. J Phys Soc Japan 46:1608-1615

Tasker PW (1979) The stability of ionic crystal surfaces. J Phys C 12:4977-4984

Templeton AS, Trainor TP, Traina SJ, Spormann AM, Brown GE (2001) Pb(II) distributions at biofilm-metal oxide interfaces. Proc Nat Acad Sci 98:11897-11902

Trainor TP, Templeton AS, Brown GE, Parks GA (2002) Application of the long-period X-ray standing wave technique to the analysis of surface reactivity: Pb(II) sorption at α-Al_2O_3/aqueous solution interfaces in the presence and absence of Se(VI). Langmuir 18:5782-5791

Vartanyants IA, Kovalchuk MV (2001) Theory and applications of X-ray standing waves in real crystals. Rep Prog Phys 64:1009-1084

Wang J, Bedzyk MJ, Thomas LP, Caffrey M (1991) Structural studies of membranes and surface layers up to 1,000 Å thick using X-ray standing waves. Nature 354:377-380

Wang J, Bedzyk MJ, Caffrey M (1992) Resonance-enhanced X-rays in thin films: A structure probe for membranes and surface layers. Science 258:775-778

Wang J, Caffrey M, Bedzyk MJ, Penner TL (2001) Direct profiling and reversibility of ion distribution at a charged membrane/aqueous interface: An X-ray standing wave study. Langmuir 17:3671-3681

Woicik JC, Nelson EJ, Heskett D, Warner J, Berman LE, Karlin BA, Vartanyants IA, Hasan MZ, Kendelewicz T, Shen ZX, and Pianetta P (2001) X-ray standing wave investigations of valence electronic structure. Phys Rev B 64:125115

Woodruff DP (1998) Normal incidence X-ray standing wave determination of adsorbate structures. Prog Surf Sci 57:1-60

Zegenhagen J (1993) Surface structure determination with X-ray standing waves. Surf Sci Rep18:199-271

Zegenhagen J, Materlik G, Uelhoff W (1990) X-ray standing wave analysis of highly perfect Cu crystals and electrodeposited submonolayers of Cd and Tl on Cu surfaces. J X-ray Sci Tech 2:214-239

5

Grazing-incidence X-ray Absorption and Emission Spectroscopy

Glenn A. Waychunas

Geochemistry Department, Earth Sciences Division, MS 70-108B
E. O. Lawrence Berkeley National Laboratory
One Cyclotron Road
Berkeley, California, 94720, U.S.A.

INTRODUCTION

Grazing-incidence X-ray methods

Although the advantages of using grazing incidence (GI) techniques in X-ray scattering and spectroscopy have been known for a long time (Yoneda and Horiuchi 1971), the use of such geometry has only been developed fully with the advent of highly collimated X-ray sources, particularly synchrotrons. It is now possible to place the entire output of a synchrotron insertion device (wiggler or undulator) onto the surface of a flat crystal or analogous sample with minimal angular divergence at the critical angle for total external reflection. This allows the application of unprecedented X-ray intensity in a region confined within a few nm of the surface. Such intensity allows for measurement of the X-ray absorption and emission spectrum with sufficient counting statistics so that extremely small surface coverages of impurities, precipitates or sorbed species can be structurally probed. This is of value to environmental geochemistry, as the majority of toxic pollutants in the environment do not occupy mineral surfaces with anything like monolayer (ML) coverages. Hence the majority of X-ray experiments to date deal with unrealistically high surface loadings, relatively thick surface precipitates, or natural samples with high levels of contaminants. This is of special significance because small quantities of surface complexes may bind to defect or otherwise unrepresentative low density sites on a surface, and hence would not behave like species sorbed at higher densities. In addition, the use of single crystal samples allows the polarization of the synchrotron radiation to be used to probe specific directions with respect to the surface plane (for K and L_1 edges), hence potentially obtaining additional crystallographic information about oriented adsorbates, precipitates or epitaxial layers. One can select the type of substrate, the plane (hkl) of the substrate, and the orientation of that plane with respect to the propagation direction and electric vector polarization of the incident synchrotron radiation. In lower symmetry cases three different polarized XAS spectra can be collected, affording three times the information obtainable from a "bulk" experiment. This allows specification of precise bonding arrangements and geometry as pair correlation functions describing different directions need contain fewer absorber-neighbor contributions. This situation is best done with the use of substrates having extremely perfect surfaces, so that the experiment is probing only a well-described set of surface topologies, as opposed to some unknown set as occurs with ground (powdered) samples. It should be noted that XAS spectra with different polarization directions can also be collected by the preparation of powder samples with high degrees of texture (see chapter by Manceau et al. in this volume). Done mainly with layered minerals, this technique is usually limited to two polarization directions by rotational disorder of the particles in the substrate plane (i.e., one in-plane spectrum and one perpendicular to plane spectrum). Besides application to structural characterization of surface complexes, GI techniques enable measurement of surface chemistry at sensitivities approached by few other methodologies. Specific trace element concentrations of as low as 10^8 atoms cm^{-2}

1529-6466/00/0049-0005$05.00

can be sensed, even in the presence of many other types of more highly concentrated impurities by the use of GI X-ray fluorescence methods. Hence GI methods are ideal for the characterization of extremely low surface concentrations of geochemically important elements provided that samples with flat surfaces are available. The gist of the situation is that GI-XAS analysis can be extremely useful in determining surface complexation toplogy in cases where bulk analysis cannot provide sufficient sensitivity or is compromised by inadequate characterization of particle surface structure. Similarly, GI-XRF methods can measure surface composition quantitatively at levels well beyond most surface analysis methods.

In this chapter the nuts and bolts of doing grazing-incidence X-ray spectroscopy are described, along with examples from the recent literature, and experiments that are being planned to further extend GI-based capabilities into important geochemical systems. Emphasis is placed on the implementation details and structural information obtained from the technique, and the latter is compared with information obtainable from other methodologies to yield a perspective on the overall complementarity of surface-structure methods. Little attention will been given herein to any detailed discussions of the fundamental aspects of X-ray absorption spectroscopy, as these have been presented previously in RIMG volumes and elsewhere (Brown and Doniach 1980; Eisenberger and Lengeler 1980; Lee et al. 1981; Hayes and Boyce 1982; Stern and Heald 1983; Brown et al. 1988; Als-Nielsen and McMorrow 2001). Similarly, there are excellent reviews available on grazing-incidence methods and reflectivity theory (Dosch 1992; Klockenkämper 1997; Stoev and Sakurai 1999; van der Lee 2000; Als-Nielson and McMorrow 2001; Yashiro et al. 2001) so that abstruse theory has been avoided here. Instead the emphasis is on the specific aspects of GI experiments that differ from "bulk" sample investigations. A set of definitions is listed below to cover most of the important and commonplace terminology and concepts connected with GI-X-ray spectroscopy.

DEFINITIONS

Absorption coefficients define the X-ray attenuation effect from a given solids. We use both mass absorption coefficients, μ_m, and linear absorption coefficients, μ, with these being related by $\mu = \mu_m \rho\, x$, where ρ is the density of a given solid and x is the X-ray path length. The mass absorption for any material can be calculated from the elemental mass absorption coefficients by the relation $\mu_m = \Sigma f_i \mu_i / \rho$, where the f_i and the μ_i are the weight fraction and linear absorption coefficients, respectively, for each element in the material.

Anharmonicity-typically refers to vibrations that lead to a shift in the center of mass of the vibrating system. However, for X-ray scattering experiments we often refer to anharmonic pair correlation functions (PCF) as well as anharmonic vibrations. Such an anharmonic PCF is asymmetric with mean and median having different values. Anharmonicity requires that a PCF must be described by an increased number of moments of the distribution, and if these can not all be extracted from an experiment (e.g., EXAFS) then the true distribution is not obtained. There is a general impression that anharmonic effects in EXAFS can usually be evaluated (i.e., "corrected") by suitable analysis, though this is untrue. However modeling may suggest permissible options when corrections are impossible.

Attenuation length is that distance where X-ray intensity is reduced by a factor of 1/e, or by 63%.

Coherence lengths for X-rays are defined as the transverse and linear coherence length, or ε_{trans} and ε_{long}, respectively. These refer to the distance over which a slightly diverging beam will produce a phase change of 2π in the direction transverse to propagation, or to the distance over which a small variation in wavelength will produce a similar phase change in the longitudinal (along propagation) direction. For example, given a wavelength variation of 1 part in 10^4, typical of synchrotron double crystal monochromators, the longitudinal coherence is $\lambda^2/\Delta\lambda = \lambda^2/(0.0001\lambda) = 10,000\ \lambda$, about a micron. The transverse coherence of a monochromatic X-ray beam is $\lambda R/\Delta d$, where R is the sample to source distance, and d is the source size in the appropriate direction. Hence a source size of $d = 1$ mm at $R = 30$ meters has a transverse coherence length of $3\times10^5\lambda$ or tens of microns.

Compton scattering is an inelastic X-ray scattering process where momentum is lost to the scattering electron. This means that the energy of the scattered X-ray is shifted to lower energy, and will not be involved in constructive or destructive interference with other scattered X-rays of the original energy. This means that Compton scattered X-rays are incoherent, and must be treated differently than other coherent and elastically scattered X-rays.

The Critical Angle for total external reflection of X-rays is given by $\alpha_c = (2\delta)^{1/2} = (4\pi\rho r_0)^{1/2}k^{-1}$, where r_0 is the electron scattering length, ρ is the electron density in the material (electrons per cubic Å) and k is the wavevector (see below). It is typically on the order of mrad.

The Cumulant expansion is a series method for approximating the moments of a distribution. It is often used to fit a non-Gaussian or anharmonic interatomic distance distribution in EXAFS spectra.

The Debye-Waller factor is an amplitude term in any scattering experiment that takes account of the movements of the scatterers about their average positions. This results in attenuation of the scattering which increases with scattering vector. For EXAFS analysis the appropriate Debye-Waller factor takes account of variations in the absorber-scatterer distance, and thus depends on how much the motion of this pair of atoms is correlated.

Dynamical theory considers all of the X-ray interactions when an X-ray beam enters a crystal. This includes all levels of scattered and rescattered beams, and can be thought of in many formulations as a field or energy-propagation theory. For strongly scattering materials or strongly interacting particles (e.g., electrons) dynamical theory must be used to obtain realistic models of scattering.

Elastic scattering refers to X-ray scattering where there is no change in energy of the scattered X-ray. It may be coherent or incoherent depending on the conditions for interference and the coherence lengths of incident radiation.

The Evanescent wave is the X-ray field that propagates into the solid surface in a reflection geometry near the critical incidence angle. It is the exponentially damped transmitted beam.

EXAFS-Extended X-ray absorption fine structure spectroscopy; fine structure above an absorption edge that contains information about local interatomic structure.

Grazing-incidence is taken to mean X-ray incidence at or below the critical angle for total external reflection. For 7 KeV X-rays on an Al_2O_3 surface, this value is ca. 0.2 degrees.

Kinematical theory is also called "two-beam" theory, as it includes consideration of only the incident and scattered X-ray beams. This is a very good approximation for thin crystals or for weakly scattering particles (e.g., most X-ray photons and neutrons), as rescattering of the initially scattered beam will be very weak. It is useful for calculating

X-ray reflectivity away from the critical angle where reflectivity is relatively weak and refraction effects are negligible.

Optical theory includes the Fresnel equations that describe the behavior of light waves at an interface between media of differing refractive index, and other relationships between the atomic scattering factors and absorption cross sections. It can be applied to X-ray reflectivity and refraction in the critical angle region.

Polarized XAS takes advantage of the direction of the electric vector in synchrotron radiation to probe specific directions in a sample (Fig. 1). For K- and L_1 edge XAS the absorption is effectively a s→p type transition with the analogous polarization

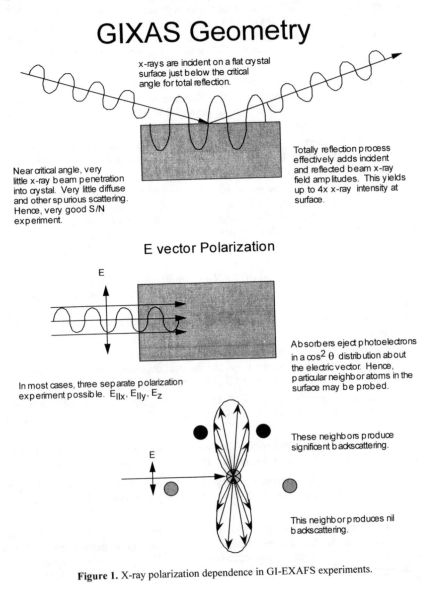

GIXAS Geometry

x-rays are incident on a flat crystal surface just below the critical angle for total reflection.

Near critical angle, very little x-ray beam penetration into crystal. Very little diffuse and other spurious scattering. Hence, very good S/N experiment.

Totally reflection process effectively adds incident and reflected beam x-ray field amplitudes. This yields up to 4x x-ray intensity at surface.

E vector Polarization

E

In most cases, three separate polarization experiment possible. E_{IIx}, E_{IIy}, E_z.

Absorbers eject photoelectrons in a $\cos^2 \theta$ distribution about the electric vector. Hence, particular neighbor atoms in the surface may be probed.

These neighbors produce significant backscattering.

E

This neighbor produces nil backscattering.

Figure 1. X-ray polarization dependence in GI-EXAFS experiments.

dependence. Hence three orthogonal directions of independent measurements are possible if the sample symmetry permits this (i.e., has orthorhombic or lower symmetry). For L_2 and L_3 edge XAS the absorption is effectively a p→d transition where there can be only limited differences due to sample orientation (Brouder 1990).

REFLEXAFS refers to the same experimental geometry as grazing-incidence-excited EXAFS, but indicates use of the reflected beam instead of a fluorescence signal to extract the absorption spectrum. EXAFS analyses via the reflected beam have been reported with results similar to ordinary transmission experiments (Barchewitz et al. 1978; Fox and Gurman 1980; Martens and Rabe 1980; Bosio et al. 1984; Alonso-Vante et al. 2000). What is called REFLEXAFS by some authors (e.g., England et al. 1999a;1999b) is identical to GI-EXAFS. Ostensibly, reflection XAS may afford high surface sensitivity without self absorption effects, though other corrections may be necessary (Heald et al. 1988).

The scattering vector is a measure of the momentum transfer in the X-ray scattering process. It is usually defined as $Q = 4\pi / \lambda \sin\theta$, where θ is the scattering angle. Also used is $k = 2\rho/l$ (see critical angle above). For EXAFS the analogous vector is defined as $k = (\pi^2 m / 2h^2 (E - E_0))^{1/2}$, where h is Planck's constant, m is the photon mass, E is the energy of the photoelectric wave, and E_0 is the energy of the edge threshold.

Standing waves refer to interacting waves that coherently interfere to annihilate all traveling wave components and produce one fixed wavefield. In X-ray standing waves the interfering beams are the incident beam and scattered beams, where the incidence is either at or below the critical angle for specular (mirror) reflection, or near a Bragg condition for Bragg scattering-based reflection.

Surface roughness is the deviation of a surface from perfect flatness. It is often defined as a RMS (root mean square) roughness as this parameter is obtained directly from reflectivity measurements.

Surface site density is the number of reactive sites per unit surface area. This differs from surface sorption density as that refers to the number of occupied surface sites per unit surface area.

Surface steps and terraces are aspects of a smooth surface. Steps represent new surface layer origination, and terraces are the areas of the layers between down and up steps to other layers.

TRXRF, TXRF or Total Reflection X-Ray Fluorescence is a method wherein the near surfaces of very flat samples can be analyzed. The incident beam is directed below the critical angle, and hence beam penetration is minimized. Similarly, noise levels are very much reduced. Easily done when setting up a GIXAS experiment.

XANES or X-ray Near Edge Structure is the fine structure very close to an X-ray absorption edge that is mainly dominated by multiple scattering effects. Also called NEXAFS for Near Edge X-ray Absorption Fine Structure.

XAS refers to X-ray Absorption Spectroscopy, which includes EXAFS, XANES, GIXAS and related methods.

The X-ray refractive index, n, is defined the same way as for visible light photons with Snell's law. However the index is slightly less than unity. This paradoxical fact, which suggests that X-rays might have to travel faster than light in vacuum when going into the denser medium during the refraction process (as the velocity in the medium must be c/n), holds because it is the velocity of the phase that effectively travels at this speed.

X-ray fluorescence is emission from an excited atom that has had an inner electron ejected by X-ray absorption. The fluorescence is due to changes in energy of electrons dropping into the inner core hole. This sets up a cascade of fluorescence as each hole is successively filled.

EARLY STUDIES AND QUANTITATIVE ISSUES

Heald et al. (1988) provided the first thorough analysis of GI EXAFS measurements performed on thin metallic films. Besides identifying quantitative issues in data analysis of either the reflected beam or fluorescent emission signal, these authors also demonstrated the utility of GI measurements for the study of "buried" interfaces—as long as the buried layer is supported on a still denser substrate material. Another observation, earlier demonstrated by Heald et al. (1984), was sub-ML sensitivity through analysis of the fluorescence emission. Two problems in data analysis are changes in reflectivity as a function of energy, especially as the absorption edge is crossed, and anomalous dispersion effects on the phase and amplitude of the EXAFS.

In the first case, changes in reflectivity with incidence angle lead to extreme sensitivity to beam position variations. This can induce far more noise than would be expected from counting statistics. Hence beam position stability is a key issue when performing GI experiments, particularly at smaller incidence angles. Related is the issue of reflectivity as a function of energy. If uncorrected, EXAFS scans over several hundred eV can show background fluctuations due only to reflectivity changes that affect the beam penetration and fluorescence signal. This problem is remedied by continuous control of the incidence angle to maintain constant beam penetration during energy scanning, but this can be a difficult proposition with very small incidence angles and thin beams (e.g., 50 μm thick). For environmental experiments when the sample is a sub-ML adsorbate on a substrate, changes in reflectivity mainly increase the elastic scattered background from the substrate. This affects the EXAFS background very little unless there is "leakage" of the scatter peak counts into the data channels in the fluorescence detector electronics. A larger problem is the change in reflectivity just at the absorption edge. The case for the Mn K edge (in MnOOH) is shown in Figure 2, where the 10 eV change in energy over the edge produces a change of penetration depth of 54 to 78 Å at 0.3 degree incident angle, and a change of 70 to 120 Å at 0.32 degrees. Alternatively, to hold the penetration depth fixed at 50 Å, the incidence angle must be changed from 0.29 to 0.26 degrees.

Correction of anomalous dispersion effects is probably unnecessary at small incidence angles and for very dilute (i.e., sub-ML samples) such as those usually studied for environmental systems. However most investigations do not consider these effects in their data analysis, assuming them to be negligible. Analysis by Heald et al. (1988) shows that phase effects should be small, affecting interatomic distances by only 0.03 Å in a worst case scenario. A similar result was obtained by Martens and Rabe (1980). But amplitude effects can be significant, leading to corrections as large as a factor of 2.5. Experimentally this effect can be approached by collecting data at different incidence angles and developing a "working curve" for amplitude correction, or the iterative procedure of Heald et al. (1988) can be employed.

GI X-RAY ABSORPTION SPECTROSCOPY

Geometric considerations

To get a rough idea of how much a grazing incidence experiment differs from a standard XAS (bulk) experiment we need to consider the geometric factors involved. The

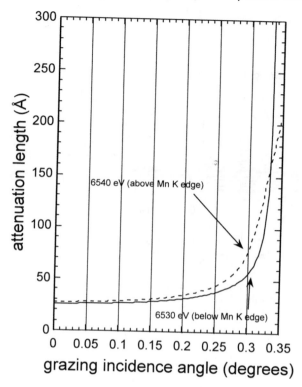

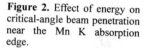

Figure 2. Effect of energy on critical-angle beam penetration near the Mn K absorption edge.

basic arrangement of a typical experiment is shown in Figure 3. Various collimating slits shape the beam so that it will not hit the front edge of the single crystal sample substrate when the substrate is positioned near the critical angle for total reflection. The substrate itself is held in a goniometer and can be moved with high precision along the incidence angle, θ, and azimuthal angle, ϕ. An adjustment of the sample χ angle might be enabled to remove any sample tilt in this direction. The detector is placed near the sample at 90 degrees to the incident beam direction. In the SSRL grazing-incidence apparatus the entire sample, goniometer and slit assembly can be rotated on another χ axis once the sample is optimally positioned, to vary the angle of substrate and X-ray polarization direction. Figure 4 shows the geometry of radiation exiting from a sample in the GI arrangement.

A straightforward analysis via X-ray line tracing requires that the emittance of the source is known, i.e., the size and shape of the emitting electron field, and the divergence of the beam of X-ray photons produced by this field. Also required is a knowledge of the polarization of the emitted X-rays, which will differ as a function of angle from the synchrotron ring plane and also within this plane. These factors will differ for insertion devices in the same synchrotron, and after exiting the source the X-ray field will be affected by downstream optics such as mirrors and monochromating crystals. For the APS and other synchrotron sources we show some approximate estimates of the effect of different source parameters on the type of samples that can be used for grazing incidence work in Figure 5. The key issue at the outset is the amount of beam divergence in the vertical plane that results in a range of X-ray incidence angles on the sample. If this spread is large compared to the critical angle the GI experiment is impossible, while if it

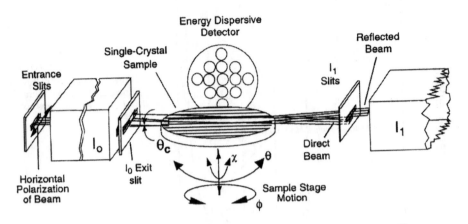

Figure 3. GI experimental arrangement used at SSRL (modified from Towle et al. 1999). The pattern on the detector face indicates the position of individual detector elements. In more recent modifications of this geometry the slits are motorized and can be continuously varied during energy scans to track the reflected beam while blocking the transmitted beam.

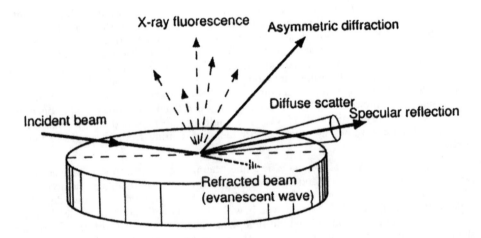

Figure 4. Beams and processes in grazing-incidence X-ray excitation (modified from Stoev and Sakurai 1999).

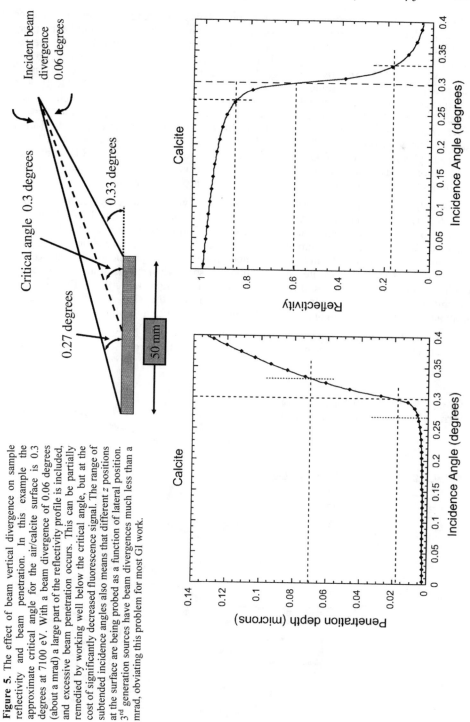

Figure 5. The effect of beam vertical divergence on sample reflectivity and beam penetration. In this example the approximate critical angle for the air/calcite surface is 0.3 degrees at 7100 eV. With a beam divergence of 0.06 degrees (about a mrad) a large part of the reflectivity profile is included, and excessive beam penetration occurs. This can be partially remedied by working well below the critical angle, but at the cost of significantly decreased fluorescence signal. The range of subtended incidence angles also means that different z positions at the surface are being probed as a function of lateral position. 3rd generation sources have beam divergences much less than a mrad, obviating this problem for most GI work.

is of the same order the experiment is impractical. This is so because a range of incidence angles produces significantly greater beam penetration into the sample, and hence more scattering from the bulk which contributes to background. Further, it is difficult then to assign the critical angle accurately, and one will have trouble optimizing the experiment, and obtaining high effective signals. Thus the vertical divergence must be small compared to the critical angle. For many synchrotron sources (e.g., SPEAR II) this divergence is about a mrad or 0.057 degrees on the bending magnet lines, which is about a third of the 7 KeV critical angle on quartz. At the APS the bending magnet vertical divergence is on the order of 10 μrad, or about 0.3% of that same critical angle. Hence much "cleaner" work with substantially better signal/noise can be done at the APS for most samples without use of additional beam collimation. In practice, at SSRL additional collimation is used to achieve reduced effective divergence, thought this is always consumptive of photons, weakening the intensity of the incident beam.

Another consideration is the beam cross section. Typical beam cross section at SSRL (SPEAR II) is on the order of several to tens of mm, while that on typical APS lines is in the tens of microns. For practical sample size, e.g., 2 cm, this means that the beam size cannot be larger than about 40 μm at a critical angle of 2 mrad. Any larger part of the beam will miss the sample (Fig. 6) and probably incur dramatic noise production. In practice this greatly reduces the utility of the GI experiment as so much incident beam is rejected by collimation (e.g., with SPEAR II with a 2 mm vertical height beam, 98% of the beam is discarded). Thus the APS beams are ideal for samples of this size, and all of the output from an insertion device can in principle be directed on a GI sample of a few mm.

Polarization dependence on beam position is generally a small concern, as the polarization is near 100% in the plane of the ring, and for larger source sizes only the

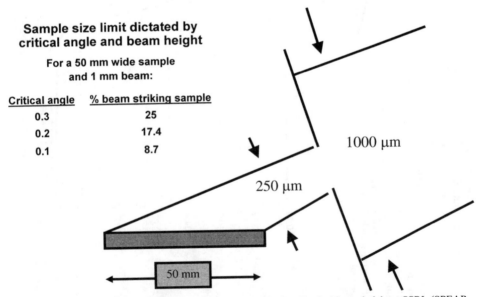

Figure 6. Effect of beam size and critical angle on sample size. Typical beam height at SSRL (SPEAR II) is 1 mm or so. In general this requires discarding 90% or more of the bam to achieve useful results. The situation is made worse due to high beam divergence which forces setup further below the critical angle. In the case of 3rd generation sources, practically all of the beam can be used on sample widths down to a few mm.

central portion would be used (or should be used). However if a focusing monochromator or mirror is being used this may mix more highly polarized photons from the center of the source beam, with less polarized photons toward the source edges. This effect may be negligible if the monochromator increases degree of polarization appreciably.

These issues lead to an approximate statement of ideal conditions for a GI experiment: vertical beam divergence less than 10% of the critical angle, and beam size 100 μm or less in the vertical direction, with near 100% polarization. Focusing should be, if possible, arranged to preserve the separate identities of vertical and horizontal divergence. The latter is accomplished by using bent flat mirrors, either separately of in Kirkpatrick-Baez pairs.

To get an idea of the enhancements produced by GI methods, let us first consider a rough comparison of bulk fluorescence XAS versus GI-XAS sensitivity for a sorption sample. To do this we can adapt the analysis originally performed by Stern and Heald (1979) and by Warburton (1986). In this context we can take into account the level of noise in the counting system, background count rates, and other noise sources to yield an effective signal. For a bulk sample we can use a 0.25 mm thick paste of quartz with sorbed Fe equivalent to a μM concentration or about 0.5 μg of Fe. The attenuation lengths of quartz and water at the Fe K edge are about 80 and 650 μm, respectively, so that the paste will have a net attenuation length of about 200 μm, and a transmission of about 30%. For the GI sample we will use a one μm film of water on a quartz substrate with surface Fe concentration equivalent to the bulk sample. If our substrate takes up the full width of a beam at SPEAR II, or about 20 mm, and is 50 mm long, then the total illuminated surface area is 1000 mm^2, which contains about 10^{16} sorption sites. One ML coverage of this surface with Fe would then amount to about 0.5 μg of Fe, so the bulk and surface samples have similar Fe concentrations in the beam. In the GI case the critical angle is about 0.25 degrees, so that the beam path though the solution film is about 230 μm, with a transmission of 79%. All of the synchrotron beam is used in the bulk case, resulting in fluorescence from all of the Fe, fluorescence from the O and Si, which will not effectively penetrate the detector windows, and elastic and inelastic scatter from the quartz and water (the Fe scattering component will be negligible). From Stern and Heald (1979) the effective signal count rate, N_{eff}, is defined as $N_{eff} = N_s /(1 + N_b / N_s)$, where N_s is the signal count rate and N_b the background count rate. A rough value for the for N_b/N_s in our bulk case is 40, based on Warburton's (1986) estimate of 20 for a 0.003 M Fe aqueous solution, and the transmission and average density of the quartz-water sample. Hence the effective signal is 1/41 the measured signal. In the GI case there is effectively no scattering from the quartz substrate, and much less water scattering. We can approximate the water scattering by assuming that the scattering is proportional to the irradiated volume weighted by the absorption profile. This is about 20 times smaller for the GI sample or $N_b/N_s = 1$. Hence the effective signal in this case is 1/2 the measured signal.

Next we consider the effect of incident X-ray intensification because of the overlap of incident and totally reflected beams. For a highly perfect substrate we will obtain about a 4 fold intensity increase. Thus assuming that each sample receives the same number of incident X-ray photons, the GI sample will have an effective count rate at the detector equal to 82 times that of the bulk sample. Waychunas and Brown (1994) made a similar comparison for several types of X-ray detectors, sample configurations and fluorescence flux rates at the detector. At that time detector counting speed dictated what could be done with bulk samples as no fluorescence detector array could count fast enough to handle unattenuated fluorescence plus background signals at high flux beamlines. Though still a problem, electronic advances have allowed count rate operation up to the several million/channel level, with large arrays being able to handle rates approaching 10^8 cps. This allows bulk detection in the hundredth mM concentration regime, and with long collection times approaching the µM regime.

However in the GI case, there is no count rate limitation, and the sensitivity extends easily into the tenth µM regime, and usefully into the hundredth µM regime. This translates in our case to 1% ML surface coverages and below. Of course the tradeoff here is in the numbers of photons deliverable to the sample surface. At SPEAR II this would only be a few percent of a full beam, thus largely mitigating the advantages of the GI geometry. At SPEAR III and at the APS all of the beam from insertion devices can be placed onto suitably sized crystals, so that the comparison would be strictly valid.

This analysis has utilized roughly equivalent amounts of sample species in bulk and GI mode, but bulk experiments can also be increased in "surface sensitivity" by using progressively smaller particle sizes. In this way for a given surface loading the amount of sample species can be increased subject to overall sample absorption. The sensitivity obtained in this way overlooks variations in particle surfaces as size decreases, and is highly dependent on X-ray energy. For example, arsenate sorbed on 3 nm ferrihydrite particles can be studied at the bulk composition of As/Fe=0.005 or so (Waychunas et al. 1993) using an 8 pole wiggler line at SSRL (SPEAR II). This works out to a surface concentration of about 1% of a ML assuming complete separation of particles, and a few percent of a ML with typical particle-particle connections. In the case of Fe^{3+} sorbed on silica (Waychunas et al. 1999) the limit of sensitivity for the bulk experiment was about 3% of a ML. However most mineral sorbates are not generally prepared as nanometer-sized particles, as at this size standard characterization methods for phase identity are difficult (e.g., XRD), and thus higher sorption densities are required to acquire workable XAS spectra.

Surface roughness and related effects

A more detailed examination of GI geometry emphasizes the effect of surface asperity on both the detected signal and noise contributions. This is an important consideration as either polished synthetic materials, or natural surfaces will deviate considerably from the ideal atomically flat interface. Effectively, increased surface roughness creates a range of local incidence angles. This produces a larger integrated volume of scattering in the substrate, and thus higher elastic and inelastic contributions to the noise. It also reduces signal due to variations in X-ray amplitude at varying incidence angles. The strong changes in beam penetration, X-ray intensity at the surface, and X-ray reflectivity as a function of grazing incidence angle are shown in Figure 7 as deduced from optical theory. Only a very small change in the grazing incidence angle can change the penetration depth by a factor of ten. Further, the evanescent intensity, or X-ray intensity of the local surface field that decays exponentially into the surface, varies dramatically with incidence angle. A detailed treatment of surface roughness using both kinematical and statistical distributions of surface features has been presented by Als-Nielsen and McMorrow (2001) and by Dosch (1992). An overview of types of surface roughness and their characterization by X-ray and neutron scattering has been given by Sinha (1996). The basic considerations are shown in Figure 8, where the effects of scattering from an uneven surface are coupled into the concept of the coherence lengths of the incident X-ray beam. The X-ray illuminated area that is considered on the surface is defined by the vertical and longitudinal coherence lengths. Hence in Figure 8 beams 0 and 4 will constructively interfere, while beams 1 and 2 will be out of phase with 0 and 4, and with each other, leading to a net reduction in scattered intensity. Beams 3 and 5 are outside the coherence space for the surface area that is illuminated, and thus these beams are incoherent with the other scattered beams and will not interfere with them. With these

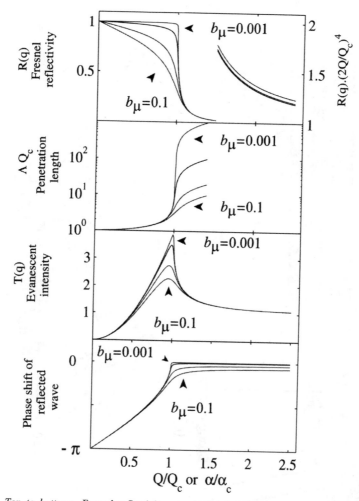

Figure 7. *Top to bottom:* Fresnel reflectivity, penetration length, evanescent wave intensity and reflected beam plane calculations for a model system. The parameter $b_\mu = 2\mu k / Q_c^2$, where k is the scattering wavevector, Q_c is the critical angle wavevector, and μ is the linear absorption coefficient. [From Elements of Modern X-ray Physics by Als-Neilson and DesMorrow, with permission from the editors at John Wiley and Sons.]

definitions, the surface length that is probed with surface X-ray scattering is thus $\varepsilon_{trans} \cot(\alpha)$, where α is the incidence angle. If the surface scattering from this region is formulated as the kinematical average, then one arrives at a particularly simple result if the roughness at the surface can be described by a Gaussian distribution (Dosch 1992), namely that the Fresnel scattered intensity is modified by a reduction factor of the form $e^{-Q^2\sigma^2}$, the Debye-Waller factor, where σ^2 is the root mean square roughness, and Q is the scattering wavevector. Typical values for σ^2 are 2 Å^2 for an "epi" polished substrate, to 20 Å^2 for a standard polished substrate, and 100 Å^2 for surfaces subject to some etching or chemical weathering. For a GI experiment near the critical angle Q is about 0.02, and thus the attentuation in the Fresnel reflectivity for 2 Å^2 to 20 Å^2 rms roughness is

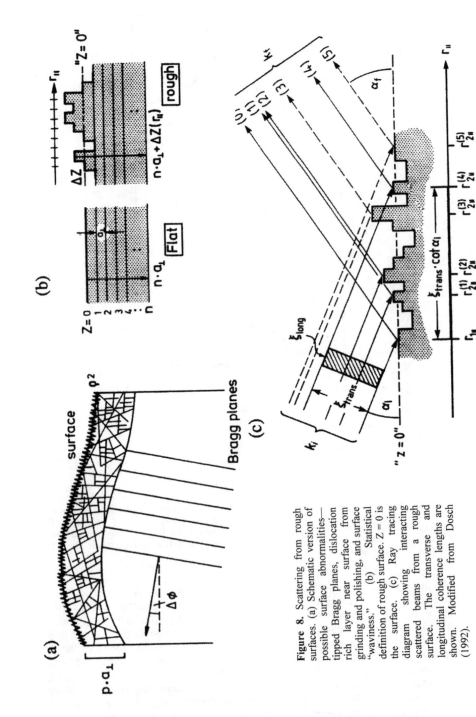

Figure 8. Scattering from rough surfaces. (a) Schematic version of possible surface abnormalities—tipped Bragg planes, dislocation rich layer near surface from grinding and polishing, and surface "waviness." (b) Statistical definition of rough surface. $Z = 0$ is the surface. (c) Ray tracing diagram showing interacting scattered beams from a rough surface. The transverse and longitudinal coherence lengths are shown. Modified from Dosch (1992).

insignificant, and is about 4% at 100 $Å^2$ roughness. Note that at higher Q, i.e., for a reflectivity experiment, the decrease in reflected intensity with roughness will be dramatic. For $Q = 0.25$, 2 $Å^2$ and 20 $Å^2$ roughness will reduce reflectivity by 12% and 71%, respectively. This must be considered in any depth profiling experiments.

The effect of surface asperity on elastic and inelastic scatter depends on the effective scattering volume in the substrate and the nature of the substrate. An amorphous substrate will have strong diffuse scattering proportional to the illuminated volume, and depending on the angle of observation. This is reduced by using a 90 degree scattering angle, but as the sample, detector and beam all have finite size, the detection is actually done over a range of angles such as 80-100 degrees. A single crystal substrate will not produce Bragg scattering to any degree, unless there is considerable beam penetration. In practice this is mainly only a problem for warped crystals. However there will be an inelastic scattering component, and a diffuse scattering component from the illuminated volume. As crystalline substrates differ in the amount of diffuse and inelastic scattering produced, this needs to be analyzed prior to any GI experiment.

When a surface has any roughness, half of the surface will lie above the mean level and shadow the rest of the surface from the incident beam. This higher material will also receive beam at large incidence angles, allowing beam penetration. For work below the critical angle, which is generally where typical GI XAS experiments are performed, the depth of penetration of the evanescent wave is about 25 Å. Hence roughness less than this will not affect the illuminated volume contributing to the noise. But for larger surface roughness, the illuminated volume will increase approximately proportional to tan α. This effect is still small for most reflectivity and GI experiments, but the shadowing effect of surface roughness may not be. One can see from Figure 8 that half or more of the surface will be affected by shadowed of the incident beam near the critical angle if the roughness equals or exceeds the evanescent wave penetration depth. For beams striking rough projections at large angles the degree of shadowing will depend on the X-ray path length and attendant absorption through the raised projections. For sufficiently rough surfaces, or surfaces with particular types of corrugated roughness, the shadowing effect can be severe. Hence some fraction of the beam does not reach the surface, and the signal is attenuated proportionally.

A more obvious but perhaps underappreciated problem with surface roughness is the existence of "defect" sites on a surface, i.e., sites that would not be exposed on a perfectly smooth surface. This type of defect is separate from classical defects like stacking faults, subgrain boundaries and dislocations, and is due just to non-uniform expression of the substrate structure in an uneven surface (Fig. 9) such as could occur with the local development of "vicinal" faces. As surface characterization methods are generally poor except in the case of a small suite of oxides and silicates, this effect has probably not been fully considered to date. For example, it is possible to imagine a low roughness (hkl) surface that is entirely terminated by small faces with other (hkl) orientations, so that the exposed surface functional groups differ both in density and orientation from what is expected.

Fresnel analysis of total external reflection and sensitivity factors

It is useful to calculate the depth of penetration of the evanescent wave as a function of incidence angle and X-ray wavelength. This has been done for quartz, hematite and calcite, using the X-ray optics calculator on the Center for X-ray Optics web page: *http://www-cxro.lbl.gov/optical-constants/* Taking hematite as an example, it is clear from Figure 10 that there is a dramatic effect from increased absorption. At 7100 eV, below the Fe K edge, the penetration function resembles those of other solids. However,

vicinal surfaces

have a surface orientation adjacent to a low-index surface, miscut angle α
example: (1121) is vicinal to (001), miscut in (111) direction
step energy $\beta = a(d\gamma/d\alpha)|_{\alpha=0}$
step-step-interaction--> Wulff plot has cusps for all vicinal surfaces with
the ratios of Miller indices being rational numbers
facetting: vicinal surface decays into terraces (surface energy $\gamma_{||}$) and
micro-facets (inclination β, surface energy γ), if
$\gamma(\alpha) > [\gamma_0 \sin(\beta\alpha) + \gamma_i \sin\alpha]\sin\beta$

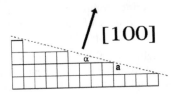

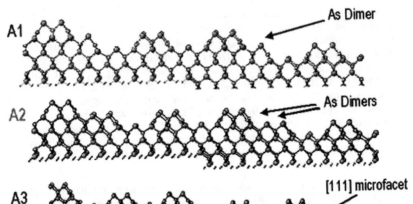

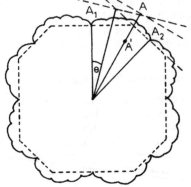

Figure 9. Development of vicinal faces on surfaces as a pathway to a more stable surface. *Top:* thermodynamic description. *Middle (A1-A3):* examples of GaAs vicinal surfaces. *Bottom left:* Wulff construction showing positions of vicinal faces. Adapted from Waychunas (2001).

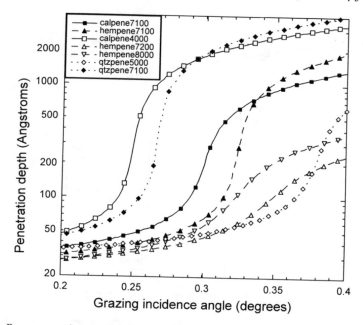

Figure 10. Beam penetration as a function of incidence angle and X-ray energy for quartz (qtzpene), hematite (hempene) and calcite (calpene).

at 7200 eV, some 88 eV above the Fe K edge, the penetration function has changed dramatically. This change allows easier control of penetration depth for depth profile experiments than below the edge. At 8000 eV, the penetration function is moving back toward the "steeper" function. The same is true for calcite between 4000 eV (below the Ca K edge) and 7100 eV. Note how little change is seen in the quartz penetration function between 5000 eV and 7100 eV, except for the angle shift, as both energies are well above the highest energy absorption edge (Si K). The effect of beam energy on the critical angle and reflectivity profile (for 0 Å surface roughness) is shown for hematite in Figure 11. Here again there is a dramatic change in the functional form when we pass through the absorption edge. Certainly if one wishes to explore the reflectivity curve in detail the use of longer wavelengths is indicated. In the case of calcite, the effect of surface roughness at a single energy is shown in Figure 12. For increasing surface roughness the higher angle part of the function sharpens, but the lower angle part softens, i.e., there is a more continuous change in reflectivity with angle. Consideration of all of these features suggests that GI experiments are best done at the lowest possible Q value. As Q decreases the penetration depth decreases slightly, and the effects of surface roughness are decreased. Perhaps most importantly, for depth profiling analysis longer wavelengths afford greater control of penetration depth due to the decreased sensitivity of penetration depth on incidence angle.

SAMPLE CELLS, SURFACE LIQUIDS AND COVER MEMBRANES

The largest problems in using GI sample cells that maintain a thin water layer over a surface is the attenuation of the incident beam, and the inelastic and elastic scatter due to the water and window films. In general, films of 1-2 μm of water will be present between

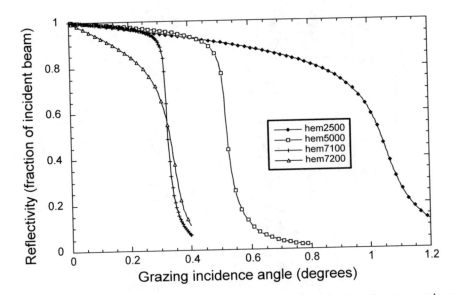

Figure 11. Reflectivity of hematite surfaces for different X-ray energies. All surfaces have rms roughness = 0.0 Å.

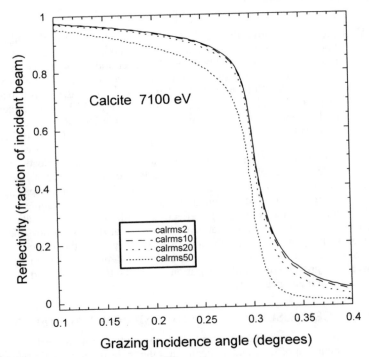

Figure 12. Effect of surface roughness on the reflectivity of calcite at 7100 eV. Lines refer to rms roughness of 2, 10, 20 and 50 Å.

sample surface and membrane, and this can be verified by X-ray reflectivity studies either off-line or on the actual sample at the synchrotron. Further, membranes of thickness less than about 6 μm (= 0.25 mil) are difficult to obtain (although kapton with a thickness of 2500 Å has been prepared), and this makes it difficult to work at K edges below that of Fe. At a critical angle of 0.20 degrees, the path length of X-rays through a 1 μm water layer will be a about 300 μm, which can create significant absorption. This is not a serious problem with X-rays of energies beyond about 7 KeV. More important is the effect of a 6 μm kapton, mylar or even polyproplyene cover film. At 0.20 degrees, the path length will be about 2 mm, equivalent to a beam attenuation of 85% at 7 KeV and 50% at 10 KeV. Polypropylene films have a lower density (of about 0.90) and the absorption at the same energies is about 40% and 20%, respectively. Recent work by the author utilizes 2 μm mylar films that enable successful work at the Fe K edge and Mn K edge, but at the cost of relative fragility.

Some considerations that must be addressed when using GI cells are illustrated in Figures 13-17. Figure 13 shows the form of idealized reflectivity functions for the three materials in a GI experiment using single crystal quartz surfaces. The "plast" refers to a polymer film membrane with density about 1.4 (like polyimide or kapton) in this plot. These curves were calculated for ideally flat interfaces, but more realistic profiles would be somewhat softened by surface roughness effects. The key consideration is that the setting of the grazing incidence angle must consider the reflectivity of each layer and interface present. Although the thickness of a membrane cover film can be measured, the surface roughness of this film is rarely known with any certainty. Further, the thickness of the underlying water film is difficult to control, and subject to evaporation through any thin film membrane. These factors are best evaluated by performing *in situ* reflectivity measurements periodically during GI-XAS work. Alternatively, samples may be studied by dosing them with water solutions starting at UHV conditions in a reaction chamber (e.g., Shirai et al. 1995). In this case the thickness of the water layer is controlled by temperature and water vapor pressure, and no membrane film comes into play.

To show a few of the experimental considerations with the film-covered cell Figure 14 shows the effect of air/film interface roughness on the overall cell reflectivity function at 7100 eV, for a sample configuration of a 6 μm film, a 1 μm water layer, and a quartz substrate. The rms roughness of the quartz/water and water/film interfaces is assumed to be 0. In the case of a relatively smooth interface with rms roughness = 10 Å, the reflectivity is established by the air/film interface, and the critical angle is that of the film. At higher angles there are finite thickness oscillations due to the thin water layer, as penetration to that layer becomes significant. Finally, at about 0.26 degrees there is a falloff in reflectivity due to the critical angle of the quartz/water interface. If the air/film interface is significantly rougher at rms roughness = 50 Å, then the reflectivity of the film is reduced enough such that the critical angle is now determined by the film/water interface at about 0.15 degrees. Note that generally the finite thickness oscillations are not observed in most experimental setups due to intrinsic roughness of the film surfaces, and hence of the air/film and film/water interfaces. What is always observed is the effect of beam intensity attenuation due to the long absorption path at small incidence angles. This is shown in Figure 15 where absorption for two energies with a realistic cell arrangement (6 μm polypropylene, 1 μm water, quartz substrate and interface roughness of 60 Å at the air/film and film/water interfaces, and 2 Å at the water/quartz interface. At 7.5 KeV, where tuning with respect to the critical angle is done for a Fe K edge experiment, we see a large drop off in reflectivity with increasing incidence angle. This is due to the air/film and film/air interfaces (the close densities of polypropylene and water yield similar reflectivity functions for both interfaces) until beam penetration and reflectivity from the water/quartz interface takes effect. Finite thickness oscillations are

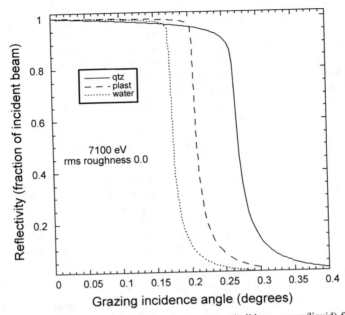

Figure 13. Reflectivity functions for idealized interfaces (vacuum/solid or vacuum/liquid) for quartz (qtz), thin film kapton (plast) and water at X-ray energy of 7100 eV. All materials are infinitely thick and have rms roughness = 0.0.

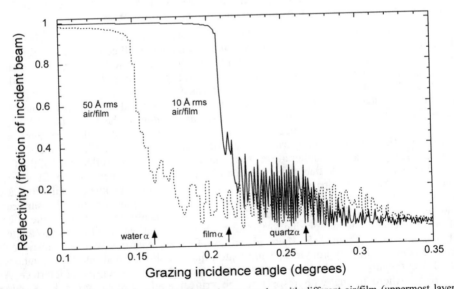

Figure 14. Reflectivity functions for two double layer samples with different air/film (uppermost layer) surface roughness mimicking the type of layers in a GIXAS cell geometry. The high frequency oscillations are finite thickness oscillations due to the second 1 μm water layer. Substrate (quartz)/water and air/film interfaces have rms roughness = 0.0. X-ray energy is 7100 eV. The arrows indicate the approximate positions of the vacuum/material interface critical reflectivity angles for the different layer materials.

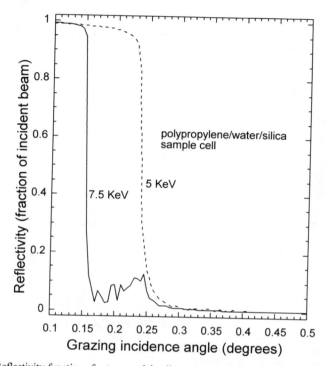

Figure 15. Reflectivity functions for two model cell samples in air. (polypropylene, density = 0.9) top layer; (water, density = 1.0) middle layer; (quartz, density = 2.6) substrate. Interface rms roughness is 60 Å for the top and bottom of the film, and 2 Å for the top of the substrate.

still present from the water layer, and the maximum reflectivity is down to 10% of the incident beam just below the water/quartz critical angle. At 5 KeV the absorption is so large that the beam does not effectively penetrate to the quartz, and the reflectivity is dominated by the air/film and film/water interfaces. The critical angle is shifted to larger angles due to the lower beam energy. This absorption effect requires that experiments for K edges below Fe either must utilize extremely thin, and possible specially manufactured films, or utilize cells wherein the water layer is controlled by H_2O vapor pressure within an enclosing chamber. Even if films can be obtained in the 1-0.5 μm thickness range, there is then the problem of finite thickness oscillations, which may make optimization of the grazing incidence angle problematic. Figure 16 shows the reflectivity for a cell using a 2 μm mylar film with a 1 μm water layer for two different energies and substrates. Note that the increased critical angle due to the greater density of the MnOOH substrate, allows for more of the lower energy X-rays to reach the substrate at it's maximum reflectivity. In this example the surfaces of the film have rms roughness of 20 nm, which practically removes any finite thickness oscillations. Figure 17 shows the similar calculations for quartz substrates with a 1 μm water cover, but no film membrane, mimicking the effect of a substrate that is dosed with water inside of a chamber. Use of this type of dosing allows GI experiments to be done down to the S K edge.

Sample heating and damage

Another experimental aspect of GI cells is the possible sample, film or water heating and damage that can be produced by a focused incident beam. With new sources having

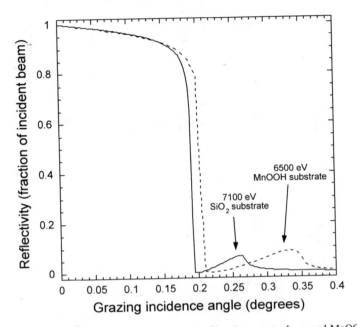

Figure 16. Reflectivity for a sample cell with 2 μm mylar film, 1 μm water layer and MnOOH substrate. Surface roughness of the mylar film is 200 Å rms.

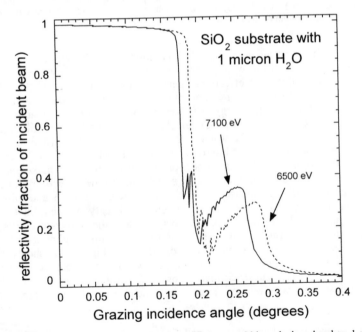

Figure 17. Reflectivity for a sample in an "*ex situ*" experiment within a dosing chamber. 1 μm of water cover but no mylar or other top film.

low emittance, such as at the APS, the entire output of an insertion device can be brought onto a single crystal surface of a few cm in length. If this beam is narrow in the lateral direction (normal to propagation) it may burn through the cover film, ionize the thin water layer over the illuminated part of the sample, or even directly damage the sample itself. Focused beams from the undulator at GSECARS beamline 13IDC obtained with use of Kirkpatrick-Baez (KB) mirrors have been shown to damage single crystal alumina substrates (Peter Eng, personal communication) with focal spots of a few microns in size. For this reason it is best to use substrate crystals that are large enough so that the beam can be spread out to minimize various possible damage modes. One way to do this with small source sizes is to use KB mirrors to defocus a beam in the lateral direction.

For chemical effects in the surface water, the only realistic possibility is to renew the water from time to time. In the case of a cell with a membrane film cover, this requires movement of the membrane and flushing, which probably will require a retuning of the sample position. For the closed chamber vapor-pressure control approach, the constant re-equilibrium of the thin layer of water on the surface should be sufficient if the vapor volume is satisfactorily large compared to the surface water volume. Beam effects on surface species themselves have also been observed, with reduction of both metals and anion complexes. If samples are studied *ex situ* in air, ozone production can cause oxidation. These effects will be most severe with third generation sources, and need to be taken quite seriously in the design of experiments.

Surface symmetry and polarized measurements

The GI approach has further dramatic advantages over bulk XAS methods that are independent of the enhanced sensitivity. These are produced by the single crystal surface which can be oriented so that the incident beam polarization can be at any angle in the plane of the surface, or perpendicular to the surface. For an absolutely perfect single crystal surface the symmetry of the uppermost atomic structure dictates how a polarized experiment can be done. Note that we say here the uppermost structure, as surface species, absorbates, and precipitates will be insensitive to deeper structure. Hence deviation of the uppermost surface atomic structure from the bulk can be significant. The most obvious example of this is if the substrate has a disordered surface layer, possibly due to polishing without proper removal of damaged material. This may effectively produce an upper layer that has reduced symmetry (or no symmetry) so that in-plane polarization variations cannot be used. If such effects are not present, and if reorganization of the surface does not occur, then the bulk symmetry will be expressed on the surface. In such case if the crystal has less than a three fold symmetry axis normal to the surface, then x,y polarization information can be obtained (Brouder 1990).

However most surfaces are not cut perfectly along the desired plane, and this gives rise to a terrace-step structure (Fig. 18). This structure alters symmetry on the surface, can create shadowing effects, and can gives rise to surface functional groups on the step edges that differ in presentation from the same groups exposed on the terraces. These aspects require that, in general, substrates must be characterized on a molecular level for GI methods to produce unambiguous results. The step-terrace structure alters symmetry by introducing a new principal direction into the surface. For an example of this consider the quartz (0001) surface (Fig. 19) with a step-terrace structure. This structure may preserve the 3-fold symmetry, or it may reduce symmetry to bare translation parallel to the step edges. In the latter case the results may be polarization dependent in the x,y plane, but in the former (or in the perfect (0001) surface) case, the 3-fold axis would remove this dependence. There is no simple analog of this situation with polycrystalline or bulk samples, as these samples have random orientations and uncharacterized individual surfaces. Hence the GI experiment not only can provide polarized information

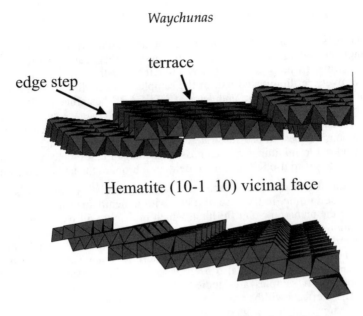

Hematite (10-1 10) vicinal face

Hematite (11-2 20) vicinal face

Figure 18. Examples of the terrace-step structure in hematite cut off a principal (0001) axis.

(10 0 -10 1) quartz vicinal face

Figure 19. Effect of terrace-step structure on symmetry at the interface for ($10\bar{1}0$) quartz.

on surface groups and sorbed species, but also yield details of surface defect regions and reduced symmetry. These types of experiments are only beginning to occur as sensitivity for surfaces complexes increases. As step edges are generally expected to be more reactive than terraces, these positions may be the initiating points for precipitation, reactions with water, and other processes. Notable recent reviews on these topics are available (Brown et al. 1999; Henderson 2002).

SELF ABSORPTION AND CONTAMINATION PROBLEMS

All GI samples measured in fluorescence mode suffer from some degree of self-absorption. This is due to the relatively long path through even a few MLs of materials. In practice, there is usually no correction for self absorption as the effects appear to be small for small amounts (sub-ML) of material on a surface. Self absorption does not markedly affect distance determinations via XAS methods, but can severely distort amplitudes. In order to make a correction for self-absorption in GI geometry the true concentration profile of the sample must be determined. For a single species this can be done by careful analysis of the fluorescence intensity as a function of incidence angle (Troger et al. 1992; Lee et al. 1999), as the reflectivity of the sample can be fit by the Fresnel function including surface roughness. The fluorescence yield can then be calculated from the knowledge of the strength of the incident and reflected beams. Alternatively, independent measurement of absorption in a transmission experiment can also be done, but this is impractical for most substrates.

An even larger problem may be contamination of the fluorescence signal when experiments are being attempted with sub-ML samples. For Fe this is a severe problem, and very careful masking and cleaning of apparatus must be done to lower stray Fe fluorescence below statistical uncertainty, i.e., to statistical baseline levels. The contamination also occurs in the detector, in detector windows, and in dust in the air. Contaminant levels now workable will become problematic as we begin to use the newer low emittance sources with faster and larger detector arrays. Ultimately a clean room type hutch may be needed for GI measurements in the sub ML regime, especially where extended defects such as step edges are being investigated.

DETECTOR LIMITATIONS

Detectors used for GI measurements today are mainly arrays of intrinsic Ge diode detectors. These typically have energy resolutions of 3-4% or so, so that at 7100 eV the resolution is about 200-300 eV, with resolution dependent on counting rate to some degree. Higher resolution detectors can be made from crystal or multilayer X-ray optic assemblies that use diffraction methods to separate the desired fluorescence energy. The disadvantage of these systems in the past has been very small solid angle for the fluorescence input signal, and use for only one wavelength. However they will outperform any other type of detector if the incident flux is high enough (Waychunas and Brown 1994). New bent crystal (transmission Laue) and multilayer designs are being developed by Bunker and associates at the APS (Zhang et al. 1999; Zhong et al. 1999) which may allow several fluorescence energies to be used with one assembly, and which afford improved solid angles. For multilayer or bent crystal diffractive optics energy resolution of about 1% or better should be achievable. What is required is complete removal of the incident beam scattering from the desired fluorescence signal. For K-edges the fluorescence is about 10% lower in energy than the absorption edge, representing the smallest energy separation that the detector must handle. However, the scattered signal can be so large than the tail of the scatter peak can be overpowering in the fluorescent region. At 1% energy resolution this should not be a problem. Unfortunately, no completely electronic detector with acceptable counting rate can yet be built with such resolution, although progress is being made with superconducting tunnel junction detectors that have intrinsic resolution near 0.1% (Hettl et al. 1999). Intrinsic Ge detector speed can be leveraged by adding array elements, and then pulling the enlarged array father away from the sample to reduce flux. This is feasible but expensive. An alternative is to utilize digital signal processing technology to achieve much higher counting rates. Both approaches are now being combined to achieve integrated counting rates of 100 Mcps or more, so that for low concentration (or coverage) samples with low scatter substrates, practically all the fluorescence in a 2π solid angle can be collected. In the case of a detector utilizing a diffractive element energy descriminator, one needs only a fast detector with no energy resolution, so that many types of elements can be used: PIN diodes, scintillation detectors, or fast ion chambers.

Another type of detector with large solid angle is that developed by Stern and Heald (1979). This is an ion chamber with a filter and soller slit assembly in front that preferentially absorbs the scattered incident X-ray beam. This method achieves good rejection of the scattered radiation, but at the cost of losses in the signal fluorescence, and thus is not practical when high scatter rates are present. Such detectors are also single elements, and as such cannot discriminate against diffracted beams from a substrate. With an array detector such is possible, although the feasibility of such data manipulation has to be evaluated in a case by case basis.

Examples of GI X-ray absorption spectroscopy

GI studies to date on environmental samples have been limited by the availability of sufficiently large single crystals of minerals, either natural or synthetic. This situation is changing as crystals of a few mm size can now be studied with low divergence sources. Table (1) lists GI studies that have been done, including a few pertinent experimental particulars.

Fe^{3+} on alpha Quartz. Fe^{3+} sorption and precipitation on quartz r ($10\bar{1}1$) and m ($10\bar{1}0$) surfaces have been studied by Waychunas et al. (1999). Bulk adsorption studies on aerosil silica show a strong effect of sorption geometry with surface density (Fig. 20). At densities equivalent to a 4% of a ML of sorbing Fe^{3+}, the EXAFS Fourier transform function shows a well-defined peak at 3.2 Å, due to Fe-Si neighbors at a distance of about 3.4 Å, and a strong beat in the first neighbor Fe-O peak due to two distance contributions at about 1.88 and 2.05 Å. Analysis of the XANES shows that the short Fe-O distance is not due to tetrahedral Fe^{3+} ions, and the best model is that of a very distorted $Fe^{3+}O_6$ octahedron bonded to surface silicate groups. At higher sorption densities (10% ML and above) the distortion in the Fe-O peak disappears, and the Fe-Si peak is replaced by a peak at about 2.7 Å which is due to Fe-Fe neighbors at a distance of 3 Å. Both changes are consistent with the formation of FeO_6 polyhedra into edge-sharing precipitates at higher surface coverages.

In the GIXAS experiments (Fig. 21), natural quartz surfaces were exposed to 10^{-4} M Fe^{3+} solutions at pH conditions chosen to suppress any Fe^{3+} polymerization in solution. Surface Fe^{3+} densities were measured at 5-10% ML via XPS after GIXAS analysis. Polymerization was observed on the surface, and distinct transform functions where found for in-plane electric vector orientation (unchanged at several azimuthal angles) compared to normal-to-plane electric vector. Analysis of these results required the development of models which took the weighted EXAFS from all Fe within a small precipitate, while factoring in polarization effects. Figure 22 shows the Fourier transforms for such EXAFS for small particles (50 total atoms) with a structure similar to hematite as a function of polarization. By iteration it was shown that a precipitate with about 20 atoms (composition Fe_5O_{15}) could explain the polarized data and the form of the transforms. The fact that similar results were obtained on both m- and r-plane faces of quartz suggests that there is no epitaxial relationship of these precipitates with the underlying structure. However the strong texture indicates a preferred growth or nucleation direction. Recent work has shown that the precipitates appear to cluster at step edges on the surface (Waychunas et al., unpublished). The experiments were done *in situ* to suppress drying of the surface, hence the polymerization was inferred to be due to the higher surface coverage.

Table 1. GI experiments of environmental relevance.

Sample	Substrate[1,2,3]	Polarization[4]		ex situ/in situ[5]	Reference
Zn	Al_2O_3 (0001)	‖ ⊥		ex situ/ in situ	Trainor et al. (2001)
	Al_2O_3 ($1\overline{1}02$)	‖[6] ⊥		ex situ/ in situ	
Oxidation	Pyrite			ex situ	England et al. (1999a)
	Chalcopyrite			ex situ	
Co(II)	Al_2O_3 (0001)	‖ ⊥		ex situ	Towle et al. (1999b)
	Al_2O_3 ($1\overline{1}02$)	‖[6] ⊥		ex situ	
Co(II)	TiO_2 (rutile) (001)	‖ ⊥		ex situ	Towle et al. (1999a)
	TiO_2 (rutile (110)	‖[6] ⊥		ex situ	
Fe(III)	Quartz ($10\overline{1}1$)	‖ ⊥		ex situ/ in situ	Waychunas et al. (1999)
	Quartz ($10\overline{1}0$)	‖ ⊥		ex situ/ in situ	
Cu(II)	Al_2O_3 (0001)	‖ ⊥		ex situ	Fitts et al. (1999)
	Al_2O_3 ($1\overline{1}02$)	‖ ⊥		ex situ	
	Quartz (0001)	‖ ⊥		ex situ	
Cr(III)	Fe_2O_3 (0001)	‖		ex situ	Grolimund et al. (1999)
Zn, Pb(II)	Chalcopyrite (polycrystalline)			ex situ	England et al. (1999b)
Pb(II)	ZnS (polycrystalline)			ex situ	Pattrick et al. (1998)
Pb(II)	Al_2O_3 (0001)	‖		ex situ	Bargar et al. (1997)
	Al_2O_3 ($1\overline{1}02$)	‖		ex situ/ in situ	
Cu(II)	Mica (0001)	‖		ex situ	Farquhar et al. (1996)
Co(II)	Al_2O_3 (0001)	‖ ⊥		in situ	Shirai et al. (1994)

1.) all Al_2O_3 is $\alpha-Al_2O_3$ (corundum or sapphire)
2.) Fe_2O_3 is $\alpha-Fe_2O_3$ or hematite
3.) "r" face on quartz or other crystals is not equivalent to the "R" plane cut often used in experiments
4.) parallel (‖) and perpendicular (⊥) symbols refer to the sample plane
5.) *ex situ* and *in situ* refer to samples run outside and inside of a membrane-covered wet cell, respectively
6.) two different azimuthal orientations examined.

In a separate set of experiments using synthetic quartz substrates, the effect of sorption from a pure Fe^{3+} solution versus one also having a near saturation level of silicic acid was examined. Surface coverages were 5-10% ML. In this case oriented Fe^{3+} complexes bonded to the r-plane surface could be detected in both polarization directions, while precipitates appeared only in the in-plane polarization. For samples without silica the precipitates were enhanced relative to the sorbed complexes, particularly in the in-plane polarization. Drying of a sample by removing it from the wet cell showed a reduction in precipitate peak amplitude, but no change in sorbed complexes. Finally, this sample was subjected to strong washing, which appeared to remove most surface precipitates (Fig. 23). One explanation of the observations is that surface polymerization does not occur in the epi samples as on a natural quartz surface. Precipitation may occur, for example, in the solution above the surface, resulting in polymers oriented along the surface. These experiments taken in total indicate dramatic changes in sorption geometry even at quite low surface densities, possibly owing to surface defect populations, or to surface catalyzed precipitation.

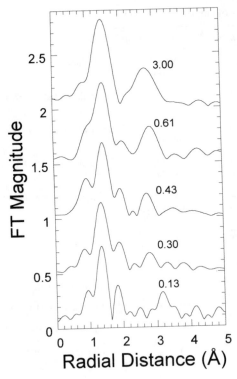

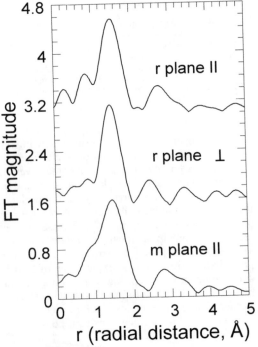

Figure 20. Fe^{3+} sorption on aerosol silica samples as function of surface sorption density. EXAFS Fourier transform functions. Indicated values are in $\mu M/M^2$. 1 ML is approximately 3 $\mu M/m^2$.

Figure 21. Fe^{3+} sorption on natural single quartz surfaces ("Herkimer diamonds"). GI-EXAFS Fourier transform functions. Parallel and perpendicular symbols refer to X-ray electric vector in the plan of the sample surface and normal to it, respectively.

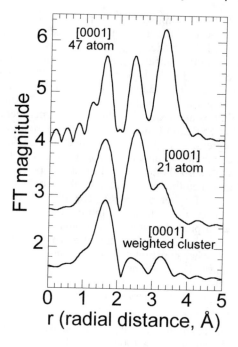

Figure 22. GI-EXAFS model Fourier transform functions for Fe^{3+} sorption on quartz as small precipitates. *Top:* 47 atom hematite-like cluster with [0001] direction normal to surface plane. *Middle:* analogous 21 atom cluster. *Bottom:* weighted 21 atom cluster, where the EXAFS for each Fe ion has been individually calculated and added as a weighted average. This gives best agreement with observations indicating oriented clusters of average 0.9 nm diameter.

Figure 23. Fe^{3+} sorption on synthetic single quartz "epi quality" surfaces. GI-EXAFS Fourier transform functions. *Bottom* two functions show effect of sorption from silica-saturated solution. "hor 90°" refers to electric vector in the plane of the quartz surface at right angles to the mirror plane. "vert" is e-vector perpendicular to the surface. *Middle* two functions contrast analogous experiment done without silica saturation. *Top* two functions show the effect of drying (i.e., "*ex situ*" experiment) on the unsaturated sorption sample. The "dry and washed" *ex situ* sample has been cleaned with a high pressure jet of DI water to remove surface precipitates. It's function is consistent with sorbed complexes with little precipitate signature.

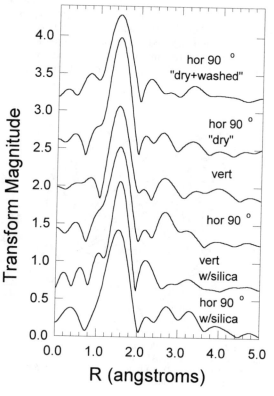

Co²⁺ on rutile, α-Al₂O₃. Towle et al. (1999) examined Co^{2+} sorption on the (110) and (001) surfaces of rutile via GI-XAS. Because of the use of the Co K edge for the measurements, polarized measurements could be made, exploiting two orthogonal directions in-plane for the lower symmetry (110) surface plus the normal-to-plane orientation; and an in-plane and normal orientation for the higher symmetry (100) plane (where all directions in the plane are symmetrically equivalent). The rutile crystals were exposed to 15 mM solutions of $Co(NO_3)_2 \cdot 6H_2O$ at near neutral pH values in CO_2-free conditions, and studied under ambient conditions approximating the *ex situ* experiments of Bargar et al. (1997). Data was collected at SSRL on beam line 6-2. The approximate critical angle for the Co K edge was 0.29 degrees, and optimal data collection was done at about 0.26 degrees grazing incidence. The maximum data range collected was k of 3-12 Å⁻¹. It was found that on both surfaces, Co^{2+} sorbed onto the surface at Ti^{4+} positions, in contrast to Co^{2+} studies on alumina that indicated the local formation of layered Co-Al hydroxides with the hydrotalcite structure (O'Day et al. 1996). The differences in behavior of the rutile surface were attributed in part to reduced solubility of rutile versus alumina. A bond valence analysis was used to predict proton transfer and the reaction stoichiometry for each interface. The geometry of the surface complexes is shown in Figure 24. This study raises the question of what constitutes a solid solution when a surface is being considered. As Co^{2+} occupies Ti^{4+} positions, this satisfies the regular definition for a solid solution. However as two oxygens for every Co^{2+} must be protonated at the surface to balance charge, the local chemistry is actually different so the appropriate solid solution endmember is $Co^{2+}(OH)_2$.

Co²⁺ oxides on α-Al₂O₃. A somewhat different type of surface study was that performed by Shirai et al. (1994) where the surface of a (0001) α-Al₂O₃ crystal was exposed for 2 hours to $Co_2(CO)_8$ by a vapor deposition process at room temperature, with excess Co carbonyl subsequently removed by evacuation while heating to 313 K. The sample was then exposed to oxygen (17 Kpa) at 300K for an additional 5 hours. A second sample was prepared in the same way, but with an additional oxygen exposure for 1 hours at 873K. Both samples were studied at grazing incidence optimized for total

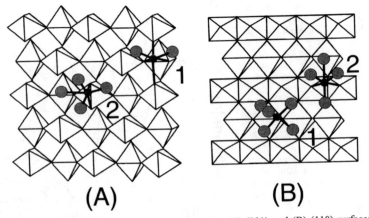

Figure 24. Possible stable Co^{2+} surface complexes on the (A) (001) and (B) (110) surfaces of rutile indicated by the GI-EXAFS data of Towle et al. (1999). In (A), site (1) is a bidentate polynuclear complex, while site (2) is a bidentate mononuclear complex. In (B), site (1) is a monodentate complex, while site (2) is a bidentate binuclear complex.

reflection. Data was collected via fluorescence with incident beam polarization both in the surface plane and normal to it, with a k-range up to 13.5 Å^{-1}. GI-EXAFS of the low temperature-equilibrated sample showed that the Co^{2+} ions all occupied "hollow" sites on the surface (Fig. 25), with no immediate second neighbor Al. This structure appears to be very similar to a hydrotalcite-like layer on the (0001) surface of the Al$_2$O$_3$. In the case of the higher temperature sample, GI-EXAFS showed occupation of several types of sites and longer range neighbor correlations. These were well-modeled by assumption of spinel-like clusters (Co$_3$O$_4$ in composition) growing epitaxially on the Al$_2$O$_3$ (0001)

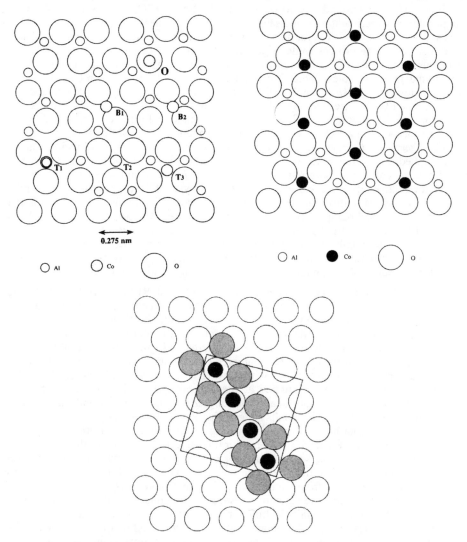

Figure 25. Co^{2+} oxide precipitation geometry on the sapphire (0001) surface. *Top left:* possible surface positions for the Co ion. *Top right:* Observed positioning for the low-temperature annealed sample. *Bottom:* observed spinel nucleus position in high-temperature annealed sample. Modified from Shirai et al. (1994).

surface with spinel (001) plane parallel to Al_2O_3 (0001) as shown in Figure 25. This is an unusual result inasmuch as one would expect the spinel (111) and Al_2O_3 (0001) planes to coincide, as these are the densest oxygen packed planes. There was thus considerable mismatch in the anion layers of layer and substrate, and it was concluded that such epitaxial growths could form a layer on the substrate. Analysis of the coordination numbers in the Fourier transforms allowed estimation of a Co_3O_4 particle size of about 0.9 nm.

Zn on α-Al₂O₃. Trainor et al. (2001) studied the sorption of aqueous Zn^{2+} on (0001) and (1-102) α-Al_2O_3 surfaces. The substrates were exposed to 30 μM $Zn(NO_3)_2$ at neutral conditions with background electrolyte of 0.01 M $NaNO_3$•CO_2 was excluded from all samples during experimentation and sample preparation. Experiments were done at SSRL on beamlines 6-2 and 4-2. Useful data out to $k = 11$ Å^{-1} was obtained using an incidence angle just below the critical angle of about 0.2 degrees. Results were compared with analogous studies of bulk XAS on powdered alumina. Zn-Al correlations at about 3.1 Å were found in all samples, indicating inner sphere complexation for all samples and surfaces. However, *ex situ* samples (those done without bulk solution present under a membrane in a GIXAS cell) showed the presence of Zn-Zn correlations at about 3.84 Å, suggesting the development of a Zn hydroxide precipitate. A further change in the *ex situ* samples was a shift in Zn-O correlation distance to about 2.07 Å, while Zn-O distances in the *in situ* samples averaged 1.98 Å. This is a clear indication of a coordination change from tetrahedral Zn complexes in the *in situ* samples to octahedral sites in the *ex situ* samples. The geometry of sorption complexes for the *in situ* samples appeared relatively similar for both kinds of faces, even though each type of face presumably presented surface oxygens with differing coordination to underlying aluminum ions (with the assumption of crystallographic perfection). This suggested that sorption may have been dominated by high affinity "defect" sites, perhaps of the types discussed earlier in this work and due to irregularity or corrugation on the surfaces. The results of the study emphasize the importance of *in situ* studies to verify any *ex situ* work, and further, the probable importance of drying reactions on surface complexation and precipitation.

Zn and Pb on CuFeS₂. England et al. (1999b) used REFLEXAFS to study the sorption of Zn^{2+} and Pb^{2+} on chalcopyrite surfaces. Polished polycrystalline $CuFeS_2$ sample were used, and exposed to 0.5 μM Zn and Pb nitrate solutions at a pH of 5.5. The samples were either examined directly after 10 minutes of solution treatment, or further reacted with sodium isopropyl [$C_3H_7OCS_2Na$] xanthate (1 gm /200 ml) solution for 10 minutes at pH 10.2 and then analyzed. Samples were washed with water prior to X-ray exposure, and mounted in a container with flowing N_2 atmosphere during REFLEXAFS measurements. Both ions were studied at the L_3 edges on station 9.3 at the SRS at Daresbury Laboratory, with data quality varying among the samples and maximum k ranging from 7.5 to 10 Å^{-1}. The results suggest that Pb^{2+} bonds directly to surface S ions, though with a first coordination sphere including oxygen or hydroxyl groups (Fig. 26). When the xanthate is present, the Pb^{2+} is coordinated only by S ions in the surface and from the xanthate molecule, so that xanthate displaces Pb-O bonds in favor of Pb-S bonds. In contrast, Zn^{2+} is initially bonded to oxygens coordinated to Fe on the $CuFeS_2$ surface, with additional hydroxyl ions in it's first shell coordination. The addition of xanthate does not disrupt the existing surface Zn-O bond, but displaces the other hydroxyls in favor of Zn-S bonds. In flotation processes it is observed that the presence of other metal species disrupts the flotation of chalcopyrite, and that this is remedied by addition of xanthate. It is thus inferred that the formation of inner sphere metal-O bonds with the surface give rise to a high density of surface hydroxyl ions that markedly affect flotation behavior.

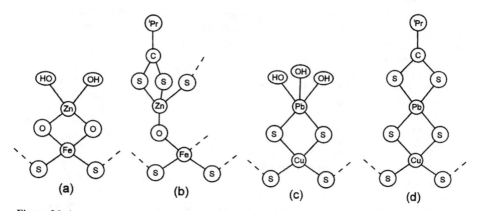

Figure 26. Arrangements of surface complexes of Zn and Pb on chalcopyrite surfaces. (a) Zn sorption. (b) Zn and xanthate. (c) Pb sorption. (d) Pb and xanthate. Modified from England et al. (1999b).

Cu^{2+} on Quartz and α-Al_2O_3. Fitts et al. (1999) studied surface complexation of Cu^{2+} aqueous species (tetragonally distorted $Cu(O,OH,H_2O)_6$ octahedra) on quartz (0001), and α-Al_2O_3 (0001) and ($1\,\overline{1}02$) surfaces. The substrates were exposed to 30 μM solution of Cu^{2+} for 2-12 hours, then studied in the *ex situ* geometry under a high humidity nitrogen flow. Data was collected at SSRL on beamlines 4-2 and 6-2 with a 13 element Ge detector system and both in-plane and normal-to-plane polarizations of the synchrotron e-vector were used on all samples. Good data was collected from 3 to 11 Å^{-1} using a 150 μM beam height and 50 mm diameter substrates. Based on interatomic distances, they found that the Cu^{2+} may occupy monodentate and bridging bidentate sites on the quartz surface, but could not detect neighbors beyond a second-shell Si at about 2.9 Å. This suggests that different types of surface sites might be occupied, leading to averaging out of longer range EXAFS signals. In the case of the Al_2O_3 substrates, Cu^{2+} formed edge-sharing bidentate clusters on both surfaces on the basis of first shell Cu-O and second shell Cu-Al interatomic distances, but analysis of longer range neighbors was inconclusive. Assumption of more distant Al neighbors with a bidentate surface geometry was not a good fit to the EXAFS at higher k values, suggesting that the surfaces may have topologies somewhat different from bulk termination. An alternative model utilized Cu-Cu longer neighbor distances as if there may be some contributions from Cu hydroxide, oxide or other Cu-rich surface precipitate phase. However such contributions did not fit the EXAFS well at lower k values, and fitted distances did not agree well with Cu model compounds.

This investigation along with those of Trainor et al. (2001) and Waychunas et al. (2002) indicate that there may be complications with studies using the *ex situ* geometry. Even though significant water can be present on a surface equilibrated with high humidity (Yan et al. 1987), fluctuations in such thin water layers may lead to dry regions which can serve as nucleation sites for precipitates. The existence of such precipitates, sorption complexes, and perhaps other species formed during drying could make *ex situ* experiments formidably difficult to analyze.

EXPERIMENTS AT SOFT X-RAY ENERGIES; ORGANICS ON SURFACES

Up to now little work has been done to study aqueous surface complexation of low Z and organic species via GIXAS methods, although chambers and instrumentation that can work at relatively low X-ray energies (perhaps down to 2 KeV) have been developed (Shirai et al. 1995; Smith et al. 1995). GI methods have particular advantages at lower energies, specifically less sensitivity to interface asperity, and use of larger (and hence more convenient experimentally) grazing incidence angles. For example, studies of S complexation as sulfate or organic sulfur species would be carried out near the S K-edge at 2.47 KeV. For a Al_2O_3 substrate this energy coincides with a critical angle for total reflection of about 1.0 degrees, and sensitivity to surface roughness on the order of 100 Å rms affects the reflectivity function only slightly below the critical angle. Hence surfaces need not have the same restrictions on polish. An additional point is the use of smaller samples. At 1 degree incidence angle, a 6 mm sample can accept all of a 100 μm incident beam. Thus in many cases samples as small as fractions of a mm can be studied with most of the available incident beam (APS), assuming that problems of sample heating from the beam itself can be handled. It is crucial that water layers be controlled in such experiments by vapor pressure control, as a 1 μm layer of water will absorb 84% of an X-ray beam at 2.5 KeV and 1 degree incidence. For work with lower energy edges, e.g., C K-edge at 284 eV the critical angle on Al_2O_3 is about 8 degrees, but the reflectivity decays quickly well before the critical angle, and a 3 degree angle yields 80% reflectivity. Unfortunately, at this angle and energy even a very thin water layer has serious absorption, e.g., 100 nm would absorb 63% of the beam. Hence water layers would need to be kept in this range and thinner, although such is conceivable. N studies would be somewhat easier, but any studies just above the O K-edge (543 eV)would be problematic due to strong absorption. Other types of related studies could be done under non-aqueous solutions, and possibly even under thin films of melts. In the latter case the melt could be produced locally via laser heating.

The most serious limits to the study of low Z species using GIXAS occur with the present detectors. Energy resolution is currently insufficient to separate incident beam and fluorescent signal, given that energy scanning ranges may be limited to 200-300 eV (7-9 Å^{-1}) by grating-optics monochromator systems. What is required is energy resolution in the 1% or better range done entirely electronically, as Bragg diffraction energy filter systems would be impractically large and expensive. Such detectors are now in development (tunneling junction detectors, see Hettl et al. 1999), but have very small solid angles of acceptance, must be cooled to liquid He temperatures, and present formidable engineering problems due to the need to juxtapose such low temperature elements with a room temperature sample. They also have relatively low counting rates (few kcps per element) at present, but this could be solved by lithographic detector production. Thus for the present, GIXAS in the 2-4 KeV range is the next relatively approachable field using existing detector technology, K-edges for their larger fluorescent yield, and windowless or very thin window detectors (that can be removed from a chamber and cleaned up periodically). Currently available detector windows made of ultrathin Be or synthetic polycrystalline diamond are useful down to 1.5 KeV or so.

TOTAL REFLECTION X-RAY ANALYSIS (TRXRF) SURFACE CHEMISTRY ANALYSIS

TRXRF is used mainly with three types of samples: thin films on the order of nm thick, fine particulates often dispersed via water suspension on a substrate, and substrates with low levels of impurities at the surface (Schwenke et al. 1999; Stoev and Sakurai 1999). As detailed earlier for grazing-incidence experiments in general, the intensity at the surface of a totally reflected X-ray beam can reach 4 times that of the incident beam. However as the incident and reflected beams sum they form a standing wave (see chapter in this volume by Fenter on X-ray standing wave measurements) with node and antinodes of intensity as a function of height above the reflection surface. This nodal structure determines just how much a signal can be enhanced for TRXRF analysis. The diagram in Figure 27 shows 4 varied TRXRF conditions with the total reflection critical angle taken as 0.1 degrees. The dry residue mode refers to fine particulates that are distributed on a flat substrate (or "carrier") surface by means of water suspension followed by evaporation. As such particles are usually in the size range up to a few μm, both a series of nodes and antinodes will overlap with the particles and the signal enhancement will be a sum of maximum (4×) and minimum (0×) contributions equivalent to 2× the incident beam intensity. In the case of a thin film at the surface of only a few nm in thickness, the layer can be superimposed with a single antinodal region of the standing wave, and so "see" an enhanced X-ray intensity of up to 4× the incident beam. In the case of a stratified thin film layer the fluorescence signal is more complex and represents the changing contributions from different parts of the film as the antinodes sweep over the film depth as a function of incidence angle. Finally, an ideal blank sample shows no fluorescence until there is increased penetration above the critical angle.

The sensitivity of TRXRF is identical to GIXAS except that much less signal is needed to produce a definite result, i.e., the appearance of a given element's "peak" in the X-ray emission spectrum, as opposed to obtaining a useful XAS spectrum for Fourier analysis. Thus where the current useful limits for GIXAS analysis are on the order of μM or somewhat less, TRXRF of a surface impurity can readily measure ng/l or 3×10^{-11} M concentrations in small particles or dried solutions (Klockenkämper and von Bohlen 2001). The use of thin film multilayer monochromators that collect a substantial solid angle of X-ray emission from a sealed tube source or from a rotating anode source have enabled commercial TRXRF instruments to achieve detectability limits of about 10^{10} atoms on a suitable carrier. Synchrotron sources have extended this to 10^8 atoms (a few femtograms), with limits mainly imposed by slow counting rates or poor energy resolution in energy dispersive detectors (Baur et al. 2001). Newer synchrotron sources are expected to extend this sensitivity into the 10^7 atom range. Light element analysis (e.g., C, N, O, Al) is relatively difficult because of window absorption, self absorption, and contamination issues. However sensitivities of 10^9 atoms have been obtained using specially designed apparatus (Streli 1997; Streli et al. 1997, 1999). Detection limits for elements excited by sealed tube sources are shown in Figure 28 (Klockenkämper and von Bohlen 1996). Theoretical calculations of detection limits have been made by Sanchez (1999, 2001).

Comparison with other methods

The utility of TRXRF elemental analysis has been compared with Auger spectroscopy, ICP-MS, neutron activation analysis (NAA), SIMS and EPR for thin film and surface analysis (Hegde et al. 1993 (Auger); Lieser et al. 1994 (NAA); Pepelnik et al. 1994; Klockenkämper et al. 2001 (NAA, ICP-MS); Calaway et al. 1994 (SIMS); Diebold 1994 (Auger, SIMS); Santos et al. 1996 (EPR)). The principle advantages of TRXRF, besides its sensitivity which rivals any technique, is the ease of sample preparation, and the resistance to various types of interferences and matrix effects. The liabilities mainly derive from the geometry, which is optimal only for nanometer thickness films with smooth surfaces, but workable for a variety of flat geometry's including fine particle suspensions (Klockenkämper 1997). Other types of samples must be ground into fine powders or have surfaces polished to optical flatness for analysis. When this cannot be done deviations in quantitative results up to a factor of about 2 are observed compared with other methods (e.g., Olsson et al. 1999).

Waychunas

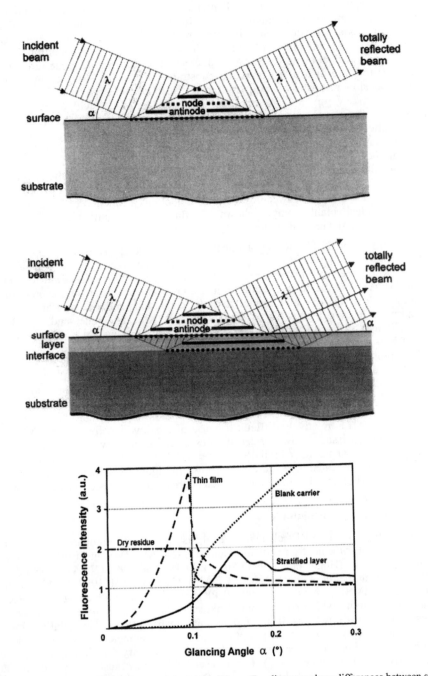

Figure 27. *Bottom:* Varied TRXRF experiment conditions. *Top* diagrams show differences between surface and thin film standing wavefield excitation conditions. [Used by permission of Elsevier Science B.V., from Klockenkamper and von Bohlen (2001), *Spectrochim Acta B,* Vol. 56, Fig. 2, p. 2007 and Fig 3, p. 2008]

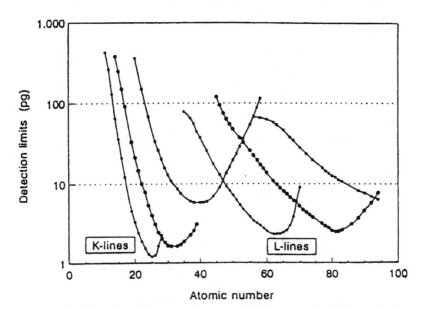

Figure 28. Detection limits for TRXRF analysis using sealed tube sources (pg = picogram). [Used by permission of Elsevier Science B.V., from Klockenkamper and von Bohlen (1996), *X-Ray Spectrom*, Vol. 25, Fig. 2, p. 160.]

Environmental Studies

TRXRF (mainly laboratory based) has been applied to the measurement of trace contaminants in river and spring waters (Prange et al. 1993; Reus et al. 1993; Dogan and Soylak 2002), coastal sediments (Battiston et al. 1993, Miesbauer 1997), oceanic waters (Schmidt et al. 1993; Haarich et al. 1993; Haffer et al. 1997), groundwaters (Vazquez et al. 2000), airborne particulates (Valkovic et al. 1996; Schmeling et al. 1997; Matsumoto et al. 2002), aerosols (Injuk and Van Grieken 1995; Stahlschmidt et al. 1997; Streit et al. 2000), geological microsamples (Ebert et al. 2000), tobacco smoke (Krivan et al. 1994), wine (Carvalho et al. 1996), biofilms (Friese et al. 1997), humic substances (Exner et al. 2000), and cancerous tissue (Benninghoff et al. 1997; von Czarnowski et al. 1997).

Garrett and Blagojevic (2001) applied synchrotron TRXRF analysis to study the surface leaching of synroc, a synthetic ceramic developed at the Australian National Science and Technology Organization (ANSTO) for the immobilization of underground nuclear waste. Leaching of a surface effectively creates a graded thin film coating whose composition as a function of depth can be determined by TRXRF collected at a range of incidence angles. Results are shown in Figure 29 where surface versus bulk contributions can be readily deduced in both a test sample and a synroc sample. Further, the leaching out of Fe in the synroc can be seen to produce both a surface Fe coating, and a Fe-poor zone a few μm deep.

Exner et al. (2000) combined laboratory TRXRF with asymmetric field-flow fractionation (AF^4) methods to analyze the trace element composition and masses of colloidal humic substances. AF^4 is a powerful technique for the separation of particles in a flow-field, and can be used with particles between 1 nm and 100 μm in size. The key aspect of the procedure is the relatively gentle handling of organic colloids with little

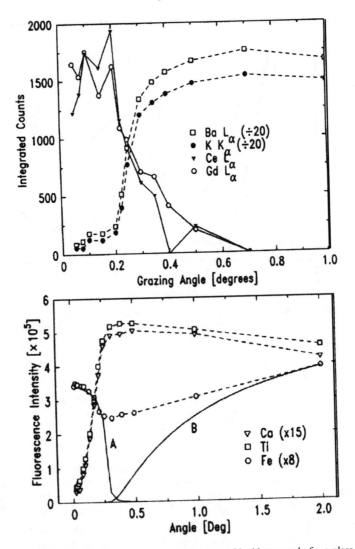

Figure 29. Variation of fluorescence emission as a function of incidence angle for a glass surface test sample (top) and a leached synroc surface (bottom). [Used by permission of Elsevier Science B.V., from Garrett and Blagojeviv (2001), *Nucl Inst Methods A*, Vol. 467-468, Figs. 1 and 2, p. 1211.]

shear forces or damage from collision with the filters or other parts of the flow system. The technique uses the simultaneous action of a carrier flow with an externally applied field, inducing differential migration within a separation channel. The sort of information obtained is shown in Figure 30 where metal content in various size colloids is shown. Typical absolute detection limits for this work were 4×10^{-14} mol for Cu, Co, Ni, Cr and Zn. The relative complexation of metals by colloids was found to decrease in the order Cu > Co > Ni > Cr > Zn via experiments with spiked comparison samples.

Chimidza et al. (2001) analyzed the mobile fraction of soil samples collected at two

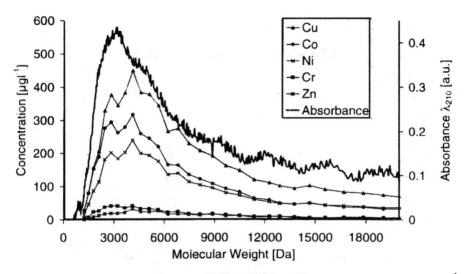

Figure 30. Soil seepage water spiked with heavy metals and analyzed with TRXRF with AF[4]. Absorbance refers to total colloid density. Modified from Exner et al. (2000).

locations in Botswana using laboratory TRXRF methods and extraction procedures. In dry countries like Botswana there is believed to be an extremely high correlation of airborne dust compositions with those in the mobile soil fraction. However, TRXRF analysis showed that the aerosol trace metal composition was more affected by bio-mass burning and automobile exhaust, than by the chemistry of the soil's mobile fraction. Sensitivity for most elements studied (Pb, Sr, Rb, Br, Zn, Mn) was on the order of ng ml^{-1}, equivalent to about 10^{-8} M solution concentration.

Cape Cod aquifer synchrotron TRXRF measurements

In a study to determine the nature of initial mineral coatings forming on quartz within an aquifer (Waychunas et al. 2002), epitaxial-quality polished quartz single crystal wafers were placed into wells at the Cape Cod USGS field site and withdrawn at intervals of six months. TRXRF measurements of these samples after one year of aquifer residence are shown in Figure 31. There are significant amounts of Ca and Fe, and detectable quantities of Mn, Ti and Cr on the wafers after this period. However the concentration of Fe at the highest levels is below 1% of one ML of surface coverage (i.e., about 10^{13} atoms). Comparison of four samples shows that Ca concentration is anticorrelated with Fe and the other transition metals (Fig. 32). The Ca signal is maximized at the critical angle for total reflection from quartz, indicating that the Ca is located immediately at the quartz surface (Fig. 33). The chemistry correlation further suggests that a Ca-rich layer forming on the quartz surface affects formation of transition metal oxides. The Fe content is large enough to obtain EXAFS spectra for each sample, the Fourier transform of each shown in Figure 34. These transforms show that Fe oxides formed in samples with the highest Ca concentration are the most crystallized, with maximum number of Fe-Fe next nearest neighbor correlations (largest second shell peak), while Fe oxides formed with high Mn has poorer crystallinity. Hence the coatings formed in synchrony with carbonate have smaller amounts of more crystalline Fe oxide, while when there is less carbonate a larger amount of poorly crystalline Fe oxide is formed.

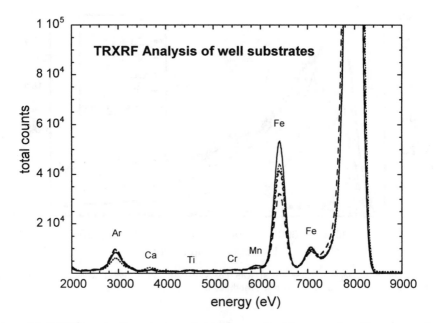

Figure 31. TRXRF spectra from quartz wafer aquifer samples (ultrapure epi-quality quartz wafers suspended in sampling wells at the Cape Cod aquifer site. (Waychunas et al., in prep.)

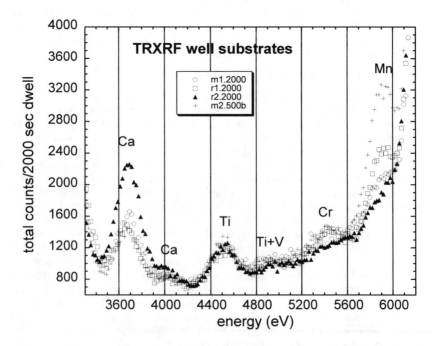

Figure 32. Detailed trace metal abundance in four different aquifer samples (Waychunas et al., in prep.).

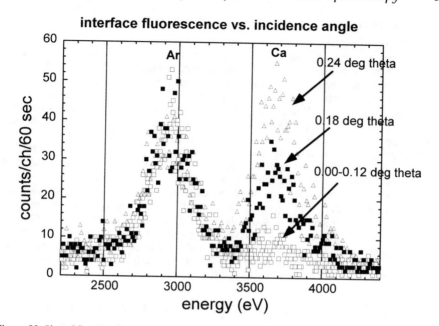

Figure 33. Size of fluorescence signal from Ca on one of the quartz wafer-aquifer samples as a function of grazing incidence angle. The critical angle at the beam energy used is 0.27 degrees. (Waychunas et al., in prep.)

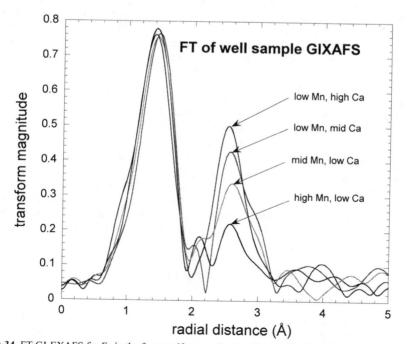

Figure 34. FT GI-EXAFS for Fe in the four aquifer samples from Figure 33. (Waychunas et al., in prep.)

Depth selective analysis

Depth profiling with TRXRF methods has been done with relatively few environmental systems to date. Measurements at various grazing-incidence angles are essentially X-ray standing wave measurements (see chapter in this volume by Fenter), and hence are ideally suited for depth selective measurements, either within a thin solid surface layer on a matrix of different density, or on a solution layer above a substrate. In general, natural samples will not be of sufficient topological quality for use in this mode, but model laboratory systems and "seeded" environmental samples should be ideal. Detailed depth-selective TRXRF analyses of the Cape Cod aquifer samples noted above have been performed (Waychunas et al. 2002), and work on the valence depth profiles of manganese oxide surfaces in contact with aqueous solution is in progress at SSRL. Examples of other types of work that can be done are studies of surface weathering and alteration, precipitation reactions at surfaces, electron transfer and redox reactions, and surface passivation layer formation. In the case of weathering, TRXRF can be used to determine if leaching reactions alter the chemistry of the surface, and the depth to which the leaching occurs. Correction procedures that compensate for the effect of increasing surface roughness as a function of leaching would need to be taken into account in such experiments. An interesting alternative to TRXRF for such analysis is the grazing-exit analog, GEXRF.

GRAZING-EXIT X-RAY FLUORESCENCE SPECTROSCOPY (GEXRF)

In this geometry the sample is irradiated at any convenient incidence angle, usually normal to the surface when an X-ray tube source is being used, but the fluorescence emission is collected only at the grazing angle for total external reflection. Interestingly, the type of information obtained from this sort of experiment is equivalent to the grazing-incidence/ large angle exit beam experiment (Becker et al. 1983), but it has some special advantages for long wavelength (low Z materials) analysis. This is because the effective source size of the sample emission is now subtended only over a few to tens of mrad, and a wavelength-dispersive (i.e., crystal analyzer) detector can be utilized. This means that energy resolution of 1% is readily achieved, and 0.2-0.3% is possible in a Rowland circle focusing instrument. With such good energy resolution it is easy to separate the fluorescence emission from lower Z elements which would be impossible with an energy dispersive detector (deBokx et al. 1997). Further, it is easy to separate inelastic X-ray scattering, diffuse scattering and other background problems.

GEXRF can also be used with high X-ray energies, and with synchrotron sources, especially wiggler insertion devices with relatively large beam divergence. If the exit angle can be scanned it is possible to collect finite thickness oscillations in the fluorescence emission due to interference effects (Fig. 35). This allows film thickness, interface roughness and density gradients to be measured as in classical reflectivity studies. Tsuji et al. (1995) have combined both grazing-incidence and exit (or takeoff angle) methods in a technique they call GIT-XRF which allows further detailed analysis of buried layers and layer profiles.

OPPORTUNITIES FOR FUTURE WORK

Redox reactions at surfaces

GI-XAS methods potentially enable study of valence changes at a surface due to electron and proton exchange reactions. Though studies of such reactions can be done in bulk samples, only for the smallest nanoparticle sizes is it possible to distinguish particle

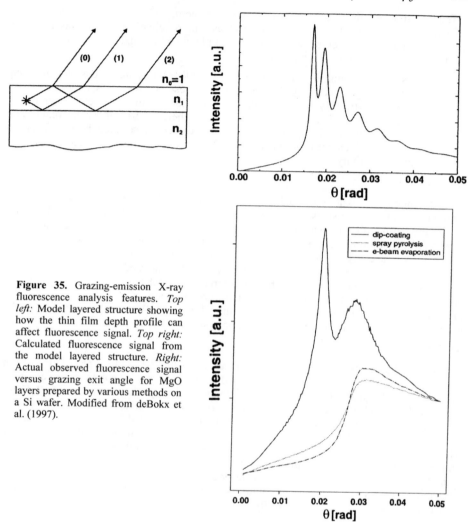

Figure 35. Grazing-emission X-ray fluorescence analysis features. *Top left:* Model layered structure showing how the thin film depth profile can affect fluorescence signal. *Top right:* Calculated fluorescence signal from the model layered structure. *Right:* Actual observed fluorescence signal versus grazing exit angle for MgO layers prepared by various methods on a Si wafer. Modified from deBokx et al. (1997).

surface from bulk valence changes. In an ideal GI-XAS experiment, the uppermost 3 nm of a surface can be preferentially examined below the critical angle for total reflection, and variation of the angle allows sampling increased thicknesses. Analysis of the XANES spectrum as a function of total probe depth can then be done to extract the valence profile of the surface. Practically, valence changes that effect the uppermost single atomic layer ought to be detectable if there is sufficient difference in the XANES spectral character between valence states.

Lower Z species GI-XAS

GI-XAS at low energies (1-4 KeV) enables probes of species that would be difficult to study in bulk samples of any grain size due to absorption effects, e.g., Al or Na in a hematite matrix. However these types of studies are rare in the literature, and have apparently not been attempted with geochemical systems. At these energies Bragg diffraction from a substrate is unlikely to be a problem, and the critical angle for total reflection is large enough such that surface flatness and roughness requirements are easier to achieve. Such experiments require special sample chambers and appropriate detectors (as discussed earlier), but are readily feasible.

GI-XAS work at still lower energies (100-1000 eV) is also possible, allowing analysis of the K edge spectra of C, N and O. For these energies bulk experiments are impractical unless extremely small nanoparticles and sample thickness can be prepared. GI-XAS experiments need to be done in chambers where control of water content and surface water layer thickness is possible. However detector resolution is insufficient at this time to separate incident beam energy and fluorescent photons so that a signal/noise problem limits sensitivity.

Defects on surfaces

Although it is possible to reach high sensitivity in bulk XAS experiments by using extremely small particle sizes with large net surface areas, such experiments will always be limited by inability to characterize the small particle surface structure and chemistry. With third generation synchrotron sources it should be possible to study surface species with GI-XAS at concentrations where defects such as vicinal faces, dislocations, grain boundaries or tilt boundaries, and growth features such as step edges and terraces can be evaluated for their reactivity and sorption complexation behavior. This can be done by preparing substrates with the desired defects or growth features, a technique that is used in materials science for preparing idealized thin film layers. With such investigations the heterogeneity known in natural samples can be systematically studied.

TRXRF analysis

The well samples examined by Waychunas et al. (2002), various types of airborne or fluvial particulates trapped on filters, and dried patches of aqueous solutions represent the range of samples that can be successfully analyzed via synchrotron or in-lab TRXRF systems. The depth profiling capability noted for valence analysis with GI-XAS can be applied for basic elemental analysis to study diffusion profiles, growth of surface precipitates, hidden layers near the surface, weathering or dissolution rates, and so forth. The main advantage of TRXRF is that it is non-destructive, relatively easy to perform, and as sensitive as most common analysis tools (ICP-AES, neutron activation, etc). Practically, TRXRF represents the opposite side of synchrotron analysis capabilities from microprobe analysis. Small regions cannot be measured, but there is much better surface sensitivity as well as element sensitivity.

CONCLUSIONS

Grazing incidence methods can provide strong complementary information when combined with bulk experiments, particularly as the substrate surface can be structurally and chemically analyzed in detail or preferentially prepared, which is impractical for small particulate surfaces. GI-XAS has been used by only a few groups for geochemical investigations, and most of the work in the literature has been done on second generation synchrotron sources where high beam divergence and source size compromise the experiment. Third generation sources will help make GI-XAS analysis a more approachable method for studying geochemical interfaces, and increase sensitivity such that surface species below 1% of a ML can be studied. This will enable idealized investigations with geologically relevant surface species concentrations.

ACKNOWLEDGMENTS

The author is very grateful for many useful discussions over the past few years that contributed to the contents of this chapter. Especially helpful have been current and past members of the Brown group at Stanford University (Gordon Brown, Jr., Steven Towle, John Bargar, Thomas Trainor), other Stanford colleagues (George Parks, George Brown, Paul Flinn), and GSECARS (U. Chicago/APS) scientists (Peter Eng, Matt Newville, Steve Sutton and Mark Rivers). Two detailed anonymous reviews greatly improved the original manuscript. Support from DOE-BES chemical, materials and geoscience program to the Earth Sciences division at LBNL is gratefully acknowledged.

REFERENCES

Alonso-Vante N, Borthen P, Fieber-Erdmann M, Strehblow H-H, Holub-Krappe E (2000) An *in situ* grazing incidence X-ray absorption study of ultra thin Ru_xSe_y cluster-like electrocatalyst layers. Electrochim Acta 45:4227-4236

Als-Nielsen J, McMorrow D (2001) Elements of Modern X-ray Physics. Wiley, New York

Aster B, von Bohlen A, Burba P (1997) Determination of metals and their species in aquatic humic substances by using total-reflection X-ray fluorescence spectrometry. Spectrochim Acta B 52:1009-1018

Authier A (2001) Dynamical Theory of X-ray Diffraction. IUCR Oxford Science Publications, Oxford University Press.

Barchewitz R, Cremonese-Visicato M, Onori G (1978) X-ray photoabsorption of solids by specular reflection. J Phys C 11:4439-4445

Bargar JR, Towle SN, Brown GE, Parks GA (1996) Outer-sphere lead (II) adsorbed at specific surface sites on single crystal α-alumina. Geochim Cosmochim Acta 60:3541-3547

Bargar JR, Towle SN, Brown GE, Parks GA (1997) Structure, composition, and reactivity of Pb(II) and Co(II) sorption products and surface functional groups on single-crystal $α-Al_2O_3$. J Coll Interface Sci 85:473-493

Barrett NT, Gibson PN, Greaves GN, Mackle P, Roberts KJ, Sacchi M (1989) ReflEXAFS investigation of the local atomic structure around Fe during the oxidation of stainless steel. J Phys D 22:542-546

Battiston GA, Gerbasi R, Degetto S, Sbrignadello G (1993) Heavy metal speciation on coastal sediments using total-reflection X-ray fluorescence spectrometry. Spectrochim Acta B 48:217-221

Baur K, Brennan S, Werho D, Moro L, Pianetta P (2001) Recent advances and perspectives in synchrotron radiation TXRF. Nucl Inst Meth A 467/468:1198-1201

Baur K, Brennan S, Burrow B, Werho D, Pianetta P (2001) Laboratory and synchrotron radiation total-reflection X-ray fluorescence: new perspectives in detection limits and data analysis. Spectrochim Acta B 56:2049-2056

Becker RS, Golovchenko JA, Patel JR (1983) X-ray evanescent-wave absorption and emission. Phys Rev Lett 50:153-156

Benninghoff L, von Czarnowski D, Denkhaus E, Lemke K (1997) Analysis of human tissues by total reflection X-ray fluorescence. Application of chemometrics for diagnostic cancer recognition. Spectrochim Acta B 52:1039-1046

Bloch JM, Sansome M, Rondelez F, Peiffer DG, Pincus P, Kim MW, Eisenberger PM (1985) Concentration profile of a dissolved polymer near the air-liquid interface: X-ray fluorescence study. Phys Rev Lett 54:1039-1042

Blum L, Abruna HD, White J, Gordon JG, Borges GL, Samant MG, Melroy OR (1986) Study of underpotentially deposited copper on gold by fluorescence detected surface EXAFS. J Chem Phys 85:6732-6738

Bosio L, Cortes R, Defrain A, Froment M (1984) EXAFS from measurements of X-ray reflectivity in passivated electrodes. J Electroanal Chem 180:265-271

Brouder C (1990) Angular dependence of X-ray absorption spectra. J Phys Cond Matter 2:701-738

Brown GS, Doniach S (1980) The principles of X-ray absorption spectroscopy. *In:* Synchrotron Radiation Research. Winick H, Doniach S (eds) Plenum Press, New York, p 353-385

Brown GE, Calas G, Waychunas GA, Petiau J (1988) X-ray absorption spectroscopy: applications in mineralogy and geochemistry. Rev Mineral 18:431-511

Brown GE, Henrich VE, Casey WH, Clark DL, Eggleston C, Felmy A, Goodman DW, Gratzel M, Maciel G, McCarthy MI, Nealson K, Sverjensky DA, Toney MF, Zachara JM (1999) Metal oxide surfaces and their interactions with aqueous solutions and microbial organisms. Chem Rev 99:77-174

Calaway WF, Coon SR, Pellin MJ, Gruen DM, Gordon M, Diebold AC, Maillot P, Banks JC, Knapp JA (1994) Characterization of Ni in Si wafers: comparison of surface analysis techniques. Surf Interface Anal 21:131-137

Carvalho ML, Barreiros MA, Costa MM, Ramos MT, Marques MI (1996) Study of heavy metals in Madeira wine by total reflection X-ray fluorescence analysis. X-ray Spectrom 25:29-32

Chimidza S, Viksna A, Lindgren ES (2001) EDXRF and TXRF analysis of aerosol particles and the mobile fraction of soil in Botswana. X-ray Spectrom 30:301-307

D'Acapito F, Mobilio S, Cikmacs P, Merlo V, Davoli I (2000) Temperature modification of the Nb oxidation at the Nb/Al interface studied by reflEXAFS. Surf Sci 468:77-84

DeBoer DKG, Leenaers AJG, Vandenhoogenhof WW (1995) Glancing-angle X-ray analysis of thin layered materials. A review. X-ray Spectrom 24:91-102

deBokx PK Kok C, Bailleul A, Wiener G, Urbach HP (1997) Grazing-emission X-ray fluorescence spectrometry: principles and applications. Spectrochim Acta B 52:829-840

Diebold AC (1994) Materials and failure analysis methods and systems used in the development and manufacture of silicon integrated circuits. J Vac Sci Technol B 12:2768-2778

Dogan M, Soylak M (2002) Determination of some trace elements in mineral spring waters by total reflection X-ray fluorescence spectrometry (TXRF). J Trace Microprobe Tech 20:261-268

Dosch H (1992) Critical Phenomena at Surfaces and Interfaces. Springer Tracts in Modern Physics 126:1-145

Ebert M, Mair V, Tessadri R, Hoffmann P, Ortner HM (2000) Total-reflection X-ray fluorescence analysis of geological microsamples. Spectrochim Acta 55:205-212

Eisenberger P, Lengeler B (1980) Extended X-ray absorption fine-structure determination of coordination numbers: limitations. Phys Rev B 22:3551-3562

England KER, Charnock JM, Pattrick RAD, Vaughan DJ (1999a) Surface oxidation studies of chalcopyrite and pyrite by glancing-angle X-ray absorption spectroscopy (REFLEXAFS). Min Mag 63:559-566

England KER, Pattrick RAD, Charnock JM, Mosselmans JFW (1999b) Zinc and Lead sorption on the surface of $CuFeS_2$ during flotation: a fluorescence REFLEXAFS study. Int J Mineral Process 57:59-71

Exner A, Theisen M, Panne U, Niesser R (2000) Combination of asymmetric flow field-flow fractionation (AF^4) and total-reflection X-ray fluorescence analysis (TXRF) for determination of heavy metals associated with colloidal humic substances. Fresenius J Anal Chem 366:254-259

Farquhar ML, Charnock JM, England KER, Vaughan DJ (1996) Adsorption of Cu(II) on the (0001) plane of mica-a REFLEXAFS and XPS study. J Coll Interface Sci 177:561-567

Fisher-Colbrie AM (1986) Grazing Incidence X-ray studies of thin amorphous layers. SSRL report 86/05

Fitts JP, Trainor TP, Grolimund D, Bargar JR, Parks GA, Brown GE. (1999) Grazing-incidence XAFS investigations of Cu(II) sorption products at alpha-Al_2O_3-water and alpha-SiO_2-water interfaces. J Synchrotron Rad 6:627-629

Fox R, Gurman SJ (1981) Theoretical calculations of EXAFS data from specular reflectivity. Phys Chem Glasses 22:32-38

Fox R, Gurman SJ (1980) EXAFS and surface EXAFS from measurements of X-ray reflectivity. J Phys C 13:L249-L253

Friese K, Mages M, Wendt-Potthoff K, Neu T (1997) Determination of heavy metals in biofilms from the river Elbe by total-reflection X-ray fluorescence spectrometry. Spectrochim Acta B 52:1019-1026

Garrett RF, Blagojevic N (2001) Grazing incidence X-ray fluorescence study of the surface leaching behavior of synroc. Nucl Inst Methods A 467/468:1209-1212

Grolimund D, Trainor TP, Fitts JP, Kendelewicz T, Liu P, Chambers S, Brown GE (1999) Identification of Cr species at the aqueous solution-hematite interface after Cr(VI)-Cr(III) reduction using GI-XAFS and Cr L-edge NEXAFS. J Synchrotron Rad 6:612-614

Haarich M, Schmidt D, Freimann P, Jacobsen A (1993) North Sea research projects ZISCH and PRISMA: Application of total-reflection X-ray spectrometry in sea water analysis. Spectrochim Acta B 48:183-192

Haffer E, Schmidt D, Freimann P, Gerwinski W (1997) Simultaneous determination of germanium, arsenic, tin and antimony with total-reflection X-ray fluorescence spectrometry using the hydride generation technique for matrix separation-first steps in the development of a new application. Spectrochim Acta B 52:935-944

Hayes TM and Boyce JB (1982) Extended X-ray absorption fine structure spectroscopy. Sol St Phys 37, 173-365

Hazemann JL, Manceau A, Sainctavit P, Malgrange C (1992) Structure of the α-$Fe_xAl_{1-x}OOH$ solid solution 1. Evidence by polarized EXAFS for an epitaxial growth of hematite-like clusters in Fe-diaspore. Phys Chem Minerals 19:25-38

Heald SM, Chen H, Tranquada JM (1988) Glancing-angle extended X-ray-absorption fine structure and reflectivity studies of interfacial regions. Phys Rev B 38:1016-1026

Hegde RI, Tobin J, Fiordalice RW, Travis EO (1993) Nucleation and growth of chemical vapor deposition TiN films on Si (100) as studied by total reflection X-ray fluorescence, atomic force microscopy, and Auger electron spectroscopy. J Vac Sci Technol A 11:1692-1695

Henderson MA (2002) The interaction of water with solid surfaces: fundamental aspects revisited. Surf. Sci. Rep. 46:1-308

Hettl P, Angloher G, Feilitzsch F, Höhne J, Jochum J, Kraus H, Mössbauer RL (1999) High-resolution X-ray spectroscopy with superconducting tunnel junctions. X-ray Spectrom 28:309-311

Injuk J, Van Grieken R, Klockenkamper R, von Bohlen A, Kump P (1997) Performance and Characteristics of two total-reflection X-ray fluorescence and a particle induced X-ray emission setup for aerosol analysis. Spectrochim Acta B 52:977-984

Injuk J, Van Grieken R (1995) Optimization of total-reflection X-ray fluorescence for aerosol analysis. Spectrochim Acta B 50:1787-1803

Klockenkämper R (1997) Total-reflection X-ray fluorescence analysis. J. Wiley and Sons, New York

Klockenkämper R, von Bohlen A (2001) Total-reflection X-ray fluorescence moving towards nanoanalysis: a survey. Spectrochim. Acta B56:2005-2018.

Klockenkämper R, Alt F, Brandt R, Jakubowski N, Messerschmidt J, von Bohlen A (2001) Results of proficiency testing with regard to sediment analysis by FAAS, ICP-MS and TXRF. J Anal At Spectrom 16:658-663

Klockenkämper R, von Bohlen A (1996) Elemental analysis of environmental samples by total reflection X-ray fluorescence—a review. X-ray Spectrom. 25:156-162

Knoth J, Prange A, Reus U, Schwenke H (1999) A formula for the background in TXRF as a function of the incidence angle and substrate material. Spectrochim Acta B 54:1513-1515

Krivan V, Schneider G, Baumann H, Reus U (1994) Multi-element characterization of tobacco smoke condensate. Fresenius J Anal Chem 348:218-225

Lagarde P, Delaunay R, Flank AM, Jupille J (1993) Site of sulfur impurities in silicate glasses and REFLEXAFS studies around the Si K-edge. Jap J Appl Phys 32:619-621

Lee PA, Citrin PH, Eisenberger PM (1981) Extended X-ray absoption fine structure-its strength and limitations as a structural tool. Rev Mod Phys 53:769-806

Lee JM, Yoo H-H, Joo M (1999) Numerical determination of a true absorption spectrum from grazing-incidence fluorescence EXAFS data. J Synchrotron Rad 6:244-246

Lieser KH, Flakowski M, Hoffman P (1994) Determination of trace elements in small water samples by total reflection X-ray fluorescence (TXRF) and by neutron activation analysis (NAA). Fresenius J Anal Chem 350:135-138

Martens G, Rabe P (1980) EXAFS studies on superfacial regions by means of total reflection. Phys Status Solidi A 58:415-424

Matsumoto E, Simabuco SM, Perez CA, Nascimento VF (2002) Atmospheric particulate analysis by synchrotron radiation total reflection (SR-TXRF). X-ray Spectrom 31:136-140

Michaelis W, Prange A (1988) Trace analysis of geological and environmental samples by total-reflection X-ray fluorescence spectrometry. Int J Radiat Appl Instrum E 2:231-245

Miesbauer H, Kock G, Fureder L (2001) Determination of trace elements in macrozoobenthos samples by total-reflection X-ray fluorescence analysis. Spectrochim Acta B 56:2203-2207

Miesbauer H (1997) Multielement determination in sediments, pore water and river water of upper Austrian rivers by total-reflection X-ray fluorescence. Spectrochim Acta B 52:1003-1008

Olsson M, Viksna A, Helja-Sisko H (1999) Multi-element analysis of fine roots of Scots Pine by total reflection X-ray fluorescence spectrometry. X-ray Spectrom 28:335-338

Pattrick RAD, Charnock JM, England KER, Mosselmans JFW, Wright K (1998) Lead sorption on the surface of ZnS with relevance to flotation: A fluorescence REFLEXAFS study. Min Eng 11:1025-1033

Pepelnik R, Prange A, Niedergesass R (1994) Comparative study of multielement determination using inductively-coupled plasma mass spectrometry, total-reflection X-ray fluorescence spectrometry, and neutron activation analysis. J Anal Atom Spectrom 9:1071-1074

Pepelnik R, Erbsloeh B, Michaelis W, Prange A (1993) Determination of trace element deposition into a forest ecosystem using total-reflection X-ray fluorescence. Spectrochim Acta B 48:223-229

Perez RD, Sanchez HJ, Rubio M (2001) Theoretical model for the calculation of interference effects in TXRF and GEXRF. X-ray Spectrom 30:292-295

Posedel D, Turkovic A, Dubcek P, Crnjak-Orel Z (2002) Grazing-incidence X-ray reflectivity on nanosized vanadium oxide and V/Ce films. Mat. Sci. Eng. B 90:154-162

Prange A, Boddeker H, Kramer K (1993) Determination of trace elements in river water using total-reflection X-ray fluorescence. Spectrochim Acta B 48:207-215

Prange A, Kramer K, Reus U (1993) Boron nitride carriers for total reflection X-ray fluorescence. Spectrochim Acta B 48:153-161

Reus U, Markert B, Hoffmeister C, Spott D, Guhr H (1993) Determination of trace metals in river water and suspended solids by TXRF spectroscopy: A methodical study on analytical performance and sample homgeneity. Fresenius J Anal Chem 347:430-435

Sanchez HJ. (2001) Detection limit calculations for the total reflection techniques of X-ray fluorescence analysis. Spectrochimica Acta B 56:2027-2036

Sanchez HJ (1999) Theoretical calculations of detection limits in total reflection XRF analysis. X-ray Spectrom 28:51-58

Santos MT, Arizmendi L, Bravo D, Dieguez E (1996) Analysis of the core in $Bi_{12}SiO_{20}$ and $Bi_{12}GeO_{20}$ crystals grown by the Czochralsk method. Mater Res Bull 31:389-396

Schmeling M (2001) Total-reflection X-ray fluorescence - a tool to obtain information about different air masses and air pollution. Spectrochim Acta B 56:2127-2136

Schmeling M, Klockenkamper R, Klockow D (1997) Application of total-reflection X-ray fluorescence spectrometry to the analysis of airborne particulate matter. Spectrochim Acta B 52:985-994

Schmidt D, Gerwinski W, Radke I, (1993) Trace metal determination by total-reflection X-ray fluorescence analysis in the open Atlantic ocean. Spectrochim Acta 48:171-182

Shirai M, Inoue T, Onishi H, Asakura K, Iwasawa Y (1994) Polarized total-reflection fluorescence EXAFS study of anisotropic structure analysis for Co oxides on α-Al_2O_3 (0001) as model surfaces for active oxidation catalysts. J Catal 145:159-165

Shirai M, Nomura M, Asakura K, Iwasawa Y (1995) Development of a chamber for *in situ* polarized total-reflection fluorescence X-ray absorption fine structure spectroscopy. Rev Sci Inst 66:5493-5498

Sinha SK (1996) Surface roughness by X-ray and neutron scattering methods. Acta Phys Polonica A 89:219-234

Smilgies D-M (2002) Geometry-independent intensity correction factors for grazing incidence diffraction. Rev Sci Inst 73:1706-1710

Smith AD, Roper MD, Padmore HA (1995) A REFLEXAFS apparatus for use with soft X-rays in the sub 4 KeV range. Nucl Inst Methods B 97:579-584

Stahlschmidt T, Schultz M, Dannecker W (1997) Application of total-reflection X-ray fluorescence for the determination of lead, calcium and zinc in size-fractionated marine aerosols. Spectrochim Acta B 52:995-1002

Steinmeyer S, Kolbesen BO (2001) Capability and limitations of the determination of sulfur in inorganic and biological matrices by total reflection X-ray fluorescence spectrometry. Spectrochim Acta 56:2165-2173

Stern EA, Heald SM (1983) Basic principles and applications of EXAFS. *In:* Handbook on Synchrotron Radiation, Vol. 1b, Koch EE (ed) North Holland, New York, p. 955-1014

Stern EA, Heald SM (1979) X-ray filter assembly for fluorescence measurements of X-ray absorption fine structure. Rev Sci Inst 50:1579-1582

Stoev KN, Sakurai K (1999) Review on grazing-incidence X-ray spectrometry and reflectometry. Spectrochim Acta B 54:41-82

Streit N, Weingartner E, Zellweger C, Schikowski M, Gaggeler HW, Baltensperger U (2000) Characterization of size-fractionated aerosol from the Jungfraujoch (3580 m asl) using total reflection X-ray fluorescence (TXRF). Int J Environ Anal Chem 76:1-16

Streli C, Kregsamer, P, Wobrauschek, P, Gatterbauer H, Pianetta P, Pahlke S, Fabry L, Palmetshofer L, Schmeling M (1999) Low Z total reflection X-ray fluorescence analysis-challenges and answers. Spectrochim Acta 54:1433-1441

Streli C, Wobrauschek P, Bauer V, Kregsamer P, Goegl R, Pianetta P, Ryon R, Pahlke S, Fabry L (1997) Total reflection X-ray fluorescence analysis of light elements with synchrotron radiation and special X-ray tubes. Spectrochim Acta B 52:861-872

Streli C (1997) Total reflection X-ray fluorescence analysis of light elements. Spectrochim Acta B 52:281-293

Schwenke H, Beaven PA, Knoth J (1999) Applications of total reflection X-ray fluorescence spectrometry in trace element and surface analysis. Fresenius J Anal Chem 365:19-27

Towle SN, Brown GE, Parks GA (1999a) Sorption of Co(II) on metal oxide surfaces: I. Identification of specific binding sites of Co(II) on (110) and (001) surfaces of TiO_2 (rutile) by grazing-incidence XAFS spectroscopy. J Coll Interface Sci 217:299-311

Towle SN, Bargar JR, Brown GE, Parks GA (1999b) Sorption of Co(II) on metal oxide surfaces. II. Identification of Co(II) (aq) adsorption sites on the (0001) and (1-102) surfaces of α-Al_2O_3 by grazing-incidence XAFS spectroscopy. J Coll Interface Sci 217:312-321

Trainor TP, Fitts JP, Templeton AS, Grolimund D, Brown GE (2002) Grazing-incidence XAFS study of aqueous Zn(II) sorption on α-Al_2O_3 single crystals. J Coll Interface Sci 244:239-244

Tröger L, Zschech E, Arvanitis D, Baberschke K (1993) Quantitative fluorescence EXAFS analysis of concentrated samples-correction of the self-absorption effect. Jap J Appl Phys 32:144-146

Tsuji K, Sato S, Hirokawa K (1995) Depth profiling using the glancing-incidence and glancing-takeoff X-ray fluorescence method. Rev Sci Inst 66:4847-4852

van der Lee A (2000) Grazing incidence specular reflectivity: theory, experiment, and applications. Sol. St. Sci. 2:257-278

Valkovic V, Dargie M, Jaksic M, Markowicz A, Tajani A, Valkovic O (1996) X-ray emission Spectroscopy applied for bulk and individual analysis of airborne particulates Nucl Inst Meth B 113:363-367

Vazquez C, de Funes SF, Casa A, Adelfang P (2000) Application of total reflection X-ray fluorescence to studies of the geographical distribution of arsenic and other toxic trace elements in ground waters of Argentina Pampa plain. J Trace Microprobe Tech 18:73-81

von Czarnowski, Denkhaus E, Lemke K (1997) Determination of trace element distribution in cancerous and normal human tissues by total reflection X-ray fluorescence analysis. Spectrochim Acta B 52:1047-1052

Warburton WK (1986) Filtered energy dispersive detector (EDD) arrays: superior detectors of EXAFS from very diluted solutions. Nucl Inst Methods A246:541-546

Waychunas GA, Brown GE (1994) Fluoresence yield XANES and EXAFS experiments: Applications to highly dilute and surface samples. Adv X-ray Anal 37:607-617

Waychunas GA, Rea BA, Fuller CC, Davis JA (1993) Surface chemistry of ferrihydrite. Part I. EXAFS studies of the geometry of coprecipitated and adsorbed arsenate. Geochim Cosmochim Acta 57:2251-2269

Waychunas GA, Davis JA, Reitmeyer R (1999) GIXAFS study of Fe^{3+} sorption and precipitation on natural quartz surfaces. J Synchrotron Rad 6:615-617

Waychunas GA (2001) Structure, aggregation and characterization of nanoparticles. Rev Mineral Geochem 44:112-166

Waychunas GA, Reitmeyer R, Davis JA (2002) Grazing incidence EXAFS characterization of Fe^{3+} sorption and precipitation on silica surfaces. (submitted to J Col Inter Sci)

Weygand M, Wetzer B, Pum D, Sleytr UB, Cuvillier N, Kjaer K, Howes PB, Lösche M (1999) Bacterial S-layer protein coupling to Lipids: X-ray reflectivity and grazing-incidence diffraction studies. Biophysical J 76:458-468

Yamiguchi H, Itoh S, Igarashi S, Naitoh K, Hasegawa R (1998) TXRF analysis of solution samples using polyester film as a disposable sample-carrier cover. Anal Sci 14:909-912

Yan B-D, Meilink SL, Warren GW, Wynblatt P (1987) Water absorption and surface conductivity measurements on alpha-alumina. IEEE Trans Components Hybrids Manufact Tech CHMT-10:247-251

Yashiro W, Ito Y, Takahasi M, Takahashi T (2001) Darwin's theory for the grazing incidence geometry. Surf Sci 490:394-408

Yoneda Y, Horiuchi T (1971) Optical flats for use in X-ray spectrochemical microanalysis. Rev Sci Inst 42:1069-1070

Zhang K, Rosenbaum G, Bunker G (1999) Design and testing of a prototype multilayer analyzer X-ray fluorescence detector. J. Synchrotron Rad 6:220-221

Zhong Z, Chapman D, Bunker B, Bunker G, Fischetti R, Segre C (1999) A bent Laue analyzer for fluorescence EXAFS detection. J Synchrotron Rad 6:212-214

6

Applications of Storage Ring Infrared Spectromicroscopy and Reflection-Absorption Spectroscopy to Geochemistry and Environmental Science

Carol J. Hirschmugl

Department of Physics and Laboratory for Surface Studies
University of Wisconsin-Milwaukee
P. O. Box 413
Milwaukee, Wisconsin, 53201, U.S.A.

ABSTRACT

Infrared radiation extracted from a storage ring affords new opportunities for scientific exploration in the areas of geochemistry, geomicrobiology and environmental science. In this review paper, the fundamental interactions between infrared light and matter are discussed, followed by an introduction to the source properties of infrared radiation emitted from relativistically accelerated electrons in a storage ring. The most important of these properties is the brightness of the source. A bright source can deliver a higher density of photons onto a small (less than 20 μm × 20 μm) sample at normal incidence than a lab based globar source, which is necessary to produce diffraction limited spatially resolved infrared images. Alternatively, this source can couple well to grazing incidence geometry for surface science experiments affording the opportunity to examine low frequency adsorbate-substrate vibrational bands. Several examples of infrared spectromicroscopy applications and surface science applications are reviewed.

INTRODUCTION

Infrared storage ring radiation (IRSR) is a bright source that affords the opportunity to examine the chemical nature of samples at diffraction-limited spatial resolution (comparable to the wavelength of light, e.g., for mid-infrared between 2-25 μm) or the chemical nature of bonding at a single crystal-adsorbate interface. Pertinent problems in geochemistry and environmental science include "geomicrobiology," i.e., how microbes interact with rocks and control chemistry of natural waters, "biomineralization," and questions about the evolution of life and related issues that can be addressed with, for example, spatially resolved chemical analysis of bone materials. In addition, time resolved infrared spectroscopy includes investigations of interest to geochemists and geologists, such as solution chemistry and reaction kinetics at interfaces. An excellent review by O'Day discusses recent advances in the understanding of molecular chemistry and reaction mechanisms at mineral surfaces and mineral-fluid interfaces (O'Day 1999).

Infrared (IR) spectroscopy is an incisive, non-destructive tool for examining interfaces and surfaces, providing valuable knowledge about chemical forces between atoms, and vibrational frequencies found in molecules and bulk solids. This tool thereby allows one to determine the flow of energy in matter, or to identify the chemical species present at the surface or interface or within the material under investigation, while, in many cases, determining their molecular structure. IR spectroscopy is a mature field, yet it is poised to make contributions to the newer directions embraced by scientists in solid-fluid interfaces: the ability of IR light to penetrate many forms of matter with low energy transfer confers access to vital information even under adverse conditions. IR radiation extracted from storage rings (radiation is emitted when swift charged particles are

1529-6466/00/0049-0006$05.00

accelerated by a magnetic field) has provided opportunities for many of these state-of-the-art experiments. For example, IR absorption measurements performed on live single cells can be used to probe the distribution of chemistry within the molecule, and monitor the changes in chemical concentration under different stages in their metabolic cycle and varying environmental stimuli (Holman et al 1999; Jamin et al. 1998).

Electromagnetic radiation, including IR light, drives the motion of electric charges in matter. If the natural time scale of any oscillations of the charges in a molecule or solid is close to the period of the electromagnetic radiation shining on the system, a condition known as resonance occurs. Like an adult timing his pushes on a child's swing to coincide with the motion of the swing, a driving force having the same frequency as the system's natural frequency efficiently couples to and excites the oscillation. Near resonance, therefore, IR light is efficiently absorbed by the system, allowing the identification of the frequencies of low-energy (1-500 meV) excitations found in the system under study. These excitations may involve nuclear motion, such as vibrating molecules, ions, radicals, or chains of atoms in extended materials.

As a practical matter, IR spectroscopy has found its widest application in identifying the chemical compounds present in an unknown sample by the virtue of frequencies of IR light the sample absorbs. Since the resonance condition occurs over a narrow range of frequencies, which differs for different compounds, the exact frequency of the absorbed light provides a characteristic signature of the molecules, ions, or radicals present in the sample. Extensive gas-phase studies have identified these "fingerprints" for a host of chemical compounds, which can be used in interpreting surface and interface data. For example, the vibrational stretching motion of a triple-bonded CO unit (such as found in CO gas) absorbs IR light at 5.70×10^{13} Hz. Similarly, CO weakly bound to a single atom on a solid surface absorbs IR light at 5.53×10^{13} Hz. The analytical capabilities of IR spectroscopy are invaluable in identifying chemical composition at surfaces and interfaces.

Einstein won a Nobel Prize in Physics (1921) for showing that the energy carried by electromagnetic radiation is directly related to the frequency of its oscillation. Thus, IR spectroscopy allows the determination of the energy of the excitations it probes, and thereby sheds light on the microscopic origin of the excitation. For example, the energies of vibrational excitations provide insight into the interatomic potentials found in molecules, including knowledge about the bond strength. Since the strength of bonding within a molecule is altered upon adsorption (e.g., bonding) to a surface, a sensitive determination of the bond strength can be invaluable in probing surface reaction paths.

The origins of IR spectroscopy date back to the 1880s. Recent advances in instrumentation, including the design of spectrometers and detectors and the development of new sources, however, provide the means to enhance significantly the capabilities of this mature field. Thus, more complex systems such as single cells, or functionally gradient material, or fluids trapped inside minerals can be accessed. Over the years there has been continual improvement in spectrometers, detectors, window materials, and data processing. Recently, however, there have been major developments in new sources of IR radiation. These sources include radiation emitted from a storage ring, lasers, and free electron lasers (FELs). FEL radiation is generated from swift charged particles that are accelerated several times by a spatially varied magnetic field over a short distance. Thus, several collinear micro-beams are generated and add in-phase to create the resulting beam. All three sources possess two important properties, brightness and pulsed time structures, which are invaluable in studies of topical interest. A high-brightness source emits radiation that originates from a point-like source, and the light is therefore confined to a well-defined, small beam (e.g., laser beams). This property is required to measure spatially resolved data with spatial resolution comparable to the wavelength of the light

(<1 μm to ten's μm). Moreover, these bright sources are pulsed in time. They provide their radiation in staccato bursts, lasting anywhere between several femtoseconds to nanoseconds (1 fs equals one quadrillionth of a second and 1 ns equals one billionth of a second). Thus, modern sources allow the possibility to study time-dependent changes in chemical composition, or structure on time scales as short as femtoseconds.

In this paper, the properties of IR spectroscopy, IR synchrotron radiation and geological and environmental molecular sciences applications are discussed. In the next section, fundamental interactions between IR light and materials are reviewed. In the following section, details of infrared storage ring spectroscopy are introduced, focusing on the benefits gained in typical experimental geometries. Finally, selected recent applications in spectromicroscopy and surface science are presented.

INTERACTIONS BETWEEN LIGHT AND MATTER

Some of the more important aspects of the interplay between light and matter that allow us to probe materials with IR radiation are reviewed in this section. Einstein explained that energy in light waves is not continuous, but is instead limited to a discrete set of values, i.e., is quantized, into small bundles called *photons*. The energy of each photon is $E = hc/\lambda$ where λ is the wavelength of the light, h is Planck's constant and c is the speed of light. Interactions between photon beams and matter offer insight into a host of fundamental processes and properties of materials systems. Upon interaction with matter, the beam may be elastically scattered (e.g., reflected or diffracted), inelastically scattered (e.g., by exciting a vibrational mode), or absorbed. Each of these mechanisms can be exploited using a photon beam of appropriate energy. If none of these interactions is strong, the light may penetrate deeply into the material, or transmit unperturbed through the material. Typical penetration depths for IR radiation range from hundreds of Å's for metals to μm's for metallic oxides, to tens of kilometers for certain insulators. The penetration depth depends on the frequency-dependent conductivity of the sample. Buried interfaces can be studied if the materials on at least one side of the interface are thinner than the IR penetration depth, rendering that layer transparent to IR radiation.

Dielectric functions, complex indices of refraction and Maxwell's equations

The *dielectric function,* $\tilde{\varepsilon}$, is a mathematical entity commonly used to describe (as a function of frequency) the interaction between light and matter. This function is closely related to the frequency-dependent index of refraction, $\tilde{n}$, that governs how light is bent by lenses. The index of refraction is separated into two parts, the real part n, which is proportional to the speed of the light as it passes through the material and the imaginary part k', which is a measure of how quickly the light is absorbed by the material. The speed of light in a material, υ, is equal to c/n, where c is the speed of light in vacuum. A strongly absorbing material that attenuates an incident beam is characterized by a large value of k'. The dielectric function $\tilde{\varepsilon}$ is also separated into two parts, the real part ε' and the complex part ε''. The dielectric function and the index of refraction are related to each other by $\tilde{\varepsilon} = \tilde{n}^2$.

Here we will examine the nature of electromagnetic propagation in a semi-infinite medium, starting with Maxwell's Equations for a non-magnetic, uniform charge density material (for more details refer to Jackson 1975):

$$\nabla \times B = -\frac{\tilde{\varepsilon}}{c}\frac{\partial E}{\partial t} + \frac{4\pi\tilde{\sigma}E}{c} \tag{1a}$$

$$\nabla \times E = \frac{1}{c}\frac{\partial B}{\partial t} \tag{1b}$$

where $\tilde{\varepsilon}$ and $\tilde{\sigma}$ are the complex, frequency dependent dielectric and conductivity functions, respectively. The conductivity is related to the absorbance of a material. When we operate on Equation (1a) with $\partial/\partial t$, and on Equation (1b) with $\nabla \times$, and add them together, we obtain a wave equation for the Electric Field:

$$\nabla^2 \mathbf{E} - \frac{\tilde{\varepsilon}}{c}\frac{\partial^2 E}{\partial t^2} = \frac{4\pi\tilde{\sigma}}{c}\frac{\partial E}{\partial t} \tag{2}$$

We propose the following plane wave solution (propagating in the x direction) for the Electric Field:

$$E = e^{i(\tilde{k}\cdot\vec{x}-\omega t)} \tag{3}$$

where the angular frequency $\omega = 2\pi\nu$. Upon substitution into Equation (2), we obtain the following expression for the magnitude of the wavenumber $\tilde{k}$ (The wavenumber is a measure of the number of waves per length and corresponds to the momentum of the wave. Note here that the wavenumber $\tilde{k}$, is frequently referred to as k, and is not the same as k', the imaginary part of the index of refraction, in terms of the dielectric function and conductivity of the material.):

$$\tilde{k}^2 = \frac{\tilde{\varepsilon}\omega^2}{c^2}\left(1 + \frac{i4\pi\tilde{\sigma}}{\tilde{\varepsilon}\omega}\right) \tag{4}$$

where $\tilde{k}$, $\tilde{\varepsilon}$ and $\tilde{\sigma}$ are all complex numbers. The first term of the wavenumber, $\tilde{k}$, represents the displacement current in the medium, while the second term is the conduction current in the medium, and only arises when there is a free electron component of the medium (metallic contribution in e.g., metals, semi-conductors and superconductors).

Now, we replace the complex dielectric function by the complex refractive index,

$$\tilde{N} = n + ik' = \sqrt{\tilde{\varepsilon}} \quad \text{and} \tag{5}$$

$$\varepsilon_1 = n^2 - k'^2$$
$$\varepsilon_2 + \frac{4\pi\sigma}{\omega} = 2nk' \tag{6}$$

The wavenumber in terms of the complex refractive index is more intuitive,

$$\tilde{k} = \frac{\tilde{N}\omega}{c} = \frac{n\omega}{c} + \frac{ik'\omega}{c} \tag{7}$$

Now substituting this result for the wavenumber back into the expression for the traveling plane wave in a semi-infinite dielectric medium for one dimension, we obtain an intuitive form for the traveling wave:

$$\mathbf{E} = e^{i\left(\frac{n\omega x}{c} - \omega t\right)}e^{-\frac{k'\omega x}{c}} \tag{8}$$

This solution for the electric field contains both traveling wave (imaginary part of the exponential function) and attenuated wave (real part of the exponential function) contributions. The traveling wave with frequency ω, travels at a reduced velocity in the medium compared to vacuum, $(n\omega/c)$.

IR absorption spectroscopy

As described in more detail below, infrared spectra are most commonly a difference between two measurements, where something has been controllably changed between the two measurements. Generally, one is searching for the correlation between what has been changed in the measurement and what spectral signatures that change gives rise to. For example, absorption features—their frequency, line-shape, and strength—can all be correlated to the fundamental origin of the absorption. Each absorption feature in a difference spectrum corresponds to an additional contribution in the dielectric function arising from the modified sample. In the applications presented below, various aspects of the derivation presented above are relevant to the variety of materials-related experiments afforded by IRSR. In all cases, we are ultimately interested in understanding some aspect of the dielectric function for bulk materials or surfaces of materials or the interface between two materials.

For IR spectroscopy, the process of interest is absorption. IR photons are absorbed by vibrations that induce *dynamic dipoles*. Dynamic dipoles are oscillations in the density of electrons or electron charge due to atomic motion (the electrons follow the motion of the nuclei). Dynamic dipoles can absorb a photon when the photon's electric field is parallel to the charge oscillation, and when the frequencies of the light and the oscillation are similar (resonance). Dynamic dipoles occur in both molecules and extended systems, and their natural frequencies are related to the masses of the displaced atoms. Vibrational modes are also referred to as *normal modes*, which are defined so that each normal mode is independent of all the other modes. Due to energy and momentum conservation (the momentum of a photon is equal to h/λ or E/c), only photons with energy and momentum matching that of an available excitation may be absorbed; light of higher or lower frequencies cannot couple to the excitation. An absorption spectrum is a plot that shows how well different frequencies of light couple to excitations in the sample. An absorption spectrum is commonly plotted as one of the following (see Fig. 1): (1) Percent transmittance T vs. frequency v, which plots the ratio of transmitted intensity to the incident intensity, or (2) Absorbance A vs. frequency v, which is related to the transmittance by $A = -\log T$ [It is conventional to convert the units for frequency v from Hz (s^{-1}) to wavenumbers (cm^{-1}) by dividing v by the speed of light c.]. Transmittance

(a) Transmittance

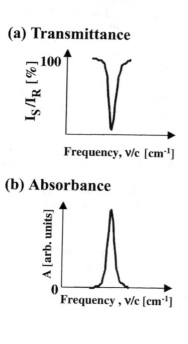

Figure 1. Infrared absorption spectra are plotted as either the percentage of (a) transmitted signal or (b) the absorbed signal versus frequency v (Conventionally, v is divided by c, where c is the speed of light). The transmittance is obtained by dividing a measured sample intensity I_S by a reference intensity I_R. Typically, a reference spectrum is obtained in reflection from a clean surface, or a gold reference mirror or in transmission through an open aperture. In these cases, the spectrum of interest, the sample spectrum, may be obtained from an adsorbate-covered surface, a sample mounted on a gold surface or a sample in transmission, respectively. In transmittance discrete absorption dips are observed at the energies of vibrational excitations. The absorbed signal A is equal to $-\log (I_S/I_R)$ and absorbance spectra exhibit discrete absorption peaks at the energies of vibrational excitations.

curves (Fig. 1a) exhibit absorption dips at energies characteristic of excitations available in the material. No additional absorption for a sample compared to its background measurement is manifested as 100% transmittance. Absorbance curves (Fig. 1b) exhibit peaks at energies where the sample has absorbed energy from the incident beam. Zero absorbance means no energy has been absorbed.

INFRARED SYNCHROTRON RADIATION PROPERTIES AND CONSIDERATIONS

IRSR sources exhibit many similar characteristics to UV-X-ray storage ring sources discussed in other chapters of this book. These sources are bright, broadband sources, with well-characterized timing and polarization structures, which generate highly reproducible measurements for challenging experiments with brightness limited geometries. In this section, calculations for IRSR power and brightness are shown. The inherent polarized emission and timing patterns are examined, and theoretical and measured reproducibilities for two brightness limited experimental geometries will be evaluated. Several comparisons will be made to a 2000 K black-body source, a standard lab-based, broadband infrared source.

IRSR power

The dipole radiation power emitted as a function of wavelength in the infrared region can be calculated using the following expression (Duncan and Williams 1983):

$$P(\lambda) = 4.38 \times 10^{14} \times I \times \theta \times BW \times (\rho/\lambda)^{1/3} \text{ photons/s} \qquad (10)$$

where I (Amps) is the beam current, θ (rads) is the horizontal collection angle, BW is the bandwidth, λ (m) is the wavelength and ρ (m) is the radius of the ring. BW is either % bandwidth or $\Delta v/v$ bandwidth. For example, 0.1% BW means ± 0.05 eV BW around 1 eV wavelength and is similar to that used for UV calculations. However in the infrared it can be more insightful to use a bandwidth based on $\Delta v/v$, where 2 cm^{-1} BW refers to a constant 2 cm^{-1} window around each frequency. Evident in Equation (10), the emitted power is directly proportional to the beam current and the opening angle. In practice, the most powerful sources have been extracted from low energy, small radius storage rings, since these facilities can maintain higher beam current conditions, and accommodate larger opening angles. The emitted power for a 1 Ampere stored beam in a storage ring that supplies a 90 mrad opening angle can be compared to the power emitted from a black body source. The most striking feature is that there is a crossover in power at approximately 100 μm, where, below this wavelength, the storage ring produces more photons/sec than the thermal source.

IRSR brightness

The brightness, or brilliance, of a source is defined as the power per source area per angle into which it emits light, which requires an accurate representation of these characteristics of the source. Next we will discuss the angle subtended by the source and the contributions to the horizontal and vertical source sizes, as functions of wavelength and storage ring extraction geometry, and give a general expression for the brightness of storage ring sources in the infrared.

The angle subtended by the source is given by the smaller of either the extraction angle or the characteristic natural opening angle, θ_{nat}, which is wavelength dependent: $\theta_{nat} = 1.66(\lambda/\rho)^{1/3}$ (Duncan and Williams 1983). Notice that the natural opening angle is proportional to wavelength to the third, which means that the angle is larger for longer wavelengths. In practice at smaller wavelengths, the natural opening angle is smaller than the extraction angle, and at larger wavelengths, the natural opening angle becomes greater than the extraction angle, which is determined by the physical limitations of the machine.

In general, there are three contributions (Hirschmugl 1994) to the horizontal and vertical source sizes: (1) the intrinsic size of the electron beam itself s_i; (2) the projected (observed) size due to the a large opening angle and hence extended source s_p; and (3) the diffraction-limited source size s_{diff}. The practical horizontal and vertical sizes are the sum of these contributions added in quadrature, thus for s_H (horizontal source size) $s_H = \sqrt{s_{Hdiff}^2 + s_{Hp}^2 + s_{Hi}^2}$. Frequently, both the horizontal and vertical sizes are dominated by that corresponding to the diffraction limit. The diffraction limit, s_{diff}, is given by the full width half maximum (FWHM) of λ/θ_{nat}. If the horizontal extraction angle is larger than the natural opening angle then s_p, the projected source size is an important contribution due to the curvature of the storage ring: the horizontal projected size is $\rho\theta_h^2/8$ and the vertical projected source size is approximately $\rho\theta_h\theta_{nat}/8$.

Recently, Murphy (Murphy 1999; Williams 1999) has determined a universal expression for the brightness for all electron storage rings, assuming that the opening angle of the storage ring matches the natural radiation opening angles, and that the intrinsic source size is smaller than that due to diffraction:

$$B(\lambda) = 3.8 \times 10^{20} \, I \times BW/\lambda^2 \text{ photons/s/mm}^2/\text{sr} \tag{12}$$

where I (Amps) is the stored current, BW (%) is bandwidth and λ (μm) is wavelength. The brightness for a storage ring source with 1 Amp stored beam and a 90 mrad collection angle can be compared to the black-body brightness. Notably, the storage ring brightness is between 3 and 4 orders of magnitude higher than the blackbody source. As shown below, for brightness limited experimental geometries one can collect up to 4 orders of magnitude more photons using a storage ring radiation source rather than a lab-based blackbody source.

IRSR polarization and pulsed patterns

IRSR has a characteristic emission pattern. Light polarized in the plane parallel to the electron orbit plane peaks in the same plane, while light polarized perpendicular to the orbit plane, peaks slightly out of the orbit plane (Green 1977; Williams 1982). Longer wavelengths are emitted at larger emission angles. The emission angle for the VUV is smaller than 10 milliradians while the emission angle for far infrared wavelengths (100 μm) is approximately 100 milliradians.

Storage ring radiation is emitted from bunched, relativistically accelerated electrons, whose bunch length is of the order of 50 to 500 picoseconds, depending on the characteristics of the storage ring. A radio frequency (RF) cavity provides the energy bursts to maintain the energy of the circulating electrons, which in turn sets the time structures that can be generated. The power of the storage ring pulse is adequate to probe time dependent phenomena that occur at timescales between the length of the pulse and the length of the time between pulses. To generate excitations that can be examined over this time scale, a mode-locked laser can be triggered by a signal from the RF cavity. The characteristics of the laser pulse are determined by the properties of the laser, and a pump pulse is generated for every probe pulse. (To learn more details about this timing structure see La Veigne et al. 1999).

Conservation of brightness: storage ring emittance and experiment throughput

An optimal optical design successfully focuses all of the available source photons onto the detector. Frequently, the limiting factor in these designs for black-body sources, is that the light emits in a 4π steradian angle, and is therefore difficult to completely collect with optical elements. However, the defining factor when designing with an IRSR source is most likely the experimental optical throughput (product of the area-angle acceptance of the experimental geometry). Ideally, the experimental optical throughput is equivalent to the source emittance (area-angle product of the source). This obeys Liouiville's theorem, which states that the brightness (power/emittance) of a source can, at best, be conserved through an optical path. Here we will introduce two brightness-limited experimental geometries, and show that the optical throughputs are well matched to the emittance of a bright storage ring source.

In Figures 2 and 3, two experimental geometries are shown; (1) grazing incidence reflection at 86° from the sample normal, and (2) micro-focusing to a 3 μm^2 spot size. Grazing incidence geometry is primarily used to examine surface related dynamics in the far-infrared, and micro-focusing geometry is used to investigate spatially varying chemical constituents which absorb in the mid- to far-infrared. The first geometry can be used for reflection absorption experiments. The reflectivity spectra for two "samples" are measured and subtracted from one another resulting in absorption features corresponding to the differences between the samples. The samples can be the same substrate prepared differently, such as a clean sample and adsorbate covered sample, or a sample with and without optical excitation. Otherwise, the substrate can be unchanged, and the reflectivity as a function of polarization (ellipsometry) can be measured and transformed into an anisotropic complex dielectric function for the material. Alternatively, the micro-focusing geometry can be used to measure spatially resolved transmitted and/or reflected signals from the sample. Recent experimental results for these applications will be given in the applications section below.

The optical throughput for an experimental geometry is equivalent to the product of the sample area and the acceptance angle. For grazing incidence geometry, the sample area is equal to the footprint of the beam at the given angle of incidence. Given a 1 mm beam size (typical beam size for diffraction limited wavelengths greater than 10 μm and no magnification or demagnification) reflected at 86° (chosen to produce the maximum

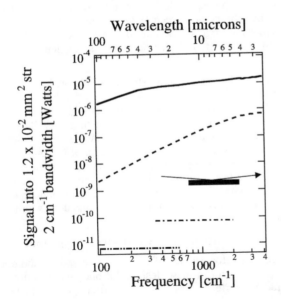

Figure 2. Broadband spectrum of a conventional 2000 K Globar IR source (short dashed line), and the spectrum of the NSLS synchrotron source (solid line) limited by an experimental throughput of 1.2×10^{-2} mm^{2}sr. This is the etendue of a grazing incidence geometry experiment to measure surface vibrational modes. The measured, background limited Noise Equivalent Power (N_{EP}) of a Cu-doped Ge photoconductive detector (dash-dot-dot) and of a Si Bolometer (dash-dot-do-dot-dash) are shown. The former detector is operated with no filter, while the latter detector has a 600 cm^{-1} filter, and both detectors are operated at liquid helium temperatures.

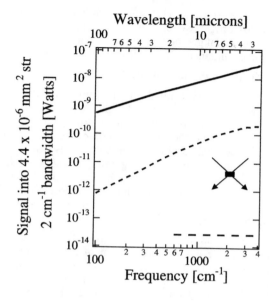

Figure 3. Broadband spectrum of a conventional 2000 K Globar IR source (short dashed line), and the spectrum of the NSLS synchrotron source (solid line) limited by an experimental throughput of 4.4×10^{-4} mm²sr. This is the etendue for a 1 μm by 1 μm sample measured with an infrared microscope. The measured, background limited Noise Equivalent Power (N_{EP}) of a Mercury Cadmium Telluride (MCT) (long dashed line) detector is shown. This detector is operated at liquid nitrogen temperatures.

E-field perpendicular to the sample) the optimal sample length is 1mm/(tan 86°) = 20 mm. In practice a longer crystal is beneficial, since the beam has an angular spread of 5° (100 mrad). The beam height is also 1 mm, however, typically a taller sample is used for easier alignment. Thus, using a 25 mm long × 5 mm high sample, the optical throughput is:

$$\text{Etendue (gr. inc.)} = (1\times10^{-3}\text{ m})(25\times10^{-3}\text{ m }(\cos86°))(0.09\text{ rad})(0.09\text{ rad})$$
$$= 1.4\times10^{-8}\text{ m}^2\text{sr} \tag{13}$$

The optical throughput for the grazing incidence geometry is well matched to the storage ring source emittance as shown here. The emittance of a source is the product of the source area times the subtended angle. The source size at a wavelength of 10 μm is diffraction limited and equal to 1 mm² for the U4IR beamline at the National Synchrotron Light Source, which has an opening angle of 90 mrad × 90 mrad. Thus, the source emittance is:

$$\text{Source emittance (gr.inc.)} = (1\times10^{-3}\text{ m})(1\times10^{-3}\text{ m})(0.09\text{ rad })(0.09\text{ rad})$$
$$= 0.8\times10^{-8}\text{ m}^2\text{sr} \tag{14}$$

In spectromicroscopy, one encounters an even smaller etendue. One can investigate a sample area of as small as 3 μm², with an angular acceptance of 0.7 rad (41°) has an etendue of:

$$\text{Etendue (mic.)} = (3\times10^{-6}\text{ m})(3\times10^{-6}\text{ m})(0.7\text{ rad })(0.7\text{ rad})$$
$$= 4.4\times10^{-12}\text{ m}^2\text{ sr} \tag{15}$$

This limited experimental throughput is advantageous for frequencies where the chemical "fingerprint" infrared region lies (20-2.5 μm or 500-4000 cm^{-1}). The storage ring source size at a wavelength of 1 μm, for a 40 mrad × 40 mrad opening angle (U2A and B and U10 A and B beamlines at Brookhaven) has an intrinsic source size of 200 μm × 400 μm and emits into a natural opening angle of 0.014 rad. Thus the emittance is:

$$\text{Source emittance (mic.)} = (2\times10^{-4}\text{ m})(4\times10^{-4}\text{ m})(0.014\text{ rad})(0.014\text{ rad})$$
$$= 15.7\times10^{-12}\text{ m}^2\text{sr} \tag{16}$$

The storage ring source emittance is larger than the optical throughput for microscopy, however, the storage ring source is clearly a better matched source for this experimental geometry compared to a blackbody source of infrared radiation which has a source emittance of $63 \times 10^{-6} \, m^2 sr$.

Diffraction limited infrared imaging

Diffraction limited infrared images have a spatial resolution equal to 1/2 the wavelength of the incident light. When the wavelength of light is comparable to an aperture it traverses, the light is diffracted—some portion of the intensity generates a pattern outside the original confined beam path. For mid-IR spectroscopy diffraction-limited spatial resolutions are between 1-8 μm for wavelengths between 2-16 μm. This limit can be achieved using storage ring-based IR microspectroscopy with a confocal microscope geometry with input and exit effective aperture sizes that are equal to or smaller than the wavelength of probing light (Carr 2001). The optimal spatial resolution is achieved when the instrument's apertures define a region having dimensions equal to the wavelength of interest.

For example, the infrared image for an absorption feature at 1000 cm^{-1} ($\lambda = 10 \, \mu m$) can achieve a spatial resolution of 5 μm if the apertures are 10 μm or smaller, with optimal results at 10 μm. However, it is frequently desirable to measure a large spectral region (4000-650 cm^{-1}, or 2.5 μm-16 μm) concurrently. Five μm apertures can typically be used with adequate signal to noise when the specimen is partially absorbing. Thus, the spatial resolution for the higher frequency absorption features (4000-1000 cm^{-1}, or 2.5-10 μm) would be 5 μm. Any infrared images for absorption bands below 1000 cm^{-1} would have a frequency dependent spatial resolution. However, one must use caution, since it is also possible that diffraction effects may alter the strength, but not typically the shape, of absorption bands for longer wavelength features (Bradley 1999).

Alternatively, one may be interested in longer wavelength features of a specimen that is smaller than the wavelength that corresponds to the absorption band frequency. In particular many absorption bands of minerals, e.g., metal-oxide and sulfides, have absorption frequencies which correspond to wavelengths of 20 μm and longer. Bradley and coworkers (Bradley 1999) have shown that they required a 12 μm aperture to successfully measure a silicate band at 9.8 μm for particles that were only a couple of microns in size. Measuring the same particles with smaller apertures suppressed the absorption band of interest. Thus, one must carefully choose the aperture sizes that are most appropriate for the given experiment.

It is also worth noting that there are efforts underway to develop near-field IR spectromicroscopy to further improve the spatial resolution perhaps to a fraction of the diffraction limit (Keilmann 1999; Talley 2000).

Spectral reproducibility and signal-to-noise

A high signal-to-noise (S/N) ratio is desirable for all spectroscopic measurements, and to achieve high S/N ratios for the experiments described here, one must obtain high quality reproducibility. For example, for the grazing incidence, reflection absorption spectroscopy experiments, we need to be able to measure the absorption caused by a small fraction of a monolayer of weakly absorbing molecules on a surface. These absorption modes are small compared with natural features in the spectra and are only observable when compared to (divided by) reference spectra which are taken on the clean substrate immediately prior to dosing. The "signal" of interest is really the *change* in the spectra. Thus, it is instructive to determine the reproducibility of the signal. For these experiments the effective S/N ratio is the relationship between the strength of the absorption feature and this reproducibility.

Here we examine the theoretically achievable limit for the spectral reproducibility using the storage ring source to investigate a sample with a limited throughput. In Figure 2, we show brightness versus frequency for the storage ring source and for a blackbody source limited by the 1.2×10^{-2} mm^{2}sr throughput of the experiment (Hirschmugl and Williams 1995). In addition, the noise equivalent power (N_{EP}) for two liquid helium cooled detectors, Cu doped Ge photoconductive detector with no filter, and Silicon Bolometer, are included. The N_{EP} corresponds to the noise level of the detectors. The signal from the storage ring is 3 orders of magnitude larger than that from the globar; however, the most striking advantage is the ratio between the N_{EP} of the detector and the signal from the storage ring. The reproducibility at 400 cm^{-1} is $(7 \times 10^{-12}/8 \times 10^{-6}) \times 100$, which equals 0.00001% in one second measuring time. While the efficiency of the optical system is probably only 1 to 10%, the typical measuring time is between 30-200 seconds, and this reproducibility is recovered. In Figure 3, a similar comparison between detector noise, storage signal and globar signals can be calculated for the diffraction limited spectromicroscopy geometry. In particular, when using a small MCT detector (1/4 mm × 1/4 mm) with the storage ring source, reproducibilities of $(2.5 \times 10^{-14}/2 \times 10^{-8}) \times 100$ at 2000 cm^{-1} or 0.00001% in one second measuring time can be obtained, ideally. The efficiency of the optical system is similar, and the typical data is collected for up to 500 seconds. Again, the signals are 2 to 3 orders of magnitude higher than one can collect with a globar source. In sum, a storage ring radiation source coupled to a limited throughput experiment produces substantially higher quality data than one can obtain with a globar source.

Status of IRSR facilities

While most storage ring facilities are optimally designed to emit X-rays or UV radiation, all synchrotrons emit IR radiation. However, since IR wavelengths are much longer than X-rays and UV radiation, substantial modifications are typically required to extract IR radiation from storage rings. In most cases, IR beamlines have been constructed on storage rings optimized for UV radiation.

There are several sources of storage ring IR radiation available worldwide. IR spectromicroscopy facilities have been established at storage rings in the US (National Synchrotron Light Source, Brookhaven, NY; Advanced Light Source, Berkeley, CA; Aladdin, Stoughton, WI) and in Europe (Synchrotron Radiation Center, Daresbury, UK; LURE, Orsay, France). There are many other synchrotrons with IR capabilities that could accommodate IR microscopes located in the US, (NIST, Gathersburg, VA) Europe (MAX, Lund, Sweden; Bessy-2, Berlin, Germany), and Japan (UVSOR, Okasaki, Japan). In addition there are many facilities that are constructing or developing IR capabilities (CAMD, Baton Rouge, LA; Delta, Dortmund, Germany; ANKA, Karlsruhe, Germany; DAPHNE, Rome, Italy; SOLEIL, France; European Synchrotron Radiation Facility, Grenoble, France; Spring-8, Japan; SRRC, Taiwan; Hefei, China; Canadian Light Source, Saskatoon, Canada; Campinas, Brazil; Singapore).

APPLICATIONS: IRSR SPECTROMICROSCOPY

Recent advances in IR spectromicroscopy have developed the technique into an incisive, routine method to obtain spatially resolved chemical maps for a wide range of samples, including samples of biological (Jamin et al. 1998), geological (Guilhaumou et al. 1998; Hofmeister 1995), and environmental molecular sciences (Holman et. al. 2000; Ghosh et al. 2000) interest. IR microspectrometers, microscopes using collinear IR/visible reflective optics coupled to an interferometer, afford the opportunity to optically choose a sample area of interest and to collect an IR spectrum for the same area. As the sample is rastered through the focus of the beam, individual spectra are collected through a fixed aperture as a function of sample position. A chemical map is created from these spectra by determining the strength of an absorption band at each measuring position. Different absorption bands are used to create additional chemical maps.

Single cell chemistry

IR spectromicroscopy affords the opportunity to examine the chemical nature of single living cells, in contrast to many standard experimental techniques where cells are typically destroyed and studied enmasse. It is well established that many constituents of cells give rise to distinct infrared absorption signatures, including for example, proteins, lipids and sugars (Stuart 1997). For example, proteins always have two absorption bands at 1650 and 1560 cm^{-1} identified as the Amide I and Amide II bands. These bands correspond to predominantly the C=O stretching vibration and a mixture of the N-H bending and C-N stretching vibrations, respectively, in the amide group of proteins (NHCO). Recently, several IRSR spectromicroscopy investigations of intracellular chemistry have utilized these basic IR absorption signatures as a starting point. There are two different foci in these studies. One approach is to obtain spatially resolved images of the intracellular constituents of live cells. The other is to investigate modified chemistry in single living cells as a function of exposure to external stimulants.

Spatially resolved distribution of proteins and lipids in living mouse cells. Jamin et al. (1998) were the first to map the distribution of functional groups, including proteins, lipids and nucleic acids, inside a single living cell with a spatial resolution of a few micrometers using IR storage ring radiation. They have studied a cell undergoing reproduction and necrosis (cell death), noting, in the latter case, the appearance of a carbonyl ester group. The cells, derived from UN2 hybridoma B cells, were cultured in a controlled environment, and then deposited on a transparent IR window (BaF$_2$) by low-speed cytospin centrifugation. These cells were measured in transmission. In Figure 4 (Fig. 3 from the paper) the optical image and two contour plots of the distribution of proteins and lipids are shown for a cell undergoing mitosis. In the optical image it is clear that the cell wall between the two daughter cells has not been fully developed. The infrared image of the proteins shows two clear maxima that are indicative of two independent nuclei. The infrared image of the lipids shows two minima at the same positions as the protein maxima. Notably, the distribution of the lipids at this stage already shows a high concentration in the central region where the contractile ring responsible for the cleavage furrow is located. While the identification of the functional groups can be challenging, it is possible that this technique will allow the possibility of new insights into intracellular chemical changes that correlate to cell modifications.

Spatially resolved distribution of proteins, lipids and carbohydrates in algae: Euglena gracilis. Euglena gracilis are a fresh water algae (optical image of a single cell is 20 μm × 20 μm—Fig. 5a) that have been examined with IR spectromicroscopy by Hirschmugl et al. (2002a). They were deposited onto a gold-coated slide, and measured using a reflection geometry. The IR images with spatial resolution of 5 μm × 5 μm are shown for the lipids and proteins for an alga cell are in Figure 5b-d. They show a low level concentration distributed over the entire cell, and a higher, overlapping, concentration located towards one end of the cell. These contour maps are consistent with the known chemistry of *Euglena gracilis* (Arnott and Walne 1967). In particular, *Euglena gracilis* is typically girdled by a pellicle (or periplast) located inside the lipid bilayer of the plasmalemma. The pellicle is composed mostly of protein with smaller concentrations of lipids and carbohydrates (Bold and Wynne 1985). Cells of *Euglena* are not known to accumulate large amounts of lipids as storage products, even if occasional phospolipid granules are present in the cytosol, most of the lipids in the cells are therefore associated with membranes, a large part of which is represented by the endoplasmic reticulum and

Cell in division

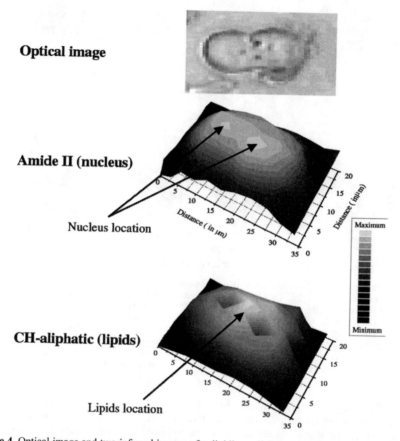

Optical image

Amide II (nucleus)

Nucleus location

Distance (in μm)

Distance (in μm)

Maximum

Minimum

CH-aliphatic (lipids)

Lipids location

Distance (in μm)

Figure 4. Optical image and two infrared images of a dividing cell (35 μm × 20 μm) obtained from IR spectromicroscopy. These chemical maps are derived from the strength of absorption bands from approximately 100 infrared transmittance spectra collected over the area of the dividing cell. The maximum in the color scale (yellow) represents the position of the maximum absorption by the cell, while the minimum (blue) represents no absorption by the cell. Strikingly, the intensity map of the amide II absorption band shows two peaks in the center of the two halves of the cell, representing the position of two separate nuclei before the cell division is complete. The intensity map of the C-H stretch absorption bands shows that lipids are concentrated at the contractil ring, where the cleavage furrow is located. [Used by permission of the National Academy of Sciences, U.S.A., from Jamin et al. (1998), *Proc Natl Acad Sci*, Vol. 95, Fig. 3, p. 4839.]

by the membrane systems of the chloroplasts (i.e., the three peripheral membranes and the thylakoids) (Gibbs 1978, 1981). Proteins are also not expected to produce a strong signal in the cytosol. The highest proportion of protein is likely to be in the chloroplasts, and specifically in the pyrenoids, where the primary photosynthetic enzyme ribulose bisphosphate carboxylase (rubisco) can be very abundant (Badger et al. 1998). Considering this physiology, it is plausible that the maxima in the protein and lipid images correspond to chloroplasts.

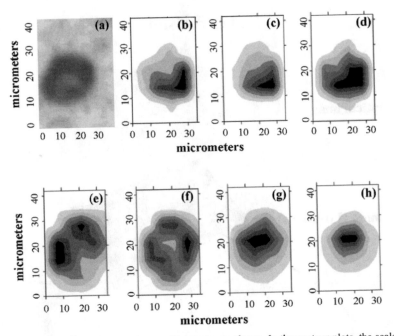

Figure 5. One optical image (a) and seven IR images are shown. In the contour plots, the scale ranges from black (maximum) to white (minimum). The IR images are derived from the peak strength as a function of position for the following peaks: (b) 1545 cm⁻¹ (N-H bending and C-N stretching for protein group—amide II), (c) 2920 cm⁻¹ (C-H asymmetric stretch for lipids), (d) 1742 cm⁻¹ (C-O stretch for phospholipids), (e) 973 cm⁻¹ (unidentified), (f) 937 cm⁻¹ (unidentified), (g) 1040 cm⁻¹ (unassigned for starch) and (h) 1014 cm⁻¹ (COH deformation for sugar). The backgrounds that were chosen for each IR image are between 1600-1500 cm⁻¹ for image (b), between 2980-2872 cm⁻¹ for image (c), between 1700-1780 cm⁻¹ for image (d), and between 3750-680 cm⁻¹ for images (e), (f), (g) and (h).

The IR images for the carbohydrate (Fig. 5g-h) absorption bands can also be reconciled with *Euglena gracilis* biochemistry. The IR image for the absorption band at 1040 cm⁻¹ (Fig. 5h) shows a maximum that is confined to a small area, neighboring the maxima in the protein. In contrast, the IR image for sugars (Fig. 5g) shows less localized distributions with a weaker gradient. *Euglena gracilis* accumulates paramylon, a β 1-3 polymer of glucose. Paramylon production is often associated with pyrenoids, but free granules of this polysaccharide are also frequently observed in the cytosol. Generally speaking, therefore, paramylon is much more localized than soluble sugars. It is very likely that the maximum in panel h of Figure 5 is the site of a concentration of paramylon located next to the pyrenoid. In comparison to the distribution of the paramylon, the soluble sugars will be diffuse and found both in the chloroplasts, due to synthesis of glucose, and the cytosol, where sugar metabolism occurs. The spectral fingerprints for the spectra at the maximum in panels h and g agree very well with the published data for starch (Giordano et al 2001), which is very similar to paramylon, and glycogen (Diem et al. 1999), respectively.

Finally, the IR images for the absorption features observed at 973 cm⁻¹ and 937 cm⁻¹ (Fig. 5e-f) show a minima in the center of the algae with a maximum around the edge of the cell. The authors propose that these absorption bands could possibly originate from a gelatinous sheath: in fact, a thick mucilaginous matrix could coat the cells, if they have become encysted or are in a palmelloid phase; moreover, even vegetative *Euglena* cells have rows of "mucilage-producing bodies" under the pellicle extruding mucilage to the exterior through canals (Bold and Wynne 1985). These investigations can also be used to examine the changes in cell chemistry as a function of exposure to external stimuli (Hirschmugl 2002b).

Human cell response to environmental organic toxins. Holman et al. (2000) have investigated the changes in intracellular phosphate and lipid concentrations in human cells as a function of exposure to polychlorinated hydrocarbons and compared their findings to quantitative results from reverse transcription polymerase chain reaction (RT-PCR). The human cells were exposed to low doses of 2,3,7,8-tetrachlorodibenzo-p-dioxin (TCDD), a potent and well-studied man-made toxin. TCDD causes harmful effects at exposure levels of hundreds to thousands of times lower doses than most chemicals of environmental concern. This polychlorinated hydrocarbon causes an increase in the production of the cytochrome P4501A1 (CYP1A1) gene, which can be monitored with RT-PCR.

Holman et al. (2000) exposed the cells to 10^{-11}, 10^{-10}, and 10^{-9} M TCDD dissolved in dimethyl sulfoxide for 20 hours and control experiments were performed on cells which were not exposed to TCDD. After this sample preparation, some cells were analyzed using the RT-PCR technique. Other cells were deposited on a gold surface and infrared spectra were collected in reflection for individual cells. These individual cells were all similar in size (20 µm). Line scans were collected across each cell and used to determined within ± 2 µm the center of each individual cell. The most absorbing, thickest part of each cell was used to collect the infrared spectra for further analysis. The data were normalized to the absorption strength of the Amide II band.

While the overall IR absorption signatures were similar to those observed before, the IR absorption signatures that showed the most significant changes as a function of exposure to TCDD include the symmetric and asymmetric stretching of phosphate functional groups. The ratio of the absorption band for symmetric PO^{2-} stretching compared to the absorption band for asymmetric PO^{2-} stretching increases with respect to the increased exposure to TCDD. In addition, the relative changes in this ratio increase in direct proportion to the logarithm of the expression of the CY1P1A gene over the two orders of magnitude difference in exposure. Other studies of malignant samples from a colorectal cancer tissue study and a lung cancer cell study observed a change in the ratio of these two absorption bands when compared to normal cells (see Fig. 6). The absorption bands did not shift as a function of exposure. The authors suggest that the hydrogen-bonding environment of the phosphate backbone in the nucleic acids is weak, and remains unchanged throughout the study. This finding is in contrast with previous studies of cancerous tissues, where the bands soften and suggest increased hydrogen bonding. The authors also examined the variation in the ratio of methyl to methylene groups, indicating differences in average hydrocarbon chain length in the lipids.

Microbial reduction of Cr^{6+} on mineral surfaces. Holman et al. (1999) have also studied the conversion of Cr(VI) to Cr(III) on a magnetite surface covered with living *A. oxydans* cells (biotic) in the presence of toluene. The mineral/microbe was exposed to chromate and toluene and then monitored with IR spectromicroscopy for changes in Cr(IV), Cr(III) and toluene absorption features at and around the living cells. After five days the spatial maps showed a significant decrease in the chromate and toluene intensities at the same position as the living cells, identified by amide absorption bands. These results suggest a close link between the biodegradation of toluene and microbial reduction of Cr(VI). Further investigations showed that there was a lack of significant reduction of the chromate species in absence of the living cells.

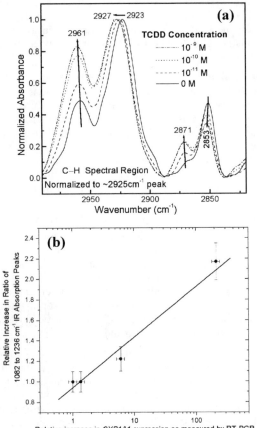

Figure 6. (a) Infrared spectra in the phosphate band region for cells treated with 0, 10^{-11}, 10^{-10}, 10^{-9} M TCDD. Spectra have been normalized to the protein amide II intensity to account for cell to cell thickness variations. (b) The ratio of the 1082-1236 cm^{-1} peak intensities as a function of TDD concentration. [Used by permission of the American Chemical Society, from Holman et al. (2000), *Environ Sci Tech,* Vol. 34, Figs. 2 & 4, p. 2514-2516.]

Spatially resolved chemistry in inhomogeneous materials

Examples of inhomogeneous materials that have distinct infrared absorption signatures include minerals, bone and chemicals that are sorbed at surfaces or are inside inclusions of geological specimen. IRSR spectromicroscopy has been useful in identifying very small samples, or revealing spatially-resolved components in functionally gradient materials. All of these experiments require diffraction limited or close to diffraction limited spatial resolution.

Fluid chemical analysis and mapping. Guilhaumou et al. (1998) have examined inclusions in geological specimens with IR spectromicroscopy, creating chemical distribution contour plots of the inclusions. These authors have studied a fluorite inclusion of Tunisia (inclusion A) using a transmission geometry. A typical IR absorbance spectrum of a 3 μm × 3 μm sample area is shown in Figure 7. The sharp absorption band at 2338 cm^{-1} is due to the anti-symmetric stretching vibrations of CO_2, and is shifted with respect to gas phase CO_2 bands. Other absorption bands include weak bands at 3592 and 3697 cm^{-1} and strong bands between 3000-2800 cm^{-1}. The former set are due to OH stretching vibrations in hydroxyl groups, and the latter set are due to symmetric and anti-symmetric CH stretching vibrations in hydrocarbons. Contour plots are derived for these three sets of bands are included in Figure 7. The contour plot for the

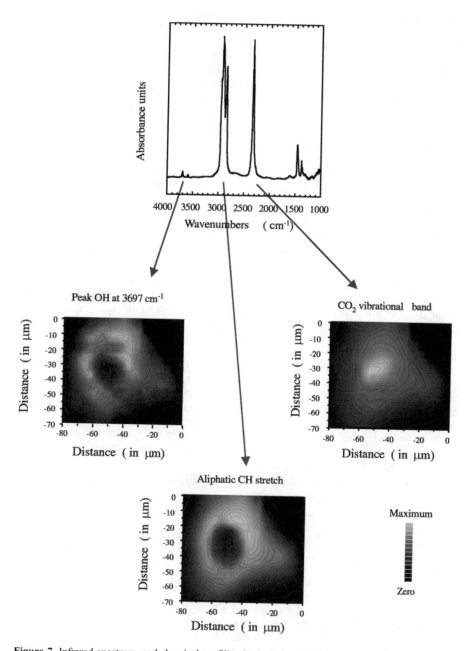

Figure 7. Infrared spectrum, and chemical profiling hydrocarbon fluid inclusion in fluorite (inclusion A). (a) Typical infrared spectrum recorded with an aperture of 3×3 μm². hemical profiling is reported for the band located at 3697 cm⁻¹ attributed to hydroxyl (b), for the CO_2 band (c), and for the aliphatic compounds (d). [Used by permission of the Society for Applied Spectroscopy, from Guilhaumou et al. (1998), *Appl Spectroscopy*, Vol. 52, Fig. 3, p. 1031.]

CO_2 band shows a concentration in the center of the inclusion. The contour plots for the 3697 cm^{-1} hydroxyl band and for the integrated intensity under the set of absorption bands between 3000-2800 cm^{-1} are very similar with higher concentration towards the outer edges of the inclusion. The authors conclude that the center of the inclusion is in a vapor phase that consists mostly of CO_2 under high pressure. In addition, the authors deduce that the surrounding material is in a liquid phase, and consists of aliphatic compounds terminated by hydroxyls. These results should motivate further studies to identify and understand the chemical nature of compounds inside inclusions.

Identification of chemical composition of interstellar grains. Bradley et al. (1999) have measured IR spectra for interplanetary dust particles (IDPs) collected in the stratosphere from asteroids and comets. The authors were concentrated on collecting the absorption signatures of glass with embedded metal and sulfides (GEMS) and comparing these signatures to emission and absorption spectra for interplanetary objects. Collecting the IR spectra from the IDPs is a challenging experiment, since the particles are very small (μm) and are frequently mixtures of the GEMS with other glassy silicates. However, regions (~ 6 μm in diameter) existed where GEMS are the only silicate present. The authors published five IR spectra focusing on the signature for the Si-O stretching band at 10 μm. The one absorption spectrum is representative of the GEMS rich IDP, with a very broad, featureless band, a peak at approximately 9.8 μm, and asymmetry on the long wavelength side. This spectrum is very similar to the emissivity of four interstellar objects. Two similar spectra are from the interstellar molecular cloud environments around two stars, Elias 16 (in Taurus) and Trapezium (in Orion). The dust around a young stellar object also emits a similar spectrum, which is recorded from DI Cephei (a T Tauri Star). Finally, a narrower emission band with a similar shape has been observed for an evolved M-type supergiant star, μ-Cephei. Similar comparisons have been made between the absorption spectrum for another IDP and the emission spectra for Comet Hale Bopp and Comet Halley. In this case the composition of the particles is a mixture of GEMS, enstatite, forsterite and other glassy silicates. The authors conclude, after introducing further corroborating evidence, that a presolar interstellar molecular cloud could consist of GEMs, and presume that the solar system was formed from this molecular cloud. The authors suggest that if indeed GEMS are astronomical amorphous silicates, that one of the long sought building blocks of the solar system has been found.

Microscale characterization and localization of hydrocarbons in sediment. Ghosh et al. (2000) have studied sediment collected from the Milwaukee Harbor with a combination of several techniques including IR spectromicroscopy. The goal of these measurements was to determine the distribution of polycyclic aromatic hydrocarbons (PAH) sorbed on or within different components of sediment, including silica, coal and wood particles. Samples were examined with scanning electron microscopy with wavelength dispersive X-ray spectroscopy (SEM/WDX) and IR spectromicroscopy to evaluate the elemental and organic components in the sediment in addition to microprobe laser desorption/laser ionization mass spectrometry. The latter technique can detect extremely low concentrations (attomoles) of PAH. In one experiment, the authors measured the reflected infrared intensity from individual silica particles. Figure 8 shows a contour map of the distribution of C-H vibrational stretch absorption bands for a 200 × 200 μm region. The contour map reveals that organic material sorbs in patches at the surface of silica particles, and since the absorption bands are predominantly between 2800-3000 cm^{-1}, the organic material is predominantly aliphatic (hydrocarbon chains). Complementary laser desorption measurements show that PAH is sorbed on silica only in co-existence with the patches of the aliphatic organic material.

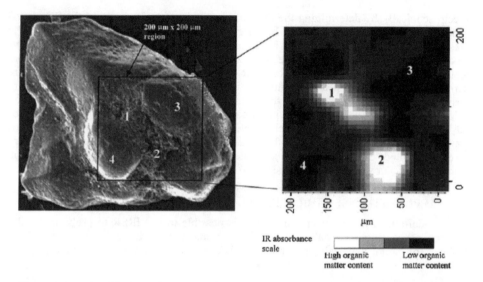

Figure 8. Scanning Electron Microscope (SEM) image of a silica particle having patches of organic matter as indicated by the white regions in the IR mapping of the C-H stretching absorbance (2800–3000 cm^{-1}) shown in the right panel. Regions 1 and 2 also showed higher abundance of aluminum and magnesium compared to spots 3 and 4 as determined by SEM-WDX. [Used by permission of the American Chemical Society, from Ghosh et al. (2000), *Environ Sci Tech*, Vol. 34, Fig. 5, p. 1733.]

The authors also examined coal and wood particles within the sediment. They found a larger concentration of PAH within these particles compared to the silica particles. They also determined that while organic carbon molecules were present uniformly throughout these particles in mixtures of aliphatic and aromatic (cyclic) forms, the PAH was concentrated at the surfaces of these particles. The conclusions derived from this powerful combination spectroscopies will allow a more successful approach to remediation of the Milwaukee Harbor. These results could also have a larger impact, since they show the feasibility of identifying chemical constituents and their distributions in sediments which are important in planning remediation efforts in different environs.

In situ analysis of mineral content and crystallinity in bone. Bone, a functionally gradient material, is composed of protein and mineral components which give rise to spectral absorptions in the mid and far-infrared spectral range. Recently, Miller et al. (2001) have initiated an investigation of cross sections of human iliac crest bones, collecting the IR absorption spectra around a human osteon. The focus of this investigation was to measure the acid phosphate content and determine mineral crystallite perfection from the IR spectra. The crystallite perfection was determined from a concurrent study of the correlation of IR absorption spectra with X-ray powder diffraction results from a series of synthetic hydroxyapatite crystals and natural bone powders of various species and ages.

The authors have examined a low frequency vibration of PO_4^{3-} (v_4) (500-650 cm^{-1}), which is accessible with a Cu doped Ge photoconductive detector coupled to the IR microscope, and has fewer components than the more commonly studied PO_4^{3-} (v_1,v_3) bands (900-1200 cm^{-1}). The combined X-ray and IR spectroscopic studies revealed that as bone matures the average crystallite size increases, and this can be monitored spectroscopically by determining the ratio of two peaks (603/563) within the PO_4^{3-} (v_4) band. Another peak near 540 cm^{-1} has been assigned to the acid phosphate, which is present in high concentrations in new bone, and decreases as the bone ages. Figure 9 shows an optical image of a human bone specimen with an osteon and a superimposed square that represents the area over which a series of IR spectra were collected. IR images, which are derived from the ratio of peak strengths within the PO_4^{3-} band as a function of measuring position, are shown in the bottom of the figure. The contour plot on the left (603/563) and (538/563) represent the variation in phosphate content and crystallinity as a function of position around the osteon. High acid content (crystallinity) is white in the right (left) contour plot. This data shows that the acid phosphate content and the crystallinity are inversely related. The authors also conclude that the most dramatic changes in osteon composition occur within 30 μm of the site of the osteon center.

APPLICATIONS: VIBRATIONAL SPECTROSCOPY AT SURFACES

In addition to IR spectromicroscopy it is possible to use IRSR to probe molecular interactions at single crystal surfaces. The chemical and physical properties of the surfaces of solid materials and the interfaces between solids and fluids play an

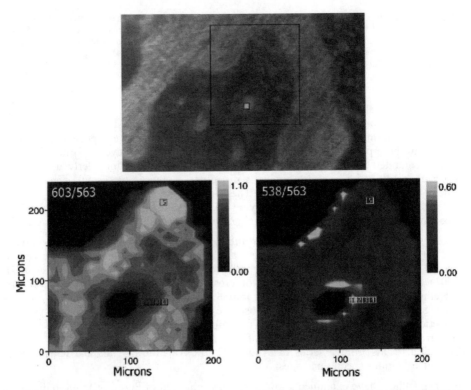

Figure 9. Optical and infrared images of human, osteopetrotic, trabecular bone. Data were collected at Beamline U4IR at the NSLS using a Spectra Tech Irμs microscope and a Cu doped Ge detector (Infrared Laboratories). Apertures were sent at 12 by 12 μm and 128 scans were collected at 4 cm^{-1} resolution. Infrared images were generated by plotting peak height ratios of (left) 603/563 cm^{-1} and (right) 538/563 cm^{-1}. [Used by permission of Elsevier Science, from Miller et al. (2001), *Biochimica at Biophysica Acta*, Vol. 1527, Fig. 8, p. 17.]

overarching role in a host of processes of interest to geochemistry and environmental science. For instance, information can be gained about interactions between adsorbates and model mineral surfaces. In particular a molecular level understanding of hydroxylation and hydration of mineral surfaces is of great interest to geochemists. Furthermore, the interaction of contaminants with mineral surfaces and hydroxylated or hydrated mineral surfaces is a rich area for further investigation.

Fe_3O_4-water interface

Recently, studies by Pilling et al. (1999) of the Fe_3O_4-water interface using IRSR have shown the feasibility of investigating the vibrational modes between the bonding adsorbate layer and the substrate, whose frequencies provide information about the adsorbate-substrate bond strength. Magnetite is a metallic transition metal oxide, requiring grazing incidence geometry for IR investigations of adsorbate species, since the substrate is both reflective and strongly absorbing. When this geometry is required, IRSR provides approximately 100 times larger density of photons for probing the interface than a standard black body source(see Fig. 2).

The Fe_3O_4 thin oriented crystal is 2000 Å thick, grown on an MgO(100) substrate. Figure 10 shows two Far-IRAS spectra for $H_2O/Fe_3O_4(100)/MgO(100)$ at coverages of 1 monolayer (bottom) and 10 monolayers (top). These data have been obtained by dividing an adsorbate covered reflectivity spectrum by a clean surface reflectivity spectrum. The experimental data have been normalized to the synchrotron beam current, and have a regular 20 cm^{-1} oscillation arising from multiple reflections in the detector. In the bottom panel, a simulated reflectivity spectrum at an angle of incidence of 80° for a three-layer model is shown. The simulation requires dielectric functions for vacuum, the adsorbate layer and substrate layer, and uses a formalism derived from Maxwell's equations to predict the reflected signal. The top layer is vacuum, the middle layer is 1 monolayer of adsorbate, with a discrete absorption at 600 cm^{-1}, and a full-width-half maximum (FWHM) of 20 cm^{-1}; the bottom layer is a thick oxide substrate, with a phonon at 400 cm^{-1}, and a FWHM of 100 cm^{-1}. There are several comparable features in the experimental results and the simulated spectrum. Notice, in both the measurement and the

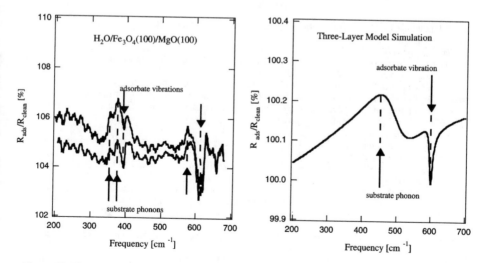

Figure 10. Three Layer simulation and Infrared Reflection Absorption Spectra for 1 and 10 monolayers of water adsorbed on Fe_3O_4. This data was taken at beamline U4IR at the NSLS.

simulation, the data lie above 100%. This result indicates that the adsorbate-substrate system is more reflective than the substrate. In addition, in the measured data the anti-absorption features correspond to Fe_3O_4 phonons at 360 cm^{-1}, 380 cm^{-1} and 570 cm^{-1}, similar to the broad feature at 450 cm^{-1} in the simulated data. Finally, the absorption features in the measured data at 400 cm^{-1} and 620 cm^{-1} are similar to the narrow absorption at 600 cm^{-1} in the simulated data. These results show promise for further adsorbate/substrate studies in the far infrared that can reveal the bonding strengths of varying adsorbate-substrate bonds.

SYNOPSIS

Storage ring IR radiation is a valuable tool for a wide range of investigations of interest in geochemistry and environmental science. IR spectroscopy, in general, has become a routine and incisive tool to provide chemical-specific information, in a wide variety of environments, and at surfaces and interfaces, all of import to geochemical systems, since IR radiation is weakly interacting with many materials, is polarization-dependent, and chargeless. As shown in this paper a bright storage ring source, coupled with stable interferometers and high quality detectors, provides the opportunity to do high resolution IR spectromicroscopy and surface-sensitive single crystal spectroscopy. Some illustrative examples presented here include *in situ* IR imaging of living cells, investigations of chemistry of living cells and measurements of the bonding vibrational mode between water and single crystal oxide surfaces. With the number of IR facilities at storage ring sources increasing dramatically, it is possible that these sources will become an indispensable tool in forefront research in geochemistry and environmental science.

ACKNOWLEDGMENTS

The author is grateful to Neil Sturchio for his guidance and advice in preparing this contribution. This work was supported in part by the following grants: NSF-CHE-9984931 and NSF-DMR-9806055.

REFERENCES

Arnott HJ, Walne PL (1967) Observations in the fine structure of the pellicle pores of *Euglena granulata*. Protoplasma. 64:330-344
Badger MR, Andrews TJ, Whitney SM, Ludwig M, Yellowlees DC, Leggat W, Price GD (1998) The diversity and co-evolution of rubisco, plastids, pyrenoids and chloroplast-based CO_2 concentrating mechanisms in algae. Can J Bot 76:1052-1071
Bold HC, Wynne MJ (1985) Introduction to the algae. 2nd edition. Prentice-Hall, Englewood Cliffs
Bradley JP, Keller LP, Snow TP, Hanner MS, Flynn GJ, Gezo JC, Clemett SJ, Brownlee DE, Bowey JE (1999) An infrared spectral match between GEMS and interstellar grains. Science 285:1716-1718
Carr GL (2001) Resolution limits for infrared microspectroscopy explored with synchrotron radiation. Rev Sci Inst 72:1613-1619
Diem M, Boydston-White S, Chiriboc L (1999) Infrared spectroscopy of cells and tissues: shining light onto a novel subject. Applied Spectroscopy 53:148A-161A
Duncan WD, Williams GP (1983) Infra-red synchrotron radiation from electron storage rings. App Optics 22:2914-2918
Ghosh U, Gillette JS, Luthy RG, and Zare R (2000) Microscale location, characterization, and association of polycyclic aromatic hydrocarbons on harbor sediment particles. Environ Sci Tech 34:1729-1736
Gibbs SP (1978) The chloroplast of *Euglena* may have evolved from symbiotic green alga. Can J Bot 56:2883-2889
Gibbs SP (1981) The chloroplast of some alga groups may have evolved from endosymbiotic eukaryotic algae. Ann N.Y. Acad Sci 361:193-208
Giordano M, Kansiz M, Heraud P, Beardall J, Wood B, McNaughton D (2001) Fourier transform infrared spectroscopy as a novel tool to investigate changes in intracellular macromolecular pools in the marine microalga *Chaetoceros muellerii* (Bacillariophyceae). J Phycol 37:271-279

Green GK (1977) Design of NSLS VUV-ring. Brookhaven National Laboratory Report 50595

Guilhaumou N, Dumas P, Carr GL, and Williams GP (1998) Synchrotron infrared microspectrometry applied to petrography in micrometer-scale range: fluid chemical analysis and mapping. Appl Spectroscopy 52:1029-1034

Hirschmugl CJ (1994) Low frequency adsorbate substrate dynamics for CO/Cu studied with infrared synchrotron radiation. PhD Dissertation, Yale University, New Haven, CT

Hirschmugl CJ, Williams GP (1995) Signal-to-noise improvements with a new far-IR rapid-scan Michelson interferometer. Rev Sci Inst 66:1487–1488

Hirschmugl CJ, Peden CH, Takasaki M, Collins MA, Chambers SA (1999) Synchrotron-based far-IRAS investigations of ice on a single crystal transition metal oxide. SPIE 3775:167-173

Hirschmugl CJ, Bunta M, Holt JB, Giordano M, Strickler R (2002a) Chemical imaging of living cells: *Euglena gracilis*. J Phycology (submitted)

Hirschmugl CJ, Bunta M, Giordano M (2002b) Synchrotron-based infrared imaging of *Euglena gracilis* single cells. *In:* Handbook of scaling methods in aquatic ecology: measurement, analysis, simulation. Seuront L and Strutton PG (eds) CRC 2002

Holman HY, Perry DL, Martin MC, Lamble GM, McKinney WR, Hunter-Cevera JC, (1999) Real-time characterization of biogeochemical reduction of Cr(VI) on basalt surfaces by SR-FTIR imaging. Geomicrobio J 16:307-324

Holman HYN, Goth-Goldstein R, Martin MC, Russell ML, McKinney WR (2000) Low dose responses to 2,3,7,8-tetrchlorodibenzo-p-dioxin in single living human cells measured by synchrotron infrared spectromicroscopy. Environ Sci Tech 34:2513-2517

Hofmeister A (1995) Infrared microspectroscopy in earth science *In:* Humecki HJ (ed) Practical Guide to Infrared Microspectroscopy. Marcel Dekker, Inc, New York, p 377-416

Jackson JD (1975) Classical Electrodynamics. John Wiley and Sons, New York

Jamin N, Dumas P, Moncuit J, Fridman W-H, Teillaud J-L, Carr GL, Williams GP, (1998) Highly resolved chemical imaging of living cells by using synchrotron infrared microspectroscopy. Proc Natl Acad Sci 95:4837-4840

Keilmann F, Knoll B, Kramer A (1999) Long-wave-infrared near-field microscopy. Physica status solidi B215:849-851

La Veigne JD, Carr GL, Lobo RPSM, Reitze DH, Tanner DB (1999) Time-resolved infrared spectroscopy on the U12IR beamline at the NSLS. SPIE 3775:128-136

Miller LM, Vairavamurthy V, Chance M, Mendelsohn R, Paschalis EP, Betts F, Boskey AL (2001) *In situ* analysis of mineral content and crystallinity in bone using infrared microspectroscopy of the v_4 PO_4^{3-} vibration. Biochim Biophys Acta 1527:11-19

Murphy J (1999) Private communication

O'Day PA (1999) Molecular environmental geochemistry. Rev of Geophys 37:249-274

Pilling M, He T, Hirschmugl CJ (1999) Far-infrared reflection absorption spectroscopy: low frequency studies on single-crystal oxide surfaces. *In:* Synchrotron Radiation Instrumentation. Pianetta P, Arthur J Brennan S (eds). Eleventh US National Conference, AIP Conference Proceedings 521:41- 46

Stuart B (1997) Biological Applications of Infrared Spectroscopy. John Wiley and Sons, New York

Talley DB, Shaw LB, Sanghera JS, Aggarwal ID, Cricenti A, Generosi R, Luce M, Margaritondo G, Gilligan JM, Tolk NH (2000) Scanning near field infrared microscopy using chalcogenide fiber tips. Materials Lett 42:6

Williams GP (1982) The national synchrotron light source in the infra-red region. Nuc Instr Meth 195:383-387

Williams GP (1999) Infrared synchrotron radiation, review of properties and prospectives. SPIE 3775:2-6

7

Quantitative Speciation of Heavy Metals in Soils and Sediments by Synchrotron X-ray Techniques

Alain Manceau[1,2], Matthew A. Marcus[2], and Nobumichi Tamura[2]

[1]*Environmental Geochemistry Group*
LGIT, University J. Fourier and CNRS
38041 Grenoble Cedex 9, France

[2]*Advanced Light Source*
Lawrence Berkeley National Laboratory
One Cyclotron Road
Berkeley, California, 94720, U.S.A.

INTRODUCTION

Human societies have, in all ages, modified the original form of metals and metalloids in their living environment for their survival and technical development. In many cases, these anthropogenic activities have resulted in the release into the environment of contaminants that pose a threat to ecosystems and public health. Examples of local and global pollution are legion worldwide, and the reader of the environmental science literature is forever faced with ever more alarming reports on hazards due to toxic metals. For example, extensive mining and associated industrial activities have introduced large amounts of metal contaminants in nature at the local, but also global, scale since anthropogenic metals are detected in remote areas including Greenland ice (Boutron et al. 1991). Industrialized countries have countless polluted sites, and the major consequence in terms of contamination by heavy metals are areas of wasteland and sources of acid and metal-rich runoff from tailings piles and waste-rock heaps, and the subsequent pollution of coastal areas. Water supplies in many areas of many countries are also extensively polluted or threatened by high concentration of metal(loid)s, sometimes from natural sources, but most often from the activities of humans (Smedley and Kinniburgh 2002). Pollution of ground and surface waters, and hence of lands, by arsenic from alluvial aquifers in the Bengal Delta plain and in Vietnam are probably the two most catastrophic actual examples of the second type, where a modification of the chemistry of deep sediment layers by intensive well drillings and pumping of drinking water has led to vast arsenic remobilization and poisoning of ecosystems (Chatterjee et al. 1995; Nickson et al. 1998; Berg et al. 2001).

Soils and sediments, being at the interface between the geosphere, the atmosphere, the biosphere and the hydrosphere, represent the major sinks for anthropogenic metals released to the environment. In industrialized countries, an estimation of background levels of trace metals is almost impossible because truly pristine ecosystems no longer exist. Indeed, in contrast to organic contaminants, which can undergo biodegradation, heavy metals remain in the environment, and no one now expects to return the earth to a pre-hazardous-substance state. Fortunately, the toxicity of metals largely depend on their forms, the rule being that the less soluble form is also the less toxic. To give two examples, arsenic is extremely toxic in its inorganic forms but relatively innocuous as sparingly soluble arsenobetaine and arsenoribosides, the main arsenic compounds that are present in marine animals and macroalgae, respectively (Beauchemin et al. 1988; Shibata and Morita 1989; Kirby et al. 2002). The semi-quantitative link between solubility and toxicity is illustrated in Table 1 with cobalt compounds. Elemental cobalt and cobalt

1529-6466/00/0049-0007$10.00

Table 1. Relation between metal concentration,
solubility, and toxicity.

Compound	Toxicity upon ingestion (mg / kg)	Solubility	[Co]
Cobalt	> 7000	2 mg/l	100%
Co oxide	> 5000	8 µg/l	71%
Co sulfate	768	60 g/l	22%
Co chloride	766	76 g/l	24%
Co nitrate	691	240 g/l	20%
Co acetate	503	237 g/l	23%

oxides have a solubility in the mg/l to µg/l range and can be ingested in significant amount without risk, whilst cobalt salts are several orders of magnitude more soluble and, hence, more toxic. The main interest of this working example comes from the inverse relationship between toxicity and metal concentration, because the two less toxic species are also those in which cobalt is the most concentrated. Therefore, the potential human and ecological impacts of hazardous heavy metals can be addressed by transforming soluble species to sparingly soluble forms, either *in situ* or in landfills after excavation. To this end, determining, as quantitatively as possible, all the forms of potentially toxic metals in soils, sediments, and solid wastes, is a key to assessing the initial chemical risk, formulating educated strategies to remediate affected areas and, eventually, to purity assessment and site monitoring.

Comprehending in full detail the environmental chemistry of an element is clearly an impossible dream, because characterizing in full the nature and proportion of all the various forms of a metal is beyond all existing analytical capabilities. We shall show in this chapter that speciation science can benefit from advanced X-ray techniques developed at 3rd generation synchrotron facilities. Whilst it is clearly impracticable to identify and quantify all species present in a bulk sample, at least the inorganic forms, and to a lesser extent the organic ones, can be characterized with unprecedented precision. This information can then be used to understand and predict the transformations between forms, to infer from such information the likely environmental consequences of a physico-chemical perturbation of the system and, in turn, to control the mobility of metal contaminants. Before discussing which analytical approach and tools are available to achieve these goals, it is essential to have in mind what are the main forms of metals in soils.

Chemical forms of metals in soils

Soils are multicomponent and open (complete equilibrium is never reached) systems in which elements can be partitioned between the solid, the gaseous, and the aqueous solution phases. Although the gaseous and liquid phases, being at the interface between the hydrosphere and the atmosphere, are the transport medium of most labile species, they generally contain a small fraction of the total amount of metals. Therefore, in the following speciation will be discussed by considering solid-phase species, thus neglecting gaseous and aqueous species. Comprehensive reviews on the two last forms can be found in monographs by Schlesinger (1991) and Ritchie and Sposito (1995).

The solid fraction is a complex heterogeneous assemblage of minerals, small organic molecules and highly polymerized organic compounds resulting from the activity of living organisms (bacteria, fungi, roots...). The minerals present consist usually of so-called primary minerals, either inherited from the parent material (quartz, titanium oxides, feldspar...) or introduced to the environment by industrial activities (e.g., zinc and nickel oxide, willemite, franklinite... released by smelters), and secondary minerals such as phyllosilicates, oxides of Fe, Al and Mn (in this paper we refer to oxides, oxyhydroxides, and hydrous oxides collectively as oxides), and sometimes carbonates, which may also have a lithogenic origin. The organic matter comprises living organisms (mesofauna and microorganisms), dead plant material (litter) and colloidal humus formed by the action of microorganisms on plant litter. These solid components are usually clustered together in the form of aggregates, thus creating a system of interconnected voids (pores) of various size filled with gases and aqueous solution. The inorganic as well as organic solid soil constituents are variable in size and composition. The finest particles, the smallest of them being colloidal in size, are the result of a more advanced weathering of rocks or a more advanced decay of plant litter. While the coarse fraction of the soil may be more important from the standpoint of soil physics and to trace the origin of the pollution, the fine materials are typically the most reactive soil portion from a chemical point of view, and hence the most important in order to assess the impact of metal contaminants to ecosystems. The mineral colloidal fraction consists predominantly of phyllosilicates and variable amounts of oxides (Fe, Mn), while the organic colloidal fraction is represented by humic substances. These nanoparticles have a large surface area per unit weight, and are characterized by a surface charge originating from surface functional groups, which attracts labile ions. Environmental physical, chemical, and biological conditions are continuously changing and, therefore, these assemblages become unstable from place to place and progressively modify the original, anthropogenic or lithogenic, forms of trace metals. In the transitory stage of their incorporation to the soil, trace metals can be present in many different forms but, with time, the more labile fractions transform into more stable forms that better correspond to the new conditions (Han et al. 2001). Some of the possible molecular-level forms of metals and some pathways of their sequestration are illustrated in Figures 1 and 2. Five principal uptake mechanisms have been identified so far, and can be conceptually described as follows (Sposito 1994; Sparks and Grundl 1998; Brown et al. 1999; Ford et al. 2001):

(1) Outer-sphere surface complexation (OSC). In this mechanism the sorbate ion keeps its hydration sphere and is retained at a charged surface within the diffuse ion swarm by electrostatic interactions. The sorbate species is screened from the sorbent metal typically by two oxygen layers, that is, is distant from it by at least 4.5 Å (Schlegel et al. 1999a). OSC frequently occurs in the interlayer space of minerals having a negative (phyllosilicates) or positive (layer double hydroxides, LDHs, Hofmeister and Von Platen 1992; Bellotto et al. 1996) permanent charge (i.e., pH-independent) arising from aliovalent isomorphic lattice substitutions. Since outer-sphere surface complex are loosely bound and can be easily replaced by an ion exchange mechanism, metals and metalloids held on exchangeable sites are highly mobile and readily available to living organisms. For many soils in the temperate region, the average cation exchange capacity (CEC) of the clay fraction as a whole is typically 50 meq/100 g. Each class of constituent has a characteristic range of CEC values, for instance, the organic matter has a CEC of about 200 meq/100 g, montmorillonite, ~100 meq/100 g, illite, ~30 meq/100 g, and kaolinite, ~8 meq/100 g, whereas Fe and Al oxides have almost no CEC. Among Mn oxides, only monoclinic birnessite exhibits CEC properties (~300 meq/100 g), but this mineral is seldom present in the environment (Usui and Mita 1995) because it is unstable and transforms to hexagonal birnessite, losing its CEC (Silvester et al. 1997). Since phyllosilicates are, together with Fe oxides, the two major constituents of the great majority of soils, and that the organic fraction is generally lower than a few percents,

Figure 1. Basic processes of adsorbate molecules or atoms at mineral-water interface (homogeneous precipitation not represented). a) physisorption; b) chemisorption; c) detachment; d) absorption or inclusion (impurity ion that has a size and charge similar to those of one of the ions in the crystal); e) occlusion (pockets of impurity that are literally trapped inside the growing crystal); f) attachment; g) hetero-nucleation (epitaxial growth); h) organo-mineral complexation; i) complexation to bacterial exopolymer and to the cell outer membrane. The photo in the background shows how mineral surface processes can be used to reduce the mobility and bioavailability of metal(loid) contaminants in the environment. The vegetation shown was established at the Barren Jales gold mine spoil (Portugal) as described in the text and shown in Figure 25. The SEM image of the bacteria is from Banfield and Nealson (1997) (credit: W.W. Barker) and the background photo from Mench et al. (2002) and Bleeker et al. (2002).

OSC essentially occur on smectitic clays (e.g., montmorillonite, bedeillite, nontronite...) and, therefore, these unstable metal species are quantitatively more abundant in clay-containing soils. The fraction of heavy metals taken up by this mechanism in soils can be greater than ten percent in contaminated acid soils (Roberts et al. 2002), but in the vast majority of soils it amounts to less than a few percent. This highly mobile pool can be easily leached or translocated to another form by increasing the pH or amending the soil with sorbent minerals (Vangronsveld et al. 1995; Mench et al. 2002).

(2) Isolated inner-sphere surface complexation (ISC). Isolated sorbate cations or oxyanions (e.g., metalloids) bond separately to the surface by sharing one or several

a

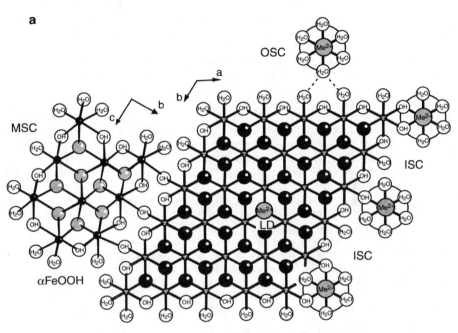

OSC

MSC

αFeOOH

ISC

LD

ISC

CoOOH

b

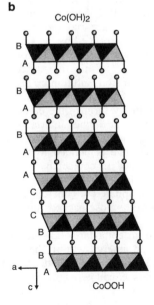

Co(OH)₂

CoOOH

Figure 2. Schematic representation of principal surface sorption processes. (a) Projection of the CoOOH structure in the *ab* plane. MSC, multinuclear surface complexation represented by an epitaxy of α–FeOOH (left); ISC, mononuclear monodentate (middle right), mononuclear bidendate (top right) and binuclear bidendate (lower right) inner-sphere complexation; OSC, outer-sphere surface complexation (top); LD, lattice diffusion (center). (b) Example of epitaxy without sharing of oxygens (Van der Waals forces). The ...AB-AB... close-packed anionic layer sequence of Co(OH)₂(s) is coherently stacked on the ...AB-BC-CA... layer sequence of CoOOH. Co(OH)₂(s) has a 1H polytypic structure, and CoOOH a 3R. Small circles are H. Adapted from Manceau et al. (1999b).

ligands (generally oxygens) with one or several cations from the sorbent. Then the sorbed ion may be incorporated progressively in the sorbent structure during crystal growth (Watson 1996). This sorption mechanism generally results from the disruption of the bulk structure of solids at their surface. Dangling chemical bonds on particle edges generate a pH-dependent macroscopic variable charge, the magnitude of which much depends on the specific surface area of the solid (the more highly divided the particles, the more reactive surface sites there are). The occurrence of this sorption mechanism in soils results in metal sorption to a far greater extent than would be expected uniquely from the CEC of a soil. Ions sorbed in this way are inherently much less reversibly released in the environment than ions sorbed by ionic exchange. The metal and metalloid polyhedra (generally octahedra or tetrahedra) usually share edge(s) and corner(s), and occasionally a face, with the sorbent metal polyhedron, yielding a characteristic metal-metal (Me-Me) distance for each type of polyhedral association. Accordingly, for a given metal-sorbent system, the crystallographic nature of the sorption site can be determined from the knowledge of the Me-Me distance using a polyhedral approach (Manceau and Combes 1988; Charlet and Manceau 1993). ISC is the primary form of metal uptake by most minerals (at least in the transitory stage), and in the last decade examination of the manner in which aqueous metal ions sorb on mineral surfaces has led to mushrooming research on model systems using synchrotron radiation. In the real world, these species are thought to be abundant but are overwhelmingly difficult to identify by conventional analytical techniques. As will be shown below, the higher brightness of 3rd generation synchrotron sources now allows the key identification of these species in unperturbed natural matrices. The first surface complex to have been positively identified in soil is tetrahedrally coordinated zinc sorbed in the interlayer space of hexagonal birnessite at vacant Mn layer crystallographic positions (Manceau et al. 2000b) (Fig. 3).

(3) Multinuclear surface complexation (MSC). Sorbate cations polymerize on the substrate and form a heteroepitaxial overgrowth, the sorbent acting as a structural template for the sorbate precipitate (Charlet and Manceau 1992; Chiarello and Sturchio 1994; Schlegel et al. 2001a). The presence of the mineral surface may reduce the extent of supersaturation necessary for precipitation to an extent determined by the similarity of the two lattice dimensions (Mc Bride, 1991). Heterogeneous nucleation is probably the most important process of crystal formation in soil systems. The mineral surface reduces the energy barrier for the nuclei of the new crystals to form from solution by providing a sterically similar, yet chemically foreign, surface for nucleation. The energy barrier arises from the fact that the small crystallites, which must initially form in the crystallization process, are more soluble than large crystals because of the higher interfacial energy between small crystals and solution. From a crystallographic standpoint, the template and the new crystal may share anions if the sorbate precipitate is bonded to the sorbent surface, or they may not if the two solids are maintained in contact by electrostatic interactions, hydrogen bonds or Van der Waals forces (Fig. 2). But in both situations, anionic frameworks of the substrate and the surface precipitate are coherently oriented. This is exemplified in Figure 2b where the ...AB-AB... close-packed anionic layer sequence of $Co(OH)_2(s)$ is coherently stacked on the ...AB-BC-CA... layer sequence of CoOOH. The multinuclear surface complex may form an epitaxial solid-solution if the sorbent and sorbate phase grow simultaneously or an intergrowth if they grow in alternating layers, leading in the former case to a mixed-solid and in the latter to a mixed-layering (Chiarello et al. 1997). The latter situation is encountered in many natural materials including manganese oxides, such as lithiophorite, which is built up of a MnO_2-$Al(OH)_3(s)$ interlayering (Wadsley 1952), and Ni-rich asbolane consisting of a MnO_2-$Ni(OH)_2(s)$ interlayering (Manceau et al. 1992), in clay minerals such as rectorite and corrensite (Brindley and Brown 1980; Beaufort et al. 1997), and in the valleriite group of

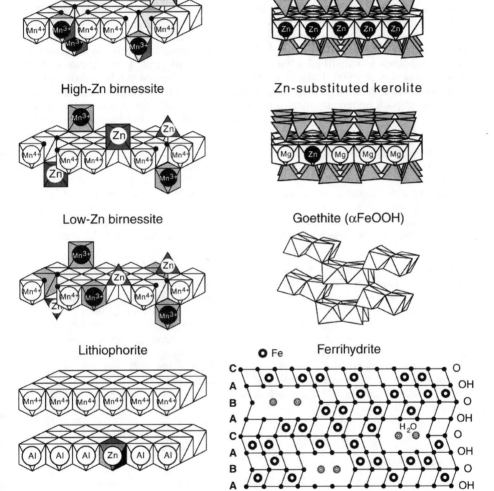

Figure 3. Idealized structures of phyllomanganate, phyllosilicate, and Fe oxyhydroxide minerals discussed in the chapter. Phyllomanganates include hexagonal birnessite (Drits et al. 1997; Silvester et al. 1997), Zn-sorbed hexagonal birnessite at high and low surface coverage (Lanson et al. 2002; Manceau et al. 2002a), and Zn-containing lithiophorite (Manceau et al. 2000c). Phyllosilicates include Zn-substituted kerolite ($Si_4(Mg_{3-x}Zn_x)O_{10}(OH)_2 \cdot nH_2O$) and Zn-kerolite (Schlegel et al. 2001a; Schlegel et al. 2001b). Oxyhydroxide includes goethite and defect-free ferrihydrite (Drits et al. 1993).

minerals where sulfide and brucitic layers alternate regularly (Organova et al. 1974). Heterogeneous nucleation is partly responsible for the widespread formation of coatings on mineral surfaces, but the role of microorganisms in this process is also often determinant (Banfield and Nealson 1997). Surface precipitates generally have a low solubility and may represent a valuable form of metal immobilization.

(4) Homogeneous precipitation (HP). Dissolved cationic species polymerize and precipitate in solution without structural link to the substrate (Espinose de la Caillerie et al. 1995; Scheidegger et al. 1997; Towle et al. 1997). The concentration of the sorbate may be lowered below that determined by the solubility of a pure sorbate hydroxide phase owing to the formation of a mixed phase incorporating dissolved species, generally from the substrate in simplified laboratory systems (Ford et al. 1999; Manceau et al. 1999b; Scheckel et al. 2000; Dähn et al. 2002a). This mechanism can be referred to as a "dissolution-induced homogeneous precipitation" (DI-HP). Homogeneous precipitates may further deposit on the substrate to form a surface coating. Since solid solutions and mixed precipitates are less soluble than pure metal phases, this mechanism can durably immobilize metals in soils.

(5) Lattice diffusion (LD). The sorbed single ion diffuses into the sorbent, filling vacancies or substituting for sorbent atoms. This phenomenon has been said to be responsible, together with DI-HP, for the progressive decrease of metal mobility in soils, that is to natural attenuation processes that convert metals over time to less detrimental forms. Direct and indirect evidence of the kinetically-controlled sequestration of metals in soil minerals by solid state diffusion has been reported for various species, including goethite, calcite, and birnessite (Brümmer et al. 1988; Stipp et al. 1992; Manceau et al. 1997).

Identifying these molecular mechanisms of trace element reactions with mineral phases is a difficult task on model compounds, but dealing with natural samples is even more challenging. The principal reasons for this include relative low metal concentrations relative to the detection limit of conventional laterally-resolved analytical and crystallographic probes; partitioning into coexisting minerals; the nanoscale size of most reactive soil particles; the difficulty of identifying the mineral species into which trace metals are bound; and the multiplicity of sorption mechanisms. In the following we shall see how these long-standing impediments to determining and quantifying at the molecular level how trace metals interact and are sequestered by soil constituents are now yielding to investigations by X-ray techniques developed at 3rd generation synchrotron facilities.

ANALYTICAL APPROACH

Electrons vs. X-rays

Since environmental materials are heterogeneous down to the nanometer scale, and the information sought is structural in essence, electrons rather than X-rays are, a priori, a better probe because electrons can be focused with magnets down to the angstrom scale. In addition, and like photons, they are absorbed, scattered and diffracted by matter, yielding the desired chemical and structural information (Buseck 1992). The unrivalled lateral resolution of transmission electron microscopy is well exemplified in the recent works by Hochella et al. (1999), Buatier et al. (2001), and Cotter-Howells et al. (1999). The first team investigated the forms of Zn and Pb in two acid mine drainage systems from former silver, gold, zinc and copper sulfide vein deposits in Montana, USA, the second the forms of Zn and Pb in smelter-affected soils from northern France, and the third the form of Pb in the outer wall of the root epidermis of *Agrostis Capillaris* grown in a soil contaminated by mine-wastes. A number of metal precipitates were identified, including gahnite ($ZnAl_2O_4$), hydrohetaerolite ($Zn_2Mn_4O_8.H_2O$), plumbo-jarosite ($Pb_{0.5}Fe_3(SO_4)_2(OH)_6$), sphalerite ($ZnS$), and pyromorphite. Interestingly, most crystallites had a nanometer size, and were never uniformly distributed but aggregated in micrometer-sized aggregates, similarly to ZnS nanocrystallites precipitated by biofilms (Labrenz et al. 2000). These independent observations support the view that highest resolution is not necessarily warranted to study most divided environmental materials, and that the micrometer scale of resolution is also appropriate, therefore the relevance of micro-X-ray diffraction (μXRD) for detecting minute precipitates. In addition to precipitated forms, ferrihydrite (hydrous ferric oxide) and phyllosilicates held significant portions of heavy metals, as attested to by elemental correlations in EDX analyses. But electron microscopy failed to identify the structural form of metals associated with these secondary minerals.

Clearly, electron diffraction as an adjunct to transmission electron microscopy is the tool needed to identify unambiguously metal precipitates in natural samples, at least when their structure is not too defective. However, as powerful as electron microprobe techniques are, none of them are at the same time element-specific and sensitive to the type and distance of neighboring atoms. The first attribute is necessary to determine the local structure around a metal contaminant, from which its containing matrix and uptake mechanism can be inferred. Electron energy loss near edge structure (ELNES) and electron energy loss structure (EELS) spectroscopy are element-specific via core-level electronic excitations, just as their X-ray equivalent, extended X-ray absorption fine structure (EXAFS) and X-ray absorption near-edge structure (XANES) spectroscopy, but the strong interaction of electrons with matter can induce amorphization or change of the oxidation state, especially in environmental samples which are often hydrated and very beam-sensitive. In the previous examples, these limitations prevented the determination of the molecular-scale environment of Zn associated with hydrous ferric oxide and phyllosilicate. Indeed, an association between an element E and a mineral M does not necessarily imply that E is chemically bound to or included in the structure of M. In soils, phyllosilicates are often coated by Fe oxides, the two phases being sometimes epitaxially related as was observed for hematite and goethite on kaolinite (Boudeulle and Muller 1988; Vodyanitskii et al. 2001), making it quite difficult to ascertain which mineral the trace metal is associated with. The answer came from the combination of micro-EXAFS and polarized-EXAFS spectroscopy, which showed that Zn is tied up at the molecular scale by both Fe oxides and phyllosilicates (Manceau et al. 2000b). Electron microscopy also suffers from relatively poor elemental detection limits and typically requires a UHV environment, which may modify the original speciation of the probed element. Finally, macroscopic quantification of the various species detected at the microscopic scale is usually impossible, as averred by Hochella et al. (1999): "The drawback in applying the methods presented here (or any other high-resolution technique) ... is one of sampling. ...we are actually looking at only an infinitesimally small portion of an otherwise highly complex, dynamic, and extensive system.... Transmission electron microscopy cannot avoid the collection of detailed information from vanishingly small amounts of material." With synchrotron radiation it is possible to probe a sample with an X-ray beam having a lateral size between about 10 mm^2 and less than one μm^2, and X-rays have a much higher penetration depth than electrons, which is an advantage to determine the proportion of metal species in the bulk.

Another useful complementarity between electrons and X-rays for the study of environmental materials is their difference of structural sensitivity to long and short range order. Electron diffraction is quite sensitive to the superstructure of minerals. This capability was used to determine the layer structure of triclinic and hexagonal birnessite and, specifically, the layer distribution of Mn^{3+} cations and octahedral vacancies which, together, account for the layer charge and, hence, the primary surface reactivity of this widespread mineral (Drits et al. 1997, 1998). Electron diffraction allowed also the two-dimensional distribution of trace elements (Ca, Zn, Pb) in the interlayer space of birnessite to be unraveled with unequalled precision (Drits et al. 2002). In contrast, electrons are less sensitive than X-rays to the defective structure of minerals, which is a strong disadvantage for the study of environmental nanoparticles because the most reactive ones are also those that contain the highest density of defects. This limitation can be exemplified with the hydrous ferric oxide, 6-line ferrihydrite, whose structure was examined recently in great details by X-ray diffraction and high resolution transmission electron microscopy (HRTEM) (Drits et al. 1993; Janney et al. 2001). Three components were necessary to simulate the XRD pattern: (i) defect-free ferrihydrite particles, consisting of ordered crystallites having a double-hexagonal (ABAC) stacking of close-packed oxygen layers (Fig. 3); (ii) defective ferrihydrite in which ABAB and ACAC structural fragments occur with equal probability and alternate completely at random; and (iii) hematite nanoparticles with mean dimension of coherent scattering domains of 10-20 Å. Of these three components, only the first was identified by HRTEM. Hematite nanoparticles were sensitive to vacuum and beam exposure and transformed to spinel (maghemite or magnetite) in the microscope (Drits et al. 1995), and the most defective second component was not detected by this technique. Therefore, the diffraction pattern calculated from HRTEM data did not match the experimental pattern (Fig. 7 in Janney et al. 2001), in contrast to the one calculated by X-ray diffraction (Fig. 13 in Drits et al. 1993).

Synchrotron-based X-ray radiation fluorescence (SXRF)

X-ray fluorescence was the first synchrotron radiation technique to take advantage of the ever-increasing brightness of X-ray sources in implementing focusing optics yielding a lateral resolution of a few μm^2 (Thompson et al. 1988; Newville et al. 1999; Sutton and Rivers 1999). The sampling depth is usually determined by the escape depth of the fluorescence X-ray of interest. For 3d transition metals, typical escape depths are on the order of 10-80 μm and, consequently, the analyzed volume depends strongly on the apparent sample thickness. The exquisite sensitivity to trace elements and high spatial resolution of this widely used technique explains its pervasive impact in environmental science. However, the difficulty of quantifying precisely matrix effects is a real limitation as it prevents determining elemental concentrations and, hence, the actual structural formula of the host phases identified by XRD. Note that in modern X-ray crystallography, crystals are still widely analyzed by wet chemistry. This drawback is less pronounced with particle-induced X-ray emission (PIXE), which achieves comparable lateral resolution, but is less sensitive than SXRF (Maxwell et al. 1995; Mesjasz-Przybylowicz and Przybylowicz 2002).

Recording elemental maps to locate trace metals within the heterogeneous matrix and to correlate trace and major elements is generally the first step towards any speciation analysis by XRD and EXAFS spectroscopy. Anti- and cross-correlations of trace elements with Fe, Mn, P and S can provide suggestive evidence for an association with Fe or Mn oxides, or with a phosphate or sulfide. Our extensive analyses of natural soils and sediments led us to the following experience. Mn is most often speciated as phyllomanganate (birnessite), and this mineral generally contains Ca (and in lesser amount Ba) in its interlayer (Taylor et al. 1964; Drits et al. 1998). Therefore, the Ca maps may not be representative of calcite ($CaCO_3$) distribution. Likewise, K is present in feldspars and micaceous materials. In general, it is hardly possible to obtain compelling evidence for trace element – phyllosilicate association by this technique because the vast majority of soil clay minerals have a dioctahedral structure, and that Fe^{3+}, Al^{3+} and Si^{4+} are present in a variety of minerals (oxides, feldspar, quartz...).

The result of µSXRF is a set of maps, one for each element or fluorescence line observed. Thus, for one run, one can end up with a three-dimensional array of size $500 \times 500 \times 10$ (height × width × number of elements) or more pixels. It is thus non-trivial to make sense of this large amount of data. The most obvious thing one can do is to make a gray-scale map for each element and compare them side-by-side. Such a set of plots is shown in Figure 4, which shows a nodule viewed in the "light" of several elements. These maps are shown in negative contrast, so that high concentrations are shown as dark areas. Letting one's eye roam around this figure lets one gather much information, for instance, that Ca is in the form of grains which appear to be unrelated to the other metals. There is an outer "rind," which appears to be enriched in Cr, Mn, Ni, Zn and Pb relative to the other areas. The maps for Ni and Mn are strikingly similar, as are those for As and Fe.

The next level of sophistication is the RGB tricolor map. In this method, one makes an image in which the R (red), G (green) and B (blue) values of each pixel are proportional to the amounts of three elements. If the elements are all the same, then R=G=B and one has a grayscale image, like one of the ones shown in Figure 4. If they are all different, one can get images that are at once informative and esthetically appealing. The overall brightness of a region is related to the sum of the concentrations, and the hue is related to the difference. An example of such a map is shown in Figure 5. There are two tricolor maps shown, one with As as green, Fe red and Mn blue, and the other showing Pb as green, Fe as red and Mn as blue (Pb was obtained by subtraction of the normalized maps taken 50 eV above and below the Pb L_3-edge to eliminate the contribution from As). At the top is a two-color (red-blue) map showing just the Mn and Fe. From that two-color map, we see that Fe is found principally in the middle of the nodule, and Mn shows up on the outside and in veins or cracks running through the Fe-rich region. The relatively pure red and blue colors suggest that Mn and Fe "avoid" each other, though there are some areas whose purple coloring shows the presence of both elements. The middle map (Pb-Fe-Mn) shows a greenish cast everywhere, indicating that Pb is widespread, but the outer surface is green or blue-green, suggesting Pb is enriched there and associated with Mn. The lower map (As-Fe-Mn) shows the Fe-rich region as yellow and orange, colors that come from mixing red and green. Thus, As is found with Fe. The outer surface is blue, showing a lack of As where the Mn is highest. There is a pocket on the bottom right of the nodule in which there is neither Fe, Mn or As, but there is Pb. While it takes many words to describe the features just noted, the practiced eye can spot them quickly. It should be noted, however, that the choice of how the R, G and B channels are scaled can make a big difference in the appearance of an image and the impression the viewer gets about features in it. For instance, suppose we pick one element to be represented by red, and some areas have more of this element than others. It is then tempting to set the levels for the red channel so that the redder areas show a range of colors and are not saturated in red. However, other areas containing less of this element may not give any visual impression of redness, so it may not be obvious that such areas contain any of the elements in question at all. Therefore, it may take more than one image to show all the details that exist in the data.

The above methods can be effective because the human visual system is a very powerful computer specialized for handling two-dimensional data in up to three channels. The tricolor map is a way of providing input to that computer in its accustomed format. However, this method provides only qualitative information about the associations between elements. The cross-correlation function is a more quantitative tool. If the

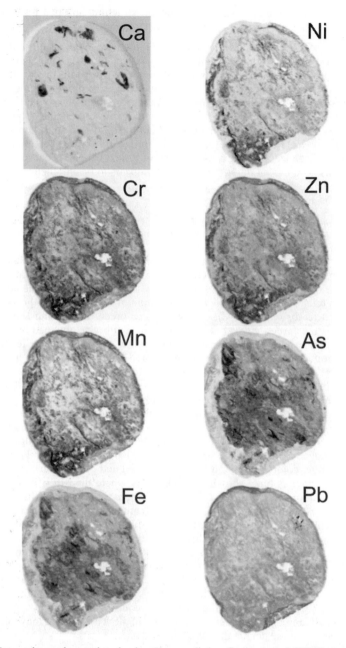

Figure 4. Gray-scale synchrotron-based micro-X-ray radiation fluorescence (µSXRF) maps in negative contrast showing the distribution of some elements in a soil nodule from the Morvan region, France (Baize and Chrétien 1994; Latrille et al. 2001; Manceau et al. 2002b). All maps except Pb were obtained by scanning the soil nodule under a monochromatic beam with an energy of 12,985 eV (Pb L_3-edge - 50 eV). The Pb map was obtained by subtraction of the normalized maps taken 50 eV above and below the Pb L_3-edge to eliminate the contribution from As. Nodule size: 3 × 3.5 mm, beam size 16 µm H × 6 µm V; step size: 16 × 16 µm; dwell time: 250 ms/point. Data were recorded on beamline 10.3.2 at the ALS (Berkeley).

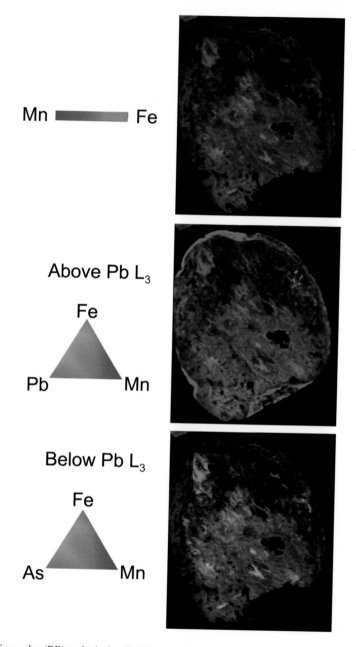

Figure 5. Two-color (RB) and tricolor (RGB) maps of the distribution of Fe, Mn, Pb and As in the nodule presented in Figure 4. Pb is present everywhere as indicated by the greenish cast of the middle map, but is enriched in the 'rind' and associated with Mn as indicated by the green or blue-green color of the outer surface. As is found with Fe as shown by the yellow and orange colors from the bottom nodule (mix of red and green). The outer surface is blue, showing a lack of As where the Mn is highest. There is a pocket on the bottom right of the nodule in which there is neither Fe, Mn or As, but there is Pb.

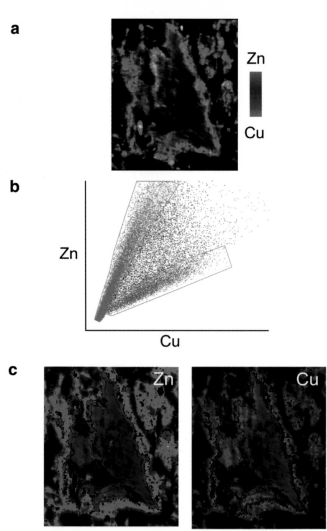

Figure 6. Demonstration of the isolation of distinct populations according to regions on a scatterplot. The sample is an organic soil used for truck farming and contaminated by sewage irrigation (Kirpichtchikova et al., unpublished). a) Bicolor map of Cu (red) and Zn (green). The large reddish (Cu-rich) area in the center is identified by light microscopy as organic matter with a reticular structure, which is visible in this image. Wherever there is Cu, there is also Zn, but the Cu/Zn ratio varies throughout the image. b) Cu vs. Zn scatterplot. There are two prominent branches and a significant number of points in a fan-shaped region between the branches. Populations were selected by defining polygons on the plot and then assigned colors: red for the branch characterized by a Zn/Cu count ratio near 1, green for the high-Zn branch (Zn/Cu around 4), and blue for the "in-between" points. The polygons used to define the "red" and "green" populations are shown in gray lines; the polygon defining the "blue" population is not shown. c) Grayscale maps for Cu and Zn colorized according to membership in the populations defined as in part b. Each pixel is colored red, green or blue depending on whether it belongs to the low-Zn/Cu, high-Zn/Cu or "in-between" populations. The brightness of each pixel is related to the Cu or Zn concentration at that location. Note that these are colorized grayscale maps and not true tricolor maps as in Figure 5. Also, these are not color-contour maps of the sort where red and blue represent different intensities.

intensities of two elements are $I_1(\vec{r})$ and $I_2(\vec{r})$, with positions $\vec{r}$ defined on a rectangular grid that covers the area of interest, then the normalized cross-correlation function is given by

$$\rho(\vec{\delta}) = \frac{\left\langle \left(I_1(\vec{r}) - \langle I_1 \rangle\right)\left(I_2(\vec{r} + \vec{\delta}) - \langle I_2 \rangle\right)\right\rangle_{\vec{r}}}{\sqrt{\left\langle \left(I_1(\vec{r}) - \langle I_1 \rangle\right)^2\right\rangle \left\langle \left(I_2(\vec{r}) - \langle I_2 \rangle\right)^2\right\rangle}} \qquad (1)$$

where $\langle \cdots \rangle$ represents an average over position on the sample. This function is 0 for a sufficiently-large sample of uncorrelated data, and equals 1 for a perfect cross-correlation (I_1 and I_2 are linearly-related) and -1 for a perfect anti-correlation. The argument $\vec{\delta}$ represents a position offset, such that if $\vec{\delta} = 0$, the values of I_1 at each position are compared with the values of I_2 at the same positions. Equation (1) is normalized so that $\rho(\vec{\delta} = 0)$ is just the Pearson correlation (r) between I_1 and I_2. If the structures in the sample are of finite size, or the size of the probe beam is more than one pixel, then $\rho(\vec{\delta})$ will be non-zero over a range of $\vec{\delta}$ corresponding to the size of the structures or the beam. We see from the above material that the autocorrelation function for one element, that is the correlation of I_1 with itself, will have a peak value of 1 at $\vec{\delta} = 0$ and fall off to 0 in a distance related to the size of the features in the sample and the size of the beam. The special feature of correlation analysis is that it looks at the whole sample at once, thus enabling one to detect patterns which are globally present, rather than just locally. Also, one can sometimes detect correlations that are too weak to be seen visually. One should bear in mind, however, that the correlation-function technique can be misleading if there are two or more distinct populations of minerals, each with its own behavior. For instance, a positive cross-correlation between two elements may be due to a small fraction of the sample, while the rest of the sample is devoid of one of the elements involved (Manceau et al. 2002b).

The obvious way to evaluate Equation (1) is to sum on each pixel, but that is very slow. A much better way is to use the convolution theorem and the fast Fourier transform (FFT). This topic, and the procedure for taking a two-dimensional FFT, are described in Press et al. (1992). This method involves the assumption that there are meaningful data at all points in a rectangular map. However, the sample is often an irregular shape surrounded by "blank" areas, such as the areas around the nodules in Figures 4 and 5. If one leaves these regions filled with zeroes, then one gets a large, illusory correlation due to the fact that sample itself is a feature which will be picked up in correlation analysis. A cure for this problem is to fill the external areas in with a value equal to the mean of the data in the region of the sample. This region can be defined by a simple threshold algorithm: a pixel is in the sample area if its value for the given element is greater than a given threshold, and is considered blank if its value is less. Our software allows the user to set the threshold and see on a binary (black/white) image what pixels are included in the sample area.

Another useful technique for understanding the data is the scatterplot, in which one plots the counts in one channel against the counts in another (Manceau et al. 2000b). Again, one uses a threshold algorithm to exclude points from the plot that come from areas not actually part of the sample. To show how this method can help us understand the data, let us assume that there is a mineral that contains Zn and Mn, and another that contains Zn and Fe. Let us suppose that both minerals occur in varying amounts throughout the sample and that Zn and Mn occur in no other forms. Now, if we plot the Zn counts on the y-axis vs. the Mn counts on x-axis, we will find a cloud of points along a diagonal representing the places in the sample in which the Zn-Mn species is dominant, and another cloud hugging the y-axis, representing the areas which have Zn and little Mn because the Zn is in the Zn-Fe mineral. One could, in principle, then use automatic classification techniques to define regions in this Zn-Mn space that would correspond to the different Zn environments. To our knowledge, this application has not been tried. Once one has defined an interesting region in this scatterplot, say the diagonal cloud, then one can work backwards and determine which points in the image fell in that cloud, and thus what areas in the sample contained Zn in the Zn-Mn form.

Figure 6 shows an example of the use of this population-segmentation method. The sample is a soil contaminated with both Zn and Cu by sewage irrigation (Kirpichtchikova et al., unpublished). Figure 6a shows a bicolor map in Cu (red) and Zn (green). Light microscopy shows that the large reddish (Cu-rich) area in the center is organic matter. There is Cu and Zn everywhere in the sample, but in varying amounts and proportions. The scatterplot for [Zn] vs. [Cu] is shown in Figure 6b. There are two distinct clouds of points with roughly constant [Zn]/[Cu] ratios in each population. The high-[Zn]/[Cu] points were colored in green and the ones with the low ratio in red. Points in between are in blue. Now, reconstructing the map from these three sets of points yields the images shown in Figure 6c. On the left, we have a grayscale map in which high-[Zn] regions are bright, and on the right a similar grayscale map for Cu. However, each pixel has been colored red, green or blue depending on which region it occupies in the scatterplot. The scatterplot separates out three regions, within the organic area, outside the organic area, and a border region. While the green region is relatively brighter in Zn than the red one, and vice versa for Cu, the two metals are present in all regions, but probably in different forms. Just looking at [Cu] or [Zn] alone does not yield such a separation. In principle, it should be possible to use automatic classification methods to define regions in scatterplots with more than two dimensions, with less subjectivity than is presently involved in the process.

Figure 7 shows the correlation maps (a) and scatterplots (b) for Fe correlated with Mn (left side) and As (right side) in the same soil nodule as was discussed above with reference to gray-scale and tricolor maps (Figs. 4 and 5). The correlation map for Fe-Mn shows a negative correlation at $\bar{y} = 0$, confirming that Fe and Mn tend to stay away from each other. The anti-correlation is not very strong, being only –0.22. However, the map shows a distinct "hole" at $\bar{y} = 0$ which is deeper (more negative) than any other feature, suggesting that the anti-correlation is real. If it were just random noise, one would expect to see similar features at other positions. However, As and Fe (right half of the figure) show a strong correlation (0.89 at the center), showing that if we know [Fe] at any point, we have a very good estimate of [As] there. The FWHM of the central feature (dip or peak) in both maps is about 100-150 μm, which is much bigger than the pixel size (16 μm) or the beam size (16 × 6 μm). There may therefore be a scale range of 16-100 μm over which the material is relatively homogeneous. The scatterplots are graphs of the counts in the Fe channel vs. the counts in Mn, and of the counts in the Fe channel vs. those in As. The Fe-Mn plot shows two, maybe three distinct "clouds" of points. One cloud is composed of points with low Fe counts, and the other has a similar appearance but with more Fe. The latter is the material near the center of the nodule, where most of the Fe resides. Within each cloud, there is a strong anti-correlation of Fe and Mn. Each cloud represents a distinct population of points in the nodule. Without the scatterplot, it would be difficult to see that there are two or more populations and that there is a strong Fe-Mn anti-correlation within each population.

We can now look at the same nodule using scatterplots displaying more relationships as shown in Figure 8. This figure shows scatterplots for As and Pb plotted against Fe and Mn. The numbers quoted by each graph are the cross-correlations at the origin of the correlation map. By definition, the correlation numbers are the Pearson *r*

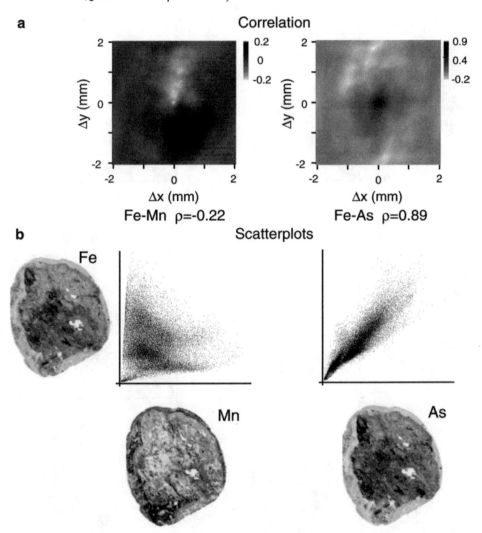

Figure 7. Cross-correlation maps (a) and scatterplots (b) for Fe-Mn and Fe-As elemental pairs in the nodule presented in Figure 4. Fe and Mn are slightly anti-correlated, and Fe and As are strongly correlated.

($\rho(0)$ in our notation) values for the scatterplots. The figure shows some obvious trends:

(1) The As is highly correlated with Fe, such that the As/Fe ratio is roughly the same everywhere in the nodule.

(2) Although Pb is strongly correlated with Mn, the Pb/Mn ratio is not constant, but instead shows two different trends. Application of the population-segmentation method showed that the upper branch represents the material in the "rind," while the lower branch, which itself seems to be split, represents the interior.

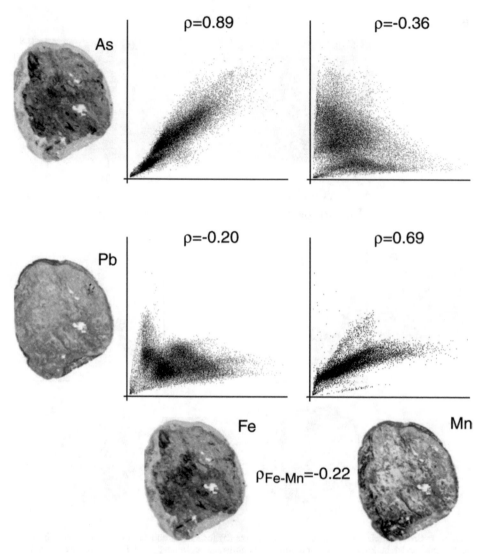

Figure 8. Scatterplots for As and Pb against Fe and Mn. As is highly correlated with Fe as suggested from the bottom tricolor map of Figure 5. As expected from the Fe-Mn-Pb tricolor map of Figure 5, there exists two dominant Pb-Mn populations, one with a high Pb/Mn ratio representing the material in the "rind," and one having a lower Pb/Mn ratio, which represents the interior. The numbers with each scatterplot are the Pearson r-coefficients ($\rho(0)$ in our notation).

(3) The positive correlations between As and Fe, and between Pb and Mn are stronger than the negative correlations (As-Mn, Pb-Fe). This suggests that the main determinant of trace-metal concentrations in this nodule is the association of the trace element with a major constituent, rather than an exclusion of the trace element from any specific material.

The fluorescence radiation is typically detected using a solid-state detector. The resolution of this type of detector is limited to 100-300 eV, the lower figure being obtainable only at very low count rates. There are also pileup and background effects such that a large signal for one element can induce a background in other spectral regions. Thus, one must be careful about choosing the excitation energy and the regions of interest. For instance, it can be difficult to detect Cr in ferromanganese nodules because of the huge Mn and Fe signals. The tail of the Mn peak may spill over into the Cr channel, thus causing a background that is proportional to the Mn signal. Thus, the Cr map may look just like the Mn map, giving a false impression of the concentration and distribution of Cr. In practice, one is often forced to make a map at an energy above the edge of the concentrated element and another below, wherein lighter elements are seen without the interfering background of the heavier one.

A variation of this method is useful when Pb and As are present in a sample. The As K_α line ($K_{\alpha 1}$ = 10,544 eV) is close enough to the Pb L_α line ($L_{\alpha 1}$ = 10,551 eV) to be inseparable with a solid-state detector. However, the Pb L_3-edge is above the As K edge. Thus, if one does a map between the As K and Pb L_3-edges, one gets only the As signal. Then, one can do a map above the Pb L_3-edge and get the Pb signal by subtraction of the signal normalized to the incident intensity. This is how the Pb and As images in Figures 4 and 5 were produced. This method of mapping just above and just below an edge and taking the difference may be useful anytime an element of interest is present in very small quantity.

It is important to be aware not only of interfering signals from K_α and L_α lines, but also from K_α and K_β lines as well. For instance, the Mn K_β line sits atop the Fe K_α line. Thus, if there is both Mn and Fe, one must record the Mn K_α line to get the Mn and the Fe K_β line for Fe. If there is also Co, then the chain grows by one link. For the $3d$ elements, the K_β line of one element overlaps the K_α line of the next. In the future, new detector technologies such as superconducting tunnel junctions may yield good enough resolution to separate these spectral features at an acceptably-high count rate. Another interesting possibility is the use of high-throughput crystal optics to select the fluorescence from the element of interest, as in Karanfil et al. (2000).

X-ray diffraction (XRD)

X-ray diffraction is the most commonly used technique to solve the atomic structure of crystalline materials, and it has provided one of the cornerstones of soil science research in determining the structure of soil inorganic constituents (Brindley and Brown 1980). The arrangement of atoms in a crystal can be determined using either single crystal diffraction or powder diffraction. Single crystal diffraction is the most straightforward and precise technique and is preferred to powder diffraction for structure refinement. However, a major limitation for its systematic use is the availability of sufficiently large good quality single crystal to obtain a workable set of reflections. For example, chalcophanite ($ZnMn_3O_7 \cdot 3H_2O$) still is the only phyllomanganate whose structure has been determined on single crystal, and that was done almost 50 years ago in the pioneering work by Wadsley (1955). X-rays produced by synchrotron radiation are several orders of magnitude brighter than those produced from X-ray tubes and rotating anodes in laboratory equipment. The high collimation and brightness of synchrotron radiation fostered the development of X-ray microfocusing optics, yielding the structure of micrometer sized crystals to be investigated, as recently achieved for 8 and 0.8 μm^3 kaolinite crystals (Neder et al. 1999). However, environmental materials are heterogeneous multi-phase systems of generally poor crystallinity, such that the quality of their diffraction patterns is dominated by sample-related peak broadening. In addition, many minerals can only be found as aggregates of nanometer-sized particles eluding the possibility of their study by single crystal diffraction techniques. For these types of sample, the powder diffraction technique has to be used. The superior collimation of the synchrotron X-ray beam, which is generally used to reduce instrumental broadening of diffraction peaks, so that peak overlap is kept to a minimum, is not critical in soil research as medium instrumental resolution is most often sufficient. The advantage of synchrotron radiation for these samples lies in the lateral resolution of the incident beam, which allows one to reduce in many cases the heterogeneity of the sample in the diffraction volume, and the high intensity of the X-ray beam, which enables the collection of diffraction patterns with excellent counting statistics. The tunability of the synchrotron radiation also permits one to avoid the fluorescence of a constituent atom in the sample (generally Fe, sometimes Mn). In the following, emphasis is placed on the use of synchrotron-based X-ray microdiffraction to study the mineral composition and distribution in soil samples, and its synergistic use with μSXRF and μEXAFS.

An X-ray microdiffraction beamline end-station (7.3.3) suitable for this type of application is being operated at the Advanced Light Source (ALS, Berkeley) (MacDowell et al. 2001). The bending magnet source of 240 (H) × 35 (V) μm size is refocused to a sub-micron size using a pair of elliptically bent mirrors in the Kirkpatrick-Baez (KB) configuration (Kirkpatrick and Baez 1948). There are several ways to focus X-rays to the micrometer scale, including zone plates, capillaries and refractive lenses (Dhez et al. 1999). However, the Kirkpatrick-Baez optics are the only ones to provide high efficiency, quality achromatic focusing, a long working distance and energy tunability over a large energy range without moving the beam focus. On the 7.3.3 beamline, the maximum beam divergence on the sample is 3.7 (H) × 1.6 (V) mrad, which translates to an instrumental resolution of $\Delta\theta/\theta$ ~3×10^{-2} to 3×10^{-3} in the 5-50° θ interval. Monochromatic beam is obtained via a 4-crystal Si monochromator placed upstream of the KB mirrors. The beamline optics was designed to meet the requirement of a stable beam position and size on the sample while changing the radiation wavelength. The energy range available on this beamline is E = 5.5-14 keV (λ = 0.885-2.25 Å). The μXRD technique greatly benefited from the concomitant development of fast and high resolution CCD (charge coupled device) detectors, making two-dimensional diffraction pattern analysis a widely used tool in crystallography and condensed matter physics during the past decade. The 7.3.3 beamline is equipped with a Bruker SMART 6000 CCD detector offering an active detection area of 9 × 9 cm, and a pixel dimension of 88 μm in the 1K × 1K mode. The resulting 2θ angular range is about 30-40° for an experimental (i.e., pattern) resolution of ~$\Delta\theta$ = 0.03°, which is amply sufficient to record the *hkl* reflections of minerals. Fluorescence signals are collected via a high purity Ge ORTEC detector coupled with a multi-channel analyzer (MCA).

CCD μXRD patterns can be recorded in reflection or transmission modes. In the former configuration, Bragg reflections are collected by inclining the sample (generally a glass or quartz mount) to Ω ~6° from the XY horizontal plane, and placing the CCD detector in the vertical plane (Fig. 9). In this geometry a higher volume of the sample is probed (better statistics) since the beam impinging the sample has an effective vertical dimension 1/sinΩ larger than the incoming beam. The Ω value determines d(*hkl*)$_{max}$, which formally equals $\lambda/[2\sin(\Omega/2)]$, that is ideally d_{max} = 18 Å for 6.3 keV (λ = 1.968 Å, E < Mn K-edge), d_{max} = 16 Å for 7.05 keV (λ = 1.759 Å, E < Fe K-edge), and d_{max} = 11 Å for 10 keV (λ = 1.240 Å). Since in fluorescence-yield EXAFS spectroscopy measurements are usually carried out by placing the detector in the horizontal plane at 90° from the X-ray beam to minimize elastically scattered X-rays in the fluorescence spectrum, and the sample vertically to 45° to both the X-ray beam and the detector, in the reflection diffraction geometry the sample has to be repositioned in going from μXRD to μEXAFS. Regions-of-interest are then re-imaged by μSXRF since fluorescence is isotropic. Higher lateral resolution (equal to the actual beam size) is obtained in

Transmission mode　　　　　　**Reflection mode**

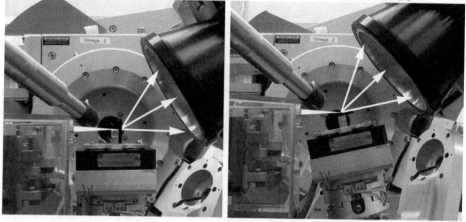

Figure 9. Side views of the sample regions of the experimental setups used on station 7.3.3 at the ALS for combined X-ray fluorescence and diffraction in transmission and reflection mode. The white wedge to the left of each frame represents the focused beam (width not to scale) coming from the vertical-focus mirror on the left. The large black object is the diffraction detector. The cylindrical object coming in from the upper left is the Si detector used for fluorescence mapping.

transmission mode with the sample placed vertically and the diffraction detector behind the specimen, but this configuration requires the availability of an X-ray transparent sample and a higher flux since the diffracted volume is significantly reduced. Another advantage of the transmission configuration is that μSXRF, μXRD and μEXAFS data can be collected *in situ* without modifying the sample orientation. Still, μXRD and μEXAFS cannot be recorded simultaneously because the wavelength is constant in the former, and variable in the latter, experiment. All patterns presented below were collected in reflection geometry.

Characteristic two-dimensional μXRD patterns from a soil sample are displayed in Figure 10a. Two types of diffraction features are almost always observed, with all possible intermediate types: sharp spots from micron to sub-micron crystallites and Debye rings of constant intensity arising from nanometer-sized particles. Since in this example the illuminated area was 14 μm (H) × 11 μm (V), leading to a diffracting volume of about $7{\times}10^3$ μm^3 (E = 6 keV), coarse grains yield spotty and discontinuous rings due to the diffraction of a finite number single crystals inside the diffracting volume. When crystallites are coarser, only a few individual crystals satisfy the Bragg condition and some *hkl* reflections are not even observed, as it is the case for the 004 reflection of anatase (2.38 Å) in Figure 10a, where only the 101 reflection (3.52 Å) is detected. In environmental materials, quartz, feldspars, carbonates, and titanium oxides are the most common minerals giving rise to point diffraction spots. Micaceous minerals and kaolinite often give moderately textured two-dimensional XRD. Texture effects can be used to differentiate mineral species having strongly overlapping XRD lines, such as kaolinite and birnessite. These two last minerals are notoriously difficult to identify unambiguously in soils because both have intense 00*l* reflections at 7.1Å. When the analyzed sample is rich in manganese, and therefore birnessite is suspected to be present, attempts to distinguish these two minerals can be made by dissolving either kaolinite with NaOH or birnessite with $NH_2OH{\cdot}HCl$ (Taylor et al. 1964; Ross et al. 1976; Tokashiki et

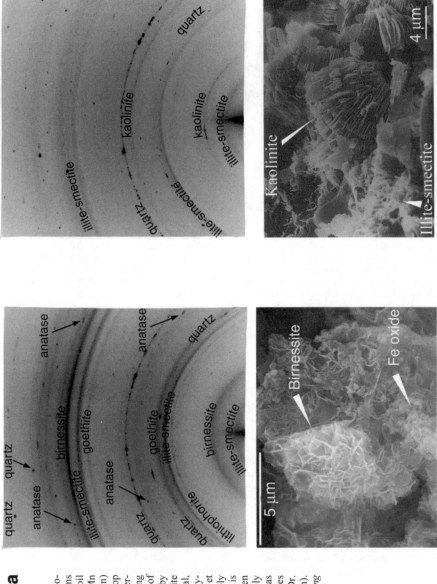

Figure 10 (a). Typical two-dimensional μXRD patterns (reflection mode) of a soil sample collected in a Fe-Mn rich region (top left pattern) and of a clayey material (top right pattern), and characteristic microstructure (scanning electron microscope image) of the soil minerals identified by μXRD. By SEM, birnessite appears as paper-like material, which often forms honey-comb-like features (Allard et al. 1999). The more rubbly material in the same image is iron oxide. Kaolinite often forms booklet-like loosely stacked assemblages, whereas illite-smectite clay aggregates are more wispy (credit: Dr. Debra Higley-Feldman). (*Figure continued on facing page.*)

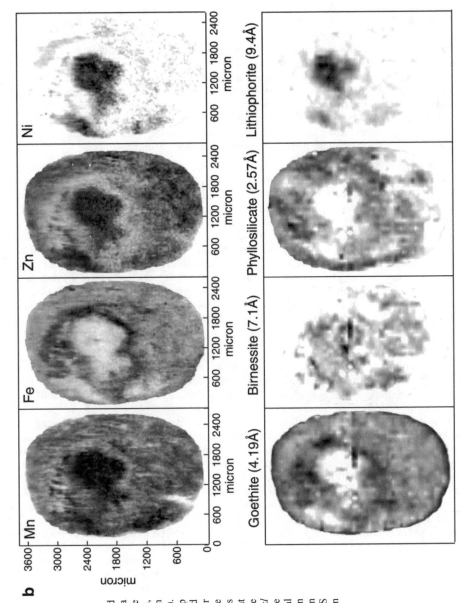

Figure 10 (b). µSXRF and µSXRD maps of a ferromanganese soil nodule from the Morvan region, France (Baize and Chrétien 1994; Latrille et al. 2001). The four images on the top are elemental maps obtained by µSXRF, and the four images on the bottom are mineral species maps obtained by integrating at each point of analysis the intensities of the relevant *hkl* reflections along the Debye rings of the two-dimensional XRD patterns. Synchrotron data were recorded on beamline 7.3.3 at the ALS (Berkeley). Adapted from Manceau et al. (2002c).

al. 1986). However, this treatment may introduce some artifacts, such as the recrystallization of poorly crystallized Mn species (Rhoton et al. 1993). Since birnessite particles are generally very small, they tend to produce continuous diffraction rings, whereas kaolinite particles tend to be large enough to produce spotty rings. Figure 10a shows examples of both minerals. Thus, μXRD offers a way to differentiate these two species *in situ*, which is without perturbing the soil matrix. Also, birnessite layers are most often randomly stacked giving rise to weak and broad 00*l* reflections as in synthetic δ-MnO$_2$, whereas kaolinite platelets or books have a large X-ray coherent size along c* resulting in relatively thin and intense 001 and 002 lines. Therefore, integrating the diffracted intensities along the Debye rings and fitting the corresponding peak profile affords a clue to differentiate these two minerals in case of mixture.

The good lateral resolution of X-ray microdiffractometers allows not only the detection of finely divided environmental particles but, more interestingly, also the mapping of their distribution in the sample from the quantitative analysis of their powder XRD pattern. Mineralogical maps of finely dispersed minerals can be obtained by rastering the sample in an XY pattern, collecting point XRD patterns at each point, and integrating the diffracted intensity along the Debye rings of X-rays. The closest equivalents to the scanning X-ray microdiffraction (μSXRD) technique in the field of electron microscopy are STEM (Scanning Transmission Electron Microscopy) and EBSD (Electron Back Scatter Diffraction). Though X-ray-based techniques have a lower spatial resolution than electrons, X-rays offer several advantages, such that the possibility of performing in parallel and without disrupting the sample, μSXRF and μEXAFS, thereby allowing on the same experimental station to spot a trace metal (μSXRF), to identify the mineral host (μSXRD), and to determine the mechanism of metal sequestration at the molecular scale (μEXAFS).

An example of the synergistic use of μSXRF and μSXRD is presented in Figure 10b. The mineral distribution maps of phyllosilicate, goethite, birnessite and lithiophorite in a soil ferromanganese nodule from the Morvan region in France (Latrille et al. 2001) were recorded in reflection diffraction mode (8° < 2θ < 59°; 2.1 Å < d < 15 Å; E = 6 keV) (Manceau et al. 2002c). The distance between the analyzed spot on the sample and CCD, and thus the 2θ scale, were precisely calibrated using the reflection peaks of quartz grains contained in the sample. Mineral abundance maps for the four designated species were produced by integrating at each point-of-analysis the diffracted intensities of the non-overlapping 020 and 200 reflections at ~4.45 Å and ~2.57 Å for phyllosilicate, the 101 and 301 reflections at 4.19 Å and 2.69 Å for goethite, the 001 reflection at 7.1-7.2 Å for birnessite, and the 001 and 002 reflections at 9.39 Å and 4.69 Å for lithiophorite. The reliability of the quantitative treatment was verified by comparing mineral maps for a same species established using different *hkl* reflections. The main interest of μSXRD lies in the comparison between μSXRF and μSXRD maps. If an element and a mineral species have the same contour map, as it is the case here for nickel and lithiophorite, then one can deduce with a good degree of certainty that this element is bound to this particular mineral phase. Unlike nickel, the Zn map does not resemble any of the four mineral species maps, nor can it be reconstructed by a combination of several. Therefore, some Zn-containing species are missing, and the most likely host phase candidate are vernadite (δ-MnO$_2$) and feroxyhite (δ-FeOOH, Fx), which yield broad lines at 2.5-2.4 Å, 2.25-2.20 Å, 1.70-1.65 Å (Fx), and 1.45-1.4 Å (Carlson and Schwertmann 1981; Varentsov et al. 1991; Drits et al. 1993). To map these species requires recording XRD patterns at high 2θ angle to get rid of peak overlaps with other mineral species in the 2.5-2.2 Å interval (birnessite, ferrihydrite, phyllosilicate…) (for more details see last section). Clearly, the technique will benefit from a better lateral resolution and, hence, from progress in microfocusing technology. Higher flux is also desirable to decrease the CCD data collection time for μSXRD measurements. On a bending magnet at the ALS, a typical collection time per point for soil samples is between several tens of seconds to a few minutes, making one μSXRD scan last for hours. Also, a smaller collection time would allow one to obtain higher resolution mineralogical maps by reducing the raster step size. Another important aspect to consider is the reduction of μSXRD data. The availability of robust user-friendly software adapted to the analysis of environmental samples is essential to make the combined use of μSXRD and μSXRF a routine tool. In the future, the larger availability of bright sources in the X-ray range will open more mapping and analysis possibilities, and this new combination of μSXRF and μSXRD should add to the arsenal of analytical methods available in mineralogy and geochemistry.

Extended X-ray absorption fine structure (EXAFS) spectroscopy

Of the three techniques covered in this chapter, EXAFS is indisputably the technique of choice for probing the speciation of metal contaminants at the molecular level. Its contemporary success in environmental science directly results from its physical characteristics (Stern and Heald 1983; Sayers and Bunker 1988; Rehr and Albers 2000). Briefly, the EXAFS experiment measures the variation of a material's X-ray absorbance as a function of the incident X-ray energy up to typically 800 eV beyond the absorption edge of a specific element. Beyond the edge, oscillations are observed, which arise from interference effects involving the photoelectron wave ejected from the absorbing atom and the fraction of the photoelectron wave backscattered by atoms surrounding the absorbing atom. Fourier transformation of the oscillatory fine structure (obtained after background subtraction) yields a radial structure function (RSF) in real space with peaks revealing the local environment of the target atom. In contrast to diffraction techniques, EXAFS does not rely on long-range order in the material and can be used to probe the local structure in both crystalline and amorphous solids. Chemical specificity is another asset, which can be used to differentiate between different components. If the probed element is concentrated, then EXAFS enable identification of the mineral host in complement to diffraction, or determination of its local structure in case of poorly-crystallized particles. If the probed element is diluted, then fluorescence-yield EXAFS spectroscopy provides the mechanism of its sequestration by the host matrix at the molecular scale (Brown et al. 1999). These two aspects are detailed successively below. A major obstacle to achieving a precise identification of metal species, however, is that EXAFS spectroscopy averages over all the atoms of a certain Z in the system under study regardless of their chemical state. Only recently, with the advent of high-brilliance synchrotron radiation sources, has it become possible to overcome this difficulty by using X-ray microprobes and principal component analysis of μEXAFS spectra. The fundamentals of this approach is described in this section, and the first working example of this new tool is given in the last section of the chapter.

Identification of the host mineral. Diffraction indisputably is the most relevant technique to identify the nature of mineral species. However, when the sorbent phase has extensive crystalline defects (stacking faults, homovalent and aliovalent substitutions, vacancies…), the size of its coherent scattering domains can be as small as a few tens of angstroms, thereby greatly decreasing the efficiency of diffraction. All structural works on poorly-crystallized environmental particles (ferrihydrite…) have shown that, even in their most disordered state, metal polyhedra (generally octahedra, less frequently tetrahedra) form coherent building blocks in which interpolyhedral bond angles (e.g., Me-(O,OH)-Me) are almost identical for a given type of bridging (face, edge, corner) and, therefore, so are Me-Me distances. This stands in strong contrast with glasses and amorphous metals, in which interpolyhedral bond angles are distributed typically over several tens of degrees resulting in a large spread of Me-Me distances. An example is amorphous silica in which SiO_4 tetrahedra are connected by their apices forming a three-dimensional network with a mean Si-O-Si bond angle of 144° as in quartz. But in contrast to the crystalline form, in the vitreous state Si-O-Si bond angles shows a large distribution extending all the way from 120° to 180°. As a result, Si-Si correlations are underrepresented in EXAFS because the Si-Si distances are too incoherent. This difference between the truly amorphous state and the long range disordered, but locally ordered, state is crucial because, even in most disordered natural compounds, EXAFS yields a metal shell contribution from which the local structure of the sorbent metal atom can be retrieved using a polyhedral approach. An illustration is given in Figure 11, which compares powder XRD patterns and RSFs of α-GeO_2 (quartz structure), amorphous germania, feroxyhite (δ-FeOOH), and hydrous ferric oxide (HFO, 2-line ferrihydrite) (GeO_2 was chosen as an example instead of SiO_2 because the Ge electron scattering amplitude is closer to that of Fe). Amorphous germania and HFO both give broad X-ray scattering bands in XRD, but little Ge-Ge interaction is detected in EXAFS whereas two prominent Fe-Fe contributions similar to the crystalline forms are at R+ΔR ~ 2.7 Å and R+ΔR ~ 3.2 Å (distances in the RSFs are uncorrected for phase shift, ΔR ~ -0.3-0.4 Å). Exhaustive analysis of the crystal structure of Fe oxides showed that the former distance

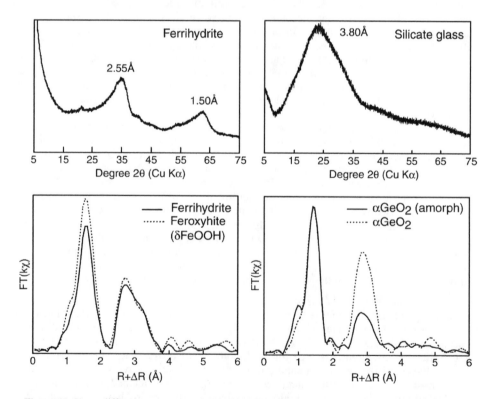

Figure 11. X-ray diffraction patterns and EXAFS radial structure functions of two poorly-crystallized compounds. Both ferrihydrite and silicate glass look "amorphous" through XRD, but the former is as well short-range ordered as feroxyhite, whereas the latter is disordered even at the local scale. GeO_2 is used as the EXAFS example instead of SiO_2 to better match the second-neighbor scattering factors for the Fe oxides.

is characteristic of edge-sharing and the latter of corner-sharing Fe octahedra. This polyhedral approach was initially developed on Fe and Mn oxides (Manceau and Combes 1988), and is now currently used to determine the nature of metal surface complexes on mineral surface inasmuch as mononuclear or polynuclear tridentate, bidentate, or monodentate surface complexes give rise to specific sorbate–sorbent metal EXAFS distances (see e.g., Spadini et al. 1994; Bargar et al. 1997, 1998).

It is clear from this discussion that EXAFS, through its ability to register metal-metal pair correlations, has the necessary sensitivity to identify host species and, accordingly, any sorbate precipitates. The uniqueness of spectra to different structural environments of the same metal atom is well illustrated with the manganate family of minerals, which encompasses an almost countless number of forms. Manganates essentially differ by their relative number of edge- to corner-sharing octahedra, Mn^{3+} to Mn^{4+} ratio, and framework structure (i.e., 3D vs. 2D structures). Figure 12 shows structures and EXAFS spectra for a number of different manganates. For each structure, the solid curve is the EXAFS spectrum for hexagonal birnessite, shown for reference, and the dashed line is that of the structure shown to the left of the curves. EXAFS spectroscopy possesses the sensitivity to even small variation in coordination geometries, interatomic distances and average oxidation state. For instance, though todorokite and hexagonal one-layer birnessite have similar proportion of edge- to corner-sharing octahedra per Mn (N_E/N_C = 4.7/2.7 and 3.5/2.5, respectively), and hence similar RSFs, their EXAFS spectra are significantly distinct, essentially because the photoelectron has different multiple-scattering paths in tectomanganate (3D) and phyllomanganate (2D) structures.

This example, wishfully extreme in choosing these overwhelmingly complex materials, clearly demonstrates that EXAFS spectroscopy possesses the intrinsic sensitivity to speciate a great number of soil minerals (mostly oxides), and it can be realized using the spectra of model compounds as fingerprints to the unknown spectrum. In practice, the main limitation resides in our capacity to obtain a good enough EXAFS spectrum of the unknown species, and our ability to interpret the experimental signal whenever the unidentified species is absent from the library of model compounds. Distinguishing phyllosilicate species by this fingerprinting approach is hopeless because these minerals are at the same time chemically and structurally quite complex since they can accommodate in their structure a large variety of cations (Al, Si, Fe, Mg....) and in various proportions (trioctahedral vs. dioctahedral, 1:1 vs. 2:1 vs. 2:1:1 structures...).

Identification of metal species. The following is an attempt to convey how the structural identity of the metal species (surface complex, sorbate precipitate...) can be recovered by EXAFS spectroscopy. For pedagogic purposes, we first consider a laboratory system aimed at showing how EXAFS spectroscopy enables to distinguish all sorption mechanisms described in the introduction (Figs. 1 and 2), then a case study taken from a real-world system is presented.

The selected model system is Zn – phyllosilicate because of its obvious relevance to soils and sediments, and the high reactivity and diversity of surface and bulk crystallographic sites of phyllosilicates. Phyllosilicates are built of layers made by the condensation of one central octahedral sheet and two tetrahedral sheets (Fig. 3). Two types of surface sites exist: permanent negative structural charge on basal planes resulting from aliovalent cationic lattice substitutions, and pH-dependent sites at layer edges resulting from the truncation of the bulk structure yielding oxygen dangling bonds. Metal cations sorbed on basal surfaces form outer-sphere (OS) surface complexes and, therefore, are easily exchanged with solute ions by varying the cationic composition of the solution (i.e., ionic strength), or may diffuse to another site (e.g., layer edges) where

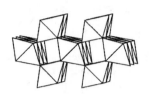

Pyrolusite (MnO₂)

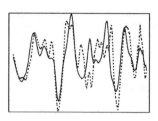

Ramsdellite (MnO₂)

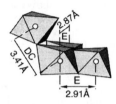

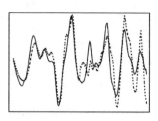

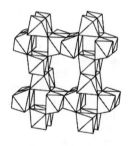

Hollandite (~MnO₂)

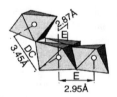

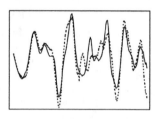

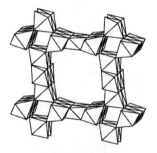

Todorokite (~MnO₂)

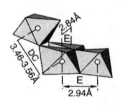

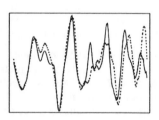

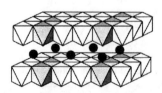

Triclinic birnessite

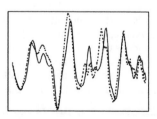

Hexagonal birnessite

$$(Mn^{2+}_{.05}\ Mn^{3+}_{.116}(Mn^{4+}_{.74}\ Mn^{3+}_{.093}\ \square_{.167})O_{1.7}(OH)_{.3})$$

Hexagonal Co-sorbed birnessite

$$(Mn^{2+}_{.03}Co^{2+}_{.06}Mn^{3+}_{.01}Co^{3+}_{.03}(Mn^{4+}_{.72}\ Mn^{3+}_{.06}Co^{3+}_{.12}\square_{.10})O_{1.72}(OH)_{.28})$$

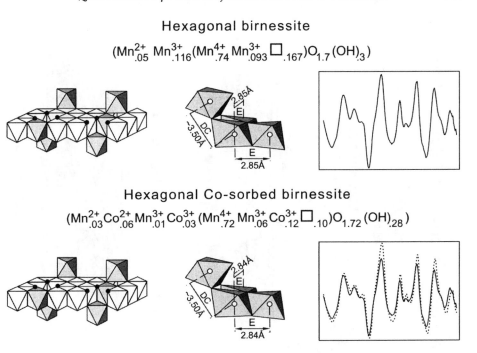

Figure 12. *(On previous page and above.)* Polyhedral representations of the structures of Mn oxides (manganates). All compounds yield distinctly different EXAFS spectra, which can be used in a spectral library to speciate Mn in unknown samples. The structure of phyllomanganates and EXAFS spectra ($k^3\chi$) are from Silvester et al. (1997), Manceau et al. (1997), and Lanson et al. (2002). In each row is shown a structural building block, a close up of a representative local Mn environment, and the EXAFS spectrum in dotted lines. The solid line in each row is the spectrum of hexagonal birnessite, shown for reference.

they are more strongly bound (Fig. 13a). Uptake on border sites is favored at near-neutral pH and high ionic strength, and is interpreted by the formation of inner-sphere (IS) complex. IS and OS mechanisms are readily differentiated by EXAFS spectroscopy since the sorbate metal is surrounded by a sorbent metal shell in the former case and not in the latter (Chen and Hayes 1999; Schlegel et al. 1999a,b; Strawn and Sparks 1999; Hyun et al. 2001; Morton et al. 2001; Schlegel et al. 2001a; Dähn et al. 2002b). At higher pH or solution cation concentration, the sorbate metal may polymerize and form various kinds of precipitates (neoformed phyllosilicate, hydrotalcite, metal hydroxide), attached or not to the surface (Scheidegger et al. 1997; Towle et al. 1997; Ford et al. 1999; Manceau et al. 1999b; Scheinost and Sparks 2000; Schlegel et al. 2001a; Dähn et al. 2002a). All these compounds have distinct EXAFS spectra, which can be used as fingerprints to speciate the metal forms in an unknown sample. However, detecting this metal pool in natural samples can be really difficult because most phyllosilicates are essentially aluminosilicates (i.e., montmorillonite, illite), and the EXAFS signal scattered by their constituent "light" elements (Si, Al, Mg) is weak and can be hidden by the intense scattering of metal species bound to Fe and Mn oxides in case of mixtures. This problem can be overcome sometimes using μEXAFS (Roberts et al. 2002), but more generally by polarized EXAFS (P-EXAFS) spectroscopy (Manceau et al. 1999a).

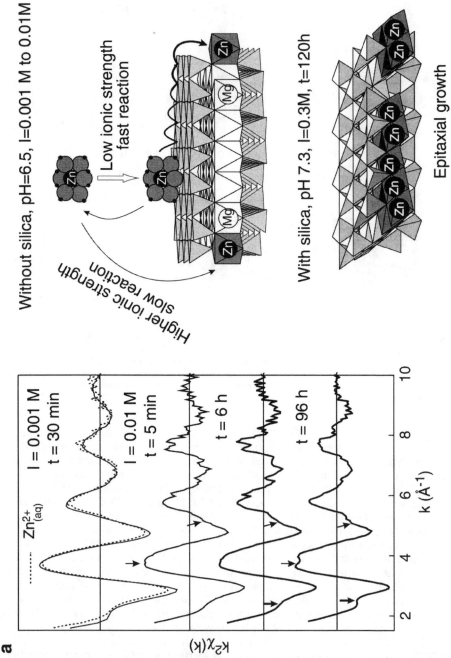

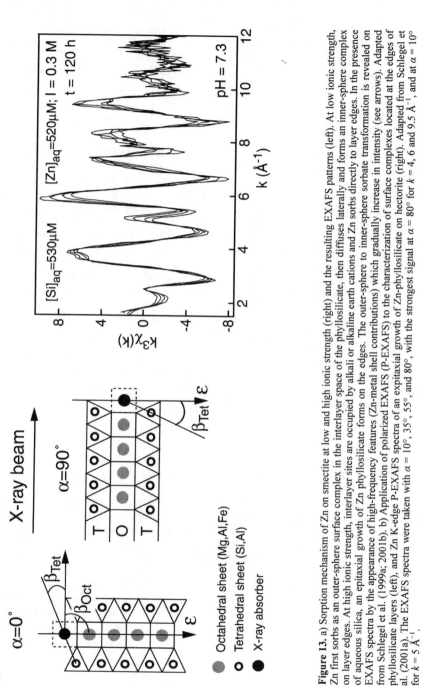

Figure 13. a) Sorption mechanism of Zn on smectite at low and high ionic strength (right) and the resulting EXAFS patterns (left). At low ionic strength, Zn first sorbs as an outer-sphere surface complex in the interlayer space of the phyllosilicate, then diffuses laterally and forms an inner-sphere complex on layer edges. At high ionic strength, interlayer sites are occupied by alkali or alkaline earth cations and Zn sorbs directly to layer edges. In the presence of aqueous silica, an epitaxial growth of Zn phyllosilicate forms on the edges. The outer-sphere to inner-sphere sorbate transformation is revealed on EXAFS spectra by the appearance of high-frequency features (Zn-metal shell contributions) which gradually increase in intensity (see arrows). Adapted from Schlegel et al. (1999a; 2001b). b) Application of polarized EXAFS (P-EXAFS) to the characterization of surface complexes located at the edges of phyllosilicate layers (left), and Zn K-edge P-EXAFS spectra of an epitaxial growth of Zn-phyllosilicate on hectorite (right). Adapted from Schlegel et al. (2001a). The EXAFS spectra were taken with $\alpha = 10°$, $35°$, $55°$, and $80°$, with the strongest signal at $\alpha = 80°$ for $k = 4$, 6 and 9.5 Å^{-1}, and at $\alpha = 10°$ for $k = 5$ Å^{-1}.

One of the important attributes of synchrotron X-ray beams is that the radiation is horizontally polarized in the plane of the electron orbit (Koningsberger and Prins 1988; Hazemann et al. 1992). This property can be used to enhance the sensitivity of EXAFS to the various metal species tied up to layered minerals in performing angular measurements on textured samples. In P-EXAFS, the amplitude of the EXAFS contribution of an atomic shell S coordinated to an element E can be written in the plane wave approximation and at the E K-edge

$$\chi_{E-S}(k,\phi) = 3 <\cos^2\phi_{E-S}> \chi_{E-S}^{iso}(k) = \sum_{j=1}^{N_{cryst}} (3\cos^2\phi_{E-j})\chi_{E-S}^{iso}(k) \qquad (2)$$

where ϕ_{E-j} is the angle between the electric field vector ε and the vector that connects the X-ray absorbing E atom to the backscattering j atom in the S shell, and χ_{E-S}^{iso} is the isotropic contribution of the S shell. The summation is made over all the N_{cryst} atoms of the S shell because, for most orientations of ε, all the $E-j$ vectors do not have the same ϕ angle. For a true powder (i.e., perfectly random crystallites), there is no angular variation, and $3 <\cos^2\phi_{E-S}> = 1$. For a textured sample, neighbouring j atoms located along the polarization direction ($\phi = 0°$) are preferentially probed, whereas atoms located in a plane perpendicular to ε ($\phi = 90°$) are not observed. It follows that P-EXAFS measurements can be used to probe the local structure of layer silicates between two different directional limits, parallel and perpendicular to the (001) plane, by varying the angle between ε and the layer plane of a single crystal or the surface of a self-supporting clay film (Fig. 13b). The contribution from the S shell is enhanced by a factor of three when the $E-S$ atomic pair lies along the direction of the X-ray polarization vector, thereby facilitating its detection. In practice, however, this factor is lower since all $E-j$ pairs of the S shell generally do not have the same ϕ angle in a given crystallite, and from one crystallite to another (basal planes are randomly distributed in the film). The actual angular dependence of EXAFS spectra can then be obtained by transforming Equation (2) in introducing angles that are independent of the distribution of orientation of the $E-j$ pairs in the S atomic shell (Manceau et al. 1998)

$$\chi_{E-S}(k,\phi) = 3\left[\cos^2\beta\sin^2\alpha + (\sin^2\beta\cos^2\alpha)/2\right]\chi_{E-S}^{iso}(k) \qquad (3)$$

or

$$\chi_{E-S}(k,\phi) = \left[1 - (1/2)(3\cos^2\beta - 1)(3\cos^2\alpha - 2)\right]\chi_{E-S}^{iso}(k) \qquad (4)$$

where α is the angle between ε and the film plane (i.e., the experimental angle), and β the angle between the vectors connecting the $E-S$ atomic pair and the perpendicular to the layer. Note that this crystallographic angle is now the same for all $E-j$ pairs of the S shell, and can be easily determined from the crystallographic structure. For $\beta = 54.7°$, $\chi_{E-S}(k,\phi) = \chi_{E-S}^{iso}(k)$, and the EXAFS amplitude is independent of the measurement α angle (β is equal to the magic angle, 54.7°). For $\beta < 54.7°$, the amplitude of $\chi_{E-S}(k,\phi)$ increases with α values, while for $\beta > 54.7°$ the amplitude decreases with increasing α (Fig. 13b). Since the amplitude of $\chi_{iso}(k)$ is obviously proportional to N_{cryst}, and that in a polarized experiment one detects an apparent number of atomic neighbors (N_{app}) which is the effective number of atoms really seen at the α angle, Equation (4) can be written

$$N_{app} = N_{cryst} - (1/2)N_{cryst}(3\cos^2\beta - 1)(3\cos^2\alpha - 2). \qquad (5)$$

In phyllosilicates, metal atoms in the octahedral sheet are surrounded by 3 (dioctahedral framework) or 6 (trioctahedral framework) nearest six-fold coordinated cations (Oct), and by 4 nearest four-fold coordinated Si,Al atoms (Tet) from the tetrahedral sheets ($\beta_{Tet} \approx 32°$, Fig. 13b). From Equation (5), the contribution of the Oct shell is cancelled ($N_{app}^{\perp}(Oct) = 0$), and that of the Tet shell selected ($N_{app}^{\perp}(Tet) = 8.6$), in the normal orientation. Conversely, when the polarization vector is parallel to the film plane, the Oct contribution is preferentially reinforced ($N_{app}^{\parallel}(Oct) = 4.5$ or 9), and the Tet contribution becomes small ($N_{app}^{\parallel}(Tet) = 1.7$). Thus, the dichroism of P-EXAFS spectra for phyllosilicate is quite strong, and this technique is a powerful adjunct to μEXAFS spectroscopy for identifying phyllosilicate-bound metal species.

Strongly textured films, with all crystallites having their (001) basal surface parallel to the film plane, can be prepared from pure smectitic clays (Manceau et al. 2000a). In soil samples, the particle packing in the film is always disrupted to some degree owing, for example, to the presence of grains, aggregates, clay coatings or gel-like neoformed clay particles (Fig. 14a). This imperfect film texture diminishes the angular dependence in an amount that depends on the orientation distribution of individual crystallites in the film, a parameter that can be measured by texture goniometry (Bunge 1981). The distribution of c^* axes of individual particles around the film normal can be introduced in Equation (5) as follows (Dittmer and Dau 1998)

$$N_{app} = N_{cryst} - (1/2)N_{cryst}(3\cos^2\beta - 1)(3\cos^2\alpha - 2)I_{ord} \tag{6}$$

with

$$I_{ord} = \frac{\int_0^{\pi/2}(3\cos^2\alpha - 1)P(\alpha)\sin\alpha d\alpha}{2\int_0^{\pi/2}P(\alpha)\sin\alpha d\alpha}. \tag{7}$$

The function I_{ord} accounts for the particle disorder, and its value is one for perfectly ordered films and zero for an isotropic sample. $P(\alpha)$ represents the profile-shape function used to model the distribution of c^* axes off the film normal.

$$P(\alpha) = \exp(-\alpha^2 / \Omega^2) \tag{8}$$

for a Gaussian distribution, where Ω is the width of the mosaic spread.

The integrals in Equation (7) have no closed form but can be evaluated numerically (Manceau and Schlegel 2001). $I_{ord}(\Omega)$ does not vary linearly with the mosaic spread: it has little sensitivity to Ω when the texture strength is high and low but a high sensitivity in the $15° < \Omega < 30°$ interval. The consequence of an imperfectly textured film on the dichroism of the Oct and Tet shells contributions in trioctahedral phyllosilicates is represented in Figure 14a. N_{app} logically converges to crystallographic values ($N_{Oct} = 6$, $N_{Tet} = 4$) when the particle disorder increases. But, more interestingly, the loss of dichroicity is more important in the normal than in the parallel orientation. For example, at $\Omega = 20°$, $N_{Tet}^{\perp}$ is 13% (7.5 vs. 8.6) and $N_{Oct}^{\parallel}$ is 7% (8.4 vs. 9), below their values at $\Omega = 0°$. Since the precision on N by EXAFS is at best 10%, one sees that $N_{Oct}^{\parallel}$ has relatively little sensitivity to Ω. At $\alpha = 90°$, not only does $N_{Tet}^{\perp}$ decrease by 13% but, more importantly, $N_{Oct}^{\perp}$ increases from 0 to 1.3 and this additional atomic contribution leads to a significant modification of EXAFS spectra (i.e., observed dichroism).

P-EXAFS and texture goniometry were employed to identify Zn-containing phyllosilicates in a smelter-affected soil (Manceau et al. 2000b). P-EXAFS spectra and P-RSFs obtained from the fine fraction of the soil had a pronounced angular dependence, indicating that the average local structure of Zn was anisotropic and also that the predominant Zn-bearing phase was oriented together with clay minerals during the film preparation (Fig. 14b). The Zn species was identified by comparing the unknown spectra with P-EXAFS spectra from layer mineral references: hydrozincite ($Zn_5(OH)_6(CO_3)_2$),

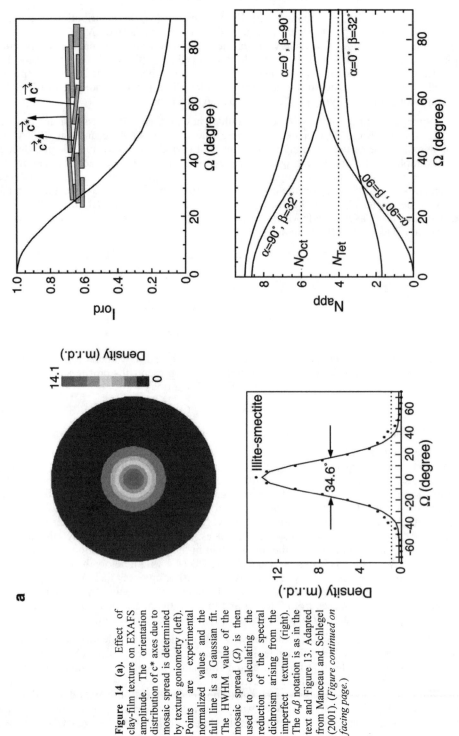

Figure 14 (a). Effect of clay-film texture on EXAFS amplitude. The orientation distribution of c* axes due to mosaic spread is determined by texture goniometry (left). Points are experimental normalized values and the full line is a Gaussian fit. The HWHM value of the mosaic spread (Ω) is then used to calculating the reduction of the spectral dichroism arising from the imperfect texture (right). The α,β notation is as in the text and Figure 13. Adapted from Manceau and Schlegel (2001). *(Figure continued on facing page.)*

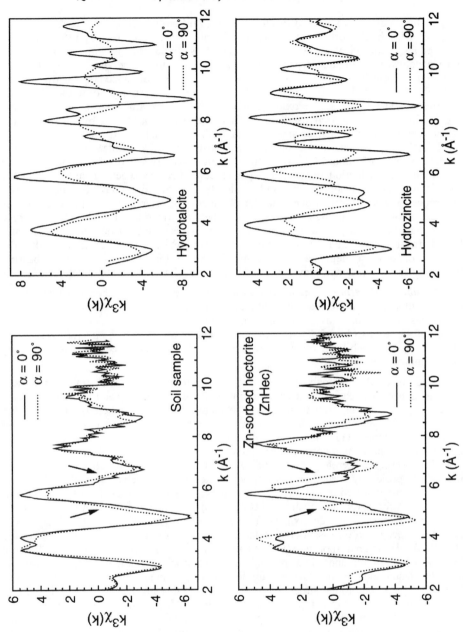

Figure 14 (b). P-EXAFS spectra of the clay fraction from a soil sample, Zn-sorbed hectorite, (Zn,Al) hydrotalcite, and hydrozincite. Adapted from Manceau et al. (2000b).

(Zn,Al)-hydrotalcite, Zn-sorbed hectorite, and Zn-kerolite. Strong resemblance was observed with the Zn-sorbed phyllosilicate reference, in which Zn forms an IS complex at layer edges (Schlegel et al. 2001b). The second oscillation of the soil EXAFS spectra have a noteworthy angular dependence which mimics that observed in the reference, though the phase shift and variation in absorbance are less pronounced (see arrows in Fig. 14b). This loss in dichroicity may be due to imperfect texture strength or to the admixture of other Zn species (e.g., Fe oxides), whose EXAFS spectra have a featureless and angular independent second oscillation. This alternative was tested by measuring the orientation distribution of illite-smectite and kaolinite particles in the soil film. The half width at half-maximum of the orientation distribution of kaolinite and illite-smectite particles were 15.5° and 17.3°, respectively (Fig. 14a). Based on the graph shown in Figure 14a, the reduction in angular dependence of EXAFS spectra resulting from the particle dispersion is about 17%, and is clearly too low to account for the observed difference between the unknown and the phyllosilicate reference. The presence in this soil sample of Zn-containing Fe oxides was further confirmed by μEXAFS. This example shows that P-EXAFS provides a direct means to assess the presence of phyllosilicate-bound metal species in natural systems. However, one should bear in mind that the absence of angular dependence does not disprove the presence of metal-bearing phyllosilicate as mentioned previously and recently evidenced by Isaure et al. (2001).

Number and nature of metal species in a mixture. In most systems of interest, the metal is attached to more than one mineral or organic molecule. In some cases, cogent arguments can be made as to which species are plausible. For instance, mineralogical maps can be used to identify likely hosts. In some cases, reference spectra exist for all plausible species, while in other cases they do not. We thus need ways of analyzing multicomponent spectra.

Traditional EXAFS analysis, involving Fourier transforming, back-transforming and filtering, generally does not work with multiple mineral species. The atomic shells from the different species overlap so that one cannot separate them out when there is a mixture (Fig. 15). Each component tends to have a complex spectrum, so the spectrum from a mixture would be even more complex, often with no resolvable shells in real space, though the EXAFS function yields a complex beat pattern rich in structural information up to about 6 Å (Fig. 16a). We therefore need other ways of proceeding in which the spectra from individual species are treated as wholes, not broken up into shells, especially when only one or two are analyzed quantitatively, thus limiting the structural analysis to ca. 2-3.5 Å. The situation for XANES is even more complex in that there is no known way to retrieve the structure of even one component from its XANES pattern, and that at high-energy XANES spectra are essentially featureless. For example, lead hydroxyapatite ($Pb_5(PO_4)_3OH$) and pyromorphite ($Pb_5(PO_4)_3Cl$) have very similar XANES spectra because lead is in both compounds divalent, and the two phosphate frameworks have about the same chemistry and topology (same multiple scattering paths) (Cotter-Howells et al. 1999). However, the pyromorphite structure has a short Pb-O distance (2.35 Å) and a Pb-Cl pair (3.11 Å) that makes this compound readily differentiable from lead hydroxyapatite by EXAFS spectroscopy (Fig. 16b). The poor sensitivity of XANES spectroscopy to metal speciation is obviously even more problematic in the case of mixtures and, therefore, in the remainder of this chapter emphasis is placed on EXAFS spectroscopy.

The framework for identifying metal species in multicomponent systems is the idea of spectra as linear combinations of component spectra (Manceau et al. 1996). This idea leads to the use of linear least-squares fitting of the unknown spectra. Unlike the method of fitting to shells, the linear-combination fit cannot produce a local minimum which is not the global minimum.

Suppose we have M spectra taken in various spots of the unknown sample or in various samples, and there are C components. Let the spectra all be interpolated to the same grid of N points in k-space (or E-space for XANES). Now, the spectra for the unknowns can be represented as sums of component spectra

$$\forall a = 1..M, i = 1..N : \chi_i^a \equiv \chi^a(k_i) = \sum_{\alpha=1}^{C} f_\alpha \chi_{comp,i}^\alpha \tag{9}$$

where f_α is the fraction of the αth component and $\chi_{comp,i}^\alpha$ is its spectrum. Here, χ represents either an EXAFS or a XANES spectrum, the latter normalized to an edge jump of unity. In what follows, Greek indices will refer to components, Roman indices starting at i to points in k- or E-space, and Roman indices starting at a to spectra of unknowns.

Now, suppose one knows from some other technique how many components there are and what they are. In that case, one can simply do a fit to Equation (9), with or without the constraint that the sum of the f_α must add up to unity. If that constraint is left off, then the resultant sum serves as a check on the relevance of the fit. If the sum of the f_α does not add up to 1, then there is something wrong with the assumptions behind the fit. An additional refinement is to add as free parameters possible energy shifts between the unknowns and the reference spectra. Since the reference spectra will normally have been taken at some other time that the unknowns, it is plausible that the unknowns will not have exactly the same energy zero. If we add the energy-shift parameters to the fit, it becomes non-linear and therefore more difficult than Equation (9). However, if the spectra were taken with due care, the shifts will be small, of order the energy step per point. In that case, one could approximate the energy-shifted spectra by Taylor expansions using numerically calculated derivatives

$$\chi^a(E_i + \delta E^a) \approx \chi_i^a + \delta E^a \chi_i^{a\prime} \tag{10}$$

where the prime represents the derivative. The fit equation is therefore

$$\chi_i^a = \sum_{\alpha=1}^{C} f_\alpha \chi_{comp,i}^\alpha + \delta E^a \chi_i^{a\prime} \tag{11}$$

which has an extra M free parameters. However, if one is willing to assume energy shifts for the unknown spectra, one must allow them for the reference spectra as well. In this case, the fit equation is non-linear even with the derivative approximation

$$\chi_i^a = \sum_{\beta=1}^{C} f_\beta (\chi_{comp,i}^\beta + \delta E^\beta \chi_{comp,i}^{\prime\beta}) + \delta E^a \chi_i^{a\prime}. \tag{12}$$

The conventional way of optimizing the fit parameters is to use a Marquardt-Levinberg algorithm, as is commonly used for fitting back-transformed atomic shells. However, the derivative terms will be approximately orthogonal to the main χ terms. Thus, if one holds the energy shifts fixed and lets the fractions vary, these fractions will converge to approximately the right answer. Then one can hold the fractions constant and vary the energy shifts. By iterating on these two sets of parameters, linear methods could be used to solve this non-linear fit. While the extra degrees of freedom represented by energy shifts can improve the fit, they represent further sources of uncertainty and must be applied carefully. Such shifts are more important for XANES than for EXAFS as the features in EXAFS are broader in energy than those of XANES. One method for minimizing the importance of energy shifts in EXAFS has been proposed by Manceau et al. (1996). In this method, one defines E_0 separately for each sample and each reference as being where the absorption, with pre-edge subtracted, is half the edge jump. Thus, one

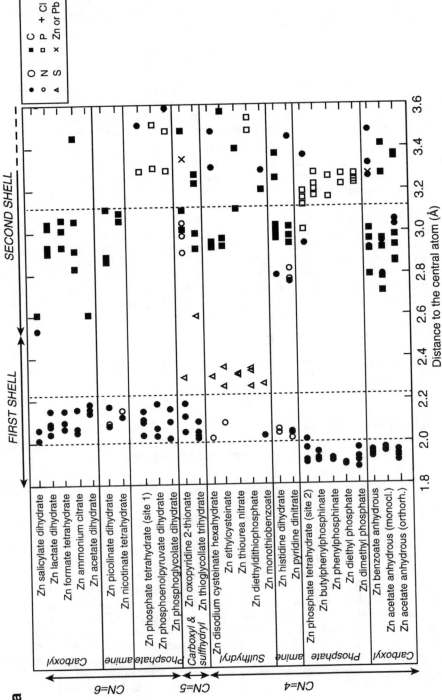

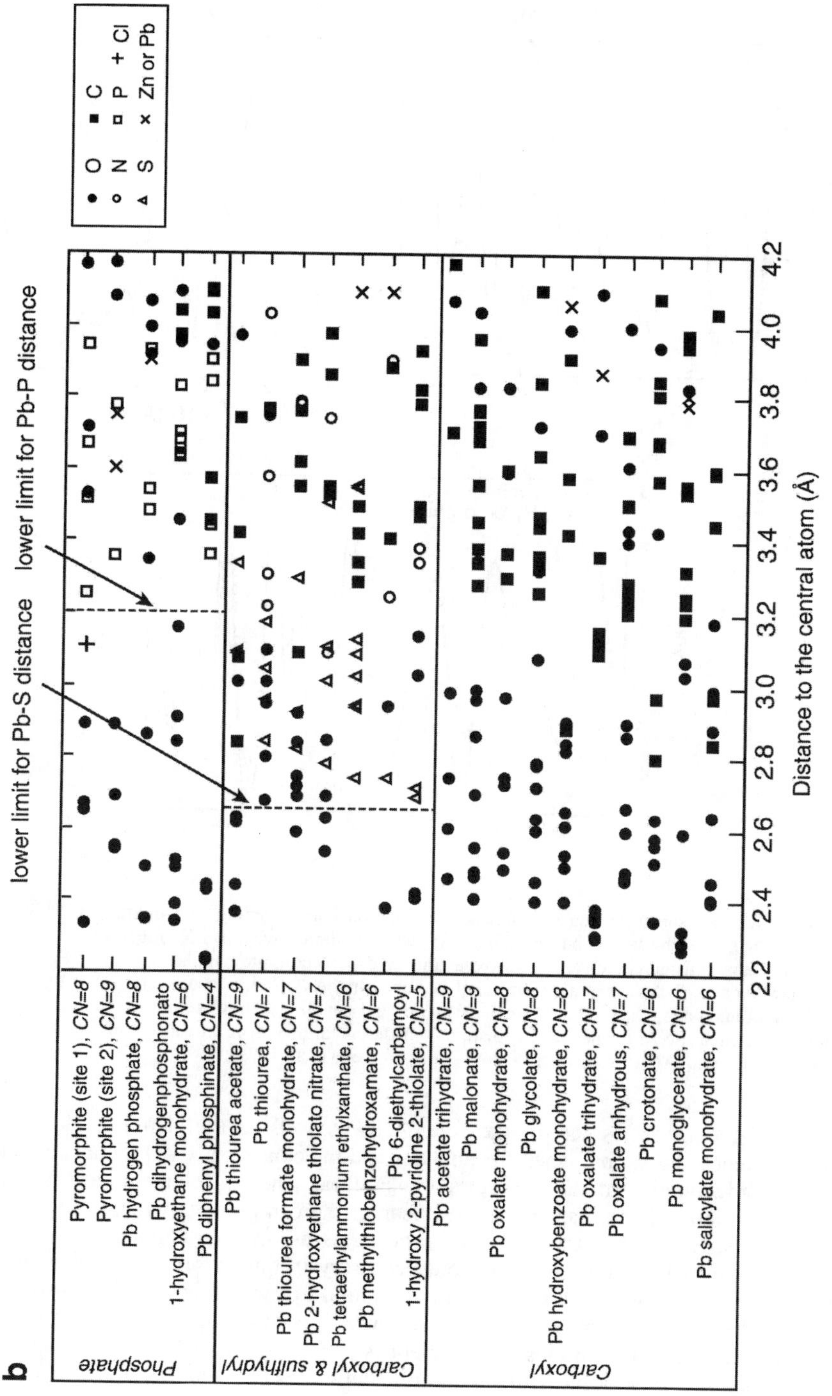

Figure 15. Distances between Zn (a) and Pb (b) and near neighbors in a variety of coordination compounds. Adapted from Sarret et al. (1998).

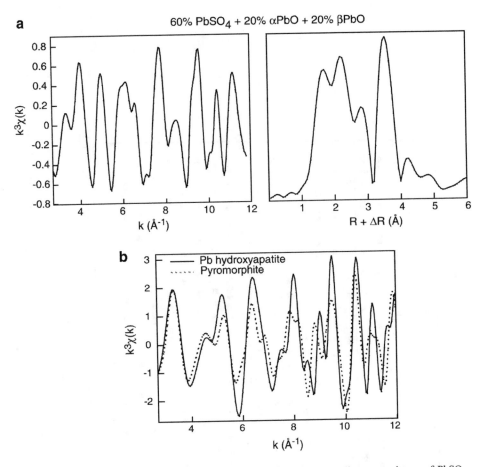

Figure 16. a) EXAFS spectrum and radial structure function corresponding to a mixture of PbSO₄, α-PbO and β–PbO. The Pb-O contributions (first RSF peak) are unresolved. b) EXAFS spectra (only 2- and 3-legged paths considered) of lead hydroxyapatite and pyromorphite simulated with FEFF 7.02 from structural data by Brückner et al (1995), and Dai and Hughes (1989), respectively ($S_0^2 = 1.0$, $\sigma = 0.08$ Å for all atomic pairs, $R_{max} = 8$ Å, $NLEG = 3$). Although lead local structure is very similar in the two compounds, their EXAFS spectra are distinctly different allowing one to differentiate the two compounds. However, XANES spectra are alike (Cotter-Howells et al. 1999).

puts in artificial edge shifts for all the references as well as the samples. However, practical experience shows that when the analysis is done consistently in this fashion, and that the edge shifts among unknown spectra are small (as when the element has the same oxidation state in all unknown species), the linear fitting can be done without additional edge adjustments. Even when the edge shifts are large, as between Cu and CuO (9eV), the edge adjustments in fitting are small, but necessary in this extreme case. This is an example of a method that seems theoretically unsound but works in practice. Still, this method is a substitute for what one really wants, which is to calibrate the beamline in a consistent manner for both samples and references (i.e., absolute calibration). Suppose, for instance, that one is working with Zn and one calibrates the beamline to the Zn edge as found for Zn metal. However one defines the edge energy for the metal, one will get consistent answers if E_0 for both sample and references is taken to be a fixed absolute energy value. Fortunately, the chemistry of the samples and references tends to be similar enough in soils so that the edge shifts are small and the half-edge-jump method is accurate enough for EXAFS, but not for XANES. Since XANES depends sensitively on both the energy calibration and the monochromator resolution, it is wisest to re-take XANES data for the references on the same beamline, preferably during the same run, as was used for the samples.

The success of this simple "fingerprint" method depends on the existence of an extensive database of reference spectra covering all relevant species. If such a database exists, one can try fitting a spectrum first with one component, then all possible pairs of components, all triples, etc. , all without adjusting E_0. Since the fits are linear and involve few parameters, they go fast enough to make such exhaustive searches practical on modern personal computers. As with any fitting process, it can be tempting to bring in many parameters and force a good fit even when the data do not really justify it. This situation commonly occurs when the data will not fit to the sum of only two or three references. Possible causes of this situation include:

(1) There really are many species in the sample. One way to untangle a complex mixture like this is to use a microprobe and examine several spots, then use PCA (Principal Component Analysis, *v.i.*) to determine how many independent components there really are. While sequential extraction seems like an attractive way of isolating or removing different components, it has been found by Ostergren et al. (1999) and Calmano et. al. (2001) that this procedure can affect the very chemistry one is trying to probe.

(2) The data are noisy or have artifacts. Certain types of drifts can mimic EXAFS.

(3) There are really only a few components, but one or more is not included in the set of references being used for the fit. Thus, the fit "tries" to make up the missing component as a sum of two or more of the included references.

The problem with having many components is that the more free parameters one has in the dataset, the more likely it is that the parameters will become highly correlated, meaning that a change in the amount or energy shift of one component can be compensated for by changes in one or more of the others. This effect can drastically stretch the error bars. However, physical constraints offer a check in that the component fractions must be nonnegative, they must add up to unity, and the energy shifts should be small. If one does an unconstrained fit and finds these conditions to be satisfied without artifice, then there are some grounds for confidence. While the fitting of two-component mixtures has been tested quantitatively (Welter et al. 1999), there has not been a good test of the validity of XANES fits to many components. Thus, a fit to seven XANES components (e.g., Calmano et al. 2001) must be regarded with suspicion, and especially in the case of a mixture of organic and inorganic metal species. For example, let us consider an EXAFS spectrum from a phase mixture (P) consisting of 80 percent ZnO + 20 percent Zn acetate dihydrate, and let us assume a 6 percent error in the determination of the amount of ZnO during the linear combination. The residual EXAFS signal not fitted by the ZnO model equals $R = P - [(0.8 + 0.8 * 0.06) \times ZnO] = P - 0.85 \times ZnO$ and, consequently, the EXAFS spectrum of the unknown second species equals $U = R/(1 - 0.85)$, since this species contributes to 15% of the total EXAFS signal. Figure 17 shows that this U spectrum is much different from Zn acetate dihydrate. Therefore, a 6 percent error in the quantification of ZnO suffices to mask the presence of 20 percent of a low X-ray scattering organic species, and may lead to a mistaken species identification.

As we have just seen, it is quite common that one does not know what the

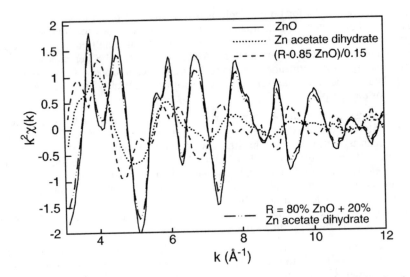

Figure 17. A demonstration of the potential for error in the analysis of mixtures containing the element of interest in organic and inorganic matrices. The simulated unknown ("R," dash-dot line) consists of a mixture of 80% ZnO (solid line) and 20% Zn actetate dihydrate (dotted line). Now, if we assume that the ZnO fraction is 85%, the inferred spectrum for what is left over is as shown in dashed lines. This spectrum is clearly quite different from the actual acetate spectrum.

components are or even how many there are. In that case, one can treat the $\chi^{\alpha}_{comp,i}$ as additional fit parameters, with constraints. The constraints are necessary because one can take a set of linear combinations of $\chi^{\alpha}_{comp,i}$ as a new set and have the same fit, albeit with different f_{α}. This situation is the domain of the Principal Components Analysis (PCA) method. This method has been used for decades in fields ranging from chemistry (Malinowski 1991; Segtnan et al. 2001) to astronomy (Mittaz et al. 1990). PCA has been applied to EXAFS and XANES in several different systems by Fay et al. (1992), Wasserman et al. (1996, 1999), and others. The idea is to take a set of M spectra and ask if they can be represented as linear combinations of a smaller set of $C \leq M$ components. If one imagines each spectrum as being a point in a multidimensional data space, then the set of spectra becomes a cloud of points. The PCA method finds the vector along which the cloud has the greatest extent, then the vector perpendicular to the first along which the cloud is longest, etc. Another way to look at it is as a form of lossy data compression. One starts with a "file" of χ^{α}_i and tries to shrink it by representing it as a sum of a smaller number of components. The new, compressed "file" now consists of a list of components and the amounts of each that make up each experimental spectrum. In this sense, the method is analogous to wavelet image compression in which small wavelet coefficients are zeroed out with little perceptible change to the original image (Press et al. 1992).

We will describe the PCA method following the treatment of Ressler et al. (2000) but with the above notation. PCA can be derived from the singular-value decomposition theorem from linear algebra, which says that any rectangular matrix can be decomposed as follows

$$A_{ia} = E_{ib}\lambda_b W_{ba} \tag{13}$$

where A is any $N \times M$ matrix, E is a column-orthogonal $N \times M$ matrix, and W is a square,

$M \times M$ orthonormal matrix. The matrix E can be considered as the set of components into which A is decomposed, the λ_b are scale factors showing the relative importance of each component, and W is a table of weights showing how much of each component is used to make up each column of A. In our case, A is the set of spectra, tabulated on a common grid and packed together so that A_{ia} is the value for the ath spectrum at the ith point. The λ_b are usually referred to as singular values or eigenvalues, as their squares are the eigenvalues of A^TA. These singular values indicate how much of a contribution to the whole dataset is made by each of the components E_{ib}, and W_{ba} is the relative amount of component b in spectrum a. As it stands, this form is not all that useful because we have not reduced the number of components. However, it generally happens that only a few of the λ_b are of significant magnitude, and the rest may be set to zero without affecting anything but noise. This step is analogous to the step in Fourier or wavelet image compression in which small-amplitude components are nulled. Now, we have a description of the experimental spectra in terms of sums of components

$$\chi_i^b \approx E_{ia}\lambda_a W_{ba} \equiv (E\Lambda W)_{ib} \tag{14}$$

where the sum on a runs from $1...C$, C being the number of components used, and b runs from $1...M$. Now, E_{ia} is the ith point in component a, and $\lambda_a W_{ba}$ is the amount of component a to use in making up the approximation to spectrum b. It should be noted at this point that the E_{ia} components are not spectra corresponding to any single species. When XANES is analyzed using PCA, the components (other than the first) do not even look like XANES spectra when plotted out. This is because these components are actually linear combinations of elementary spectra. For instance, if there were only two components, one of them would look like the average of the spectra of two species, while the other would resemble the difference. Thus, these E_{ia} are sometimes called "abstract components." For EXAFS, the components typically do look like EXAFS wiggles. Wasserman et al. (1996) point out that since components are orthogonal, the EXAFS shells tend to appear separated and may be analyzed with normal EXAFS methods.

How many components are needed to reproduce the observed spectra? To determine this, we can use Equation (14) to create simulated spectra and ask whether these simulated spectra agree with the actual ones to within statistics. We can also look at the individual E_{ia} and see if they look like signal or noise. For EXAFS, for instance, one could see if Fourier transforming them yields peaks in the range of plausible interatomic distances. This problem is similar to that found in other fitting methods where one must decide how many parameters to use in the fit. By reconstructing the spectra using only the number of components needed to represent that which is not noise, one is essentially doing a form of filtering or signal averaging over the set of spectra. This filtering is what we are doing by making the sum in Equation (14) run only from $1...C$ instead of $1...M$.

We have seen that the abstract components do not resemble real spectra. What we want to find out is which real references might go into making up the abstract components. The target transformation is a tool commonly used for this purpose. The idea is to take a given reference and remove everything from it that does not look like a feature found in any of the unknown spectra. If this procedure leaves the reference spectrum unaffected, then that reference is a good candidate for inclusion in the description of the set of unknowns. If the target transformation alters the reference spectrum significantly, then that reference must include some details or features not found in any combination of the unknowns and therefore should not be considered as a possible component. The target transformation is done as follows: Let the candidate reference spectrum be χ_i^{ref}, and perform the operation

$$\tilde{\chi}^{ref} = EE^T \chi^{ref} \tag{15}$$

which is a projection of χ^{ref} onto the subspace spanned by the vectors in E.

Of course, one would like an objective way of determining whether the target-transformed candidate really resembles its original version closely enough to be able to call it a match. Malinowski (1978) used the theory of errors to come up with a number he calls *SPOIL* which measures the degree to which replacing an abstract component with the candidate would increase the fit error. This is a non-negative dimensionless number for which values < 1.5 are considered excellent, 1.5-3 good, 3-4.5 fair, 4.5-6 poor, and > 6 unacceptable. The *SPOIL* function, re-cast from its original form into the present notation, is given by

$$SPOIL = \sqrt{\frac{N(M-C)\sum_i (\tilde{\chi}_i^{ref} - \chi_i^{ref})^2}{(N-C)\sum_{\alpha=C+1}^{M} \lambda_\alpha^2 \sum_{a=1}^{C}\left[\lambda_a^{-1}\sum_i E_i^a \chi_i^{ref}\right]^2} - 1} \quad . \tag{16}$$

The sums on i are over points in each spectrum. If the quantity in the radical is negative, as can happen when the fit is excellent, then *SPOIL* is set to 0. The sum in the numerator represents how poorly the target-transformed spectrum matches the input, and the first sum in the denominator is the amount of noise or error represented by the components we did not use in the reconstruction.

Now, let us suppose we have a set of references that we believe describe all the experimental spectra. We can finish off the analysis by doing the least-squares fit as described above. The paper by Ressler et al. (2000) shows this step-by-step process very nicely as applied to their study of Mn-bearing particles in automotive exhaust. This least-squares fitting step affords the opportunity to make the small E_0 corrections that can improve the fit quality significantly, though caution is always advocated when fitting XANES spectra.

Since we can improve the fit as much as we want by adding more components, we need a way of telling when we have the right number. Malinowski (1977) gave a criterion that he called the "indicator" or "*IND*". The *IND* function is given by

$$IND = \left(\frac{\sum_{\alpha=C+1}^{M} \lambda_\alpha^2}{N(M-C)^5}\right)^{1/2} \tag{17}$$

where we vary C, the number of considered components. The value of C for which *IND* is a minimum is the number of components to use. This function was developed empirically and does not have full theoretical support, but does seem to work whenever tested by using simulated data or when using real data and screening plausible references with the target transformation. Note that some papers will quote formulae like this one with λ_α instead of λ_α^2. This is because in such papers the eigenvalues quoted are those of the covariance matrix $\sum \chi_i^a \chi_i^b$, which are the squares of the singular values λ_α used here.

As a demonstration of the PCA method, we have analyzed simulated data produced by summing together measured spectra from three compounds (γ-MnOOH, MnS and MnCO$_3$) in various proportions and adding noise. The spectra are shown in Figure 18a. Visual inspection shows that there is more than one component. If one plots the spectra without the vertical offsets shown, one finds that there are no values of the abscissa for which all the curves meet, as would be the case if there were only two components.

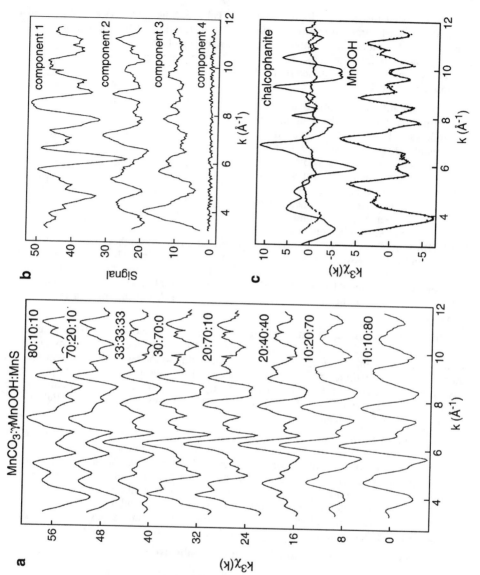

Figure 18. Principal component analysis (PCA) of EXAFS spectra. a) linear combinations of $MnCO_3$, γ–MnOOH and MnS reference spectra used as unknowns. b) First four components calculated by PCA of the "unknowns" shown in a). c) Target transformation on γ-MnOOH and chalcophanite. The reconstructed and reference spectra for γ-MnOOH are identical, which indicates that this species is contained in the series of unknowns. In contrast, the two chalcophanite ($ZnMn_3O_7 \cdot 3H_2O$) spectra are very different since this species is absent.

Therefore, visual inspection tells one that there are at least three components, but does not yield more information. The PCA method yields numbers for the singular values λ_a and the indicator IND as shown in Table 2.

The indicator is a minimum at $C = 3$, suggesting that there are three significant components. If we look at those components (Fig. 18b), we see that the first three look like EXAFS and the other five like noise. Next, we did a target transformation on each of the references from which the data were generated as well as two other Mn compounds which were not present, chalcophanite and $Mn(OH)_2$. Figure 18c shows the results for one of the references that were used to make the data and one that was not. It is obvious which one is which. The *SPOIL* values bear out the conclusion reached by visual inspection, being < 1 for the three pure references, 26.8 for $Mn(OH)_2$, and 41.5 for Mn chalcophanite. In actual practice, the distinctions are not always as clear as in this case, especially for XANES, but the method is still quite powerful.

Table 2. Singular values λ_a and IND values as a function of component number for sample problem.

C	λ_a	IND
1	70.4	0.413
2	38.0	0.427
3	36.8	0.145
4	4.22	0.217
5	3.96	0.365
6	3.62	0.777
7	3.31	2.885
8	2.88	

In some cases, the target data are missing over some range of k or E. In that case, one can use iterative target factor analysis (Malinowski 1991). The idea is to start by guessing values for the missing points, perhaps zero to start with, do the target transformation, then put the resulting predicted values back in instead of the missing values. This procedure has the side benefit of "filling in" missing data points in the target, which can then be used elsewhere. If this is done, then the *SPOIL* formula needs to be modified slightly; see Malinowski (1978) for details.

It is important to keep in mind that PCA is sensitive only to variations among the sample spectra. Thus, if a set of samples all contains the same fraction of some species, then PCA will not pick up that common species as a separate component. The working example given at the end of this chapter is of such a system. This particular system is approximately a pseudobinary, in which one end-member is a mixture of two species that we may refer to as A and B, and the other is a mixture of B and a third species C. The amount of B in this series of samples is approximately constant, so PCA only picks up two components, which may be thought of as A+(constant)*B+C and A−C.

Tips, tricks and cautions for hard X-ray microprobe users

Beam instability. The use of EXAFS on heterogeneous samples, especially with a microfocused beam, tends to exacerbate certain problems that are well known from other systems. One simple example of such a problem is the sensitivity to beam motion, which results from putting a small beam on an equally small particle. The beam tends to move on the sample during data acquisition, for a variety of reasons. These motions can cause artifacts in the data. For example, during the course of a synchrotron fill, the source point may move, and the decreasing power incident on the optics may also cause position shifts. Vibrations either of the sample or the optics result in increased noise and an increased effective spot size. If the position drifts on a time scale comparable to the length of time required to scan over an EXAFS oscillation, one could get artifacts, which are indistinguishable from EXAFS, except that they do not repeat from scan to scan. If the monochromator produces a beam whose position or angle depends on energy, then there will be position shifts on the sample, which will repeat from scan to scan. As the beam moves on and off a particle, one may get a strongly curved background that could distort the EXAFS signal.

There are various strategies for avoiding these effects, some of which are common to bulk and non-environmental EXAFS. For instance, if the beamline is capable of "quick-EXAFS," then each scan will be taken so quickly that slow drifts will not cause artifacts (Gaillard et al. 2001). Some beamlines (MacDowell et al. 2001) create an image of the source on a set of slits, which in turn becomes a fixed "virtual source." Attention to mechanics and temperature stability can pay off in terms of beam-position stability. If one is looking at a particle whose fluorescent yield or transmission is very different from that of its surroundings, then it pays to put the beam accurately on an extremum of yield or transmission. That way, small motions only cause second-order perturbations in the signal. This procedure also minimizes the effect of vibrations.

Beam damage. The power density in a microfocus beam is much greater than in a "normal" beamline fed by the same source, and with the advent of extremely intense X-ray beams from third generation sources, observation of damage to samples has become more common. In organic materials, radiation damage has been shown to cause the breaking of disulphide bonds, the decarboxylation of glutumatic and aspartic acids, and the removal of hydroxyl groups from tyrosines and methylthio groups from methionine residues (Burmeister 2000; Ravelli and McSweeney 2000; Weik et al. 2000). A way of delaying the onset of radiation damage is to cryocool the sample in order to decrease the mobility of the free radicals (mainly electrons) formed in the primary radiation damage process (Powers 1982). However, this problem is not trivial because there are many variables, and in many cases collecting the data with a reduced total photon dose is the best option to alleviate the problem because radiation damage rates can only be slowed.

In X-ray absorption experiments of environmental samples, photoreduction and photooxidation of the target element are the two phenomena most commonly observed. The reduction of Mn(IV,III) to Mn(II) in soils has been investigated in detail by studying the influence of the sample, the exposure time, and the photon density (Ross et al. 2001). The drop in energy of the main absorption edge varied from a sample to another. This change was as high as 1.8 eV in the Supersoil sample ($[C]$ = 11.3 wt. %) irradiated for 180 minutes with a photon density of 4.8×10^3 ph s^{-1} μm^{-2}, and equal to 1.3 eV in the Hickory sample ($[C]$ = 15.9 wt. %) exposed for the same duration, but with a beam two orders of magnitude more intense (4.4×10^5 ph s^{-1} μm^{-2}). In general, the shift in energy depended not only on the intensity and exposure time, but also above all on the sample. In this study, the sources of electrons for the photoreduction likely were organic molecules, which in soil samples are generally intimately associated with inorganic constituents. The materials used in sample preparation are additional potential sources of electrons for reduction. Zavarin (1999) found reduction of selenate sorbed to calcite after extended exposure to a synchrotron X-ray beam. The inferred source of electrons causing reduction to selenite was the adhesive on Kapton or Mylar tapes used to mount the sample. Ross et al. (2001) tried to get rid of this effect in using a polypropylene film with no adhesive but, as they acknowledged, contributions of electrons from sample-holding materials are difficult to suppress and generally cannot be completely ruled out.

In polarized experiments performed on self-supporting films of the SHCa-1 hectorite from the Source Clay Minerals Repository of the Clay Minerals Society ($[MnO]$ = 0.008 wt. %), we observed that Mn(II) present in trace amounts (0.03 weight %) was oxidized to Mn(III) when measurements were carried out on the undulator beamline ID26 of the ESRF (photon density $\sim 10^8$ ph s^{-1} μm^{-2}), and remained in reduced form when measurements were carried out on the bending magnet BM32 (photon density $\sim 10^7$ ph s^{-1} μm^{-2}). Figure 19 shows fluorescence-yield Mn K-edge XANES spectra recorded at α = 35° and 55° during the photooxidation reaction, together with the polarized XANES spectrum of the unoxidized sample (all scans lasted about 25 minutes). In the unperturbed state, the edge crest of Mn(II) is split as systematically observed for 3d transition metals in the octahedral layer of phyllosilicates (Dyar et al. 2001). The two maximum absorption

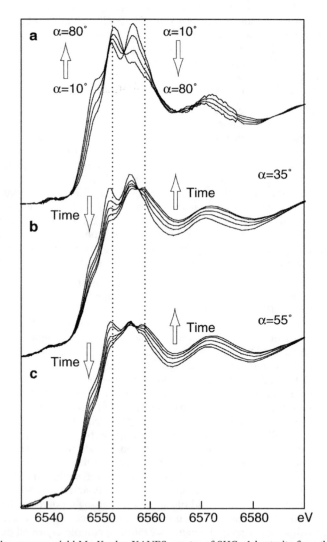

Figure 19. Fluorescence-yield Mn K-edge XANES spectra of SHCa-1 hectorite from the Source Clay Minerals Repository of the Clay Minerals Society ([MnO] = 0.008 wt. %). a) Polarized spectra from a self-supporting film recorded on a bending magnet beamline at ESRF (BM32, Grenoble). In phyllosilicates, metal octahedra are flattened, and this distortion likely accounts for the observed angular dependence (Manceau et al. 1998; Dyar et al. 2001). b-c) XANES scans recorded at fixed polarization angle (α = 35° and 55°) and increasing time on an undulator beamline at ESRF (ID26). Mn(II) is gradually photooxidized to Mn(III).

features occur as sharp peaks at ca. 6552 eV and 6557 eV. During the Mn(II) oxidation, a shoulder at ca. 6559 eV, characteristic of Mn(III) (Bargar et al. 2000), appeared and increased with time, while the amplitudes of the two Mn(II) maxima decreased. Isosbestic points (constant absorbance) at ca. 6558 eV indicate that two components, i.e., Mn(II) and Mn(III), dominate the system. The absence of Mn(IV) suggests that the oxidation occurs via a one-electron transfer reaction. Close comparison of XANES spectra collected on BM32 and ID26 shows that Mn(II) oxidation already started during the first scan on the undulator beamline, meaning that the collection time was not short enough to prevent an experimental artifact. Since the SHCa-1 hectorite contains very low amounts of ferric iron ($[Fe_2O_3]$ = 0.02 wt. %), its color is originally pale, and the radiation damage could be monitored optically by the formation of a dark smudge as X-rays impinged the sample. Sample darkening induced by long exposure to an undulator X-ray source was also observed within the beam footprint by Schlegel et al. (2002) in reflectivity measurements of quartz. In this experiment, the authors likely formed color centers as in smoky quartz (Rossman 1994).

Sample preparation and characteristics. Sample preparation is something that needs to be carefully considered. As we will see below, EXAFS spectra can be distorted if the sample is too thick or too concentrated. Also, much of the information one wants is related to the morphology of intact soils or sediments. It is therefore useful to be able to make a uniform, thin section and to fix its structure. A typical way of doing this involves infiltration of the sample with a liquid casting resin, which then hardens. The sample may then be sliced into sections, generally 10-50 µm thick. Unfortunately, this procedure may affect the structure and chemistry of the sample, as recently pointed out by Strawn et al. (2002). In a study on the speciation of arsenic and selenium in soil, these authors showed that added As(III) was partly oxidized to As(V) after the impregnation with LR White resin (SPI Supplies, West Chester, PA) and thin sectioning steps. Therefore, care should be taken in selecting a casting resin that contains as small as possible a proportion of free acids and solvents so as to minimize the potential chemical impact on the sample. But even with due care, it is still possible for species in the soil to diffuse through the resin before it sets. The chemical purity of the resin and of the glass mount is sometimes an issue. For instance, some glass slides contain arsenic, and it is generally preferable to use fused quartz slides (Fig. 20, As map). Scotchcast[TM] epoxy from 3M is of high purity, but it has an extremely high viscosity and complete diffusion throughout a porous sample is typically not possible, but multiple embeddings of the same sample usually gives the desired result.

Another point to consider is that the penetration depth for fluorescent X-rays depends on their energy. Thus, the map for Ca will represent material closer to the surface than that of Zn (Fig. 20). This effect can distort the results of correlation analysis because the maps for different elements will represent different volumes. Since the incident beam is usually not normal to the surface, features deep inside the sample will appear out of registry with those nearer the surface. The net effect is as if the sample is viewed from the direction from which the beam comes, rather than straight on, as is usually the case for optical microscopy.

It is important to know where the X-ray beam strikes the sample so that features seen in fluorescence maps may be correlated with known structure as seen by optical microscopy. Ideally, one would like an optical microscope that views the sample from the direction of the beam so that the optical and fluorescence images maintain a consistent registry regardless of surface topography. This condition may be arranged by having the beam pass through a hole in a small prism or a mirror, which reflects an image of the sample to a microscope (Perrakis et al. 1999). If the microscope is properly lined up, the

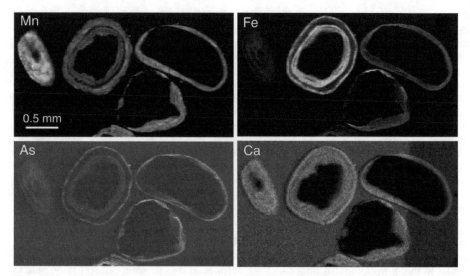

Figure 20. µSXRF maps of (Fe,Mn)-coated gravels used to filter waste waters. The sample was prepared as a 30 µm micropolished thin section mounted on a glass slide. Arsenic is selectively associated with Fe oxides, but the slide contains As impurity which yields a background signal. Calcium is also present both in the (Fe,Mn) layers and glass substrate. In the Ca map, the quartz core regions are darker than the substrate areas because the Ca Kα radiation is absorbed by the quartz. This map also shows shading and highlighting effects because the sample protrudes 30 µm from the surface of the slide. The right hand darker strip originates from an adhesive tape stuck on the forward face. Data collected on beamline 10.3.2 at the ALS (Berkeley).

beam will always appear in the same place regardless of motion of the sample. However, a simpler system in which the sample is viewed from the side is often used, in which case the registry between sample and beam depends on the longitudinal position of the sample along the beam direction. In that case, it becomes important to align the beam and microscope for each sample.

In addition to changing the registry between the beam and the optical image, sample topography can have some other effects. The next paragraphs discuss "thickness" and "hole" effects, which are sensitive to topography. Also, topography can produce "highlighting" or "shading" effects in the fluorescence maps of low-Z elements. For instance, suppose there is a bump on the sample surface and the beam comes in from the left as viewed looking straight at the sample. Then, when the probe beam is on the left side of the bump, the bump itself blocks some of the fluorescence light from reaching the detector, resulting in a dark border on that side. Similarly, there will be a bright border on the right due to the reduced angle of incidence of the beam on the surface on the right-hand slope of the bump. The net effect is as if the sample were illuminated from the right—the opposite of the actual illumination direction. This seeming discrepancy is common to scanning-beam microscopies and is well known in scanning electron microscopy. This effect is clearly visible in the calcium maps of Figures 4 and 20.

The effects known as "hole effect" and "thickness effect" (also, "self-absorption" and "over-absorption") are well known from other sorts of EXAFS studies (Goulon et al. 1981), but tend to be difficult to quantify and manage when doing environmental studies, especially with a microbeam. The hole effect is seen when doing a transmission spectrum on an inhomogeneous sample. As a simple example, imagine that half the beam goes through a part of the sample for which the number of absorption lengths (μt) is 2 above the edge. For this simple example, let us neglect the pre-edge absorption. Assume that the other half of the beam goes through a blank area with no absorption. Now, the total transmission above the edge is $[1 + \exp(-2)]/2 \approx 0.57$, whereas that below the edge is 1. The effective absorption is the negative log of this quantity, which is also about 0.57. Now, let there be an EXAFS oscillation of 10%, so that the absorption in the sample goes from 2 to 2.2. The transmission becomes 0.56, and the change in the effective absorption is only 0.0218, out of an edge jump of 0.57. Thus, we have a measured amplitude for this EXAFS wiggle of 3.8% instead of the real value of 10%. What is happening here is that the hole produces a non-linear relation between the measured absorbance and the real one. This relation displays a saturating behavior, so that small changes at the top end (where the EXAFS is) are underrepresented relative to the whole edge jump. Figure 21 shows this phenomenon graphically. In this figure, the actual EXAFS is shown on the bottom, and its mapping to the measured EXAFS at the right is via the smooth curve in the middle. Figure 22 shows some experimental microprobe data of a ferromanganese crust (Hlawatsch et al. 2001) taken in transmission at three spots on areas of the same composition but differing thickness. We see from these data how the EXAFS amplitude can be reduced, and the pre-

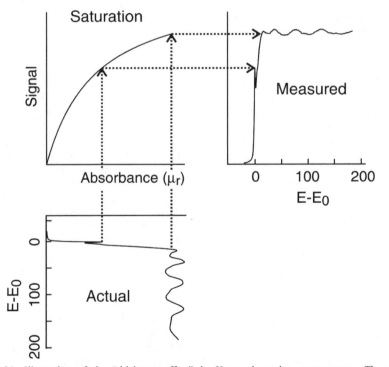

Figure 21. Illustration of the "thickness effect" in X-ray absorption spectroscopy. The actual absorbance vs. energy is shown at the bottom. Due to these thickness effects, the measured signal (right) is related to the actual absorbance via a sub-linear transfer curve (Saturation). Two specific points along the curves are picked out with dotted lines and arrows, showing how the pre-edge features are raised relative to the edge. Notice also that the EXAFS amplitude in the Measured curve is reduced compared to its actual value. The Actual curve is transmission data for a Ti foil and the Measured curve is the fluorescence data for the same sample (6 μm, 45° incidence and exit angle). The Saturation curve comes from a fit between the Actual and Measured curves.

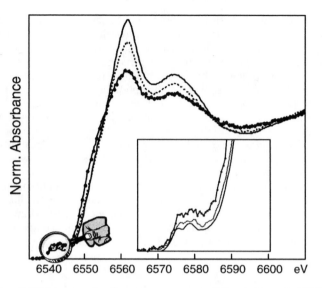

Figure 22. Micro-XANES data of a ferromanganese crust (Hlawatsch et al. 2001) recorded in transmission mode in spots of the same composition but differing thickness. The pre-edge is enhanced and the EXAFS amplitude is reduced by hole effect.

edge enhanced (Manceau and Gates 1997), by these effects. As will be explained below, similar effects can occur in fluorescence as well as transmission.

With this simple example as background, let us consider a more general situation. Consider a beam incident on a particle such that the thickness of material is not uniform within the beam. Let the total transmission of the beam be $T(\mu(E))$ where $\mu(E)$ is the absorption coefficient at a particular energy E. This absorption coefficient has two parts, a resonant part μ_r which is 0 below the edge and which is modulated by the EXAFS, and a non-resonant or background part μ_b that is continuous and slowly varying through the edge. The EXAFS oscillations cause μ_r to be replaced by $\mu_r(1 + \chi)$ where χ is the EXAFS wiggle. What one plots is the negative logarithm of the transmission $P = -\ln(T)$. In analyzing the EXAFS, one subtracts the pre-edge background and normalizes to the smooth part of the resonant absorption. If this could be done perfectly, the result would be the effective EXAFS wiggle

$$\tilde{\chi} = \frac{\ln\left[T(\mu_b)/T(\mu_b + \mu_r(1 + \chi))\right]}{\ln\left[T(\mu_b)/T(\mu_b + \mu_r)\right]} \tag{18}$$

where the denominator represents the post-edge background normalization. We can check this formula by noting that if the sample is uniform and of thickness t, then the transmission $T(\mu)$ is given by $\exp(-\mu t)$ and the above equation evaluates to $\tilde{\chi} = \chi$. If the EXAFS wiggle is small, we can expand Equation (18) with respect to χ and come up with

$$\tilde{\chi} = \chi \frac{\mu_r \left.\dfrac{d\ln T}{d\mu}\right|_{\mu = \mu_b + \mu_r}}{\ln[T(\mu_b + \mu_r)] - \ln[T(\mu_b)]} \equiv S_t \chi \tag{19}$$

where S_t is the ratio of the amplitude of the measured EXAFS wiggle to the actual signal.

Thus, if S_t is 0.8, then the coordination number will appear to be 80% of what it really is.

The above formalism is quite general, and can be used to study the effects of doing transmission EXAFS on non-uniform samples and particles of all sorts. However, the mathematics quickly become forbidding if one looks at a realistic situation such as a particle illuminated by a Gaussian beam. Such situations require numerical methods. For a simple, analytically accessible example, imagine a uniform foil with a hole in it such that a fraction f of the beam goes through the hole. Then, we have

$$T(\mu) = f + (1 - f)\exp(-\mu t) \tag{20}$$

where t is the thickness of the hole. Turning the crank on Equation (19) yields, after some algebra

$$S_t = -\frac{\mu_r t(1 - \tilde{f})\exp(-\mu_r t)}{[\tilde{f} + (1 - \tilde{f})\exp(-\mu_r t)]\ln[\tilde{f} + (1 - \tilde{f})\exp(-\mu_r t)]} \tag{21}$$

where $\tilde{f}$ is an "effective" hole fraction which reduces to f for $\mu_b = 0$ and is given by

$$\tilde{f} = f/[f + (1 - f)\exp(-\mu_b t)] \tag{22}$$

which is always larger than f. It is easy to show that S_t is 1 for $\tilde{f} = 0$ or $\mu_r t = 0$ and decreases monotonically with both $\tilde{f}$ and $\mu_r t$.

Figure 23 shows a contour map of S_t as a function of $\tilde{f}$ and $\mu_r t$. For a thick sample, even a small fraction of beam skimming by can affect the EXAFS amplitude. Similarly, if there are harmonics in the beam, they go through the sample with less absorption than the fundamental, leading to the same effect as if there were a hole (Stern and Kim 1981). While the sample may not have holes in it, it is common to encounter particles smaller than the beam. Suppose one is dealing with a primary mineral such as magnetite (Fe_3O_4),

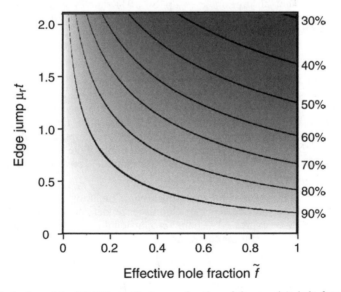

Figure 23. Reduction of the EXAFS amplitude as a function of the sample's hole fraction and the absorption edge jump. Percentages correspond to the ratio of the amplitude of the measured EXAFS wiggle to the actual signal.

to give a simple example. At the edge, $1/\mu_r$ is about 8.7 μm. Therefore, if we have a particle of 5 μm size, we would want < 30% of the beam to miss the particle in order to keep 90% of the EXAFS amplitude, if the data were being taken in transmission.

Since these calculations require one to evaluate the absorption coefficients (absorbances) of materials at various energies, we will now make a short diversion into how these coefficients may be calculated. There are several standard sources of tables, some of which are on the World Wide Web. Two such sets of tables are those by Henke et al. (1993) and McMaster (1969). The latter is embodied in a FORTRAN program called mucal. An approximate method, good to 20% or so between 5 and 100 keV, is to use the polynomial fits due to Gerward (1981), which is the source of the numbers quoted here. All of these methods yield the mass absorption coefficient $\mu\,\rho|_{E,Z}$ (dimension cm^2/g) for a given element Z at a given energy E. To get the absorbance of a sample of a given composition, the formula is

$$\mu(E) = \rho\sum_i (\mu/\rho)\big|_{E,Z_i} \tag{23}$$

where i is the index of elements and ρ is the density of the sample in g/cm^3. The result is the absorbance in cm^{-1}. To get the resonant part, subtract the absorbance evaluated just below the edge from that just above. The non-resonant part is approximately the absorbance just below the edge. For more accuracy, extrapolate the non-resonant part to the energy of interest.

For small particles, and in many other cases, one will use fluorescence instead of transmission. However, saturation effects such as we saw for transmission mode also occur in fluorescence as well (Troger et al. 1992; Castaner and Prieto 1997). The classic case here is that of a thick piece of pure metal such as Cu. In this material, the ratio of the resonant to non-resonant absorption is about 85:15. This means that for every 100 incident photons, 85 of them create K-holes and thus could stimulate fluorescence. Now, suppose the resonant absorption goes up by 10% due to EXAFS. Now the ratio is 93.5:15, or about 86.1:13.9. Thus, the resonant process accounts for 86.1% of the total, which means the fluorescence intensity only goes up by 1.3% instead of 10%. This example shows that, again, the response saturates as a function of the absorption one wants to measure.

This effect is known variously as "self-absorption," "thickness effect" or "over-absorption." An analogous effect in photochemistry is known as "saturation of the action spectrum." The first term is somewhat of a misnomer because it suggests that the problem has to do with re-absorption of the fluorescence radiation, by analogy with certain effects in optical spectroscopy. Actually, the absorption of the fluorescence is independent of the incident energy, hence does not contribute to any non-linearity. What is important is that the penetration depth for the incident radiation depends on the quantity one wants to measure.

In the same spirit as the previous discussion on transmission, let us look at the general theory of thickness effects in fluorescence detection mode. The fluorescence signal received may be expressed as being proportional to the resonant absorption times a factor that would be constant for a thin sample

$$I_f = \mu_r F(\mu_r, \mu_b, \mu_f) \tag{24}$$

where μ_f is the absorption coefficient for the fluorescence radiation. Now, the background below the edge is 0, so the EXAFS modulation is given by

$$I_f(1 + S_f\chi) \approx \mu_r(1 + \chi)F(\mu_r(1 + \chi), \mu_b, \mu_f) \tag{25}$$

to first order in χ. Expanding this, we find a simpler formula than for transmission

$$S_f = 1 + \frac{\partial \ln F}{\partial \ln \mu_r}.$$ (26)

Now let us do a simple example. Consider an infinitely-thick material oriented at an angle θ to the beam (0 = grazing) and ϕ to the detector, which is considered to subtend a small angle. The beam intensity after the beam has gone a distance z into the sample is $\exp(-(\mu_r + \mu_b)z)$, and the fraction of the fluorescence not absorbed by the sample is $\exp(-\tilde{\mu}_f)$, where $\tilde{\mu}_f$ is an effective absorption given by $\mu_f \sin\theta / \sin\phi$. Integrating along the beam path, we find that up to constant factors, the fractional fluorescence yield is

$$F \propto \int_0^\infty dz e^{-z(\mu_r + \mu_b + \tilde{\mu}_f)} = \frac{1}{\mu_r + \mu_b + \tilde{\mu}_f}.$$ (27)

Carrying out the math, we find that

$$S_f = \frac{\mu_b + \tilde{\mu}_f}{\mu_b + \tilde{\mu}_f + \mu_r}$$ (28)

which shows that if we want to retain 90% of the EXAFS amplitude in fluorescence detection mode, the resonant absorption must be < 10% of the total, including fluorescence. This treatment extends to electron-yield detection if one substitutes the appropriate mean free path for $\tilde{\mu}_f$. This also shows that by collecting the fluorescence at grazing angles to the sample, one can raise $\tilde{\mu}_f$ to the point where the thickness effect drops out and a correct EXAFS spectrum is obtained (Brewe et al. 1992). However, this method only works for smooth, flat samples, which are not always available, and it throws away a lot of signal. Still, it may be useful for certain standard minerals which can be polished and which are so concentrated one does not mind losing signal.

If the sample is of finite thickness, the formula gets somewhat more complex

$$S_f = 1 - \frac{\mu_r}{\mu_t} + \frac{\mu_r}{\mu_t} \frac{\mu_t t \csc\theta}{\exp(\mu_t t \csc\theta) - 1}$$ (29)

where t is the sample thickness and $\mu_t \equiv \mu_r + \mu_b + \tilde{\mu}_f$ is the total absorption along the incident direction. For $\mu_t t \csc\theta \gg 1$ we recover the bulk Equation (28), and for $\mu_t t \csc\theta \ll 1$ we get $S_f = 1$ as expected, because for diluted samples (regardless their thickness) the fluorescence intensity is proportional to the resonant absorption. Figure 24 shows a contour map of this formula, plotted as a function of the edge jump $\Delta\mu t = \mu_t t \csc\theta$ (proportional to the sample thickness) and the resonant fraction μ_r/μ_t (proportional to the metal concentration). We see two distinct limits, corresponding to the two branches of the contours: thick sample, in which the resonant fraction has the most effect on S_f, and thin sample, where the edge jump matters most.

Now for some examples. One of these has already been shown as Figure 21. The "measured" curve in Figure 21 is an experimental fluorescence curve for Ti foil (6 μm, 45° incidence and exit angle), while the "actual" curve is transmission data for the same sample. The smooth mapping curve is a fit to Equation (29). The first environmental example is Zn in a primary mineral, sphalerite (ZnS), measured at 45° incident and exit angles. For this case, the resonant fraction is 59% ($\mu_b = 160$ cm^{-1}, $\mu_r + \mu_b = 700$ cm^{-1}, $\mu_f = \tilde{\mu}_f = 220$ cm^{-1}, so $\mu_t = 920$ cm^{-1}, assuming $\rho = 4.102$). Suppose we want the amplitude to be good to 10%, so S_f must be at least 90%. Referring to Figure 24, we see that with $\mu_r/\mu_t = 59\%$, we are in the thin-sample (high [Me]) branch of the curve, and that

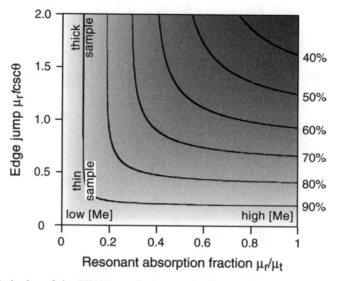

Figure 24. Reduction of the EXAFS amplitude as a function of the resonant absorption fraction (proportional to the metal concentration) and the absorption edge jump (proportional to the sample thickness and metal concentration). Percentages correspond to the ratio of the amplitude of the measured EXAFS wiggle to the actual signal.

the edge jump $\mu_t t \csc\theta$ must be 0.2 or less. Plugging in the values above, we find a maximum sample thickness of $0.2 \times \sin(45°)/920\ \mathrm{cm}^{-1} = 1.5\ \mu\mathrm{m}$. Fortunately, we do not need the beam to be as small as the particle in order to do fluorescence. A bigger beam would make for better stability and still yield the correct spectrum provided there are no larger particles or particles containing Zn in a different form in that beam. This example shows the importance of mapping inhomogeneous samples before measuring them.

The opposite limit is represented by a dilute mixture. For this example, let us take 1 cation % of Zn in kaolinite (used as an example of a light matrix). The empirical formula of this sample is $\mathrm{Al_4Si_4O_{10}(OH)_8{:}Zn_{.08}}$. Now, the Zn resonant absorption is < 5% of the total, so one is free to use any thickness. Further, non-uniformities in concentration over the beam spot would have no effect, provided that the Zn environment is uniform.

What can be done in the case of a concentrated sample? Some possibilities include:

(1) Look for a particle in fluorescence mode that is small enough to be in the thin-sample limit. In a fluorescence map, we do not see the thickness directly, but only the width across the beam, so one may have to make assumptions about the shape of the particle in order to decide if it is small enough. With a micron-size beam, one would look for a resolution-limited spot on a fluorescence map.

(2) Use electron-yield detection. In the hard X-ray range, the escape depth for electrons is much less than the absorption depth for X-rays, so there is little thickness effect because the electrons are effectively probing a part of the sample that is thin compared with the absorption depth. This method can be useful for standards and other highly-concentrated samples. However, if the sample is illuminated at grazing incidence, there can be an over-absorption effect as well because then the X-rays do not penetrate much farther than where the electrons

come from. Thus, the amount of material probed depends on the X-ray absorption one wants to measure, just as with fluorescence. Also, fluorescence photons are detected by photoemission, yielding a contribution to the signal that is proportional to the fluorescence and hence affected by over-absorption (Schroeder et al. 1997). Thus, it is commonly observed that the EXAFS amplitudes seen in electron yield are slightly smaller than those measured by transmission.

(3) Use grazing-exit detection. This method, as described above, requires a very flat sample and so is not suited to most natural materials.

(4) Measure a thin section in transmission. This works reasonably well for samples in which one wants to look at the matrix, and the matrix is uniform on the scale of the beam size.

(5) If the effect is not too severe and the various quantities are known, then the spectrum may be corrected so the distortions are removed (Pfalzer et al. 1999). However, this correction occurs at the cost of some signal-to-noise. Also, this procedure requires knowing the various quantities in the above equations. For standards and identified primary minerals, the composition is known, so that the resonant fraction is known. Then, if there is transmission data, the edge jump may be measured. This level of knowledge occurs less often than one might like. If transmission data are available, it is sometimes possible to do an approximate correction based on the amplitude of the transmission EXAFS signal at low k (higher S/N ratio), which is used to correct the fluorescence spectrum. This procedure is used when the transmission signal is not good enough to use on its own. Unfortunately, hole effects can cause errors in this type of correction.

The above discussion brings us to another set of problems typically encountered in environmental science, mainly those encountered when trying to look at small particles. One of these has already been described, that is beam stability. This problem can only be addressed by proper engineering of the beamline. If the particle is concentrated and has a thickness comparable to the absorption length, then thickness effect in fluorescence mode, and hole effect in transmission mode, will cause problems. The hole effect arises because the particle will never be uniform in thickness across the beam. As a limiting case, if the beam spills over the particle, then there will be a hole effect just as described as our first example of thickness effects. The measured transmission edge jump will be smaller than the real edge jump in the thick part of the particle, so it becomes impossible to estimate the real thickness, which is what is needed for correction. The particle will appear to be thinner than it really is, so there will be saturation effects but no warning that these may occur. The beam from a typical Kirkpatrick-Baez setup has a halo which is several times larger than the nominal (FWHM) beam diameter and which carries a large fraction of the energy in the beam. Therefore, even if the FWHM of the beam is less than the particle size, a good part of the beam will pass by the particle. We see that saturation effects may be important and uncorrectable when looking at particles. Some of the measures described above may help avoid these problems. Fortunately, such concentrated particles in soils are often composed of single, simple minerals whose spectra may be recognized, even in distorted form, by comparison with standards

If the sample is thick, the particle may be buried in a matrix which contains the same element one wants to probe in the particle (Roberts et al. 2002). In that case, the spectrum will be a weighted sum of those from the matrix and the particle. In this case, making the sample thin might help by reducing the ratio of the matrix contribution to that of the particle. Another possibility is to record μEXAFS spectra in different spots in order to vary the proportion of each species, and to analyze the set of spectra by PCA, as illustrated below.

APPLICATION TO HEAVY METAL SPECIATION

In the remaining section of this chapter, synchrotron radiation-based studies on the speciation of heavy metals in soils and sediments are reviewed, and specifically those employing EXAFS spectroscopy since this technique has the desired attributes to provide the forms of metals in the host solid fraction at the molecular scale. For reasons discussed in previous sections, XANES spectroscopy, if not used as an adjunct to EXAFS spectroscopy, does not uniquely speciate metals in complex matrices, in particular those containing secondary forms of the metal (sorption complex, substitutional forms...). Therefore, such measurements are beyond the scope of this review. Several strategies have been developed to determine the proportion of metal species from EXAFS data in the case of mixture, encompassing approaches that use linear combinations of EXAFS spectra (Manceau et al. 1996) and shell-by-shell fitting (Bostick et al. 2001). As discussed previously, the latter approach is inappropriate because of the presence of overlapping and multi-atom shells in multiphase systems matrices, which cannot be discriminated by the traditional Fourier filtering technique. A survey of the literature shows that the most comprehensive studies are those in which an array of analytical techniques (PIXE, SXRF, XRD...), and physical (size and densimetric separation) and chemical fractionation (selective dissolution), were combined. Therefore, emphasis will be placed on multi-approach and multi-scale work. However, it should be bear in mind that chemical extractions can cause artifacts, as convincingly demonstrated by Ostergren et al. (1999).

In general, inorganic contaminants enter soil as a result of mining, metallurgy, use of fossil fuels, and application of soil amendments, that is in primary forms which are not in equilibrium with the physico-chemical conditions of their new resting place. Weathering leads to the liberation of labile and potentially toxic metastable forms, and determining their nature and properties is fundamental to assess and reduce risks to living organisms. Therefore, the real scientific challenge relies upon the identification of these secondary species, and main focus here will be placed on this hellishly difficult task. Also, primary forms are generally best identified by traditional mineralogical techniques, including electron microscopy. Finally, since heavy metals are persistent (other than radioisotopic forms), and because their negative effects in soil may be long-lasting, their mobility can be mitigated by sequestration in sparingly soluble forms. Therefore, understanding how trace metals are naturally sequestered in soils provides a solid scientific basis for formulating educated strategies to clean-up severely impacted areas and for maintaining soil quality. Recent results obtained in this emerging field shall be presented in the second part of this section.

Speciation of metal(loid)s in the contaminated environment

Arsenic. More than 300 arsenate and associated minerals have been identified (Escobar-Gonzalez and Monhemius 1988). Inevitably, some of the arsenic contained in these minerals enters any industrial circuit, and concentrations of As in soils and waters can become elevated due to mineral dissolution. The original National Priority List (USA) identified approximately 1000 sites in the United States (USA) that posed environmental health risks (Nriagu 1994; Allen et al. 1995) with arsenic cited as the second most common inorganic constituent after lead (Database 2001). The more common oxidation states of arsenic are III and V, and the predominant form is influenced by pH and redox potential. In aqueous solutions of neutral pH, arsenate is present predominantly as $H_2AsO_4^-$, while arsenite is uncharged as H_3AsO_3 (Cullen and Reomer 1989). Both forms are often present in either reduced or oxidized environments due to their relatively slow redox transformations (Masscheleyn et al. 1991), but seasonal variations of the mean As oxidation state has been observed by XANES in wetlands undergoing flooding and drying periods (La Force et al. 2000). The reduced form of arsenite is more soluble and toxic than the oxidized form, and both have strong affinities for iron oxides, though they behave oppositely with regard to the influence of pH. In the pH range of 3 to 10, adsorption of arsenate on iron oxide generally decreases with increasing pH, while arsenite adsorption increases with pH, with maximum adsorption at approximately pH 9.0 (Manning et al. 1998; Jain and Loeppert 2000; Goldberg and Johnston 2001).

Tailings from uranium and gold mines represents one of the main source of As pollution to the environment (Roussel et al. 2000; Woo and Choi 2001), with As being contained generally in arsenopyrite and pyrite. In mining, especially during bioleaching of gold from arsenopyrite ore, soluble arsenic (mostly arsenate) levels in waste waters can reach 30 g/l (Rawlings and Silver 1995). The speciation of arsenic in aqueous colloidal particles and secondary phases from acid rock drainage solutions and precipitates have been extensively investigated recently using EXAFS spectroscopy. In the Freiberg (Germany) mining area, Zänker et al. (2002) showed that dissolved As was mainly complexed to nanometer-sized jarosite - schwertmannite particles. These colloids ultimately aggregated and precipitated leading to the incorporation of arsenate in the jarosite ($K(Fe_3(SO_4)_2(OH)_6)$) structure and/or to the formation of very small scorodite ($FeAsO_4 \cdot 2H_2O$) clusters as occlusions within jarosite. In keeping with these results, three secondary As species were positively identified in the solid fraction of Californian mine wastes: As sorbed on ferric oxyhydroxides, scorodite, and As^{5+}-containing jarosite (Foster et al. 1997; Savage et al. 2000). All these species are stable in the acidic conditions generated by the oxidation of sulfides (Singer and Stumm 1970), but none of them is a relevant candidate for As on-site immobilization. Jarosite, which is a common weathering product in acidic environments (Alpers et al. 1989), is not stable above pH ~3 (Fig. 8 in Savage 2000) and is expected to dissolve releasing associated As to water. Scorodite's solubility strongly depends on pH with a minimum at pH ~ 4.0 (50 µg/L), which still exceeds drinking water standards, and a sharp increase of its solubility for lower and higher pH values (Krause and Ettel 1988). Lastly, though arsenate forms a strong inner-sphere surface complex on ferric (oxyhydr)oxides (Waychunas et al. 1993; Manceau 1995), there is always the risk that a surface complex is desorbed inasmuch as the capacity of these minerals to retain As depends on the solution composition (possible presence of strong ligands), the bacterial activity, and the seasonal pH and redox conditions within the wetland or mine spoil. The speciation of arsenic in Californian reservoir sediments used to remove As from water supplied to the city of Los Angeles has been also investigated by XANES and EXAFS (Kneebone et al. 2002). In the uppermost sediment (0-25 cm), As was present as a mix of As(III) and As(V), whereas only oxygen-coordinated As(III) was detected (no sulfide) in the deeper sediments (to 44 cm). A metal shell at 3.42 Å was identified and interpreted as arising from Fe atoms in an (oxyhydr)oxide phase. This study represents the first strong suggestive evidence of the uptake of As(III) by ferric oxides in nature, and, if it is confirmed, it validates the relevance to the field of all the solution chemistry (see e.g., Wilkie and Hering 1996; Jain and Loeppert 2000; O'Reilly et al. 2001) and spectroscopic (Fendorf et al. 1997; Manning et al. 1998; Goldberg and Johnston 2001) investigations realized in the last ten years on the As(III) - ferric oxide model system.

Many sites contaminated by arsenic are too toxic to allow the spontaneous colonization and growth of even hyperaccumulating plants, and present remediation technologies focus on *in situ* inactivation procedures rather than excavation which is costly, disruptive and requires a source of clean soil and a repository emplacement (Van der Lelie et al. 2001). All mineralogical and geochemical studies indicate that there is no single amendment that durably immobilizes arsenic in oxidizing to mildly reducing environments and the current trend consists of using co-amendments and seeking synergistic effects between the added reactive phases to produce the maximum reduction in metal(oid) bioavailability. An example is the on-going remediation and revegetation program of the Barren Jales gold mine spoil in Portugal (Fig. 25, Bleeker et al. 2002; Mench et al. 2002). Vegetation establishment on the treated spoil was successful with *Holcus lanatus* the 1st year and *Pinus pinaster* the 2nd year after *in situ* As inactivation by the addition of steel shot, beringite, and municipal compost. The role of steel shot (elemental Fe) is to provide a source of Fe^{2+} and Fe^{3+} ions, which can immobilize the labile As fraction in a similar manner to Fe^{2+} released by sulfide minerals in mine spoils. But as steel shot is devoid of sulfur, it does not have the adverse effect of lowering the pH. Beringite is used for its alkaline and sorptive properties and the compost to remedy to the lack of organic matter in the spoil. This real world example demonstrates how a firm knowledge of chemical reactions with mineral phases can help to control the mobility of trace contaminants in the environment.

Lead. Lead contamination of the environment is synonymous with civilization. A great number of studies have been conducted on this element because of its detrimental effect on intelligence, behavior and educational and social attainment (Needleman 1983; Needleman et al. 1990). Lead was the first heavy metal whose speciation in contaminated soil was studied by EXAFS (Cotter-Howells et al. 1994), and its has received considerable attention over the last six years (Manceau et al. 1996; O'Day et al. 1998; Morin et al. 1999; Ostergren et al. 1999; Welter et al. 1999; O'Day et al. 2000; Calmano et al. 2001; Hansel et al. 2001; Morin et al. 2001). Five secondary forms of lead have been identified from this breadth of work, including Pb^{2+}-organic complex, Pb^{2+}-sorbed ferric and manganese oxides, plumbo-jarosite, and lead phosphates, and the factors controlling their occurrence, formation, and stability have been also examined. As shown below, the basic research conducted on this element with synchrotron radiation already has important practical consequences as it provides a solid scientific basis to improve the effectiveness and reduce the cost of existing technologies to lead-contamination remedial.

Lead is known to be fairly immobile in organic soils, with a mean residence time

Figure 25. On-going remediation and revegetation program of the Barren Jales gold mine spoil (Portugal). Vegetation establishment on the treated spoil was successful with *Holcus lanatus* the 1st year and *Pinus pinaster* the 2nd year after *in situ* As inactivation by the co-amendment of steel shots, beringite, and municipal compost. Adapted from Mench et al. (2002) and Bleeker et al. (2002).

evaluated from hundreds to thousands of years (Benninger et al. 1975; Heinrichs and Mayer 1977, 1980). Its low mobility and bioavailability had been interpreted in terms of the formation of stable organo-lead species (Stevenson 1979; Bizri et al. 1984; Taylor et al. 1995), but no direct structural evidence was available until EXAFS spectroscopy was applied. Manceau et al. (1996) examined lead structural chemistry in a garden soil contaminated by tetraalkyllead species used as additive in gasoline constituents. Since alkyllead species have a life-time of a few hours, lead is rapidly bound to soil constituents, and in the present case to organic molecules that are abundant in garden soils. Therefore, this system was particularly suited to determine the nature of complexing functional groups because, in contrast to most situations, lead speciation was unique. Confirmation of the degradation of the initial lead tetraethyl species was obtained by comparing EXAFS spectra of the organic compound and the soil sample (Fig. 26). The unknown spectrum did not resemble that of lead carboxylate complex, and good modeling was obtained with a mixture of 60% Pb salicylate and 40% Pb catechol functional groups. In spite of the generally higher proportion of carboxylate groups in organic matter, lead was preferentially chelated to functional groups bonded to aromatic rings forming a bidentate complex. At the root surface of *Phalaris arundinacea*, a reed canary-grass commonly found in submerged areas of wetlands, lead was inferred to be complexed by organic molecules present in ferric (oxyhydr)oxide – biofilm coatings (Hansel et al. 2001) (Fig. 27). Organo-lead species were also identified in wooded (pH 6.5) and tilled (pH 7.5) soils affected by smelting activities, but the simultaneous presence of Pb^{2+}-sorbed ferric and manganese oxides prevented identifying the nature of organic ligands (Morin et al. 1999). Lead adsorbed to iron oxyhydroxide surfaces was also found in the carbonate-rich (pH 6.8) Hamms tailings of Leadville, Colorado (USA) (Ostergren et al. 1999). This Pb species is likely widespread in nature owing to the abundance of iron (oxyhydr)oxides and the near-neutral pH of many lands since lead is sorbed on these minerals at pH > ~5 (Müller and Sigg 1992; Ostergren et al. 2000). The identification of this Pb adsorbate clearly demonstrates the value of EXAFS spectroscopy to reveal sorption complexes as mineral surface sorbates in natural systems are not yielding to identification by any other currently available analytical tool. Sulfide-rich mine spoils (pH 2-6) have a completely different geochemistry, and Fe-rich (hydroxy)sulfates represent the main oxidation products of pyrite. As for As-contaminated spoils, jarosite is generally the main sink of lead in these environments, sometimes in association with anglesite ($PbSO_4$) (Ostergren et al. 1999). This sulfate mineral, which often forms as efflorescences, is however highly soluble and lead is more strongly retained in plumbo-jarosite.

Lead phosphates are the most insoluble forms of Pb in soils under a wide variety of environmental conditions (Nriagu 1974; Lindsay 1979). Of this family of minerals, pyromorphite ($Pb_5(PO_4)_3Cl$) is the most stable ($K_{sp} = 10^{-84.4}$), and experimental evidence supports the hypothesis that lead species would be converted to pyromorphite by a dissolution – precipitation mechanism, provided the soil is not chloride- or phosphate-limited (Lower et al. 1998; Ryan et al. 2001). Pyromorphite has been reported in mine-waste tailings, garden soil, soil contaminated with Pb ore (PbS) adjacent to phosphoric acid plant, and soil developed over a geochemical anomaly (Cotter-Howells and Thornton 1991; Cotter-Howells et al. 1994; Ruby et al. 1994; Buatier et al. 2001; Morin et al. 2001). Pyromorphite was also positively identified with electron diffraction and EXAFS spectroscopy in the outer wall of the epidermis of *Agrostis capillaris*, a plant used in phytostabilization (Vangronsveld and Cunningham 1998; Cotter-Howells et al. 1999). However, the role of pyromorphite precipitation on the root surface remains unclear and, as averred by Cotter-Howells (1999), it could be a physiological detoxification mechanism or simply a passive chemical reaction between the heavy metal

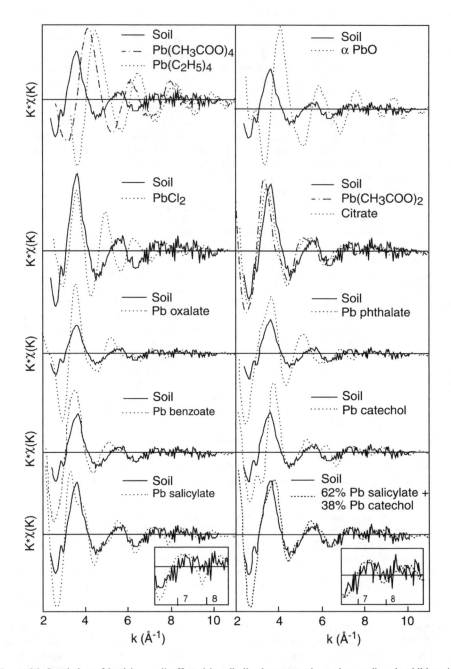

Figure 26. Speciation of lead in a soil affected by alkyllead compounds used as antiknock additives in gasoline. The soil EXAFS spectrum (solid) is compared to model spectra from a library of lead compounds. This comparison identifies salicylate and catechol functional groups as the predominant complexing chelates for divalent Pb. Adapted from Manceau et al. (1996). (*Figure continued on facing page.*)

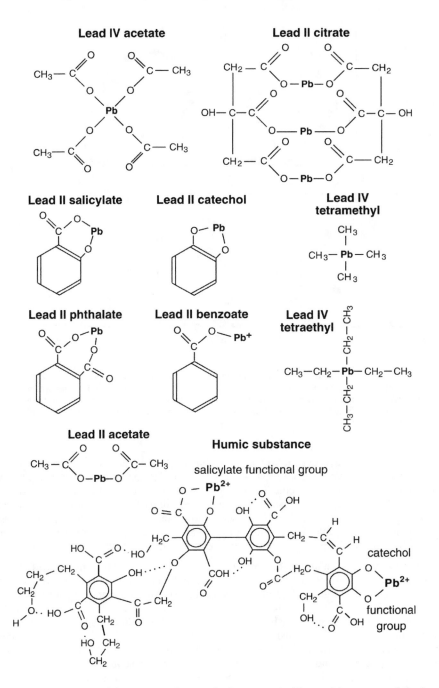

Figure 26 continued. Model structures of organolead compounds. The model structure of the humic substance is adapted from Morel (1983).

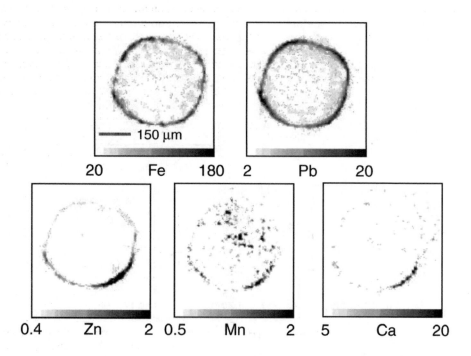

Figure 27. Metal distributions on the root surface of *Phalaris arundinacea* obtained by X-ray fluorescence microtomography. Pb and Fe are concentrated in the surficial rind on the epidermis and have similar distribution patterns. These images have been converted to negative-contrast grayscale from their original forms published by Hansel et al. (2001). Mn, Zn, and Ca are correlated and exist as discrete aggregates.

and phosphate. Lead sequestration in phosphates undoubtedly represents an attractive mechanism to remove Pb from waste waters by precipitation from apatite, $Ca_5(PO_4)_3OH$ (Ma et al. 1994), or as a remediation technology to stabilize Pb in soils and waste materials (Ruby et al. 1994; Eusden et al. 2002). Similarly to arsenic, all the fundamental knowledge gained recently on lead speciation clearly demonstrates how a better understanding of underlying molecular scale mechanisms of trace element immobilization can help provide more confident responses to environmental concerns.

Mercury. Hg is undoubtedly one of the major contaminants introduced into the environment because of its toxicity and the relatively easy uptake of its chemical compounds, mostly organic, by biota (Darbieu 1993). Anthropogenic Hg sources include mercury, gold, and silver mining, and agricultural and industrial uses. For example, amalgamation with mercury has been used as a method of gold and silver extraction since Roman times, and is still extensively used in some regions, such as in the Amazon basin, though its emission now tends to decline with the introduction of cyanidation processing technology (Lacerda and Salomons 1998). EXAFS spectroscopy has been employed to probe the identities and proportions of mercury species present in different mine waste piles of the California Coast Range (Kim et al. 2000). Cinnabar (HgS, hexagonal), and to a lesser extent metacinnabar (HgS, cubic), were by far the two predominant species identified in all samples. Only minor amounts of more soluble sulfate and chloride species were identified, with the Hg-Cl species being exclusively encountered near hot-spring mercury deposits, where chloride levels are elevated. However, since mercury forms highly soluble species with organic sulfur and chloride, and can be converted to volatile elemental Hg and methylated species by bacteria, these labile and pervasive species are hardly detected and largely overlooked upon analyzing uniquely the bulk fraction of Hg-containing rocks and wastes. Therefore, it seems important in further studies on mercury to combine structural and biogeochemical investigations to better comprehend its fate in terrestrial systems.

Zinc. Zinc fertilizers, sewage sludges and atmospheric dust of industrial origin are the principal sources of Zn accumulation in soils (Adriano 1986; Robson 1993). While Zn is an essential metal, being an important constituent of cells, and specifically a cofactor of many enzymes, it can become toxic at elevated levels. Most acute cases of land contamination by zinc result from smelting activities. More than 80 percent of the current primary zinc production capacity in the world is now based on the less polluting Roast-Leach-Electrowinning (RLE) process (Wickham, 1990), which is a wet chemical or hydrometallurgical process. This process replaced the former pyrometallurgical Imperial Smelting Process (ISP), which has produced in the past considerable amounts of dusts and exhaust fumes rich in zinc and lead, leading to important contamination of the local environment. Zinc speciation in neutral to near-neutral soils (pH 5.5 or greater) highly affected by more than one century of smelting activities in northern France and Belgium has been investigated in detail by a combination of techniques including size and densitometric fractionation, conventional XRD, µSXRF, and powder and polarized EXAFS (Manceau et al. 2000b). Franklinite ($ZnFe_2O_4$), willemite (Zn_2SiO_4), hemimorphite ($Zn_4Si_2O_7(OH)_2 \cdot H_2O$), and Zn-containing magnetite ($[Fe,Zn]Fe_2O_4$) were identified in dense soil fractions, and represented the main contaminant source term. In the fine soil matrix, Zn was shown to be primarily bound to phyllosilicates and, to a lesser extent, to Mn and Fe (oxyhydr)oxides. The Fe and Mn populations were heterogeneously distributed in the clay matrix material. Zn-containing Fe-rich grains (feroxyhite) were typically 10 to 20 µm in size, and manganese-rich hexagonal birnessite spherules much larger at 300 µm (Fig. 28a,b). Application of µEXAFS within these regions revealed the identity and structure of the zinc species. The zinc-containing manganese compound was birnessite, in which the zinc was demonstrated to be adsorbed in the interlayer space above and below vacant sites with four-fold coordination (Fig. 3). Fe µEXAFS spectra collected on the Fe grains resembled the mineral feroxyhite (δ-FeOOH), and zinc was assumed to be substituted for Fe in the structure (no Zn-µEXAFS collected). Finally, the percentage of each secondary species in the bulk was quantified by fitting powder EXAFS spectra with a linear combination of reference spectra corresponding to Zn species identified individually (Fig. 28c). Roberts et al. (2002) interrogated the form of Zn in a strongly acidic soil (pH 3.2-3.9) also contaminated by smelting operation. Franklinite and sphalerite (ZnS) were identified by bulk EXAFS, and Zn adsorbed to Fe and Mn oxides were detected by microspectroscopy. The main difference with the former study was the absence of Zn-containing phyllosilicate, and the high amount of loosely bound Zn (presumably outer-sphere Zn complexes) easily leached by 0.1 M NaNO₃ at soil pH in a stirred-flow reaction chamber. The difference of speciation is well explained by the difference of soil pH in the two studies because Zn is not quantitatively adsorbed on Fe oxides below pH 5, and it is highly soluble with respect to phyllosilicate at low pH. For example, Manceau et al. (2000b) calculated that the saturation concentration of dissolved Zn^{2+} with respect to trioctahedral phyllosilicates is as high as ca. 10^{-2} M at pH 4. These two complementary studies illustrate how metal species in complex natural matrix are now yielding to identification by X-ray techniques developed at 3rd generation synchrotron facilities in

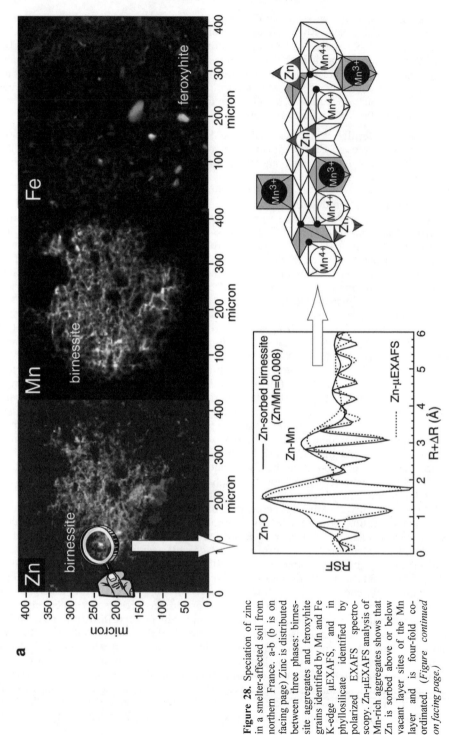

Figure 28. Speciation of zinc in a smelter-affected soil from northern France. a-b (b is on facing page) Zinc is distributed between three phases: birnessite aggregates and feroxyhite grains identified by Mn and Fe K-edge μEXAFS, and in phyllosilicate identified by polarized EXAFS spectroscopy. Zn-μEXAFS analysis of Mn-rich aggregates shows that Zn is sorbed above or below vacant layer sites of the Mn layer and is four-fold coordinated. *(Figure continued on facing page.)*

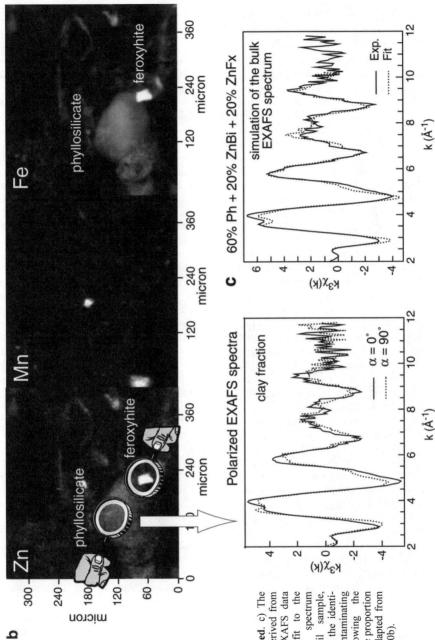

Figure 28 continued. c) The three Zn species derived from μEXAFS and P-EXAFS data provide a good fit to the powder-EXAFS spectrum from a bulk soil sample, thereby confirming the identification of the contaminating species, and allowing the quantification of the proportion of each species. Adapted from Manceau et al. (2000b).

minimally disturbed environmental samples. They also demonstrate that lower spatial resolution is required to assess and quantify the representativity of observations made at high resolution with X-ray microprobes.

In industrialized countries, mining-related activities, such as smelting and dumping of mining material into lakes, rivers, channels and harbors used to carry the ore, have contributed to the pollution of aquatic systems. Using analytical electron microscopy and bulk EXAFS spectroscopy, Zn was shown to be speciated as sphalerite, smithsonite ($ZnCO_3$), zinc carbonate hydroxide, and zinc phosphate in the sediments of Lake DePue (IL, USA) (Webb et al. 2000; Gaillard et al. 2001). This sediment was extremely rich in biological cells and structures (bacteria, algae) to which zinc-iron-phosphate particles were intimately associated. Alluvial sediments are periodically dredged to clean and maintain the depth of navigational units worldwide. Land disposal of these dredge sediments may affect the surrounding environment due to the presence of metal contaminants. Due to the anoxic conditions of the immersed sediments, metals are generally in sulfidic forms (Luther et al. 1980; Parkman et al. 1996; O'Day et al. 1998, 2000; Gaillard et al. 2001). Upland disposals cause changes in the redox potential, pH, and bacterial activity, and these physico-chemical changes enhance the leachability and bioavailability of trace metals. Increasing environmental awareness is leading to more stringent regulations for the disposal of dredged sediments. In an attempt to evaluate the impact of this practice on the environment, the nature of secondary Zn phases formed after the deposition of a sediment (dredged nearby from previous smelter-affected soil) were scrutinized by a whole set of complementary techniques (Isaure et al. 2001, 2002). This detailed mineralogical and geochemical analysis was carried out by optical and scanning electron microscopy, XRD, µPIXE, µSXRF, and powder and polarized EXAFS spectroscopy on bulk, and on physically and chemically fractionated, samples. In each class of samples, Zn species were identified by PCA and their proportions were evaluated quantitatively using linear combinations of model compound spectra. The coarse grain fraction, essentially of anthropogenic origin, was highly heterogeneous in color, chemistry, and mineral composition. Three Zn-bearing primary constituents were identified which were, in decreasing order of abundance, sphalerite (ZnS), willemite (Zn_2SiO_4), and zincite (ZnO). Their weathering under oxidizing conditions after the sediment disposal led to the formation of Zn-containing Fe (oxyhydr)oxides and phyllosilicates. The preferred formation of phyllosilicate Zn species over metal hydroxides (e.g., hydrotalcite) or hydroxycarbonate phases (e.g., hydrozincite, $Zn_5(OH)_6(CO_3)_2$) in the sediment matrix, but also in the soils of this region, was consistent with the composition of pore waters, which were supersaturated with respect to Zn phyllosilicate precipitation (Manceau et al. 1999b; Manceau et al. 2000b). More generally, and with the exception of strongly acidic matrices, Zn-phyllosilicate is thought to be abundant in nature because quartz and amorphous silica are ubiquitous (Fig. 10a), and that silicic acid in low-temperature terrestrial waters typically ranges from 10 to 80 ppm (Davies and DeWiest 1966), the lower and upper ends of which are controlled by quartz (~11 ppm, Rimstidt 1997) and amorphous silica (~116 ppm, Rimstidt and Barnes 1980) solubility, respectively.

In addition to lead, the form of zinc in the Fe plaque of *Phalaris arundinacea* was also examined by fluorescence microtomography and EXAFS spectroscopy (Fig. 27, Hansel et al. 2001). Unlike lead, which was clearly associated with Fe (oxyhydr)oxides and inferred to be complexed to organic molecules, zinc was clearly correlated to manganese. On the basis of Mn-O (2.20 Å) and Zn-O (1.96 Å, 2.06 Å, 2.08 Å, 2.15 Å) EXAFS distances, Mn was assumed to be speciated as rhodochrosite ($MnCO_3$) and Zn as hydrozincite ($Zn_5(OH)_6(CO_3)_2$). While this study documented for the first time the combined use of fluorescence microtomography and EXAFS spectroscopy, and demonstrated the interest of this approach for providing *in situ* environmentally relevant information on the forms of metals in the rhizosphere, caution with the structural interpretation is warranted because, as explained previously, the analysis of the first coordination shell is not differential enough to distinguish metal species (Fig. 15). For instance, d(Zn-O) = 1.96 Å is only characteristic of four-fold coordinated zinc and d(Zn-O) ≥ 2.06 Å of six-fold coordinated zinc (Sarret et al. 1998), and these two Zn coordinations and three sets of distances (EXAFS does not allow one to differentiate two shells at 2.06 Å and 2.08 Å) are also encountered in Zn-sorbed birnessite (Lanson et al. 2002; Manceau et al. 2002a). A similar approach was used by Bostick et al. (2001) to speciate Zn in contaminated wetland. In this study the authors conclude on the basis of a shell-by-shell fitting analysis that "four principal phases, ZnO, Zn on (hydr)oxide, $ZnCO_3$, and ZnS, were found in seasonally flooded wetland," but it is unclear which species is really present in this natural system because no other metal species was included as a reference phase, and the conclusions are therefore susceptible to misinterpretation. In addition, ZnO is unlikely to form in a wetland system resulting from mining activities; so far, its environmental occurrence was always due to contamination by smelting plants (Sobanska et al. 1999; Thiry and Van Oort 1999).

Speciation of trace metal(loid)s in uncontaminated soils

Much of the literature on heavy-metal-bearing soils and sediments has been devoted to the speciation of anthropogenic metal(oid)s in contaminated matrices, but few papers focus on their crystal chemistry when they are present in trace amounts. This section reviews this topic and supplements and enhances the existing literature by describing the forms of arsenic, selenium, nickel, and zinc in two natural soils. It also attempts to illustrate with one example (Zn) how the novel synergistic use of μSXRF, μSXRD, and μEXAFS provides a quantitative analytical tool to speciate dilute multi-component metals in heterogeneous environmental materials.

Arsenic and selenium in acid sulfate soil. The forms of As (13.7 mg kg^{-1} soil) and Se (6.8 mg kg^{-1} soil) in the upland soils from the Western San Joaquin Valley (Panoche Hills, California, USA, Gewter series) have been investigated in detail using μSXRF, μXANES (Se and As), and μEXAFS (Fe) (Strawn et al. 2002) (Fig. 29). These soils are developed on shale parent materials, and the arsenic and selenium initially contained in pyrite inclusions are released to the soil matrix following the oxidation of sulphide minerals under the very oxidizing conditions of these unsaturated soils. In these soils nearly all of the As, and most of the Se, are associated with orange aggregates, and secondarily with yellow aggregates, which are generally smaller than 100 microns. μEXAFS spectroscopy indicated that the yellow aggregates consist of jarosite, and the orange aggregates of ferrihydrite and goethite. The presence of jarosite and goethite was confirmed recently by μXRD performed on mixed orange-yellow aggregates, but this technique identified feroxyhite (δ-FeOOH) instead of ferrihydrite. It is likely that the ferrihydrite species inferred from μEXAFS actually corresponds to feroxyhite because these two poorly-crystallized Fe oxides have a similar short-range structure (Manceau and Drits 1993) and, therefore, EXAFS spectroscopy has relatively little sensitivity to either species in the case of phase mixture. The lower affinity of As and Se for the iron sulfate mineral (i.e., jarosite) compared to Fe oxides was confirmed from the strong anti-correlation between (K,S) and (As,Se) obtained from μSXRF maps using K and S as indicators for jarosite. In keeping with the result of Strawn et al. (2002), Dudas et al. (1988) found concentrations of As in the Fe oxide fraction at levels of 10-20 times those found in the jarosite fraction in acidic soils from Alberta, Canada, formed from shale materials. As and Se oxidations states in the soils from the California Coast Range were determined by μXANES microspectroscopy after taking care of the effect of the resin

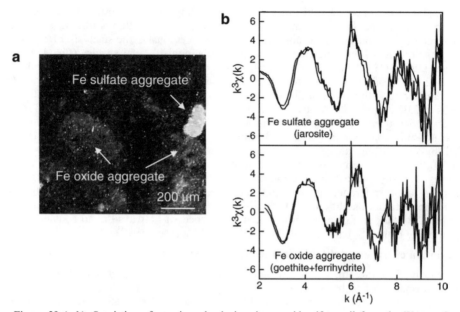

Figure 29 (a-b). Speciation of arsenic and selenium in an acid sulfate soil from the Western San Joaquin Valley (Panoche Hills, California, USA, Gewter series). a) Fe oxide and Fe sulfate aggregates observed in reflected light microscopy. b) µEXAFS spectroscopy shows that Fe is speciated as jarosite ($K(Fe_3(SO_4)_2(OH)_6)$ in the sulfate aggregates and as goethite (α–FeOOH) and poorly-crystallized oxide (modeled here by ferrihydrite) in the oxide aggregates. Adapted from Strawn et al. (2002). (*Figures c-e on facing page.*)

impregnation and thin sectioning steps on As and Se oxidation during the sample preparation. As was found to be exclusively five-valent, which is the most common form of As in aerobic environments. Since As is negatively charged in pyrite, and that As(III) is more mobile in acidic soils than As(V), some of the arsenic likely was leached out of the soil as As(III) during pyrite oxidation. Selenium was present as both Se(IV) and Se(VI) oxidation states, this later form being thought to be stabilized in the jarosite structure by isomorphic substitution of selenate for sulfate. This inferred immobilization mechanism parallels that identified by Savage et al. (2000) in mine tailings from California, where arsenate tetrahedra were shown to substitute for SO_4 in jarosite.

Nickel and zinc in soil ferromanganese nodules. One of the most efficient and durable process responsible for trace metal sequestration in soils is the formation of ferromanganese micronodules, which often have been compared to the well-known oceanic Mn nodules (Glasby et al. 1979; White and Dixon 1996; Han et al. 2001) (Fig. 30a). Although soil micronodules are the premier reservoir for many trace elements in soils, the crystal chemistry of the sequestered elements remains unknown. Chemical analyses of individual nodules from Sicilia and New Zealand showed that some elements, such as Co, Ce, Ba, Pb, Ni, are several times to more than one order of magnitude enriched in the concretions relative to the soil matrix, whereas others are less (e.g., Zn) or even depleted (Childs 1975; Palumbo et al. 2001) (Fig. 30b). Since nodules are formed *in situ* within the soil matrix by local enrichment of Fe and Mn oxides, the observed difference of metal enrichment in the concretions relative to the soil matrix likely results from a difference of metal affinity for soil minerals. The micronodule analyzed here by

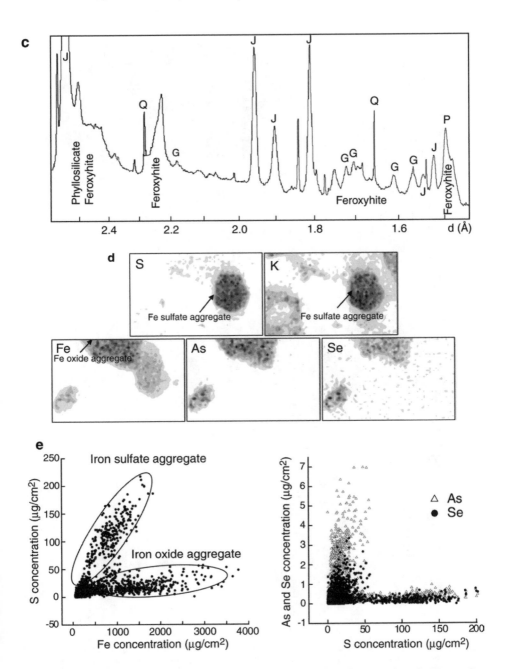

Figure 29 continued (c-e). Speciation of arsenic and selenium in an acid sulfate soil from the Western San Joaquin Valley (Panoche Hills, California, USA, Gewter series). c) μXRD applied to mixed Fe aggregates identifies jarosite, goethite, and feroxyhite (δ–FeOOH). d-e) μSXRF elemental maps and scatterplots. As and Se are essentially associated with Fe in the oxide aggregates. K and S are used as indicators for jarosite. Adapted from Strawn et al. (2002).

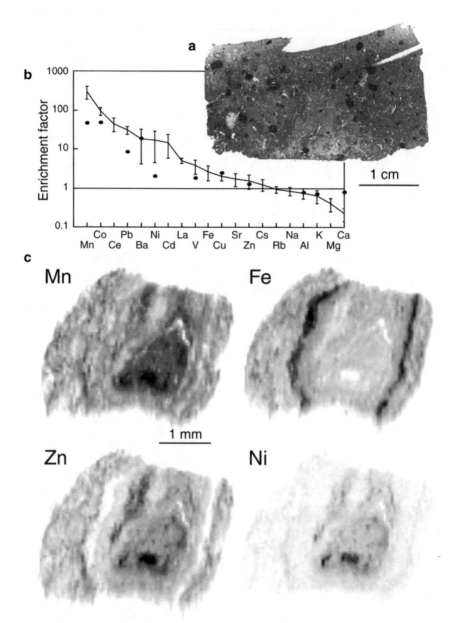

Figure 30. Speciation of Ni and Zn in a ferromanganese nodule from the Morvan region, France (Baize and Chrétien 1994; Latrille et al. 2001; Manceau et al. 2002b). a) Photomicrograph of a soil thin section (30 μm thick) containing embedded ferromanganese concretions. b) Average enrichment factor of elements in soil nodules. Adapted from Palumbo et al. (2001) and Childs (1975). c) μSXRF elemental maps recorded on beamline 7.3.3 at the ALS (Berkeley). Beam energy: 10 keV, illuminated area 14 μm (H) × 11 μm (V), step size 20 × 20 μm, dwell time 3s/pt.

the three X-ray microprobe techniques was collected in the plowed layer (0-25 cm) of an agricultural field of the Morvan region in France used for livestock breeding activities. Soils in this area are developed on the Sinemurian geological setting and are known to contain several percent per soil's weight of ferromanganese pellets generally termed "shot," "lead shot," or "buckshot" by agriculturists (Wheeting 1936; Baize and Chrétien 1994). These soils frequently contain several hundreds ppm of metals (Ni, Zn, Pb...), but which are preferentially concentrated in micronodules, and, hence, in a form that is highly immobile and therefore hardly accessible to living organisms (Baize and Chrétien 1994; Latrille et al. 2001). Nodules from these soils are sub-rounded in shape, a few millimeters in diameter, and reddish-brown to dark brown in color. Inspection of thin sections under the microscope shows that the concretions have grains of primary minerals, chiefly quartz, cemented by a ocherous-brown or dark non-birefringent substance consisting Fe-Mn oxides. Analysis of 10 individual nodules using inductively coupled plasma mass spectroscopy (ICP-MS) indicated that they contain [Fe] = 121.5 (σ = 53.4), [Mn] = 104.7 (σ = 32.0), [Zn] = 4.63 (σ = 1.83), [Ni] = 0.63 (σ = 0.38) mg/kg. The σ variability is relatively high and comparable to that measured for Sicilian nodules (Palumbo et al. 2001).

Several nodules were studied, and results reported here correspond to one that was dark in color, indicating a high Mn content. Scanning X-ray microfluorescence (μSXRF) and microdiffraction (μSXRD) first were used to identify the host solid phase by mapping the distributions of elements and some solid species, respectively. μSXRF showed that Fe, Mn, Ni, and Zn are unevenly distributed at the micrometer scale (Fig. 30c). Fe and Mn have no detectable correlation (ρ_{Fe-Mn} = -0.02), and Ni and Zn are both strongly correlated with Mn (ρ_{Mn-Ni} = 0.58, ρ_{Mn-Zn} = 0.86) and not with Fe (ρ_{Fe-Ni} = -0.29, ρ_{Fe-Zn} = -0.22). Visual inspection of the contour maps shows that the degree of correlation varies laterally: Highest Zn and Ni amounts are observed in the Mn-rich core; the outer regions also contain significant amounts of Mn and Zn but are depleted in Ni. The variability in composition from nodule to nodule as measured by ICP, and within a given nodule as measured by μSXRF, demonstrates that elemental correlations of environmental materials should be interpreted cautiously. The overall likeness of the Mn and Zn maps, and the partial Ni-Mn association, suggest that Mn is present in at least two forms, with only one containing Ni. Point μXRD measurements in the core region showed the major presence of finely-divided lithiophorite, a MnO_2 - $Al(OH)_3$ mixed-layer phyllomanganate (Fig. 3), and minor phyllosilicate, together with coarse quartz, feldspar and anatase grains (Fig. 31). Finely-divided goethite admixed with phyllosilicate and birnessite were detected in Fe-rich areas forming two strips on each side of the Mn core. Lithiophorite and goethite contour maps were imaged by scanning the nodule under the micro-focused monochromatic beam, recording point XRD patterns in reflection mode at a resolution of 20 × 20 μm, and integrating diffracted intensity along the 001 (lithiophorite) and 301 (goethite) Debye rings at 9.39 Å and 2.69 Å, respectively. The comparison of μSXRF and μSXRD maps clearly shows that nickel and lithiophorite have the same distribution, therefore indicating that Ni is exclusively bound to this particular Mn phase. The goethite map does not match the Zn and Fe elemental maps, which means that this constituent is devoid of zinc and also that iron is not uniquely speciated as goethite. The missing Fe- and Zn-containing phases were identified by moving the CCD camera to higher 2θ angle in order to detect most poorly-crystalline compounds. Turbostratic birnessite (also termed vernadite, δ-MnO_2) and feroxyhite (δ-FeOOH) were identified in the outer region (Fig. 31). In some rare sites, broad humps with maxima at 2.45-2.40 Å, 2.25-2.20 Å, and 1.42 Å characteristic of vernadite were observed, but in most areas Fe and Mn were present in similar amount and an additional hump at 1.70-1.65 Å typical of feroxyhite was also present (Fig. 31). The intimate association of

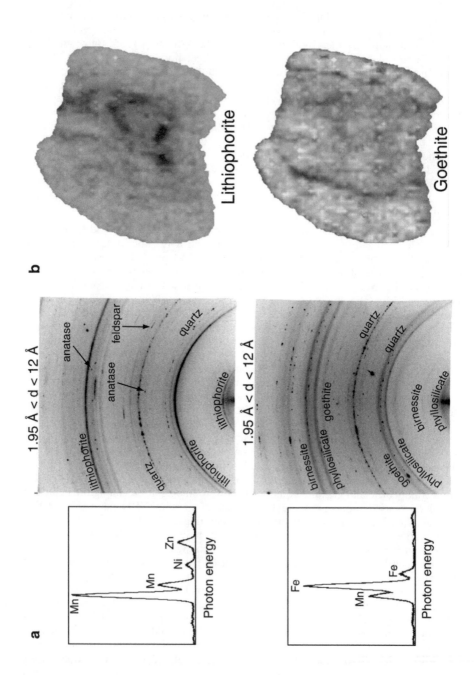

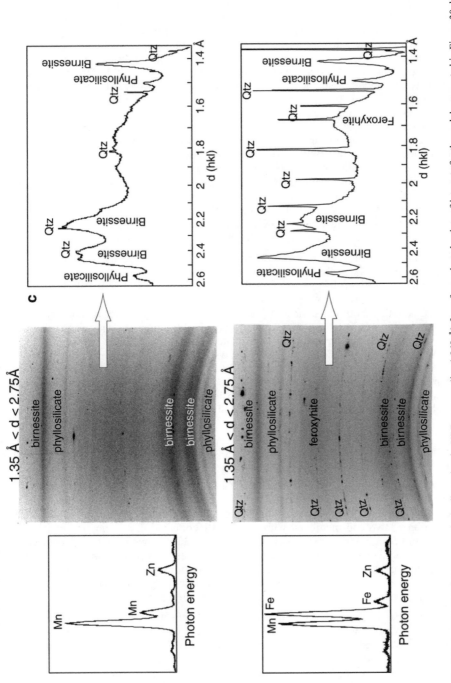

Figure 31. a) X-ray fluorescence spectra and two-dimensional μXRD patterns ($\lambda = 1.968$ Å) from four selected points-of-interest for the nodule presented in Figure 30. b) μSXRD (negative contrast) maps of lithiophorite and goethite. c) One-dimensional μXRD patterns obtained by integrating intensities of 2D patterns at constant Bragg angle.

feroxyhite and vernadite is widespread in low-temperature formations (Chukhrov and Gorshkov 1981; Varentsov et al. 1991; Vodyanitskii et al. 2001), and may have a bacterial or a chemical origin. Microorganisms that precipitate iron and manganese are nearly universal in their distribution, and their role in the geochemical cycling of Fe and Mn has been acknowledged for nearly two (Fe) and one (Mn) centuries (Thiel 1925). The intermeshed occurrence of Fe and Mn oxides in "Fe-vernadite" or "Mn-feroxyhite" solid phases may also result from the catalytic oxidation of Fe^{2+} to Fe^{3+} by Mn^{4+} present in vernadite (Golden et al. 1988). Regardless of their origin, these Fe and Mn nanoparticles act as the main cementing agent of coarse grains. Under reducing conditions, Fe and Mn are dispersed throughout the soil matrix as water fills soil pores, and when the soil dries and oxygen levels gradually increase, Fe and Mn precipitates in surface of soil skeletal grains (White and Dixon 1996). Point µXRD analyses indicated that phyllosilicate, vernadite and feroxyhite are relatively homogeneously distributed within the examined nodule down to a spatial extent of 14×11 µm (the resolution of the XRD microprobe, Manceau et al. 2002c), and presumably also at lower scale owing to the very fine-grained crystalline nature of these materials. Note that this nodule may very well contain also two-line ferrihydrite because the two XRD broad humps of this hydrous ferric oxide at ~2.55 Å and ~1.50 Å are masked by the 130, 200 and 060 reflections of phyllosilicates.

The nature and proportion of the Zn host phases, and the binding mechanism of Zn and Ni at the molecular scale, were determined by EXAFS spectroscopy. The uniqueness of the sequestration mechanism of nickel inferred from µSXRF - µSXRD was first verified by Ni K-edge µEXAFS measurements from selected points-of-interest in this nodule and another from the same soil (Manceau et al. 2002c). Nickel was demonstrated to substitute for Mn^{3+} in the manganese layer of lithiophorite. The comparison of effective ionic radii for octahedrally coordinated Ni^{2+} (0.70 Å), Mn^{3+} (0.65 Å), Mn^{4+} (0.54 Å), Al^{3+} (0.53 Å), and Li^+ (0.74 Å) shows that the Ni^{2+} and Mn^{3+} sizes differ by only 7%, with the Mn ion being slightly smaller (Shannon and Prewitt 1969). Therefore, the Mn^{3+} crystallographic site is presumably the most favorable energetically (Davies et al. 2000), and the affinity of nickel for this particular site provides a rationale to explain the higher partitioning of this element in the soil nodules over the soil matrix mentioned previously (Childs 1975; Palumbo et al. 2001). Since lithiophorite is common in soils, and in particular in nodules (Taylor 1968; Uzochukwu and Dixon 1986; Golden et al. 1993), the Ni species identified in this work may correspond to a major sequestration form of Ni in Earth near-surface environments.

Five points-of-interest were selected for Zn K-edge µEXAFS measurements in such a manner as to vary the proportions of lithiophorite and Mn-feroxyhite on the basis of µXRD analysis (Fig. 32a). As a first approximation, the five spectra look like a linear combination of two end-members, with spectrum 1 collected in a lithiophorite-rich spot and spectrum 5 in a Mn-feroxyhite-rich spot. Figure 32b shows in dots a fluorescence-yield spectrum taken of a powdered pellet of ten nodules, on ESRF Beamline BM32 and with a beam size big enough to ensure that the spectrum represents an average over all nodules. We see that this spectrum does not seem to belong to the same pseudobinary series as the five micro-EXAFS spectra, a point to which we shall return below.

Let us first consider only the micro-EXAFS spectra. PCA was done on this set of spectra and revealed two components, with *IND* values of 0.694, 0.649, 1.303 and 5.034 for the first four eigenvalues. The reconstruction of the data from just two components were relatively good, with the worst example shown in Figure 32c (#4). Target testing against a variety of references turned up three good candidates (^{VI}Zn-containing lithiophorite (Zn in Al(OH)$_3$ layer), low-Zn birnessite (Zn 4-coordinate), ^{IV}Zn-sorbed ferrihydrite, and one marginal (low ^{VI}Zn-containing kerolite with Zn primarily

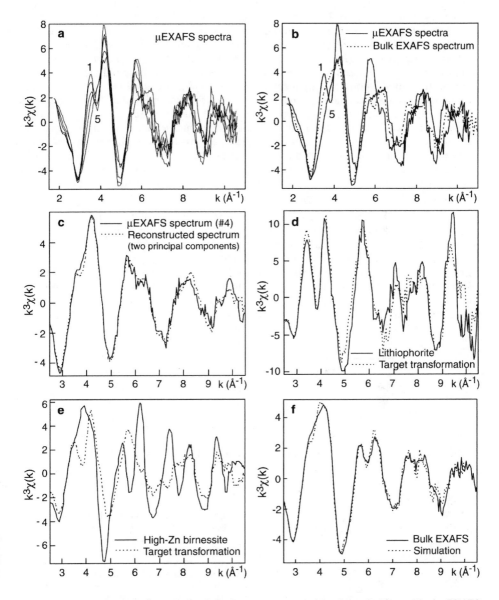

Figure 32. Zn K-edge EXAFS analysis of the ferromanganese nodule shown in Figure 30. a) µEXAFS spectra collected in five points-of-interest having variable Mn/Fe ratio and, hence, variable proportions of lithiophorite and Mn-feroxyhite. b) Comparison of the bulk EXAFS spectrum to the two end-member µEXAFS spectra. c-e) PCA analysis, showing (c) the worst fit of any of the unknowns to a sum of the first two abstract components, and (d-e) the results of a target transformation done on lithiophorite and high-Zn birnessite reference . f) Three-component least-squares fitting of the bulk EXAFS spectrum (solid line) to a linear combination of reference spectra (dotted line). The best simulation was obtained for 41% high-Zn birnessite + 34% low-Zn birnessite + 23% low-Zn kerolite (estimated accuracy 10% of the total Zn).

surrounded by Mg in the octahedral sheet) (Fig. 3). The *SPOIL* values for these references were 2.10, 2.56, 3.13, and 4.51, respectively. High-Zn birnessite (Zn mostly 6-coordinate, Fig. 3) and high-Zn kerolite (Zn surrounded by Zn in the octahedral sheet, Fig. 3) were poorer candidates than their low-Zn equivalents, with *SPOIL* values of 7.12 and 9.79, respectively. The results of target transformation on the lithiophorite and high-Zn birnessite spectra are shown in Figure 32d-e. Next, we fitted the individual spectra to two of the three "good" components, but all combinations failed and only three-component fits gave satisfactorily results. We also tried adding in the "bad" components but, as expected, their addition did not improve the fit. Adding in the "marginal" component (low-Zn kerolite) improved the fit a little, but the data are still well-explained by using just the three species, lithiophorite (Li), birnessite (Bi) and ferrihydrite (Fh), all in low-Zn version. The results of the three-species fit are shown in Table 3. The entry for Sum is the sum of all the fractions, which would ideally add up to 1. The Sum-sq. error is the sum of the squares of the residuals, normalized to the sum of the squares of the data values.

Since there are three predominant species, we can represent the data on a ternary plot, as shown in Figure 33. We see that the points lie roughly on a line, as expected from the pseudobinary nature of the system as shown by the PCA results. The data are approximately consistent with a model in which the samples are all mixtures of just two substances. One of these is a mixture of Fh and Li, and the other is a mixture of Fh and Bi. Both end-members have about the same amount of ferrihydrite. We thus consider Fh

Table 3. Linear-combination fit results
for the micro-EXAFS spectra.

	#1	#2	#3	#4	#5
Low-Zn Bi	0.175	0.164	0.379	0.290	0.469
Ferrihydrite	0.529	0.431	0.395	0.487	0.477
Lithiophorite	0.311	0.311	0.168	0.109	0
Sum	1.01	0.91	0.94	0.88	0.95
Sum-sq. error	0.108	0.140	0.119	0.106	0.099

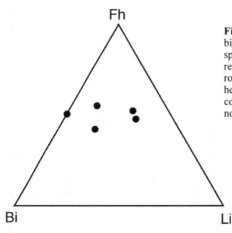

Figure 33. Proportion of ferrihydrite (Fh), birnessite (Bi) and lithiophorite (Li) Zn species measured by μEXAFS spectroscopy represented on a ternary plot. The points lie roughly on a line because ferrihydrite, used here as a proxy for poorly-crystallized Fe compounds, is present everywhere in the nodule (background Zn species).

to be a "background" species that is present everywhere and does not participate in the chemical and mineralogical development of the nodule. Here the ferrihydrite reference is used as a proxy for the general class of defective Fe oxides (ferrihydrite *sensu stricto*, feroxyhite...) because previous EXAFS investigation showed that these compounds have a similar local structure (Manceau and Drits 1993). Poorly crystalline Fe oxides are likely present throughout the sample, and the present fits suggest that they sorb about the same fraction of Zn everywhere.

Now, going back to the bulk, we obtained an excellent fit using three different components: high-Zn birnessite, low-Zn birnessite, and low-Zn kerolite, in proportions of 41%, 34% and 23%, respectively (estimated accuracy 10% of the total Zn) (Fig. 32f). At first sight, it is surprising that the bulk spectrum should turn up with two species (phyllosilicate and high-Zn birnessite) not found by micro-EXAFS. The most plausible explanation is that the sampling of the particular nodule used for micro-EXAFS was not representative of all the nodules. This interpretation is supported by chemical analyses, which showed that the nodules from this soil are heterogeneous in composition: [Mn] = 104.7 (σ = 32.0), [Zn] = 4.63 (σ = 1.83) mg/kg. In particular, the nodule on which micro-EXAFS measurements were carried out was arbitrarily chosen for its dark color (high Fe and Mn content). The speciation of Zn is logically expected to depend on the Mn, Fe, and Zn contents. This story should be viewed as a cautionary tale about the importance of representative sampling. We see from this example that by using μEXAFS to generate a set of related spectra and then analyzing them with PCA, we can garner much information, which would not be accessible with single spectra. However, in order to get information that truly represents all of what is found in a soil, it is still very important to look at a bulk average. Finally, the multiplicity of Zn species in soil nodules, and specifically its association with phyllosilicates, provide a clue to the lower partitioning of Zn relative to Ni in the soil nodules over the soil matrix (Fig. 30). Since ferromanganese nodules are differentiated *in situ*, and differ essentially from the soil matrix by their much higher concentration in manganese, they have typically the same quantity of phyllosilicates, and roughly of Fe oxides, as the soil matrix. Therefore, these two Zn-bearing species are almost uniformly distributed in the soil.

CONCLUDING REMARKS

For numerous questions related to the speciation of metal(loid) contaminants in natural and waste matrices, the combination of X-ray fluorescence, diffraction and absorption presented in this chapter offers a unique access to the problem. X-ray microscopy cannot compete with the atomic resolution offered by electron microscopy, but it offers a number of unique features. The chemical and structural information obtained by μSXRF and μSXRD can be used to identify the host solid phase by mapping the distributions of elements and solid species, respectively. Then the molecular-scale binding mechanism of trace elements by the host phase can be unraveled by μEXAFS spectroscopy. All these techniques can be applied with minimum preparation, minimizing any possible alteration of the sample. However, caution should be taken to not modify the initial form of the metal species by photon-assisted oxidation or reduction. This problem can be circumvented by decreasing the exposure time, photon density, or temperature. The polarization of the synchrotron radiation can be used to analyze anisotropic materials, which is important since many environmental minerals have a layered structure.

In many cases, the most critical problem to the application of these techniques is the access to the installations. This situation should improve in the coming years with the ongoing or planned development of new X-ray microscopes, with ever better resolution and intensity, at existing and new 3rd generation synchrotron facilities (ALS, ANKA, APS, SLS, SSLS, Diamond, Soleil...). Fledgling attempts have only just begun and, given the rapid development of this field in response to the growing societal demand for site remediation and green technologies, we expect that the ability to speciate metals in natural matrix will become routine as improvements in X-ray microprobe design continue and accessibility to beamlines increases.

ACKNOWLEDGMENTS

We thank Paul Fenter for his thorough review of the manuscript. The first author also thanks V.A. Drits, J.L. Hazemann, B. Lanson, and M. Mench for their many inspiring discussions on various aspects of this work, which seeded more than a decade of fruitful and friendly collaboration. This paper was written while the first author was on sabbatical leave in the Environmental Geochemistry Group, Division of Ecosystem Sciences of the University of California at Berkeley, and at the Advanced Light Source of the Lawrence Berkeley National Laboratory. We thank the ALS (Berkeley), ESRF (Grenoble) and LURE (Orsay) for their continuous provision of beamtime over the years. This work was in part supported by the Laboratory Directed Research and Development Program of the ALS, and by the Director, Office of Energy Research, Office of Basic Energy Sciences, Materials Sciences Division of the U.S. Department of Energy, under Contract No. DE-AC03-76SF00098.

REFERENCES

Adriano DC (1986) Trace Elements in the Terrestrial Environment. Springer Verlag, New York
Allard T, Ildefonse P, Beaucaire C, Calas G (1999) Structural chemistry of uranium associated with Si, Al, Fe gels in a granitic uranium mine. Chem Geol 158:81-103
Allen HE, Huang CP, Bailey GW, Bowers AR (1995) Metal Speciation and Contamination of Soil. Lewis, Ann Arbor, MI
Alpers CN, Nordstrom DK, Ball JW (1989) Solubility of jarosite solid solutions precipitated from acid mine waters, Iron Mountain, California, U.S.A. Sci Geol Bull Strasbourg 42:281–298
Baize D, Chrétien J (1994) Les couvertures pédologiques de la plate-forme sinémurienne en Bourgogne: particularités morphologiques et pédo-géochimiques. Etude Gestion Sols 2:7-27
Banfield JF, Nealson KH (1997) Geomicrobiology: Interactions between microbes and minerals. Reviews in Mineralogy. Vol 35. Mineralogical Society of America, Washington DC
Bargar JR, Brown GE, Parks GA (1997) Surface complexation of Pb(II) at oxide-water interfaces: II. XAFS and bond-valence determination of mononuclear and polynuclear Pb(II) sorption products on iron oxides. Geochim Cosmochim Acta 61:2639-2652
Bargar JR, Brown GE, Parks GA (1998) Surface complexation of Pb(II) at oxide-water interfaces: III. XAFS determination of Pb(II) and Pb(II)-chloro adsorption complexes on goethite and alumina. Geochim Cosmochim Acta 62:193-207
Bargar JR, Tebo BM, Villinski JE (2000) *In situ* characterization of Mn(II) oxidation by spores of the marine Bacillus sp. strain SG-1. Geochim Cosmochim Acta 64:2775-2778
Beauchemin D, Bednas ME, Berman SS, Sui KWM, Sturgeon RE (1988) Identification and quantification of arsenic species in a dogfish muscle reference material for trace elements. Anal Chem 60:2209-2212
Beaufort D, Baronnet A, Lanson B, Meunier A (1997) Corrensite: A single phase or a mixed-layer phyllosilicate in the saponite-to-chlorite conversion series? A case study of Sancerre-Couy deep drill hole (France). Am Mineral 82:109-124
Bellotto M, Rebours B, Clause O, Lynch J, Bazin D, Elkaim E (1996) A reexamination of hydrotalcite crystal chemistry. J Phys Chem 100:8527-8534
Benninger LK, Lewis DM, Turekian KK (1975) The use of natural Pb-210 as a heavy metal tracer in the river-estuarine system *In* Marine chemistry and the coastal environment, Vol 18. Church TM (ed) Am Chem Soc Symp, Washington DC, p 201-210
Berg M, Tran HC, Nguyen TC, Pham HV, Schertenleib R, Giger W (2001) Arsenic contamination of groundwater and drinking water in Vietnam: A human health threat. Env Sci Tech 35:2621-2626

Bizri Y, Cromer M, Scharff JP, Guillet B, Roullier J (1984) Constantes de stabilité de complexes organo-minéraux. Interactions des ions plombeux avec les composés organiques hydrosolubles des eaux gravitaires de podzol. Geochim Cosmochim Acta 48:227-234

Bleeker PM, Assuncao AGL, Teiga PM, deKoe T, Verkleij JAC (2002) Revegetation of the acidic, As contaminated Jales mine spoil tips using a combination of spoil amendments and tolerant grasses. Sci Tot Env: in press

Bostick BM, Hansel CM, La Force MJ, Fendorf S (2001) Seasonal fluctuations in zinc speciation within a contaminated wetland. Env Sci Tech 35:3823-3829

Boudeulle M, Muller JP (1988) Structural characteristics of hematite and goethite and their relationships with kaolinite in a laterite from Cameroon . A TEM study. Bull Mineral 111:149-166.

Boutron CF, Görlasch U, Candelone JP, Bolshov MA, Delmas RJ (1991) Decrease in anthropogenic lead, cadmium and zinc in Greenland snows since late 1960s. Nature 353:153-156

Brewe DL, Bouldin CE, Pease DM, Budnick JI, Tan Z (1992) Silicon photodiode detector for a glancing rmergence angle EXAFS technique. Rev Sci Instr 63:3298-3302

Brindley GW, Brown G (1980) Crystal structures of clay minerals and their X-ray identification. Mineralogical Society, London

Brown GE, Foster AL, Ostergren JD (1999) Mineral surfaces and bioavailability of heavy metals: A molecular-scale perspective. Proc Natl Acad Sci USA 96:3388-3395

Brückner S, Lusvardi G, Menabue L, Saladini M (1995) Crystal structure of lead hydroxyapatite from powder X-ray diffraction data. Inorg Chem 236:209-212

Brümmer GW, Gerth J, Tiller KG (1988) Reaction kinetics of the adsorption and desorption of nickel, zinc and cadmium by goethite: I. Adsorption and diffusion of metals. J Soil Sci 39:37-52

Buatier MD, Sobanska S, Elsass F (2001) TEM-EDX investigation on Zn- and Pb-contaminated soils. Appl Geoch 16:1165-1177

Bunge HJ (1981) Textures in Materials Science. Butterworths, London

Burmeister WP (2000) Structural changes in a cryo-cooled protein crystal owing to radiation damage. Acta Cryst D56:328-341

Buseck PR (1992) Minerals and reactions at the atomic scale: Transmission electron microscopy. Reviews in Mineralogy. Vol 27. Mineralogical Society of America, Washington DC

Calmano W, Mangold S, Welter E (2001) An XAFS investigation of the artefacts caused by sequential extraction analyses of Pb-contaminated soils. Fres J Anal Chem 371:823-830

Carlson L, Schwertmann U (1981) Natural ferrihydrites in surface deposits from Finland and their association with silica. Geochim Cosmochim Acta 45:421-429

Castaner R, Prieto C (1997) Fluorescence detection of extended X-ray absorption fine structure in thin films. J Phys III 7:337-349

Charlet L, Manceau A (1992) X-ray absorption spectroscopic study of the sorption of Cr(III) at the oxide/water interface. II Adsorption, coprecipitation and surface precipitation on ferric hydrous oxides. J Coll Interf Sci 148:25-442

Charlet L, Manceau A (1993) Structure, formation and reactivity of hydrous oxide particles: Insights from X-ray absorption spectroscopy *In* Environmental Particles, Vol Vol. 2. Buffle J, Van Leeuwen HP (eds) Lewis Publ., Chelsea, Michigan, p 117-164

Chatterjee A, Das D, Mandal BK, Chowdhury TR, Samanta G, Chakraborti D (1995) Arsenic in ground water in six districts of West Bengal, India: the biggest arsenic calamity in the world. Part I. Arsenic species in drinking water and urine of the affected people. Analyst 120:643-650

Chen CC, Hayes KF (1999) X-ray absorption spectroscopy investigation of aqueous Co(II) and Sr(II) sorption at clay-water interfaces. Geochim Cosmochim Acta 63:3205-3215

Chiarello RP, Sturchio NC (1994) Epitaxial growth of otavite on calcite observed *in situ* by synchrotron X-ray scattering. Geochim Cosmochim Acta 58:5633-5638

Chiarello RP, Sturchio NC, Grace JD, Geissbuhler P, Sorensen LB, Cheng L, Xu S (1997) Otavite-calcite solid-solution formation at the calcite-water interface studied *in situ* by synchrotron X-ray scattering. Geochim Cosmochim Acta 61:1467-1474

Childs CW (1975) Composition of iron-manganese concretions from some New Zealand soils. Geoderma 13:141-152

Chukhrov FV, Gorshkov AI (1981) Iron and manganese oxide minerals in soils. Trans Royal Soc, Edinburgh 72:195-200

Cotter-Howells J, Thornton I (1991) Sources and pathways of environmental lead to children in a Derbyshire mining village. Env Geochem Health 12:127-135

Cotter-Howells JD, Champness PE, Charnock JM (1999) Mineralogy of lead-phosphorus grains in the roots of Agrostis capillaris L. by ATEM and EXAFS. Miner Mag 63:777-789

Cotter-Howells JD, Champness PE, Charnock JM, Pattrick RAD (1994) Identification of pyromorphite in mine-waste contaminated soils by ATEM and EXAFS. Eur J Soil Sci 45:393-402

Cullen WR, Reomer KJ (1989) Arsenic speciation in the environment. Chem Rev 89:713-764

Dähn R, Scheidegger AM, Manceau A, Schlegel M, Baeyens B, Bradbury H, Morales M (2002a) Neoformation of Ni phyllosilicate upon Ni uptake on montmorillonite. A kinetics study by powder and polarized EXAFS. Geochim Cosmochim Acta 66:2335-2347

Dähn R, Scheidegger AM, Manceau A, Schlegel ML, Baeyens B, Bradbury MH, Chateigner D (2002b) Structural evidence for the sorption of metal ion on the edges of montmorillonite layers. A polarized EXAFS study. Geochim Cosmochim Acta: in press

Dai Y, Hughes JM (1989) Crystal structure refinements of vanadinite and pyromorphite. Can Mineral 27:189-192

Darbieu MH (1993) Mercury in Our Environment. Rev Roum Chim 38:219-227

Database C (2001). U.S. Environmental Protection Agency, Office of Solid Waste and Emergency Response, Washington DC

Davies RA, Islam MS, Chadwick AV, Rush GE (2000) Cation dopant sites in the $CaZrO_3$ proton conductor: a combined EXAFS and computer simulation study. Sol State Ionics 130:115-122

Davies SN, DeWiest RCM (1966) Hydrogeology. Wiley & Sons, New York

Dhez P, Chevallier P, Lucatorto TB, Tarrio C (1999) Instrumental aspects of X-ray microbeams in the range above 1 keV. Rev Sci Instr 70:1907-1920

Dittmer J, Dau H (1998) Theory of the linear dichroism in the extended X-ray absorption fine structure (EXAFS) of partially vectorially ordered systems. J Phys Chem B102:8196-8200

Drits VA, Gorshkov AI, Sakharov BA, Salyn AL, Manceau A, Sivtsov AB (1995) Ferrihydrite and its phase transitions during heating in the oxidizing and reducing environments. Lith Min Res 1:68–75 (translated from Litologiya (1995) 1:76–84)

Drits VA, Lanson B, Bougerol-Chaillout C, Gorshkov AI, Manceau A (2002) Structure of heavy metal sorbed birnessite. Part 2. Results from electron diffraction. Am Mineral: in press

Drits VA, Lanson B, Gorshkov AI, Manceau A (1998) Sub- and super-structure of four-layer Ca-exchanged birnessite. Am Mineral 83:97-118

Drits VA, Sakharov BA, Salyn AL, Manceau A (1993) Structural model for ferrihydrite. Clay Miner 28:185-208

Drits VA, Silvester E, Gorshkov AI, Manceau A (1997) The structure of synthetic monoclinic Na-rich birnessite and hexagonal birnessite. Part 1. Results from X-ray diffraction and selected area electron diffraction. Am Mineral 82:946-961

Dudas MJ, Warren CJ, Spiers GA (1988) Chemistry of arsenic in acid sulphate soils of northern Alberta. Comm Soil Sci Plant Ana 19:887-895

Dyar MD, Delaney JS, Sutton SR (2001) Fe XANES spectra of iron-rich micas. Eur J Mineral 13:1079-1098

Escobar-Gonzalez VL, Monhemius AJ (1988) The mineralogy of arsenates relating to arsenic impurity control *In* Arsenic metallurgy fundamentals and applications. Reddy RG, Hendrix JL, Queneau PB (eds) The Minerals, Metals and Materials Society, Warrendale, PA, USA, p 405-453

Espinose de la Caillerie JBD, Kermarec M, Clause O (1995) Impregnation of gamma-alumina with Ni(II) or Co(II) ions at neutral pH: Hydrotalcite-type coprecipitate formation and characterization. J Am Chem Soc 117:11471-11481

Eusden JD, Gallagher L, Eighmy TT, Crannell BS, Krzanowski JR, Butler LG, Cartledge FK, Emery EF, Shaw EL, Francis CA (2002) Petrographic and spectroscopic characterization of phosphate-stabilized mine tailings from Leadville, Colorado. Waste Manag 22:117-135

Fay MJ, Proctor A, Hoffmann DP, Houalla M, Hercules DM (1992) Determination of the Mo Surface Environment of Mo/TiO_2 Catalysts by EXAFS, XANES and PCA. Mikrochim Acta 109:281-293

Fendorf S, Eick MJ, Grossl P, Sparks DL (1997) Arsenate and Chromate Retention Mechanisms on Goethite. 1. Surface Structure. Env Sci Tech 31:315-320

Ford RG, Scheinost AC, Scheckel KG, Sparks DL (1999) The link between clay mineral weathering and the stabilization of Ni surface precipitates. Env Sci Tech 33:3140-3144

Ford RG, Scheinost AC, Sparks DL (2001) Frontiers in metal sorption/precipitation mechanisms on soil mineral surfaces. Adv Agron 74:41-62

Foster AL, Brown GE, Tingle TN, Parks GA (1997) Quantitative arsenic speciation in mine tailings using X-ray absorption spectroscopy. Am Mineral 83:553-568

Gaillard JF, Webb SM, Quintana JPG (2001) Quick X-ray absorption spectroscopy for determining metal speciation in environmental samples. J Synch Rad 8:928-930

Gerward L (1981) Analytical approximations for X-ray attenuation coefficients. Nucl Instr Meth Phys Res 181:11-14

Glasby GP, Rankin PC, Meylan MA (1979) Manganiferous soil concretions from Hawaii. Paci. Sci. 33:103-115

Goldberg S, Johnston CT (2001) Mechanisms of arsenic adsorption on amorphous oxides evaluated using macroscopic measurements, vibrational spectroscopy, and surface complexation modeling. J Coll Interf Sci 234:204-216

Golden DC, Chen CC, Dixon JB, Tokashki Y (1988) Pseudomorphic replacement of manganese oxides by iron oxide minerals. Geoderma 42:199-211

Golden DC, Dixon JB, Kanehiro Y (1993) The manganese oxide mineral, lithiophorite, in an oxisol from Hawaï. Aust J Soil Res 31:51-66

Goulon J, Goulon-Ginet C, Cortes R, Dubois JM (1981) On experimental attenuation factors of the amplitude of the EXAFS oscillations in absorption, reflectivity and luminescence measurements. J Phys 42:539-548

Han FX, Banin A, Triplett GB (2001) Redistribution of heavy metals in arid-zone soils under a wetting-drying cycle soil moisture regime. Soil Sci 166:18-28

Hansel CM, Fendorf S, Sutton S, Newville M (2001) Characterization of Fe plaque and associated metals on the roots of mine-waste impacted aquatic plants. Env Sci Tech 35:3863-3868

Hazemann JL, Manceau A, Sainctavit P, Malgrange C (1992) Structure of the α-Fe$_x$Al$_{1-x}$OOH solid solution. I. Evidence by polarized EXAFS for an epitaxial growth of hematite-like clusters in diaspore. Phys Chem Miner 19:25-38

Heinrichs H, Mayer R (1977) Distribution and cycling of major and trace elements in two Central European forest ecosystems. J Env Qual 6:402-407

Heinrichs H, Mayer R (1980) The role of forest vegetation in the biogeochemical cycle of heavy metals. J Env Qual 9:111

Henke BL, Gullikson EM, Davis JC (1993) X-ray interactions: Photoabsorption, scattering, transmission and reflection at E=50-30000eV, Z=1-92. Atom Data Nucl Data Tab 54:181-342

Hlawatsch S, Kersten M, Garbe-Schönberg CD, Lechtenberg F, Manceau A, Tamura N, Kulik DA, Harff J, Suess E (2001) Trace metal fluxes to ferromanganese nodules from the western Baltic Sea as a record for long-term environmental changes. Chem Geol 182:697-710

Hochella MF, Moore JN, Golla U, Putnis A (1999) A TEM study of samples from acid mine drainage systems: Metal-mineral association with implications for transport. Geochim Cosmochim Acta 63:3395-3006

Hofmeister W, Von Platen H (1992) Crystal chemistry and atomic order in brucite-related double-layer structures. Crystallography Review 3:3-29

Hyun SP, Cho YH, Hahn PS, Kim SJ (2001) Sorption mechanism of U(VI) on a reference montmorillonite: Binding to the internal and external surfaces. J Radio Nucl Chem 250:55-62

Isaure MP, Laboudigue A, Manceau A, Sarret G, Tiffreau C, Trocellier P (2001) Characterisation of zinc in slags issued from a contaminated sediment by coupling µPIXE, µRBS, µEXAFS and powder EXAFS spectroscopy. Nucl Instr Meth Phys Res 181:598-602

Isaure MP, Laboudigue A, Manceau A, Sarret G, Tiffreau C, Trocellier P, Hazemann JL, Chateigner D (2002) Quantitative Zn speciation in a contaminated dredged sediment by µPIXE, µSXRF, EXAFS spectroscopy and principal component analysis. Geochim Cosmochim Acta 66:1549-1567

Jain A, Loeppert RH (2000) Effect of competing anions on the adsorption of arsenate and arsenite by ferrihydrite. J Env Qual 29:1422-1430

Janney DE, Cowley JM, Buseck PR (2001) Structure of synthetic 6-line ferrihydrite by electron nanodiffraction. Am Mineral 86:327-335

Karanfil C, Zhong Z, Chapman LD, Fischetti R, Bunker GB, Segre CU, Bunker BA (2000) A bent Laue analyzer detection system for dilute fluorescence XAFS. Synchrotron Radiation Instrumentation, Eleventh U.S. National Conference, 178-182

Kim CS, Brown GE, Rytuba JJ (2000) Characterization and speciation of mercury-bearing mine waste using X-ray absorption spectroscopy. Sci Tot Env 261:157-168

Kirby J, Maher W, Chariton A, Krikowa F (2002) Arsenic concentrations and speciation in a temperate mangrove ecosystem, NSW, Australia. Appl Org Chem 16:192-201

Kirkpatrick P, Baez AV (1948) Formation of optical images by X-rays. J Opt Soc Am 38:766

Kneebone PE, O'Day PA, Jones N, Hering JG (2002) Deposition and fate of arsenic in iron- and arsenic-enriched reservoir sediments. Env Sci Tech 36:381-386

Koningsberger DC, Prins R (1988) X-ray Absorption. Principles, Applications, Techniques of EXAFS, SEXAFS and XANES. John Wiley & Sons, New York

Krause E, Ettel V (1988) Solubility and stability of scorodite, FeAsO$_4$.2H$_2$O: new data and further discussion. Am Mineral 73:850-854

La Force MJ, Hansel CM, Fendorf S (2000) Arsenic speciation, seasonal transformations, and co-distribution with iron in a mine waste-influenced palustrine emergent wetland. Env Sci Tech 34:3937-3943

Labrenz M, Druschel GK, Thomsen-Ebert T, Gilbert B, Welch SA, Kemner KM, Logan GA, Summons RE, De Stasio G, Bond PL, Lai B, Kelly SD, Banfield JF (2000) Formation of sphalerite (ZnS) deposits in natural biofilms of sulfate-reducing bacteria. Science 290:1744-1747

Lacerda LD, Salomons W (1998) Mercury from gold and silver mining: a chemical time bomb? Springer, Berlin

Lanson B, Drits VA, Gaillot AC, Silvester E, Plançon A, Manceau A (2002) Structure of heavy metal sorbed birnessite. Part 1. Results from X-ray diffraction. Am Mineral: in press

Latrille C, Elsass F, van Oort F, Denaix L (2001) Physical speciation of trace metals in Fe-Mn concretions from a rendzic lithosol developed on Sinemurian limestones (France). Geoderma 100:127-146

Lindsay WL (1979) Chemical equilibria in soils. John Wiley & Sons, New York

Lower S, Maurice PA, Traina SJ (1998) Simultaneous dissolution of hydroxylapatite and precipitation of hydroxypyromorphite: Direct evidence of homogenoeus nucleation. Geochim Cosmochim Acta 62:1773-1780

Luther GWI, Meyerson AL, Krajewski J, Hires RI (1980) Metal sulfides in estuarine sediments. J Sed Petrol 50:1117-1120.

Ma QY, Logan TJ, Traina SJ, Ryan JA (1994) Effects of aqueous Al, Cd, Cu, Fe(II), Ni, and Zn on Pb immobilization by hydroxyapatite. Env Sci Tech 28:1219-1228

MacDowell AA, Celestre RS, Tamura N, Spolenak R, Valek BC, Brown WL, Brawman JC, Padmore HA, Batterman BW, Patel JR (2001) Submicron X-ray diffraction. N Instr Meth A 468:936-943

Malinowski ER (1977) Determination of the number of factors and the experimental error in a data matrix. Anal Chem 49:612-617

Malinowski ER (1978) Theory of error for target factor analysis with applications to mass spectrometry and nuclear magnetic resonance spectrometry. Anal Chim Acta 103:359-354

Malinowski ER (1991) Factor Analysis in Chemistry. John Wiley, New York,

Manceau A (1995) The mechanism of anion adsorption on Fe oxides: evidence for the bonding of arsenate tetrahedra on free $Fe(O,OH)_6$ edges. Geochim Cosmochim Acta 59:3647-3653

Manceau A, Boisset MC, Sarret G, Hazemann JL, Mench M, Cambier P, Prost R (1996) Direct determination of lead speciation in contaminated soils by EXAFS spectroscopy. Env Sci Tech 30:1540-1552

Manceau A, Chateigner D, Gates WP (1998) Polarized EXAFS, distance-valence least-squares modeling (DVLS) and quantitative texture analysis approaches to the structural refinement of Garfield nontronite. Phys Chem Miner 25:347-365

Manceau A, Combes JM (1988) Structure of Mn and Fe oxides and oxyhydroxides: A topological approach by EXAFS. Phys Chem Miner 15:283-295

Manceau A, Drits VA (1993) Local structure of ferrihydrite and feroxyhite by EXAFS spectroscopy. Clay Miner 28:165-184

Manceau A, Drits VA, Silvester E, Bartoli C, Lanson B (1997) Structural mechanism of Co(II) oxidation by the phyllomanganate, Na-buserite. Am Mineral 82:1150-1175

Manceau A, Gates W (1997) Surface structural model for ferrihydrite. Clays Clay Miner 43:448-460

Manceau A, Gorshkov AI, Drits VA (1992) Structural chemistry of Mn, Fe, Co, and Ni in Mn hydrous oxides. II. Information from EXAFS spectroscopy, electron and X-ray diffraction. Am Mineral 77:1144-1157

Manceau A, Lanson B, Drits VA (2002a) Structure of heavy metal sorbed birnessite. Part III. Results from powder and polarized extended X-ray absorption fine structure spectroscopy. Geochim Cosmochim Acta 66:2639-2663

Manceau A, Lanson B, Drits VA, Chateigner D, Gates WP, Wu J, Huo DF, Stucki JW (2000a) Oxidation-reduction mechanism of iron in dioctahedral smectites. 1. Structural chemistry of oxidized nontronite references. Am Mineral 85:133-152

Manceau A, Lanson B, Schlegel ML, Hargé JC, Musso M, Eybert-Bérard L, Hazemann JL, Chateigner D, Lamble GM (2000b) Quantitative Zn speciation in smelter-contaminated soils by EXAFS spectroscopy. Am J Sci 300:289-343.

Manceau A, Schlegel M, Chateigner D, Lanson B, Bartoli C, Gates WP (1999a) Application of polarized EXAFS to fine-grained layered minerals *In* Synchrotron X-ray Methods in Clay Science. Schulze D, Bertsch P, Stucki J (eds) Clay Mineral Society of America, p 68-114

Manceau A, Schlegel M, Nagy KL, Charlet L (1999b) Evidence for the formation of trioctahedral clay upon sorption of Co^{2+} on quartz. J Coll Interf Sci 220:181-197

Manceau A, Schlegel ML (2001) Texture effect on polarized EXAFS amplitude. Phys Chem Miner 28:52-56.

Manceau A, Schlegel ML, Musso M, Sole VA, Gauthier C, Petit PE, Trolard F (2000c) Crystal chemistry of trace elements in natural and synthetic goethite. Geochim Cosmochim. Acta 64:3643-3661

Manceau A, Tamura N, Celestre RS, MacDowell AA, Sposito G, Padmore HA (2002b) Molecular-scale speciation of Zn and Ni in soil ferromanganese nodules. Env Sci Technol: accepted

Manceau A, Tamura N, Marcus MA, MacDowell AA, Celestre RS, Sublett RE, Sposito G, Padmore HA (2002c) Deciphering Ni sequestration in soil ferromanganese nodules by combining X-ray fluorescence, absorption and diffraction at micrometer scales of resolution. Am Mineral 87:1494-1499

Manning, Fendorf, Goldberg (1998) Surface structures and stability of arsenic(III) on goethite: spectroscopic evidence for inner-sphere complexes. Env Sci Tech 32:2383-2388

Masscheleyn PH, Delaune RD, Patrick WH (1991) Effect of redox potential and pH on arsenic speciation and solubility in a contaminated soil. J Env Qual 25:1414-1419

Maxwell JA, Teesdale WJ, Campbell JJ (1995) The Guelph PIXE software pacakage II. Nucl Instr Meth Phys Res B, 95:407-421.

McMaster WH, Kerr Del Grande N, Mallett JH, Hubbell JH (1969) Compilation of X-ray Cross Sections. US National Technical Information Service, Springfield

Mench M, Bussière S, Boisson J, Castaing E, Vangronsveld J, Ruttens A, De Koe T, Bleeker P, Assunçao A, Manceau A (2002) Progress in remediation and revegetation of the Barren Jales gold mine spoil after *in situ* inactivation. Plant Soil: in press

Mesjasz-Przybylowicz J, Przybylowicz WJ (2002) Micro-PIXE in plant sciences: Present status and perspectives. Nucl Instr Meth Phys Res B, 189:470-481.

Mittaz JPD, Penston MV, Snijders MAJ (1990) Ultraviolet variability of NGC4151: A study using principal component analysis. Mon Not R Astr Soc 242:370-378

Morel FMM (1983) Principles of Aquatic Chemistry. Wiley-Interscience, New York

Morin G, Juillot F, Ildefonse P, Calas G, Samama JC, Chevallier P, Brown GE (2001) Mineralogy of lead in a soil developed on a Pb-mineralized sandstone (Largentière, France). Am Mineral 86:92-104

Morin G, Ostergren JD, Juillot F, Ildefonse P, Calas G, Brown GE (1999) XAFS determination of the chemical form of lead in smelter-contaminated soils and mine tailings: Importance of adsorption processes. Am Mineral 84:420-434

Morton JD, Semrau JD, Hayes KF (2001) An X-ray absorption spectroscopy study of the structure and reversibility of copper adsorbed to montmorillonite clay. Geochim Cosmochim Acta 65:2709-2722

Müller B, Sigg L (1992) Adsorption of lead(II) on the goethite surface: Voltammetric evaluation of surface complexation parameters. J Coll Interf Sci 148:517-532

Neder RB, Burghammer M, Grasl T, Schulz H, Bram A, Fiedler S (1999) Refinement of the kaolinite structure from single-crystal synchrotron data. Clays Clay Miner 47:487-494

Needleman HL (1983) Low level lead exposure and neuropsychological performance *In* Lead versus health - Sources and effects of low level lead exposure. Rutter M, Russell JR (eds) Wiley, Chichester, p 229-248

Needleman HL, Schell A, Bellinger D, Leviton A, Allred EN (1990) The long-term effects of exposure to low doses of lead in childhood: an 11-year follow-up report. N Engl J Med 322:83-88

Newville M, Sutton S, Rivers M, Eng P (1999) Micro-beam X-ray absorption and fluorescence spectroscopies at GSECARS: APS beamline 13ID. J Synch Rad 6:353-355

Nickson R, McArthur J, Burgess W, Ahmed KM, Ravenscroft P, Rahman M (1998) Arsenic poisoning of Bangladesh groundwater. Nature 395:338.

Nriagu JO (1994) Arsenic in the Environment, Part I: Cycling and Characterization. John-Wiley & Sons, New York

Nriagu JO (1974) Lead orthophosphate - IV Formation and stability in the environment. Geochim Cosmochim Acta 38:887-898

O'Day PA, Carroll SA, Randall S, Martinelli RE, Anderson SL, Jelinski J, Knezovich JP (2000) Metal speciation and bioavailability in contaminated estuary sediments, Alameda Naval Air Station, California. Env Sci Tech 34:3665-3673

O'Day PA, Carroll SA, Waychunas GA (1998) Rock-water interactions controlling zinc, cadmium, and lead concentrations in surface waters and sediments, U.S. Tri-State Mining District. I. Molecular identification using X-ray absorption spectroscopy. Env Sci Tech 32:943-955

O'Reilly S, Strawn DG, Sparks DL (2001) Residence time effects on arsenate adsorption/desorption mechanisms on goethite. Soil Sci Soc Am J 65:67-77

Organova NI, Drits VA, Dmitrik AL (1974) Selected area electron diffraction study of a type II "valleriite-like" mineral. Am Mineral 59:190-200

Ostergren JD, Brown GE, Parks GA, Tingle TN (1999) Quantitative lead speciation in selected mine tailings from Leadville, CO. Env Sci Tech 33:1627-1636

Ostergren JD, Trainor TP, Bargar JR, Brown GE, Parks GA (2000) Inorganic ligand effects on Pb(II) sorption to goethite (alpha-FeOOH). I. Carbonate. J Coll Interf Sci 225:466-482

Palumbo B, Bellanca A, Neri R, Roe MJ (2001) Trace metal partitioning in Fe-Mn nodules from Sicilian soils, Italy. Chem Geol 173:257-269

Parkman RH, Curtis CD, Vaughan DJ, Charnock JM (1996) Metal fixation and mobilisation in the sediments of Afon Goch estuary - Dulas Bay, Anglesey. Appl Geoch 11:203-210

Perrakis A, Cipriani F, Castagna JC, Claustre L, Burghammer M, Riekel C, Cusack S (1999) Protein microcrystals and the design of a microdiffractometer: current experience and plans at EMBL and ESRF/ID13. Acta Cryst D55:1765-1770

Pfalzer P, Urbach JP, Klemm M, Horn S, denBoer ML, Frenkel AI, Kirkland JP (1999) Elimination of self-absorption in fluorescence hard-X-ray absorption spectra. Phys Re B60:9335-9339

Powers L (1982) X-ray absorption spectroscopy. Application to biological molecules. Biochim Biophys Acta 638:1-38

Press WH, Teukolsky SA, Vetterling WT, Flannery BP (1992) Numerical Recipes. Cambridge University Press

Ravelli RGB, McSweeney SM (2000) The "fingerprint" that X-rays can leave on structures. Str Folding Design 8:315-328

Rawlings DE, Silver S (1995) Mining with microbes. Bio Tech 13:773-778

Rehr JJ, Albers RC (2000) Theoretical approaches to x-ray absorption fine structure. Rev Modern Phys 72:621-654

Ressler T, Wong J, Roos J, Smith I (2000) Quantitative speciation of Mn-bearing particulates emitted from autos burning methylcyclopentadienyl manganese tricarbonyl- (MMT-) added gasolines using XANES spectroscopy. Env Sci Tech 34:950-958

Rhoton FE, Bigham JM, Schulze DG (1993) Properties of iron-manganese nodules from a sequence of eroded fragipan soils. Soil Sci Soc Am J 57:1386-1392

Rimstidt JD (1997) Quartz solubility at low temperatures. Geochim Cosmochim Acta 61:2553-2558

Rimstidt JD, Barnes HL (1980) The kinetics of silica-water reactions. Geochim Cosmochim Acta 44:1683-1699

Ritchie GSP, Sposito G (1995) Speciation in soils *In* Chemical speciation in the environment. Ure AM, Davidson CM (eds) Blackie Academic & Professional, Glasgow, p 201-233

Roberts DR, Scheinost AC, Sparks DL (2002) Zinc speciation in a smelter-contaminated soil profile using bulk and microscopic techniques. Env Sci Tech 36:1742-1750

Robson AD (1993) Zinc in Soils and Plants. Klumer Academic Publishers, Australia

Ross DS, Hales HC, Sea-McCarthy GC, Lanzirotti A (2001) Sensitivity of soil manganese oxides: XANES spectroscopy may cause reduction. Soil Sci Soc Am J 65:744-752

Ross SJ, Franzmeier DP, Roth CB (1976) Mineralogy and chemistry of manganese oxides in some Indiana soils. Soil Sci Soc Am J 40:137-143

Rossman GR (1994) Colored varieties of the silica minerals *In* Silica, Vol 29. Heaney PJ, Prewitt CT, Gibbs GV (eds) Mineralogical Society of America, Washington DC, p 433-467

Roussel C, Bril H, Fernandez A (2000) Arsenic speciation: Involvement in evaluation of environmental impact caused by mine wastes. J Env Qual 29:182-188

Ruby MV, Davis A, Nicholson A (1994) *In situ* formation of lead phosphates in soils as a method to immobilize lead. Env Sci Tech 28:646-654

Ryan JA, Zhang P, Hesterberg DA, Chou J, Sayers DE (2001) Formation of chloropyromorphite in lead-contaminated soil amended with hydroxyapatite. Env Sci Tech 35:3798-3803

Sarret G, Manceau A, Spadini L, Roux JC, Hazemann JL, Soldo Y, Eybert-Bérard L, Menthonnex JJ (1998) Structural determination of Zn and Pb binding sites in Penicillium chrysogenum cell wall by EXAFS spectroscopy. Env Sci Tech 32:1648-1655.

Savage KS, Tingle TN, O'Day PA, Waychunas GA, Bird DK (2000) Arsenic speciation in pyrite and secondary weathering phases, Mother Lode Gold District, Tuolumne County, California. Appl Geochem 15:1219-1244

Sayers DE, Bunker G (1988) EXAFS data analysis *In* X-ray absorption: Principles, Applications, Techniques of EXAFS, SEXAFS and XANES. Koningsberger DC, Prins R (eds) John-Wiley & Sons, New York

Scheckel KG, Scheinost AC, Ford RG, Sparks DL (2000) Stability of layered Ni hydroxide surface precipiates - A dissolution kinetics study. Geochim Cosmochim Acta 64:2727-2735

Scheidegger AM, Lamble GM, Sparks DL (1997) Spectroscopic evidence for the formation of mixed-cation hydroxide phases upon metal sorption on clays and aluminum oxides. J Coll Interf Sci 186:118-128

Scheinost AC, Sparks DL (2000) Formation of layered single- and double-metal hydroxide precipitates at the mineral/water interface: A multiple-scattering XAFS analysis. J Coll Interf Sci 223:167-178

Schlegel M, Charlet L, Manceau A (1999a) Sorption of metal ions on clay minerals. II. Mechanism of Co sorption on hectorite at high and low ionic strength, and impact on the sorbent stability. J Coll Interf Sci 220:392-405

Schlegel M, Manceau A, Chateigner D, Charlet L (1999b) Sorption of metal ions on clay minerals. I. Polarized EXAFS evidence for the adsorption of Co on the edges of hectorite particles. J Coll Interf Sci 215:140-158

Schlegel ML, Manceau A, Charlet L, Chateigner D, Hazemann JL (2001a) Sorption of metal ions on clay minerals. III. Nucleation and epitaxial growth of Zn phyllosilicate on the edges of hectorite. Geochim Cosmochim Acta 65:4155-4170

Schlegel ML, Manceau A, Charlet L, Hazemann JL (2001b) Adsorption mechanism of Zn on hectorite as a function of time, pH, and ionic strength. Am J Sci 301:798-830

Schlegel ML, Nagy KL, Fenter P, Sturchio NC (2002) Structures of quartz (1010)- and (1011)-water interfaces determined by X-ray reflectivity and atomic force microscopy of natural growth surfaces. Geochim Cosmochim Acta: in press

Schlesinger WH (1991) Biogeochemistry. Academic Press, San Diego

Schroeder SLM, Moggridge GD, Lambert RM, Rayment T (1997) "Self-absorption" effects in grazing-incidence total electron-yield XAS. J Phys IV 7:91-96

Segtnan VH, Sasic S, Isaksson T, Ozaki Y (2001) Studies on the Structure of Water Using Two-Dimensional Near-Infrared Correlation Spectroscopy and Principal Component Analysis. Anal Chem 73:3153-3161

Shannon RD, Prewitt CT (1969) Effective ionic radii in oxides and fluorides. Acta Cryst B25:925-945

Shibata Y, Morita M (1989) Exchange of comments on identification and quantitation of arsenic species in a dogfish muscle reference material for trace elements. Anal Chem 61:2116-2118

Silvester E, Manceau A, Drits VA (1997) The structure of synthetic monoclinic Na-rich birnessite and hexagonal birnessite. Part 2. Results from chemical studies and EXAFS spectroscopy. Am Mineral 82:962-978

Singer PC, Stumm W (1970) Acidic mine drainage: the rate determining step. Science 167:1121-1123

Smedley PL, Kinniburgh DG (2002) A review of the source, behaviour and distribution of arsenic in natural waters. Appl Geochem 17:517-568

Sobanska S, Ricq N, Laboudigue A, Guillermo R, Brémard C, Laureyns J, Merlin JC, Wignacourt JP (1999) Microchemical investigations of dust by a lead smelter. Env Sci Tech 33:1334-1339

Spadini L, Manceau A, Schindler PW, Charlet L (1994) Structure and stability of Cd^{2+} surface complexes on ferric oxides. I. Results from EXAFS spectroscopy. J Coll Interf Sci 168:73-86.

Sparks DL, Grundl TJ (1998) Mineral-water interfacial reactions: Kinetics and mechanisms. Am Chem Soc, Washington DC

Sposito G (1994) Chemical equilibria and kinetics in soils. Oxford University Press, New York

Stern EA, Heald SM (1983) Basic principles and applications of EXAFS *In:* Handbook of Synchrotron Radiation. Koch E (ed) North-Holland, Amsterdam, New York, p 955-1014

Stern EA, Kim K (1981) Thickness effect on the extended X-ray absorption fine structure amplitude. Phys Rev B23:3781-3787

Stevenson FJ (1979) Lead-organic matter interactions in a mollisol. Soil Biol Biochem 11:493-499

Stipp SL, Hochella MF, Parks GA, Leckie JO (1992) Cd^{2+} uptake by calcite, solid-state diffusion, and the formation of solid-solution: Interface processes observed with near-surface sensitive techniques (XPS, LEED, and AES). Geochim Cosmochim Acta 56:1941-1954

Strawn D, Doner H, Zavarin M, McHugo S (2002) Microscale investigation into the geochemistry of arsenic, selenium and iron in soil developed in pyritic shale materials. Geoderma 108:237-257

Strawn DG, Sparks DL (1999) The use of XAFS to distinguish between inner- and outer-sphere lead adsorption complexes on montmorillonite. J Coll Interf Sci 216:257-269

Sutton SR, Rivers ML (1999) Hard X-ray synchrotron microprobe techniques and applications *In* Synchrotron X-ray Methods in Clay Science. Schulze D, Bertsch P, Stucki J (eds) The Clay Minerals Society of America, p 146-163

Taylor R, Xiu H, Mehadi A, Shuford J, Tadesse W (1995) Fractionation of residual cadmium, copper, nickel lead and zinc in previously sludge-amended soil. Comm Soil Sci Plant Ana 26:2193-2204

Taylor RM (1968) The association of manganese and cobalt in soils- further observations. J Soil Sci 19:77-80

Taylor RM, McKenzie RM, Norrish K (1964) The mineralogy and chemistry of manganese in some Australian soils. Aust J Soil Res 2:235-248

Thiel GA (1925) Manganese precipitated by microorganisms. Econ Geol XX:301-310

Thiry M, Van Oort F (1999) Les phases minérales majeures et mineures d'une friche industrielle de métallurgie des métaux non-ferreux: état d'altération, évolution géochimique et devenir des métaux polluants du site de Mortagne-du-Nord. *In:* Spéciation des métaux dans le sol. Les Cahiers des Clubs CRIN, Paris

Thompson AC, Underwood JH, Wu Y, Giauque RD, Jones KW, Rivers ML (1988) Elemental measurements with an X-ray microprobe of biological and geological samples with femtogram sensitivity. Nucl Instr Meth Phys Res A266:318

Tokashiki Y, Dixon JB, Golden DC (1986) Manganese oxide analysis in soils by combined X-ray diffraction and selective dissolution methods. Soil Sci Soc Am J 50:1079-1084

Towle SN, Bargar JR, Brown GE, Parks GA (1997) Surface Precipitation of Co(II)(aq) on Al_2O_3. J Coll Interf Sci 187:62-82

Troger L, Arvanitis D, Baberschke K, Michaeis H, Grimm U, Zschech E (1992) Full correction of the self-absorption in soft-fluorescence extended x-ray absorption fine structure. Phys Rev B46:3283-3289

Usui A, Mita N (1995) Geochemistry and mineralogy of a modern buserite deposit from a hot spring in Hokkaido, Japan. Clays Clay Miner 43:116-127

Uzochukwu GA, Dixon JB (1986) Manganese oxide minerals in nodules of two soils of Texas and Alabama. Soil Sci Soc Am J 50:1358-1363

Van der Lelie D, Schwitzguébel JP, Glass DJ, Vangronsveld J, Baker A (2001) Assessing phytoremediation's progress in the United States and Europe. Env Sci Tech 35:446A-452A

Vangronsveld J, Cunningham S (1998) *In situ* Inactivation and Phytorestoration of Metal-Contaminated Soils. Landes Biosciences, Georgetown, TX, USA

Vangronsveld J, Van Assche F, Clijsters H (1995) Reclamation of a bare industrial area contaminated by non-ferrous metals: *In situ* metal immobilization and revegetation. Env Poll 87:51-59

Varentsov IM, Drits VA, Gorshkov AI, Sivtsov AV, Sakharov BA (1991) Me-Fe oxyhydroxide crusts from Krylov Seamount (Eastern Atlantic): Mineralogy, geochemistry and genesis. Mar Geol 96:53-70

Vodyanitskii YN, Lesovaya SN, Sivtsov AV (2001) Iron minerals in soils on red-colored deposits. Eurasian Soil Sci 34:774-782

Wadsley AD (1952) The structure of lithiophorite, $(Al,Li)MnO_2(OH)_2$. Acta Cryst 5:676-680

Wadsley AD (1955) The crystal structure of chalcophanite, $ZnMn_3O_7.3H_2O$. Acta Cryst 8:1165-1172

Wasserman SR, Allen PG, Shuh DK, Bucher JJ, Edelstein NM (1999) EXAFS and principal component analysis: a new shell game. J Synch Rad 6:284-286

Wasserman SR, Winans RE, MacBeth R (1996) Iron species in Argonne premium coal samples: An investigation using X-ray absorption spectroscopy. Energy Fuels 10:392-400

Watson EB (1996) Surface enrichment and trace-element uptake during crystal growth. Geochim Cosmochim Acta 60:5013-5020

Waychunas GA, Rea BA, Fuller CC, Davis JA (1993) Surface chemistry of ferrihydrite: Part 1. EXAFS studies of the geometry of coprecipitated and adsorbed arsenate. Geochim Cosmochim Acta 57:2251-2269

Webb SM, Leppard GG, Gaillard JF (2000) Zinc speciation in a contaminated aquatic environment: Characterization of environmental particles by analytical electron microscopy. Env Sci Tech 34:1926-1933

Weik M, Ravelli RBG, Kryger G, McSweeney S, Raves ML, Harel M, Gros P, Silman I, Kroon J, Sussman JL (2000) Specific chemical and structural damage to proteins produced by synchrotron radiation. Proc Natl Acad Sci USA 97:623-628

Welter E, Calmano W, Mangold S, Troger L (1999) Chemical speciation of heavy metals in soils by use of XAFS spectroscopy and electron microscopical techniques. Fresenius J Anal Chem 364:238-244

Wheeting LC (1936) Shot soils of Western Washington. Soil Sci 41:35-45

White GN, Dixon JB (1996) Iron and manganese distribution in nodules from a young Texas vertisol. Soil Sci Soc Amer J 60:1254-1262

Wilkie JA, Hering JG (1996) Adsorption of arsenic onto hydrous ferric oxide: Effects of adsorbate/adsorbent ratios and co-occurring solutes. Coll Surf A107:97-110

Woo NC, Choi MJ (2001) Arsenic and metal contamination of water resources from mining wastes in Korea. Env Geol 40:305-311

Zänker H, Moll H, Richter W, Brendler V, Hennig C, Reich T, Kluge A, Huttig G (2002) The colloid chemistry of acid rock drainage solution from an abandoned Zn-Pb-Ag mine. Appl Geochem 17:633-648

Zavarin M (1999) Sorptive properties of synthetic and soil carbonates for selenium, nickel, and manganese. PhD Dissertation, University of California, Berkeley

8

Microfluorescence and Microtomography Analyses of Heterogeneous Earth and Environmental Materials

Stephen R. Sutton[1,2], Paul M. Bertsch[3], Matthew Newville[2], Mark Rivers[1,2], Antonio Lanzirotti[2] and Peter Eng[2]

[1]*Department of Geophysical Sciences and*
[2]*Consortium for Advanced Radiation Sources*
University of Chicago
Chicago, Illinois, 60637, U.S.A

[3]*Savannah River Ecology Laboratory*
University of Georgia
Aiken, South Carolina, 29802, U.S.A.

INTRODUCTION

Analytical techniques with high sensitivity and high spatial resolution are crucial for understanding the chemical properties of complex earth materials and environmental samples, and these so-called "microprobes" have become workhorses of the geochemical community as well as important tools for environmental scientists. These microanalytical instruments are based on various forms of sample excitation and detection. They are complementary in terms of spatial resolution, element sensitivity, energy deposition and non-destructiveness.

Several techniques fall in the class of methods employing charged particle excitation of X-ray fluorescence, including electron microprobe analysis (EMPA) and particle-induced X-ray emission (PIXE). EMPA is capable of μm-sized spots with minimum detection limits near 100 mg kg^{-1}. PIXE is well suited for analyses of relatively light elements with 10 mg kg^{-1} sensitivity and μm-sized spots. The relatively large energy deposited by the charged particle beam can complicate the analysis of volatile elements or induce valence state changes of redox sensitive elements. Sensitivity of these technologies is a relatively smooth function of atomic number.

Other techniques are based on sample sputtering followed by mass spectrometry of the vaporized products, including secondary ion mass spectrometry (SIMS) and laser ablation inductively coupled plasma mass spectrometry (LA-ICP-MS). Beam sizes are in the few to tens of μm range. Elemental sensitivities for SIMS are highly variable depending on ion yield, and quantification can be difficult because of matrix effects in the ion production process. SIMS and LA-ICP-MS have very high sensitivities for some elements and low sensitivity for others. These and other microanalytical techniques used in earth science research are described in Potts et al. (1995).

The subject of this chapter is synchrotron X-ray fluorescence (SXRF) microprobe analysis (Horowitz and Howell 1972) and microtomography. Unlike other microanalytical techniques, SXRF uses photons for excitation. Spot sizes are in the μm range with sensitivities in the sub-mg kg^{-1} range. Like EMPA and PIXE, SXRF sensitivities are smooth functions of atomic number (Smith and Rivers 1995). Quantification is comparatively straightforward for SXRF because the physics of photon interactions with matter is well understood. Another advantage is that a vacuum sample chamber is unnecessary allowing more versatility in sample state, including analyses of liquids and hydrated solids.

1529-6466/00/0049-0008$10.00

The key to the practicality of the X-ray microprobe is the availability of synchrotron X-ray sources. The high brightness and brilliance of these sources allows small, intense X-ray beams to be produced leading to high sensitivity and spatial resolution. The sensitivity of the technique is further enhanced by the polarization properties of the synchrotron radiation that allows the geometry of the experiment to be optimized for highest signal to noise ratios. In this way, the capabilities of an X-ray microprobe are enhanced by orders of magnitude when used with a synchrotron source as compared to what is achievable with a laboratory tube source.

A unique capability of the synchrotron X-ray microprobe is microbeam applications of X-ray absorption fine structure spectroscopy (XAFS) for the determination of chemical speciation. Thus, not only can one obtain the concentration of a trace element but also its oxidation state, coordination number, and the identity of nearest neighbors.

X-ray computed microtomography (CMT) has great potential for application to research problems in earth and environmental sciences. It provides the ability to examine the internal structure of objects which cannot be sectioned, either because the objects are too valuable (e.g., some meteorites and fossils), because the objects are too fragile to be sectioned (e.g., interplanetary dust particles and soil aggregates), or simply because physical sectioning is too time consuming. Synchrotron radiation makes it possible to extend the CMT technique to μm spatial resolution with reasonable data acquisition times. In addition, element specific imaging is possible either by collecting transmission tomograms above and below an absorption edge or by using fluorescence techniques.

In this chapter, we discuss synchrotron X-ray microprobe and microtomography instrumentation, practical aspects of μXRF, μXAFS and microtomography, and examples of applications to earth and environmental materials. Other discussions of synchrotron-based microprobe and microtomography techniques can be found in the following references: Dunsmuir et al. 1991; Janssens et al. 1993; Sutton et al. 1994; Schulze and Bertsch 1995; Smith and Rivers 1995; Jones et al. 1997; Sutton and Rivers 1999; Rivers et al. 1999; Bertsch and Hunter 2001; Baker 2002.

X-RAY MICROPROBE ANALYSIS

Hard X-ray microprobes in the U.S.

Hard X-ray microprobes are currently in use for earth and environmental science research primarily at three synchrotrons in the US: NSLS, APS and ALS (Table 1). At the NSLS (*http://nslsweb.nsls.bnl.gov*), the X26A Microprobe Facility (*http://www.-bnl.gov/x26a*) is the only high-energy X-ray microprobe and resides on a dedicated beam line. At the APS (*http://www.aps.anl.gov*), there are three microprobes on undulator (vide infra) beam lines. GeoSoilEnviroCARS (13-ID; *http://gsecars.org*) is dedicated to earth and environmental science research with 95% of the beam time allocated to the scientific community through a proposal based system but the microprobe receives only 20% of the available beam time (sharing with other techniques such as high pressure crystallography). The 20-ID microprobe (PNC-CAT; *http://www.pnc.aps.anl.gov*) is similar in configuration to the 13-ID microprobe and the general scientific community can compete for 25% of the total available time (through the APS Independent Investigator program). The 2-ID microprobe (SRI-CAT; *http://www.aps.anl.gov/sricat*) is focused on instrumentation development and some of their time is also available through the APS Independent Investigator program. At the ALS (*http://www-als.lbl.gov*), the 10.3.1 and 10.3.2 microprobes have environmental science components and some of this time is open to the community. There is also a Kirkpatrick-Baez (KB) mirror based microprobe at CAMD (*http://www.camd.lsu.edu*) utilized primarily for materials science.

Table 1. Hard X-ray microprobes at United States synchrotron facilities.

Beam Line*	Managing Agent(s)	Scientific Program	Source	Microbeam Apparatus	Beam Size (μm)	Flux [†]
NSLS X26A	Univ. of Chicago; Univ. of Georgia; Brookhaven Nat. Lab.	Earth and environmental science	Bending Magnet (2.8 GeV)	KB mirrors	10	2×10^8
APS 13-ID	GeoSoilEnviroCARS/ Univ. of Chicago	Earth and environmental science	Undulator (7 GeV)	KB mirrors	1	4×10^{11}
APS 20-ID	Univ. of Washington; Pacific Northwest Nat. Lab.; Simon Fraser Univ.	Materials science and environmental science	Undulator (7 GeV)	KB mirrors	1	4×10^{11}
APS 2-ID	Argonne National Laboratory	High resolution imaging	Undulator (7 GeV)	Zone plate	0.1	1×10^8
ALS 10.3.2	Lawrence Berkeley National Laboratory	Materials science and environmental science	Bending Magnet (1.9 GeV)	KB mirrors	5	3×10^9

[†] Flux – maximum delivered to sample at stated beam size; ph/s/0.01%BW

These instruments are complementary in terms of spatial resolution, sensitivity, flexibility and availability. The APS undulator source, in combination with KB mirrors, is an excellent X-ray microprobe source, delivering ~1 μm X-ray beams with fluxes that are factors of ~10^3 greater than those on other sources. The APS zone plate microprobe achieves smaller spot sizes at the expense of sensitivity and chromaticity. The choice of the optimum instrument for a particular experiment depends on the specific requirements. On the other hand, because of the unique characteristics of the APS insertion device, the APS microprobes must share time with other techniques, the resource is limited and only those experiments that require the undulator typically receive beam time. In addition, these microprobes are historically oversubscribed. The NSLS and ALS microprobes have the advantage of residing on dedicated beam lines so that more beam time is typically available for particular experiments. These microprobes have poorer sensitivity and spatial resolution than the APS instruments but are very well suited for many experiments.

X-ray fluorescence process

The X-ray microprobe relies on the X-ray fluorescence process for identification and quantification of specific elements. At photon energies below ~0.1 MeV, the interactions between photons and matter are dominated by the photoelectric effect. In this process, a photon having energy greater than the binding energy of the electron it encounters gives up its energy to this electron ejecting it from the atom. Atomic recoil takes place to conserve momentum. As we will see later, another important aspect of this process is the transition of the core electron to an outer, bound state of the atom. Such bound state transitions are initiated by interaction with a photon with energy less than the binding energy of the electron. This emptying of the core level leaves the atom in an unstable excited state and the subsequent de-excitation involves filling of the electron hole by an electron in a higher level. The excess energy of the electron is emitted as a photon in the process (fluorescence), where the photon energy is equal to the energy difference between the two bound states involved in the transition. Because these energy levels are distinct for each atom, precise measurement of the photon energies can be used to identify the emitting atom and the intensity of the fluorescence is proportional to the number of such atoms.

The probability of fluorescence being generated by this process depends on the atomic number of the material and the energy of the photons. The dependencies are roughly Z^4 and E^{-3} although the exact powers have been shown by theory and experiment to be non-integral (Evans 1955). So, for a given material, a plot of photoelectric absorption coefficient versus energy shows a rapid decrease with increasing energy. Superimposed on this E-cubed dependence are the absorption edges which are discontinuous jumps where the energy becomes greater than the binding energy of some electrons so the total number of electrons that the photon can energetically eject goes up (and the probability of interaction is therefore higher).

X-ray fluorescence analysis has a long history as a standard technique for analysis of mm to cm specimens using a hot-cathode X-ray source. However, the X-ray flux from a hot-cathode source is too divergent (isotropic emission) to permit efficient focusing. Thus, the niche of the synchrotron X-ray beam is in microbeam applications of the XRF technique, the inherent collimation and polarization of synchrotron radiation is well suited to use in an XRF microprobe. Details of synchrotron radiation generation are given in an accompanying chapter (Sham and Rivers, this volume).

Synchrotron X-ray microprobe instrumentation

The incident X-ray beam path of a typical microprobe configuration (Figs. 1 and 2) consists of the following components (running downstream from the X-ray source): monochromator, intensity monitor, microbeam optics and sample stage. The sample stage is mounted either normal to the incoming beam or in a 45°/45° geometry. In the former configuration, viewing and detection are conducted from behind the sample, requiring samples and mounting substrates to be X-ray thin and visually transparent. In the latter configuration, a visible light microscope can be mounted normal to the sample surface for front surface viewing and the X-ray detector can be positioned at 90° to the incident beam and within the horizontal synchrotron plane. This arrangement is highly preferable because it allows thick samples to be analyzed and minimizes background from inelastically scattered radiation, a consequence of the beam polarization. These components are described briefly below.

Monochromator. This device consists of a diffracting crystal that transmits a single energy X-ray from the wide energy range of the beam striking it. This is useful in allowing control of the excitation energy for the measurement (see below). The energy (wavelength) of the transmitted beam is given by Bragg's law:

$$n\lambda = 2d \sin\theta \tag{1}$$

where n is an integer, λ is the wavelength (Å), d is the interplanar distance (Å), and θ is the angle (radians) of the crystal lattice planes relative to the incoming beam. A typical monochromator crystal is the (111) plane of silicon that has a lattice spacing $2d = 6.2712$ Å. When operated over the angular range of 5 to 45 degrees, this crystal transmits first order X-rays over the energy range of 3 to 23 keV. The energies of these beams correspond to the absorption K-edges of elements with atomic number between 17 and 44 (chlorine to ruthenium) and L-edges for elements with atomic number above 44. The bandwidth (dE/E) is ~10^{-4}, i.e., ~1 eV. The crystal also transmits higher harmonics ($n > 1$) but at reduced intensity (intensity scales as the quadrature sum ratio of the Miller

Schematic for X-ray Microprobe using KB Mirrors
(top view)

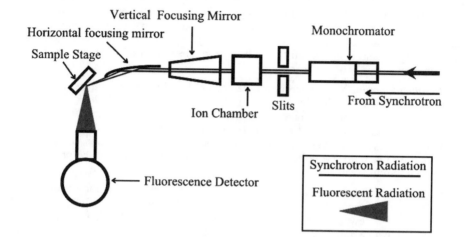

Figure 1. Schematic showing the components of an X-ray microprobe based on Kirkpatrick-Baez microfocusing mirrors.

Figure 2. Photograph of the X-ray microprobe at GeoSoilEnviroCARS Sector 13 (Advanced Photon Source, Argonne National Laboratory).

indices). The higher harmonics are typically rejected by the high-energy cut-off of the microfocusing system (see below). An advantage of the Si (111) reflection is that the second harmonic is forbidden.

Intensity monitor. A beam intensity monitor is used for normalization purposes. In most cases this device is an ion chamber consisting of the two parallel metal plates, one grounded and the other at high voltage, separated by approximately a cm gap residing in an inert gas chamber with thin entrance and exit windows. The X-ray beam traverses this gas path with some probability of creating electron-hole pairs thus generating an electrical current proportional to the beam intensity. The gas is typically a noble gas (helium or argon) and is optimized to result in beam absorption of a few percent for the typical operating energy of the beam.

Microbeam optics. X-ray microbeams can be produced using techniques that rely on collimation, refraction, diffraction, or reflection (e.g., Ice 1996). The most straightforward devices are collimators, apertures that merely allow a small portion of the synchrotron X-ray beam to be transmitted. The devices are achromatic (i.e., lacking energy dispersion) but the vast majority of the usable beam is absorbed so that beam intensities are low. Hard X-ray beams of a few micrometers in size can be produced in this way (e.g., Sutton et al. 1994).

Refractive lens take advantage of the slight differences in X-ray refractive indices of various materials compared to air (Snigirev et al. 1996; Ohishi et al. 2001). Focused beams down to a few micrometers in size have been produced using an aluminum parabolic stack with a gain of 1000 (Lengeler et al. 2001).

Diffraction-based focusing devices include Fresnel zone plate (FZP) multilayer structures. Zone plates operating in the hard X-ray regime have achieved sub-micrometer beams (Yun et al. 1999; Kagoshima et al. 2001; Suzuki et al. 2001) at the expense of achromaticity.

Reflective devices include capillaries, Bragg-Fresnel lens, and mirrors. Capillaries, typically tapered, rely on internal reflection to concentrate photons into small spots defined by the downstream aperture (Bilderback and Thiel 1995; Heald et al. 1997; Dhez et al. 1999; Bilderback and Huang 2001). Sub-micrometer beams have been achieved whereas the small working distances tend to be the main limitation. Bragg-Fresnel lens (Hayakawa et al. 1989; Chevalier et al. 1995; Snigirev et al. 1995; Dhez et al. 1999) are similar to FZPs but operate in reflection mode.

Kirkpatrick-Baez (KB) microfocusing mirrors (Kirkpatrick and Baez 1948; Bonse et al. 1992; Eng et al. 1995, 1998; Yang et al. 1995) have become in wide use. The KB mirrors have the advantages of achromaticity (all energies are focused to the same spot), achievable gains in flux density in excess of 10^5, and long working distances (centimeters) to accommodate an array of ancillary instruments (microscopes, detectors, etc.). A KB mirror system consists of two mirrors, one in the horizontal plane and one in the vertical plane where the mirrors have an elliptical shape to perform point (source)-to-point (sample) focusing. The elliptical shape can be achieved statically (Thompson et al. 1992) or dynamically (Eng et al. 1998). One such system used at the APS (Eng et al. 1998) focuses the APS undulator X-ray beam from 350 μm down to 1 μm (flux density gain of about 10^5). The mirrors are highly polished, single crystal silicon coated with several hundred Å of Rh. The flat surfaces are dynamically bent to elliptical shapes using a mechanical bender. The "double-bounce" focusing system provides excellent harmonic rejection capabilities for μXAFS applications.

Sample stage. A sample manipulation stage is required to position samples and the region of interest in the X-ray beam and also for moving the sample along x and y coordinates to produce 2D elemental maps. Motion in three orthogonal axes are required to bring the analysis volume to the single point in space defined by the intersection of the X-ray beam and the focal plane of the visible microscope. Positioning resolution of ~0.1 μm or better is usually required.

X-ray fluorescence detectors. Various complementary detectors are useful for collecting X-ray fluorescence spectra. These detectors vary in terms of count rate capability and energy resolution and there is usually a trade-off between these two characteristics. That is, one can achieve highest count rate capabilities with low energy resolution, and vice versus. The optimum detector for a particular experiment depends on the requirements of the experiment. In most studies of earth and environmental samples, chemical compositions tend to be complex and a necessary detector requirement is sufficient energy resolution to differentiate the fluorescence peak of interest from all the other peaks in the spectrum. Normally, this requires a resolution near 100 eV. For this reason, the solid state, energy-dispersive (EDS) detector is the workhorse in microfluorescence analysis. The EDS detector typically consists of a biased, single crystal of Ge or Li-drifted Si. X-rays traversing this crystal produce electron-hole pairs proportional to the X-ray energy and the associated current pulse is amplified and stored in a multi-channel analyzer (MCA). The XRF spectrum is the histogrammed pulse-height distribution of these pulses. Achievable count-rate depends on the selected electronic shaping time (time constant of the pulse height determination circuitry). Longer shaping times result in higher energy resolution but greater dead time and therefore lower achievable maximum count rate. The rule of thumb is to use the shortest shaping time needed to achieve the required energy resolution. A common variant of the solid-state EDS detector is the multi-element array that consists of multiple (up to 100), detecting crystals packed in a single detector snout and each with its own dedicated electronics chain. A common commercial array detector is a germanium array produced by Canberra Industries, Inc. (Meriden, CT), coupled with digital signal processing electronics, such as those offered by X-ray Instrumentation Associates (Newark, CA).

Another useful fluorescence detector is the wavelength dispersive spectrometer. Commercial spectrometers are available, such as the WDX-600 by Oxford Instruments (Fremont, CA). This scanning spectrometer employs diffraction in a Johansson geometry with the sample, bent crystal analyzer, and detector residing on a 210 mm Rowland circle. The detected wavelength (energy) is defined by the crystal lattice spacing and the angle the analyzing crystal presents to the sample. The WDX is equipped with two LiF cuts (200 and 220) with a 420 also available. The combination of these LiF crystals provides a usable energy range from 1 to 17 keV. The Rowland geometry is maintained mechanically with the analyzer driven along a linear rail with coupled angular adjustment and the detectors are moved on an arc to intercept the diffracted beam. Tandem proportional counters are used, a P-10 flow counter in front followed by a sealed 2 atm Xe counter. The energy resolution at 6 keV is ~10 eV, about a factor of 10 better than that of a solid-state detector. There are two disadvantages of the WDS. First, the collection solid angle of the WDS is substantially smaller than that of a solid-state array detector, thus its sensitivity is lower. Second, only a single energy is detected at a given time, thus applications requiring the detection of multiple elements are hampered by the time required to scan the device to the desired energies.

Optical viewing microscope. A conventional optical microscope is needed to view the sample/beam position relationship. A handy instrument is the Optem 7 motorized transmission tube equipped with a long working distance objective (e.g., Mitutoyo M Plan Apo 10×, 33.5 mm working distance) and color CCD camera (e.g., Panasonic GP-

KR222). This system produces a field of view of ~1 mm with 7× motorized zoom and focusing capability via a joystick-controller.

Auxiliary apparatus. A useful device for performing analyses at elevated temperature is the Linkam 1500 heating stage. This device was designed for diffraction experiments at Daresbury but has been used for fluid inclusion analyses up to 500 °C (Mavrogenes et al. 2002). It is also occasionally desirable to perform analyses in helium or vacuum to reduce absorption of low energy fluorescent X-rays and allow analyses on lighter elements. The approach to implement this environment varies from a gas tight enclosure encompassing the entire microprobe instrument to a gas-filled bag enclosing only the beam path from the sample to the detector.

Microprobe platform. Typically the entire microprobe apparatus is installed on an optical breadboard (e.g., Newport RG series) driven by a multi-axis lift table. This configuration allows the entire internally aligned instrument to be positioned precisely (better than 10 μm resolution) as a unit with respect to the incoming X-ray beam.

Instrument control. The main requirement for a control system is the ability to acquire X-ray fluorescence spectra (detector acquisition on, off, save, clear) in coordination with a scanning routine manipulating the sample stage and monochromator energy.

Sample preparation

Samples in any form (solid, liquid, gas) can be analyzed and, with hard X-ray microprobes, the technique is particularly well suited for studies of wet materials because no vacuum chamber is required. In practice, most samples consist of individual particles mounted on thin plastic films or conventional thin sections with pure silica as the substrate.

The third dimension in the analysis (depth) is significant because the synchrotron X-rays penetrate deeply in most materials. This has several ramifications. First, buried volumes, such as fluid inclusions, can be analyzed. Second, one needs to know the sample thickness to correct for absorption effects (see quantification discussion below). Third, it is important to minimize the mounting substrate because this material is a source of scattered radiation that contributes to the spectral background. Fourth, the sampling depth depends on the energy of the fluorescence X-ray and therefore is element dependent. This fact can be used to advantage by selecting a sample thickness to achieve optimum sensitivity for the suite of elements of interest.

Choice of excitation energy

Microfluorescence analyses can be made using white synchrotron radiation as the excitation source, but this is rarely done for several reasons. First, the scattered photons produce a huge background level at all energies in the spectrum resulting in reduced elemental sensitivity. Second, the white beam has high power and can cause radiation damage problems whereas many of the photons in the beam have energies that are not useful for exciting the elements of interest. For these reasons, monochromatic beams are normally utilized as the excitation source.

Care must be taken in selecting the optimum excitation energy for a particular experiment. Obviously, the energy should be high enough to excite the absorption edges of interest but "escape" peaks from the detector are common obstacles. This is particularly true in using Ge detectors. An escape peak is a detector artifact that is produced by processes whereby an incoming fluorescent X-ray ionizes an atom within the detector crystal and a secondary X-ray is produced. This secondary X-ray has some probability of escaping the detector crystal without being absorbed. As a result, the energy deposited in the crystal is artificially low by the energy of this secondary X-ray. For example, an incoming Sr K_α X-ray (14.1 keV) ionizing a Ge atom (9.9 keV K_α fluorescence energy) will produce a spectral artifact peak at 4.2 keV. This peak can interfere with primary fluorescence peaks from Ca (K_β) and Cs (L_α). Earth and environmental materials are typically chemically complex, producing robust fluorescence spectra, and each fluorescent peak whose energy is greater than the Ge K absorption edge has the potential for producing an escape peak.

Even more problematic in practice is the fact that the Compton scatter and associated escape features tend to be the dominant features in an SXRF spectra. The Compton peak is broad with a maximum occurring at lower energy than the elastic peak where the so-called "Compton shift" is energy dependent and given by:

$$\Delta E/E_0 = \gamma/(1-\gamma) \tag{2}$$

where $\gamma = E_0(1-\cos\theta)/m_0 c^2$, E_0 = incident energy, θ = observation angle relative to incoming beam, and $m_0 c^2$ = rest mass of the electron = 0.511 MeV. Thus, the Compton energy shift increases with increasing energy and has a maximum into 180° (backscatter). For example, the energy shift of 20 keV photons (a typical XRF excitation energy) scattered into 90° is –0.815 keV. The Compton peak has a maximum at this energy but the peak itself is broadened substantially due to the motion of the electrons. It produces a broad escape feature that interferes with a wide range of the spectrum at lower energy. However, unlike the escape peaks from primary fluorescence peaks, the Compton scatter peak and its escape move with the excitation energy. Consequently, it is often necessary to adjust the excitation energy to optimize the position of the Compton peak and its escape to minimize interference with fluorescence peaks of interest. The excitation energy should be high enough so that high-energy peaks do not reside directly on intense portions of the Compton scatter peak. Of course, this flexibility is not available in µXAFS applications.

Detector optimization

As mentioned above, the optimum detector is typically the one that provides the highest count rate at the minimum required resolution. Thus, for analysis of a single element that is the highest concentration detectable element in the sample, a low energy resolution, high solid angle detector, such as a Lytle detector, is optimum. At the other end of the sensitivity scale, analysis of a trace element in a chemically complex material usually requires a multi-element array solid-state detector. In cases where peak overlap requires higher resolution than offered by the solid state detector, such as analysis of trace Au in the presence of major As or trace Cs in the presence of major Ti, a high resolution spectrometer is optimum. Ideally, all these detector types could be installed permanently in the apparatus to allow simultaneous collection. Space limitations are an obstacle to this approach but another important consideration is the desire to have each detector at the minimum Compton scattering point (in plane at 90° to beam) for highest sensitivity.

Data collection

The basic µXRF measurement is a spot analysis where the beam is positioned on the area of interest and a fluorescence spectrum is accumulated. Efforts must be made to optimize the detector configuration for a particular measurement. In particular, with a synchrotron excitation source, it is common for a solid-state detector to be easily saturated by fluorescence from major elements in the sample as well as from scattered radiation. Consequently, the count rate in the detector needs to be optimized by varying one or more of the following: incident beam intensity, detector collection solid angle (distance and/or apertures), and fluorescence beam filtering. In practice, one normally wants to work at dead times around 30% to minimize errors due to dead time corrections and avoid producing pileup peaks in the spectrum (a pileup peak occurs when two X-rays enter the detector within the sampling time of the electronics producing an current pulse equal to the sum of the two individual photons).

To achieve this level of dead time, one should first attempt to reduce or eliminate unwanted signals because this approach leaves the intensities of the peaks of interest unaffected. There are two principal possibilities: (1) add filters to the detector to reduce the intensities of low energy peaks that are typically from major elements. Kapton film works well for calcium K fluorescence and aluminum for iron; (2) reduce the incident beam energy to be below the absorption edges of uninteresting elements producing intense high-energy peaks. Other adjustments include varying the tune of the monochromator second crystal and adjusting the sample-detector distance, effectively varying the detector collection solid angle. Both of these approaches modify the intensities of all peaks. The required measurement dwell time depends on the precision desired. For 1% precision, one needs to count long enough to accumulate ~10,000 counts in the peak of interest.

µXAFS spectra can be obtained by scanning the monochromator energy through the absorption edge of interest. Normally such measurements are made in fluorescence mode with the region of interest (ROI) saved for the element's alpha fluorescence peak, either K or L. As for the XRF measurement, it is important to adjust the detector dead time conditions with the monochromator energy above the absorption edge, i.e., ~energy of maximum fluorescence intensity).

Scanning the sample on a grid and recording a spectrum at each position can produce a two-dimensional map. The solid-state detector is very useful in this application because of its ability to collect peaks for multiple elements simultaneously, thus a single scan produces maps for multiple elements that are exactly registered in 2-D space. A disadvantage is the significant time overhead (~second) in reading out the detector. This detector dead time limits the scanning speed to a couple of seconds per point. Such maps can be produced by saving the full spectrum at each point or the total counts in "regions-of-interest" defined about peaks of interest in the XRF spectrum. The ROI approach is most commonly used and works well for well-defined peaks, i.e., intense peaks. The disadvantage of the full spectrum approach is the need to perform subsequent peak fitting on a very large number of SXRF spectra. The ROI image data allow spectral interferences to be handled by array manipulation. For example, an ROI image having contributions from peaks of two different elements may be deconvoluted by using other ROIs for those elements and knowledge of peak ratios, such as known alpha/beta intensity ratios.

Two-dimensional maps of oxidation state can be obtained by collecting consecutive elemental maps with the monochromator energy set to excite different proportions of the chemical species. This approach takes advantage of the fact that the absorption edge energy for a particular element increases with increasing valence. The shifts are of the order of a few electron volts per unit charge in valence. Under certain circumstances, pre-edge features of first row transition elements can also be used as diagnostic indicators of valence states.

Fluorescence tomography, described in a separate section of this chapter, is a variant of the two-dimensional map where one of the scanned dimensions is sample angle. The resulting data can be used to produce elemental maps of planes through solid objects without slicing.

Data analysis

Element concentrations from XRF spectra. The μXRF spectrum consists of various peaks due to fluorescence, scattering, and detector artifacts such as escape and pileup peaks. The basic procedure for obtaining quantitative compositional information from XRF spectra is to fit a background spectrum, subtract the background, fit the remaining peaks to obtain net peak areas, and use these net areas to compute concentrations, where the concentration calculations include information on the analysis conditions and physical state and major element composition of the sample.

Background fitting. The background in an XRF spectrum derives principally from elastic and inelastic (Compton) scattering of the incident X-ray beam. The elastic peak is at the incident beam energy with a width equal to the detector resolution at that energy. In fact, the energy of this peak can be used to determine the incident beam energy with a precision of ~10 eV. This Compton profile is asymmetric and characterized by a long tail to lower energies. The escape peaks associated with the scatter peaks have the same shape and occur at energies that are 9.9 keV lower in energy (for Ge solid state detectors). Fitting of the scatter peaks themselves is typically not attempted because one attempts to adjust the incident beam energy to keep fluorescence peaks of interest away from the intense portions of these features. In this case, the background under the peaks of interest is a smooth, slowly varying function of energy and a simple spline or polynomial fit to selected portions of the spectrum can suffice. However, the presence of the scatter peaks and their escapes often prevents the entire spectrum from being fit *in toto* and fitting various segments of the spectrum is usually more successful.

Peak fitting. Once the background has been fit and subtracted, the next step is to fit the remaining peaks. These peaks include fluorescence, escape and pileup peaks. MCA software with fluorescence peak libraries is very useful in identifying the various peaks in a complex spectrum. Peak overlaps in energy-dispersive XRF spectra are common and known alpha/beta ratios can be used as constraints to deconvolute these overlaps during the fitting process (Fig. 3). Gaussian peak shapes are typically appropriate.

Concentration calculations. In a geometry where the incident angle equals the detection angle (θ), the fluorescence intensity, I, measured by a detector for a given fluorescence transition (e.g., Fe K$_\alpha$) from a sample of arbitrary thickness, S, excited by a monochromatic beam of intensity, I_0, is given by

$$= I_0 \gamma \phi w C \Omega \tau \left[\frac{1 - e^{-(\mu_0 + \mu_f)R}}{\mu_0 + \mu_f} \right] \tag{3}$$

where γ is the detector efficiency at the fluorescence energy, ϕ is the attenuation factor for filters on the fluorescence X-ray beam (e.g., air path, detector windows), w is the fluorescence yield, C is the element concentration, Ω is the fraction of a sphere intercepted by the detector, τ is the photoelectric absorption coefficient, R is the beam path length ($= S/\sin\theta$) and μ_0, and μ_f, are the sample attenuation coefficients for the incoming beam and fluorescence beam, respectively. The detector filter attenuation factor ϕ is the product of the attenuations of the individual filters,

$$\phi = \prod_i e^{-\mu_i t_i} \tag{4}$$

where μ_i are the attenuation coefficient for the ith fluorescence beam filter and t_i is the filter thickness.

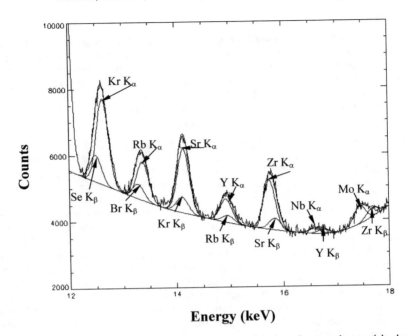

Energy (keV)

Figure 3. High energy region of the μXRF spectrum for a 3 μm interplanetary dust particle showing peaks due to trace elements at the ~10 ag level. Shown are the raw spectrum, background fit, individual fitted peaks (labeled) and the overall fitted spectrum. The K fluorescence spectra for elements with atomic number near 40 are complicated by the overlap between the K_α peak of element Z and the K_β peak of element Z-2. The odd-even abundance effect in the chondritic (solar) composition is apparent. Krypton derives from air in the analysis environment.

An example of this calculation is given in Table 2. In this example, we calculate the measured Sr K_α fluorescence intensity from a 100 μm thick slice of calcite containing 100 ppm Sr. The sample is excited with 20 keV X-rays in a 45°-in/45°-out geometry. The detector is a single, 1 cm diameter Ge crystal 10 cm from the sample. The computed ratio of I/I_0 is 3×10^{-8}. For a typical incident flux from a microfocused undulator beam of $\sim 3 \times 10^{11}$ photons/sec, this translates into a measured intensity of about 1000 counts per second.

Although it is possible in principle to theoretically calculate the element concentrations on the basis of this equation (taking into account the incident beam intensity, instrument geometry, absorption effects, detector efficiencies, etc.) (see e.g., Bos and Vrielink 1998; Jenkins 1999; Hansteen et al. 2000), this is almost never done because of the inherent uncertainties in the numerous parameters. Instead various forms of standardization are utilized in an attempt to reduce the number of dependent variables. There are two general approaches: (1) use a standard for each element of interest and (2) use an internal standard element.

In the "standard" approach, one collects an XRF spectrum for a standard material with the element of interest in known concentration C_{stand}. The National Institute of Standards and Technology (NIST) offers several materials useful for this purpose including the SRM 1832/1833 thin films. Then, to a first approximation, the concentration in the unknown C_{unk} is determined by the relationship,

Table 2. Fluorescence intensity example (refer to Eqn. 3).

Symbol	Parameter	Units	Value
E_0	Incident energy	keV	20
E_f	Fluorescence energy	keV	14.1
θ	Incident and detection angles	deg	45
z	Sample to detector distance	cm	10
γ	Detector efficiency		0.98 [†]
ϕ	Detector filter attenuation		0.96 [‡]
w	Fluorescence yield		0.7 [§]
C	Strontium concentration	wt fraction	1.00×10^{-4}
Ω	Detector subtension		0.000625
τ	Photoelectric cross section	1/cm	180.0253
μ_0	Attenuation coefficient for incident energy	1/cm	68.0752
μ_f	Attenuation coefficient for fluorescence energy	1/cm	182.5727
S	Sample thickness	cm	0.01
I/I_0	Ratio of measured fluorescence to incident intensity		2.87×10^{-8}

[†] Canberra LEGe
[‡] 10 cm air + 11 mil Be
[§] Evans 1955

$$C_{unk} = C_{stand} \left(I_{unk} / I_{stand} \right) \tag{5}$$

where I_{unk} and I_{stand} are the net peak areas per second for the unknown and standard, respectively. However, this assumes the standard and unknown are identical in terms of major element composition, thickness, and density, which is rarely the case. So this equation must be corrected by the difference in element sensitivities from these differences in matrix.

$$C_{unk} = C_{stand} \left(I_{unk} / I_{stand} \right) \left(S_{stand} / S_{unk} \right) \tag{6}$$

where S_{stand} / S_{unk} is the relative sensitivity (counts/second/mg kg^{-1}) for the standard compared to the unknown. Values for S_{stand} / S_{unk} can be obtained theoretically with good reliability because the interactions between photons and matter are well understood. A program used by us for this purpose is NRLXRF (Criss et al. 1978), a package developed by the Naval Research Laboratory for conventional XRF. This program works particularly well for homogeneous slab geometries (appropriate for many earth and environmental science applications, e.g., thin sections) and includes corrections for secondary fluorescence, thickness, major element composition, density, and analysis angles. Its prediction mode generates relative fluorescence intensity values in units of photons per second per square steradian (X-ray flux produced by a 1 steradian incident beam and detected by a 1 steradian detector).

The NRLXRF program makes theoretical predictions of fluorescence intensities based on formulae from various published sources. Primary and secondary intensity contributions use the formulae of Gillam and Heal (1952). Wavelengths for absorption edges and emission lines derive from Bearden (1964, 1967) and Bearden and Burr (1967). Photoelectric mass absorption coefficients are calculated from the parameters of

McMaster et al. (1969). Fluorescence yields (including Auger and Coster-Kronig transitions) are from Bambynek et al. (1972). Sample thickness corrections use the formula of Compton (1929).

The main complication in the standards approach for μXRF analysis is that trace element standards that are homogeneous at the micrometer scale are uncommon. For this reason, the internal element reference approach has become more widely utilized. In this method, the concentration of one of the abundant elements in the XRF spectrum is determined independently, e.g., by stoichiometry or electron microprobe, and the sensitivity for this element can be computed directly. The sensitivities for all other elements are then computed theoretically relative to this internal reference. Again, we find the NRLXRF program useful for this purpose after making a few modifications, to include detector filtering, for example. The input parameters for this type of computation are:

- Incident X-ray energy
- Major element composition
- Thickness (g/cm^2)
- Absorbers between the sample and detector (thickness and composition)
- Peaks to be used for concentration calculations and the net areas of those peaks
- Reference element peak and the concentration of that element.

Figure 4 is an example of the relative K_α fluorescence intensities obtained using the prediction mode of NRLXRF. In this example, we consider a 10 μm thick quartz sample containing trace elements with atomic number between 15 and 44 (P to Nb). The assumed analytical conditions were excitation energy of 20 keV and detector filtering consisting of 10 cm helium and 1 mil Be (detector window). The predicted relative sensitivity is a smooth function of atomic number (with the exception of a small dip associated with the Si absorption edge). With functions of this type, one can compute the concentrations of all elements if the concentration of one element is known independently and can be used as a normalizing point.

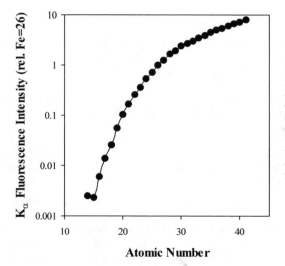

Figure 4. Predicted fluorescence intensity relative to Fe detected from a 10 μm thick quartz sample containing trace elements with atomic numbers between 13 and 44 (P to Nb) at 1 ppm concentration. The excitation energy is 20 keV with sample-to-detector filtering consisting of 10 cm of helium and 1 mil of Be (detector window).

Figure 5 shows how the fluorescence intensity varies with sample thickness assuming uniform depth distribution. In the "thin/negligible absorption" regime, the fluorescence intensity increases linearly with thickness. In the "infinitely thick" regime, the fluorescence intensity saturates. The highest sensitivity will be achieved with a thickness near where absorption becomes significant. At greater thickness, more scatter background is contributed compared to additional fluorescence. Figure 6 shows the effect of fluorescence intensity on incident beam energy. The highest sensitivity is achieved with the incident beam energy just above the absorption edge of interest.

μXAFS analysis. μXAFS data analysis is identical to large sample XAFS data analysis (see chapters in this volume by Waychunas and Manceau et al.). This section describes aspects of the measurements that are unique to microbeam applications. Perhaps the most important issue is spatial stability between the sample and X-ray beam. Such instabilities can be caused by instrumentation vibration or beam motion. In the limits where the sample is small compared to the beam or large and homogeneous compared to the beam, positional drifts at the beam size scale are manageable. In the former case, the beam fully illuminates the sample at all times whereas, in the latter case, the beam excites the

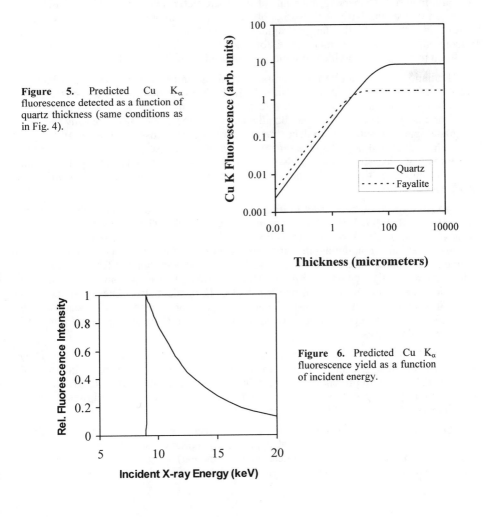

Figure 5. Predicted Cu K$_\alpha$ fluorescence detected as a function of quartz thickness (same conditions as in Fig. 4).

Figure 6. Predicted Cu K$_\alpha$ fluorescence yield as a function of incident energy.

same volume of material. In both cases, the fluorescence intensity is constant. However, complications arise when the sample and beam are comparable in size. Consider the hypothetical case of a two-dimensional Gaussian beam with a full width at half-maximum (FWHM) equal to the diameter of a uniform disk (the sample). Compared to the centered situation, a lateral shift of 20% FWHM leads to a decrease in the fluorescence intensity of 8% because part of the beam misses the sample. Such relative motion will lead to significant fluorescence oscillation during the XAFS spectrum collection. For a 3 μm FWHM X-ray beam, this motion corresponds to 0.6 μm. To maintain a stable fluorescence intensity at the 1% level (minimum requirement for μEXAFS) requires stability at the 5% FWHM level, or 0.1 μm in this example. Thus, it is important prior to a μXAFS measurement to verify that the fluctuation in "above-edge" fluorescence intensity is statistical in nature, i.e., follows a square root of the intensity behavior.

Another important consideration is the desire to obtain μXAFS spectra on individual components in a complex material such as a soil. The spatial resolution required depends on the scale of heterogeneity in the sample. Measurements on aggregates substantially larger than this heterogeneity scale can produce XAFS spectra no different from that of the bulk material. Dispersing materials at the micrometer level can be challenging. One approach that works fairly well is to apply a very small amount of the sample to a pure silica slide, "tap off" the larger fragments, then search the electrostatically-adhered microparticles under the microscope for analysis candidates. A micromanipulator is useful in transferring individual fragments to individual analysis holders.

Radiation effects can initiate chemistry within a sample and can potentially be more problematic in μXAFS analyses because of the high power density in the focused beam. Thus, it is important to test for such effects by collecting a time sequence of spectra to look for changes. These effects can be alleviated to some extent by defocusing the beam (degrade spatial resolution), reduce the incident beam intensity either by filtering or monochromator detuning (degrade sensitivity), or in some cases, eliminating contact with air. Experience has shown the following elements to be radiation sensitivity, undergoing redox changes, particularly in the presence of water: $Cr^{6+} \rightarrow Cr^{3+}$, $Mn^{2+} \rightarrow Mn^{4+}$, and $As^{3+} \rightarrow As^{5+}$.

μXAFS spectra can be used to quantify the oxidation state of multivalent elements (e.g., Bajt et al. 1993, 1995; Delaney et al. 1996a,b, 1998, 1999; Dyar et al. 1998, 2001; Mosbah et al. 1999). First row transition metals have been the focus of recent work because they are ubiquitous in earth and environmental materials. Algorithms for extracting oxidation state rely on energy and/or intensity variations of features in the pre-edge and near-edge region of XAFS spectra (XANES). The well-known example is chromium that can occur either in the trivalent or hexavalent state. These cations can be easily distinguished by the intensity of a pre-edge peak, prominent in Cr^{6+} spectra and the intensity of this peak can be used to quantify the C^{6+}/total Cr ratio (e.g., Bajt et al. 1993; Peterson et al. 1997).

Vanadium shows a similar pre-edge peak to that of chromium (Fig. 7) and obeys similar systematics, i.e., increasing intensity and energy with increasing oxidation state (Wong et al. 1984). In principle, it is possible to infer oxygen fugacity of igneous systems by calibrating the pre-edge peak systematics with samples produced under known conditions (Schreiber 1987; Delaney et al. 2000).

Oxidation state maps can be produced for samples containing an element in multiple states (Sutton et al. 1995). The basic approach in oxidation state mapping is to make multiple, 2-dimensional X-ray fluorescence (elemental) maps of the specimen using monochromatic radiation where the monochromatic energy for each map is chosen to

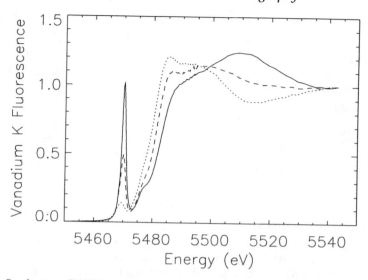

Figure 7. Synchrotron μXANES spectra of V for glasses produced at different $\log(fO_2) = 0$ (solid), −6.8 (dash), −9 (dot). The effective oxidation states determined by optical spectrometry (H. Schreiber, VMI) were 4.7, 4.0, and 3.2, respectively. The systematic change in energy and intensity of the pre-edge peak near 5470 eV is the basis of the oxidation state determination.

preferentially excite particular oxidation state components of the element of interest. The distributions of individual oxidation states are then determined by deconvolution of these maps. This procedure assumes pure samples of the multiple species are available for measurement as standards and has the following steps:

1. Measure XANES spectra for pure standards of each of the n components.

2. Define n monochromator energies, E_i at which these components are most distinguishable, e.g., white line energies, plus an above-edge energy, E_0 and a below edge energy, E_b

3. Normalize each standard XANES spectrum using the intensities at E_b and E_0.

4. Define the $n \times n$ matrix, **S** to contain the normalized intensities for the standards at each of the E_i.

5. Measure n complete 2-dimensional (*columns* × *rows*) fluorescence intensity maps M_i, one map obtained using each E_i.

6. Measure 2D maps at the energies, E_b and E_0

7. Normalize the intensities in each map M_i using the maps collected at E_b and E_0.

8. Define a 3-dimensional array Φ of dimensions $n \times$ *columns* × *rows* containing these normalized fluorescence intensities.

9. At each map pixel (i.e., discrete value of *column* and *row*),

   • define a 1-dimensional matrix **F** containing the normalized intensities in the n maps

$$\mathbf{F} = \Phi(*,column,row) \qquad (7)$$

   • calculate the 1-dimensional component fraction matrix **X** by

$$\mathbf{X} = \mathbf{F}\,\mathbf{S}^{-1} \qquad (8)$$

• define a 3-dimensional, valence state array V of dimensions $n \times columns \times rows$ as

$$V(i, column, row) = \mathbf{X}(i) \qquad \text{for } 1 \leq i \leq n \qquad (9)$$

The $*$ in the equation of $\mathbf{F}$ refers to all values of that index. The distribution of a particular valence state (e.g., component i, where $1 \leq i \leq n$) is then given by the two-dimensional array $V(i, *, *)$.

Current capabilities

Hard X-ray microprobes utilizing KB microfocusing mirrors currently have μXRF detection limits near 1 mg kg^{-1} in 30 picogram samples which translates into ~100 ag (= 10^{-16}g) or ~10^6 atoms. For μXANES, a concentration of ~10 mg kg^{-1} is required and 100-1000 mg kg^{-1} for μEXAFS.

MICROTOMOGRAPHY

Microtomography facilities in the US

There are three, high energy microtomography instruments at the Advanced Photon Source: GSECARS (13-BM), SRI-CAT (2-BM), and PNC-CAT (20-BM). At the NSLS, beamlines X27A and X2B have dedicated computed microtomography (CMT) stations. At the ALS, a new superbend tomography station (8.3.2) is scheduled to come online in 2003. The CAMD port 7A is a CMT station.

Basic principles

Synchrotron based computed microtomography is an extension of conventional medical CAT scans to high spatial resolution. The higher spatial resolution is achievable because of the high collimation and intensity of the X-ray beam. There are four basic types of microtomography: transmission, edge, fluorescence and diffraction. In all cases, the approach is to measure projections of some parameter through the object, collect these projections viewing from different angles, and then reconstruct the dataset to obtain the internal distribution of the parameter.

In transmission CT, a "flood field" approach is used and the parameter is the linear attenuation coefficient σ that describes the attenuation of a beam of intensity I_0 to an intensity I over a linear distance x where,

$$I/I_0 = e^{-\sigma x} \qquad (10)$$

The transmission CT measurement is actually a line integral along l through the sample which is spatially heterogeneous in σ (Kinney et al. 1994), so that

$$\ln(I/I_0) = \int \sigma(x, y, z) \, dl \qquad (11)$$

Two-dimensional maps of I/I_0 at various sample viewing angles are used to solve this equation for $\sigma(x,y,z)$.

Element specificity can be achieved either by the edge or fluorescence methods. In the former, two tomograms are collected, one with the incident energy above the absorption edge of the element of interest and one below the edge. These two images are then subtracted and the resulting image shows the distribution of that element only. This approach is applicable to only one element per tomogram pair and the concentration normally must be in the weight percent range. In the fluorescence approach, the parameter collected is the fluorescence from the element of interest and a "pencil-beam" approach must be taken because the energy-dispersive detectors lack position sensitivity.

Advantages include higher elemental sensitivity and the ability to collect images for multiple elements simultaneously. The disadvantages are that it is more time consuming and self-absorption effects complicate data processing.

In diffraction CT, the measured parameter is a diffraction feature from the sample and information on strain and texture can be obtained in this way.

Instrumentation

This section describes the tomography instrumentation at GSECARS (APS) as representative of the typical setup. For transmission CT, a bending magnet source is typically used to provide a wide fan for flood field applications. The APS bending magnet source has a critical energy of 20 keV, and thus provides high flux at photon energies up to 100 keV, making it well suited to imaging a wide range of earth materials up to several cm in size. The GSECARS bending magnet beamline provides a 2.5 mrad fan of radiation, permitting, in principle, studies of objects up to 125 mm in diameter in the experimental station 50 m from the source. The vertical beam size is about 5 mm maximum, due to the narrow opening angle of the synchrotron radiation. The monochromator currently in use is a narrow gap, Si(111) double crystal monochromator providing photons out to 90 keV with beam sizes up to 18mm wide and 4mm high. The APS undulator source provides an extremely bright X-ray beam, more than 1000 times brighter than the bending magnet. The undulator beam size is less than 1mm vertical by 3mm horizontal in the experimental station 50m from the source. The APS undulator has a first harmonic energy tunable from approximately 3.5keV to 13 keV, and provides higher harmonics to well beyond 50 keV. The undulator brightness is generally not required for most absorption CT experiments, but is very useful for producing a focused monochromatic beam for fluorescence CT experiments.

The transmitted X-rays are converted to visible light with a single crystal YAG scintillator (Fig. 8). The scintillator is imaged with either a microscope objective (5X to 20X) or a zoom lens. The field of view can be adjusted between 3mm and 50mm. The image is projected onto a CCD detector. Two different Princeton Instruments (Roper Scientific, San Diego, CA) CCD detectors are used: 1) Pentamax camera: Kodak

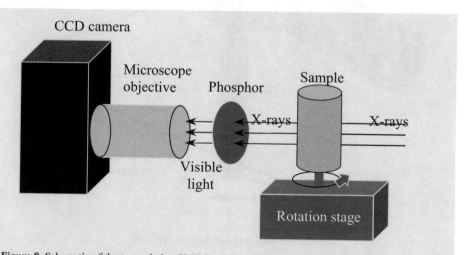

Figure 8. Schematic of the transmission CMT flood field apparatus at GSECARS Sector 13 at the APS (adapted from Rivers et al. 1999).

~1300×1200 chip, 7 μm pixels, 1 and 5 MHz A/D, 12 bits; 2) ST-138 camera: EEV ~1200×1100 chip, 25 μm pixels, 430 kHz and 1 MHz A/D, 16 and 14 bits.

For fluorescence CT on the undulator, the conventional microprobe apparatus is used with the addition of a rotation stage about the vertical axis (Fig. 9). Fluorescence X-rays are collected with a 16-element Ge array.

Sample preparation and experiment setup

An attractive aspect of microtomography is the nearly complete lack of need for sample preparation. Virtually any object can be imaged. The main considerations are in optimizing the sample size (attenuation/absorption), incident energy and detector for the particular application. In some cases, these cannot be freely adjusted. For example, precious samples cannot generally be physically modified, the absorption edge used in edge tomography dictates the energy, and a fluorescence experiment usually requires the energy dispersive array. In transmission tomography on samples of adjustable size (e.g., imaging of microfractures in rock cores), the setup can be optimized to produce highest resolution and contrast.

As an example, consider an experiment designed to image the pores in a sandstone with a mean grain size of 300 μm. A specimen of about 3 mm diameter is needed to

Fluorescence Microtomography Apparatus

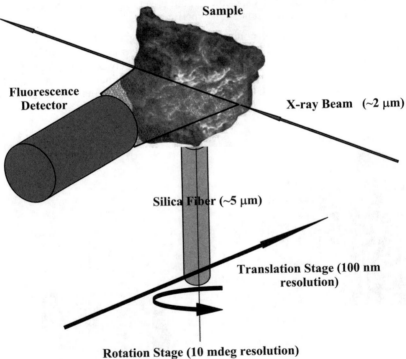

Figure 9. Schematic of the fluorescence CMT apparatus at GSECARS Sector 13 at the APS.

have a sample with about 100 grains in a cross section. At this size, X-rays of energy about 30 keV should be used where the sample thickness is equal to the 1/e absorption length (Fig. 10). The detector objective and extension tube length should be selected to achieve a field of view of ~5mm on the CCD. (Contrast can be enhanced by backfilling air filled cavities with solutions containing a high concentration of a high atomic number element, e.g., CsCl.) If one wants to image the distribution of Sr in this sample using edge tomography, the energy is fixed to be near 16.1 keV (Sr K absorption edge) and Figure 10 indicates a sample size slightly less than 1 mm is optimum. On the other hand, the Sr content is likely to be low and fluorescence tomography may be more successful. In this case, the specimen size is dictated by the need to have Sr K$_\alpha$ fluorescence X-rays (14.1 keV) traverse the sample without major absorption. Figure 10 shows the absorption length to be about 0.5 mm which means the sample should be no more than a couple of hundred μm in diameter to minimize absorption corrections. A good excitation energy to use for this analysis is around 20 keV, high enough above the Sr K fluorescence peak to minimize Compton scattered background (see microprobe discussion above).

Data collection

The GSECARS transmission CT data collection software uses a layered approach, in which each layer performs a specific function well. The lowest layer is the Princeton Instruments WinView program, which collects a single data frame, and allows the user to specify the exposure time, binning factors, region-of-interest and other camera related parameters. The next level is a Visual Basic program called TOMO, which collects a single CT data set. This program rotates the object and translates it out of the beam to collect flat field images. Data collection is typically done from 0 to 180 degrees in 2 passes, where first the 0, 1, 2, ...degree projections are collected, and then

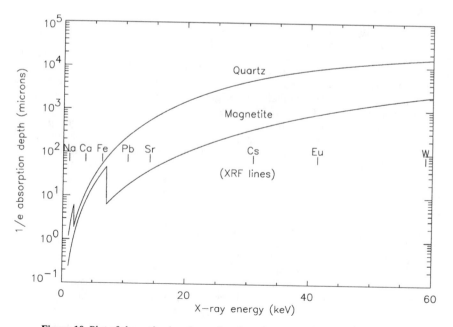

Figure 10. Plot of absorption length as a function of energy for quartz and magnetite.

the 0.5, 1.5, ... projections are collected. We find that this permits useful data to be collected even if something goes wrong near the end of the run, and it also permits detecting unwanted sample motions since closely spaced projections are collected far apart in time. TOMO collects white field images every N projections, where N is typically 50. The exposure times are typically 1-3 seconds on the bending magnet beamline with total data collection times in the range of 20 to 40 minutes for 360 projections. The final software layer can consist of an IDL or Visual Basic program that automates a more complex experiment. This is typically done to collect multiple vertical sections when imaging objects that are taller than the beam, to change energy or other parameters between data sets.

The system is aligned by imaging a pin that is placed near the edge of the field of view. The pin is imaged at 0, 90, 180 and 270 degree rotations. The 0 and 180 degree views allow one to measure and correct the tilt of the rotation axis with respect to the incident X-ray beam. The 90 and 270 degree views allow one to measure and correct the tilt of the rotation axis with respect to the CCD columns.

Collection of fluorescence CT data on the undulator beamline is analogous to making a 2D elemental map with the energy dispersive array detector where the two axes are translation and rotation. The translation step size is chosen to be comparable to the beam size (typically 5 μm), the translation length must reach air on both sides of the sample at all angles (typically 0.5 mm) and the rotation step size is then defined to be "π/number of translation steps" (typically 2°). Count times of 1 second are common but, because the energy dispersive detector system readout has ~2 second overhead, a single slice dataset can take several hours to collect, making 3D imaging impractical. Regions of interest are defined in the XRF spectrum for each element of interest and the total counts in these regions are saved at each point thereby producing simultaneously tomograms for multiple elements.

Data processing

The processing of transmission CT data consists of several steps: 1) preprocessing, 2) sinogram creation, and 3) reconstruction. The first preprocessing step is dark current subtraction. Exposure times are so short that actual dark current is not measurable. However, there is a constant digitization offset (~50 counts for the Pentamax detector) that is subtracted from each pixel in both the flat field and data frames. The next step is zinger removal. Zingers are anomalously bright pixels, typically caused by scattered X-rays directly striking the CCD chip. If these zingers are not removed they will cause ring artifacts if they are present in a flat field frame, or linear streaks if they are present in a data frame. A double correlation technique is used to detect zingers in the flat field images, which is possible because there are multiple flat field images that should only differ slightly due to beam decay and slight motions of the beam or monochromator. For the data frames, a spatial filter is used to detect and correct the zingers. The next step is to normalize each data frame to the flat field image, using interpolation between flat fields to correct for drift.

The next step in the data processing is to construct the corrected sinogram. This involves first taking the logarithm of the data (for absorption but not fluorescence CT) relative to air on each side of the object. The next step is centering the rotation axis in the projections. We determine the rotation axis by fitting a sinusoid to the center-of-gravity of each row in the sinogram, and the symmetry center of the fitted sinusoid is the rotation axis. The image is padded with extra columns of air on one side or the other to put the rotation axis in the center column. The final optional step is to reduce ring artifacts by detecting and correcting anomalous columns (bright or dark) in the sinogram.

The final step is tomographic reconstruction, presently done using filtered backprojection using the IDL Riemann function. Reconstruction times are a few hours for a 512×512×512 3-D data set using a single 450 MHz Pentium PC running Windows NT. Visualization is currently performed in IDL on the local PC at the beamline.

An example of this process is shown in Figure 11 that demonstrates the imaging of inclusions within an opaque diamond (courtesy of J. Parise and W. Yang, SUNY-SB) showing a) the radiogram with sample at 0 degrees, b) the radiogram with sample at 90 degrees, c) the sinogram for a single horizontal slice within the diamond and d) the reconstructed image of that single slice. The reconstruction d) shows the presence of two inclusions (dark) and several voids (light). The traces of the inclusions can be seen to trace out sine waves in the sinogram c).

Processing of fluorescence tomographic data is similar to that for transmission except that the sinogram is derived directly from the dataset (i.e., no white field, zinger subtraction, dark current correction, etc.). Reconstruction leads to a separate image for each of the regions of interest collected. In some cases, a single ROI will have contributions from more than one peak, e.g., peak overlaps of alpha peak for one element and beta peak for another. Such overlaps can be deconvoluted using these images and additional constraints such as knowledge of the ratios of alpha and beta fluorescence from a single element.

Current capabilities

Collection times for transmission CT are typically 20-60 minutes for a 658×658×517 pixel 3-D data set with resolutions near 1 μm. The recent addition of a vertical focusing mirror at the GSECARS bending magnet beam line has increased the usable flux by more than 100 and allows rapid data collection in the 1-2 minute range. This makes feasible the implementation of near real-time computed tomography of transient systems, such as fluid flow in sediments. Fluorescence CT collections times are more time consuming because of the need for the pencil beam approach. Typical collection times are a few hours for a single 2D slice with element sensitivities in the 10 mg kg^{-1} range and spatial resolutions of a few μm.

APPLICATIONS

Synchrotron-based microanalytical techniques offer distinct advantages over other analytical techniques by allowing analyses in situ, an important example being the ability to determine chemical speciation of a wide variety of toxic elements in moist soils, waste-forms, and biological specimens with little or no chemical pretreatment at detection limits that typically exceed those of conventional methods by several orders-of-magnitude. The availability of these quantitative, high-sensitivity, high-spatial resolution, microanalytical techniques have led to major advances in our understanding of fluid/mineral interactions and geochemical siting and behavior of contaminants. In particular, μXAFS allows one to quantify and map oxidation state ratios in heterogeneous earth materials and individual mineral grains. Such information is crucial in understanding the toxicity, mobility and containment of contaminating metals in the environment, mechanisms of trace element partitioning, and paths of strategic metal enrichment in nature. Examples of applications of these techniques are given below in the sections that follow. These examples were chosen to demonstrate the range of scientific applicability including microparticle chemistry, mini-soil column studies of redox reactions, speciation of metals in fluid inclusions at elevated temperature, fluid flow physics, chemical reactions between radionuclides in solution and mineral surfaces, and genesis of carbonates.

a) Radiogram
(0 degrees)

b) Radiogram
(90 degrees)

c) Sinogram
(slice 264)

d) Reconstruction
(slice 264)

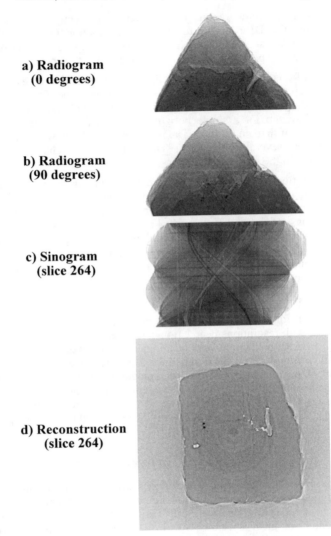

Figure 11. Example of the transmission CMT process for imaging inclusions within an opaque diamond (specimen courtesy of J. Parise and W. Yang, SUNY-Stony Brook) showing a) radiogram with sample at 0 degrees, b) radiogram with sample at 90 degrees, c) sinogram for a single horizontal slice within the diamond and d) the reconstructed image of the single slice shown in c). The reconstruction d) shows the presence of two inclusions (dark) and several voids (light). The inclusions can be seen to trace-out sine waves in the sinogram c).

Microbial processes and biomineralization

Interest in environmental microbiology and geomicrobiology has arisen in the past decade because of the importance of microbial-mineral interactions at controlling critical biogeochemical processes and because of the growing interest in bioremediation of contaminated sites. Soft X-ray microscopy and spectromicroscopy at a spatial resolution of ~50 nm have recently been applied to examine mineral surfaces and microbial-mineral interactions including the detailed characterization of biomineralized phases and these novel applications have been recently reviewed (Bertsch and Hunter 2001). Applications of hard X-ray μXRF and/or XAFS have included the investigation of biogenic sphalerite (ZnS) grains in biofilm deposits of sulfate reducing bacteria (Labrenz et al. 2000), the reduction of U(VI) by dissimilatory metal reducing bacteria (DMRB) in the presence of Fe-oxyhydroxide phases (Fredrickson et al. 2000), and preliminary work on the distribution of Cr in single cells of bacteria (Kemner et al. 2001), as well as on the coordination environment of U sorbed to bacterial cell walls (Kelly et al. 2001; Kemner et al. 2001).

Biomineralized sphalerite formed in biofilm deposited by sulfate reducing bacteria (SRB) was examined by a combination of electron microscopic/spectroscopic and synchrotron based energy dispersive X-ray photoelectron emission microscopy (PEEM) and μXRF techniques (Labrenz et al. 2000). Images generated with SEM and TEM revealed that the bacterial cells were closely associated with spherical ~10 μm aggregates comprised of randomly oriented ~3 nm crystallites, which were demonstrated to contain abundant Zn and S by both EDX and XRF (Fig. 12). Additionally, μXRF of the sphalerite aggregates demonstrated elevated As and Se and depleted Fe concentrations (Table 3). This investigation demonstrated the important role of SRB at concentrating metals and metalloids, with the Zn concentrations in the biofilm estimated to be 10^6 times that of the bulk fluid. This level of metal and metalloid sequestration in SRB biofilms suggests that microbial processes may be critical in the functioning of natural or engineered wetlands in environmental remediation processes.

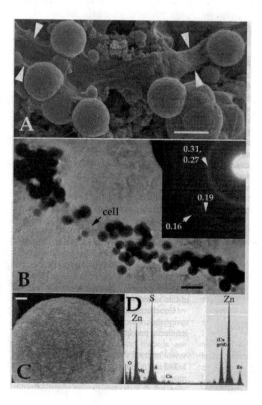

Figure 12. (A) SEM image of a chain of filamentous cells and associated micrometer-scale mineral aggregates. Bar, 1 μm. (B) TEM image of a chain of cells (bar, 1μm) and aggregates (dark contrast). (Inset) Selected area electron diffraction pattern from aggregates with interplanar spacings (in nanometers). (C) High-resolution SEM image showing that the aggregates in (A) and (B) consist of few-nanometer-diameter particles. Bar, 0.1 μm. (D) Energy-dispersive X-ray spectrum. (Reprinted with permission from Labrenz et al. 2000. Copyright 2000 American Association for the Advancement of Science.)

Table 3. Impurity content in ZnS (wt %) determined by synchrotron-based microprobe analyses (Labrenz et al. 2000).

Ion	1 μm diameter aggregate	Average of four analysis points on other μm-scale aggregates
Mn	0.12	0.103 ± 0.024
Fe	0.20	0.174 ± 0.025
As	0.0096	0.009 ± 0.0032
Se	0.0042	0.0043 ± 0.0018
Ca	0.67	0.902 ± 0.295
K	2.37	4.67 ± 1.28

Microbial reduction of U(VI), Tc(VII), and Cr(VI) coupled to the oxidation of organic compounds or H_2 by dissimilatory metal reducing bacteria (DMRB) and sulfur reducing bacteria (SRB) has been demonstrated to be effective at accelerating bio-immobilization of these contaminants under a wide range of conditions commonly found in subsurface environments. A number of mechanistic studies have established the role of electron donors and natural humic substances acting as an electron shuttle on microbial reduction of metals and radionuclides. Additionally, a number of studies have begun to examine the biogeochemical dynamics regulating contaminant metal reduction in the presence of Fe-and Mn-oxyhydroxide phases commonly found in soils and sediments that can compete with U as the primary terminal electron acceptor. μXANES analysis of ~100-200 μm dark colored biogenic grains or aggregates formed in suspensions of the DMRB, *Shewanella putrefaciens,* containing uranyl acetate revealed nearly complete reduction of U(VI) to U(IV) after 7 to 14 days of growth, even in the presence of goethite (Fredrickson et al 2000). Treatments with heat-killed cells demonstrated total lack of U reduction based on the energy of the L_{III} absorption edge (Bertsch et al. 1994) providing strong evidence for the utilization of UO_2^{2+} as a terminal electron acceptor by the DMRB. Another observation was that the uraninite (UO_2)-like phase formed on reduction was readily oxidized after 12 hr. exposure to air, suggesting that processes controlling the kinetics of reoxidation of the biogenic uraninite need to be examined relative to the utilization of in situ bioimmobilization as a long-term remediation strategy. Another interesting observation was that the scattering resonance to the high energy side of the main absorption feature, indicative of scattering along the O-U-O linear group commonly found in actinides having axial O's, i.e., UO_2^{2+} was absent for the reoxidized suspensions. This phenomenon had been observed previously in sediments subjected to reduction/oxidation cycles (Duff et al. 1999c). One explanation is that the reoxidized U exits in a constrained coordination environment with organic ligands, for example, which results in a shortening of the equatorial O and lengthening of the axial O, similar to U(VI) uranates.

Other recent preliminary studies have demonstrated the utility of X-ray microprobes to interrogate the distribution and chemical speciation of transition metals and other elements in single cells of *Pseudomonas fluorescens* adhered to Kapton film and the DMRB, *Shewanella putrefaciens* adhered to an iron oxide film (Kemner et al. 2001). While preliminary, this study demonstrates the utility of synchrotron X-ray microprobes to examine single fully hydrated bacterial cells. The Ca K_α map of *P. fluorecens* adhered to Kapton film can be used to define the location of the bacterium cell on the Kapton. The Cr K_α map of the same bacterium exposed to a Cr(VI) solution demonstrates a different distribution than the Ca, indicating that the Cr is not concentrated within the cell (Fig. 13).

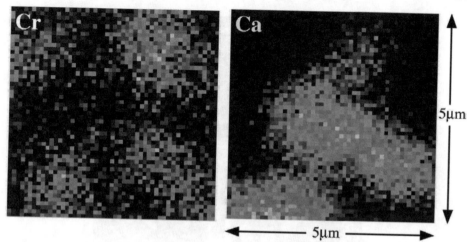

Figure 13. Ca and Cr maps of a hydrated *P. fluorescens* bacterium exposed to a 1000 mgL^{-1} Cr(VI) solution and adhered to a kapton film. (Reprinted from Kemner et al. 2001 with permission from Argonne National Laboratory.)

Plant rhizosphere processes

Metal sequestration by plaques. Steep redox gradients in wetland systems result in significant influences on the biogeochemical cycling of nutrients and contaminant metals and metalloids. One particularly important redox boundary exists in the vicinity of wetland plant roots, many of which are known to deliver oxygen to the rhizosphere from aerial tissues. One important result of this redox gradient in the rhizosphere is the formation of Fe-plaques on root surfaces. It has been hypothesized that these Fe-plaques may play a critical role in the regulation of potentially toxic metals to wetland plants by providing a reactive surface for the sequestration of soluble metals, thus, serving as an exterior barrier to metal uptake. Understanding the role of metal and metalloid interactions in the rhizosphere of wetland plants is important as natural and engineered wetlands are being used increasingly in the remediation of metal and metalloid contaminated environments, such as areas surrounding ore and coal mines. In a recent study, intact roots of a common wetland plant, *Phalaris arundinacea*, collected from a wetland receiving drainage from a century old Ag mine were examined by X-ray microprobe and microtomography and XAFS spectroscopy to provide insights into root-mediated metal sequestration mechanisms (Hansel et al. 2001). Metal distribution patterns of intact roots generated by X-ray fluorescence microtomography revealed a particularly vivid co-association of Fe and Pb accumulated on a rind formed on the outer part of the root, or epidermis, consistent with the notion that the root bound Fe-oxyhydroxide phases were acting as a sink for contaminant metals (Fig. 14). Conversely, the Mn and Zn distribution patterns were dissimilar to those for Fe and Pb, being localized in specific regions on the epidermis as well as in the root interior (Fig. 14). The Fe XANES analysis coupled to Raman spectroscopic analysis of the roots revealed that the Fe-plaque was comprised primarily of the poorly ordered phase ferrihydrite and goethite, both of which are highly surface active phases for Pb, as well as minor quantities of siderite. Based on the characterization of the Fe-plaque material and the elemental distribution maps, it would be tempting to conclude that the Pb was sorbed to the Fe-oxyhydroxide phases; however the Pb L$_{III}$ EXAFS revealed a different story.

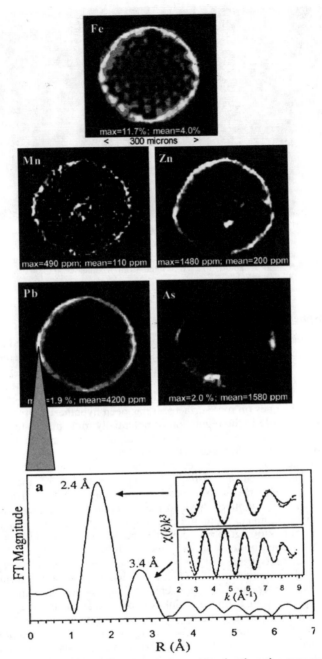

Figure 14. Fluorescence microtomographic images of a cross section through a grass root showing the internal distributions of Mn, Fe, Zn, As and Pb. White is the highest element concentration in each image. Lead is well correlated with the Fe-rich plaque whereas the other metals are not. (*bottom*) Fourier transformed EXAFS spectra of Pb on root surface (*left*) from the Pb rich Fe-plaque region of the root (*right*). [Reprinted with permission from Hansel et al. 2001, Figs. 1 and 3a. Copyright 2001 American Chemical Society].

Analysis of the Pb EXAFS data from the Fe-plaque revealed a Pb-O interatomic distance of 2.4 Å, which is inconsistent with published data for Pb sorbed to Fe-oxides (~2.27 Å). Furthermore, optimized fitting of the second shell EXAFS function was obtained for C or N at a distance of 3.4 Å, which is longer than previous reports for Pb bound humic material (Hansel et al. 2001). Based on fits that included a sample where Pb was bound to microbial biofilm it was concluded that the Pb was bound to a microbial biofilm-like material apparently intimately associated with the Fe-plaque. This finding demonstrates the important role of microorganisms in rhizosphere processes and the preferential binding of Pb to extracellular polymeric substances even in the presence of a highly reactive high surface area Fe-oxide phase.

Arsenic contamination of sediments throughout the world has stimulated study of the potential for human exposure to arsenic. The likelihood of human exposure largely arises from processes that mobilize arsenic from the soils and sediments into groundwater. Recently, the EPA adopted a drastically lower arsenic standard for drinking water (10 ppb), underscoring the potential detrimental impact of arsenic exposure. Natural biogeochemical processes that stabilize arsenic contamination within sediments may provide a means of limiting redistribution by groundwater transport.

Recent research is focused on understanding the cycling of arsenic within the Wells G&H wetland, Woburn, MA. This 16-acre riverine wetland, a Superfund site that gained notoriety in *A Civil Action*, is a reservoir of approximately 10 tons of arsenic within the upper 50 cm of the sediment profile. Most of the arsenic is sequestered in the wetland peat sediments with relatively little in the groundwater, as revealed by geochemical characterization of the site. In contrast to the wetland peat, riverbed sediments in the wetland (5 feet away) have higher concentrations of aqueous (mobile) arsenic despite lower solid phase concentrations. This research is currently exploring the differences between the two environments, which might explain these contrasting geochemical behaviors for arsenic. The most notable between the riverbed and wetland systems is the presence of vegetation (wetland) or lack thereof (riverbed).

This study focuses on cattails (Typha spp.), an invasive wetland plant species with a dense root structure that is found throughout the U.S and much of the world. Cattail roots may release oxygen during root respiration in anoxic sediments, precipitating an iron plaque on the roots (Armstrong 1979, Taylor et al. 1984). The adsorption of arsenic onto these plaques has been hypothesized to be a mechanism of arsenic immobilization in sediments (Otte et al. 1995). The root zone, approximately 30 cm in depth, coincides with the highest concentrations of arsenic in the Wells G & H wetland. The hypothesis is that the metabolic activity of the wetland plants (i.e., near-root (rhizosphere) oxygenation and exudation of organic complexes) may help to explain the sequestration of arsenic in the Wells G & H wetland.

Geochemical analyses of the contaminated sediments in the root zone using sequential chemical extractions showed that greater than half of the arsenic is strongly adsorbed (Keon et al. 2000, 2001). A mixture of arsenic oxidation states and associations was observed and supported by bulk XANES and EXAFS data collected at the SSRL. Arsenic in the upper 40 cm of the wetland, which contains the peak corresponding to maximum deposition, appears to be controlled by iron phases, with a small contribution from sulfidic phases. The results suggest that iron oxide phases may be present in the otherwise reducing wetland sediments as a substrate onto which arsenic can adsorb, perhaps due to cattail root plaque formation.

The tools of XRF microtomography and μXAFS are ideally suited for this exploration, as the necessity of root preservation is bypassed and the potential for analysis-induced alterations in solid-phase associations (as in extractions) is reduced. Direct cross-sectional elemental mapping of roots on the μm scale offers an extraordinary opportunity to determine how wetland plants in arsenic-rich sediments affect the cycling of the arsenic and if they act to internally or externally biostabilize the arsenic. Hansel et al. (2001, 2002) conducted the first studies of metal distribution in wetland plant roots, demonstrating that the technique might be successful in metal-contaminated wetland sediments.

Initial findings (Keon et al. 2002) demonstrate that arsenic distribution is affected by the activity of cattails in the upper wetland sediments. These XRF tomographic images of intact cattail roots provide the first direct evidence supporting this interpretation. Analysis of the tomographic images reveals a very strong correlation between arsenic and iron (R > 0.95) in the ~30 μm plaque. Furthermore, oxidation state tomograms (Fig. 15) show higher As^{3+}/As^{5+} ratios (more reduced arsenic) in more mature roots and a tendency for As^{3+} to occur on the interior of the iron plaque. These results suggest that either arsenic is being incorporated in the plaque as As^{5+} and that aging leads to reduction to the more toxic and less mobile form of arsenic, or differential adsorption strengths allow the As^{3+} to travel further into the plaque before being precipitated. The long-term implications are profound suggesting that both oxidation states of As are being effectively sequestered by the plants.

Selenium reduction by decomposing roots. The 2-D oxidation state method was used to map selenium oxidation states in water-saturated sediment containing decomposing roots of Scirpus, a common wetland plant (Tokunaga et al. 1994). Selenium was introduced homogeneously as selenate (Se^{6+}). Se K XANES spectra were measured for pure standards of selenium metal (Se^0), selenite (Se^{4+}) and selenate (Se^{6+}). Monochromatic energies were selected for elemental mapping to maximize the fluorescence intensity from the different species. The resulting oxidation state maps clearly showed that soluble Se^{6+} was reduced to the less mobile Se^{4+} and insoluble Se^0 in the regions of high microbial activity, i.e., immediately adjacent to the decaying roots (Fig. 16).

Fungal infection. In another application, Mn oxidation state maps of wheat roots growing in agar amended with Mn^{2+} and infected with the take-all fungus provided direct evidence for the accumulation of insoluble Mn^{4+} (Schulze et al. 1995a,b). These results supported the hypothesis that the oxidation of Mn^{2+} to Mn^{4+} is intimately involved in the fungal infection process. The procedure also is applicable to oxidation state microtomography where one of the linear scan axes is replaced by a rotation.

Biota collected from contaminated environments

A major obstacle in developing realistic ecological risk assessment or for designing environmentally sound chemical and biological remediation strategies has been the ability to characterize the chemical speciation of metals and metalloids in complex environmental samples, free from artifacts introduced by indirect chemical extraction or by sample preparation methods. Furthermore, major challenges in environmental toxicology and ecological risk assessment include the identification of appropriate biological receptors and tissues to be analyzed, the evaluation of principal toxicant(s) in mixed waste systems containing several metals and metalloids, and the assessment of the bioavailable fraction of the total contaminant concentration in soil. μXRF and XAFS techniques are unique tools for examining spatially heterogeneous elemental distributions in biota and accretionary biomineralized tissues, which may be useful for providing evidence implicating the specific contaminant(s) that may be involved in a biological effect under chronic exposure or in recording the environmental geochemical signature of contaminant exposure.

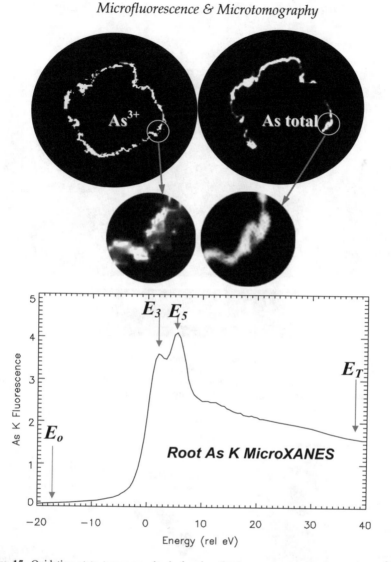

Figure 15. Oxidation state tomogram (*top*) showing the heterogeneous distribution of As³⁺ in the plaque. The μXANES spectrum (*bottom*) shows the presence of both As³⁺ and As⁵⁺. The As³⁺ image was collected with incident energy E_3 that preferentially excites that cation (Keon et al. 2002).

There has been interest in recent years at utilizing indigenous organisms as indicators of environmental contamination. It has been suggested that turtles could potentially be very attractive indicator species of metal contamination because they are extremely long-lived organisms, appear to be relatively tolerant to a range of pollutants, and because their shell is comprised of bone (apatite), which is a well-known target organ for a number of transition and heavier elements. Furthermore, turtle shells exhibit growth annuli in the shell for both the bone and protein (keratin) coating, within which periodic growth deposition bands are discernible (Fig. 17), thus providing the potential to examine historical information related to metal, metalloid, and radionuclide exposure.

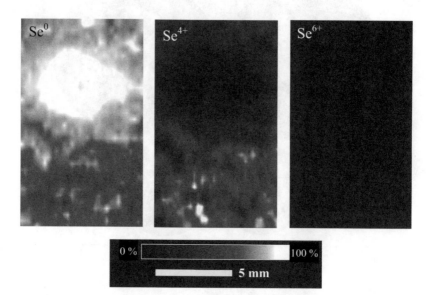

Figure 16. Selenium oxidation state maps (8×17 mm) in water-saturated sediment containing a decomposing root (top center of each image) of Scirpus initially containing Se⁶⁺. *From left to right*, Se⁰, Se⁴⁺, and Se⁶⁺. Soluble Se⁶⁺ was reduced to the less mobile Se⁴⁺ and insoluble Se⁰ in the regions of high microbial activity, i.e., immediately adjacent to the decaying roots (Tokunaga et al. 1994).

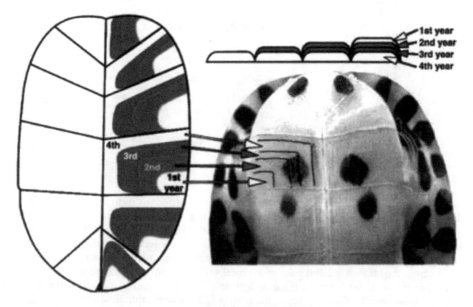

Figure 17. Photomicrograph of a turtle shell and an illustration of the growth annuli.

Turtle (*Trachemys scripta*) shell sections of individuals collected from a contaminated wetland adjacent to a former nuclear materials processing facility (Steed Pond) and a wetland receiving outfall from a coal fly ash basin (Ash Basin), both on the U.S. Department of Energy's Savannah River Site near Aiken, SC, as well as from a pristine wetland at a managed wildlife refuge (control site) were prepared with a diamond blade sectioning device. The sections were mounted on Kapton tape and the bone and protein coating examined for evidence of metal and metalloid accumulation via 2-D scans by XRF spectroscopy and, on appropriate regions of elevated contaminant concentration, by spatially resolved XANES spectroscopy (Hunter et al. 1997; Bertsch and Hunter 2001). The Steed Pond site was heavily contaminated with U and Ni and also received lesser quantities of other contaminant metals such as Cr, Cu, Cd, and Pb, resulting from decades of discharge of metallurgical waste associated with a nuclear materials fabrication facility (Sowder et al., in press). The Ash Basin site received outfall from a coal combustion ash basin and was contaminated with a range of metals and metalloids commonly concentrated in coal, such as As, Se, Cu, Cr, Ni, and Cd (Rowe et al. 1996; Hopkins et al. 2002a).

The micro-SXRF analysis of turtle shell sections from both contaminated sites revealed that several metals and/or metalloids were bioaccumulated and deposited in the bone, whereas no evidence for these metals were evident for the control samples (Fig. 18). Bone sections from individuals collected from the Steed Pond site had a characteristic signature of Ni but not U, despite the fact that total soil U concentrations exceed that of Ni, demonstrating a clear difference in the bioavailability of these metals. Bone sections from individuals collected at the Ash Basin site were elevated for a number of metals and metalloids but were particularly enriched in As and Se (Fig. 18). Age dependent accumulation patterns were observed for bone samples of individuals collected from both sites, although significant differences in the nature of these patterns were discernible (Fig. 19). As, Se, and Ni concentrations tended to decrease exponentially for turtles collected from the Ash Basin site, while this pattern was not evident for individuals collected at the Steed Pond site. The hypothesis is that the accumulation pattern for the turtles inhabiting the Ash Basin site is related to dietary exposure preference, which shifts from being primarily carnivorous for juveniles to herbivorous for adults. A plausible explanation for the different accumulation pattern observed for the turtles from Steed Pond is that this wetland is very large (~7 ha) compared to the wetland at the Ash Basin site (< 0.5 ha) and the contamination is heterogeneously distributed in the Steed Pond Site (Sowder et al., in press). Thus, since turtles home range is relatively large, the distribution pattern comprised of periodic spikes in the bone Ni concentration most likely represents periods related to feeding in the vicinity of highly contaminated sediments. Controlled contaminant exposure experiments in mesocosms are currently underway to test these hypotheses.

Contaminant metals and metalloids were also found to be localized to varying degrees in the ~50-100 μm bands associated with the keratin-based epithelial layer of the shells, which are also deposited in discrete annuli. Se was particularly enriched in the keratin layers of turtles collected from the Ash Basin site (Fig. 20). Contaminant association with the keratin could conceivably result from surface sorption to functional groups associated with the protein or *via* physiological incorporation of bioaccumulated metals and metalloids. μXANES spectra generated in regions of elevated Se concentration within the keratin of a bone fragment demonstrate a coordination environment similar to a Se-methionine standard suggesting that the Se exists predominantly in a coordination environment resembling Se-substituted methionine Se(−II), a functional group generally associated with S containing structural proteins (Fig. 20). These data demonstrate that the Se associated with the epithelial layer has been bioaccumulated and metabolically incorporated into the proteins associated with the shell (*vide infra*).

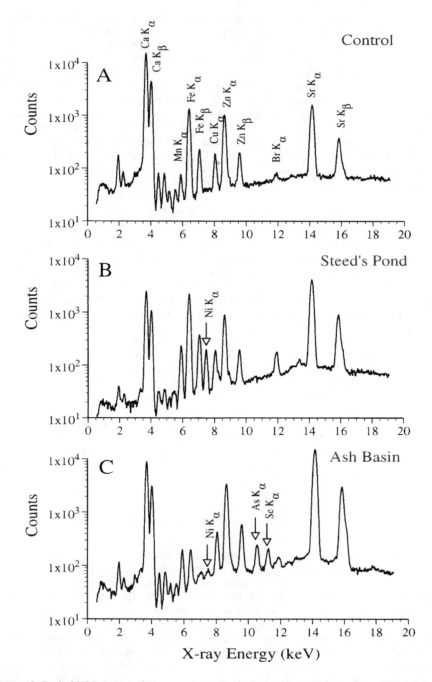

Figure 18. Typical XRF spectra of 10 μm regions of bone from turtle shell fragments collected from two contaminated sites and the control site (Reprinted with permission of EDP Sciences from Hunter et al. 1997, Fig. 1.)

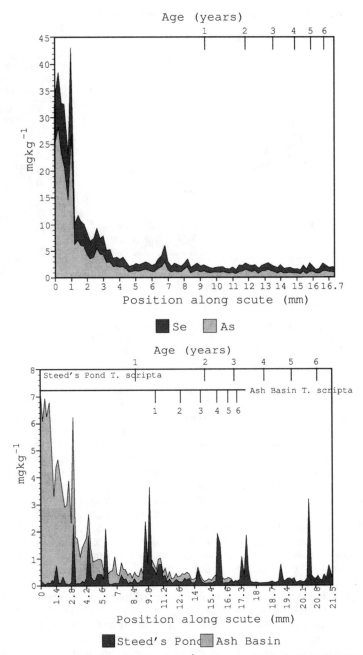

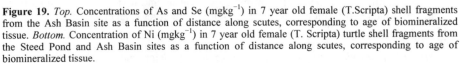

Figure 19. *Top.* Concentrations of As and Se (mgkg⁻¹) in 7 year old female (T.Scripta) shell fragments from the Ash Basin site as a function of distance along scutes, corresponding to age of biomineralized tissue. *Bottom.* Concentration of Ni (mgkg⁻¹) in 7 year old female (T. Scripta) turtle shell fragments from the Steed Pond and Ash Basin sites as a function of distance along scutes, corresponding to age of biomineralized tissue.

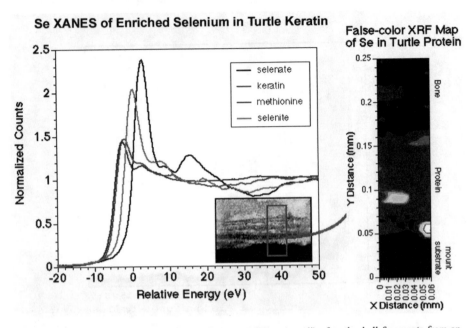

Figure 20. Photomicrograph of the keratin epidermal layers (annuli) of turtle shell fragments from an individual collected at the Ash Basin site, a Se map, and Se K_α XANES spectra of the epidermal layers and inorganic and organo-Se compounds (Hunter and Bertsch, unpublished).

A variety of other aquatic and semi-aquatic organisms inhabiting the wetland environments around the Ash Basin site have been investigated for chronic effects associated with exposure to multiple contaminant metals and metalloids (Rowe et al. 1996; Hopkins et al. 2002a,b; Jackson et al. 2002). An earlier study, examining tadpoles (*Rana catesbeiana*) inhabiting the Ash Basin site demonstrated a 90% greater incidence of oral deformities compared to a control site. These oral deformities were found to have significant effects on the long-term survival and reproduction of the tadpoles due to the influences on grazing and growth rates. Total elemental concentrations of tadpoles collected from the Ash Basin site revealed elevated concentrations of As, Ba, Cd, Cr, and Se; however, µXRF maps generated on the oral lesions revealed significantly elevated Se concentrations compared to the mouth structures of control tadpoles (Fig. 21). µXANES collected at the Se K edge in the Se rich regions of the oral toxic lesions indicate that the Se is incorporated into the proteinaceous mouth structure. Preliminary EXAFS of these regions suggest that Se has both Se and S in the first shell suggesting that Se could be substituting into cysteine-rich structural proteins, consistent with results of size exclusion chromatography/ICP-MS analysis of similar extracted tissue, which shows abundant low molecular wt. Se-rich peptides (Jackson et al. 2002). While these data cannot unequivocally demonstrate that Se is the primary toxin in this mixed waste system, it does provide strong evidence for the role of Se in the oral deformities and supports the emerging hypothesis that the significant biomagnification and toxicity of Se results from the indiscriminant substitution of Se for S in protein structures (Fan et al. 2002).

Vadose zone processes

Plutonium sorption onto Yucca Mountain tuff. A major challenge in environmental

Malformed Mouth Structures of Tadpoles from Coal Fly Ash Basins

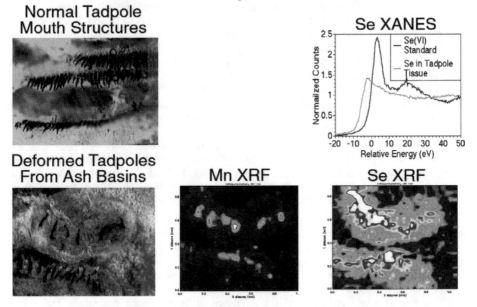

Figure 21. Photomicrograph of normal (control) and deformed (Ash Basin) tadpole mouth structures, Mn and Se elemental maps of the deformed mouth and Se K_α XANES spectra of a region in the deformed mouth structure having elevated Se concentration (Hunter, Bertsch, and Rowe, unpublished).

chemistry and geochemistry is defining reactive mineral phases involved in the sorption and sequestration of contaminant metals and radionuclides under field conditions. While decades of research on the sorption of solutes to a range of model mineral phases has provided important information on reactive surface functional groups and their coordination with a variety of contaminants, the extension of this information to complex mineral assemblages found in soils and sediments has been less than satisfying (Bertsch and Seaman 1999). The ability to interrogate rock and soil thin sections on a microscale via spatially-resolved XRF and XAFS has facilitated the examination of contaminant sorption and/or partitioning to complex mineral assemblages. In recent studies, the spatial distribution and chemical speciation of Pu sorbed to rock thin sections prepared from deep-core samples (450m) taken from the proposed long term high level radioactive waste repository at Yucca Mountain, NV were examined with μXRF and μXAFS techniques (Duff et al. 1999a,b). Thin sections were equilibrated with a groundwater simulate containing low concentrations of Pu(V). Studies on the redox and speciation chemistry of Pu in synthetic and natural waters suggest Pu can exist in four oxidation states (III, IV, V, VI) simultaneously (Choppin et al. 1997). Pu(V) was chosen for these experiments since it has been shown to be the primary valence state of Pu in neutral to basic oxic natural waters, such as synthetic Yucca Mountain groundwater (Boust et al. 1996).

Extensive elemental distribution maps generated via μSXRF revealed that Pu was heterogeneously distributed and co-associated with a number of trace elements, being highly localized in ~150 μm² regions associated with a low abundance mineral assemblage comprised of the 2:1 phyllosilicate mineral, smectite and the Mn-oxide, rancieite [(Ca, Mn^{2+})O•$4Mn^{4+}O_2$•$3H_2O$] (Fig. 22). In contrast to predictions based on previous batch adsorption experiments, Pu concentrations elsewhere within the thin section, i.e., the zeolitic rich matrix and relatively abundant Fe-oxide grains, were generally below detection. Spatially resolved Pu L_{III} edge XANES and EXAFS were collected on regions having the highest Pu concentrations (~2000 mg kg^{-1} in a 10 × 10 μm spot) (Duff et al. 1999a,b). The XANES spectra collected shortly following the sorption experiments revealed spatially variable Pu oxidation states, with edge positions indicative Pu(V) (region 1) and (VI) (region 2) (Duff et al. 1999a). From the edge position alone it is not possible to conclude if the Pu sorbed in region 1 was predominately Pu(V) or a mixture of Pu (IV, V, and VI). μEXAFS of region 1 revealed six oxygen atoms in the first shell at 2.25 Å, consistent with both Pu(IV) and Pu(V) compounds. The absence of the post-edge shoulder on the main absorption peak in the XANES spectra, and the lack of a shorter Pu-O bonds (at distances shorter than ~2.0 Å) indicative of axial O suggest that there is no Pu(VI) associated with this region. Another notable observation was the presence of Mn and the lack of Pu present in the second shell, suggesting that most of the Pu accumulating in these localized regions was adsorbed to mineral surfaces rather than being present as a surface precipitate (Fig. 23).

Following a two-year storage period, μXANES and EXAFS measurements were repeated and the data revealed that Pu associated with all the Pu-enriched areas previously examined had been fully reduced to Pu(IV) (Duff et al. 2001). The μEXAFS spectra once again revealed oxygen in the first shell and Mn in the second, suggesting a surface sorption mechanism rather than precipitation. These data suggest that the ranciete in these samples is facilitating the reduction of the Pu(V) over time. This observation is inconsistent with the Ce LIII XANES of rancieite from Yucca Mountain Tuff, that revealed that all of the Ce is in the Ce(IV) rather than the Ce(III) form, suggesting a surface mediated oxidation reaction. This indicates that surface controlled redox reactions associated with rancieite in YM tuff are complex and metal specific. The observation of some localized Pu(VI) in the samples initially most likely represented disproportionation

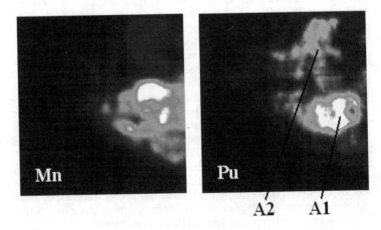

Figure 22. (left) X-ray fluorescence maps of a Pu-doped piece of Yucca Mountain tuff. Shown are the intensities of the fluorescence lines for Mn and Pu in the region of a Mn-rich mineral phase. Each map is 150 × 150 μm, and was made with 5 μm step sizes (adapted from Newville et al. 1999).

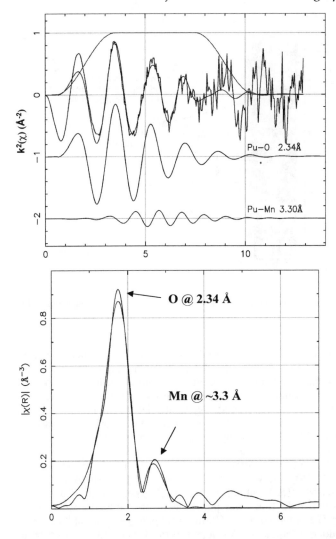

Figure 23. μXAFS results for Pu on tuff. *Top:* The k^2-weighted Pu-EXAFS spectra for one Pu-enriched region on the tuff. *Bottom:* the complementary Fourier-transformed data of Pu-EXAFS spectra showing the pseudo-radial distribution of atoms around the Pu at the region examined (from Duff et al. 2001).

of Pu(V) to Pu(IV) and (VI) at short equilibration times followed by the longer term reduction of the Pu(VI). Additional work will be required to fully elucidate the speciation dynamics in this system, but these studies clearly demonstrate the utility of μXRF and XAFS for examining metal and radionuclide sorption mechanisms in minimally disturbed environmental samples containing complex mineral assemblages.

Chromium diffusion and reduction in soil microsites. The X-ray microprobe is useful for in-situ studies of redox reactions within soil columns. T. Tokunaga (LBNL) has pioneered this approach for study of heterogeneous reactions at small scales in soils and has developed mini-soil columns with X-ray transparent windows that incorporate electrodes for redox potential measurement (Fig. 24).

Anoxic microsites within soil aggregates may be responsible for redox transformations and subsequent heterogeneous spatial distribution of reduced

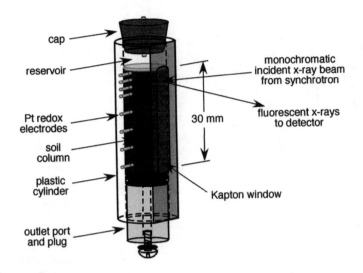

Figure 24. Mini-soil column used for in-situ μXRF and μXANES studies of heterogeneous redox reactions (courtesy of T. Tokunaga and J. Wan, LBNL).

precipitates. These processes have important consequences in biogeochemical cycling of major (e.g., N, S, Fe) and trace (e.g., Cr, As, Se, U) elements. μXANES measurements are made in order to understand whether or not, and under what conditions, it is possible to establish redox zonation in synthetic soil aggregates for Cr (Tokunaga et al. 1994, 1997, 1998, 2002). Variables being tested include aggregate size, organic matter content, concentrations of Cr(VI), and redox potentials. Total elemental distribution maps and μXANES information on Cr(III,VI) from soil aggregates incubated under the various test conditions were obtained to assess the significance of aggregate microsite-dependent reduction of Cr(VI) to Cr(III), under conditions limited by oxygen and Cr(VI) diffusion. Some initial results (Fig. 25) show the remarkable interplay between Cr(VI) concentration and organic carbon (OC) content. At high Cr(VI) concentration and no OC (profiles A), the diffusion and reduction front penetrates deeply into the soil. At high concentrations of both Cr(VI) and OC (profiles C), the front penetrates about half as deep and an extremely sharp reduction boundary develops (10 mm). With low concentrations of both, diffusion occurs only to relatively shallow depths (profiles B). Microbial activity was significantly higher in the more reducing soils, while total microbial biomass was similar in all of the soils. The small fraction of Cr(VI) remaining unreduced resides along external surfaces of aggregates, leaving it potentially available to future transport down the soil profile.

Fluid transport

Transient film flow on rough fracture surfaces. The X-ray microprobe has been valuable in making in situ measurements of water film hydraulic properties. The nature of water flow in unsaturated fractured rock is generally complex and incompletely understood. Various modes of fast, preferential flow through the vadose zone have been identified previously including macropore flow, fracture flow, flow fingering in soils and along fractures, and funneled flow. One set of experiments (Tokunaga et al. 2000) has focused on film flow and developing new methods for measuring relations between

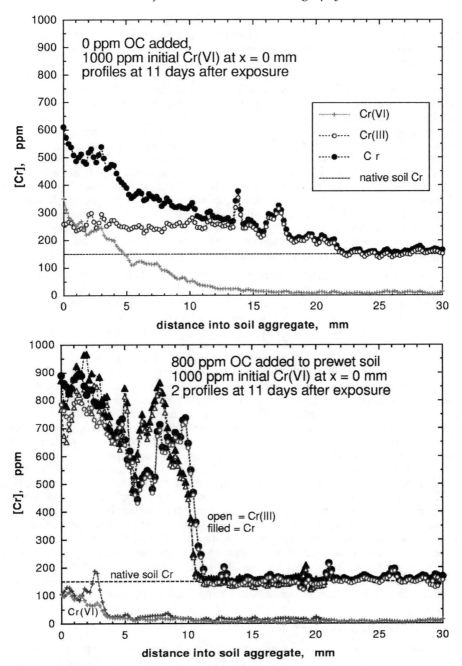

Figure 25. μXANES profiles of Cr(VI), Cr(III) and total Cr in soil aggregate microcosms at 11 days after contamination. These results illustrate the counter-balancing influences of OC amendment (0, 800 ppm, *top and bottom*, respectively) and boundary Cr(VI) concentration (260 ppm *top*, 1000 ppm *bottom*) in controlling Cr transport distances (adapted from Tokunaga et al. 2002).

average film thickness, matric potential, and film hydraulic diffusivity. The basic approach is to use the X-ray fluorescence intensity of a tracer to infer water thickness in a spatially resolved manner. By considering films on fracture surfaces as analogues to water in partially saturated porous media, the film hydraulic diffusivity and equation for transient film flow are obtained from their porous medium counterparts, the hydraulic diffusivity and the Richards equation. A modified version of a suction plate was developed for purposes of measuring film hydraulic properties on rough surfaces, specifically designed to measure equilibrium relations between matric potential and average film thickness, and film hydraulic diffusivities on surfaces of impermeable solids through X-ray fluorescence of a solute tracer. The tracer used in these experiments was selenate (SeO_4^{2-}), added as a sodium salt at a concentration of 242 mM. The choice of SeO_4^{2-} was based on its very low affinity to most mineral surfaces (Neal and Sposito 1989). Experiments on roughened glass showed that the average film thickness dependence on matric potential can be approximated as a power function. It was also shown that the film hydraulic diffusivity increases with increased film thickness (and with increased matric potential). Fast film flow (average velocities greater than 3×10^{-7} m s^{-1} under unit gradient conditions) was observed for average film thickness greater than 2 μm and matric potentials greater than -1 kPa (Tokunaga et al. 2000).

Dependence of physical properties on outflow conditions. To correctly describe transport of contaminant species it is essential to understand the interplay of advection, mechanical dispersion and diffusion, and their dependency on soil water distribution. As the relative importance of each transport component strongly depends on the soil water flow rate, the study of dynamic flow systems is extremely important. X-ray computed tomography (CT) offers an advantageous possibility to non-invasively investigate these dynamic processes on the pore scale (Hopmans et al. 1992; Clausnitzer and Hopmans 1999). Recent experiments carried out by Wildenschild et al. (1996, 2002) support the notion that the flow rate may vary notably between different types of experiments and therefore influence the estimation of the unsaturated hydraulic properties. The main objective of this study is to use microtomography to observe spatial and temporal distribution of air and water phases during drainage experiments (Wildenschild et al. 2002). Two different types of experiments were performed: (1) one-step experiments for which one relatively high pneumatic pressure is imposed on the sample to induce outflow and (2) multi-step outflow experiments where a varying number of smaller pressure increments are applied. Initial results are encouraging (Fig. 26). The figure shows two slices through the center of a packed-sand core in identical position, and scanned after (a) the core has been drained slowly to a final capillary pressure of 490 cm using a multi-step approach (left), and (b) the core has been drained very quickly using just one large pressure step of 490 cm. Clearly, there are major differences in overall amount of water retained in the core and the water distribution is also very different. It is expected that these experiments will lead to quantitative interpretations of flow-rate dependent phenomena and, potentially, verification of Lattice Boltzmann numerical simulations of flow and transport in unsaturated porous media.

Hydrothermal fluids and seawater

Copper speciation in vapor-phase fluid inclusions. The X-ray microprobe is valuable in studying the chemistry of intact fluid inclusions. Much of the work focuses on determining the chemical compositions of inclusions (Mavrogenes et al. 1995; Hayashi and Iida 2001; Menez et al. 2001; Philippot et a. 2001; Vanko et al. 2001) but chemical speciation studies using μXAFS have also become fruitful (Anderson et al. 1995, 1998; Mavrogenes et al. 2002).

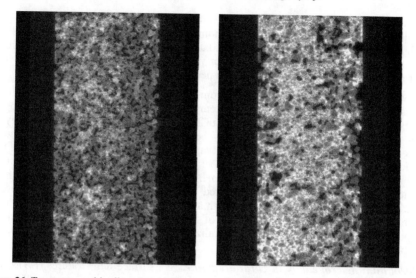

Figure 26. Two tomographic slices through the center of a packed-sand core in identical position. Left: after the core has been drained slowly to a final capillary pressure of 490 cm using a multi-step approach. Right: after the core has been drained very quickly using one large pressure step of 490 cm. Note the difference in overall water content and distribution (pore space is dark).

Hydrothermal ore deposits are the world's principal source of copper, gold, silver, lead, zinc, and uranium. The deposits form when metal complexes, transported in solution until cooling, decompression, or chemical reaction, cause precipitation and concentration. To understand the formation of these deposits it is necessary to determine both the amount and type of the metal complexes involved. A major impediment to advancing our understanding of these hydrothermal processes is the difficulty of undertaking spectroscopic experiments above the critical point of water. Thermodynamic modeling and predictive exploration is hindered by the lack of data in the geologically relevant temperature range of 400-700°C. XAFS measurements on hydrothermal solutions have great potential in this area but experiments have generally been limited to below 350°C (e.g., Mosselmans et al. 1996; Seward et al. 1996). These PT limitations can be overcome by using synthetic fluid inclusions to trap brine and vapor under known pressure, temperature, and chemical conditions (Loucks and Mavrogenes 1999). The application of XAFS to single fluid inclusions has been demonstrated (Anderson et al. 1995, 1998).

High Cu concentrations in magmatic-related hydrothermal vapors have been reported on a number of occasions (e.g., Lowenstern et al. 1991; Heinrich et al. 1992). PIXE analyses of coexisting vapor and brine fluid inclusions from the Mole Granite, NSW, Australia, found preferential partitioning of Cu into the vapor in a ratio of approximately 10:1 in contrast to Cl which is around 1:10 (Heinrich et al. 1992).

The coordination of copper in vapor phase inclusions coexisting with brine from the Mole Granite, NSW, Australia has been studied (Mavrogenes et al. 2002). X-ray fluorescence maps indicate that copper is concentrated in the vapor inclusions and at room temperature uniformly distributed in the condensed fluid. Opaque precipitates in these inclusions do not contain copper. Through a comparison with XAFS spectra on known copper compounds at temperatures up to 325°C (Fulton et al. 2000), the absorption spectra (Fig. 27) identify the stable complexes as $[Cu(OH_2)_6]^{2+}$ at 25°C, $[CuCl_2]^-$ at 200°C, and $[CuCl(OH_2)]$ at the homogenization temperature of around 400°C. The change in copper coordination and oxidation state is fully reversible. These results are the first direct spectroscopic evidence for vapor-phase Cu speciation and suggest that copper is transported in the vapor phase as a neutral chloride complex.

Copper in vapor-phase inclusions from the Mole Granite is dissolved in the condensed liquid phase as Cu^{2+} at room temperature. With heating a linear copper(I) complex is formed which appears to be $[CuCl_2]^-$. It has been suggested previously that the presence of S in these inclusions may be important in forming a volatile copper species to partition Cu into a low salinity vapor in preference to a chloride-rich brine (Heinrich et al. 1992). This work indicates that S is probably not important as a ligand but could be involved in changing the solution pH, and hence Cu volatility, by promoting the formation of the neutral and poorly solvated complex $[CuCl_2]H$. A linear dichloro Cu species is stable at high temperatures in both a brine and vapor and partitions depending upon the acidity of the solution. Estimates of fluid acidity at the time of boiling may be an important indicator of the type of mineralization (porphyry, epithermal, or both) that might be expected.

Strontium heterogeneity in coral aragonite. Sea surface temperatures (SSTs) have been inferred previously from the Sr/Ca ratios of coral aragonite (Beck et al. 1992; Evans et al. 1998). However microanalytical studies have indicated that Sr in some coral skeletons is more heterogeneously distributed than expected from SST data (Allison 1996; Hart and Cohen 1996). Synchrotron X-ray fluorescence (XRF) and X-ray absorption-spectroscopy (XAS) has been used to study the microdistribution and speciation of Sr in skeletons of *Porites lobata* coral (Allison et al. 2001). An XRF map on a small region (200 × 400 μm) of coral with an X-ray spot and pixel size of ~5 × 5 μm shows variability in the Sr/Ca ratio far in excess of that expected from the variations in sea temperature possible during the growth of this region. Furthermore, XAS measurements (Fig. 28) on a region of the coral with Sr concentration well above its bulk solubility limit in aragonite shows the Sr to still have primarily Ca as its second neighbor, indicating that Sr is still in the aragonite phase, and not in a Sr-rich phase such as strontianite ($SrCO_3$). In contrast, Greegor et al. (1997) found Sr in their coral specimens to be associated with strontianite. These results suggest that the Sr in coral skeleton is in disequilibrium with the aragonite, and that further study of the calcification processes in corals is needed to resolve the source of the Sr heterogeneity and to evaluate the use of the Sr/Ca ratio as a paleothermometer.

Extraterrestrial materials

Chronology and formation of the carbonate in a Martian meteorite. It has been suggested that carbonate globules in the ALH84001 meteorite from Mars (an igneous rock collected from Antarctica) contain evidence consistent with the development of bacterial life early in the history of Mars (McKay et al. 1996). Magnetite associated with this carbonate has been reported to be of the size and shape produced by terrestrial bacteria (Thomas-Keprta et al. 2000). μXRF measurements (Flynn et al. 2002; Wadhwa et al. 2002) have provided valuable information on the formation of the carbonates and these results are aiding the interpretation of radiochronometric data. The typical ALH84001 carbonate globule consists of four regions (Fig. 29): a core of Fe-rich carbonate, a thin magnetite-rich band, a rim of Mn-rich carbonate, and another thin magnetite-rich band.

Two carbonate globules were removed from ALH84001, embedded in epoxy, and ~4 μm thick slices was prepared by ultramicrotomy. The samples, which showed the same

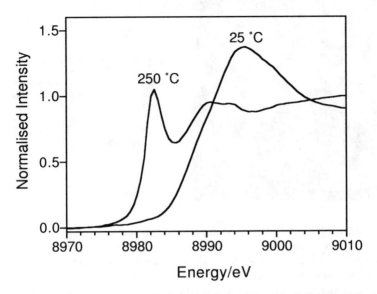

Figure 27. Copper K XANES spectra for a single fluid inclusion (30 μm) in the Mole granite showing the reversible speciation change from Cu(II) at ambient conditions to Cu(I) at elevated temperature (adapted from Mavrogenes et al. 2002)

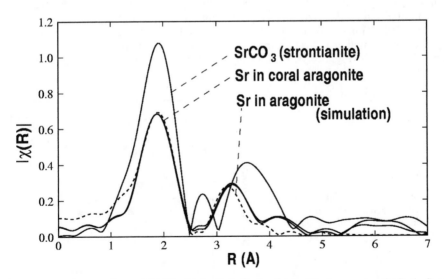

Figure 28. Fourier Transform of XAFS χ(k) for Sr in *Poritas lobata* (blue), pure strontianite (red), and a simulation of XAFS for Sr substituted in aragonite (black). The first neighbors for the two phases are similar (9 O at ~2.5 Å, 6 C at ~3.0 Å). The second neighbors (Sr or Ca) are easily distinguished, and the coral can easily be seen to be dominated by Sr substituted into aragonite (adapted from Allison et al. 2002)

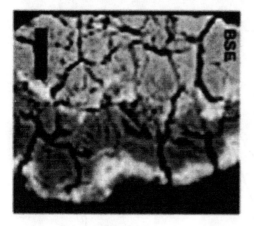

Figure 29. SEM backscattered electron image of a polished section of a carbonate from ALH84001 showing the four distinct zones: the Fe-rich core carbonate (dark, top), a thin magnetite band (bright), the rim carbonate (dark), and another magnetite band (bright, bottom) (courtesy of L. Keller, NASA-JSC; Flynn et al. 2002).

four zones as in Figure 29, measured about 40 × 40 μm. They were placed on ~8 μm thick Kapton films for analysis. X-ray spectra were collected with 15 minute dwell times, in one μm steps (using a ~3 μm analysis beam spot) along a line traversing the entire sample from the core carbonate to the outer magnetite-rich band.

Element abundances for Ca, Ti, V, Cr, Mn, Fe and Sr were measured. The highest Cr concentration was observed in the inner magnetite band. The two inner magnetite band analysis spots that have minimal carbonate contamination have Cr ~0.14%, while the underlying core carbonate has a mean Cr concentration of ~0.03 %. More striking is the Cr/Fe ratio, which is ~0.014 in the inner magnetite band, but almost an order-of-magnitude lower, ~0.003, in the high-Fe core carbonate. This indicates that the magnetite band is unlikely to be derived by a simple thermal decomposition of the core carbonate (Brearley 1998; Golden et al. 2001), which would preserve the Cr/Fe ratio in the magnetite band. Another indication that the rim and core carbonate are unrelated are the distinct Sr/Ca ratios (Fig. 30). Since Ca and Sr generally have rather similar geochemical

Figure 30. Ca vs. Sr in carbonate globule components determined by X-ray microprobe analysis (Flynn et al. 2002)

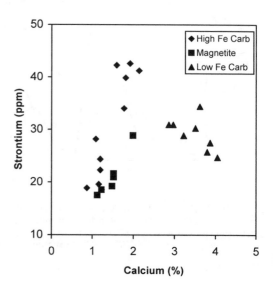

behavior, the distinct Sr/Ca ratios suggest that the rim carbonate did not form as a reaction product of the core carbonate or from a fluid of similar Sr/Ca.

The age of the carbonate has been inferred from Rb-Sr radiometric dating of leachates to be 3.9 Ga (Borg et al. 1999). μXRF direct measurement of Rb contents in carbonate were 0.4 to 0.5 ± 0.2 ppm in rim carbonate in one sample (Wadhwa et al. 2002) and below the detection limit of ~0.6 ppm in all spots in a second sample (Flynn et al. 2002). These results are much lower than the 1.3 to 6.4 ppm Rb reported by Borg et al. (1999) but consistent with the Rb concentrations reported by ion microprobe measurements (Eiler et al. 2002). Thus, the leached Rb used for radiometric dating may not have been hosted in rim carbonate but may have derived from some minor, yet-to-be-identified, Rb-rich phase, such as highly fractionated mesostasis material.

Chemical heterogeneity of interplanetary dust particles. Interplanetary dust particles (IDPs) are collected from the Earth's stratosphere by NASA planes, curated and allocated for study (Zolensky et al. 1994). The chemical heterogeneity of IDPs has been investigated to (1) obtain correlated trace element and noble gas analyses on sub-fragments of cluster particles and (2) evaluate the degree of stratospheric contamination in whole IDPs. Cluster particles are a subset of the IDPs that are so friable that they break into multiple fragments upon impact with the collection surface. While the IDPs are in space, each particle is exposed to the solar wind, and most IDPs contain significant quantities of solar noble gases. Low He concentrations (Nier and Schlutter 1993) unusually high ^{3}He/^{4}He ratios (Nier and Schlutter 1993) and high D/H ratios (Messenger 2000) have been reported for "cluster" IDPs. These results suggest the weak, porous IDPs which fragment on collection probably sample a different type of parent body, which better preserves the isotopic signature of interstellar material than the stronger, more common IDPs. Coupled trace element measurements allow us to use the Zn thermometer (Flynn and Sutton 1992) to determine if the low He contents are due to atmospheric heating or short space exposure time. Particles with "normal" Zn and low He are of particular interest because it is likely they are associated with cometary dust trails which the Earth transects (short space exposure) (Flynn and Sutton 1998). The X-ray microprobe allows high sensitivity (ppm) analyses on these sub-fragments down to 1 μm.

Eight individual sub-fragments have been analyzed from a single cluster particle (L2011-AD21; Flynn and Sutton 2002). These sub-fragments have sizes in the 3-5 μm range, i.e., total individual masses in the 10-100 pg range. Figure 31 shows the abundance patterns for some of the detected elements for each of the siblings, as well as the average of the eight. The average abundances fall within a factor of 3 of the carbonaceous chondrite (CI) composition, a typical observation for IDPs, and each of the elements has a total range of about an order of magnitude. The most notable features are the coupled depletions in Zn and Ge. Such coupled depletions in these two elements have been observed before in whole IDPs and attributed to atmospheric entry heating (Flynn and Sutton 1992, 1998; Greshake et al. 1998). The wide range in Zn and Ge contents, coupled with the much smaller range of variations for the other elements, demonstrates the heterogeneous nature of the heating process at the micrometer scale and will be extremely valuable in interpreting low He cluster sub-fragments (Thomas et al. 1996; Messenger and Walker 1998).

Internal microdistribution of volatile elements in dust particles. Fluorescence microtomography can be used to image the internal element distributions in interplanetary dust particles (IDPs) collected from the stratosphere (Sutton et al. 2000). L2036H19 is a ~30 μm compact IDP, with a chondritic EDX spectrum. As expected based on the diverse mineralogy of individual IDPs reported in TEM examinations, each element exhibits a high degree of heterogeneity in L2036H19. Figure 32 shows the Fe,

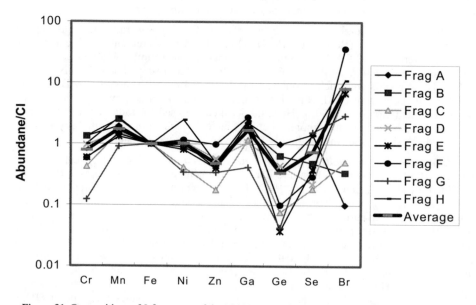

Figure 31. Compositions of 8 fragments of the L2011-AD21 cluster particle (Flynn and Sutton 2002).

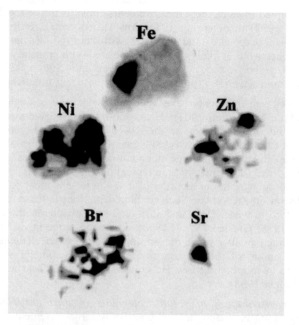

Figure 32. X-ray fluorescence microtomogram of a 30 μm interplanetary dust particle. The distributions (black represents a high concentration) of the volatile elements Zn and Br suggest an indigenous origin for these elements rather than stratospheric contamination by aerosols (Sutton et al. 2000).

Ni, Zn, Br, and Sr distributions in a single "slice" through L2036H19. Iron and Ni are not strictly correlated in this particle. In particular, the particle includes a high Fe region that is not enriched in Ni. The Fe/Ni ratio in this region is about a factor of four higher than that in the remainder of the particle. One possible interpretation is that this few μm region is dominated by pyrrhotite (Orgueil pyrrhotite Fe/Ni = 50 ≅ 3×CI; Greshake et al. 1998). Figure 32 also shows that Sr is concentrated in a single region near the center of this slice. The Fe and Zn contents are quite low in this spot. While the mineralogy of L2036H19 has not yet been determined, the high Sr is probably a Ca-rich mineral such as Ca-carbonate. The Zn is concentrated in two regions, both of which are adjacent to Ni-rich regions. The size is consistent with the large (μm sized) sphalerite grains reported by TEM examination of some other IDPs. However, in this preliminary study, Se was not measured, so presently neither S nor Se analyses exist to confirm that these regions of high Zn are sulfides.

The distribution of Br is particularly important since IDPs, on average, contain unusually high concentrations of Br (average ~30×CI) compared to the other volatile elements (which are enriched to ~3×CI). Chlorine contamination has been detected on the surface of Al-oxide spheres recovered from the Earth's stratosphere (Mackinnon and Mogk 1985), and Rost et al. (1999) have reported enrichments of F, Cl, and Br on the surface of an Al-oxide sphere collected from the stratosphere. The Br in this slice of L2036H19 is concentrated in several regions located throughout the IDP. No obvious Br-rich rim was detected.

FUTURE DIRECTIONS

Spatial resolution currently achievable with hard X-ray microbeam techniques, down to the 100 nm range, is quite satisfactory. In many cases, spatial resolution at the beam size cannot be realized because of the properties of the sample, notable thickness (because of the penetrating nature of the X-rays, 100 nm resolution can only be achieved in samples <100 nm thick). Improvements in sensitivity can be achieved in two principal ways, increase in incident flux or improvement in detection efficiency. Major increases in incident flux will be impractical in many cases because of the radiation sensitivity of the earth and environmental samples of interest. Improvements in fluorescence detection efficiency will be a more fruitful avenue because current detection schemes use very small solid angles. Energy dispersive detectors that intercept large solid angles would be a major advance in sensitivity enhancement.

Because of the interest in analyzing intact specimens with the minimum of sample preparation, fluorescence microtomography will likely be a technique used more frequently in the future. Improvements in data collection speed will be needed to make this technique more routinely applicable.

ACKNOWLEDGMENTS

The authors wish to thank the following individuals for their contributions to the applications section of this chapter: N. Allison (University of Brighton), A. Berry (Australian National University), N. K. Blute (Massachusetts Institute of Technology), M. Duff (Westinghouse Savannah River Company), A. Finch (University of St. Andrews), G. Flynn (SUNY-Plattsburgh), C. Hansel (Stanford University), D. Hunter (Westinghouse Savannah River Company), K. Kemner (Argonne National Laboratory), J. Mavrogenes (Australian National University), T. Tokunaga (Lawrence Berkeley National Laboratory), M. Wadhwa (Field Museum), J. Wan (Lawrence Berkeley National Laboratory), and D. Wildenschild (Technical University of Denmark). The chapter benefited from the thoughtful reviews of P. Fenter and an anonymous reviewer. This work was supported in part by the following grants: NSF (EAR-9906456), DOE (DE-FG02-94ER14466, DE-FG02-92ER14244). PMB was partially supported by contract DE-FC09-96SR-18546 between the U.S. Department of Energy and The University of Georgia.

REFERENCES

Allison N (1996) Comparative determinations of trace and minor elements in coral aragonite by ion microprobe analysis, with preliminary results from Phuket, South Thailand. Geochim Cosmochim Acta 60:3457-3470

Allison N, Finch AA, Sutton SR, Newville M (2001) Strontium heterogeneity and speciation in coral aragonite: Implications for the strontium paleothermometer. Geochim Cosmochim Acta 65:2669-2676

Anderson AJ, Mayanovic RA, Bajt S (1995) Determination of the local-structure and speciation of zinc in individual hypersaline fluid inclusions by micro-XAFS. Can Mineral 33:499-508

Anderson AJ, Mayanovic RA, Bajt S (1998) A microbeam XAFS study of aqueous chlorozinc complexing to 430 degrees C in fluid inclusions from the Knaumuhle granitic pegmatite, Saxonian Granulite Massif, Germany. Can Mineral 36:511-524

Armstrong W (1979) Aeration in higher plants. Advances in Botanical Research 7:226-332

Bajt S, Clark SB, Sutton SR, Rivers ML, Smith JV (1993) Synchrotron X-ray microprobe determination of chromate content using X-ray absorption near-edge structure. Anal Chem 65:1800-1804

Bajt S, Sutton SR, Delaney JS (1995) X-ray microprobe analysis of iron oxidation-states in silicates and oxides using X-ray absorption near-edge structure (XANES). Geochim Cosmochim Acta 58:5209-5214

Baker DR (2002) The Hard X-ray Microprobe. *In*: Synchrotron Radiation: Earth, Environmental, and Material Sciences Applications. Short Course Series Vol. 30. Henderson GS and Baker DR (eds) Mineralogical Society of Canada, Ottawa, Canada, p 99-130

Bambynek W, Crasemann B, Fink RW, Freund HU, Mark H, Swift CD, Price RE, Rao PV (1972) X-ray fluorescence yields, Auger and Coster-Kronig transition probabilities. Rev Mod Phys 44:716-813

Bearden JA (1964) X-ray wavelengths. U. S. Atomic Energy Commission Report NY0-10586

Bearden JA (1967) X-ray wavelengths. Rev Mod Phys 39:78-124

Bearden JA, Burr AF (1967) Reevaluation of X-ray atomic energy levels. Rev Mod Phys 39:125-142

Beck JW, Edwards R, Ito E, Taylor F, Recy J, Rougerie F, Joannot P, Henin C (1992) Sea surface temperature from coral skeletal strontium/calcium ratios. Science 257:644-647

Bertsch PM, Hunter DB (2001) Applications of synchrotron-based X-ray microprobes. Chem Rev 101:1809-1842

Bertsch PM, Seaman JC (1999) Characterization of complex mineral assemblages: Implications for contaminant transport and environmental remediation. Proc Nat Acad Sci 96:3350-3357

Bertsch PM, Hunter DB, Sutton SR, Bajt S, Rivers ML (1994) *In situ* chemical speciation of uranium in soils and sediments by micro X-ray absorption spectroscopy. Envir Sci Techno 28:980-984

Bilderback DH, Huang R (2001) X-ray tests of microfocusing mono-capillary optic for protein crystallography. Nucl Instrum Methods Phys Res A 467-468:954-967

Bilderback DH, Thiel DJ (1995) Microbeam generation with capillary optics. Rev Sci Instrum 66:2059-2063

Bonse U, Riekel C, Snigirev AA (1992) Kirkpatrick-Baez microprobe on the basis of two linear single crystal Bragg-Fresnel lenses. Rev Sci Instrum 63:622-624

Borg LE, Connelly JN, Nyquist LE, Shih CY, Wiesmann H, Reese Y (1999) The age of the carbonates in martian meteorite ALH84001. Science 286:90-94

Bos M, Vrielink JAM (1998) Constraints, iteration schemes and convergence criteria for concentration calculations in X-ray florescence spectrometry with the use of fundamental parameter methods. Anal Chim Acta 373:298-302

Boust D, Mitchell PI, Garcia K, Condren O, Vintro LL, LeClerc G (1996) A comparative-study of the speciation and behavior of plutonium in the marine-environment of 2 reprocessing plants. Radiochim Acta 74:203-210

Brearley AJ (1998) Magnetite in ALH84001: Product of the Decomposition of Ferroan Carbonate. Lunar Planet Sci XXIX:1451

Chevallier P, Dhez P, Legrand F, Idir M, Soullie G, Mirone A, Erko A, Snigirev A, Snigireva I, Suvorov A, Freund A, Engström P, Nielsen JA, Grübel A (1995) First test of the scanning X-ray microprobe with Bragg-Fresnel multilayer lens at ESRF beam line. Nucl Instrum Methods Phys Res A 354:584-587

Choppin G R, Bond AH, Hromadka PM (1997) Redox speciation of plutonium. J Radio Nucl Chem 219:203-210

Clausnitzer V, Jopmans JW (1999) Estimation of phase-volume fractions from tomographic measurements in two-phase systems. Advances in Water Resources. 22:577-584

Compton AH (1929) The efficiency of production of fluorescent X-rays. Phil Mag 8:961-977

Criss JW, Birks LS, Gilfrich JV (1978) Versatile X-ray analysis program combining fundamental parameters and empirical corrections. Anal Chem 50:33-37

Delaney JS, Bajt S, Dyar MD, Sutton SR, McKay G, Roeder P (1996a) Comparison of quantitative synchrotron microXANES (SmX) FeIII/(FeII+FeIII) results for amphibole and silicate glass with independent measurements. Lunar Planet Sci XXVII:299-300

Delaney JS, Bajt S, Sutton SR, Dyar MD (1996b) In situ microanalysis of $Fe^{3+}/\Sigma Fe$ ratios in amphibole by X-ray absorption near edge structure (XANES) spectroscopy. In Mineral Spectroscopy, Roger Burns Memorial Volume 5:165-171

Delaney JS, Dyar MD, Sutton SR, Bajt S (1998) Redox ratios with relevant resolution: Solving an old problem by using the synchrotron microXANES probe. Geology 26:139-142

Delaney JS, Jones JH, Sutton SR, Simon S, Grossman L (1999) *In situ* microanalysis of vanadium, chromium, and iron oxidation states in extraterrestrial samples by synchrotron microXANES (SmX) spectroscopy. Meteorit Planet Sci 34:A32

Delaney JS, Sutton SR, Newville M, Jones JH, Hanson B, Dyar MD, Schreiber H (2000) Synchrotron micro-XANES measurements of vanadium oxidation state in glasses as a function of oxygen fugacity: Experimental calibration of data relevant to partition coefficient determination. Lunar Planet Sci XXXI:1806

Dhez P, Chevallier P, Lucatorto TB, Tarrio C (1999) Instrumental aspects of X-ray microbeams in the range above 1 keV. Rev Sci Instrum 70:1907-1920

Duff MC, Newville M, Hunter DB, Sutton SR, Triay IR, Vaniman DT, Bertsch PM, Eng P, Rivers ML (2001) Heterogeneous plutonium sorption on Yucca Mountain tuff. APS Forefront (ANL/APS/TB-42). 18-21

Duff MC, Hunter DB, Bertsch PM, Amrhein C (1999c) Factors influencing uranium reduction and solubility in evaporation pond sediments. Biogeochemistry 45:94-114

Duff MC, Hunter DB, Triay IR, Bertsch PM, Reed DT, Sutton SR, Shea-McCarthy G, Kitten J, Eng P, Chipera SJ, Vaniman DT (1999a) Mineral associations and average oxidation states of sorbed Pu on tuff. Environ Sci Technol 33:2163-2169

Duff MC, Newville M, Hunter DB, Bertsch PM, Sutton SR, Triay IR, Vaniman DT, Eng P, Rivers ML (1999b) Micro-XAS studies with sorbed plutonium on tuff. J Synch Rad 6:350-352

Dunsmuir JH, Ferguson SR, D'Amico KL, Flannery BP, Deckman HW (1991) X-ray microtomography: Quantitative three-dimensional X-ray microscopy. Rev of Progress in Quantitative Nondestructive Evaluation 10A:443-449

Dyar MD, Delaney JS, Sutton SR (2001) Fe XANES spectra of iron-rich micas. Eur Jour Mineral 13:1079-1098

Dyar MD, Delaney JS, Sutton SR, Schaefer MW (1998) Fe^{3+} distribution in oxidized olivine: A synchrotron micro-XANES study. Am Mineral 83:1361-1365

Eiler JM, Valley JW, Graham CM, Fournelle J (2002) Two populations of carbonate in ALH84001: Geochemical evidence for discrimination and genesis. Geochim Cosmochim Acta 66:1285-1303

Eng PJ, Newville M, Rivers ML, Sutton SR (1998) Dynamically figured Kirkpatrick Baez micro-focusing optics. In X-ray Microfocusing: Applications and Technique, I. McNulty, ed., SPIE Proc. 3449:145

Eng PJ, Rivers M, Yang BX, Schildkamp W (1995) Micro-focusing 4keV to 65keV X-rays with bent Kirkpatrick-Baez mirrors. In X-ray Microbeam Technology and Applications, W. Yun, ed., Proc., 2516:41

Evans RD (1955) The Atomic Nucleus. McGraw-Hill Book Company, Inc., New York

Evans MN, Fairbanks RG, Rubenstone JL (1998) A proxy index of ENSO teleconnections. Nature 394:732-733

Fan TW-M, Teh SJ, Hinton DE, Higashi RM (2002) Selenium biotransformations into proteinaceous forms by foodweb organisms of selenium-laden drainage waters in California. Aquatic Toxicology 57:65-84

Flynn GJ, Sutton SR (1992) Trace Elements in Chondritic Stratospheric Particles: Zinc Depletion as a Possible Indicator of Atmospheric Entry Heating. Proc. Of Lunar and Planetary Science 22, Lunar and Planetary Institute, Houston, 171-184

Flynn GJ, Sutton SR (1998) Trace Element Contents of L2011 Cluster Fragments: Implications for Comet Schwassmann-Wachmann 3 as a Source of L2011 Cluster Interplanetary Dust Particles. Meteoritics and Planetary Science 33:A49-A50

Flynn GJ, Sutton SR, Keller LP (2002) X-ray microprobe measurements of the chemical compositions of ALH84001 carbonate globules. Lunar Planet Sci XXXIII:1648

Flynn GJ, Sutton SR (2002) An assessment of the chemical heterogeneity of individual ~3 μm fragments from an L2011 cluster IDP. Supplement to Met & Planet. Sci. 37:5012

Fredrickson JK, Zachara JM, Kennedy DW, Duff MC, Gorby YA, Li Shu-Mei W, Krupka KM (2000) Reduction of U(VI) in goethite (α-FeOOH) suspension by a dissimilatory metal-reducing bacterium. Geochim. Cosmochim. Acta 64:3085-309

Fulton JL, Hoffman MM, Darab JG (2000) An X-ray absorption fine structure study of copper(I) chloride coordination structure in water up to 325 degrees. Chem Phys Lett 330:300-308

Gillam E, Heal HT (1952) Some problems in the analysis of steels by X-ray fluorescence. British J Appl Phys 3:353-358

Golden DC, Ming DW, Schwandt CS, Lauer HV, Socki RA, Morris RV, Lofgren GE, McKay GA (2001) A simple inorganic process for formation of carbonates, magnetite and sulfides in Martian meteorite ALH84001. Am Min 86:370-375

Greegor RB, Pingitore NE Jr, Lytle FW (1997) Strontianite in coral skeletal aragonite. Science 275:1452-1454

Greshake A, Klock W, Arndt P, Maetz M, Flynn GJ, Bajt S, Bischoff A (1998) Heating experiments simulating atmospheric entry heating of micrometeorites: Clues to their parent body sources. Meteoritics and Planet Sci 33:267-290

Hansel CM, Fendorf S, Sutton S, Newville M (2001) Characterization of Fe plaque and associated metals on the roots of mine-waste impacted aquatic plants. Environ Sci Technol 35:3863-3868

Hansel CM, LaForce MJ, Fendorf S, Sutton S (2002) Spatial and temporal association of As and Fe species on aquatic plant roots. Environ Sci Technol 36(9):1988-1994

Hansteen TH, Sachs PM, Lechtenberg F (2000) Synchrotron-XRF microprobe analysis of silicate reference standards using fundamental-parameter quantification. Eur J Mineral 12:25-31

Hart SR, Cohen AL (1996) An ion probe study of annual cycles of Sr/Ca and other trace elements in corals. Geochim Cosmochim Acta 60:3075-3084

Hayakawa S, Iida A, Aoki S, Gohshi Y (1989) Development of a scanning X-ray microprobe with synchrotron radiation. Rev Sci Instrum 60:2452-2455

Hayashi K, Iida A (2001) Preliminary study on the chemical mapping of individual fluid inclusion by synchrotron X-ray fluorescence microprobe. Resource Geol 51:259-262

Heald SM, Brewe DL, Barg B, Kim KH, Brown FC, Stern EA (1997) Micro-XAS using tapered capillary concentrating optics. J de Physique IV 7 (C2):297-301

Heinrich CA, Ryan CG, Mernagh TP, Eadington PJ (1992) Segregation of ore metals between magmatic brine and vapor – A fluid inclusion study using PIXE microanalysis. Econ Geol 87:1566-1583

Hopkins WA, Roe JH, Snodgrass JW, Staub BP, Jackson BP, Congdon JD (2002b) Effects of chronic dietary exposure to trace elements on banded water snakes (*Nerodia Fasciata*). Environ Tox Chem 21:906-913

Hopkins WA, Snodgrass JW, Roe JH, Staub BP, Jackson BP, Congdon JD (2002a) Effects of food ration on survival and sublethal responses of lake chubsuckers (*Erimyzon sucetta*) exposed to coal combustion wastes. Aquatic Toxicology 57:191-202

Hopmans JW, Vogel T, Koblik PD (1992) X-ray tomography of soil water distribution in one-step outflow experiments. Soil Sci Soc Am J 56(2):355-362

Horowitz P, Howell JA (1972) A scanning X-ray microscope using synchrotron radiation. Science 178:608-611

Hunter DB, Bertsch PM, Kemner KM, Clark SB (1997) Distribution and chemical speciation of metals and metalloids in biota collected from contaminated environments by spatially resolved XRF, XANES, and EXAFS. Journal de Physique IV France 7:767-771

Ice GE (1996) Microbeam forming methods for synchrotron radiation. X-ray Spectrom 26:315-326

Jackson BP, Shaw-Allen P, Hopkins WA, Bertsch PM (2002) Trace element speciation in largemouth bass (*Micropterus salmoides*) from a fly ash settling basin by liquid chromatography-ICP-MS. Analytical and Bioanalytical Chemistry, in press

Janssens K, Vincze L, Adams F, Jones KW (1993) Synchrotron-radiation-induced X-ray-microanalysis. Anal Chim Acta 283:98-110

Jenkins R (1999) X-ray Fluorescence Spectrometry, Second Edition. John Wiley and Sons, New York, NY

Jones KW, Berry WJ, Borsay DJ, Cline HT, Conner WC, Fullmer CS (1997) Applications of synchrotron radiation-induced X-ray emission (SRIXE). X-ray Spectrometry 26 (6):350-358

Kagoshima Y, Takai K, Ibuki T, Yokoyama Y, Hashida T, Yokoyama K, Takeda S, Urakawa M, Miyamoto N, Tsusake Y, Matsui J, Aino M (2001) Scanning hard X-ray microscope with tantalum phase zone plate at the Hyogo-Bl (BL24XU) of SPring-8. Nucl Instrum Methods Phys Res A 467-468:872-876

Kelly SD, Kemner KM, Fein JB, Fowle DA, Boyanov MNI, Bunker BA, Yee N (2001) XAFS Study of U Sorption to Bacterial Cell Wall. Proc. 6th International Conference on the Biogeochemistry of Trace Elements; Guelph, Ontario. p 19

Kemner K, Lai B, Maser J, Cai Z, Kelly S, Legnini D, Ilinski P, Rodrigues W, Nealson K, Schneegurt M, Kulpa C Jr, Fredrickson J, Gorby Y, Zachara J (2001) High energy X-ray microprobe investigations of metal-microbe and mineral-microbe interactions. Proc. 6th International Conference on the Biogeochemistry of Trace Elements; Guelph, Ontario. p 17

Keon NE, Swartz CH, Brabander DJB, Harvey C, Hemond HF (2000) Evaluation of arsenic mobility in sediments using a validated extraction method. EOS Transactions 81:F526

Keon NE, Swartz CH, Brabander DJB, Myneni SCB, Hemond HF (2001) Validation of an arsenic sequential extraction method for evaluating mobility in sediments. Environ Sci Technol 35:2778-2784

Keon NE, Swartz CH, Brabander DJB, Myneni SCB, Sutton S, Newville M, Hemond HF (2002) Combined XAS and chemical extraction investigation of arsenic distribution in sediment phases and in cattail roots. Amer Chem Soc National Meeting, GEOC-127

Kinney JH, Haupt DL, Nichols MC, Breunig TM, Marshall GW Jr., Marshall SJ (1994) The X-ray tomographic microscope: three-dimensional perspectives of evolving microstructures. Nucl Instrum Methods A 347:480-486

Kirkpatrick P, Baez AV (1948) Formation of optical images by X-rays. J Opt Soc Am 38:766-774

Labrenz M, Druschel GK, Thomsen-Ebert T, Gilbert B, Welch S, Kemner K, Logan G, Summons R, De Stasio G, Bond P, Lai B, Kelly SD, Banfield JF (2000) Formation of sphalerite (ZnS) deposits in natural biofilms of sulfate-reducing bacteria. Science 290:1744-1747

Lengeler B, Schroer CG, Benner B, Güzler TF, Kuhlmann M, Tümmler J, Simionovici AS, Drakopoulos M, Snigirev A, Snigireva I (2001) Parabolic refractive X-ray lenses: a breakthrough in X-ray optics. Nucl Instrum Methods Phys Res A 467-468:944-950

Loucks RR, Mavrogenes JA (1999) Gold solubility in supercritical hydrothermal brines measured in synthetic fluid inclusions. Science 284:2159-2163

Lowenstern JB, Mahood GA, Rivers ML, Sutton SR (1991) Evidence for extreme partitioning of copper into a magmatic vapor-phase. Science 252:1405-1409

Mackinnon IDR, Mogk DW (1985) Surface sulfur measurements on stratospheric particles. Geophys Res Lett 12:93-96

Mavrogenes JA, Berry AJ, Newville M, Sutton SR (2002) Copper speciation in vapor-phase fluid inclusions from the Mole Granite, Australia. Amer Min 87:1360-1364

Mavrogenes JA, Bodnar RJ, Anderson AJ, Bajt S, Sutton SR, Rivers ML (1995) Assessment of the uncertainties and limitations of quantitative elemental analysis of individual fluid inclusions using synchrotron X-ray-fluorescence (SXRF). Geochim Cosmochim Acta 59:3987-3995

McKay DS, Gibson EK, Thomas-Keprta KL, Vali H, Romanek CS, Clemett SJ, Chillier XDF, Maechling CR, Zare RN (1996) Search for past life on Mars: Possible relic biogenic activity in Martian meteorite ALH84001. Science 273:924-930

McMaster WH, del Grande NK, Mallet JH, Hubbell JH (1969) Compilation of X-ray cross sections. Lawrence Livermore National Laboratory Report UCRL-50174 sec II revision I

Menez B, Simionovici A, Philippot P, Bohic S, Gibert F, Chukalina M (2001) X-ray fluorescence micro-tomography of an individual fluid inclusion using a third generation synchrotron light source. Nucl Instrum Methods Phys Res B 181:749-754

Messenger S (2000) Identification of molecular-cloud material in interplanetary dust particles. Nature 404:968-971

Messenger S, Walker RM (1998) Possible association of isotopically anomalous cluster IDPs with comet Schwassmann-Wachmann 3. Lunar and Planetary Science XXIX, 1906

Mosbah M, Curaud JP, Metrich N, Wu Z, Delaney JS, San Miguel A (1999) Micro-XANES with synchrotron radiation: a complementary tool of micro-PIXE and micro-SXRF for the determination of oxidation state of elements. Application to geological materials. Nucl Instrum Methods Phys Res B 158:214-220

Mosselmans JFW, Schofield PF, Charnock JM, Garner CD, Pattrick RAD, Vaughan DJ (1996) X-ray absorption studies of metal complexes in aqueous solution at elevated temperatures. Chem Geol 127:339-350

Neal RH, Sposito G (1989) Selenate adsorption on alluvial soils. Soil Sci Soc Am J 53:70-74

Newville M, Sutton S, Rivers M, Eng P (1999) Micro-beam X-ray absorption and fluorescence spectroscopies at GSECARS: APS beamline 13ID. J Synchrotron Rad 6:353-355

Nier AO, Schlutter DJ (1993) The thermal history of interplanetary dust particles collected in the Earth's stratosphere. Meteoritics 28:675-681

Ohishi Y, Baron AQR, Ishii M, Ishikawa T and Shimomura O (2001) Refractive X-ray lens for high pressure experiments at SPring-8. Nucl Instrum Methods Phys Res A 467-468:962-965

Otte ML, Kearns CC, Doyle MO (1995) Accumulation of arsenic and zinc in the rhizosphere of wetland plants. Bulletin of Environmental Contamination and Toxicology 55:154-161

Peterson ML, Brown GE Jr. and Parks GA (1997) Quantitative Determination of Chromium Valence in Environmental Samples Using XAFS Spectroscopy. Mat Res Soc Symp Proc 432:75

Philippot P, Menez B, Drakopoulos M, Simionovici A, Snigirev A, Snigireva I (2001) Mapping trace-metal (Cu, Zn, As) distribution in a single fluid inclusion using a third generation synchrotron light source. Chem Geol 173:151-158

Potts PJ, Bowles JFW, Reed SJB, Cave MR (eds) (1995) Microprobe Techniques in the Earth Sciences. Chapman and Hall, London

Rivers M, Sutton SR, Eng P, Newville M (1999) Applications of microfluorescence in earth sciences. Proceedings of 47th Denver X-ray Conference 47:122

Rost D, Stephan T, Jessberger EK (1999) Surface analysis of stratospheric dust particles. Meteoritics and Planet. Sci 34:637-646

Rowe, CL, Kinney OM, Fiori AP, Congdon JD (1996) Oral deformities in tadpoles (Rana catesbeiana) associated with coal ash deposition: Effects on grazing ability and growth. Freshwater Biology 36:723-730

Schreiber HD (1987) An electrochemical series of redox couples in silicate melts: A review and applications to geochemistry. Jour Geophys Res 92:9225-9232

Schulze D, McCay-Buis T, Sutton SR, Huber DM (1995a) Manganese oxidation states in gauemannomyces infested wheat rhizospheres probed by micro xanes spectroscopy. Phytopathology 85:990-994

Schulze D, Sutton SR, Bajt S (1995b) Determination of manganese oxidation state in soils using X-ray absorption near-edge structure (XANES) spectroscopy. Soil Sci Soc Amer J 59:1540-1548

Schulze DG, Bertsch PM (1995) Synchrotron X-ray techniques in soil, plant and environmental research. Adv Agron 55:1-66

Seward TM, Henderson CMB, Charnock JM, Dobson BR (1996) An X-ray absorption (EXAFS) spectroscopic study of aquated $Ag+$ in hydrothermal solutions to 350 degrees C. Geochim Cosmochim Acta 60:2273-2282

Smith JV and Rivers ML (1995) Synchrotron X-ray microanalysis. *In:* Microprobe Techniques in the Earth Sciences. Potts PJ, Bowles JFW, Reed SJB, Cave MR (eds). Chapman and Hall (London, UK), p 163-233

Snigirev A, Kohn V, Snigireva I, Lengeler B (1996) A compound refractive lens for focusing high-energy X-rays. Nature 384:49-51

Snigirev A, Snigireva I, Egström P, Lequien S, Suvorov A, Hartman Ya, Chevallier P, Idir M, Legrand F, Soullie G, Engrand S (1995) Testing of submicrometer fluorescence microprobe based on Bragg-Fresnel crystal optics at the ESRF. Rev Sci Instrum 66:1461-1463

Sowder AG, Bertsch PM, Morris PM (2002) Uranium and nickel speciation and availability in riparian sediments: Impact of aging, source term, and geochemical controls. J Environ Qual, in press

Sutton S, Bajt S, Delaney J, Schulze D, Tokunaga T (1995) Synchrotron X-ray fluorescence microprobe: quantification and mapping of mixed valence state samples using micro-XANES. Rev Sci Instrum 66:1464-1467

Sutton SR, Rivers ML (1999) Hard X-ray synchrotron microprobe techniques and applications. *In:* Synchrotron Methods in Clay Science. CMS Workshop Lectures Vol. 9. Schulze DG, Stucki JW, Bertsch PM. (eds). The Clay Mineral Society, Boulder CO, p 146-163

Sutton SR, Flynn G, Rivers M, Newville M, Eng P (2000) X-ray fluorescence microtomography of individual interplanetary dust particles. Lunar Planet Sci XXXI:1857

Sutton SR, Rivers ML, Bajt S, Jones KW, Smith JV (1994) Synchrotron X-ray-fluorescence microprobe-a microanalytical instrument for trace element studies in geochemistry, cosmochemistry, and the soil and environmental sciences. Nucl Instrum Methods Phys Res A 347:412-416

Suzuki Y, Awaji M, Kohmura Y, Takeuchi A, Takano H, Kamijo N, Tamura S, Yasumoto M, Handa K (2001) X-ray microbeam with sputtered-sliced Fresnel zone plate at SPring-8 undulator beamline. Nucl Instrum Methods Phys Res A 467-468:951-953

Taylor GJ, Crowder AA (1984) Formation and morphology of an iron plaque on the roots of Typha latifolia L. grown in solution culture. American Journal of Botany 71:666-675

Thomas KL, Blanford GE, Clemett SJ, Flynn GJ, Keller LP, Klöck W, Maechling CR, McKay DS, Messenger S, Nier AO, Schlutter DJ, Sutton SR, Warren JL, Zare RN (1996) An asteroidal breccia: The anatomy of a cluster IDP. Geochim Cosmochim Acta 59:2797-2815

Thomas-Keprta KL, Bazylinski DA, Kirschvink JL, Clemett SJ, McKay DS, Wentworth SJ, Vali H, Gibson EK, Romanek CS (2000) Elongated prismatic magnetite crystals in ALH84001 carbonate globules: Potential Martian magnetofossils. Geochim Cosmochim Acta 64:4049-4081

Thompson AC, Chapman KL, Ice GE, Sparks CJ, Yun W, Lai B, Legnini D, Vicarro PJ, Rivers ML, Bilderback DH, Thiel DJ (1992) Focusing optics for a synchrotron-based X-ray microprobe. Nucl. Instrum. Methods A 319 (1-3):320-325

Tokunaga T, Sutton SR, Bajt S (1994) Mapping of selenium concentrations in soil aggregates with synchrotron X-ray fluorescence microprobe. Soil Science 158:421-433

Tokunaga TK, Brown GE Jr., Pickering IJ, Sutton SR, Bajt S (1997) Selenium redox reactions and transport between ponded waters and sediments. Environ Sci Technol 31:1419-1425

Tokunaga TK, Sutton SR, Bajt S, Nuessle P, Shea-McCarthy G (1998) Selenium diffusion and reduction at the water-sediment boundary: Micro-XANES spectroscopy of reactive transport. Environ Sci Technol 32:1092-1098

Tokunaga TK, Wan J, Sutton SR (2000) Transient film flow on rough fracture surfaces. Water Resources Research 36:1737-1746

Tokunaga TK, Wan J, Hazen TC, Schwartz E, Firestone MK, Sutton SR, Newville M, Olsen KR, Lanzirotti A and Rao W (2002) Distribution of chromium contamination and microbial activity in soil aggregates. J Environmental Quality, submitted

Vanko DA, Bonnin-Mosbah M, Philippot P, Roedder E, Sutton SR (2001) Fluid inclusions in quartz from oceanic hydrothermal specimens and the Bingham Utah porphyry-Cu deposit: a study with PIXE and SXRF. Chem Geol 173:227-238

Wadhwa M, Sutton SR, Flynn GJ, Newville M (2002) Microdistributions of Rb and Sr in ALH84001 carbonates: Chronological implications for secondary alteration on mars. Lunar Planet. Sci. XXXIII:1362

Wildenschild D, Hopmans JW, Vaz CMP, Rivers ML, Rikard D (2002) Using X-ray computerized tomography in hydrology: systems, resolutions, limitations, Journal of Hydrology 267:285-297

Wildenschild D, Jensen KH, Hollenbeck KJ, Illangasekare TH, Znidarcic D, Sonnenborg T, Butts MB (1996) A two-stage procedure for determining unsaturated hydraulic characteristics using a syringe pump and outflow observations. Soil Sci Soc Am J 61:347-359

Wong J, Lytle FW, Messmer RP, Maylotte DH (1984) K-edge absorption spectra of selected vanadium compounds. Phys Rev B 30:5596-5610

Yang BX, Rivers M, Schildkamp W, Eng PJ (1995) GeoCARS microfocusing Kirkpartrick-Baez mirror bender development. Rev Sci Instrum 66:2278-2280

Yun W, Lai B, Cai Z, Maser J, Legnini D, Gluskin E, Chen Z, Krasnoperova AA, Vladimirsky Y, Cerrina F, Di Fabrizio E, Gentili M (1999) Nanometer focusing of hard x rays by phase zone plates. Rev Sci Instrum 70:2238-2241

Zolensky ME, Wilson TL, Rietmeijer FJM, Flynn GJ (eds) (1994) Analysis of Interplanetary Dust, AIP Press, NY

9

Soft X-ray Spectroscopy and Spectromicroscopy Studies of Organic Molecules in the Environment

Satish C. B. Myneni

Department of Geosciences
Princeton University
Princeton, New Jersey, 08544, U.S.A.
and
Earth Sciences Division
Earnest Orlando Lawrence Berkeley National Laboratory
Berkeley, California, 94720, U.S.A.

INTRODUCTION

Organic molecules are found everywhere and play an important role in almost all biogeochemical processes occurring on the surface of the Earth (Aiken et al. 1985; Thurman 1985; Schwartzenbach et al. 1993; Senesi and Miano 1994). They are found in soluble and insoluble phases, coatings on mineral and colloidal particles, and in gas phase molecules in soils, sediments and aquatic systems. The activities of macro- and micro-fauna and flora release organic molecules of various sizes and composition. Photochemical reactions in the atmosphere also add certain small chain molecules to the organic carbon content in the environment. Significant compositional variations occur in natural organic molecules, which include small chain carboxylic acids, alcohols and amino acids; and polymeric, polyfunctional and polydisperse macromolecules such as humic and fulvic acids. The behavior of small chain molecules and their influence on different geochemical reactions is well understood. However, understanding of the chemistry of biopolymers and their role in different biogeochemical processes in the environment is poor, which may be attributed to the unavailability of instrumentation to examine the chemistry of natural organic molecules in their pristine state.

Two important properties that dictate the behavior of natural organic molecules and biopolymers in the environment are: functional group chemistry and (macro)molecular structure (Schnitzer 1991). Evaluation of these two properties is complicated by the compositional and structural heterogeneity of the naturally occurring organic molecules, and their ability to form intramolecular and intermolecular H-bonds, which further modify their structure and chemical reactivity. These two properties are interrelated and one influences the other. The chemical composition of natural waters (pH, ionic composition and concentration, redox conditions) and soil and sediment particle surface chemistry (composition, coordination environment, number of reactive groups) also modify their behavior. It is not well understood how each of these environmental variables influences the behavior of organic molecules in nature. One of the challenges that geochemists face is to gain a better understanding of the relation between chemical composition and structure, and the biogeochemical behavior of naturally occurring organic molecules. While conventional laboratory spectroscopy and microscopy techniques suffer from lack of sensitivity, element- and group-specific X-ray spectroscopy and spectromicroscopy methods using synchrotron radiation can allow the examination of these two fundamental properties of natural organic molecules in their original state.

Following are some of the various direct and indirect macroscopic and molecular methods researchers have been using routinely to evaluate the properties of organic molecules:

1529-6466/00/0049-0009$10.00

Functional group chemistry: Nuclear magnetic resonance spectroscopy, vibrational spectroscopy, electron spin resonance spectroscopy, UV-VIS absorption and fluorescence spectroscopy, pyrolysis mass spectrometry, size exclusion chromatography, and potentiometric titrations

Macromolecular structure: Electron microscopy, atomic force microscopy, scanning tunneling microscopy and light scattering.

In order to use these routine laboratory techniques, organic molecules have to be isolated and concentrated from their surrounding geological matrices. Very often samples are subjected to harsh chemical treatment before they are characterized (Oades 1989; Christensen 1992; Stevenson 1994). During isolation, some of the molecules are preferentially concentrated, while others are lost or perhaps altered. Hence, several researchers question the validity of information obtained from studies conducted on organic molecule isolates. Of the chemical characterization techniques, NMR spectroscopy has provided most of the information available today about the functional group chemistry of natural organic molecules. However, only C, N, and P functional groups can be examined using NMR spectroscopy. This method also has limitations when it comes to the direct probing of molecules in geological matrices. Although vibrational, UV-VIS photoabsorption and luminescence spectroscopy can provide direct information on the functional group chemistry of organic molecules without any special sample preparation, they lack the sensitivity necessary to differentiate different forms of carbon. Evaluation of the macromolecular and aggregate structure of organic molecules also suffers from similar problems. Although electron microscopy provides atomic scale information, the samples have to be examined under extremely dry ultrahigh vacuum conditions. Further, the damage caused by the electron beam to organic molecules is not desirable. For the scanning force microscopes, the samples have to be placed on atomically flat surfaces for obtaining nanometer scale resolution.

Despite these serious limitations, researchers have made significant contributions to the understanding of the chemistry of natural organic molecules and their role in environmental processes. Application of *in situ* X-ray spectroscopy and spectromicroscopy methods will facilitate a big leap in the current understanding of the natural organic molecule chemistry. This chapter discusses how organic molecule functional group chemistry and macromolecular structures can be probed using X-rays. Researchers have begun using X-ray spectroscopy and spectromicroscopy methods in the last few years for studying the chemistry of natural organic molecules. Several research facilities optimized to examine environmental samples are being built at the synchrotrons and this will have a major impact on this field. This will only be a beginning for the exploration of the chemistry of natural organic molecules.

SOFT X-RAY SPECTROSCOPY & CHEMICAL BONDING IN ORGANIC MOLECULES

Electrons are important players in all chemical reactions. The number of electrons and their location (energy levels) in the molecule governs its behavior in the environment, and the electron and X-ray spectroscopy methods can provide information on the status of electrons in molecules. The electron energy levels in molecules and their respective electronic spectra can be well understood by examining the electronic states and spectra of simple mono and diatomic molecules.

In single atoms, such as in H and noble gases, energy states of electrons can be determined from the Coulomb potential of the nucleus. Some of these states are empty (Rydberg states) and differences in their energy states becomes smaller and smaller with distance away from the nucleus and finally converge to the vacuum level (corresponding to the ionization energy). The continuum states lie above the ionization energy and have no relation to the atom (Fig. 1). When electrons in single atoms are excited, the electrons jump from occupied atomic orbitals to the empty Rydberg orbitals. Such transitions can occur when incident energy is equal to the difference between the energy states of these orbitals. The electronic spectra (absorption as a function of energy) show distinct sharp peaks that correspond to the electronic transitions to these discrete levels (Fig. 1).

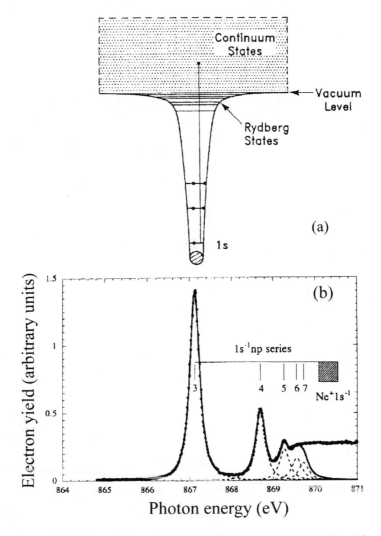

Figure 1. The atomic orbitals, and the potential well of a single atom are shown in (a), and the core hole experimental NEXAFS spectrum of a single atomic species is shown in (b). The NEXAFS spectrum is for Ne. The first available empty orbitals are of $3s$ and $3p$ character, and the most intense first peak at ~ 867.2 eV in the spectrum corresponds to $1s$ electronic transitions into $3p$ states. The second and higher peaks correspond to the higher energy empty orbitals of $4p$ and $5p$ character [(a) Used by permission of Springer-Verlag from Stohr (1992); (b) after Coreno et al. 1999).

When two atoms are brought closer, such as in the diatomic molecules O_2 and N_2, significant overlap of the atomic orbitals of these two atoms forms molecular orbitals. Electrons from these two atoms become delocalized and move into these molecular orbitals (Fig. 2). The linear combinations of atomic orbitals results in formation of both bonding and antibonding orbitals, which are below and above the energy levels of the atomic orbitals of individual atoms, respectively. The energy levels of the bonding and

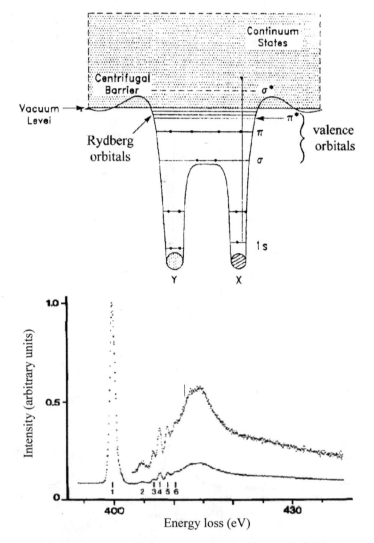

Figure 2. The atomic and molecular orbitals, and the potential well for a diatomic molecule are shown at the top, and a representative corehole NEXAFS spectrum is shown at the bottom (for N in NO). The most intense low-energy peak in the experimental spectrum of NO corresponds to the electronic transitions of N $1s \rightarrow \pi^*$ orbitals, and peak 2 corresponds to the N $1s \rightarrow \sigma^*$ orbitals, and all other higher energy peaks correspond to the Rydberg orbitals [(a) Used with permission of Springer-Verlag from Stohr (1992); (b) Used with permission of Elsevier Science from Wight and Brion (1974a)].

antibonding orbitals and their energy differences are affected by the intensity of the overlap of the atomic orbitals. In addition, changes in the type of atom and its energy states also modify the energy levels of molecular orbitals (e.g., O_2 versus CO). The electronic spectra of these molecules can provide direct information on orbital energy levels, and thus the strength of interactions between atoms. Since the innermost electrons (core electrons), such as the $1s$ electrons, are localized and are not affected significantly by the overlap of the outermost (valence) orbitals, the core electron spectrum (transitions of core electrons) can be used as a ruler to measure the bonding interactions between different atoms. In the case of CO, the energy states of valence orbitals can be probed by exciting electrons in C and electrons in O. Similarly, electrons in bonding orbitals can also be excited and the energy gap between the occupied and unoccupied orbitals can be measured (UV-VIS spectroscopy).

The same approach can be used to understand the chemical bonding in polyatomic molecules and polymers. The electronic spectrum of a polymer can be described as the sum of the spectra of individual diatomic components (Fig. 3). Since the orbital

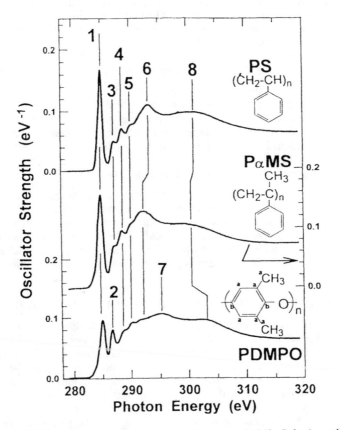

Figure 3. C NEXAFS spectrum of polymers. PS: polystyrene, PαMS: Poly (α-methyl styrene), PDMPO: poly(dimethyl p-phenylene oxide). The peaks 1 through 8 correspond to the C $1s$ transitions into orbitals of C=C π^* (peak 1), [b]C=C π^* (peak 2), C-H σ^* (peak 3), C=C π^* (peak 4), ionization threshold (peak 5), C-C σ^* (peak 6), [b]C-C σ^* and C-O σ^* (peak 7), and C=C σ^* (peak 8). [Reprinted with permission of Elsevier Science from Kikuma and Tonner (1996)]

composition and energy states of an atom of interest in a molecule are modified primarily by the energy states of its first neighbors, such a building block picture is valid for describing the electronic structure of polymers (Stohr 1992; Kikuma and Tonner 1996; Brown et al. 1999). However, it should be remembered that it is not simply the interactions of diatomic molecules and their respective orbitals that contribute to the overall spectrum of this central atom; the complete local coordination around this atom of interest and the nature of the hybridized orbitals of this atomic group should be considered. In case of molecules that can exhibit delocalized electrons (e.g., molecules with a conjugated π system), coordination environment beyond the first neighbors should also be considered.

The core electrons in atoms and molecules can be excited using high-energy electrons (electron energy-loss spectroscopy, EELS) or X-ray photons (X-ray absorption spectroscopy, XAS). A typical energy-loss or X-ray absorption spectrum of a molecule exhibits intense peak(s) that correspond to the bound state electronic transitions at low energies (also referred to as the absorption edge or white line) followed by the scattering related features a few eV above the bound state transitions (Fig. 4). The part of the X-ray absorption spectrum that corresponds to the bound state transitions is referred to as near edge X-ray absorption fine structure (NEXAFS) or X-ray absorption near edge structure (XANES), and the spectrum well above the bound state transitions is called extended X-ray absorption fine structure (EXAFS). The corresponding EELS are referred to as energy-loss near-edge structure (ELNES), and extended energy-loss fine structure (EXELFS) for the bound state and above bound state scattering features, respectively.

Since the electronic transitions are sensitive to the local coordination environment, electronic spectra can be used to identify the functional groups of organic molecules and their coordination environment. One of the important questions is how sensitive are these electronic spectra to the variations in the local coordination environment around atoms? Is it possible to identify all the functional groups in molecules? Are these spectral features useful for identifying the coordination state of organic molecules in unknown samples? These questions are addressed in the next few sections.

A few years ago, all these techniques and this energy range was used for probing samples under ultrahigh vacuum conditions only. However, the availability of high-flux and high brightness synchrotron sources and the recent innovations in sample chamber designs completely altered this scenario, and made studies at atmospheric pressure conditions possible. This will bring a new suite of element and functional group-specific X-ray spectroscopy and spectromicroscopy methods to studies of solutions and organic molecule chemistry.

INSTRUMENTATION FOR STUDYING ORGANIC MOLECULES

Electronic structure of organic molecules has been traditionally investigated using the EELS facilities (Hitchcock and Mancini 1994; Egerton 1997). The EELS studies have provided the very first information on the electronic structure of gas phase organic molecules and of molecules on surfaces. But poor detection limits and the electron beam damage to organic molecules have limited the applications of this technique in geochemistry and environmental interfacial chemistry. When analyzed for radiation doses, studies on synthetic polymers indicate that there is approximately a 500-fold advantage for XAS conducted at room temperature over the EELS conducted at 100 K for a 20 eV NEXAFS scan (Rightor et al. 1997). In addition, the ultrahigh vacuum sample environment is disadvantageous for probing wet samples. However, EELS can be obtained from subnanometer regions of a sample in an electron microscope, which is impossible to achieve with any other technique at this stage. On the other hand, high-

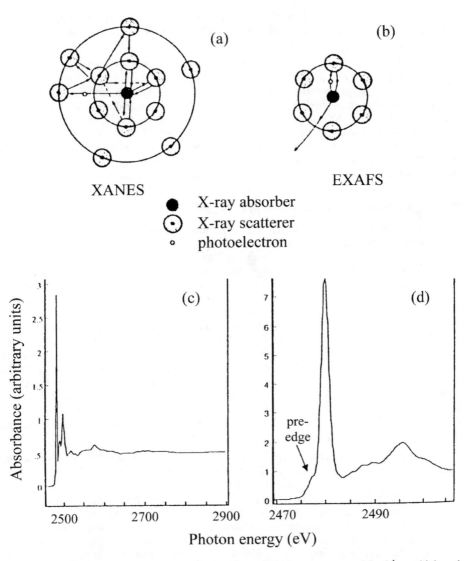

Figure 4. XANES (or NEXAFS) and EXAFS regions of XAS spectra are explained by multiple and single scattering of photoelectrons by the neighboring atoms of the absorber, respectively (a, b: modified from Brown et al. 1988; Myneni 2000). The EXAFS and XANES spectral contributions of the S X-ray absorption spectrum are shown in (c) and (d), respectively (Myneni 2000). The spectrum in (c) is for alunite and the spectrum in (d) is for jarosite.

resolution electronic spectra can be collected for wet samples using X-ray spectroscopy methods. In addition, electronic spectra can also be collected from small regions of a sample (as low as 30 nm for light elements in organic molecules) using X-ray spectromicroscopy facilities. Sample alterations are also minimum or absent when electronic excitations are conducted using X-ray photons. This is because; the depth of penetration of the X-ray beam is much greater and the energy is distributed over a larger volume than that of the incident electron beam. Different X-ray spectroscopy methods (NEXAFS, XES, XPS) are also available for studying the chemistry of organic molecules under wet conditions.

The chemistry of organic molecule functional groups can be probed using their $1s$ (K absorption edges), $2s$ and $2p$ (L_1, and L_2 and L_3 absorption edges, respectively) electronic transitions. The L edges are useful in the case of P, S and Cl, and complement the K edge spectra obtained at much higher energies. The $1s$ electronic transitions of C, N, O, P, S, and Cl are at ~ 280, 400, 530, 2140, 2475, 2820 eV, respectively; and the $2s$ and $2p$ transitions of P, S and Cl are at 190,135; 230,163; and 270,200 eV, respectively. All these energies cannot be accessed at any single synchrotron beamline. Photon energies below 1200 eV are accessible at regular soft X-ray beamlines, and those above 2000 eV are accessible at some soft X-ray and intermediate energy beamlines. Photons having energies < 4000 eV are generally referred to as soft X-rays.

Probing the 1s electronic transitions of C, N and O

The $1s$ states of C, N, O and the $2s$ and $2p$ transitions of P, S, and Cl can be accessed at the same beamlines because of the proximity of their absorption edge energies (130-600 eV). All soft X-ray synchrotron facilities and a majority of the high-energy synchrotron facilities have beamlines optimized for conducting NEXAFS spectroscopy at these energies. These facilities are commonly equipped with grating monochromators that can provide a theoretical spectral resolution better than 0.1 eV (Stohr 1992; Attwood 2000). Since soft X-ray photons at these energies are strongly absorbed by air, all beamlines that can access these energies are maintained under high vacuum and samples experience approximately 10^{-6} torr or less. For this reason, only gas phase molecules or dry solids can be examined at these facilities. Samples can be examined in transmission (for extremely thin films < 100 nm), and also by using sample fluorescence yield and electron yield (partial and total electron yield). The transmission and fluorescence methods are bulk-sensitive (several tens of nanometers), and the electron yield detection is more surface-sensitive (top few nanometers or less). Stohr (1992) discussed details of data collection procedures and the types of detectors for these soft X-ray energies, and the reader can refer to this publication for further details.

Probing the 1s electronic transitions of light elements in aqueous systems at the atmospheric pressure. Applications of core electron spectroscopy to molecules in liquid state and at relatively high pressures (a few torr) were begun by Seigbahn and his group members (Siegbahn 1985). The apparatus built by this group consists of an excitation source in ultrahigh vacuum, a sample chamber maintained at elevated pressures (close to 10 millitorr), and an electron detection system for collecting the released electrons from the sample (Fig. 5). The pressure in the sample chamber is maintained at several millitorr using a differential pumping system. The liquid samples can be pumped continuously into the sample chamber, or the solution is exposed to the beam using a rotating drum, which is partly immersed in a sample reservoir and exposes fresh sample to the beam continuously upon rotation of the drum. The sample can be excited with either the monochromatic electrons or X-rays. The vapors produced in the sample chamber can be removed continuously by pumping and also by using a liquid nitrogen cold trap. Since electron detection is used, this method is extremely sensitive to the liquid surface (a few tens of Å) and researchers should be cautious with sample surface contamination. In addition to electron yield detection, researchers can use photon detection in this setup.

Synchrotrons are ideal for probing the core electronic transitions of light elements in ambient conditions because of the availability of intense and bright X-ray beams. Recently numerous Siegbahn-type chambers, with several modifications, have been built

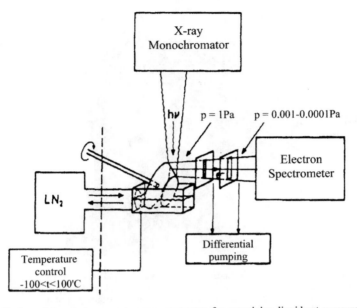

Figure 5. Design of the electron spectroscopy apparatus for examining liquids at pressures close to several millitorr (from Siegbahn 1985).

at the synchrotrons. At the Advanced Light Source (ALS; Lawrence Berkeley National Laboratory, LBNL) researchers built 3 different types of endstations that can facilitate studies on liquids and their interfaces, wet pastes, and solid-vacuum interfaces. All these chambers are used in studying the chemistry of liquid water and their interfaces, and interested readers can refer to Myneni et al. (2002), Bluhm et al. (2002), and Wilson et al. (2002) for more details on these tools. For a majority of these investigations, it is necessary to have high-flux and bright synchrotron beams. Synchrotron endstations that can operate at a few torr of pressure at other synchrotron sources are available and studies on monolayers of organic films on water surfaces have been conducted. A new synchrotron beamline optimized for examining environmental samples of interest in this energy range is under construction at the ALS and this facility will be operational by the beginning of 2003. Similar facilities are also planned in Europe at the Swiss Light Source.

This author, in collaboration with scientists at the ALS, built a special apparatus, Soft X-ray Endstation for Environmental Research (SXEER) to obtain high flux beams to examine dilute organic molecules in aqueous solutions and at mineral-water interfaces under atmospheric pressure conditions (Fig. 6). In addition, this endstation is also useful for examining the L-, and other higher-edges of heavy elements in different sample environments. The SXEER endstation has been operational at the ALS for the past four years. The SXEER is differentially pumped and the vacuum condition in the chamber increases from 10^{-9} torr to 10^{-5} torr without any windows (Fig. 6). This windowless environment minimizes photon losses. The pressure can also be increased to a few torr by using a series of differential pumps aided with strong turbo pumps. In SXEER, this high vacuum environment is separated from the sample environment, which is maintained at the atmospheric pressure conditions, by a thin silicon nitride, diamond, graphite or a carbon polymer window. The sample chamber is filled with He to minimize the photon absorption by air. The liquid or wet samples are placed in a tube of diameter ~ 2-3 mm,

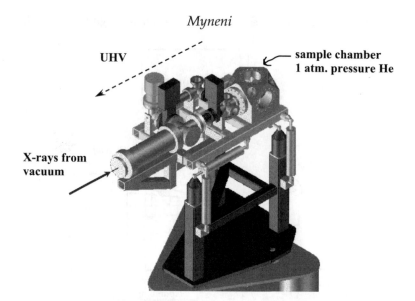

Figure 6. Design of the Soft X-ray Endstation for Environmental Research (SXEER), optimized to examine the chemistry of light elements under ambient conditions. The samples are placed at 1 atm. pressure in the sample chamber, which is isolated from the high vacuum part of the station with thin X-ray transparent windows (Myneni, unpublished data).

and the tube holds the samples through capillary forces. The samples are exposed to the X-ray beam directly and their NEXAFS spectra can be collected using sample fluorescence. Liquid samples sensitive to the X-ray beam can be examined using a flow-through cell, in which fresh samples are exposed to the X-ray beam continuously. Using this chamber the author and his collaborators examined the coordination environment of H_2O molecules in liquid water and in a variety of other environments (e.g., ion solvation, pH influence) by studying at the oxygen edge (Myneni et al. 2002).

A similar synchrotron endstation, optimized to conduct X-ray photoelectron spectroscopy (XPS) at the solid-vacuum, and liquid-vacuum interfaces at pressures up to 5 torr ($0°C$), has been built at the ALS by Salmeron's group (Bluhm et al. 2002), which is similar to the endstation built by Siegbahn's group. Using this high-pressure XPS chamber, researchers have examined the chemical state of H_2O on ice surfaces at different temperatures and pressures in the stability region of water. Similar experiments at soft X-ray energies have been conducted using the water-jet experiments (Saykally and his group at the University of California at Berkeley; Wilson et al. 2002). Nordgren and his group members have been accessing these energies by placing liquid samples in gas cells, completely isolated from the vacuum environment of the beamline using thin silicon nitride windows. They have successfully collected the X-ray emission spectra at the O K-edge for liquid water (personal communication). The core hole spectra of these light element absorption edges can also be accessed at transmission X-ray microscopy beamlines but only for concentrated samples (details discussed later).

Probing the 1s electronic transitions of P, S and Cl

The 1s transitions of P, S and Cl (energy range: 2100-3500 eV) are typically conducted at beamlines equipped with crystal monochromators. Since crystal monochromators that can provide the optimal spectral resolution, stability and structureless incident beam are uncommon for the 2-4 KeV energy range, and since grating monochromators can not be used in this range, it is a difficult range in which to work (Myneni 2000). For this reason, many synchrotron beamlines are not optimized for this energy range. The monochromator crystals that allow for the study of the absorption edges of these elements are Si, Ge, InSb and Yb_{66}. The energy resolution obtained with these monochromator crystals is ~ 0.5-0.7 eV. Although NEXAFS spectroscopy studies can be conducted under ambient conditions at these energies more easily when compared to the very soft XAS investigation discussed above, such studies can only be conducted at high flux facilities because of the strong absorption of X-ray photons by air. Although the core hole broadening for these absorption edges is around 0.4 eV (Brown et al. 1988; Stohr 1992), researchers have recorded spectra with a resolution better than 0.4 eV. Further, spectral variations as low as 0.1 eV can be recorded accurately at these energies. EXAFS spectroscopy studies can be conducted for these elements, but dilute samples cannot be examined at the undulator or wiggler beamlines (which typically provide the highest photon flux) because of the spectral normalization problems. High flux bend magnet beamlines are ideal for EXAFS spectroscopy studies because of the smooth incident beam. For S samples containing large concentrations of Fe, the higher order harmonics component of the beam at S absorption edge can cause significant Fe fluorescence and increases the background and decreases the sensitivity.

Liquid or wet samples are typically probed either in fluorescence or in transmission. Fluorescence detection offers better sensitivity and sample handling is much easier. Currently Stern-Heald-Lytle type detectors and photodiodes are used for detecting sample fluorescence. Availability of multielement solid state detectors would improve the sensitivity and would allow the examination of these elements at micromolar concentrations.

Detection limits

As mentioned earlier, the C, N, O functional group K edges can be probed either in vacuum or under ambient conditions using different specialized chambers. Using the vacuum beamlines researchers have examined the chemical state of organic molecules adsorbed on metal and mineral surfaces well below monolayer coverages (Stohr 1992; Brown et al. 1999). For aqueous solutions, the author and his group have collected NEXAFS spectra down to 10 mM, and concentrations as low as a few mM can be measured using the SXEER and one of the highest flux beamlines at the ALS. Although aqueous samples and wet pastes have been examined at the X-ray microscopy beamlines, the flux at these beamlines is about 3 orders of magnitude less than that obtained with the SXEER (when attached to a similar vacuum beamline). As such, the SXEER can be used to study lower sample concentrations. The detection limits at the other endstations described earlier were not evaluated. The fluorescence yields at the L edges are much lower and thus the detection limits would be much lower than that of K-edges. Using a high flux wiggler or an undulator beamline optimized for P, S, and Cl X-ray studies (photon flux $> 10^{10}$ photons/sec), NEXAFS spectra can be collected for less than 500 µM in aqueous solution.

Soft X-ray spectromicroscopy facilities

Imaging with X-rays can be conducted using the differences in:

- mass absorption between different parts of the sample,
- characteristic excitations of different elements and also different functional groups of the same element, and
- polarization dependent absorption.

Recently, a summary of spectromicroscopy facilities has been reviewed by several researchers (Bertsch and Hunter 2001), including the chapter by Sutton et al. (this volume), and the readers may refer to this chapter for more details. A brief introduction to these facilities and their applications to organic molecule studies has been presented here.

With soft X-ray microscopy (or spectromicroscopy), NEXAFS spectra for different elements can be conducted in transmission mode at a spatial resolution of ~ 30 nm. Thus the high spatial resolution and spectral sensitivity are advantages when compared to visible light microscopes or spectromicroscopy methods that use infrared radiation. However, visible and infrared light causes much less or no damage to the samples when compared to X-rays. According to this author, not all natural samples are sensitive to the X-ray beam and is dependent upon the chemical states of the molecules. Of all the microscopes, the electron microscopes produce the highest spatial resolution. However, the vacuum environment of samples and beam damage in the presence of the electron beam are disadvantageous for probing organic molecule functional groups. Recently cryogenically cooled sample stages for X-ray and electron microscopes have been introduced. Although sample cooling drastically reduces beam damage, it is not clear whether the hydration and solvation state of wet samples are well preserved. During ice formation in wet samples, it is possible that concentration and pH gradients may be established and may modify the sample state completely.

For imaging with X-rays, the X-ray beam can be focused on a small spot using zone plates, Kirkpatrick-Baez mirrors, or tapered capillaries (Attwood 2000; Bertsch and Hunter 2001). With the exception of photoemission electron microscopes (PEEM, details discussed later), all soft X-ray microscopes are based on focusing with zone plates and uses sample transmission to collect high-resolution images (Jacobsen et al. 1992; Ade 1998). The spatial resolution depends on the outer zone width of the zone plates. There are two types of transmission X-ray microscopes: conventional X-ray microscopes (pioneered by Schmahl and his collaborators at the University of Gottingen) and scanning X-ray microscopes (pioneered by Kirz and his group at the State University of New York, Stony Brook). Currently the highest resolution in an X-ray microscope is 25-30 nm obtained with the conventional transmission X-ray microscope installed at the ALS by Meyer-Ilse. In this microscope, a selected area of a sample is illuminated by X-rays using a pinhole and a condenser zone plate. The transmitted X-ray beam is enlarged and imaged onto a CCD detector using an objective zone plate. The chromatic aberrations of the condenser zone plate can be used to get monochromatic X-rays to the sample (Fig. 7). The radiation dose at the sample in these microscopes is about an order of magnitude greater than the scanning microscopes, and this is due to the poor efficiency of the objective zone plates.

In the scanning transmission X-ray microscopes (STXM), the beam is focused on the sample using a condenser zone plate and the transmitted beam is directly collected using a photodiode, or a similar detector (Fig. 7). In STXM, a 2-D image is collected by scanning the sample stage at a fixed photon energy, set by the grating and by keeping the zone plate at the appropriate focal distance from sample. A sequence of images can be collected at different energies by changing the photon energy and by moving the zone plate accordingly. This stack of images provides the absorption contrast and a NEXAFS spectrum in the third dimension at any selected location on the image. Currently Jacobsen at Stony Brook, and Warwick, Ade and Hitchcock at the Advanced Light Source are optimizing the STXM to improve spatial and spectral resolution. Similar STXMs are also in operation/planned in Europe at BESSY and Swiss Light Source.

For imaging at the P, S, and Cl K-absorption edges and for other heavy element absorption edges, researchers use pinholes (at several synchrotrons), tapered capillaries (GN George, personal communication; Steve Heald and colleagues at the APS; Bilderback et al. 1994), Kirkpatrick-Baez mirrors (at several synchrotrons), and also zone

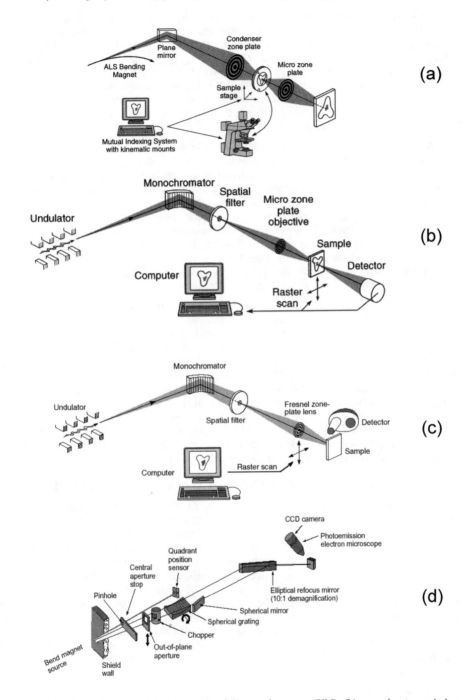

Figure 7. Soft X-ray microscopes. (a): conventional X-ray microscope (XM); (b): scanning transmission X-ray microscope (STXM); (c) Scanning Photoemission Microscope (SPEM); and (d) Photoemission Electron Microscope (PEEM). (taken from *www.als.lbl.gov*).

plates (at the Advanced Photon Source, European Synchrotron Radiation Facility). Currently research facilities optimized to examine wet samples under ambient conditions with a high spatial resolution (300 nm), similar to that achieved at the STXMs optimized for lighter elements, are not available. Theoretically, the highest resolution imaging can be conducted below 2500 eV. Since fluorescence detection is possible much more easily at the P, S and Cl absorption edges, molecules containing these atoms can be detected at lower concentration using spectromicroscopy methods. High resolution imaging with hard X-rays (> 3000 eV) is more difficult to achieve because of diminished absorption and phase contrast at higher energies.

All these microscopes allow imaging of organic molecules in soil or sediment samples without any special treatment. The samples have to be thin enough so the X-ray beam can penetrate through the sample (~ few hundred nanometers optimum thickness, varies with the incident photon energy and sample composition and density). The author and his group tried to obtain images by using fluorescence detection. With the existing detectors, however, it is not feasible to obtain images for thick samples. Although the origin of the signal is not well understood, electron/ion yield type spectra can be collected by applying a voltage to the order sorting aperture in front of the sample (Hitchcock, pers. comm.). However, the spectra are noisier than the transmitted spectra. At present these difficulties limit the examination of molecules present on thick particle surfaces. Development of better detection schemes will allow imaging with fluorescence detection in future.

For imaging organic molecules in vacuum at soft X-ray energies, there are two types of microscopes: the scanning photoemission microscope (SPEM) and the photoemission electron microscope (PEEM) (Fig. 7). The operation of SPEM is similar to the STXM, but the sample surface is probed by examining the released photoelectron (using an electron analyzer), or the photoabsorption (by measuring the sample current). The spatial resolution obtained with the SPEM is dependent on the zone plate used. In PEEM, the sample is illuminated with the X-rays, and the released electrons are collected and imaged using a series of lenses. Although atomic resolution is theoretically possible with this microscope, the current PEEMs can produce a spatial resolution of about 50 nm (Tonner et al. 1994). High resolution PEEM is currently under construction at the ALS and the expected resolution is close to a nanometer. Although samples have to be dried, organic molecules on thick sample surfaces can be examined using these microscopes.

FUNCTIONAL GROUP CHEMISTRY OF ORGANIC MOLECULES

In naturally occurring organic molecules N, O, P and S functional groups are the most common moieties. Recent studies also indicate that organochlorines are present at high concentrations in the organic fraction of soils and natural waters, and warrant the consideration of these molecules and other organohalogens in all biogeochemical reactions. Although metalloproteins and enzymes derived from parent biological material also enter into the dead organic fraction of soils and sediments, their stability and residence time in the natural systems is not well understood. Stable organometallic compounds of As, Se, Hg and other heavy elements also constitute a significant fraction of these respective elements in different environments. Researchers have been using metal NEXAFS and EXAFS to understand the coordination chemistry of organometallic systems and metalloproteins, and a review of these studies is presented elsewhere. Hence the focus of this discussion is limited to the chemistry of dominant functional groups of organic molecules present in the environment.

C functional groups

Carbon constitutes the primary building block of all organic molecules, and several spectroscopic methods are available for characterizing the chemistry of different C functional groups in organic molecules. Of these methods, NMR spectroscopy has provided unique information on the chemical states of different C functional groups in organic macromolecules, which is difficult to obtain with any other technique. Applications of XAS have changed that view, and this method has shown to provide *in situ* chemical information on untreated natural samples. Although NMR spectroscopy is the most sensitive method for characterizing different types of C, NEXAFS spectroscopy is the only technique that can offer at least the same level of information as NMR when direct chemical characterization is necessary for natural samples. Researchers have been examining the electronic states of C in several different gas phase molecules and solids using EELS and XPS for over three decades. At the advent of synchrotron methods and NEXAFS spectroscopy, researchers have started examining the reactions of molecules on surfaces at submonolayer coverages without causing damage to the sorbed species or the substrate. Probing of the chemistry of C in geological matrices has started only recently.

In natural systems, C exists in several different aliphatic and aromatic forms (Thurman 1985; Wilson 1987; Stevenson 1994; Orlov 1995). A variety of substituents can be expected, and these can include a series of organic moieties, halogens, chalcogens, and metal atoms. The NEXAFS spectra of each group are expected to change with the type(s) of substituents in the molecule. With a combination of XAS studies at the absorption edges of C and other common elements in organic molecules (N, O, etc. absorption edges), unequivocal information on the types of functional groups and their relative concentrations in natural organic molecules can be obtained. However, the NEXAFS technique cannot provide the size and structure of each type of these molecules, but only the local coordination environment around different atoms.

X-ray spectra and chemical state of model organic compounds. An attempt has been made to present a survey of the published electronic spectra of a series of model organic molecules, small and long chain organic polymers and biopolymers with their band assignments in Table 1 (in Appendix). Fortunately several researchers have examined a series of C containing compounds using the EELS and NEXAFS spectroscopic techniques, and this information is useful in identifying the coordination environment of C in naturally occurring organic molecules. A summary of these observations is presented here.

The EELS and NEXAFS spectra of different gas phase molecules and solids exhibit distinct spectral features corresponding to the C $1s \rightarrow \pi^*$ transitions of different C moieties, and these features can be used to identify the presence of respective functional groups. A majority of the π^* transitions occur between 284 and 291 eV, with some overlap with the σ^* transitions above 287 eV (with some exceptions, Table 1—in Appendix). Molecules containing C=C and C≡C exhibit C $1s \rightarrow \pi^*$ transitions around 285 eV; C=N and C≡N around 286.3; C=O of ketonic group around 286.8 eV; COO in carboxylate groups around 288.6 eV; and CO$_3$ in carbonate around 290.5 eV. Their corresponding σ^* transitions are relatively broad (especially in condensed phases) and are present at energies above ~ 293 eV and do not interfere with these assignments. Variations in the coordination environments of these different groups may shift the energies of these different transitions (Table 1—in Appendix). Although the C=N and C=O (ketones) moieties exhibit significant spectral overlap, it is possible to use the NEXAFS spectra of N and O edges to identify the presence of these individual groups unambiguously. If the elemental composition of these molecules is known, phenolic groups can be identified unequivocally from one of its $1s \rightarrow \pi^*$ transitions at ~ 287.2 eV (Table 1—in Appendix).

Similarly, the $1s \rightarrow \sigma^*$ transitions of certain C moieties are also useful for their identification. Some of these are C-H in alkanes and several other common organic molecules that exhibit distinct transitions around 287.8 eV, and C-OH in alcohols that show features around 291 eV. However, the 1s→σ* transitions in large polyfunctional polymeric organic molecules may not be clearly identified if they are above 292 eV. This is because of broader peak widths for these transitions, and significant overlap of bands corresponding to several transitions in polymers. However, all the functional groups and electronic transitions of smaller molecules, either in solution or on surfaces, can be identified. In addition, the energies of the 1s→σ* transitions vary linearly with bond length. This is useful for identifying the local coordination variation around functional groups of interest, provided the transitions are identified unambiguously (Fig. 8, Stohr 1992).

The NEXAFS spectrum of a large molecule can be deconvoluted and the individual contributions from different functional groups can be examined directly using the building block approach (discussed earlier; Stohr 1992; Kikuma and Tonner 1996; and several articles referred to in the next section). As mentioned earlier, researchers should be cautious with this kind of interpretation for molecules that contain conjugate π system of electrons. A discussion on spectral variations in molecules with conjugate double bonds is presented for the N functional groups, and the readers can refer to this section for more details. In addition, the vibrational fine structure associated with the core hole transitions, especially for the intense and narrow 1s→π* transitions, causes broadening of the spectra. It is possible that these individual features can be mistaken for different functional groups during peak deconvolution. Theoretical analysis can be used to resolve this issue to some extent and to identify the expected number of transitions for each of the moieties (Fig. 9; Urquhart et al. 2000). Chemical composition (e.g., elemental analysis) of the molecules is also useful in spectral interpretation.

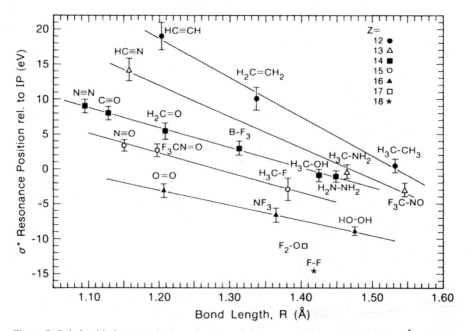

Figure 8. Relationship between the bond length, and the energy difference between the σ* resonance and the ionization potential (IP). Z = sum of the atomic numbers of the bonded pair. [Used with permission of Springer-Verlag from Stohr (1991), Fig 8.6, p. 250.]

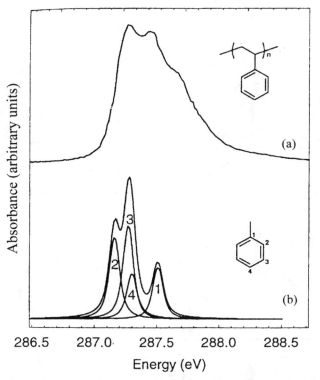

Figure 9. Experimental C 1*s* NEXAFS spectrum of polystyrene (a), and the simulated C 1*s* spectrum of toluene (b). The numbered peaks in (b) correspond to the π^* transitions of different C atoms in the molecule, and spectral broadening and additional spectral features in the experimental spectrum are attributed to the vibrational fine structure. [Reprinted with permission from Elsevier Science from Urquhart et al. (2000), Fig. 3].

X-ray spectra and chemical state of organic compounds in the environment. Preliminary NEXAFS spectroscopic studies conducted at the C edge focused on understanding the coordination environment of C in synthetic polymers. Using C NEXAFS spectra, researchers have identified the coordination environments of different groups, and correlated this information to the physical and chemical properties of synthetic polymers. Interested readers may refer to a series of articles published by Ade, Stohr, Tonner, Hitchcock and their coworkers (Ade et al. 1992, 1995; Ade 1998; Rightor et al. 1997; Rightor et al. 2002; Hitchcock et al. 1989; Urquhart et al. 1997; Kikuma and Tonner 1996). Their spectral interpretations are useful for identifying different components of the NEXAFS spectra of natural organic molecules.

Initial studies of C NEXAFS spectra of natural organic molecules in aqueous solutions were performed using STXM at the ALS and at the National Synchrotron Light Source. Extremely high concentrations of organic molecules (~M or greater) were used in all these investigations. One of the detailed investigations conducted on the NEXAFS spectroscopy of a series of amino acids under ambient conditions indicates that the carboxylic groups of all amino acids exhibit a distinct feature corresponding to the $1s \rightarrow \pi^*$ transitions at 288.6 eV, with significant variations for the other C groups in the amino acids (Fig. 10; Kaznacheyev et al. 2002). These studies were also complemented by

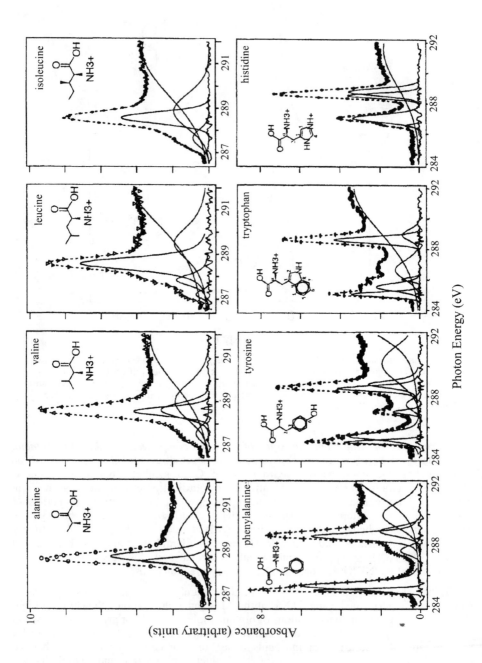

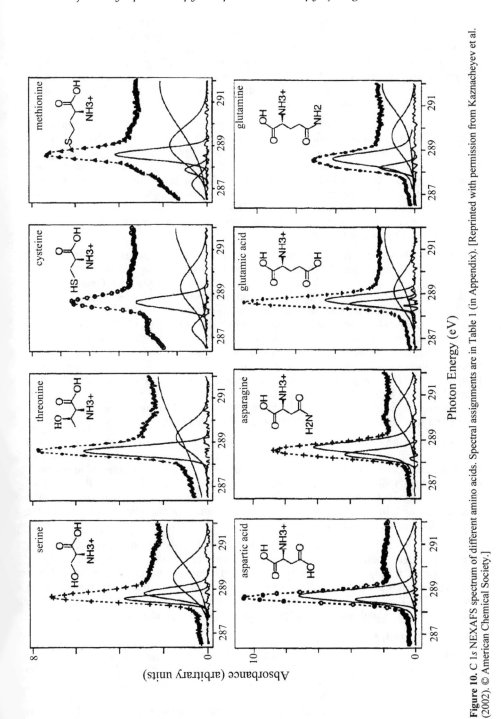

Figure 10. C 1s NEXAFS spectrum of different amino acids. Spectral assignments are in Table 1 (in Appendix). [Reprinted with permission from Kaznacheyev et al. (2002). © American Chemical Society.]

theoretical interpretations. Amino acids containing aromatic groups showed a distinct feature at 285 eV, in addition to the peaks at 288.6 eV corresponding to the carboxylic groups. When one of the amino acids (glycine) was titrated to evaluate the influence of pH on its functional group chemistry and respective NEXAFS spectra, the molecule did not exhibit any variations for the $1s{\rightarrow}\pi^*$ transitions of the carboxylic groups. However, variations were observed in the $1s{\rightarrow}\sigma^*$ transitions. Studies are in progress to understand the nature of other functional groups in amino acids.

Researchers have also used wet sample specimens in STXM to measure the NEXAFS spectra of bovine serum albumin and DNA (Ade et al. 1992; Fig. 11). These studies indicate that the C $1s{\rightarrow}\pi^*$ transitions of these samples are distinctly different with a greater signal for the CN moieties in DNA. Variations in these spectral features have been used to image the distribution of protein and DNA in cells (Fig. 11). This author has evaluated the variations in the C NEXAFS spectra of microorganisms to obtain a better understanding of their cell wall chemistry. His studies indicated that the surfaces of microorganisms exhibit distinct spectral differences when compared to that of the interior (SCB Myneni, unpublished results).

Initial studies of naturally occurring organic molecules, such as humic substances and soil organic molecules, have been conducted by this author using the prototype STXM under atmospheric conditions and with the vacuum facilities at the ALS (Warwick et al. 1998; Myneni and Martinez, 1999). The spectra of isolated humic substances did not exhibit spectral variations with the vacuum conditions of the sample. These studies also indicate that distinct $1s{\rightarrow}\pi^*$ transitions corresponding to the C=C, C=O, COOH, and the $1s{\rightarrow}\sigma^*$ transitions corresponding to the C-H groups can be identified from the NEXAFS spectra of humic substances (Fig. 12). The ratio of aromatic to carboxylic groups, estimated based on their relative spectral intensities, indicates that humic acids are enriched in unsaturated C. The humic substances isolated from different sources also exhibit similar spectral features indicating that the nature of C moieties may be similar. However, their spectral intensities indicate that the relative concentrations of different groups are different. These data sets agree with the ^{13}C NMR spectra of isolated humic substances.

A comparison of the NEXAFS spectra of isolated humic substances and pristine soil organic molecules for a pine ultisol indicate that the undisturbed soil samples indicate much greater concentrations of carboxylic acids when compared to that of aromatic C (Fig. 13, details discussed later). Similar behavior is observed for N, P, S and Cl functional groups (discussed later). These studies suggest that chemical extraction procedures used in the isolation of humic substances preferentially concentrate the aromatic C over the carboxylic groups.

Recently several other groups have begun investigating the functional group chemistry of natural organic molecules using the STXM (Joerge et al. 2000; Scheinost et al. 2001). Scheinost et al. (2001) examined a series of isolated humic acids and fulvic acids, and their studies indicated that the carboxylic groups are more dominant in fulvic acid than in humic acid (Fig. 14). Based on the peak positions, they have identified the presence of aromatic, carboxylic, O-alkyl, and phenolic groups in these isolated molecules.

N functional groups

Nitrogen exists in a variety of oxidation states in inorganic and organic forms and plays a critical role in the biogeochemical cycling of several major and trace elements in soils and aquatic systems, particularly the C cycle. NMR spectroscopy is the method that has provided the most comprehensive information on N forms in the organic fractions of soils and sediments. NMR data indicate that the natural organic matter primarily contains amines with a small concentration of pyrroles, indoles and pyridine functional groups

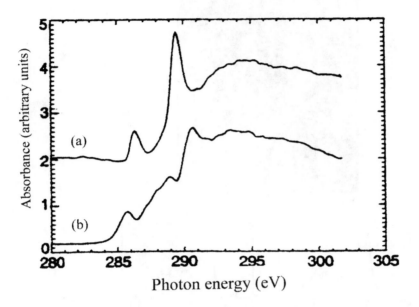

Figure 11. C 1*s* NEXAFS spectrum of bovine serum albumin (a) and DNA (b) showing differences for the C=N bonds between 286 and 288 eV (modified from Ade at al. 1992).

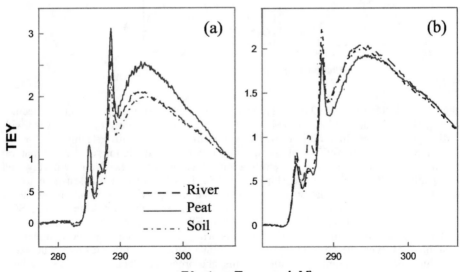

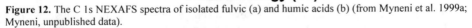

Figure 12. The C 1s NEXAFS spectra of isolated fulvic (a) and humic acids (b) (from Myneni et al. 1999a; Myneni, unpublished data).

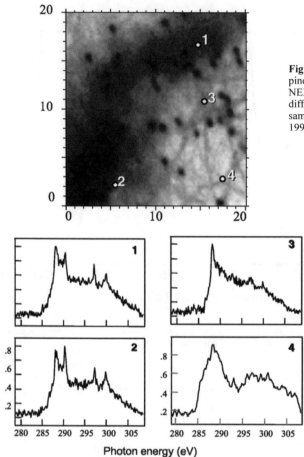

Figure 13. STXM image of pine Ultisol, and the C NEXAFS spectra from different locations in the sample (from Myneni et al. 1999a; Warwick et al. 1998).

(Stevenson 1994; Orlov 1995; Bortiatynski et al. 1996). However, organic molecules from geological matrices have to be isolated and concentrated before being analyzed with NMR spectroscopy. In addition, NMR is not the most sensitive technique for the characterization of N functional groups because of the depletion of ^{15}N in natural systems, which also limits the identification of N heterocycles (Vairavamurthy and Wang, 2002). Researchers have shown that N NEXAFS spectra are sensitive to the variations in local coordination environment, and can provide information on the N groups in natural organic molecules (Fig. 15; Vairavamurthy and Wang 2002; SCB Myneni, unpublished results). In addition, NEXAFS spectroscopy allows direct probing of the chemistry of N in soils and sediments without any special sample treatment or isolation. However, the concentration of N is much smaller in organics associated with the soluble organic fraction in aquatic systems, and organic molecules have to be isolated before they can be characterized using these methods.

Electronic structure analysis of N in several inorganic and organic species has been conducted using EELS and NEXAFS spectroscopy (Table 2—in the Appendix). A majority of the earlier EELS studies focused on small chain gas phase molecules, and

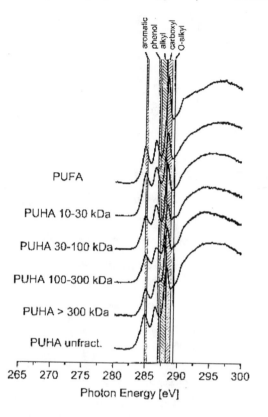

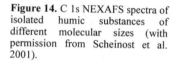

Figure 14. C 1s NEXAFS spectra of isolated humic substances of different molecular sizes (with permission from Scheinost et al. 2001).

thin solid films containing N. Organic macromolecules and polymers were not examined until recently. In the author's opinion, the peak assignments in the case of several molecules are not complete. Theoretical investigations are required in several cases. A survey of the NEXAFS spectra of several different molecules is presented in Table 2 (in Appendix) with the assignments proposed by different researchers. Most commonly, researchers use the π* and Rydberg transitions of N_2 for calibration, and it is expected that the energies reported in one study may well be correlated with those reported in other studies. Because of inconsistencies in calibration, the reader should be cautious while examining absolute energies of different transitions.

X-ray spectra and chemical state of model organonitrogen compounds. The author tried to make some general observations after comparing different molecular structures and their reported electronic transitions in Table 2 (in Appendix). Molecules with distinct C=N (linear, 5 and 6 membered rings) and C≡N bonds exhibit intense $1s \rightarrow \pi^*$ transitions in the range of 398.5-400.5 eV. Molecules such as hydrogen cyanide, pyridine, acridine, imidazole, dicyanoimidazole, and triazine exhibit distinct features in this energy range. With changes in the substituents in these molecules, the π^* transition could shift to different energies based on the relative electron withdrawing tendency of the entering and leaving groups.

Molecules with conjugate π system exhibit new spectral features that correspond to transitions into these π^* orbitals. The intensity and the energy of such features vary with the nature of the chemical interactions, symmetry and the orbital composition. For

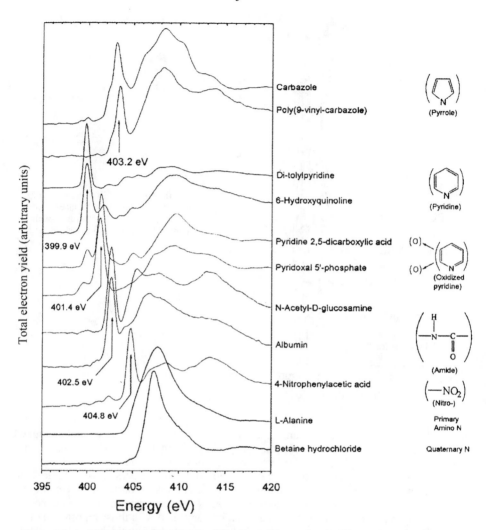

Figure 15. 1*s* NEXAFS spectra of N in different model compounds. The spectra are collected in electron yield mode and thus are sensitive to their surface structure (reprinted with permission from Vairavamurthy and Wang 2002. © American Chemical Society).

example, imidazole exhibits two types of N; one with C=N character and the other with C_2NH in a 5-membered ring (Fig. 16, Table 2—in Appendix). The NEXAFS spectrum of this molecule exhibits two distinct transitions that correspond to both types of N: ~ 400 eV for C=N, and 401.5 eV for C_2NH. 4,5-dicyanoimidazole exhibits three peaks at 399.4, 401.3, and 403 eV, and Hitchcock and Ishii (1987) assigned them to the N 1*s* electronic transitions to the π^* orbitals of C=N character, and also into delocalized π system. The low energy peak may be assigned to the transitions in to the π^* orbitals of C=N character, but the other two high-energy peaks may correspond to the valence orbitals of amine groups (details discussed later; Fig. 16). Acrylonitrile ($H_2C=CH-C\equiv N$) exhibits two distinct, intense features below ~ 398 eV. The low energy feature is assigned to the

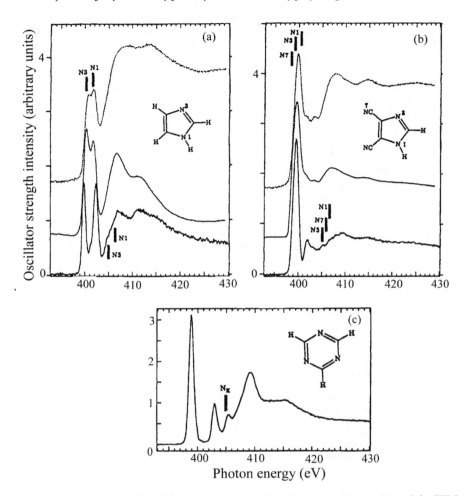

Figure 16. 1s EELS and NEXAFS spectra of imidazole (a); 4,5 dicyanoimidazole (b), and the EELS spectra of s-trazine (c). Different curves in (a) and (b) correspond to fluorescence yield, electron yield, and EELS (from top to bottom). The hatched lines in each figure are the ionization energies for different nitrogen atoms. The spectral assignments are given in Table 2 (in Appendix). (Reprinted with permission from Apen et al. 1993. © American Chemical Society.)

$1s \rightarrow \pi^*$ transitions of C=C, and the high-energy transition is attributed to the C=/≡N. Pyrroles also exhibit similar features; details are discussed below.

Before other N heterocycles are discussed, it is important to evaluate the spectral features of amines from first principles. The electronic state and spectral features of amines in organic molecules can be well understood by looking at the spectral features of simple molecules such as NH_3. The gas phase NH_3 exhibits two distinct intense transitions: 400.6 and 402.2 eV, which correspond to the N 1s transitions to the $3s_{a1}$ and $3p_e$ energy states, respectively, and both of these transitions exhibit significant Rydberg character (Fig. 17; Table 2—in Appendix). As the NH_3 molecule adsorbs to metal surfaces, these transitions become weaker and the Rydberg orbitals shift to higher energies. However, these features are not lost because of interactions at the interfaces. Similar features are also noticed in the case of a series of amines: mono-, di-, and tri-methyl amines, which showed features at 400.8, 402; 401, 402.3; and 403 (401.8) eV, respectively (Fig. 17). In these molecules the coordination around N is changing from C-NH_2, C_2-NH to C_3N, respectively. These studies indicate that the presence of amine groups exhibit features around 401 and 402 eV. Metal surface adsorption of glycine, which contains the C-NH_2 group, clearly showed distinct features at these energies (Hasselstrom et al. 1998). Examinations of other organic molecules that contain amine and amide groups also exhibit such transitions. The NH groups in imidazole, 4,5-dicyanoimidazole, albumin and N-acetyl-D-glucosamine exhibit features at 401.5, 401.1 (assigned to a broad feature at 400.3 eV), 402.5 and 402.5, respectively. However, it is not clear whether they have any π^* character or they are purely of σ^* character. Theoretical studies are necessary to understand the orbital composition of these molecules, which can lead to the identification off different electronic transitions.

The N heterocycles with saturated C in the ring exhibit interesting spectral variations. The EELS of pyrrole, in the gas phase, shows distinct features at 402.3 (sharp, intense) and 405.9 eV (Fig. 18). As the double bond character is lost in pyrrole, such as in pyrrolidine, this molecule exhibits spectral features at 401 (sharp, weak), 402.5 and 404.5 eV. However, the intensity of the feature at 402.5 eV is reduced significantly (Fig. 18). Similarly when an NH group is in a 6-membered ring that contains only saturated C, such

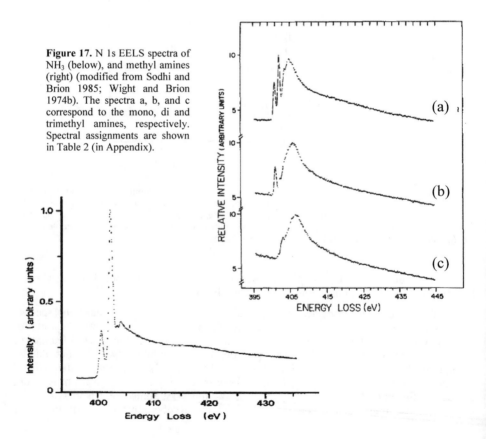

Figure 17. N 1s EELS spectra of NH_3 (below), and methyl amines (right) (modified from Sodhi and Brion 1985; Wight and Brion 1974b). The spectra a, b, and c correspond to the mono, di and trimethyl amines, respectively. Spectral assignments are shown in Table 2 (in Appendix).

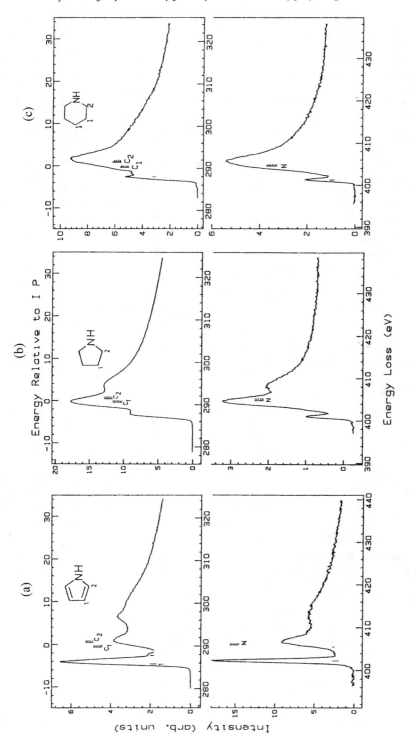

Figure 18. 1s EELS spectra of C (top) and N (bottom) in a series of heterocyclic gas phase molecules (modified from Newbury et al. 1986). (a) pyrrole, (b) pyrrolidine, and (c) piperdine. The hatched lines in each figure are the ionization energies, determined using the XPS.

as in piperdine, it exhibits a distinct feature at 401.2 and a broad peak at 405.6 eV. It should be noted that pyridine, with unsaturated C in the ring, exhibits distinct features at 398.8 eV that corresponds to the $1s{\to}\pi^*$ transitions of C=N. These molecular spectra suggest that as the delocalization of the electron density is minimized around an NH group, the molecules exhibit a distinct small feature around 401 eV. Carbazole (pyrrole between two benzene rings), on the other hand, exhibits a weak peak at 403.2 eV with a shoulder at 402.5 eV. The variations between pyrrole and carbazole are attributed to the electron delocalization tendency of the benzene rings. The oxidized forms of N, such as nitrosyls and nitrates exhibit intense $1s{\to}\pi^*$ transitions around 405 eV and can be identified unambiguously.

X-ray spectra and chemical state of natural organonitrogen compounds. NEXAFS spectroscopy has been used to identify functional groups of N in petroleum hydrocarbons, humic substances, and other biological macromolecules (Mitra-Kirtley et al. 1992, 1993; Vairavamurthy and Wang, 2002, SCB Myneni, unpublished data). All these studies indicate that several different forms of N can be identified unambiguously.

When nonmetallated porphyrines (protoporphyrin–IX, 5,10,15,20- tetraphenyl-porphine) are examined at the N-edge, these molecules clearly exhibit intense peaks at 399, 401.1 eV and 2-3 overlapping weak features between 402 and 405 eV (Vairavamurthy and Wang, 2002). Although Vairavamurthy and Wang have not assigned these features, this author interprets and assigns the peak at 399 eV to the N $1s{\to}\pi^*$ of the C=N in the 5 membered ring, and the peaks at 401.1 and other features correspond to the pyrrole group in the porphyrin molecule (Fig. 19). When metals occupy the center of the porphyrin ring, for example Ni^{2+} in nickel octaethylporphyrin, its N NEXAFS spectrum exhibits peaks at $\sim$ 400, 402 and 403.2 eV (Mitra-Kirtley et al. 1993a). When Ni enters into the ring, it replaces H from the amine group of the pyrrole ring, and accordingly the peak at 401.1 eV disappears. Similar spectral variations are also noticed in different models discussed above.

The N NEXAFS spectra of humic acids and sediment samples have been reported for the first time by Vairavamuthy and Wang (2002), and their spectra exhibited 4 peaks around 400, 401.3, 402.5, and 404.8 eV (Fig. 20). They have assigned these peaks to pyridine, oxidized pyridine, amide groups and nitro compounds, respectively. Although this author found similar spectral features for humic substances isolated from soils and river waters, the spectral interpretation is different for the following reasons: the peak around 399 eV cannot be assigned to pyridine unambiguously (discussed earlier), and the peaks around 401.3 and 402.5 can be produced by several different forms of N (e.g., aliphatic and aromatic amines, amides). Theoretical studies of model molecules are necessary to make unambiguous assignments of these different peaks.

X-ray spectra and chemical state of aqueous organonitrogen compounds. Studies on aqueous molecules have been started in the last three years and to the author's knowledge only two groups have conducted these investigations so far: Jacobsen's group at the State University of New York, and the author's group. Jacobsen's group used the STXM and collected NEXAFS spectra of amino acids ($\sim$ 1 M) in aqueous solutions as a function of pH (personal communication). This author and his group have examined amino acids and biomacromolecules such as siderophores at $\sim$ 25 mM concentration using the SXEER. Their studies indicate that the N NEXAFS spectra can be collected for molecules at a concentration of $\sim$ 5 mM using this synchrotron endstation in fluorescence mode. They have examined the protonation and deprotonation and metal complexation reactions of these molecules, which showed distinct changes in the NEXAFS spectra. Theoretical investigations are in progress to identify these different transitions unambiguously (results not published). This group has also conducted experimental and

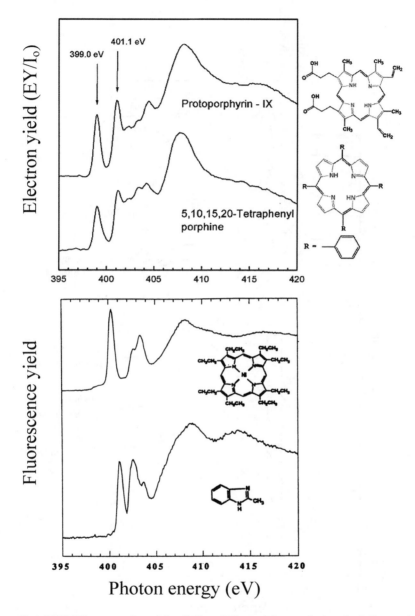

Figure 19. 1s NEXAFS spectra of metal free (left), and Ni containing porphyrines (top). Spectra shown on the top are collected in electron yield mode, and the spectra shown on the bottom are collected using fluorescence. Spectrum of 2-methylbenzimidazole is shown for a comparison (lower spectrum in bottom figure). One of the distinct features between the metalated and metal-free porphyrine is the absence of peak at 401.1 eV in the former. [Used with permission from Vairavamurthy and Wang 2002. and Mitra-Kirtley et al. 1993. © American Chemical Society.]

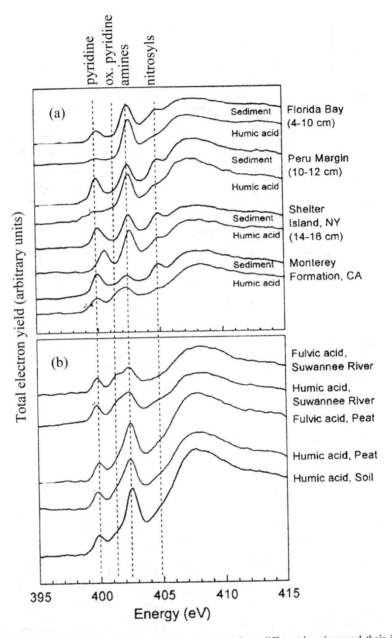

Figure 20. NEXAFS spectra of N in sediments collected from different locations and their isolated humic substances (a), and humic substances isolated by the International Humic Substance Society (b). The spectra are collected in electron yield mode and are sensitive to surface coordination (especially for the sediment samples). [Used with permission from Vairavamurthy and Wang 2002, © American Chemical Society].

theoretical vibrational spectroscopy studies on some of these molecules, and the band assignments were difficult for biomacromolecules (Edwards et al., to be submitted). The N NEXAFS spectral features and complementary metal EXAFS spectroscopic data (in the case of metal organic complexes) provided unique and unambiguous element and group specific information on these molecules (Edwards and Myneni, unpublished data). The same synchrotron endstation can also be used to examine the organonitrogen compound interactions at mineral-water interfaces, and the forms of organonitrogen present at dilute concentrations in soils, sediments and aquatic systems without any sample treatment.

O functional groups

O functional groups are probed using variations in the C NMR or C NEXAFS spectral features. Although O NMR is commonly used for studying inorganic systems, the low natural abundance (0.037%) and the relatively small magnetic moment of ^{17}O produce broad spectra, and thus it is difficult to distinguish the spectral features of organic moieties in different coordination environments. The electronic structure of O can also be probed using the electronic excitations at the O absorption edge. Although O NEXAFS can provide useful information, absorption features of water, and oxygen and hydroxyls in the mineral fraction limits the identification of different groups (Myneni et al. 2002).

A survey of different O containing organic molecules indicate that peaks corresponding to the O $1s \rightarrow \pi^*$ transitions of ketones and aldehydes are around 531 eV; carboxylic acids, esters, and amides are around 532 eV (Table 3—in Appendix). There is some overlap of these bands based on the substituents around the O functional group in the molecule. Whereas alcohols and ethers exhibit their $1s \rightarrow \sigma^*$ (C-O) transitions around 534 and 535.5 eV respectively. The energies of these different functional groups are different, and their O NEXAFS spectra can be used in their identification. The O NEXAFS spectra of isolated humic substances from aquatic and soils systems also exhibit distinct peaks that correspond to ketones, carboxylic acids and alcohols (SCB Myneni, unpublished data).

P functional groups

Phosphorus is one of the essential elements required for all biological systems, and it exists in several different inorganic forms, with only a few varieties of organic forms in the environment. The inorganic species include different forms of orthophosphates in different protonation states, and polyphosphates. Organophosphorus exists in the form of organophosphates that include inosital compounds and phosphate esters, and phosphonates (Stevenson 1994). Among these compounds, organophosphates are the most dominant forms and may constitute more than 50% of the total organophosphorus in soil system (Anderson 1980). Although ^{31}P NMR spectroscopy has been used for examining the coordination environment of P in organic molecules, the presence of paramagnetic ions such as Fe and Mn significantly affect the NMR spectral signatures (Cade-Menun and Preston 1996). In addition, chemical extraction methods used to isolate P from soils and sediments before their characterization by NMR or other techniques are shown to alter P-speciation (Cade-Menun and Preston, 1996; Cade-Menun and Lavkulich, 1997). For example, extraction with NaOH resulted in the conversion of phosphate diesters to monoesters. In addition, P concentrations in organic molecule isolates from soils should be higher for their characterization using NMR. Although NEXAFS spectroscopy can provide direct information on the chemical state of P in organic molecules, the low photon flux in the vicinity of the P absorption edge at several synchrotron beamlines limits the applications of this technique to concentrated samples.

For this reason there have not been many synchrotron studies at the P edge. To the author's knowledge, only a couple of P NEXAFS spectroscopic studies have been conducted on the natural organic molecules (Myneni and Martinez 1999; Vairavamurthy 1999).

X-ray spectra of model inorganic and organic phosphorus compounds. The valence orbitals of phosphate in tetrahedral symmetry are of type a_1^* and t_2^*, and have P s and p character, respectively. The P $1s$ NEXAFS spectra of PO_4^{3-} exhibit an intense absorption edge that corresponds to the electronic transitions from the $1s \rightarrow t_2^*$ orbitals (Fig. 21, 28). The $1s \rightarrow a_1^*$ transitions are forbidden because of the s character of these orbitals. As the phosphate protonates, the P $1s$ spectra exhibit distinct variations, which are similar to those discussed later for S (Fig. 21). However, the splitting of the triply degenerate t_2^* orbitals in the case of protonated phosphate is not as intense as that of protonated sulfate. A high-energy feature that appears as a shoulder on the main absorption edge disappears with phosphate protonation (compare PO_4^{3-}, $H_2PO_4^-$). The origins of these features are not well understood. Phosphate polymerization also has the same effect as that of phosphate protonation on the main absorption edge. However, the scattering features approximately 10 eV above the absorption edge vary significantly in the case of polyphosphates (Fig. 22).

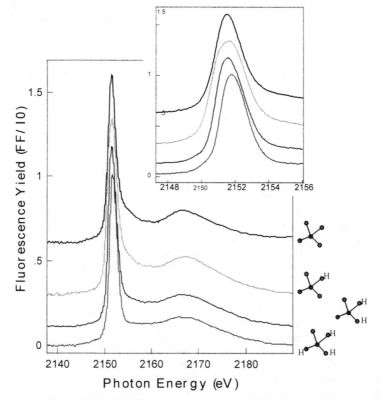

Figure 21. P $1s$ NEXAFS spectra of inorganic phosphate in different protonation states. The inset shows the close up of the absorption edge (from Myneni and Martinez 1999; Myneni, unpublished results).

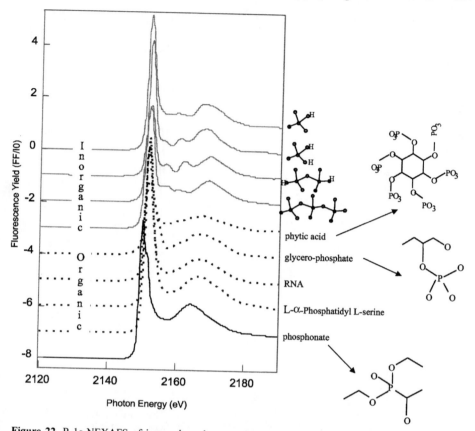

Figure 22. P $1s$ NEXAFS of inorganic and organophosphates, and phosphonate. The corresponding structures are shown on the right (from Myneni and Martinez 1999; Myneni, unpublished results).

In organophosphates, the phosphate group is connected to C through O, and the white line of organophosphate is broader than that of unprotonated phosphate. This could be caused by the splitting of the t_2^* orbitals of the phosphate group associated with the symmetry variations (compare spectral features of organosulfates, discussed later). However, the P NEXAFS spectra of different organophosphate molecules are similar. Only glycerophosphate showed distinct differences at the main absorption edge. When compared to the NEXAFS spectra of inorganic polyphosphates, the spectral features of organophosphate compounds are smoother above the white line. This suggests that multiple scattering of the released photoelectron in the organophosphate group is much weaker. In phosphonate, P is connected to C directly ($C-PO_3$), and the absorption edge of P shifts to lower energy by about 2 eV (Fig. 22).

Although the occurrence of ethylphosphines and phenylphosphates in natural environments has not been reported, the NEXAFS spectra of these molecules exhibit significant changes with variations in coordination environment around P and these spectral features may be useful in understanding the chemical state of P in different organic molecules (Fig. 23; Engeman et al. 1999). These NEXAFS studies suggest that the electronegativity of atoms in the first shell around P influences the energy of the

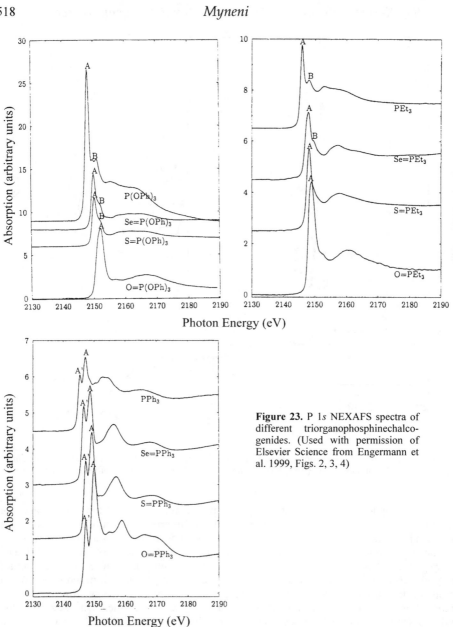

Figure 23. P 1*s* NEXAFS spectra of different triorganophosphinechalcogenides. (Used with permission of Elsevier Science from Engermann et al. 1999, Figs. 2, 3, 4)

white line, which shifts to higher energy with an increase in the electronegativity of the neighboring atoms. Presence of an aromatic ring in the second and higher shells splits the white line, and this is attributed to the delocalization of π-electrons of the ring and resulting shortening of the P-C bond. A summary of these model system studies indicates that P NEXAFS spectra are useful for identifying inorganic and organophosphorus compounds, but not variations in different organophosphates from the NEXAFS alone. The NEXAFS spectra can be used for identifying phosphonate unambiguously.

X-ray spectra of phosphorus in natural organic molecules. The P NEXAFS spectroscopic studies conducted on isolated humic substances and soil organic molecules indicate that the primary form of P is phosphate and phosphonate (Fig. 24; Myneni and Martinez 1999). When compared to humic substances from soil systems, the fluvial humic substances exhibit phosphonate as one of the important components. However, phosphonate constitutes only a minor fraction of total P in humic substances. Soil samples also exhibit features that correspond to polyphosphate. Another study conducted using NEXAFS spectroscopy at the P absorption edge suggested that marine sediments and humic materials do not exhibit phosphonate and the P NEXAFS spectra of these samples more closely resembled that of hydroxyapatite (Vairavamurthy 1999).

A summary of these investigations on natural organic molecules suggests that the spectral contributions of phosphonate and polyphosphate are separable from those of ortho-, and organophosphates. The spectral similarities between organo- (except glycerophosphate), and protonated phosphate forms prevent their unambiguous *in situ* identification in complex natural samples. Recently P NEXAFS spectroscopy studies have been conducted on aerosol particles, which indicated the presence of P in +V and +III oxidation states. However, the reported spectra are too noisy to identify P in the +III oxidation state unambiguously (Tohno et al. 2001).

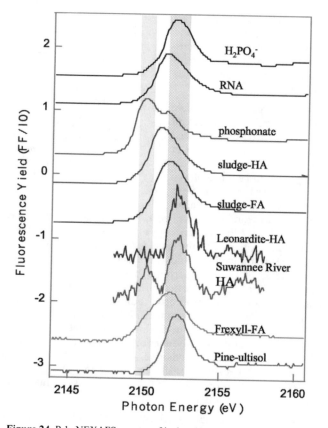

Figure 24. P $1s$ NEXAFS spectra of isolated humic substances and soils.

S functional groups

In natural organic molecules S exists in a variety of forms and in several different oxidation states (Fig. 25). In reduced form, S exists as mono and disulfides that are important components of amino acids. Reduced S species are also common ligands around reactive metal centers in enzymes and proteins, and play a critical role in biological function. The hydration and protonation state of reduced S forms modify the H-bonding environment around these moieties and thus affect protein folding and reactivity of these functional groups. The reduced S groups also exhibit extremely high binding affinity for soft metals, such as Cd and Hg, and organic molecules containing reduced S forms are used to isolate and collect toxic metals from waste matrices and contaminated water. In oxidized form, S exists as sulfoxides, sulfonates and sulfate esters that are common in all plant cells and biosurfactants released by several species of microorganisms. Some of the oxidized S forms are also used as surfactants because of their high solubility.

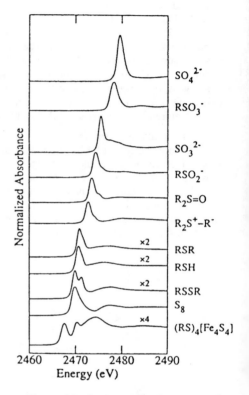

Figure 25. S 1s NEXAFS spectra of different inorganic and organosulfur compounds. 'R' represents an organic molecule. (modified from Pickering et al. 1998, Myneni 2000)

Unlike other organic molecule functional groups, S moieties in natural organic molecules cannot be characterized by NMR, UV-VIS absorption or fluorescence spectroscopy methods. X-ray spectroscopy has proven to be a useful tool for examining different forms of S in simple and complex organic macromolecules (Fig. 25). Researchers have been using synchrotron X-ray spectroscopy methods (specifically NEXAFS spectroscopy) to examine the chemistry of S in humic substances and metalloproteins for the past ten years. Even though there are a few synchrotron beamlines optimized for studying S chemistry, there has been an explosive growth of XAS studies at the S absorption edge in the past few years. This may promote the construction of several new facilities for probing S (and P and Cl) in the near future. While S NEXAFS spectroscopy studies are conducted routinely these days, in the author's opinion, the spectral interpretation is still in its nascent stages. Further, the overlapping spectral features of S in different coordination environments limit the spectral interpretation of complex polyfunctional natural organic macromolecules. In addition, the S NEXAFS spectra are sensitive to small variations in local coordination environment, which exhibit spectral shifts or completely new spectral features and complicate spectral interpretation (Fig. 25, details discussed later). This has led to incorrect band assignments for S in macromolecules.

A detailed discussion of available synchrotron methods for studying S is presented in one of the recent reviews, and the focus of this previous review is on the coordination chemistry of inorganic and organic sulfates (Myneni 2000). Information on synchrotron facilities for conducting S XAS, spectral collection and interpretation and other relevant information is also presented there, and interested researchers can refer to this article for more details. Different model molecules and energies are used by experimentalists for calibration, and this has resulted in significant variations in the reported absolute energies of organosulfur compounds. Readers should refer to the original references to obtain information on spectral calibration.

X-ray spectra and electronic structure of organosulfur compounds. Sulfur in natural organic molecules can exist in the form of reduced and oxidized forms, and their electronic structures are distinctly different. Correlations between the energy of the S absorption edge of small chain molecules (energies of individual peaks in the case of polyfunctional organic macromolecules) and their oxidation states can be used to identify different S functional groups in humic substances, petroleum hydrocarbons, and coals (Vairavamurthy et al. 1993; Morra et al. 1995; Vairavamurthy 1998; Xia et al. 1998; Sarret et al. 1999). Although such a correlation between oxidation states and energies of absorption edges works well for heavy elements such as Cr, Mn, Fe, As, and Se, in the author's opinion such a correlation may not be entirely suited for defining the chemical state of light elements such as C, N, P, S and Cl. This is because the light element spectra are sensitive to small variations in local coordination environment and symmetry. For instance, when the energy and peak area of the white line are compared, there is a linear correlation between the two with a significant scatter for several molecules indicating that the spectra are very sensitive to local coordination environment (Sarret et al. 1999; Fig. 26). The high spectral resolution obtained at these light element absorption edges (when compared to the spectral resolution for heavy elements) is useful when examining small variations in coordination environment. Small core hole broadening for light elements is also an added advantage. As discussed later, sulfur in a given oxidation state can exhibit spectra at different energies, and these may not be assigned to variations in oxidation state alone. Orbital composition and symmetry also shifts these molecular spectra.

The energies and intensities of electronic transitions are modified by the overlap of the atomic orbitals and the symmetry and composition of molecular orbitals. A detailed examination of the electronic structure of different organosulfur compounds and their complexes in different coordination environments would be useful to interpret their NEXAFS spectra. This information is also useful in the interpretation of the NEXAFS spectra of unknown compounds in the natural systems.

X-ray spectra and electronic structure of oxidized organosulfur compounds. For many years, researchers have been working on the electronic structure and bonding interactions of simple molecules, such as sulfate, and trying to establish a correlation between the two (Nefedov and Formichev 1968; Dehmer 1972; Sekiyama et al. 1986; Tyson et al. 1989). The electronic structure of organosulfates, sulfonates and other oxidized S forms can be well understood by examining the chemical state of sulfate and sulfite in different coordination environments.

The tetrahedral sulfate ion (T_d) exhibits two unoccupied antibonding molecular orbitals: a_1^* and t_2^* of $3s$ and $3p$ character, respectively (Sekiyama et al. 1986; Tyson et al. 1989; Hawthorne et al., 2000; Myneni 2000). The $1s \rightarrow a_1^*$ transitions are dipole-forbidden (because of the s-character of a_1^* orbitals), and the intense features in the NEXAFS spectra of sulfate correspond to the $1s \rightarrow t_2^*$ orbitals (Fig. 27). The high-energy features above these intense bound state transitions correspond to the Rydberg-type transitions of $3d$ character or continuum state transitions (Table 4—in Appendix).

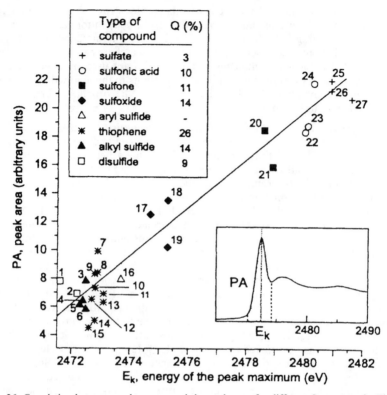

Figure 26. Correlation between peak energy and the peak area for different S compounds. The inset shows a S NEXAFS spectrum, and how the peak area is measured for the plot. [Reprinted from Sarret et al. 1999, Fig. 4, with permission from Elsevier Science.]

Complexation of sulfate, such as in $-H_2C-O-SO_3$, modifies the tetrahedral symmetry of sulfate to C_{3v} or other low symmetry groups. This leads to the splitting of empty triply degenerate t_2^* molecular orbitals into either a_2^* and e^*, or its degeneracy may be completely eliminated. In inorganic sulfate salts, the chemical bonding of cations with sulfates modifies the symmetry of the sulfate group, but these interactions are not strong enough to split the degenerate orbitals significantly (Okude et al. 2000; Myneni 2000). However, protonation (e.g., $H-O-SO_3$), direct coordination of organic molecules to sulfate (e.g., $-H_2C-O-SO_3$), and linkages between two sulfates (such as in $O_3S-O-O-SO_3^{2-}$) produces dramatic splitting of these t_2^* orbitals when compared to the metal complexed sulfates (Myneni 2000). The S-O bond distances also exhibit distinct variations for the above molecules. The S-O bond distances are 1.475 Å in uncomplexed sulfate, and they do not change significantly with variations in metal complexation in solids (Hawthorne et al. 2000). However, protonation and organic molecule covalent interactions of sulfate produced significant changes in S-O bond distances: the S-OH/C bond distance exceeds 1.55 Å, and the uncomplexed S-O bond distance is ~ 1.46 Å. Such distortions in the sulfate polyhedron cause significant changes in the electronic structure of sulfate. Since the strength of interactions modifies the intensity of splitting, the NEXAFS spectra can provide direct information on the nature of these interactions. Among the protonated-, organo-, and persulfate molecules,

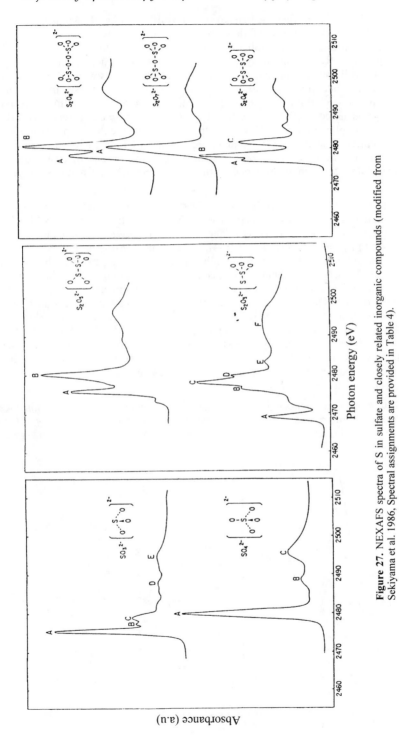

Figure 27. NEXAFS spectra of S in sulfate and closely related inorganic compounds (modified from Sekiyama et al. 1986, Spectral assignments are provided in Table 4).

the amount of splitting is intermediate for organosulfates (Fig. 28, Myneni 2000). The experimental NEXAFS spectra of these molecules exhibit two intense peaks, which correspond to the $1s$ electronic transitions to the a_2^* and e^* orbitals (Fig. 27, 28; Table 4—in Appendix). Reduction in symmetry and splitting of degenerate orbitals also lowers the overall spectral intensity of these covalent molecules when compared to that of uncomplexed sulfate (Myneni, unpublished results). In these molecules, the transitions to the a_2^* orbitals are much more intense when compared to the forbidden a_1^* orbitals (discussed above), due to the significant sulfur $3p$ character in a_2^* orbitals. In addition, complexation of sulfate with transition metals exhibits pre-edge features that vary with the chemical state and coordination environment of the complexing transition metal (Okude et al. 2000; Myneni 2000). Interactions of transition metals with organosulfates and their influence on the spectral S NEXAFS features are yet to be probed.

Sulfite ion in C_{3v} symmetry has two valence molecular orbitals, e^* and a_1^*, which have S p and s character, respectively. The NEXAFS spectra of sulfite exhibit intense electronic transitions that correspond to the $1s$ to the empty e^* orbitals, and the orbital composition of a_1^* orbitals forbids the $1s \rightarrow a_1^*$ electronic transitions (Fig. 25). Transition metal complexation of sulfite may also exhibit spectral features similar to that of sulfates (Okude et al. 2000, Myneni 2000). The spectral variations are intense if the complexing atom is another S. For instance, thiosulfate (S-SO_3) contains an S-S bond and a SO_3 group, and the NEXAFS spectra of this molecule exhibit electronic transitions that correspond to disulfide at low-energy and a sulfite group at higher energy (Fig. 27). In molecules of similar structural environment, such as $S_2O_3^{2-}$ (S-SO_3), $S_2O_5^{2-}$ (O_2S-SO_3), $S_2O_6^{2-}$ (O_3S-SO_3), the electronic transitions from S $1s$ to the orbitals representing disulfide bond shifts gradually to higher energy, which suggests that the disulfide bond is not identical in these molecules, but exhibits an increase in the apparent charge state of S (Fig 27; Table 4—in Appendix).

Sulfite groups are connected to C through S in sulfonates (-C-SO_3), and this does not influence the symmetry of sulfite group. However, the higher electronegativity of C when compared to S modifies the electronic state of S in sulfonates and thus the energy levels

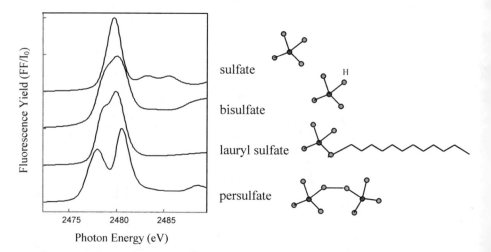

Figure 28. NEXAFS spectra of S in covalently bound inorganic and organosulfates (from Myneni and Martinez 1999; Myneni, unpublished results).

of its molecular orbitals. Accordingly the NEXAFS spectra of sulfonate shifts to higher energy when compared to that of sulfite (Fig. 29). It is interesting to note that the sulfonate molecule does not exhibit any intense transition corresponding to the C-S σ* orbital, such as in reduced S compounds (discussed later). It is possible that this transition is weak and occurs in the vicinity of e^* transitions corresponding to the SO_3 group (Fig. 30). Theoretical studies are necessary to identify these different electronic transitions. The influence of metal complexation of the sulfonate group on its spectral features has not been investigated.

The sulfonate groups connected to aliphatic and aromatic C exhibit minor variations in their spectral features (Fig. 30). Although the absorption edges, corresponding to the electronic transitions from the S $1s{\rightarrow}e$* orbitals, are at the same energy for these molecules, the sulfonate group connected to aromatic C exhibits a distinct low-energy shoulder approximately 2 eV below the main absorption edge (Fig. 30, Vairavamurthy

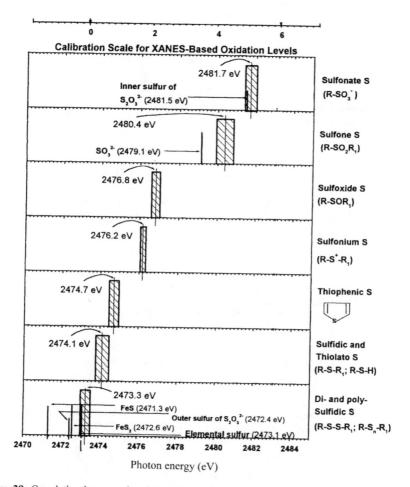

Photon energy (eV)

Figure 29. Correlation between the absorption edge energies of S in different compounds and the oxidation states estimated based on the NEXAFS spectra [reprinted with permission of Elsevier Science from Vairavamurthy 1998].

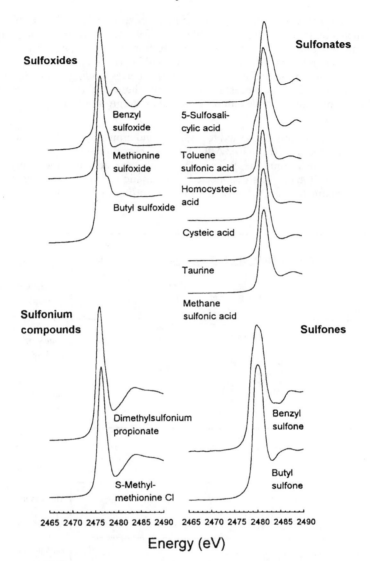

Figure 30. S NEXAFS spectra of different oxidized S species (reprinted with permission of Elsevier Science from Vairavamurthy 1998).

(1995); Myneni, unpublished spectra). Although these variations were not reported earlier, the author hypothesizes that this pre-edge feature corresponds to the electronic transitions of S $1s$ to π^* orbitals of the ring. It is not clear how these pre-edge features change with the substituents on ring.

Other relatively oxidized forms of sulfur are sulfones (-C_2-SO_2) and sulfoxides (-C_2-S=O), which are also common in biological systems. The spectra of sulfones and sulfoxides exhibit an intense feature with a shoulder at higher energy. These transitions correspond to the $1s \rightarrow \pi^*$ and $1s \rightarrow \sigma^*$ of S=O and C-S bonds, respectively (Fig. 30).

Sulfur in sulfones is in a coordination environment that is structurally similar to that of sulfate and sulfonate. Since the differences in interactions between C and O produces different spectral features, the main absorption edge shifts to about 1.3 eV below that of sulfonate and about 2.5 eV below that of sulfate (Figs. 25 and 30; Pickering et al. 1998; Vairavamurthy, 1998; Sarret et al. 1999). The coordination environment of sulfoxide is similar to that of sulfite, and for the reasons discussed above, the absorption edge of sulfoxide is approximately 2.5 eV lower than that of sulfite.

The L-edge spectra, corresponding to the $2s$ and $2p$ ($2p_{1/2}$ and $2p_{3/2}$) states occur at approximately 231, 163.5, and 162.5 eV, respectively. These edges are termed L_1, L_2 and L_3, respectively. The electronic transitions associated with the L_1 edges are similar to those of K edges, and perhaps may not offer new information other than better spectral resolution. On the other hand, the L_2 and L_3 edges correspond to the electronic transitions of $2p$ electrons to the empty $4s$ and $3d$ states, and offer complementary information to the K-edges described above. The low fluorescence yields at the L edges limits the detection of S using this technique. The L-edge NEXAFS spectroscopy studies indicate that the L edges of oxidized forms are at higher energy when compared to the reduced S forms, and the L-edges are very sensitive to the local coordination environment around S (Fig. 31; Kasrai et al. 1994; Sarret et al. 1999; Myneni 2000). Although Kasrai et al. (1994) and Sarret et al. (1999) reported three distinct peaks in oxidized and reduced S forms (marked A, B and C; Fig. 31), they did not make any specific assignments. However, they used variations in the positions of these features to identify different S forms in unknown geological matrices. According to Hitchcock et al. (1990), these peaks correspond to more than one type of electronic transition.

X-ray spectra and electronic structure of reduced organosulfur compounds. The spectral interpretation for oxidized S species is in a better state because of significant separation of their spectra in a majority of molecules. However, all of the reduced forms exhibit strong electronic transitions in a small energy range, and spectral variations above the absorption edge have to be used for their identification.

In reduced aliphatic S compounds, such as methionine and cysteine, the S NEXAFS spectra are almost identical, with absorption edges at the same energy (Fig. 25). These transitions are assigned to the S $1s \rightarrow \sigma^*$ transitions of the C-S bond. However, in cyclic polymethylene sulfides of composition $(CH_2)_nS$ (where n = 2,3,4,5), the $1s \rightarrow \sigma^*$ transitions exhibit significant variations (Fig. 32; the maximum shift is ~ 1.2 eV; Derzanaud-Dandine et al. 2001). The spectral variations between these molecules are attributed to variations in the molecular strain of cyclic compounds. All of these molecules exhibit an intense peak with a high-energy shoulder, and these peaks are assigned to the $1s \rightarrow \sigma^*$ transitions for the main edge, and the $1s \rightarrow \pi^*$ (CH_2) type transitions for the shoulder. The $1s \rightarrow \pi^*$ (CH_2) type transitions are assigned to several organosulfide molecules by various researchers following the assignments of Hitchcock et al. (1986a) for thiophene and thiolane. However, these assignments are valid for molecules with some unsaturated C-C bonds.

In cystine, another common amino acid, S exhibits S-S and C-S linkages, and its NEXAFS spectra exhibit two distinct peaks separated by about 1.5 eV (Fig. 25). These two peaks correspond to the S $1s \rightarrow \sigma^*$ transitions of S-S and C-S bonds at low and high energies, respectively. Examination of a series of dimethyl polysulfides also exhibits the same spectral features (Fig. 33; Derzanaud-Dandine et al. 1998). Dimethyl monosulfide exhibits an intense feature at 2473.1 eV, and dimethyl disulfide and dimethyl trisulfide exhibit a doublet that corresponds to the $1s \rightarrow \sigma^*$ transitions of S-S and C-S at 2472.2 and 2473.5 eV. Using low-energy electron transmission spectroscopy and NEXAFS spectroscopy, Derzanaud-Dandine et al. (1998) suggest that the HOMO-LUMO energy gap decreases with the S atom chain length.

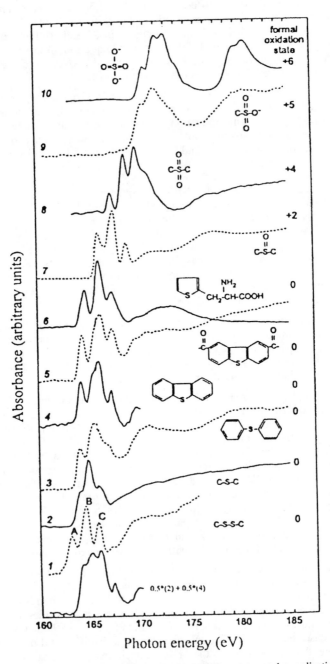

Figure 31. L edge NEXAFS spectra of S in different oxidation states and coordination environments (modified from Sarret et al. 1999). The bottom spectrum is a combination of spectra (2) and (4).

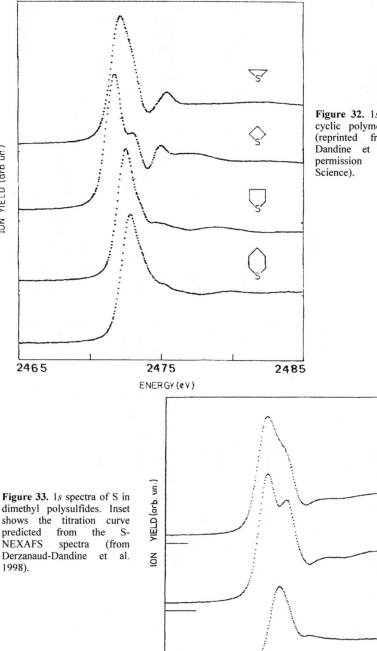

Figure 32. 1*s* spectra of S in cyclic polymethylene sulfides (reprinted from Derzanaud-Dandine et al. 2001 with permission from Elsevier Science).

Figure 33. 1*s* spectra of S in dimethyl polysulfides. Inset shows the titration curve predicted from the S-NEXAFS spectra (from Derzanaud-Dandine et al. 1998).

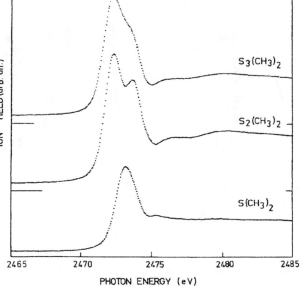

The electronic structure of S in several amino acids is sensitive to the variations in protonation state of the molecule. Cysteine has different pKa (protonation or deprotonation constant) values at ~ 1.9, 8.3 and 10.3 that correspond to the deprotonation of carboxylate, thiol and amino groups, respectively. Pickering et al. (1999) have shown that the NEXAFS spectra of cysteine can be used to evaluate variations in its protonation state (Fig. 34). As the molecule deprotonates at the thiol group, the absorption edge of S shifts to lower energy by approximately 2 eV. Based on these spectral variations, a pKa value of 8.6 has been assigned for the sulfhydryl groups. These spectral changes are directly related to variations in the coordination environment of S. The NEXAFS spectra also indicate that cysteine experiences significant chemical changes well below the pKa of the thiol group (SCB Myneni, unpublished data). Using the NEXAFS spectra of sulfhydryl groups in proteins and other biological macromolecules, it is possible to establish their protonation states and their variation with changes in solution chemistry accurately. It is not clear whether NEXAFS spectra are sensitive to changes in the hydration state of these groups.

Among S containing heterocycles, thiophenes and related compounds are common in the natural organic molecules. The NEXAFS spectra of thiophene exhibits an intense transition corresponding to the π^* transition about 0.5 eV above the σ^* transitions of organic monosulfides, such as methionine and cysteine. Examination of the NEXAFS spectra of a series of S containing heterocycles indicates that the S $1s \rightarrow \pi^*$ type transitions of the C=S (thione group) are the lowest energy transitions and occur about 1.5 eV below that of S $1s \rightarrow \sigma^*$ transitions of C-S bonds (Fig. 35; Fleming et al. 2001). Similar results are also shown by the polarization studies of organosulfur molecules adsorbed on single crystal surfaces.

Organosulfur compounds at the interfaces. Several studies have been conducted for the interactions of reduced S species on metal surfaces; and a small sample of these investigations is presented here to show the type of information obtained for organosulfur

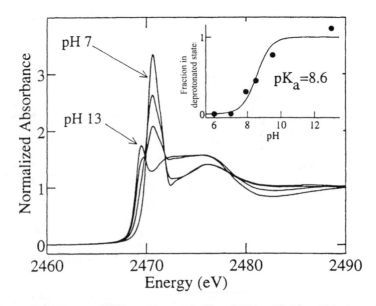

Figure 34. 1s NEXAFS spectra of S in cysteine as a function of pH (modified from Pickering et al. 1998).

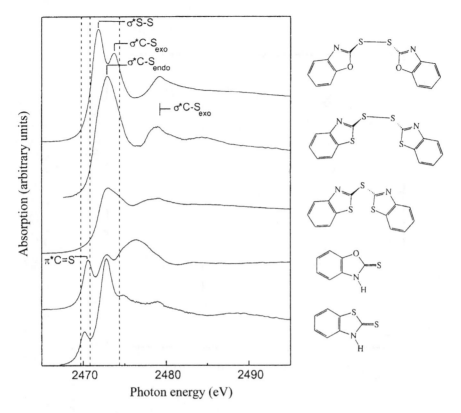

Figure 35. S1s NEXAFS spectra of sulfur heterocycles (modified from Fleming et al. 2001).

reactions with surfaces. Interactions of dimethyl sulfide on the Ni(100) surface indicates that the molecule lies flat at submonolayer coverages (Yagi et al. 2001). The polarization dependent S NEXAFS spectra indicate that the $1s \rightarrow \sigma^*$ transitions of C-S bonds are intense when the E vector of the incident photon is parallel to the crystal surface (or the incident beam is perpendicular to the surface; Fig. 36). The peak corresponding to the $1s \rightarrow \sigma^*$ transitions of C-S bonds almost disappears at grazing incident angles. Reactions of chlorothiophene with Cu(111) surface also shows the same result (Fig. 36; Milligan et al. 1999). Several researchers have examined the molecular chemistry of thiophene on metal surfaces and a great deal of controversy exists on the band assignments. Polarization dependent studies of this system indicate a small peak corresponding to the $1s \rightarrow \sigma^*$ transitions of C-S bonds at normal incidence. However, at grazing incidence the peak shifts to lower energy, by almost an eV, and the peak intensity increases. This peak is attributed to the $1s \rightarrow \pi^*$ transitions of the ring. These results indicate that thiophene is lying flat on the surface. In the case of 2-mercaptobenzoxazole adsorption on the Pt(111) surface, a combination of C, N, O, and S NEXAFS spectroscopic spectra suggests that the angle between the surface and the plane of the molecule is about 48^0 (Carravetta et al. 1999). The S L-edge spectrum of this molecule does not exhibit significant variation with changes in angle of the incident beam. To the author's knowledge NEXAFS studies have not been conducted for the interactions of organosulfur species with mineral surfaces using the NEXAFS spectroscopy.

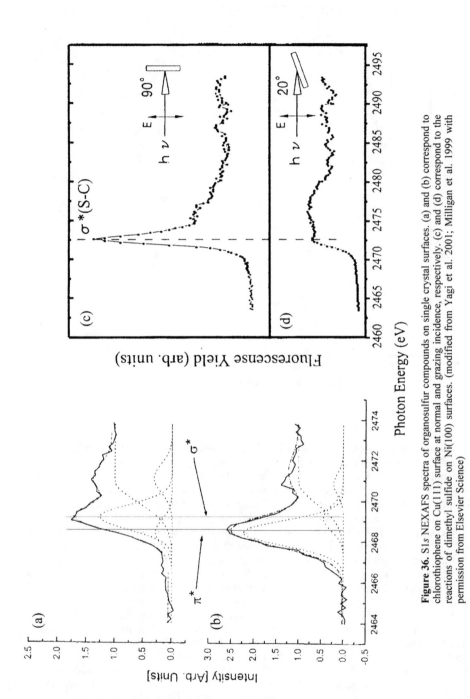

Figure 36. S1*s* NEXAFS spectra of organosulfur compounds on single crystal surfaces. (a) and (b) correspond to chlorothiophene on Cu(111) surface at normal and grazing incidence, respectively. (c) and (d) correspond to the reactions of dimethyl sulfide on Ni(100) surfaces. (modified from Yagi et al. 2001; Milligan et al. 1999 with permission from Elsevier Science)

X-ray spectra and chemical state of organosulfur in natural organic molecules.
Sulfur exists in several different forms in biological material, including sulfides, sulfonates, and organosulfates (Thurman 1985; Stevenson 1994). Several organosulfates, because of their high solubility and ability to form micelles, are used in industrial and household applications and find their way into the surface and subsurface waters. Reduced organosulfur forms are used to scavenge toxic metals from waste streams. Organosulfur compounds are also found in atmospheric particles, and are considered to play a critical role in the water adsorption and light scattering properties of aerosols (Warneck 1988).

NEXAFS spectra of S have been used to identify the oxidation state and coordination environment of S in small and large biological macromolecules (George 1993; Frank et al. 1994, 1999; Rose et al. 1998; Prange et al. 1999, 2001, 2002; Glaser et al. 2001; Pickering et al. 2001; Auxulabehere-Mallart et al. 2001; Rompel et al. 2001), organic macromolecules in soils and aquatic systems (Morra et al. 1997; Vairavamurthy et al. 1997; Xia et al. 1998; Myneni and Martinez; 1999; Hundal et al. 2000; Beauchemin et al. 2002), and in coals and petroleum hydrocarbons (Spiro et al. 1988; George and Gorbaty 1989; Gorbaty et al. 1990, 1992; Kelemen et al. 1991; George et al. 1991; Waldo et al. 1991; Calkins et al. 1992; Karai et al. 1994; Sarret et al. 1999; Sugawara et al. 2001; Olivella et al. 2002). The primary focus of geochemists has been on the chemistry of S in humic substances in soils and sediments, coals and petroleum hydrocarbons.

NEXAFS spectroscopy studies at the S absorption edge indicate that humic substances contain both reduced and oxidized S species. The humic substances from terrestrial environments consist of higher concentrations of oxidized sulfur than petroleum hydrocarbons. Morra et al. (1997) suggested that isolated aquatic, soil and peat humic substances contain high concentrations of sulfonates and sulfonic acids with some ester bound sulfates. Their studies suggest the presence of reduced S species, though they have not identified them. A detailed investigation of S in marine sediments and their humic substance isolates has been conducted by Vairavamurthy et al. (1997) using the NEXAFS spectroscopy (Fig. 37). They suggested that organic mono, di- and polysulfides, sulfonates and organosulfates are the dominant forms of S in humic substances. Sulfoxide is also common in all of their samples, but it is present at small concentrations. Their spectral fits indicate the presence of thiophene in some of the samples. Xia et al. (1998) identified thiols and thiophenes as the dominant reduced forms in humic substance isolates and their concentration decreased in the order: aquatic samples > organic soil sample > mineral soil sample. The concentration of sulfate showed the opposite trend. However, the study by Myneni and Martinez (1999) indicated that organo sulfates exhibit spectral features, which overlap the features of sulfonate, and it is difficult to distinguish these groups unambiguously. Based on these spectral features, they suggested that sulfonate may be present as a minor species in a majority of the humic substances. Principal component analysis also supports this observation (Beauchemin et al. 2002).

In contrast to the S forms identified in isolated humic materials, S NEXAFS spectroscopy studies of pristine soil samples showed distinctly different ratios for the oxidized to reduced sulfur species (Fig. 38; Myneni and Martinez 1999). A detailed examination of the pine Ultisol from Puerto Rico showed that the percentage of oxidized S forms is higher than that of reduced forms. The NEXAFS spectroscopy studies indicated that the oxidized forms of sulfur (sulfonate, organo-sulfate, inorganic sulfates) are preferentially lost and the reduced-sulfur forms are retained during humic substance extraction.

Cl functional groups

Among halogenated organic compounds, chlorinated compounds are the most common anthropogenic pollutants and tend to persist in the environment with longer half-

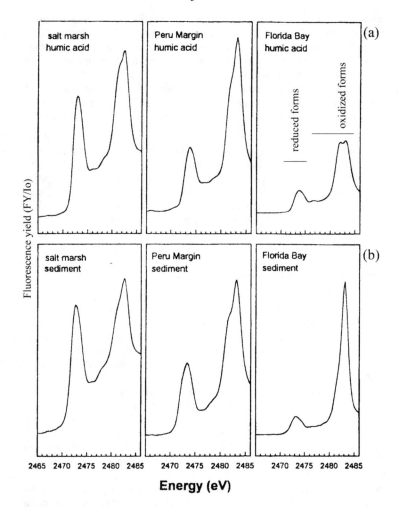

Figure 37. NEXAFS spectra of S in sediments (a) and their humic acids (b) (Used with permission from Vairavamurthy et al. 1997. © American Chemical Society).

lives. Their high environmental stability and volatility enhances their transport in the environment, and threatens to modify the stratospheric ozone concentrations. Their reactivity and ability to fractionate into lipids makes them highly toxic to the biota. Traces of these compounds are found in the atmosphere, surface and subsurface waters of all continents, far from all human inhabitants, including the Arctic and Antarctic regions. Recent investigations indicate that organohalogens are also produced by both natural biotic and abiotic processes in the environment, and that organochlorines are the dominant forms of Cl in the organic fraction of soils and sediments (Gribble 1995; Myneni 2002).

Researchers have been using gas and liquid chromatography and mass spectrometry to examine the identity and concentration of small chain halogenated compounds in the environment. Spectroscopic techniques, such as vibrational and UV absorbance and

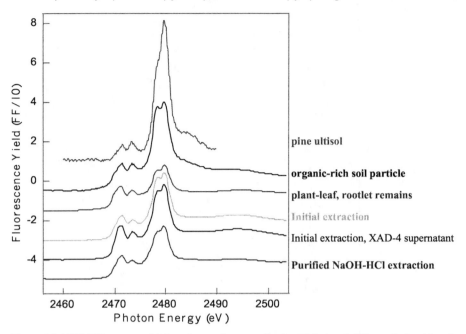

Figure 38. NEXAFS spectra of S in soil organic matter in pine Ultisol and different isolated humic substance fractions. (from Myneni and Martinez 1999; Myneni, unpublished results).

luminescence spectroscopy, have been used to examine the coordination chemistry of simpler halogenated compounds. However, these methods are not useful for the direct examination of organohalogens associated with organic macromolecules in the environment, such as humic substances. X-ray spectroscopy appears to be the only technique that can provide information on the coordination state of organohalogens associated with macromolecules in soils and natural waters. However, X-ray spectroscopy methods were not applied to explore the chemistry of natural or manmade organochlorines until recently. As discussed later, XAS can provide direct information on the chemical state and coordination environment of Cl in organic contaminants and its reactions with geologic media. However, as discussed later there are certain limitations to the type of information obtained using X-ray spectroscopic methods alone.

This discussion is focused on the research conducted at the Cl absorption edge, since very little information is available on the XAS of other organohalogens. Applications of X-ray spectroscopy to other halocarbons will be discussed briefly later.

X-ray spectra and electronic state of Cl in inorganic and organic molecules. X-ray spectroscopy studies at the Cl K-absorption edge indicate that the X-ray absorption spectra are sensitive to the Cl local coordination environment, and the spectra can be used to identify strong covalent and weak H-bonding interactions of Cl atoms (Fig. 39; Table 5—in Appendix; Myneni 2002). An examination of the NEXAFS spectra of Cl in inorganic molecules suggests that the energy of the K-absorption edge, corresponding to the transitions of Cl $1s$ electrons to vacant molecular orbitals of Cl $3p$ character or its unoccupied $4p$ states, shifts to higher energy with an increase in Cl oxidation state. The intensity of the absorption edge also increases with an increase in the oxidation state of Cl, caused by the increase in vacancies of molecular orbitals. Chlorine complexation with

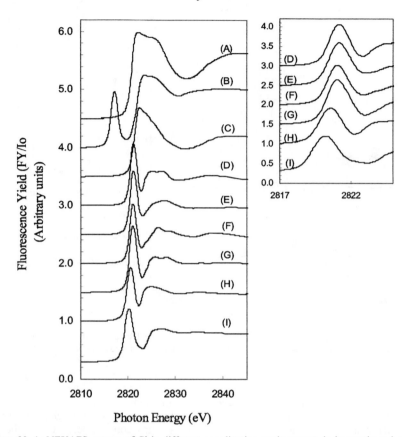

Figure 39. 1*s* NEXAFS spectra of Cl in different coordination environments in inorganic and organic compounds (from Myneni 2002). The inset shows the close up of the absorption edge of organo-Cl compounds. (A) aqueous Cl⁻, (B) solid $FeCl_3.6H_2O$, (C) solid glycine.HCl, (D) monochlorodimedone, (E) chlorophenol red, (F) 2-chlorobenzoic acid, (G) tetrachlorophenol, (H) chlorodecane, and (I) trichloroacetic acid. The pre-edge feature in Fe-chloride salt, which is well below the main Cl-edge, represents the electronic transitions to the hybridized d-orbitals of the transition metal (Fig. 40 for more details). In glycine.HCl, Cl⁻ exhibits intense H-bonding environment with organic molecule functional groups (Fig. 41 for more details). The HCl gas spectra are shown in Fig. 42 for comparison.

transition metals, specifically in the case of Cl⁻, exhibits distinct pre-edge features (Fig. 40). These pre-edge features arise from the transitions of Cl 1*s* electrons to the hybridized *d*-orbitals of metals that have some Cl 3*p* character (Shadle et al. 1994, 1995). The energy and intensity of these pre-edge features change with the number of electrons in the *d*-shell of the transition metal, and the atomic overlap of the Cl *p* orbitals. The energies of the pre-edge features and their intensities are useful in identifying the coordination state of inorganic transition metal chloride complexes in solution and at mineral-water interfaces. Although such features are expected in the case of transition metal complexes of ClO_3^- and ClO_4^-, they are expected to be weaker because of the secondary effects associated with transition metal *d*-electron interactions with the hybridized Cl-O molecular orbitals.

The NEXAFS spectra of Cl are also highly sensitive to the H-bonding environment around Cl, especially for the Cl⁻ ion (Fig. 41, Huggins and Huffman 1995; Myneni

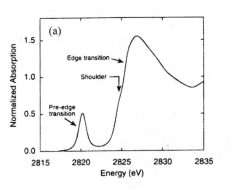

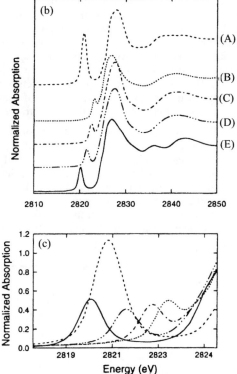

Figure 40. 1*s* NEXAFS spectra of Cl⁻ in transition metal-chloro complexes. (a) Cl NEXAFS spectrum of a transition metal-chloro complex showing the pre-edge and the main edge region (modified from Shadle et al. 1994). The energy of the main absorption edge is very close to that of uncomplexed aqueous Cl⁻. (b) and (c) shows the Cl NEXAFS spectra of different transition metal chlorides: (A) Fe(III), (B) Fe(II), (C) Co(II), (D) Ni(II), (E) Cu(II) (modified with permission from Shadle et al. 1995. © American Chemical Society).

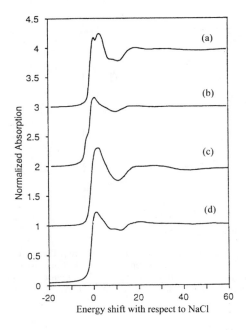

Figure 41. 1*s* NEXAFS spectra of Cl⁻ in organic molecules with different states of H-bonding interactions: (a) aniline.HCl, (b) pyridine.HCl, (c) tetracycline.HCl, (d) semicarbazide (modified with permission from Elsevier Science from Huggins and Huffman 1995). Aqueous Cl⁻ is shown in Fig. 39 for a comparison.

2002). Variations in the H-bonding interactions of Cl⁻ with different functional groups in solid organic molecules exhibit significant variations in their Cl NEXAFS spectra. Although the energy of the absorption edge of Cl⁻ (inflection point in the Cl NEXAFS spectrum) in these molecules is similar, their spectra exhibit significant spectral variations in the NEXAFS region. The origins of these spectral variations are expected to be similar to those observed in the case of O NEXAFS spectra of liquid H_2O (Myneni et al. 2002), which are related to the symmetry and the intensity of H-bonding interactions. A detailed understanding of the origin of these features, and a correlation between the coordination environment and these spectral features is not yet available.

The electronic structure of Cl in organic molecules can be well understood by examining the chemical state of Cl in simple gas phase molecules such as Cl_2 and HCl (Fronzoni and Decleva 1998; Fronzoni et al. 1998, 1999). These molecules exhibit a similar type of bonding environment with minor variations in their NEXAFS spectra, and a detailed experimental and theoretical $1s$ and $2p$ NEXAFS spectral analysis of Cl in these molecules have been conducted recently (Fig. 42). These studies indicate that the Cl $1s$ spectra exhibit an intense transition from the Cl $1s$ to σ* orbitals at low energy, and to sharp Rydberg orbitals of $4p$ and $5p$ character at high energy. The low energy Rydberg

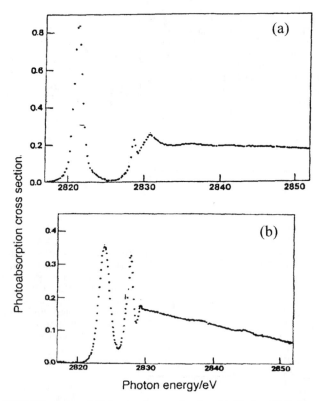

Figure 42. $1s$ NEXAFS spectra of Cl in gas phase molecules: (a) Cl_2 (modified from Fronzoni et al. 1999); (b) HCl (Modified after Fronzoni et al., 1998). The peaks represent the electronic transitions from the Cl $1s \rightarrow \sigma^*$ at low energy (lowest energy peak), and the Rydberg orbitals at high energies [the experimental spectra are originally adapted from Bodeur et al. (1990) Z. Phys. D: At. Mol. Clusters. 17:291].

transitions are just below the ionization energy of Cl. Although HCl produces the same features, the primary differences in the NEXAFS spectra of these molecules are: i) $1s \rightarrow \sigma^*$ transitions in the case of Cl_2 are at a lower energy (2821.3 eV) when compared to the HCl (2823.9 eV), ii) the energy difference between the bound-state and lowest Rydberg transition decreases from 8.0 eV in Cl_2 to approximately 3.0 eV in HCl, and iii) the intensity of transitions into the lowest energy $4p$ states is stronger in HCl. This suggests that the bonding is much stronger in HCl than in Cl_2, and the σ^* orbitals are shifted to higher energy and interact with Rydberg transitions. The strong interactions between the σ^* orbitals and the Rydberg orbitals allows the latter to shift to energies lower than the ionization potential. The Cl $2p$ spectra of Cl_2 and HCl exhibit relatively broad features corresponding to the $2p_{1/2}$ and $2p_{3/2}$ transitions to the valence σ^* orbitals at low energies and to the distinct sharp Rydberg transitions at high energies (Fig. 43). Several peaks corresponding to the Rydberg transitions exist, and it is difficult to make unequivocal assignments for these transitions. Theoretical calculations conducted using density functional theory reproduced the relative energy shifts between different transitions, but not the absolute energies. The relative shifts of $2p$ to σ^* orbitals between Cl_2 and HCl are similar to those reported for the K-edge transitions.

Electronic transitions similar to those reported for the molecules discussed above are expected for gas phase organochlorine compounds (Figs. 39, 44; Perera et al. 1987, 1991; Lindle et al. 1991; Fronzoni and Decleva 1998; Huffman and Huggins 1995; Myneni 2002). In the case of aliphatic organochlorine molecules, the strength of the interactions between Cl and C, and the coordination environment around the linking C atom will influence the energy levels of the σ^* orbitals. These bound state transitions exhibit intense features when compared to the spectral features of the gas phase molecules discussed earlier. The Rydberg transitions, which are distinct in small chain gas phase molecules, are broader and of much lower intensity when these molecules are present in a

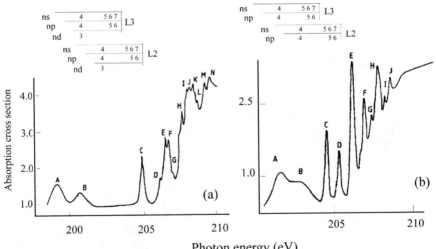

Figure 43. $2p$ (L_2, L_3) NEXAFS spectra of Cl in gas phase molecules: (a) Cl_2 (modified from Fronzoni et al. 1999); (b) HCl (modified from Fronzoni et al. 1998). The broad low energy peaks, marked "A" and "B", in both figures represent the electronic transitions to σ^* valence unoccupied molecular orbital. All other high energy peaks (C+) correspond to the Rydberg state transitions for both $2p_{1/2}$ and $2p_{3/2}$ series (the experimental spectra are originally adapted from Ninomiya et al. (1981) J Phys B: 14:1777).

Myneni

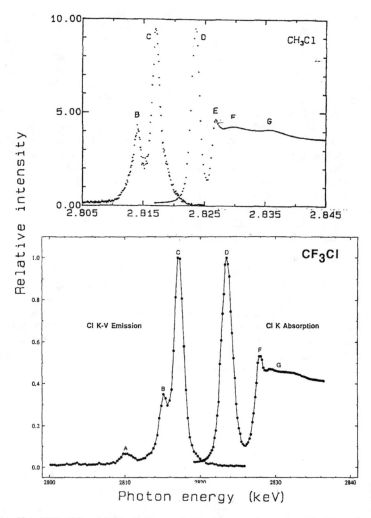

Figure 44. Cl 1s NEXAFS and XES of chloro and chlorofluoromethanes (modified from Perera et al. 1985; Lindle et al. 1991). The peaks D, E, F, and G correspond to the absorption spectrum, and A, B and C correspond to the emission spectrum.

condensed phase. The energies of the bound-state and Rydberg transitions are useful for identifying the coordination state of Cl in organic molecules (Myneni 2002).

In the case of aliphatic organochlorine compounds, the coordination state of Cl influences the energies of the $1s \rightarrow \sigma^*$ and π^* transitions, and the high-energy Rydberg and multiple scattering features (Fig. 39). In small chain aliphatic chlorinated alkanes, the energy of the $1s \rightarrow \sigma^*$ transition decreases with an increase in the Cl content of the molecule. Fluorinated chloroalkanes also exhibit the same behavior. The Cl NEXAFS spectra of chloromethane and chlorodecane are identical, which suggests that the chain length does not influence the Cl spectral features. These molecular spectra are also similar to that of polyvinyl chloride (Myneni 2002). These studies clearly indicate that the local coordination around Cl atoms affect its NEXAFS spectra, and not the chain length and the number of Cl atoms in the entire molecule.

In the case of aromatic organochlorine compounds, Cl connected to a benzene ring with or without COOH and OH groups exhibits $1s{\rightarrow}\sigma^*$ and π^* transitions at the same energy, while the higher energy Rydberg state transitions or the multiple scattering features vary significantly (Fig. 39). Unlike the polychlorinated aliphatic molecules, an increase in Cl concentration in the ring has no influence on the energies of $1s{\rightarrow}\sigma^*$ and π^* transitions (Myneni 2002). However, the number of Cl atoms in the molecule does affect the high-energy features. Theoretical studies have not been conducted on these molecules to identify the origin of these high-energy features. However, polarized NEXAFS spectroscopy studies of a single crystal of dichloroanthracene indicate that the contribution to the most intense peak in the Cl $1s$ NEXAFS spectra comes from the $1s{\rightarrow}\sigma^*$ transitions, and the transitions to the π^* transitions exhibit much weaker spectral features and occurs at a lower energy (by about 0.4 eV) when compared with the σ^* transitions (Fig. 45; Smith et al. 1994).

To the author's knowledge, Cl $2p$ spectra of organochlorine molecules have not been reported so far. The author believes that the $2p$ spectra would be useful to identify the coordination chemistry of Cl in organic molecules, and would complement the K-edge spectra. However, low fluorescence yields associated with the $2p$ state transitions limit the detection of these molecules at low concentrations.

Electronic state of Cl in natural organic molecules. NEXAFS spectroscopy has been used successfully to examine the chemical state of Cl in biomacromolecules and

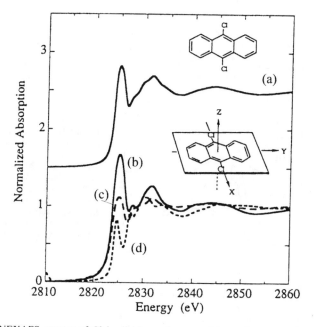

Figure 45. $1s$ NEXAFS spectra of Cl in dichloroanthracene. (a) powder spectrum; (b), (c), and (d) represent polarized S spectra of single crystal collected along the X, Y and Z axes of the molecule, respectively (shown in the inset). [Used with permission from Smith et al. 1994. © American Chemical Society.].

plant decay products directly (Myneni 2002). It has been difficult to characterize the Cl functional groups in natural organic macromolecules in their original state using the conventional techniques. Traditionally researchers have been isolating organic molecules using a variety of extraction procedures, reacting them with different chemicals before analyzing them using gas and liquid chromatography techniques. However, the coordination chemistry of Cl could be altered during such procedures, and certain types of organic molecules may be preferentially isolated.

The NEXAFS spectroscopic studies conducted on Cl functional groups of humic substances isolated from soils, sediments and river waters showed an intense feature at 2821.0 eV (Fig. 46; Myneni 2002). Nonlinear least squares fitting of these sample spectra indicate that a majority of Cl is present in the form of aromatic organochlorine compounds, with a small fraction of Cl in the aliphatic fraction. Since the number of Cl atoms and the local coordination environment around Cl in an aromatic molecule does not influence the energy of the intense $1s \rightarrow \sigma^*$ and π^* transitions, variations in weak high-energy features have to be used in the identification of chemical state of Cl. This has limited the identification of aromatic Cl forms in humic substance samples. Based on the variations in high-energy spectral features, the forms of aromatic Cl in these samples are identified as mono and dichlorophenols. These phenolic compounds are less volatile and could not be extracted using isolation procedures, and thus are not considered to be part of small chain compounds. Instead, these molecules are considered part of the large biopolymers such as lignin, or other macromolecules. Studies using $2p$ electronic

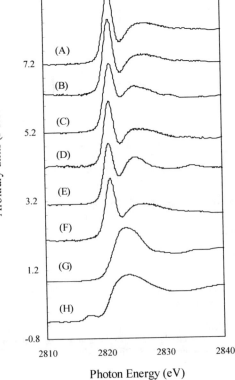

Figure 46. 1*s* NEXAFS spectra of Cl in humic substances isolated from river water, soils, peat and lignite (taken from Myneni 2002). HA: humic acid HA, FA: fulvic acid. Different spectra represent (A) Suwannee River FA, (B) Suwannee River HA, (C) soil FA, (D) soil HA, (E) Lake Fryxell FA, (F) peat HA, (G) peat FA, and (H) Leonardite HA.

transitions and X-ray emission may provide complementary information to the K-edge studies described above and aid the interpretation of different functional groups.

NEXAFS spectroscopy can also provide useful information on the nature of Cl in different fractions of soils (Myneni 2002). Detailed investigation of organochlorines associated with different fractions of soil components indicates that organochlorine compounds are primarily associated with the organic-rich fraction, while the mineral fraction is enriched in inorganic Cl$^-$. A summary of these investigations indicates that organochlorine compounds are the dominant forms of Cl in the organic matrix (Fig. 47). Using NEXAFS spectroscopy the chemical state of Cl in fresh plant material and its alteration during plant weathering has been identified (Myneni 2002). These studies suggest that the form of Cl in fresh green plant leaves and other living material is primarily hydrated Cl$^-$, which converts to organo-Cl during the plant decay process. During leaf senescence, and when the leaves accumulate on soil surface at the end of the autumn season, the leaves exhibit Cl$^-$ with minor variations in its spectral features. These changes are attributed to the dehydration of hydrated Cl$^-$ and formation of strongly H-bonded Cl$^-$ with the organic molecule functional groups (Figs. 41 and 47). These spectral variations are distinct in leaves accumulated on dry soil surfaces. Haloperoxidase enzymes produced by common soil fungi and certain bacteria may convert inorganic Cl$^-$ of the leaf to organochlorine using organic molecules in leaves as substrates for the reaction. Prior to the application of X-ray spectroscopy methods, these alterations had been widely speculated, and the extraction and analytical procedures were highly

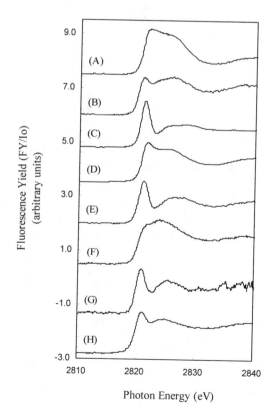

Figure 47. Cl K-edge NEXAFS spectra of live, senescent and highly humified plant material (taken from Myneni 2002). (A) green pine needles, (B) senile reddish brown pine needles attached to the tree, (C) senile leaves from soil surface, (D) moss collected from pine tree bark surface, (E) top soil in the vicinity of a pine tree, (F) ground pine ultisol, (G) organic-rich aggregate near pine tree, and (H) plant remains (leaves, rootlets, wood) of pine.

criticized. Researchers have now begun examining the chemical transformations of plant material in the presence of enzymes common in soils using soft X-ray spectroscopy methods.

The forms of Cl in coals collected from different locations have also been examined using NEXAFS spectroscopy (Huggins and Huffman 1995). These studies indicate that the chemical state of Cl in these ancient plant derived organic matrices is primarily in the form of inorganic Cl⁻, without any detectable organochlorine compounds. However, a recent spectromicroscopy study conducted at the Cl L₁ edge exhibited spectral features very similar to the Cl K-edge spectra described above. Although Cody et al. (1995) have attributed these spectral features to the inorganic Cl species, they have not examined the 2s spectra of Cl in inorganic and organic forms to make definitive spectral assignments. Since the transitions of Cl $1s$ and $2s$ electrons are similar, this author tentatively assigns these $2s$ features to the organochlorine forms. All these studies suggest that the maturation stage and thermal history of coals and buried plant debris are important in the understanding of the preservation of organochlorine compounds formed during humification. These XAS studies will help to resolve the uncertainties related to the origin and stability of the naturally formed organohalogens in the environment, and in the evaluation of the fate of anthropogenic chlorinated compounds.

EXAFS spectroscopic investigation of organochlorine compounds. Conducting EXAFS spectroscopy at the Cl absorption edge necessitates a high flux photon source that can produce a smooth incident beam without any structure. Although there are several beamlines in the USA, Europe and Japan where one can access the Cl absorption edge, high flux beamlines where Cl EXAFS spectroscopy studies can be conducted on dilute samples are not available. Further, contamination of Ar in air within the sample compartment also limits the EXAFS studies under ambient conditions to < 3200 eV. For these reasons, the majority of previous Cl EXAFS spectroscopy studies focused on concentrated model compounds or Cl containing adsorbates on metal surfaces in vacuum (Huggins and Huffman 1995; Myneni 2002).

Cl EXAFS spectroscopy studies indicate that the radial structure functions for inorganic and organochlorine compounds are different, and scattering from neighboring C atoms in organochlorine compounds is strong enough to identify their coordination environment unambiguously (Fig. 48; Huggins and Huffman 1995; Myneni 2002). EXAFS spectroscopy studies of aqueous Cl⁻ indicate that Cl-H, and Cl-O bond distances are around 2.6 and 3.0-3.2 Å. The EXAFS spectroscopy studies on organochlorines have indicated that the C-Cl bond distances in several organochlorines is around 1.75 Å (Fig. 48). Aliphatic and aromatic compounds have shown distinct variations in their radial structure functions, suggesting that Cl EXAFS may be useful for identifying the variations in Cl local coordination environment (Fig. 48). Availability of high flux beamlines will allow the examination of speciation and chemical reactivity of Cl at dilute concentrations in geological matrices.

X-ray spectroscopy studies on other organohalogen compounds. To the author's knowledge, X-ray spectroscopy studies have not yet been conducted on naturally occurring organobromines, organoiodine and organofluorine compounds. Some of these studies have recently begun in the author's laboratory and at a few other research facilities. However, X-ray spectroscopy studies have been conducted on synthetic organobromine and organoiodine compounds. Bromine NEXAFS spectroscopy studies indicate that Br doped polyacetylene shows distinct differences between inorganic Br⁻ and organobromine (Oyanagi et al. 1987). Bromine EXAFS spectroscopy studies indicate that the Br-C bond distances in organobromines are in the range of 1.9 to 2.06 Å (Fujikawa et al. 1987; Epple et al. 1997; Elizabe et al. 1998; Dau et al. 1999). The

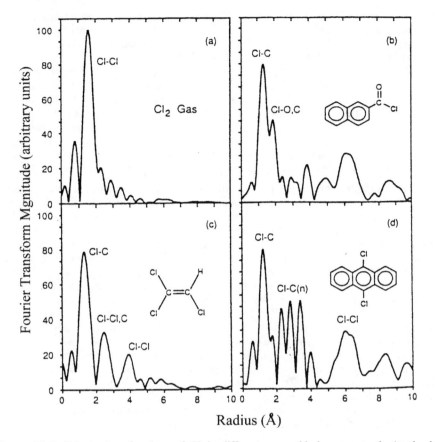

Figure 48. Radial structure functions of Cl in different organochlorine compounds (used with permission from Elsevier Science from Huggins and Huffman 1995).

corehole broadening at the Br and I edges limits the spectral resolution obtained at these element K-edges. Although the resolution is sufficient to resolve the spectral contributions of inorganic Br$^-$ and organobromine compounds, small variations in the coordination chemistry of organobromines and their corresponding energy shifts may not be resolved. This is worse at the K edge of iodine. Preliminary K-edge studies conducted on inorganic and organoiodine compound K-edges exhibited distinct variations in the spectra, but they are too subtle to identify the local coordination variation in organoiodine compounds (Feiters et al. 2001). However, the high resolution L edge spectra can provide detailed and complementary information on the coordination environments of organobromines and organoiodines in soil and sediment matrices.

APPLICATIONS OF SPECTROMICROSCOPY FOR STUDIES ON ORGANIC MOLECULE AGGREGATES

Several spectromicroscopy studies have been conducted to examine the chemical heterogeneity of organic molecules in biological samples (Ade et al. 1992; Zhang et al.

1996; Loo et al. 2000; Hitchcock et al. 2002; SCB Myneni, unpublished results), synthetic organic materials (Ade et al. 1995, 1997; Ade 1998; Warwick et al. 1998; Neuhausler et al. 1999), and naturally occurring organic molecules (Botto et al. 1994; Cody et al. 1995a,b, 1998; Myneni et al. 1999a,b; Rothe et al. 2000; Scheinost et al. 2001; Plaschke et al. 2002; Russell et al. 2002). Although high spectral and spatial resolution can be obtained with the X-ray spectromicroscopy methods and allows the probing of samples under ambient conditions, beam damage may be an important issue. As discussed earlier, not all samples are radiation sensitive, and researchers may take different approaches to minimize sample damage. These include, rapid imaging with sensitive detectors, low incident beam intensity, and collection of images at wavelengths where samples are more stable. Some special sample preparation methods have been proposed for preserving and imaging biological tissue (Loo et al. 2001).

X-ray spectromicroscopy has facilitated the direct examination of aggregate structures of aqueous humic substances and their association with mineral oxides and silicates (Myneni et al. 1999). Macroscopic studies conducted on isolated humic substances have indicated that their macromolecular structures vary as a function of solution pH, ionic strength and the chemical composition of aqueous solutions and humic substances (e.g., humic versus fulvic acids). The electron microscopy studies of isolated humics (for air dried humic materials) indicated that they coil at low pH values and form linear long chain-like structures at high pH values. X-ray spectromicroscopy has provided the first direct evidence on the nature of humic substance aggregation processes in aqueous solutions (Fig. 49; Myneni et al. 1999). These studies indicate that humic substances exhibit more than one type of structure in aqueous solutions in contrast to the observations made from earlier electron microscopy studies. These X-ray studies also indicate that complexing cations and the presence of minerals can significantly influence aggregate structures of humic substances (Myneni et al. 1999; Rothe et al. 2000; Plaschke et al. 2002). Aggregation and macromolecular structure of humic substances are also examined using AFM at a higher resolution than with X-ray microscopy; however, these are limited to humic substances sorbed on flat mineral surfaces (Dejanovic and Maurice 2001).

Using X-ray spectromicroscopy studies, it is feasible to examine the C, N and O functional groups of organic molecules in soil samples from regions as small as 100 nm. These methods allow the direct probing of the chemistry of organomineral interactions at the nanoscale. Using X-ray spectromicroscopy methods, researchers have examined the chemistry of organic molecules in soil samples and their isolates (Myneni et al. 1999a, Fig. 13). When compared to the C NEXAFS spectra of isolated humic substances, organic molecules in unaltered soil samples exhibit low concentrations of aromatic C, and a high concentration of carboxylic, ketonic C, and aliphatic C (discussed above). These studies suggest that isolation procedures preferentially extract certain forms of C. Similar results are also shown for S forms. As discussed earlier, fluorescence detection would assist in the probing of organic molecules present on thick soil particles surfaces, which is currently not possible.

Using X-ray microscopy, precipitation and crystallization processes, and the influence of solution chemistry on these processes have been examined (Rieger et al. 2000). Since an image can be collected in a fraction of a second, crystallization process can be examined at least on the time scale of seconds. Such fast scanning is helpful for imaging X-ray sensitive samples. Soft X-ray spectromicroscopy has been used to explore the interfacial chemistry of microorganisms, trace and major element geochemistry and formation of biofilms (Tonner et al. 1999; Hitchcock et al. 2002; SCB Myneni, unpublished results). These studies suggest that functional group specific

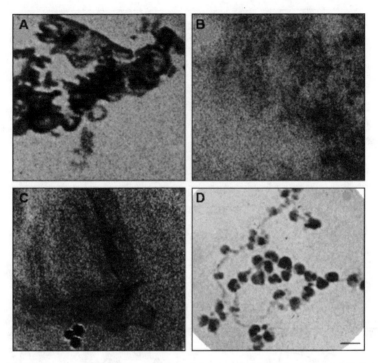

Figure 49. X-ray microscopy images of isolated humic substances in water and their aggregate structures as a function of pH (A and B), and in the presence of different soluble metal ions in water (C and D). (A) pH = 3.0; (B) pH = 9.0; (C) pH = 4.0, 0.018 M CaCl$_2$; (D) pH 4.0, 1 mM FeCl$_3$. Scale bar is 500 nm (from Myneni et al. 1999b).

information can be obtained from biolfilms at a resolution better than 100 nm for samples in water. These investigations can be conducted directly for samples in water since X-ray transmission is greater below the O-absorption edge, where imaging can be conducted using the absorption intensity at the C and N edges. In addition, the L-edges of transition metals can be used to examine the interactions of heavy elements and toxic metals with biofilms.

The high spatial resolution and spectral sensitivity of X-ray spectromicroscopy has also been used in probing the chemistry of organic molecules in aerosols (Russell et al. 2002). Since the X-ray studies can be conducted under atmospheric pressure conditions, it is possible to preserve or minimize the losses of the highly volatile organic molecules. Further, X-ray spectromicroscopy offers better spectral sensitivity than commonly used infrared spectromicroscopy techniques. Russell et al. (2002) used the chemical sensitivity of X-ray spectroscopy to map different functionalities of organic molecules and was able to correlate them with the inorganic constituents, such as K and Ca (Fig. 50). For the first time, this study provided information on the surface active carboxylic acids on aerosol particles. This *in situ* information is useful in predicting the water adsorption characteristics, growth and the light scattering characteristics of aerosols, which significantly influences micro and global climates.

Myneni

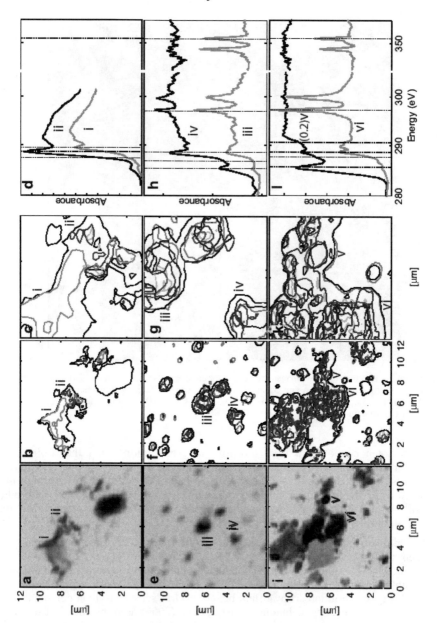

Figure 50. STXM images and C NEXAFS spectra of marine boundary layer particles collected from St. Croix (taken from Russell et al. 2002). The particles shown in (a) are collected at 310 m, and (e) and (i) are collected from 30 m from the sea level. Maps of different functional groups are shown in b and c (for a), f and g (for e), and j and k (for i). The C NEXAFS spectra of different particles are shown in d, h, and l, which correspond to different marked regions in images a, e, and i. The peaks around 290 eV correspond to C, around 300 eV correspond to the K L-edges, and at 350 eV correspond to the Ca L edges. The images shown on the left (a, e, i) are collected at 300 eV.

SUMMARY AND FUTURE DIRECTIONS

(1) Soft X-ray spectroscopy and spectromicroscopy methods can provide high-resolution spectral and spatial information on the functional group chemistry and aggregate structures of organic molecules in aqueous solutions and on particle surfaces. The detection limits are around a few mM for C, N, and O functional groups, and about 0.5 mM for the P, S and Cl functional groups. Development of better detection schemes for spectroscopy and spectromicroscopy techniques would allow the examination of molecules at much lower concentrations, and in a variety of samples.

(2) Spectral assignments and interpretation for the functional group chemistry of several aqueous and biological macromolecules is still in its early stages. Theoretical investigations of these systems together with the calculated electronic spectra would help in the interpretation of the coordination environment of organic molecule functional groups.

(3) Number of synchrotron facilities available for examining the chemistry of organic molecules in aqueous solutions or in wet state is small. Only a selected group of researchers have been using the X-ray spectroscopy methods for this reason. Construction of new synchrotron facilities optimized for examining environmental samples would further these investigations.

(4) Details on other complementary techniques, such as small and wide angle X-ray scattering, X-ray Raman scattering, and X-ray emission, and synchrotron infrared spectromicroscopy methods are not discussed here. However, these methods can provide complementary information on the macromolecular structure and functional group chemistry of organic molecules at different resolution in different sample environments.

(5) Fundamental information on the composition and the structure of naturally occurring organic molecules and their interactions with geologic media would help in the prediction of biogeochemical cycling of different major and minor elements and toxic chemicals in the environment. *In situ* molecular information is critical in identifying the role of organic molecules in different biogeochemical processes.

ACKNOWLEDGMENTS

The author would like to thank N. Sturchio, M. Hay, and K. Jyothi for providing comments on the manuscript and D. Edwards for making some of the tables. The author would like to acknowledge the help of the scientists and other staff members at the Stanford Synchrotron Light Source and Advanced Light Source in helping with studies on organic molecules. The author is supported by the funding from BES, DOE (Geosciences), NSF (Chemical sciences), and Princeton University during the preparation of this manuscript.

REFERENCES

Ade H, Zhang X, Cameron S, Costello C, Kirz J, Williams S (1992) Chemical contrast in X-ray microscopy and spatially resolved XANES spectroscopy of organic specimens. Science 258:972-975

Ade H, Smith AP, Cameron S, Cieslinski R, Mitchell G, Hsiao B, Rightor E (1995) X-ray microscopy in polymer science: prospects of a "new" imaging technique. Polymer 36:1843-1848

Ade H (1998) X-ray spectromicroscopy. *In:* Experimental Methods in the Physical Sciences. Vol 32. Samson JAR, Ederer DL (eds) Academic Press, San Diego. p 225-261

Aiken GR, McKnight DM, Wershaw RL, MacCarthy P (1985) Humic Substances in Soil Sediment and Water: Geochemistry, isolation, and characterization. Wiley, New York

Amemiya K, Kitajima Y, Yonamoto Y, Terada S, Tsukabayashi H, Yokoyama T, Ohta T (1999) Oxygen K-edge X-ray-absorption fine-structure study of surface methoxy species on Cu(111) and Ni(111). Phys Rev B 59:2307-2312

Anderson G (1980) Assessing organic phosphorus in soils. *In:* The role of phosphorus in agriculture. FE Khasawnech et al. (ed) ASA, Madison, WI. 411-431

Anxolabehere-Mallart E, Glaser T, Frank P, Aliverti A, Zanetti G, Hedman B, Hodgson KO, Solomon EI (2001) Sulfur K-edge X-ray absorption spectroscopy of 2Fe-2S Ferredoxin: Covalency of the oxidized and reduced 2 Fe forms and comparison to model complexes. J Am Chem Soc 123:5444-5452

Apen E, Hitchcock AP, Gland JL (1993) Experimental studies of the core excitation of imidazole, 4,5-dicyanoimidazole, and s-triazine. J Phys Chem 97:6859-6866

Attwood DT (2000) Soft X-rays and Extreme Ultraviolet Radiation: Principles and Applications. Cambridge, New York

Beauchermin S, Hesterberg D, Beauchermin M (2002) Principal component analysis approach for modeling sulfur K-XANES spectra of humic acids. Soil Soc Am J 66:83-91

Beck M, Stiel H, Leupold D, Winter B, Pop D, Vogt U, Spitz C (2001) Evaluation of the energetic position of the lowest excited singlet state of β-carotene by NEXAFS and photoemission spectroscopy. Biochim Biophys Acta 1506:260-267

Benitez A, Moore JH, Tossell JA (1988) The correlation between electron transmission and inner shell electron excitation spectra. J Chem Phys 88:6691-6698

Bertsch PM, Hunter DB (2001) Applications of synchrotron-based X-ray microprobes. Chem Rev 101:1809-1842

Bilderback DH, Thiel DJ, Pahl R, Brister KE (1994) X-ray applications with glass capillary optics. J Sync. Radiation 1:37-42

Bluhm H, Ogletree DF, Fadley CS, Hussain Z, Salmeron M (2002) The premelting of ice studied with photoelectron spectroscopy. J Phys Condens Matter 14:L227-L233

Boese J, Osanna A, Jacobsen C, Kirz J (1997) Carbon K-edge XANES spectroscopy of amino acid and peptides. J Electron Spectrosc Relat Phenom 85:9-15

Bonnin-Mosbah M, Metrich N, Susini J, Salome M, Massare D, Menez B (2002) Micro X-ray absorption near edge structure at the sulfur and iron K-edges in natural silicate glasses. Spectrochim Acta B 57:711-725

Bortiatynski JM, Hatcher PG, Knicker H (1996) NMR techniques (C, N, and H) in studies of humic substances. *In:* Humic and Fulvic Acids. Gaffney JS, Marley NA, Clark SB (eds) ACS Publication, Washington, 651:57-77

Botto RE, Cody GD, Kirz J, Ade H, Behal S, Disko M (1994) Selective chemical mapping of coal microheterogeneity by scanning transmission X-ray microscopy. Energy & Fuels 8:151-154

Brown GE Jr, Calas G, Waychunas GA, Petiau J (1988) X-ray absorption spectroscopy and its applications in mineralogy and geochemistry. Rev Min 18:431-512

Brown GE Jr, Henrich VE, Casey WH, Clark DL, Eggleston C, Felmy A, Goodman DW, Gratzel M, Maciel H, McCarthy MI, Nealson KH, Sverjensky DA, Toney MF, Zachara JM (1999) Metal oxide surfaces and their interactions with aqueous solutions and microbial organisms. Chem Rev 99:77-174

Cade-Menun BJ, Preston CM (1996) A comparison of soil extraction procedures for ^{31}P NMR spectroscopy. Soil Sci 161:770-785

Cade-Menun BJ, Lavkulich LM (1997) A comparison of methods to determine total, organic, and available phosphorus on forest soils. Commun Soil Sci Plant Anal 28:651-663

Calkins WH, Torres_ordonez RJ, Jung B, Gorbaty ML, George GN, Kelemen SR (1992) Comparison of pyrolytic and X-ray spectroscopic methods for determining organic sulfur species in coal. Energy Fuels 6:411-413

Carravetta V, Contini G, Plashkevych O, Agren H, Polzonetti G (1999) Orientational probing of multilayer 2-mercaptobenzoxazole through NEXAFS: An experimental and theoretical study. J Phys Chem A 103:4641-4648

Christensen BT (1992) Physical fractionation of soil and organic matter in primary particle size and density separates. Adv Soil Sci 20:1-90

Cody GD, Botto RE, Ade H, Behal S, Disko M, Wirick S (1995a) C NEXAFS microanalysis and scanning X-ray microscopy of microheterogeneities in a high-volatile A bituminous coal. Energy Fuels 9:75-83

Cody GD, Botto RE, Ade H, Behal S, Disko M, Wirick S (1995b) Inner-shell spectroscopy and imaging of a subbituminous coal: *In situ* analysis of organic and inorganic microstructures using C (1s)-, Ca(2p)-, and Cl(2s)-NEXAFS. Energy Fuels 9:525-533

Cody GD, Ade H, Wirick S, Mitchell GD, Davis A (1998) Determination of chemical-structural changes in vitrinite accompanying luminescence alteration using C-NEXAFS analysis. Org Geochem 28:441-445

Contini G, Ciccioli A, Laffon C, Parent P, Polzonetti G (1998) NEXAFS study of 2-mercaptobenzoxazole adsorbed on Pt(111): multilayer and monolayer. Surf Sci 412/413:158-165

Contini G, DiCastro V, Stranges S, Richter R, Alagia M (2000) Gas-phase photoemission study of 2-mercaptobenzoxazole. J Phys Chem A 104:9675-9680

Coreno M, Avaldi L, Camilloni R, Prince KC, de Simone M, Karvonen J, Colle R, Simonucci S (1999) Measurement and *ab initio* calculation of the Ne photoabsorption spectrum in the region of the K edge. Phys Rev A 59:2494-2497

Dau H, Dittmer J, Epple M, Hanss J, Kiss E, Rehder D, Schuzke C, Vilter H (1999) Bromine K-edge EXAFS studies of bromide binding to bromoperoxidase from *Ascophyllum nodosum*. FEBS Letters 457:237-240

Dezarnaud C, Tronc M, Hitchcock AP (1990) Inner shell spectroscopy of the carbon sulfur bond. Chem Phys 142:455-462

Dezarnaud-Dandine C, Bournel F, Mangeney C, Tronc M, Jones D, Modelli A (1998) σ* resonances in electron transmission (ETS) and X-ray absorption (XAS) spectroscopies of dimethyl(poly)sulphides $(CH_3)_2S_x$ (x = 1,2,3). J Phys B: At Mol Opt Phys 31:L497-L502

Dezarnaud-Dandine C, Bournel F, Mangeney C, Tronc M, Modelli A, Jones D (2001) Empty levels probed by XAS and ETS in cyclic polymethylene sulfides $(CH_2)_nS$, n = 2,3,4,5 Chem Phys 265:105-112

Egerton R (1997) Electron energy-loss spectroscopy. Phys World 10:47-51

Elizabe L, Yeo L, Harris KD (1998) Conformational properties of guest molecules in constrained solid state environments: Bromine K-edge X-ray absorption spectroscopy of 2-bromoalkane/urea inclusions. Chem Mater 10:1220-1226

Engemann C, Franke R, Hormes J, Lauterbach C, Hartmann E, Clade J, Jansen M (1999) X-ray absorption near-edge spectroscopy (XANES) at the phosphorous K-edge of triorganophophinechalcogenides. Chem Phys 243:61-75

Epple M, Troger L, Hildebrandt (1997) Quantitative reaction kinetics in the liquid state by *in situ* XAFS. J Chem Soc, Faraday Trans, 93:3035-3037

Eustatiu IG, Huo B, Urquhart SG, Hitchcock AP (1998) Isomeric sensitivity of the C 1s spectra of xylenes. J Electron Spectrosc Relat Phenom 94:243-252

Feiters MC, Kupper FC, Kroneck PMH, Meyer-Klaucke (2001) Investigation of the iodine compounds involved in iodine uptake, iodine accumilation, and iodovolatilization in brown algae. Hasy Laboratory Annual Reports.

Flemmig B, Modrow H, Hallmeier KH, Hormes J, Reinhold J, Rudiger S (2001) Sulfur in different chemical surroundings – S K XANES spectra of sulfur-containing heterocycles and their quantum-chemically supported interpretation. Chem Phys 270:405-413

Francis JT, Hitchcock AP (1992) Inner-shell spectroscopy of para-benzoquinone, hydroquinone, and phenol-distinguishing quenoid and benzenoid structures. J Phys Chem 96:6598-6610

Francis JT, Hitchcock AP (1994) Distinguishing keto and enol structures by inner-shell spectroscopy. J Phys Chem 98:3650-3657

Francis JT, Enkvist C, Lunell S, Hitchcock AP (1994) Studies of X 1s to π* triplet states of carbon monoxide, benzene, ethylene, and acetylene. Can J Phys 72:879-884

Frank P, Hedman B, Carlson RMK, Hodgson KO (1994) Interaction of vanadium and sulfate in blood cells from the tunicate *Ascidia ceratodes*: Observations using X-ray absorption edge structure and EPR spectroscopies. Inorg Chem 33:3794-3803

Frank P, Hedman B, Hodgson KO (1999) Sulfur allocation and vanadium-sulfate interactions in whole blood cells from the tunicate *Ascidia ceratodes*, investigated using X-ray absorption spectroscopy. Inorg Chem 38:260-270

Fronzoni G, Decleva P (1998) Ab-initio CI calculations of the C 1s and Cl 1s and 2p core excitation spectra of the freon molecules: CCl_4, $CFCl_3$, CF_2Cl_2, and CF_3Cl. Chem Phys 237:21-42

Fronzoni G, Stener M, Decleva P, De Alti G (1998) Theoretical study of the Cl 1s and 2p near edge photoabsorption spectra of HCl by accurate ab-initio configuration interaction and density functional approaches. Chem Phys 232:9-23

Fronzoni G, Stener M, Decleva P (1999) Theoretical study of the excited and continuum states in the NEXAFS regions of Cl_2. Phys Chem Chem Phys 1:1405-1414

Fujikawa T, Oizumi H, Oyanagi H, Tokumoto M, Kuroda H (1986) Short-range order full multiple scattering approach to the polarized K-edge XANES of Br-doped in Trans-polyacetylene. J Phys Soc. Japan 55:4090-4102

George GN, Gorbaty ML (1989) Sulfur K-edge X-ray absorption spectroscopy of petroleum asphaltenes and model compounds. J Am Chem Soc 111:3182-3186

George GN, Gorbaty ML, Kelemen SR, Sansone M (1991) Direct determination and quantification of sulfur forms in coals from the Argonne Premium Sample Program. Energy & Fuels 5:93-97

George GN (1993) X-ray absorption spectroscopy of light elements in biological systems. Cur Opin Struct Biol 3:780-784

Glaser T, Rose K, Shadle SE, Hedman B, Hodgson KO, Solomon EI (2001) S K-edge X-ray absorption studies of tetranuclear iron-sulfur clusters: μ-sulfide bonding and its contribution to electron delocalization. J Am Chem Soc 123:442-454

Gorbaty ML, George GN, Kelemen SR (1990) Direct determination and quantification of sulfur forms in heavy petroleum and coals: 2. The sulfur K edge X-ray absorption spectroscopy approach. Fuel 69:945-950

Gorbaty ML, Kelemen SR, George GN, Kwiatek PJ (1992) Characterization and thermal reactivity of oxidized organic sulphur forms in coals. Fuel 71:1255-1264

Gurevich AB, Bent BE, Teplyakov AV, Chen JG (1999) A NEXAFS investigation of the formation and decompostion of CuO and Cu₂O thin films on Cu(100). Surf Sci 442:L971-L976

Harrick NJ (1987) Internal reflection spectroscopy. Harrick Scientific Corporation, New York

Hasselstrom J, Karis O, Weinelt M, Wassdahl N, Nilsson A, Nyberg M, Pettersson LGM, Samant MG, Stohr J (1998) The adsorption structure of glycine adsorbed on Cu(110); comparison with formate and acetate/Cu(110). Surf Sci 407:221-236

Hawthorne FC, Krivovichev SV, Burns PC (2000) The crystal chemistry of sulfate minerals. Rev Min Geochem 40:1-113

Hempelmann A, Piancastelli MN, Heiser F, Gessner O, Rudel A, Becker U (1999) Resonant photofragmentation of methanol at the carbon and oxygen K-edge by high-resolution ion-yield spectroscopy. J Phys B 32:2677-2689

Hennig C, Hallmeier KH, Szargan R (1998) XANES investigation of chemical states of nitrogen in polyaniline. Synt Metals 92:161-166

Hibble SJ, Walton RI, Feaviour MR, Smith AD (1999) Sulfur-sulfur bonding in the amorphous sulfides WS3, WS5, and Re2S7 from sulfur K-edge EXAFS studies. J Chem Soc, Dalton Trans 2877-2883

Hitchcock AP, Brion CE (1978a) Inner shell excitation of CH_3F, CH_3Cl, CH_3Br AND CH_3I by 2.5 KeV electron-impact. J Electron Spec Rel Phen 13:193-218

Hitchcock AP, Brion CE (1978b) Inner-shell excitation and EXAFS-type phenomena in chloromethanes. J Electron Spec Rel Phen 14:417-441

Hitchcock AP, Mancini DC (1994) Bibliography and database of inner shell excitation spectra of gas phase atoms and molecules. J Electron Spectrosc Rel Phenom 67:1-132

Hitchcock AP, Pocock M, Brion CE, Banna MS, Frost DC, McDowell CA, Wallbank B (1978) Inner shell excitation and ionization of monohalobenzenes. J Electron Spec Rel Phen 13:345-360

Hitchcock AP, Brion CE (1979a) Inner shell electron energy loss studies of HCN and C_2N_2. Chem Phys 37:319-331

Hitchcock AP, Brion, CE (1979b) K-shell excitation of HCN by electron energy loss spectroscopy. J Electron Spec Rel Phen 15:201-206

Hitchcock AP, Brion CE (1980a) Inner-shell excitation of formaldehyde, acetaldehyde, and acetone studied by electron impact. J Electron Spec Rel Phen 19:231-250

Hitchcock AP, Brion CE (1980b) K-Shell excitation spectra of CO, N_2, and O_2. J Electron Spec Rel Phen 18:1-21

Hitchcock AP, Beaulieu S, Steel T, Stohr J, Sette F (1984) Carbon K-Shell electron energy loss spectra of 1- and 2-butenes, trans-1-3-butadiene, and perfluoro-2-butene. Carbon-carbon bond lengths from continuum shape resonances. J Chem Phys 80:3927-3935

Hitchcock AP, Horseley JA, Stohr J (1986a) Inner-shell excitation of thiophene and thiolane: Gas, solid, and monolayer states. J Chem Phys 85:4835-4848

Hitchcock AP, Newbury DC, Ishii I, Stohr J, Horsley JA, Redwing RD, Johnson AL, Sette F (1986b) Carbon K-shell excitation of gaseous and condensed cyclic hydrocarbons - C_3H_6, C_4H_8, C_5H_8, C_5H_{10}, C_6H_{10}, C_6H_{12}, and C_8H_8. J Chem Phys 85:4849-4862

Hitchcock AP, Ishii I (1987) Carbon K-shell excitation spectra of linear and branched alkanes J Electron Spec Rel Phen 42:11-26

Hitchcock AP, Tourillon G, Braun W (1989a) Inner-Shell excitation studies of conducting organic polymers: selenophene, 3-methyl selenophene, and their polymers. Can J Chem 67:1819-1827

Hitchcock AP, Tourillon G, Garrett R, Lazarz N (1989b) C 1s excitation of azulene and polyazulene studied by electron energy loss spectroscopy and X-ray absorption spectroscopy. J Phys Chem 93:7624-7628

Hitchcock AP, Tronc M, Modelli A (1989c) Electron transmission and inner-shell electron energy loss spectroscopy of CH_3CN, CH_3NC, CH_3SCN, and CH_3NCS. J Phys Chem 93:3068-3077

Hitchcock AP, Tourillon G, Garrett R, Williams GP, Mahatsekake C, Andrieu C (1990) Inner-shell excitation of gas phase and polymer thin film 3-alkylthiophenes by electron energy loss and X-ray photoabsorption spectroscopy. J Phys Chem 94:2327-2333

Hitchcock AP, Dewitte RS, Vanesbroeck JM, Aebi P, French CL, Oakley RT, Westwood NPC (1991) A valence shell and inner-shell electronic and photoelectron spectroscopic study of the frontier orbitals of 2,1,3-benzothiadiazole, $C_6H_4SN_2$, 1,3,2,4-Benzodithiadiazine, $C_6H_4S_2N_2$, and 1,3,5,2,4-benzotrithiadiazepine, $C_6H_4S_3N_2$. J Electron Spec Rel Phen 57:165-187

Hitchcock AP, Urquhart SG, Rightor EG (1992) Inner-shell spectroscopy of benzaldehyde, terephthaldehyde, ethyl benzoate, terephthaloyl chloride, and phosgene – models for core excitation of poly(ethylene-terephthalate). J Phys Chem 96:8736-8750

Hitchcock AP, Wen AT, Glass SW, Spencer JT, Dowben PA (1993) Inner-shell excitation of boranes and carboranes. J Phys Chem 97:8171-8181

Horsley JA, Stohr J, Koestner RJ (1985a) Structure and bonding of chemisorbed ethylene and ethylidyne on Pt(111) from near edge X-ray absorption fine structure spectroscopy and multiple scattering calculations. J Chem Phys 83:3146-3153

Horsley JA, Stohr J, Hitchcock AP, Newbury DC, Johnson AL, Sette F (1985b) Resonances in the K shell excitation spectra of benzene and pyridine: Gas phase, solid, and chemisorbed states. J Chem Phys 83:6099-6107

Huang SX, Fischer DA, Gland JL (1996) Aniline adsorption, hydrogenation, and hydrogenolysis on the Ni(100) surface. J Phys Chem 100:10223-10234

Huffman GP, Mitra S, Huggins FE, Shah NS, Vaidya N, Lu F (1991) Quantitative analysis of all major forms of sulfur in coal by X-ray absorption fine structure spectroscopy. Energy & Fuels 5:574-581

Huffman GP, Shah NS, Huggins FE, Stock LM, Chatterjee K, Kilbane JJ, Chou M, Buchanan DH (1995) Sulfur speciation of desulfurized coals by XANES spectroscopy. Fuel 74:549-555

Huggins FE, Huffman GP (1995) Chlorine in coal: an EXAFS spectroscopic investigation. Fuel 74:556-569

Hundal LS, Carmo AM, Bleam WL, Thompson ML (2000) Sulfur in biosolids-derived fulvic acidic characterization by XANES spectroscopy and selective dissolution approaches. Environ Sci Technol 34:5184-5188

Ikenaga E, Kudara K, Kusaba K, Isari K, Sardar SA, Wada S, Mase K, Sekitani T, Tanaka K (2001) Photon-stimulated ion desorption for PMMA thin film in the oxygen K-edge region studied by Auger electron-photoion coincidence spectroscopy. J Electron Spec Rel Phen 114-116:585-590

Ikeura-Sekiguchi H, Sekiguchi T, Kitajima Y, Baba Y (2001) Inner shell excitation and dissociation of condensed formamide. App Surf Sci 169:282-286

Ishii I, Hitchcock AP (1987) A quantitative experimental study of the core excited electronic states of formamide, formic acid, formylfluoride. J Chem Phys 87:830-839

Ishii I, McLaren R, Hitchcock AP, Robin MB (1987) Inner-shell excitations in weak-bond molecules. J Chem Phys 87:4344-4360

Ishii I, Hitchcock AP (1988) The Oscillator strengths for C 1s and O1s excitation of some saturated and unsaturated organic alcohols, acids and esters. J Electron Spec Phen 46:55-84

Ishii I, McLaren R, Hitchcock AP, Jordan KD, Choi Y, Robin MB (1988) The σ^* molecular orbitals of perfluoroalkanes as studied by inner-shell electron energy-loss and electron transmission spectroscopies. Can J Chem 66:2104-2121

Jacobsen C, Kirz J, Williams S (1992) Resolution in soft X-ray microscopes. Ultramicrosc 47:55-79

Jacobsen C, Wirick S, Flynn G, Zimba C (2000) Soft X-ray spectroscopy from image sequences with sub-100 nm spatial resolution. J Microscopy 197:173-184

Jordan-Sweet JL, Kovac CA, Goldberg MJ, Morar JF (1988) Polymer/ metal interfaces studied by carbon near edge X-ray absorption fine structure spectroscopy. J Chem Phys 89:2482-2489

Jugnet Y, Himpsel FJ, Avouris P, Koch EE (1984) High resolution C 1s and O 1s core-excitation spectroscopy of chemisorbed, physisorbed, and free CO. Phys Rev Let 53:198-202

Kasrai M, Bancroft GM, Brunner RW, Jonasson RG, Brown JR, Tan KH, Feg X (1994) Sulfur speciation in bitumens and asphaltenes by X-ray absorption fine structure spectroscopy. Geochim Cosmochim Acta 58:2865-2872

Kaznacheyev K, Osana A, Jacobsen C, Plashkevych O, Vahtras O, Agren H, Carravetta V, Hitchcock AP (2002) Innershell absorption spectroscopy of amino acids. J Phys Chem A 106:3153-3168

Kelemen SR, Gorbaty ML, George GN, Kwiatek PJ, Sansone M (1991) Thermal reactivity of sulfur forms in coal. Fuel 70:396-402

Kikuma J, Tonner BP (1996) XANES spectra of a variety of widely used organic polymers at the C K-edge. J Electron Spect Rel Phen 82:53-60

Koningsberger DC, Prins R (eds) (1988) X-ray absorption: Principles, applications, techniques of EXAFS, SEXAFS, and XANES. John Wiley & Sons, New York

Lindle DW, Cowan PL, Jach T, LaVilla RE, Deslattes RD, Perera RCC (1991) Polarized X-ray emission studies of methyl chloride and chlorofluoromethanes. Phys Rev A 43:2353-2362

Loo BW, Sauerwald IM, Hitchcock AP, Rothman SS (2001) A new sample preparation method for biological soft X-ray microscopy: nitrogen based contrast and radiation tolerance properties of glycol methacrylate-embedded and sectioned tissue. J Microsc 204:69-86

Matsui F, Yeom HW, Matsuda I, Ohta T (2000) Adsorption and reaction of acetylene and ethylene on the Si(001)2×1 surface. Phys Rev B 62:5036-5044

McLaren R, Clark SAC, Ishii I, Hitchcock AP (1987) Absolute oscillator strengths from K-shell electron energy-loss spectra of the fluoroethenes and 1,3-perfluorobutadiene. Phys Rev A 36:1683-1701

Medlin JW, Shrrill AB, Chen JG, Barteau MA (2001) Experimental and theoretical probes of the structure of oxametalacycle intermediates derived from 1-epoxy-3-butene on Ag(110). J Phys Chem B 105:3769-3775

Meyer-Ilse W (1999) X-ray microscopy and microanalysis. *In:* Encyclopedia of Applied Physics. VCH Publishers, New York, p. 323-341.

Milligan PK, Murphy B, Lennon D, Cowie BCC, Kadodwala M (1999) Probing the adsorption structure of a multifunctional organic molecule: a NIXSW and NEXAFS study of 3-chlorothiophene in Cu(111). Surf Sci 430:45-54

Mills JD, Sheehy JA, Ferret TA, Southworth SH, Mayer R, Lindle DW, Langhoff PW (1997) Nondipole resonant X-ray raman spectroscopy: Polarized inelastic scattering at the K-edge of Cl_2. Phys Rev Lett 79:383-386

Mitra-Kirtley S, Mullins OC, Chen J, van Elp J, George SJ, Chen CT, Halloran T, Cramer SP (1992) Nitrogen chemical structure in DNA and related molecules by X-ray absorption spectroscopy Biochim Biophys Acta 1132:249-254

Mitra-Kirtley S, Mullins OC, Elp JV, George SJ, Chen J, Cramer SP (1993) Determination of the nitrogen chemical structures in petroleum asphaltenes using XANES spectroscopy. J Am Chem Soc 115:252-258

Mochizuki Y, Agren H, Pettersson LGM, Carravetta V (1999) A theoretical study of sulphur K-shell X-ray absorption of cysteine. Chem Phys Let 309:241-248

Morra MJ, Fendorf SE, Brown PD (1997) Speciation of sulfur in humic and fulvic acids using X-ray absorption near-edge structure (XANES) spectroscopy. Geochim Cosmochim Acta 61:683-688

Myer-Ilse W, Warwick T, Attwood D (2000) X-ray microscopy. Proceedings of the Sixth International Conference, Berkeley, California (1999), American Institute of Physics, New York

Myneni SCB, Martinez GA (1999) P and S functional group chemistry of humic substances. SSRL Activity Reports-1998, 364-368

Myneni SCB, Warwick TA, Martinez GA, Meigs G (1999a) C-functional group chemistry of humic substances and their spatial variation in soils. ALS activity reports.

Myneni SCB, Brown JT, Martinez GA, Meyer-Ilse W (1999b) Imaging of humic substance macromolecular structures in water and soils. Science 286:1335-1337

Myneni SCB (2000) X-ray and vibrational spectroscopy of sulfate in earth materials. Rev Min 40:113-172

Myneni SCB (2002) Formation of stable chlorinated hydrocarbons in weathering plant material. Science 295:1039-1041

Myneni SCB, Luo Y, Naslund LA, Ojamae L, Ogasawara H, Pelmenshikov A, Vaterlain P, Heske C, Pettersson LGM, Nilsson A (2002) Spectroscopic probing of local hydrogen-bonding structures in liquid water. J Phys 14:L213-L219

Nakamura M, Sasanuma M, Sato S, Watanabe M, Yamashita H, Iguchi Y, Ejiri A, Nakai S, Yamaguchi S, Sagawa T, Nakai Y, Oshio T (1969) Absorption structure near the K edge of the nitrogen molecule. Phys Rev 178:80-82

Namjesnik-Dejanovic K, Maurice PA (2001) Conformational and aggregate structures of sorbed natural organic matter on muscovite and hematite. Geochim Cosmochim Acta 65:1047-1057

Naves de Brito A, Svensson S, Correia N, Keane MP, Agren H, Sairanen OP, Kivimaki A, Aksela S (1992) X-ray induced electron yield spectrum of thin films of 1,3-trans-butadiene and 1,3,5-trans-hexatriene. J Electron Spect Rel Phen 59:293-305

Neuhahausler U, Abend S, Jacobsen C, Lagaly G (1999) Soft X-ray spectromicroscopy of solid-stabilized emulsions. Colloid Polym Sci 277:719-726

Newbury DC, Ishii I, Hitchcock AP (1986) Inner Shell electron energy loss spectroscopy of some heterocyclic molecules. Can J Chem 64:1145-1155

Nyberg M, Hasselstrom J, Karis O, Wassdahl N, Weinelt M, Nilsson A (2000) The electronic structure and surface chemistry of glycine adsorbed on Cu(110). J Chem Phys 112:5420-5427

Oades JM (1989) An introduction to organic matter in mineral soils. *In:* Minerals in Soil Environments, 2nd edition. Dixon JB, Weed SB (eds) Soil Sci. Soc. Am., Madison, 89-159

Okude N, Nagoshi M, Noro H, Baba Y, Yamamoto H, Sasaki TA (1999) P and S K-edge XANES of transition-metal phosphates and sulfates. J Electron Spec Rel Phen 101-103:607-610

Olivella MA, Palacios JM, Vairavamurthy A, del Rio JC, de las Heras FXC (2002) A study of sulfur functionalities in fossil fuels using destructive- (ASTM and Py-GC-MS) and non-destructive –(SEM-EDX, XANES and XPS) techniques. Fuel 81:405-411

Orlov DS (1995) Humic Substances of Soils and General Theory of Humification. Balkema, Rotterdam

Outka DA, Stohr J, Madix RJ, Rotermund HH, Hermsmeier B, Solomon J (1987) NEXAFS studies of complex alcohols and carboxylic acids on the Si(111)(7×7) surface. Surf Sci 185:53-74

Oyanagi H, Tokumoto M, Ishiguro T, Shirakawa H, Nemoto H, Matsusita T, Ito M, Kuroda H (1984) Polarized X-ray absorption spectra of bromine doped polyacetylene; evidence for highly-oriented polybromine ions. J Phys Soc Japan 53:4044-4053

Pauling L (1960) The Nature of the chemical bond. Cornell Univ. Press, Ithaca, NY

Perera RCC, Barth J, LaVilla RE, Deslattes RD, Henins A (1985) Multivacancy effects in the X-ray spectra of CH_3Cl. Phys Rev A 32:1489-1494

Perera RCC, LaVilla RE, Gibbs GV (1987) Cl-Kb emission of chlorofluoromethanes and comparison with semiempirical and ab initio MO calculations. J Chem Phys 86:4824-4830

Perera RCC, Cowan PL, Lindle DW, La Villa RE, Jach T, Deslattes RD (1991) Molecular-orbital studies via satellite-free X-ray fluorescence: Cl K absorption and K-valence-level emission spectra of chlorofluoromethanes. Phys Rev A 43:3609-3618

Pickering IJ, Prince RC, Divers T, George GN (1998) Sulfur K-edge X-ray absorption spectroscopy for determining the chemical speciation of sulfur in biological systems. FEBS Letters 441:11-14

Pickering IJ, George GN, Yu EY, Brune DC, Tuschak C, Overmann J, Beatty JT, Prince RC (2001) Analysis of sulfur biochemistry of sulfur bacteria using X-ray absorption spectroscopy. Biochemistry 40:8138-8145

Plaschke M, Rothe J, Schafer T, Denecke MA, Dardenne K, Pompe S, Heise K (2002) Combined AFM and STXM *in situ* study of the influence of Eu(III) on the agglomeration of humic acid. Colloids Surf A 197:245-256

Plashkevych O, Yang L, Vahtras O, Agren H, Pettersson LGM (1997) Substituted benzenes as building blocks in near edge X-ray absorption spectra. Chem Phys 222:125-137

Polzonetti G, Carravetta V, Russo MV, Contini G, Parent P, Laffon C (1999) Phenylacetylene chemisorbed on Pt (111), reactivity and molecular orientation as probed by NEXAFS. Comparison with condensed multilayer and polyphenylacetylene. J Electron Spec Rel Phen 99:175-187

Prange A, Arzberger I, Engemann C, Modrow H, Schumann O, Truper HG, Steudel R, Dahl C, Hormes J (1999) *In situ* analysis of sulfur in the sulfur globules of phototrophic sulfur bacteria by X-ray absorption near edge spectroscopy. Biochim Biophys Acta 1428:446-454

Prange A, Kuhlsen N, Birzele B, Arzberger I, Hormes J, Antes S, and Kohler P (2001) Sulfur in wheat gluten: In situ analysis by X-ray absorption near edge structure (XANES) spectroscopy. Eur Food Res Technol 212:570-575

Prange A, Chausvistre R, Modrow H, Hormes J, Truper HG, Dahl C (2002) Quantitative speciation of sulfur in bacterial sulfur globules: X-ray absorption spectroscopy reveals at least three different species of sulfur. Microbiology 148:267-276

Ramm M, Ata M, Gross T, Unger W (2000) X-ray photoelectron spectroscopy and near-edge X-ray-absorption fine structure of C-60 polymer films. App Phys A 70:387-390

Reeve DW, Tan Z (1998) The study of carbon-chlorine bonds in bleached pulp with X-ray photoelectron spectroscopy. J Wood Chem Technol 18:417-426

Ressler T, Wong J, Roos J, Smith I (2000) Quantitative speciation of Mn-bearing particles emitted from autos burning (methylcyclopentadienyl) manganese tricarbonyl-added gasolines using XANES spectroscopy. Environ Sci Technol 34:950-958

Revel R, Bazin D, Parent P, Laffon C (2001) NO adsorption on $ZnAl_2O_4/Al_2O_3$ powder: a NEXAFS study at the nitrogen K edge. Catal Let 74:189-192

Rightor EG, Hitchcock AP, Ade H, Leapman RD, Urquhart SG, Smith AP, Mitchell G, Fischer D, Shin HJ, Warwick T (1997) Spectromicroscopy of poly(ethylene terephthalate): Comparison of spectra ans radiation damage rates in X-ray absorption and electron energy loss. J Phys Chem B 101:1950-1960

Rightor EG, Urquhart SG, Hitchcock AP, Ade HW, Smith AP, Mitchell GE, Priester RD, Aneja A, Appel G, Wilkes G, Lidy WE (2002) Identification and quantification of urea precipitates in flexible polyurethane foam formulations by X-ray spectromicroscopy. Macromolecules 35:5873-5882

Robin MB, Ishii I, McLaren R, Hitchcock AP (1988) Fluoination Effects on the Inner-Shell Spectra of Unsaturated Molecules. J Electron Spec Rel Phen 47:53-92

Rompel A, Andrews JC, Cinco RM, Wemple MW, Christou G, Law NA, Pecoraro VL, Sauer K, Yachandra VK, Klein MP (1997) Chlorine K-edge X-ray absorption spectroscopy as a probe of chlorine-manganese bonding: model systems with relevance to the oxygen evolving complex in photosystem II. J Am Chem Soc 119:4465-4470

Rompel A, Cinco RM, Latimer MJ, McDermott AE, Guiles RD, Quintanilha A, Krauss RM, Sauer K Yachandra V, Klein MP (1998) Sulfur K-edge X-ray absorption spectroscopy:A spectroscopic tool to examine the redox state of S-containing metabolites *in-vivo*. Proc Natl Acad Sci 95:6122-6127

Rompel A, Cinco RM, Robblee JH, Latimer MJ, McFarlane KL, Huang J, Walters MA, Yachandra VK (2001) S K- and Mo L-edge X-ray absorption spectroscopy to determine metal-ligand charge distribution in molybdenum-sulfur compounds. J Synchrotron Rad 8:1006-1008

Rose Williams K, Hedman B, Hodgson KO, Solomon EI (1997) Ligand K-edge X-ray absorption spectroscopic studies: Metal-ligand covalency in transition metal thiolates. Inorg Chim Acta 263:315-321

Rose K, Shadle SE, Eidness MK, Kurtz Jr. DM, Scott RA, Hedman B, Hodgson KO, Solomon EI (1998) Investigation of Iron-sulfur covalency in Rubredoxins and a model system using sulfur K-edge X-ray absorption spectroscopy. J Am Chem Soc 120:10743-10747

Rothe J, Denecke MA, Dardenne K (2000) Soft X-ray spectromicroscopy investigation of the interaction of aquatic humic acid and clay colloids. J Col Interf Sci 231:91-97

Ruhl E, Wen AT, Hitchcock AP (1991a) Inner-shell excitation of 5-$C_5H_5Co(CO)_2$ and related compounds studied by gas phase electron energy loss spectroscopy. J Electron Spec Rel Phen 57:137-164

Ruhl E, Hitchcock AP (1991b) Oxygen K-Shell excitation spectroscopy of hydrogen peroxide. Chem Phys 154:323-329

Rusell LM, Maria SF, Myneni SCB (2002) Mapping organic coatings on atmospheric particles. Geophys Res Let 29:26.1-26.4

Sarret G, Connan J, Kasrai M, Bancroft GM, Charrie-Duhaut A, Lemoine S, Adam P, Albrecht P, Eybert-Berard L (1999) Chemical forms of sulfur in geological and archaeological asphaltenes from Middle East, France, and Spain determined by sulfur K-, and L-edge X-ray absorption near-edge structure spectroscopy. Geochim Cosmochim Acta 63:3767-3779

Scheinost AC, Kretzschmar R, Christl I, Jacobsen Ch (2001) Carbon group chemistry of humic and fulvic acid: A comparison of C-1s NEXAFS and ^{13}C-NMR spectroscopies. *In:* Humic Substances: Structures, Modes and Functions. Ghabbour EA, Davies G (eds) The Royal Society of Chemistry, Cambridge p 39-47

Schnitzer M (1991) Soil organic matter- the next 75 years. Soil Science 151:41-58

Schulze DG, Stucki JW, Bertsch PM (eds) (1999) Synchrotron X-ray methods in clay science. The Clay Minerals Society, Boulder

Schwarz WHE, Chang TC, Connerade JP (1977) Core-electron excitation in NO_2. Chem Phys Let 49:207-212

Schwartzenbach RP, Gschwend PM, Imboden DM (1993) Environmental Organic Chemistry. Wiley, New York

Sekiyama H, Kosugi N, Kuroda H, Ohta T (1986) Sulfur K-edge absorption spectra of Na_2SO_4, Na_2SO_3, $Na_2S_2O_3$, and $Na_2S_2O_x$ (x = 5-8). Bull Chem Soc Jpn 59:575-579

Senesi N, Miano TM (1994) Humic Substances in the Global Environment and Implications on Human Health. Elsevier, Amsterdam.

Sette F, Stohr J, Hitchcock AP (1984) Determination of intramolecular bond lengths in gas phase molecules from K shell shape resonances. J Chem Phys 81:4906-4914

Shadle SE, Hedman B, Hodgson KO, Solomon EI (1994) Ligand K-edge X-ray absorption spectroscopy as a probe of ligand-metal bonding: charge donation and covalency in copper-chloride systems. Inorg Chem 33:4235-4244

Shadle SE, Hedman B, Hodgson KO, Solomon EI (1995) Ligand K-edge X-ray absorption spectroscopic studies: Metal-ligand covalency in a series of transition metatetrachlorides. J Am Chem Soc 117:2259-2272

Sham TK, Yang BX, Kirz J, Tse JS (1989) K-edge near-edge X-ray-absorption fine structure of oxygen- and carbon-containing molecules in the gas phase. Phys Rev A 40:652-669

Siegbahn H (1985) Electron spectroscopy for chemical analysis of liquids and solutions. J Phys Chem 89:897-909

Smith TA, DeWitt JG, Hedman B, Hodgson KO (1994) Sulfur and chlorine K-edge X-ray absorption spectroscopic studies of photographic materials. J Am Chem Soc 116:3836-3847

Sodhi RNS, Brion CE (1985a) High resolution carbon 1s and valence shell electronic excitation spectra of trans-1,3-butadiene and allene studied by electron energy loss spectroscopy. J Electron Spec Rel Phen 37:1-21

Sodhi RNS, Brion CE (1985b) Inner shell electron energy loss spectra of the methyl amines and ammonia. J Electron Spec Rel Phen 36:187-201

Spiro CL, Wong J, Lytle FW, Greegor RB, Maylotte DH, Lamson SH (1984) X-ray absorption spectroscopic investigation of sulfur sites in coal: Organic sulfur identification. Science 226:48-50

Sposito G (1989) The chemistry of soils. Oxford University Press, New York

Stevenson FJ (1994) Humus Chemistry. John Wiley, New York

Stöhr J (1992) NEXAFS Spectroscopy. Springer-Verlag, Berlin

Stohr J, Baberschke K, Jaeger R, Treichler R, Brennan S (1981) Orientation of chemisorbed molecules from surface-absorption fine-structure measurements: CO and NO on Ni(100). Phys Rev Lett 47:381-384

Stohr J, Jaeger R (1982) Absorption-edge resonances, core-hole screening, and orientations of chemisorbed molecules: CO, NO, and N_2 on Ni(100). Phys Rev B 26:4111-4131

Sugawara K, Enda Y, Sugawara T, Shirai M (2001) XANES analysis of sulfur form change during pyrolysis of coals. J Synchrotron Rad 8:955-957

Sze KH, Brion CE (1991) Inner-shell and valence-shell electronic excitation of cyclopropane and ethylene oxide by high resolution electron energy loss spectroscopy. J Electron Spectrosc Relat Phenom 57:117-135

Thurman EM (1985) Organic Geochemistry of Natural Waters. Dordrecht, Boston

Tohno S, Kawai J, Kitajima Y (2001) Identification of the chemical states of phosphorous in atmospheric aerosols by XANES spectrometry. J Synch Rad 8:958-960

Tonner BP, Droubay T, Denlinger J, Myer-Ilse W, Warwick T, Rothe J, Kneedler E, Pecher K, Nealson K, Grundl T (1999) Soft X-ray microscopy and imaging of interfacial chemistry in environmental specimens. Surf Interface Anal 27:247-258

Tyson TA, Roe AL, Frank P, Hodgson KO, Hedman B (1989) Polarized experimental and theoretical K-edge X-ray absorption studies of SO_4^{2-}, ClO_3^-, $S_2O_3^{2-}$, and $S_2O_6^{2-}$. Phys Rev B 39:6305-6315

Urquhart SG, Hitchcock AP, Priester RD, Rightor EG (1995) Analysis of polyureathanes using core excitation spectroscopy. Part II: Inner shell spectra of ether, urea, and carbamate model compounds. J Polymer Sci B 33:1603-1620

Urquhart SG, Hitchcock AP, Smith AP, Ade HW, Lidy W, Rightor EG, Mitchell GE (1999). NEXAFS spectromicroscopy of polymers: overview and quantitative analysis of polyurethane polymers. J Electron Spec Rel Phen 100:119-135

Urquhart S, Ade H, Rafailovich M, Sokolov JS, Zhang Y (2000) Chemical and vibronic effects in the high-resolution near-edge X-ray absorption fine structure spectra of polystyrene isotopomers. Chem Phys Lett 322:412-418

Urquhart SG, Ade HW (2002) Trends in the carbonyl core (C 1s, O 1s) to π^* transition in the near edge X-ray absorption fine structure spectra of organic molecules. J Phys Chem B 106:8531-8538

Vairavamurthy A, Manowitz B, Zhou W, Jeon Y (1994) Determination of hydrogen sulfide oxidation products by sulfur K-edge X-ray absorption near edge structure spectroscopy. In: Environmental Geochemistry of Sulfide Oxidation. Alpers CN, Blowes DW (eds) Am Chem Soc Symp Series, Vol 550, Am Chem Soc, Washington, DC, p 412-430

Vairavamurthy A, Maletic D, Wang S, Manowitz B, Eglinton T, Lyons T (1997) Characterization of sulfur-containing functional groups in sedimentary humic substances by X-ray absorption near-edge structure spectroscopy. Energy Fuels 11:546-553

Vairavamurthy A (1998) Using X-ray absorption to probe sulfur oxidation states in complex molecules. Spectrochim Acta 54A:2009-2017

Vairavamurthy A (1999) Speciation of phosphorus in sediments and humic materials as revealed by phosphorus K-edge XANES spectroscopy. NSLS Users Report.

Vairavamurthy A, Wang S (2002) Organic nitrogen in geomacromolecules: Insights on speciation and transformation with K-edge XANES spectroscopy. Env Sci Technol 36:3050-3056

Wako S, Sano M, Ohno Y, Matsushima T, Tanaka S, Kamada M (2000) Orientation of oxygen ad-molecules on stepped platinum (112). Surf Sci 461:L537-L542

Waldo GS, Carlson RMK, Moldowan JM, Peters KE, Penner-Hahn JE (1991) Sulfur speciation in heavy petroleums: Information from X-ray absorption near-edge structure. Geochim Cosmochim Acta 55:801-814

Warwick T, Ade H, Cerasari S, Denlinger J, Franck K, Gracia A, Hayakawa A, Hitchcock A, Kikuma J, Kortright J, Meigs G, Moronne M, Myneni SCB, Rightor E, Rotenberg E, Seal S, Shin H-J, Steele R, Tyliszczak T, Tonner B (1998) A scanning transmission X-ray microscope for materials science spectromicroscopy at the Advanced Light Source. Rev Sci Instr 69:2964-2973

Wasserman SR, Allen PG, Shuh DK, Bucher JJ (1999) EXAFS and principal component analysis: A new shell game. J Sync Radiation 6:284-284

van der Wiel MJ, El-Sherbini TM (1970) K Shell Excitation of nitrogen and carbon monoxide by electron impact. Chem Phys Let 7:161-164

Wen AT, Hitchcock AP, Werstiuk NH, Nguyen N, Leigh WJ (1990) Studies of electronic excited states of substituted norbornenes by UV absorption, electron energy loss, and HeI photoelectron spectroscopy. Can J Chem 68:1967-1973

Wight GR, Brion CE, Vanderwiel MJ (1972/73) K-shell energy loss spectra of 2.5 keV electrons in N_2 and CO. J Electron Spec Rel Phen 1:457-469

Wight GR, Brion CE (1974a) K-shell excitations in NO and O_2 by 2.5 keV electron impact. J Electron Spec Rel Phen 4:313-325

Wight GR, Brion CE (1974b) K-shell excitation of CH_4, NH_3, H_2O, CH_3OH, CH_3OCH_3 and CH_3NH_2 by 2.5 keV electron impact. J Electron Spec Rel Phen 4:25-42

Wight GR, Brion CE (1974c) K-Shell energy loss spectra of 2.5 keV electrons in CO_2 and N_2O. J Electron Spec Rel Phen 3:191-205

Wilson MA (1987) NMR techniques and applications in geochemistry and soil chemistry. Pergamon, Oxford

Wilson KR, Cavalleri M, Rude BS, Schaller RD, Nilsson A, Pettersson LGM, Goldman N, Catalano T, Bozek JD, Saykally RJ (2002) Characterization of hydrogen bond acceptor molecules at the water surface using near-edge X-ray absorption fine structure spectroscopy and density functional theory. J Phys Condens Matter 14:L221-L226

Xia K, Weesner F, Bleam W, Bloom PR, Skyllberg UL, Helmke PA (1998) XANES studies of oxidation states of sulfur in aquatic and soil humic substances. Soil Sci Soc Am J 62:1240-1246

Yagi S, Matano A, Kutluk G, Shirota N, Hasimoto E, Taniguchi M (2001) Molecular adsorption of (CH3)2S on Ni(100) studies by S K-edge NEXAFS and XPS. Surf Sci 482-485:73-76

Yokoyama T, Imanishi A, Terada S, Namba H, Kitajima Y, Ohta T (1995) Electronic properties of SO_2 adsorbed on Ni(100) studied by UPS and O K-edge NEXAFS. Surf Sci 334:88-94

Zhang X, Balhorn R, Mazrimas J, Kirz J (1996) Mapping and measuring DNA to protein ratios in mammalian sperm head by XANES imaging. J Struct Biol 116:335-344

On the following pages ....

APPENDIX — TABLES 1-5

Table 1. C $1s$ NEXAFS of different inorganic and organic molecules. Sample state (**Phase**): g = gas, s = solid, SL = several layers, ML = monolayer; TF = thin films. Method (**Met.**): N = NEXAFS, E = EELS. (Prepared by Edwards and Myneni).

Molecule	Peak position (eV)				Phase	Met.	Ref.	Notes
Alkanes	σ^*(C-H)	σ^*(C-C)	Rydberg					
methane			287,288,289.4,289.8		g	E	56	(a)
ethane	287.9	290.8	286.9,289.3		g	E	11	(b)(≠)
propane	287.8	291.6	287,289.4		g	E	11	(b)
n-butane	287.8	291.7	287.1,291.7		g	E	11	(b)
n-pentane	287.6	291.4	287,289.1		g	E	11	(b)
n-hexane	287.4	291.2	286.8,288.9		g	E	11	(b)
isobutane	287.8	292.4	287,288.6,289.9		g	E	11	(b)
isopentane	287.7	292.2	286.9,289.3		g	E	11	(b)
neopentane	287.7	293	286.8,288.8		g	E	11	(b)
methanethiol	288.5		290.5,291.3		g	E	20	(b)

Molecule	Peak position (eV)				Phase	Met.	Ref.	Notes
Cycloalkenes	π^*(C=C)	σ^*(C-H)	σ^*(C-C)	σ^*(C=C)				
cyclohexene	285.3	287.7	292	298.5	g	E	12	(b)
cyclopentene	285	287.6	290.6	295.6	g	E	12	

Molecule	Peak position (eV)			Phase	Met.	Ref.	Notes
Cycloalkanes	σ^*(C-H)	σ^*(C-C)					
cyclopropane	287.7	289.0,293.2		g	E	12	(b)
cyclobutane	287.4	289.9,291.6		g	E	12	(b)
	288.2	290,292		TF/Pt(111)	N	12	(b)
cyclopentane	287.6	290.5		g	E	12	(b)
cyclohexane	287.7	291.9		g	E	12	(b)
	288.2	292.2		TF/Pt(111)	N	12	(b)

(a) Low energy peaks could be σ^* C-H   (b) σ^* C-H is assigned to π^* CH$_2$ in the Ref.   (≠) assigned by the author

Molecule	Peak position (eV)			Phase	Met.	Ref.	Notes
Alcohols	π*(C=C)	σ*(C-H)	σ*(C-O); **Rydberg**				
methanol		288.1	289.4; 290.3,291.3	g	E	56	(b)
ethanol			288.9; 286.9,293	g	N	37	
propanol		289.3	292.7; 287(σ*(O-H)),288	g	E	24	
t-butanol		289.8	291.5; 287,289.8	g	E	27	
butanol			291.2; 286.9,289.9,295(σ*(C-C))	g	N	37	
propenol	284.8	288.4, 289.4	292.8; 287.4,290.8	g	E	24	(b)
propargyl alcohol	285.8	288.9,290	293.4; 290.7	g	E	24	(b)
Ethers			σ*(C-O); **Rydberg**				
dimethyl ether			288.5, 289.4, 291.1 (all assigned to Rydberg)	g	E	56	
diethyl ether			293; 286.5,287.9	g	N	37	
polyol			289.1; 294 (σ*(C-C)),287.4(Ryd./σ*(C-H)),288.1(Ryd./σ*(C-H))	l	N	64	
Aromatic groups	π*(C=C) & others		σ*(C=C)/(C-C)/(C-O)/**Rydberg**				
benzene	285.2		287.2,288,288.9 (all Ryd.)	g	E	8	
	285.2,288.9		287.2,293.5(σ*(?))	s	N	45	
phenol	285.4,287.4(COH),289.1,291		295,302 (σ*(C-C), (C-O))	s	N	21	
	285.2,287.1(COH),289.2,291		294.5,302 (σ*(C-C), (C-O))	g	E	21	
o-xylene	285.4,285.8		287.1,293.6,288.1(π*(C-H))	g	E	65	
m-xylene	285.4		287.1,287.4,288.2(π*(C-H)), 293.8(σ*(?))	g	E	65	
p-xylene	285.2, 285.4		287.1,287.3;288,288.3(π*(C-H)), 293.8(σ*(?))	g	E	65	
hydroquinone	285.2,287.25(COH),289.4,291.8		295.5,303 (σ*(C-C), (C-O))	g	E	21	
p-benzoquinone	283.8,286,288.3(COH),290.7		293.6,296.2,303 (σ*(C-C), (C-O))	g	E	21	
2-trifluoromethylnorbornene	284.9		287.9(π*(C-H₂));298,304(σ*(C-C))	g	E	39	
norbornene	284.9		287.5(π*(C-H₂)); 299.3(σ*(C-C))	g	E	39	
2-methylnorbornene	285.2		287.9(π*(C-H₂)),299.2(σ*(C-C))	g	E	39	

(b) σ* C-H is assigned to π* CH₂ in the Ref.

Molecule	π^* (C=C) & others	σ^*(C=C)/(C-C)/(C-O)/Rydberg	Phase	Met.	Ref.	Notes
Aromatic groups						
2-mercaptobenzoxazole	285.5,286.9, 287.4 288.4 (π^*(C=S))		g	N	49	
	285.6, 287.1,287.5, 288.4 (π^*(C=S))	298.2	SL/Pt(111)	N	28	
nido-2,3-Diethyl-2,3-carborane	285.8	291.8(σ^*(C-C)),287.1(σ^* (C-H)),288(σ^* (C-H))	g	E	1	
closo-o-carborane	288.7	294.3(σ^*(C-C));290.8,291.6,292.4	g	E	1	
nido-1,2-Dicarbaundecaborane	284.9	288.9,290.7,293(σ^*(C-C))	g	E	1	
phenylacetylene	285.5(C=C),286(C≡C), 288.7,290.4	287.3(σ^*(C-H)),294	SL/Pt(111)	N	52	
aniline	285.3,286.7, 289.0,290.4	294.3(σ^*(C-C)),301	SL/Ni(100)	N	53	
	285,286.2,288.8,290	294.1(σ^*(C-C)),301	ML/Ni(100)	N	53	
azulene	284.4,285.3,287.3,288.9	291.9,294.2,298.2,303.4,all (σ^*(C-C))	g	E	62	
naphthalene	285,285.8,287.2,288.4	293.6,300.2,305 all (σ^*(C-C))	g	E	40	
PMPO Poly(dimethylphenyleneoxide)	285.6,287.2,289.2,290.6	296(σ^*(C-C))	s	N	41	
PMDA-ODA polypyromellitimido oxydianiline	(C=C): 284.8, 285.2, 286.6, 289.2; (C=O): 287.4	291.9(σ^*(C-O)),303.1(σ^*(C=O)),295.4(σ^*(C=C))	s	N	41	
Haloaromatics (X=halogen)	π^* (C=C)	σ^* (C-X)				
fluorobenzene	285.3,287.5,289.2		g	E	8,16	
	285.3,287.6,289.3	290	g	N	30	
chlorobenzene	285.1,286.3,288.8,290.3		g	E	8,16	
bromobenzene	285.1,286,289,289.7		g	E	8,16	
iodobenzene	285.1,285.8,289.9,289.5		g	E	8,16	
1,4-difluorobenzene	285.3,287.6,289.9	289.9,291.3	g	N	30	
1,3,5-trifluorobenzene	285.5,287.7,290.2	290.3,291.4	g	N	30	
1,2,4,5-tetrafluorobenzene	285.4,287.8,289.6	289.7,290.7	g	N	30	
pentafluorobenzene	285.5,287.8,289.1	289,290.9	g	N	30	

Molecule	Peak position (eV)			Phase	Met.	Ref.	Notes
Haloaromatics (X=halogen)	π*(C=C)	σ*(C-X)					
hexafluorobenzene	287.9	289,291		g	N	30	
octafluoronaphthalene	285.4,286.2,287.4(σ*(C-F))	σ*(C-C): 299.2,304		g	E	40	
Haloalkanes (X= halogen)	σ*(C-H); (C-C); Rydberg						
methyl fluoride		289.0; -; 290.5,290.9,291.3,292.3		g	E	2	
methyl chloride	288.5,288.3	287.3; -289.4,289.8,290.2,290.9		g	E	2	
methyl bromide	288.1,288.4	286.5; -289.2,289.5,289.7,290.6		g	E	2	
methyl iodide	287.6,288	285.7; -288.7,289,289.4,290		g	E	2	
methylene dichloride		288.8,289,289.9,290.8,291.4,292.1,292.4,292.8 (all Rydberg states)		g	E	3	
trichloromethane		289.3,290.1,291.2,292.7,293.3 (all Rydberg states)		g	E	3	
tetrachloromethane		290.9,294.5 (all Rydberg states)		g	E	3	
	σ*(C-C)	σ*(C-X); Rydberg					
perfluoroethane	300.5	295.4, 297.7		g	E	25	
perfluoropropane	298.5,300.8	293,295.3,297.6		g	E	25	
perfluorobutane	298.9,300.8	292.8,295.4,297.5		g	E	25	
perfluoropentane	298.8,302	292.7,295.2, 297.3		g	E	25	
perfluorohexane	298.6,301	292.7,295.3,297.3		g	E	25	
perfluoroneopentane	300.7	290.4,295.7,298.1		g	E	25	
perfluorocyclopropane	291.1,298	292.8,294.1		g	E	25	
perfluorocyclobutane	291.2,297.4	293.1,295		g	E	25	
perfluorocyclopentane	291.7,299	293.2, 295.1		g	E	25	
perfluorocyclohexane	292.2,298.3	293.2, 295.1		g	E	25	
Alkenes and alkynes	π*(C=C)	σ*(C-C)/(C=C); Rydberg					
ethylene	284.6	286.9,287.8,288.2 (all Rydberg states)		g/Pt(111)	N	15,44	
	284.7	301; 287.4,287.8,289.3		g	E	26,40	
	284.6	293		g/Pt(111)	E	44	

Molecule	π* (C=C)	σ* (C-C)/(C=C); Rydberg	Phase	Met.	Ref.	Notes
Alkenes and alkynes						
ethylene	284.5	301; π*(CH₂); 287.7, σ*(CH₂): 288.7	SL/Si(001)	N	19	
acetylene	285.8		g	E	61	
	286.4	310; σ*(CH): 289.2	SL/Si(001)	N	19	
1,3-*trans*-butadiene	284.3,284.8	296,304; 287.6,288.8,289.6	g	E	26,40	
	284.4,285.1	297.8,288.5 (mixed Rydb and π*)	g	N	31	
1,3,5-*trans*-hexatriene	283.8, 284.1, 284.5		g	N	31	
allene	285.4, 286.0	287.6,288.5,288.8,289.1,289.6	g	E	33	
2-butyne	286.1	295.3; 287.7,288.5,290.5	g	E	40	
hexafluoro-2-butyne	285.7	294.3,298; 289.1,290.6	g	E	40	
1-butene	284.9	301.0	g	E	50	
cis-2-butene	285.0	301.0	g	E	50	
trans-2-butene	285.0	301.0	g	E	50	
Haloalkenes & alkynes (X= halogen)	π* (C=C)	σ* (C-X)/(C-C); Rydberg states				
ethylene fluoride	284.6,286.9	286.2,288.8,292.2 (all Rydberg states)	g	N	15	
1,1'-difluoro ethylene	285,287.1	289.7; 290.4,291.9	g	E	26,40	
	284.9,289.2	287.7,292.2 (all Rydberg states)	g	N	15	
cis-difluoro ethylene	285.4,289.6	292.2; 287.5,288.4,290.5	g	E	26,40	
	284.9,287.8	289.8 (all Rydberg states)	g	N	15	
trifluoro ethylene	287.3	289.1; 290.1,291.2	g	E	26,40	
	287.7,289.9	292.1 (all Rydberg states)	g	N	15	
tetrafluoro ethylene	287.6,289.7	288.9,291.6; 290.7,292.4	g	E	26,40	
	290.1	291.9 (all Rydberg states)	g	N	15	
octafluorocyclopentene	290.1	291.8; 292.5,294.4	g	E	26,40	
perfluoro-2-butene	287.4,291.7	292.8,289.8, 299 (C-C)	g	E	40	
1,3-perfluorobutadiene	288.1,290.7	298.3,295.9(C-C); 292.4,293.6,294.5	g	E	50	
	287.8,289,289.5	291.5,292.7,298.5(C-C)	g	E	26,40	

Molecule	Peak position (eV)			Phase	Met.	Ref.	Notes
Nitriles	π^*(C≡N)	σ^*(C-H); Rydbergs	σ^*(C≡N)				
hydrogen cyanide	286.4	290.6; 2885,289.1,290.9	307.9	g	E	4	
cyanogen	286.3	290.5,291.7,293.1 (all Ryd.)	306.3	g	E	42	
acetonitrile	286.9	289.1; 288.1,289.8,291.5	308	g	E	38	
methyl thiocyanide	286.4,287.2	291; 288.1,291	309.4	g	E	38	
methyl isocyanide	286.6,287.7	289.7; 291.3	307.8	g	E	38	
methyl isothiocyanide	287.2,287.9	288.9; 290.9,292.1	310.5	g	E	38	
Aldehydes	π^*(C=C); π^*(C=O)	σ^*(C=O); σ^*(C=C)/(C-C)					
benzaldehyde	285.1,286,289.2,290.3;287.7	295.4; 293.6,301		g	E	5	
terephthaladehyde	284.7,285.8,288.8,290.3;288.2	294.7; 293.6,302		g	E	5	
formaldehyde	286	300.9; Rydb: 290.2,291.2		g	E	40,46	
formyl fluoride	288.2	303; Rydb: 292.7,294		g	E	40	
acetaldehyde	286.3	301; Rydb:287.2,288.3,289.2,289.5		g	E	60	
Ketones	π^*(C=C)	π^*(C=O)	σ^*(C=O); Rydberg				
cyclohexanone		286.6	302; 287.8,289.8	g	E	22	
	π^*(C=C)	π^*(C=O)	σ^*(C=O)/(C-O); Rydberg				
1,2-cyclohexanedione	284.3,287.6	286.2	302,298; 288.4,289.5	g	E	22	
1,3-cyclohexanedione		286.7	302; 288.4	g	E	22	
1,4-cyclohexanedione		286.6	301; 288.5,289.7	g	E	22	
formyl fluoride		288.2	303; 292.7	g	E	23	
acetone		287.1	303.6; 288.8,291.2	g	N	37	
N,N'-diphenyl urea	285.1,286.4	289.5	303	s	N	59	
ethyl N-phenyl carbamate	285.3,286.8	290	304	s	N	59	
2,4-toluene di(methyl carbamate)	285.3,286.7	289.9	304	s	N	59	

Molecule	Peak position (eV)				Phase	Met.	Ref.	Notes
Ketones	π*(C=C)	π*(C=O)	σ*(C=O)/(C-O); Rydberg					
4,4' methylene bis(ethyl N-phenyl carbamate)	285.2,286.6	289.9	303		s	N	59	
polycarbonate		290.4			s	N	63	
polyurethane		289.9			s	N	63	
polyurea		289.5			s	N	63	
poly(ethylene succinate)		288.6			s	N	63	
poly(methyl methacrylate)		288.5			s	N	63	
nylon-6		288.2			s	N	63	
poly(ethylene terephthalate)		288.2			s	N	63	
	π*(C=C)	π*(C=O)	σ*(C=O)	σ*(C=C)				
poly(vinyl methyl ketone)		286.6			s	N	63,41	
toluene diisocyanate urethane	285.2,285.4, 286.7	289.9		304	l	N	64	
toluene diisocyanate urea	285.2, 285.4,286.6	289.5		302	l	N	64	
carbonyl fluoride		290.9	305		g	E	40	(c)
hexafluoroacetone		286.2, 292.4	306.4		g	E	40	(d)
Amides	π*(C=O)	σ*(C=O)	σ*(C-N)					
formamide	288.1	303	296.9		g	E	23	(e)
Amines	σ*(C-N)/ **Rydbergs**							
methylamine	287.7,288.8,289.5,290.2,290.6,291.8,				g	E	34	
dimethylamine	287.6,287.9,288.3,288.7,289.5,290,290.4,293.2				g	E	34	
trimethylamine	287.8,288.1,288.8,289.1,289.5,290.1,294.3				g	E	34	
Carboxylic acids	π*(C=O)	σ*(C=O)	σ*(C=C)					
formic acid	288.2	303			g	E	23	(f)
	288.6	301.4			g/Si(111)	N	43	
acetic acid	288.7	303.1			g	E	40	(g)

(c) Rydb:294.6,296.9   (d) Rydb:293.5,295.9   (e) Rydb:290.7,292.9   (f) Ryd:292,293.2   (g) Ryd:287.4,290.3

Molecule	Peak position (eV)				Phase	Met.	Ref.	Notes
Carboxylic acids	$\pi^*(C=C)$	$\pi^*(C=O)$	$\sigma^*(C=O)$	$\sigma^*(C=C)$				
acrylic acid	284.3	288.2	304	304	g	E	24	(h)
		288.5	302		g/Si(111)	N	43	
ethyl benzoate	285,287.5	288.2	296.5	303	g	E	5	(i)
propanoic acid		288.5	297,303		g	E	24	(j)
		288.5	301		g/Si(111)	N	43	
propiolic acid	284.9,285.9, all (C≡C)	288.9	303	312 (C≡C)	g	E	24	(k)
	284.8,286	288.9	304		g/Si(111)	N	43	
trifluoroacetic acid		288.5,293.6	303.2		g	E	40	(l)
Amino acids	$\pi^*(C=O)$	$\sigma^*(C-N)$	$\sigma^*(C=O)$	$\sigma^*(C-H)$				
glycine	288.6	290.8	302	289.4	s	N	7, 35	(m)
	288.3		298,302	291.2	g/Cu(110)	N	17, 18	
alanine	288.6	289.6		287.6	s	N	7	
valine	288.6	289		287.5	s	N	7	
leucine	288.6	289		287.2,287.9	s	N	7	
isoleucine	288.6	288.8		287.6	s	N	7	
serine	288.6	289.6			s	N	7	
threonine	288.6	289.3		287.6	s	N	7	
cysteine	288.6	289.3			s	N	7	
methionine	288.6	289.3		287.2	s	N	7	
aspartic acid	288.5, 288.7	290.1			s	N	7	
asparagine	288.3,288.6	289.8			s	N	7	
glutamic acid	288.5, 288.7	289.1			s	N	7	
glutamine	288.2,288.6	289.7			s	N	7	
arginine	288.6	289.2			s	N	7, 35	
lysine	288.7	289.5		287.8	s	N	7	
proline	288.7	289.3		287.6	s	N	7	

(h) Ryd:288.2,290.6 (i) Ryd:289.6,290.7 (j) Ryd:287.2,290.2 (k) Ryd:287.2,290.2 (l) Ryd:290.8,294.8 (l) Ryd:295.6,297.4 (m) σ^*(C-C): 294.3

Molecule	Peak position (eV)				Phase	Met.	Ref.	Notes
Amino acids	π*(C=O)	σ*(C-N)	σ*(C=O)	σ*(C-H)				
phenylalanine	288.6	289		287.6	s	N	7,35	
tyrosine	288.5	288.8		287.5	s	N	7,35	
tryptophan	288.6	288.9		286.6	s	N	7,35	
histidine	288.7				s	N	7,35	
Acyl halides	π*(C=O)	σ*(C=O)	σ*(C-Cl)					
phosgene	288.7	304.5	290.7,292.6		g	E	5	
terephthaloyl chloride	287.4,288.7	303	290.3,293.1		g	E	5	
Heterocyclic aromatics	π*(C=C)	σ*(C=C)	σ*(C-C)	σ*(C-X)				
selenophene (X=Se)	285.3,287.3	299.5	293.2,295	286.7	g	E	6	(n)
3-methylselenophene (X=Se)	285.3,287.4	298.8	292.8,295	286.7	g	E	6	(o)
polyselenophene (X=Se)	285.3		293.5,296.8,302	286.9	TF	N	6	
poly-3-methylselenophene (X=Se)	285.2		293.5,296.1,302	287.1	TF	N	6	
1,3,2,4-benzodithiadiazine (X=S,N)	284.4,285.3	301	294	287.5(C-S),293(C-N)	g	E	9	
1,3,5,2,4-benzotrithiadiazipene (X=S)	284.3,285.1	301	294.3	285.6,287.6	g	E	9	
2,1,3-benzothiadiazole (X=N)	284.4,284.8	301.4	293.4	286.9	g	E	9	
thiophene (X=S)	285.4,288.2	301	295.2	287.1	g	E	13	
ethylthiophene (X=S)	285.4,287.9	301	293.5,294.6	287.2	g	E	13	
butylthiophene (X=S)	285.6,287.8	301	292.7,294.9	287.2	g	E	13	
hexylthiophene (X=S)	285.5,287.7	300	291.9,294.7	287.2	g	E	13	
octylthiophene (X=S)	285.5,287.6	300	292.1,295.3	287.2	g	E	13	
decylthiophene (X=S)	285.4,287.9	301	292,294.2	287.3	g	E	13	
imidazole (X=N)	286.5,288.5		292.5,297.5		s	N	14	(p)
	286.7,289.5		292.2,298		g	E	14	(p)
4,5-Dicyanoimidazole (X=N)	286.6,288.9		293.4,300		s	N	14	(p)
	286.6,290.6		293.4,300.7		g	E	14	(p)

(n) Ryd: 289.5,290.4   (o) Ryd: 289.9   (p) σ*(C-C) coupled with σ*(C-N)

Molecule	Peak position (eV)				Phase	Met.	Ref.	Notes
	π*(C=C)	σ*(C=C)	σ*(C-C)	σ*(C-X)				
Heterocyclic aromatics								
s-triazine (X=N)	286.6,290.1			296.5,303	g	E	14	
furan (X=O)	285.6,286.5,288.5,289.3		297	291.4	g	E	29	
pyrrole (X=N)	285.6,286.3,288.1	300.5	296.8	291.3	g	E	29	
tetrahydrofuran (X=O)			294.9	291	g	E	29	(q)
pyrrolidine (X=N)			294.6	290.7	g	E	29	(r)
tetrahydropyran (X=O)			292.3	292.3	g	E	29	(s)
piperidine (X=N)			292.2	292.2	g	E	29	(t)
cyclopentadienyl (Cp)	284.7	297	291		g	E	32	
dicyclopentadienyl	285.1	294.3	290.6		g	E	32	
Co(Cp)2	286.8	297.7	291.9		g	E	32	
288.2,288.8,295.4 are of some σ character. Identified several peaks which they assign to Rydberg states								
ethylene oxide (X=O)	285.8		296.6		g	E	36	
	π*(C=C)	σ*(C=C)/(C=N)		σ*(C-X)				
1-epoxy-3-butene (X=O)	285.8		296.6	289.4,290.6	s/Ag(110)	N	10	
pyridine	285.3,289.2	294.2,300.1			g	E	45	
	285.5,289.1	294.1,301			s	N	45	
Miscellaneous	π*(C=C)	σ*(C-C)	σ*(C-F)	σ*(C-O)				
trifluoromethyl hypofluorite			298	300	g	E	27	
bis-trimethylfluoro peroxide			297.7	299.8	g	E	27	
di(t-butyl) peroxide		293.5,300		291.3	g	E	27	
β-carotene	284.8,286.4				TF	N	48	
Buckminster fullerene (C60)	284.9				s	N	51	
Inorganic Carbon	π*(C=O)	Rydberg		σ*(C=O)				
CO	286.9	292.3,293.3,294.8			g	E	54	
	287.3	292.5,293.4,294.8		303.9	g	E	55	

(q) π^*CH_2: 287.6,289.2   (r) π^*CH_2: 287.7,289.3   (s) π^*CH_2: 287.7,288.4   (t) π^*CH_2: 288,288.8

Molecule	Peak position (eV)				Phase	Met.	Ref.	Notes
	π^* (C=O)	σ^* (C=O)	Rydberg					
Inorganic Carbon								
CO	287.5	303.0			ML/Ni(100)	N	57,58	
	287.3	301.1,306.6	292.8		g	N	37	
	287.3				solid	N	47	
	π^* (C=O)	σ^* (C=O)	Rydberg	π^* (C-S)				
CO_2	290.7	313.8	292.7,295,297		g	N	37	
OCS	288.4	312.0	291.4,293.5,294.4	305.0	g	N	37	

Cited References in Table 1:

(1) Hitchcock et al. (1993)
(2) Hitchcock and Brion (1978a)
(3) Hitchcock and Brion (1978b)
(4) Hitchcock and Brion (1979a)
(5) Hitchcock et al. (1992)
(6) Hitchcock et al. (1989a)
(7) Kaznacheyev et al. (2002)
(8) Hitchcock et al. (1978)
(9) Hitchcock et al. (1991)
(10) Medlin et al. (2001)
(11) Hitchcock et al. (1987)
(12) Hitchcock et al. (1986)
(13) Hitchcock et al. (1990)
(14) Apen et al. (1993)
(15) Beckmann et al. (1985)
(16) Benitez et al. (1988)

(17) Nyberg et al. (2000)
(18) Hasselstrom et al. (1998)
(19) Matsui et al. (2000)
(20) Dezarnaud et al. (1990)
(21) Francis and Hitchcock (1992)
(22) Francis and Hitchcock (1994)
(23) Ishii and Hitchcock (1987)
(24) Ishii and Hitchcock (1988)
(25) Ishii et al. (1988)
(26) McLaren et al. (1987)
(27) Ishii et al. (1987)
(28) Contini et al. (1998)
(29) Newbury et al. (1986)
(30) Plashkevych et al. (1997)
(31) Naves de Brito et al. (1992)
(32) Ruhl et al. (1991)

(33) Sodhi and Brion (1985a)
(34) Sodhi and Brion (1985b)
(35) Boese et al. (1997)
(36) Sze and Brion (1991)
(37) Sham et al. (1989)
(38) Hitchcock et al. (1989c)
(39) Wen et al. (1990)
(40) Robin et al. (1988)
(41) Jordan-Sweet et al. (1988)
(42) Hitchcock and Brion (1979a)
(43) Outka et al. (1987)
(44) Horsley et al. (1985)
(45) Horsley et al. (1985)
(46) Sette et al. (1984)
(47) Jugnetet al. (1984)
(48) Beck et al. (2001)

(49) Contini et al. (2000)
(50) Hitchcock et al. (1984)
(51) Ramm et al. (2000)
(52) Polzonetti et al. (1999)
(53) Huang et al. (1996)
(54) Wight et al. (1972/73)
(55) Hitchcock and Brion (1980b)
(56) Wight and Brion (1974)
(57) Stohr and Jaeger (1982)
(58) Stohr et al. (1981)
(59) Urquhart et al. (1999)
(60) Hitchcock and Brion (1980a)
(61) Francis et al. (1994)
(62) Hitchcock et al. (1989b)
(63) Urquhart and Ade (2002)
(64) Rightor et al. (2002)
(65) Eustatiu et al. (1998)

Table 2. N 1s NEXAFS spectral assignments of inorganic and organic molecules. Sample state (*Phase*): g = gas, s = solid. Method (*Met.*): N = NEXAFS, E = EELS. (Prepared by Edwards and Myneni). σ* for single bonds are at low energy when compared to those of double and triple bonds.

Molecule	Peak position (eV)			Phase	Met.	Ref.	Comments
N₂	π* (N≡N)	Rydberg					
N₂	400.8	405.6, 406.7		g	N	1	
	400.62	406.1, 407		g	E	2	
	400.96	406.13, 407.01		g	E	3	(a)
Nitriles	π* (N=C)/(N≡C)	Rydberg	σ* (N=C)/(N≡C)/(N-C)				
hydrogen cyanide (HCN)	399.7	401.8, 402.5		g	E	4	(a)
Cyanogens (C₂N₂)	398.9	403.4, 404.7	419.3	g	E	5	
Acetonitrile (CH₃CN)	399.9	401.1, 403.2	422.5	g	E	6	
methyl thiocyanide (CH₃SCN)	399.1, 400	402.2, 403.4	422.2	g	E	6	(b)
methyl isocyanide (CH₃NC)	401.2	404.4, 405.5	406.7, 418	g	E	6	
methyl isothiocyanide (CH₃NCS)	399.5, 400.4	403.3, 405.1	407.2, 419	g	E	6	(b)
Amides	π* (C=O)	σ*(N-C)	σ*(HCN)				
Formamide (HCONH₂)	401.9	410.5	402.9	g	E	7	(c)
	400.7	405.2	402.6	cond. g	N	8	
Heterocyclic Aromatics	π* (N=C)	σ* (N-C)/(N-H)					
Pyrrole	402.3	406.7		g	E	9	(b)
pyrrolidine		401, 404.5		g	E	9	(d) (e)
protoporphyrin IX	399.0	401.1		s	N	10	(e) (f)
Imidazole	400.4	406.8, 411.5		s	N	11	(g)
	399.9	406.8, 411.4		g	E	11	(g)
4,5-Dicyanoimidazole	399.4, 401.3, 403	406.9, 414.2		s	N	11	(g)
	399.5, 402, 403.5	409, 414.9		g	E	11	(h)

Molecule	Peak position (eV)			Phase	Met.	Ref.	Comments
	π* (N=C)	σ* (N-C)/(N-H)					
2-methylbenzimidazole	~401.0	402.5,408.5		s	N	12	(f)
	398.8,402.7	408,414.3		g	E	13	
Pyridine	398.8,403.3	408,414.3		s	N	12	
	400.1	408.2		g/Pt(111)	N		
	π* (N=C)	σ* (N-C)/(N-H)/(N=C)					
Triazine	398.8,402.9	409.2,415.3		g	E	11	
piperidine		401.2,405.6		g	E	9	(d) (f)
acridine	~399	Several peaks but assignments not done		s	N	12	(f)
phenanthridine	~399	- same as above -		s	N	12	(f)
Carbazole	~403.2	- same as above -		s	N	12	(f)
2-phenylindole	~402.5	- same as above -		s	N	12	(f)
2,6-Di-p-tolyl-pyridine	~400	- same as above -		s	N	12	(f)
poly(4-vinyl-pyridine-co-styrene)	~400	- same as above -		s	N	12	(f)
6-hyroxyquinoline	399.9	- same as above -		s	N	10	
pyridoxal 5'-phosphate	401.4	- same as above -		s	N	10	
Albumin	402.5	- same as above -		s	N	10	
polyaniline	397.4,400.4	402.1		s	N	14	
Thiadiazine	π* (N=S)	σ* (N-S)	σ* (N-C)				
1,3,2,4-benzodithiadiazine	397.8	401.1	407.4	g	E	15	
1,3,5,2,4-benzotrithiadiazipene	398.6	401.6		g	E	15	
2,1,3-benzothiadiazole	398.6	401.3	409.5	g	E	15	
Molecules with Nitro Groups	π* (N=O)	σ* (N-O)/(N-C)					
4-nitrophenylacetic acid	404.8	Peaks present, not assigned		s	N	10	(e)

Molecule	Peak position (eV)		Phase	Met.	Ref.	Comments
Amines	σ^* (N-H)	σ^* (N-C)				
Methylamine	400.8,402,403.6	404.8	g	E	16	(i)
Dimethylamine	401,402.3,403.2	405.8	g	E	16	(i)
Trimethylamine	403 (401.8)	406.5	g	E	16	(i)
Ammonia	400.6,402.2,403.5		g	E	17	(i)
Miscellaneous Molecules	π^* (N=O)	σ^* (N-O)				
NO	399.7	404.7	g	E	18	
NO	400.9, 402.7	410.2	g/Al$_2$O$_3$	N	19	
NO	401.5		g/Ni(100)	N	20,21	

Cited References in Table 2:

(1) Nakamura et al. 1969
(2) Wight et al. (1972/73)
(3) Hitchcock and Brion (1980b)
(4) Hitchcock and Brion (1979b)
(5) Hitchcock and Brion (1979a)
(6) Hitchcock et al. 1989
(7) Ishii and Hitchcock (1987)
(8) Ikeura-Sekiguchi et al. (2001)
(9) Newbury et al. (1986)
(10) Vairavamurthy and Wang (2002)
(11) Hitchcock and Ishii (1987)
(12) Mitra-Kirtley et al. (1993)
(13) Horsley et al. (1985)
(14) Hennig et al. (1998)
(15) Hitchcock et al. (1991)
(16) Sodhi and Brion (1985)
(17) Wight and Brion (1974b)
(18) Wight and Brion (1974a)
(19) Revel et al. (2001)
(20) Stohr and Jaeger (1982)
(21) Stohr et al. (1981)

Comments in Table 2:

(a) Shape resonance around 420 eV
(b) Transitions into delocalized π^* (?)
(c) Transitions into delocalized π^* C=O (?)
(d) No π system in the molecule.
(e) Spectral assignment is done by the author
(f) Absolute peak energies not available
(g) Broad π^* transition
(h) π^* transition assigned to different N atoms in molecule
(i) Some of the σ^* (N-H) transitions are assigned to Rydberg states

Table 3. O 1s NEXAFS of different inorganic and organic molecules. Sample state (*Phase*): g = gas, s = solid, TF = thin film. Method (*Met.*), N = NEXAFS, E = EELS, A = Auger electron-photoion coincidence, H = High-resolution coincident partial ion yield (Prepared by Edwards and Myneni).

Molecule	Peak Position (eV)			Phase	Met.	Ref.	Comments
	π^* (O=C)	σ^* (O=C)	Rydberg				
Aldehyde							
benzaldehyde	531	544	533.6,536.5	g	E	1	
terephthalaldehyde	530.6	544	533.5,536.1	g	E	1	
phosgene	531.5, 534.4	540.5,546		g	E	1	
terephthaloyl chloride	531.3, 534.4	545, 549		g	E	1	
formaldehyde	530.8	544	535.5,536.1	g	E	2,3	
acetaldehyde	531.1	544.5	535.1, 535.8	g	E	3	
formyl fluoride	532.1	547	536.6,537.6	g	E	2	
carbonyl fluoride	532.7	547.6	535.4,538.1	g	E	2	
	π^* (O=C)	σ^* (O=C)					
Ester							
ethyl benzoate	531.5,534.3	540.8,546.3		g	E	1	Possible typo for σ^*
methyl formate	532.1,535.1	541, 547		g	E	1	
ethyl carbamate	532.8,536.5	540, 548		g	E	4	
ethyl N-phenyl carbamate	532.5,535.7	540, 548		g	E	4	
N-methyl N-phenyl carbamate	532.5, 535.8	540, 548		g	E	4	
benzyl carbamate	532.62,535.9	540, 549		g	E	4	
poly-methylmethacrylate	531.5, 534.3	~546		TF	A	5	
	π^* (O=C)	σ^* (O=C)					
Ketone							
cyclohexanone	531.2	545.8		g	E	6	
1,2-cyclohexanedione	530.9,534.4	545.3		g	E	6	
1,3-cyclohexanedione	531.1,533.9	546		g	E	6	
1,4-cyclohexanedione	531.1	545		g	E	6	
p-benzoquinone	529.9,533.2	542.7, 554		g	E	7	

Molecule	Peak Position (eV)			Phase	Met.	Ref.	Comments
	π*(O=C)	**σ*(O=C)**					
acetone	531.3	545.3		g	E	3	
	530.4	545.1		g	N	8	
hexafluoroacetone	531	548.2		g	E	2	
Alcohols	**σ*(O-H)**	**σ*(O-C)**	**Rydberg**				
methanol			534.0,537.1,538.5	g	E	9	Peak at 534 may
		539.9	534.1, 537.1	g	H	10	correspond to σ*(O-H)
ethanol		536.4	532.6	g	N	8	
propanol	533.9	537.2	536	g	E	11	
propenol	533.9	538	537	g	E	11	
propargyl alcohol	534.1	539.2	535.6,536.8	g	E	11	
phenol	534.9	539.4	537.2 (?σ*)	g	E	7	Delocalized π*
hydroquinone	534.6	539.6, 556	537.2 (?σ*)	g	E	7	Delocalized π*
t-butanol	533.9	537.7	535.9	g	E	12	
Amide	**π*(O=C)**	**σ*(O=C); Rydberg**					
formamide	531.5	547; 533.8,535.9		g	E	13	
urea	532.5	549; 534.4		g	E	4	
N-phenyl urea	532.5	548; 534.6		g	E	4	
N, N'-diphenyl urea	532.5	548		g	E	4	
Ether	**σ*(O-C)**	**Rydberg**					
dimethyl ether		535.5,538.6		g	E	9	All assigned to Ryd.
diethyl ether		534.8,535.6		g	N	8	
diisopropyl ether	538, 546	534,536.1		g	E	4	
	538, 544	534.2,535.8		g	E	4	

Molecule	Peak Position (eV)		Phase	Met.	Ref.	Comments
	π^* (O=C)	σ^* (O=C); Rydberg				
Carboxylic acids						
formic acid	532.1	547; 535.3,538.3	g	E	13	
formyl fluoride	532.1	547; 536.6,539.2	g	E	13	Ryd. coupled with σ^* O-H
propanoic acid	532.1, 535.4	547; 537.7	g	E	11	Same as above
acrylic acid	531.5, 535.1	546; 537.9	g	E	11	Same as above
propiolic acid	531.5, 535.2	546; 537.8	g	E	11	Same as above
methyl formate	532.1, 535.1	547; 539.8	g	E	11	
acetic acid	532, 535.2	546.3; 537.6	g	E	2	Same as above
trifluoroacetic acid	531.9, 535.4	547; 538.9	g	E	2	Same as above
	σ^* (O-O)	σ^* (O-H); Rydberg				
Peroxides						
hydrogen peroxide	533	535.5; 536.8,538.3	g	E	14	
di (t-butyl) peroxide	533	538.2 (C-O); 536.5	g	E	12	
	π^* (O=X)	σ^* (O-X); Rydberg				
Inorganic molecules						
oxygen difluoride (X=F)	534.4	533.7,534.8; 541.6,542.6,544	g	E	12	
CO (X=C)	533.1	550.5; 541.1	g	E	15	
	530.9,534.1	550.9; 538.8,539.8	g	N	8	
	533.5	538.8, 539.8	g	E	16	
CO₂ (X=C)	535.4	538.7, 539.9, 540.8	g	E	17	All Rydbergs
	535	542.5,559.9; 539.3	g	E	18	All Rydbergs
	533.1		g	N	8	
OCS		550.2, 558; 539.6	g	N	8	
NO₂ (X=N)		531.7, 533.5,537.4,539.3	g	N	19	All Rydbergs
NO (X=N)	532.7, 540.2	546.3 (?)	g	E	20	
N₂O (X=N)	534.6	536.5,538.8,540	g	E	18	All Rydbergs

Molecule	Peak Position (eV)		Phase	Met.	Ref.	Comments
	π^* (O=X)	σ^* (O-X)				
O$_2$	530.8	539.2, 541.9	g	E	20	
	530.8	539.2, 541.2, 541.8	g	E	16	
	530	535	g	N	21	Pt(112) surface
	529.8	533.3	g	N	22	Cu(111) surface
H$_2$O (X=H)		534, 536	g	N	23	
		535, 537.5, 541	s	N	23	
		535, 537, 542	l	N	23	
		534,535.9,537.1,538.5	g	E	9	All Rydbergs
CuO		530.8, 535	s	N	24	
Cu$_2$O		533.5	s	N	24	
SO$_2$(X=S)	530.7	534.9	g	N	25	Ni(100) surface
Heterocyclic molecules	π^* (O=C)	σ^* (C-O)				
furan	535.3	539.4	g	E	26	
tetrahydrofuran		537.4	g	E	26	531: O$_2$ contamination (?) 536: π^* CH$_2$
		535.6,540.1,543.7	g	N	8	
	π^* (O=C)	σ^* (C-O)				
tetrahydropyran		538.3	g	E	26	531: O$_2$ contamination (?) 536.6: π^* CH$_2$
ethylene oxide		540.1,542.4	g	E	27	Rydbergs: 535.2,535.4,537.1
p-dioxane		536.6; 539.3	g	N	8	

Cited References in Table 3:
(1) Hitchcock et al. 1992, (2) Robin et al. 1988, (3) Hitchcock and Brion 1980, (4) Urquhart et al. 1995, (5) Ikenaga et al. 2001, (6) Francis and Hitchcock 1994, (7) Francis and Hitchcock 1992, (8) Sham et al. 1989, (9) Wight and Brion 1974, (10) Hempelmann et al. 1999, (11) Ishii and Hitchcock 1988, (12) Ishii et al. 1987, (13) Ishii and Hitchcock 1987, (14) Ruhl and Hitchcock 1991, (15) van der Wiel et al. 1970, (16) Hitchcock and Brion 1980, (17) Wight et al. 1972/73, (18) Wight and Brion 1974, (19) Schwartz et al. 1977, (20) Wight and Brion 1974, (21) Wako et al 2000, (22) Amemiya et al. 1999, (23) Myneni et al. 2002, (24) Gurevich et al. 1999, (25) Tokoyama et al. 1995, (26) Newbury et al. 1986, (27) Sze and Brion 1991.

Myneni

Table 4. Assignments of the absorption spectra of S K-edge of Na$_2$SO$_4$, Na$_2$SO$_3$, Na$_2$S$_2$O$_3$, Na$_2$S$_2$O$_5$ Na$_2$S$_2$O$_6$, Na$_2$S$_2$O$_7$, and Na$_2$S$_2$O$_8$ (from Sekiyama et al. 1986). The alphabetical labels shown in the third column correspond to the spectral features marked in Figure 27.

Molecule	Energy (eV)	Label	Transition
SO$_4^{2-}$	2479.9	A	S $1s \rightarrow t_2^*$
	2488.7	B	d-type shape resonance
	2495.6	C	d-type shape resonance
SO$_3^{2-}$	2475.5	A	S $1s \rightarrow e^*$
	2477.5	B	S $1s \rightarrow a_1^*$
	2478.9	C	d-type shape resonance
	2487.6	D	d-type shape resonance
	2494.6	E	d-type shape resonance
S$_2$O$_3^{2-}$	2469.2	A	terminal S $1s \rightarrow a_1^*$ (terminal, central S $3p\sigma$)
	2476.4	B	central S $1s \rightarrow a_1^*$ (terminal, central S $3p\sigma$)
	2478.0	C	S $1s \rightarrow e^*$ (central S $3p\pi$)
	2479.8	D	S $1s \rightarrow a_1^*$ (central S $3s$ and $3p\pi$)
	2483.2	E	d-type shape resonance
	2493.2	F	d-type shape resonance
S$_2$O$_5^{2-}$	2475.8	A	S $1s \rightarrow 3p$ (-SO$_2$)
	2480.2	B	S $1s \rightarrow 3p$ (-SO$_3$)
	2489.2	C	d-type shape resonance
	2495.9	D	d-type shape resonance
S$_2$O$_6^{2-}$	2476.5	A	S $1s \rightarrow 3p\ \sigma$ (S-S)
	2477.8	B	S $1s \rightarrow 3p\ \pi$ (S-O)
	2481.7	C	S $1s \rightarrow 3p\ \sigma$ (S-O)
S$_2$O$_7^{2-}$	2480.5	A	S $1s \rightarrow 3p$
S$_2$O$_8^{2-}$	2478.2	A	S $1s \rightarrow 3p\ \sigma$
	2480.7	B	S $1s \rightarrow 3p\ \pi$

Table 5. The energies of the K-absorption edges of Cl in different coordination environments and oxidation states (Modified after Huggins and Huffman 1995).

Functional Groups	Compound	Peak Position (eV)[*]
Aryl Chlorides	9,10-dichloroanthracene	-0.8
	1,2,4,5-tetrachlorobenzene	-0.7
	2,6-dichlorophenol	-0.5
Acyl chlorides	diphenylacetyl chloride	-1.3
	2-naphthoyl chloride	-1.0
Alkyl chlorides	1,2-bis(chloromethyl) benzene	-1.9
	1,4-bis(chloromethyl) benzene	-1.6
	poly(vinyl chloride)	-1.2
	perchloroethylene	-0.9
Organic hydrochlorides (H-bonded Cl⁻)	aniline hydrochloride	+1.1 to +2.1[a]
	pyridine hydrochloride	
	semicarbazide hydrochloride	
	tetracycline hydrochloride	
Gases and vapors	Cl_2	-3.8[b]
	CCl_4, CH_3Cl, $C_2H_4Cl_2$, C_2HCl_3	-2.3 to -0.9
Alkali chlorides	NaCl, KCl, RbCl, CsCl	+0.4 to +2.2
Alkaline earth chlorides	$CaCl_2$, $SrCl_2$	+2.0 to +1.7
	$CaCl_2.2H_2O$, $CaCl_2.6H_2O$, $SrCl_2.6H_2O$	+2.3 to +2.5
Other inorganic chlorides	Hg_2Cl_2, $HgCl_2$, $CuCl_2$	-2.6 to -3.0[c]
Chloride solutions	0.2-6 M NaCl solutions, 1.0M HCl, sea water (Long Island, NY)	+1.5 to +2.0
Oxychlorides	CuOCl, $Ca(OCl)_2$	-2.6[d]
Other chlorine compounds	$KClO_3$, H_2PtCl_6	+5.8 to +4.2

* All spectra are calibrated against the most intense peak in NaCl. Experimental uncertainty is ± 0.2 eV.

(a) Cl K-edge NEXAFS spectra of these samples exhibit fine spectral variations, which vary with Cl local coordination environment (Fig. 41). It is difficult to assign the absorption edges for these samples.

(b) The most intense transition in the NEXAFS spectra of Cl correspond to the 1s→σ* transitions, and these transitions are different from that of NaCl (Fig. 42).

(c) Transition metal chlorides exhibit pre-edges that correspond to the transitions into hybridized d-orbitals of the metal. These are different from that of NaCl.

(d) Electronic transitions in oxychlorides are different from that of NaCl.